Statistics for Biology and Health

Statistics for Biology and Health

Bacchieri/Cioppa: Fundamentals of Clinical Research
Borchers/Buckland/Zucchini: Estimating Animal Abundance: Closed Populations
Burzykowski/Molenberghs/Buyse: The Evaluation of Surrogate Endpoints
Everitt/Rabe-Hesketh: Analyzing Medical Data Using S-PLUS
Ewens/Grant: Statistical Methods in Bioinformatics: An Introduction, 2nd ed.
Gentleman/Carey/Huber/Irizarry/Dudoit: Bioinformatics and Computational Biology Solutions Using R and Bioconductor
Hougaard: Analysis of Multivariate Survival Data
Keyfitz/Caswell: Applied Mathematical Demography, 3rd ed.
Klein/Moeschberger: Survival Analysis: Techniques for Censored and Truncated Data, 2nd ed.
Kleinbaum/Klein: Logistic Regression: A Self-Learning Text, 2nd ed.
Kleinbaum/Klein: Survival Analysis: A Self-Learning Text, 2nd ed.
Lange: Mathematical and Statistical Methods for Genetic Analysis, 2nd ed.
Manton/Singer/Suzman: Forecasting the Health of Elderly Populations
Martinussen/Scheike: Dynamic Regression Models for Survival Data
Moyé: Multiple Analyses in Clinical Trials: Fundamentals for Investigators
Nielsen: Statistical Methods in Molecular Evolution
Parmigiani/Garrett/Irizarry/Zeger: The Analysis of Gene Expression Data: Methods and Software
Proschan/Lan/Wittes: Statistical Monitoring of Clinical Trials: A Unified Approach
Siegmund/Yakir: The Statistics of Gene Mapping
Simon/Korn/McShane/Radmacher/Wright/Zhao: Design and Analysis of DNA Microarray Investigations
Sorensen/Gianola: Likelihood, Bayesian, and MCMC Methods in Quantitative Genetics
Stallard/Manton/Cohen: Forecasting Product Liability Claims: Epidemiology and Modeling in the Manville Asbestos Case
Sun: The Statistical Analysis of Interval-censored Failure Time Data
Therneau/Grambsch: Modeling Survival Data: Extending the Cox Model
Ting: Dose Finding in Drug Development
Vittinghoff/Glidden/Shiboski/McCulloch: Regression Methods in Biostatistics: Linear, Logistic, Survival, and Repeated Measures Models
Wu/Ma/Casella: Statistical Genetics of Quantitative Traits: Linkage, Map and QTL
Zhang/Singer: Recursive Partitioning in the Health Sciences
Zuur/Ieno/Smith: Analysing Ecological Data

Alain F. Zuur
Elena N. Ieno
Graham M. Smith

Analysing Ecological Data

Alain F. Zuur
Highland Statistics Ltd.
Newburgh AB41 6FN
UNITED KINGDOM
highstat@highstat.com

Elena N. Ieno
Highland Statistics Ltd.
Newburgh AB41 6FN
UNITED KINGDOM
bio@highstat.com

Graham M. Smith
School of Science and the Environment
Bath Spa University
Bath BA2 9BN
UNITED KINGDOM
g.m.smith@bathspa.ac.uk

Library of Congress Control Number: 2006933720

ISBN-10: 0-387-45967-7
ISBN-13: 978-0-387-45967-7

e-ISBN-10: 0-387-45972-3
e-ISBN-13: 978-0-387-45972-1

Printed on acid-free paper.

Printed in the United States of America

9 8 7 6 5 4 3 2 1

springer.com

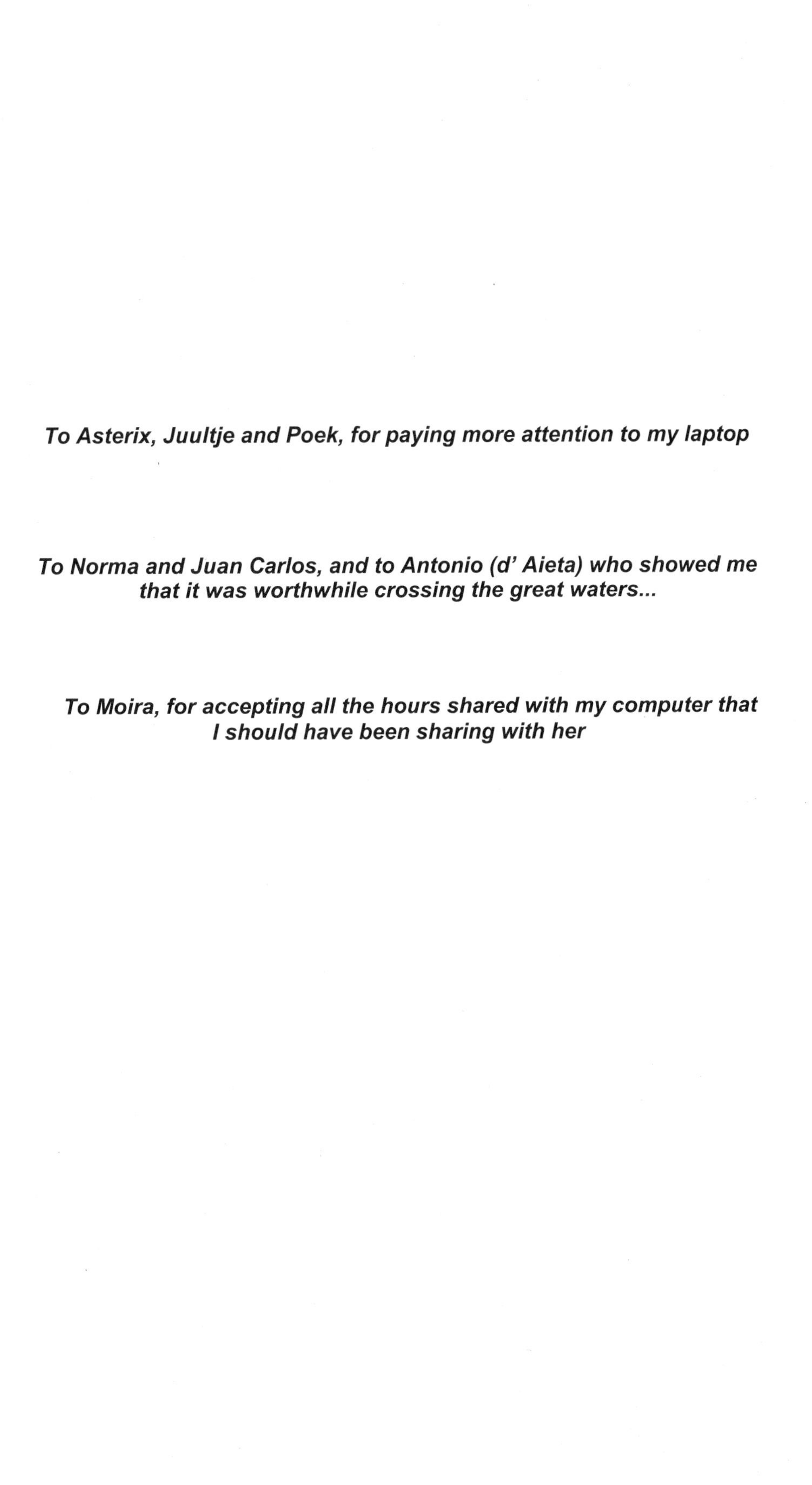

To Asterix, Juultje and Poek, for paying more attention to my laptop

To Norma and Juan Carlos, and to Antonio (d' Aieta) who showed me that it was worthwhile crossing the great waters...

To Moira, for accepting all the hours shared with my computer that I should have been sharing with her

Preface

'*Which test should I apply?*' During the many years of working with ecologists, biologists and other environmental scientists, this is probably the question that the authors of this book hear the most often. The answer is always the same and along the lines of 'What are your underlying questions?', 'What do you want to show?'. The answers to these questions provide the starting point for a detailed discussion on the ecological background and purpose of the study. This then gives the basis for deciding on the most appropriate analytical approach. Therefore, a better starting point for an ecologist is to avoid the phrase 'test' and think in terms of 'analysis'. A test refers to something simple and unified that gives a clear answer in the form of a p-value: something rarely appropriate for ecological data. In practice, one has to apply a data exploration, check assumptions, validate the models, perhaps apply a series of methods, and most importantly, interpret the results in terms of the underlying ecology and the ecological questions being investigated.

Ecology is a quantitative science trying to answer difficult questions about the complex world we live in. Most ecologists are aware of these complexities, but few are fully equipped with the statistical sophistication and understanding to deal with them.

Even data gathered from apparently simple ecological research can require a level of statistical awareness rarely taught at the undergraduate or even the postgraduate level. There is little enough time to teach the essentials of ecology, let alone finding the time to teach 'advanced' statistics. Hopefully, for post graduates moving into academia there will be some advanced statistical support available, but many ecologist end up working in government, a voluntary organisation or consultancy where statistical support is minimal.

Although, the authors of this book believe that a quantitative approach is at the core of being a good ecologist, they also appreciate how challenging many ecologists find statistics. This book is therefore aimed at three levels of reader.

At one level it is aimed at making ecologists aware of how important it is to design scientifically robust ecological experiments or monitoring programmes, and the importance of selecting the best analytical technique. For these readers we hope the book, in particular the case studies, will encourage them to develop their personal statistical skills, or convince them they need statistical support.

On the next level it is aimed at the statistically literate ecologist, who may not be fully aware of the techniques we discuss, or when to use them. Hopefully, we have explained things well enough for these readers to feel confident enough to use some of the techniques we describe. Often these techniques are presented in a fairly impenetrable manner, even for the statistically aware ecologist, and we have tried to make our presentation as 'ecologist friendly' as possible.

Finally, we hope the book will be of value to statisticians, whether they have a background in ecology or statistics. Ecological data can be particularly challenging to analyse, and we hope that providing an insight into our approach, together with the detailed case studies, will be of value to statistician readers, regardless of their background and expertise.

Overall, however, we hope this book will contribute in some small way to improving the collection and analysis of ecological data and improve the quality of environmental decision making.

After reading this book, you should be able to apply the following process: 'These are my questions', 'This is my statistical approach', 'Here is proof that I did it all correct (model validation)', 'This is what the data show' and 'Here is the ecological interpretation'.

Acknowledgement

A large part of the material in this book has been used by the first two authors as course material for MSc and PhD students, post-docs, scientists, both as academic and non-academic courses. We are greatly indepted to all 1200–1500 course participants who helped improve the material between 2000 and 2005 by asking questions and commenting on the material.

We would also like to thank a series of persons who commented on parts of this book: Ian Jolliffe, Anatoly Saveliev, Barry O'Neill, Neil Campbell, Graham Pierce, Ian Tuck, Alex Douglas, Pam Sikkink, Toby Marthews, Adrian Bowman, and six anonymous reviewers and the copy-editor. Their criticisms, comments, help and suggestions have greatly improved this book.

The first author would like to thank Rob Fryer and FRS Marine Laboratory for providing the flexibility to start the foundation of this book.

We would also like to thank the people and organizations who donated data for the theory chapters. The acknowledgement for the unpublished squid data (donated by Graham Pierce, University of Aberdeen) used in Chapters 4 and 7 is as follows. Data collection was financed by the European Commission under the following projects: FAR MA 1.146, AIR1-CT92-0573, FAIR CT 1520, Study Project 96/081, Study project 97/107, Study Project 99/063, and Q5CA-2002-00962. We would like to thank Roy Mendelssohn (NOAA/NMFS) for giving us a copy of the data used in Mendelssohn and Schwing (2002). The raw data are summaries calculated from the COADS dataset. The COADS references are Slutz et al. (1985) and Woodruff et al. (1987). We thank Jaap van der Meer (NIOZ) for allowing us to use the Balgzand data, The Bahamas National Trust and Greenforce Andros Island Marine Study for providing the Bahamas fisheries dataset, Chris Elphick (University of Connecticut) for the sparrow data, and Hrafnkell Eiríksson (Marine Research Institute, Reykjavik) for the Icelandic Nephrops time series. The public domain SRTM data used in Chapter 19 were taken from the U.S. Geological Survey, EROS Data Center, Sioux Falls, SD. We thank Steve Hare (University of Washington) for allowing us to use the 100 biological and physical time series from the North Pacific Ocean in Chapter 17. A small part of Chapter 13 is based on Zuur (1999, unpublished PhD thesis), which was partly financed by the EU project DYNAMO (FAIR-CT95-0710).

A big ‘thank you’ is also due to the large number of folks who wrote R (www.r-project.org) and its many libraries. We made a lot of use of the lattice, regression, GLM, GAM (mgcv) and mixed modelling libraries (nlme). This thank you is probably also on behalf of the readers of this book as everything we did can be done in R.

Finally, we would like to thank John Kimmel for giving us the opportunity to write this book, and his support during the entire process. On to the next book.

Alain F. Zuur
Elena N. Ieno
Graham M. Smith

February 2007

Contents

Contributors

BASTIDA, R.
Departamento de Ciencias Marinas
Universidad Nacional de Mar del Plata
Consejo Nacional de Investigaciones Científicas y Técnicas
Casilla de Correo 43
(7600) Mar del Plata
Argentina

BASUALDO, M.
Fac. Cs. Veterinarias
UNCPBA, Campus Universitario
-7000-Tandil
Argentina

BUDGEY, R.
Central Science Laboratory
Sand Hutton
York, YO41 1LZ
United Kingdom

CABRAL, H.
Universidade de Lisboa
Faculdade de Ciências, Instituto de Oceanografia,
Campo Grande
1749-016 Lisboa
Portugal

CAMPBELL, N.
School of Biological Sciences
University of Aberdeen
Zoology Building
Tillydrone Avenue
Aberdeen, AB24 2TZ
United Kingdom
Current address:
FRS Marine Laboratory
375 Victoria Road.
Aberdeen, AB11 9DB
United Kingdom

CHIZHIKOVA, N.A.
Faculty of Ecology
Kazan State University
18, Kremlevskaja Street
Kazan, 420008
Russia

CLAUDE, J.
Université de Montpellier 2
ISE-M, UMR 5554 CNRS, cc 64
2, place Eugène Bataillon
34095 Montpellier cedex 5
France

DEVINE, J.A.
Memorial University
Department of Biology
4 Clark Place
St. John's NL, A1C 5S7
Canada
Current address:
Middle Depth Fisheries & Acoustics
National Institute of Water and Atmospheric Research Ltd
Private Bag 14-901 Kilbirnie
Wellington 6241
New Zealand

DIJKEMA, K.S.
Wageningen IMARES, Department Texel
Institute for Marine Resources & Ecosystem Studies
P.O. Box 167
1790 AD Den Burg, Texel
The Netherlands

ELPHICK, C.S.
Department of Ecology and Evolutionary Biology
University of Connecticut
75 N. Eagleville Road, U-43
Storrs, CT 06269-3043
USA

ERZINI, K.
CCMAR
Universidade do Algarve
8005-139 Faro
Portugal

GALLEGO, A.
FRS Marine Laboratory
375 Victoria Road
Aberdeen, AB11 9DB
United Kingdom

GUEVARA, S.
Instituto de Ecología, A.C.
km 2.5 Carretera antigua a Coatepec 351, Congregación El Haya
Xalapa 91070, Veracruz
Mexico

HAY, S.
FRS Marine Laboratory
375 Victoria Road
Aberdeen, AB11 9DB
United Kingdom

IENO, E.N.
Highland Statistics LTD.
6 Laverock Road
Newburgh, AB41 6FN
United Kingdom

JANSSEN, G.M.
Directorate-General of Public Works and Water Management RWS-RIKZ
P.O. Box 207
9750 AE Haren
The Netherlands

JOLLIFFE, I.T.
30 Woodvale Road
Gurnard, Cowes
Isle of Wight, PO31 8EG
United Kingdom

KRIJGSVELD, K.L.
Bureau Waardenburg
P.O. Box 365
4100 AJ Culemborg
The Netherlands

LABORDE, J.
Instituto de Ecología, A.C.
km 2.5 Carretera antigua a Coatepec 351, Congregación El Haya
Xalapa 91070, Veracruz
Mexico

LEARMONTH, J.A.
School of Biological Sciences
University of Aberdeen
Zoology Building
Tillydrone Avenue
Aberdeen, AB24 2TZ
United Kingdom

LIRA-NORIEGA, A.
Instituto de Ecología, A.C.
km 2.5 Carretera antigua a Coatepec 351, Congregación El Haya
Xalapa 91070, Veracruz
Mexico
Current address:
Instituto de Ecología, U.N.A.M.
Ciudad Universitaria, Circuito Exterior S/N. anexo a Jardín Botánico
México D.F. 04510
Apdo. Postal 70-275
Mexico

LYKKE, A.M.
Department of Systematic Botany
Universitetsparken, bygn. 1540
8000 Aarhus C
Denmark

LYUBINA, O.E.
Faculty of Ecology, Kazan State University
18, Kremlevskaja street
Kazan, 420008
Russia

MACKENZIE, K.
School of Biological Sciences
University of Aberdeen
Zoology Building
Tillydrone Avenue
Aberdeen, AB24 2TZ
United Kingdom

MARTIN, J.P.
Centro de Investigaciones de Puerto Deseado
Universidad Nacional de la Patagonia Austral
CC 238 (9050), Puerto Deseado
Argentina

MBOW, C.
Institut des Sciences de l'Environnement, Faculté des Sciences et Techniques
Université Cheikh Anta Diop
BP 5005 Dakar-Fann
Senegal

MEESTERS, H.W.G.
Wageningen IMARES, Department Texel,
Institute for Marine Resources & Ecosystem Studies
P.O. Box 167
1790 AD Den Burg, Texel
The Netherlands

MULDER, S.
RIKZ
P.O. Box 207
9750 AE Haren
The Netherlands

MUKHARAMOVA, S.S.
Faculty of Ecology
Kazan State University
18, Kremlevskaja Street
Kazan, 420008
Russia

PAN, M.
FRS Marine Laboratory
375 Victoria Road.
Aberdeen. AB11 9DB
United Kingdom

PIERCE, G.J.
School of Biological Sciences
University of Aberdeen
Zoology Building
Tillydrone Avenue
Aberdeen, AB24 2TZ
United Kingdom

REED, J.M.
Department of Biology
Tufts University
Medford, MA 02155
USA

ROGOVA, T.V.
Faculty of Ecology
Kazan State University,
18 Kremlevskaja Street
Kazan, 420008
Russia

SAMBOU, B.
Institut des Sciences de l'Environnement, Faculté des Sciences et Techniques
Université Cheikh Anta Diop
BP 5005 Dakar-Fann
Senegal

SÁNCHEZ-RÍOS, G.
Instituto de Ecología, A.C.
km 2.5 Carretera antigua a Coatepec 351, Congregación El Haya
Xalapa 91070, Veracruz
Mexico

SANTOS, M.B.
Instituto Español de Oceanografía
Centro Oceanográfico de Vigo
P.O. Box 1552
36200, Vigo
Spain

SAVELIEV, A.A.
Faculty of Ecology
Kazan State University.
18 Kremlevskaja Street
Kazan, 420008
Russia

SIKKINK, P.G.
College of Forestry and Conservation
The University of Montana
Missoula, MT 59812
USA

SMITH, G.M.
School of Science and Environment
Bath Spa University
Newton Park, Newton St Loe
Bath BA2 9BN
United Kingdom

TRASSENS, M.
Departamento de Ciencias Marinas
Universidad Nacional de Mar del Plata
Consejo Nacional de Investigaciones Científicas y Técnicas
Casilla de Correo 43, (7600) Mar del Plata
Argentina

TRENDAFILOV, N.
Department of Statistics
The Open University
Walton Hall
Milton Keynes
MK7 6AA
United Kingdom

TUCK, I.
FRS Marine Laboratory
P.O. Box 101
375 Victoria Road
Aberdeen, AB11 9DB
United Kingdom
Current address:
NIWA
269 Khyber Pass Road
Newmarket, Auckland 1023
New Zealand

VAN DUIN, W.E.
Wageningen IMARES, Department Texel
Institute for Marine Resources & Ecosystem Studies
P.O. Box 167
1790 AD Den Burg, Texel
The Netherlands

ZUUR, A.F.
Highland Statistics LTD.
6 Laverock Road
Newburgh, AB41 6FN
United Kingdom

1 Introduction

This book is based on material we have successfully used teaching statistics to undergraduates, postgraduates, post-docs and senior scientists working in the field possibly best described as the 'environmental sciences'. The required background for these courses, and therefore this book, is some level of 'familiarity with basic statistics'. This is a very loose phrase, but you should feel reasonably comfortable with concepts such as normal distribution, p-value, correlation, regression and hypothesis testing. Any first-year university undergraduate statistics module should have covered these topics to the depth required.

The book is in two main parts. In the first part, statistical theory is explained in an applied context, and in the second part 17 case study chapters are presented. You may find it useful to start by reading the second part first, identify which chapters are relevant, and then read the corresponding theory chapters from part one. In fact, we anticipate this being the way most people will approach the book: by finding the case study that best matches their own ecological question and then using it and the matching theory chapters to guide their analysis.

1.1 Part 1: Applied statistical theory

In the first part, we discuss several techniques used in analysing ecological data. This part is then divided into six sections:

- Data exploration.
- Regression methods: linear regression, generalised linear modelling (GLM) generalised additive modelling (GAM), linear mixed modelling, generalised least squares (GLS) and multinomial logistic regression.
- Classification and regression tree models.
- Multivariate methods: principal component analysis, redundancy analysis, correspondence analysis, canonical correspondence analysis, principal coordinate analysis, discriminant analysis and (non-metric) multidimensional scaling.
- Time series techniques: auto- and cross-correlations, auto-regressive moving average models, deseasonalising, random walk trends, dynamic factor analysis (DFA), min/max auto-correlation factor analysis (MAFA), chronological clustering.
- Spatial statistics: spatial correlation analysis, regression extensions with spatial neigbourhood structure, variogram analysis and kriging.

The main statistical content of the book starts in Chapter 4 with a detailed discussion of data exploration techniques. This is the essential step in any analysis! In Chapter 5 we move onto a detailed discussion about linear regression and explain how it is the basis for more advanced methods like GLM, GAM, mixed modelling, GLS and multinomial logistic regression. Our experience from courses shows that students find it easier to understand GLM and GAM if you can demonstrate how these methods are an extension of something with which they are already familiar.

In Chapters 6 and 7 we explain GLM and GAM. The GAM chapter is the first chapter where we suggest some readers may wish to skip some of the more mathematical parts, in their first reading of the text. We could have omitted these more mathematical parts entirely, but as most books on GAM can be mathematically difficult for many biological and environmental scientists, we chose to include the key mathematical elements in this text. You would not be the first person to use a GAM in their PhD thesis, and then face a grilling in the viva (oral PhD defence) on 'smoothing splines', 'degrees of freedom', and 'cross-validation'. The optional sections in the GAM chapter will allow you to provide some answers.

Chapter 8 contains mixed modelling and tools to impose an auto-correlation structure on the data using generalised least squares. Our experience suggests that most datasets in biological and environmental studies require mixed modelling and generalised least squares techniques. However, the subject is complicated and few books approach it in a biological or environmental context. In Chapter 9 we discuss regression and classification tree models, which like mixed modelling, are techniques well suited to the difficult datasets often found in ecological and environmental studies.

In Chapters 10 to 15 we discuss multivariate analysis. Although there is only limited controversy over the application of univariate methods, there is considerable debate about the use of multivariate methods. Different authors give equally strong, but very different opinions on when and how a particular multivariate technique should be used. Who is right depends on the underlying ecological questions, and the characteristics of the data. And deciding on the best method and approach is the most important decision you will need to make. Sometimes, more than one method can be applied on your data, and it is here that most of our students start to panic. If two analyses show the same results, then this adds confidence to any real-world decisions based on them, but what if two methods give different results? One option is to cheat and only use the results that give the answer you were hoping for, because you believe this increases your chances of publication or getting a consultancy contract renewed. Alternatively, you can try and understand why the two methods are giving different results. Obviously, we strongly advocate the second approach, as working out why you are getting different results can greatly improve you understanding of the underlying ecological processes under study.

In Chapters 16 and 17 we discuss time series techniques and show how auto-correlation can be added to regression and smoothing models. We also present auto-regressive integrated moving average models with exogenous variables

(ARIMAX) and deseasonaling time series. In Chapter 17 we discuss methods to estimate common trends (MAFA and dynamic factor analysis).

In Chapter 18, spatial statistics is discussed. Variograms, kriging and adding spatial correlation to regression and smoothing techniques are the key techniques.

Using several methods to analyse the same dataset seems to go against the convention of deciding your analysis methods in advance of collecting your data. This is still good advice as it ensures that data are collected in a manner suitable for the chosen analytical technique and prevents 'fishing' through your data until you find a test that happens to give you a nice '*p*' value. However, although this approach may well be appropriate for designed experiments, for the types of studies discussed in the case study chapters, it is the ecological question and the structure of the collected data that dictate the analytical approach. The ecological question helps decide on the type of analysis, e.g., univariate, classification, multivariate, time series or spatial methods, and the quality of the data decides the specific method (how many variables, how many zeros, how many observations, do we have linear or non-linear relationships), which can only be addressed by a detailed data exploration. At this point you should decide on the best methodological approach and stick with it. One complication to this approach is that some univariate methods are extensions of other univariate methods and the violation of certain statistical assumptions may force you to switch between methods (e.g., smoothing methods if the residuals show non-linear patterns, or generalised linear modelling if there is violation of homogeneity for count data). This is when it becomes crucial to understand the background provided in the first part of the book

Returning to the question set earlier, in some scientific fields, it is the general belief that you need to formulate your hypothesis in advance and specify every step of the statistical analysis before doing anything else. Although we agree with specifying the hypothesis in advance, deciding on the statistical methods before seeing the data is a luxury we do not have in most ecological and environmental studies. Most of the time you are given a dataset that has been collected by someone else, at some time in the distant past, and with no specific design or question in mind. Even when you are involved in the early stages of an ecological experiment, survey or monitoring programme, it is highly likely that the generated data are so noisy that the pre-specified method ends up unsuitable and you are forced to look at alternatives, As long as you mention in your report, paper or thesis what you did, what worked, and what did not work, we believe this is still a valid approach, and all too often the only approach available.

1.2 Part 2: The case studies

In part two we present 17 case study chapters. The data for each case study are available online (www.highstat.com). Each chapter provides a short data introduction, a data exploration, and full statistical analysis, and they are between 14 and 26 pages long. The aims of the case studies are to:

- Illustrate most of the statistical methods discussed in the theoretical chapters using *real* datasets, where the emphasise is on 'real'. All datasets have been used for PHD theses (9 of the 17 case study chapters), consultancy contracts or scientific papers. These are not toy datasets!
- Provide a range of procedures that can be used to guide your own analysis. Find a chapter with a dataset similar to your own and use the same steps presented in that chapter to lead you through your own analysis.
- Show you, step by step, the decisions that were made during the data analysis. What techniques best suit this particular dataset and the ecological question being asked? How can the data be modified so that a standard statistical technique can be applied? What should you look for when deciding on a certain method; why additive modelling and not linear regression?

Of course the approaches presented in these chapters reflect our own approach to data analysis. Ask another statistician and they may well suggest something different. However, if there is a strong pattern in your data, then regardless of approach it should be detected.

The case studies have been selected to cover a wide diversity of ecological subjects: marine benthos, fisheries, dolphins, turtles, birds, plants, trees, and insects. We have also tried to use data from a range of continents, and the case studies come from Europe, North America, South America, Asia and Africa. We would have liked to include a wider range of data and continents, but finding datasets that were available for publication was difficult. We are therefore especially grateful to the owners of the data for letting us include them in the book and making them available to our readers.

The case study chapters are divided into four groups and follow the same structure as the theory sections:

1. Case study chapters using univariate techniques.
2. Case study chapters using multivariate techniques.
3. Chapters using time series methods.
4. Chapters using spatial statistics.

Each chapter illustrates a particular range of methods, or a specific decision process (for example the choice between regression, additive modelling, GLM or GAM). We also have some chapters where a series of methods are applied.

Univariate case study chapters

The first case study investigates zooplankton measured at two locations in Scotland. The aim of the chapter is to show how to decide among the application of linear regression, additive modelling, generalised linear modelling (with a Poisson distribution) or generalised additive modelling. We also discuss how we used smoothing methods to gain an insight into the required sample size: essential information when designing a cost-effective, but still scientifically robust study.

In the second case study, we go further south, to an estuary in Portugal, where we look at flatfish and habitat relationships. The first time we looked at these data, it was a "nothing works" experience. Only after a transformation to presence-absence data, were we able to obtain meaningful results. The methods used in this chapter are logistic regression models (both GLM and GAM) and classification tree models.

We then present two case studies using (additive) mixed modelling techniques. The first one is about honeybee pollination in sunflower commercial hybrid seed production (Argentina). We also apply mixed modelling to investigate the abundance of Californian wetland birds (USA) in relation to straw management of rice fields. This chapter shows that if you wrongly ignore auto-correlation in your data, all the parameters can end up as significant. But by including an auto-correlation structure on the error component, the same parameters end up as only borderline significant!

In the fifth and sixth case studies, we apply classification methods. First, we use bird data from a Dutch study. The birds were recorded by radar and surveyed on the ground by biologists. The aim of the analysis was to investigate whether radar can be used to identify bird species, and this was investigated using classification trees. In the other classification chapter, we are back to fish and demonstrate the application of neural networks. The aim of this chapter is to identify discrete populations or stocks of horse mackerel in the northeast Atlantic by using a neural network to analyse parasite presence and abundance data. This chapter is different from the other case studies in that neural networks are not discussed in the preceding theory chapters. However, it is a popular method in some ecological fields and this chapter gives a valuable insight into its value and application.

Multivariate case study chapters

In the seventh case study, the first multivariate one, we look at plant species from the western Montana landscape (USA). Classic multivariate methods like non-metric multidimensional scaling (NMDS) and the Mantel test are used. We also apply GLS on a univariate diversity index and take into account the auto-correlation structure. Several auto-correlation structures are investigated.

In the next case study, we analyse marine benthic data from the Netherlands. This country lies below sea level, and one way of defending the country from the sea is to pump sand onto the beach. This process is also called beach re-nourishment or beach re-charge, and this chapter investigates the impact this process might have on the beach living animals. Several univariate and multivariate methods are considered: GAM, redundancy analysis and variance partitioning. Results of this work were used by the Dutch authorities to improve their coastal management procedures.

In case study nine, Argentinean zoobenthic data are used to show how easy it is to cheat with classic multivariate methods like NMDS and the Mantel test. We also discuss a special transformation to visualise Chord distances in redundancy analysis (RDA).

In the tenth case study, aspects of principal component analysis (PCA) are discussed and illustrated using fatty acid data from stranded dolphins. Covariance vs. correlation, how many PCs and normalisation constraints are all discussed and a biplot is interpreted. An alternative to PCA that gives simplified interpretations is also included.

In the next chapter, the focus is on morphometric data analysis. PCA, together with landmark data analysis, is applied on skull measurements of turtles.

The twelfth case study, explores Senegalese savanna tree distribution and management using satellite images, and RDA, plus additive modelling to verify the RDA results.

In the last multivariate chapter, Mexican plant data are analysed using canonical correspondence analysis. The role of an invasive species is investigated.

Time series case study chapters

In the first time series case study, Portuguese fisheries landing trends are analysed. The main aim of this chapter is to estimate common trends using DFA and MAFA. In another time series case study chapter, groundfish research survey data from the northwest Atlantic are analysed and the effects of time lags are explored using MAFA and DFA. The techniques are used to highlight the importance of scale and to explore the complex dynamics of a system that has experienced drastic change. In the third time series case study, effects of sea level rise on Dutch salt marsh plant species are analysed using additive mixed modelling and mixed modelling methods that allow for auto-correlation. And we conclude the time series chapters with endangered Hawaiian birds and estimate common trends by applying a form of intervention analysis to detect the effect of management actions.

Spatial case study chapter

The only spatial chapter is on Russian tree data and demonstrates various spatial analysis and interpolation techniques for tree data (including variography, kriging, and regression models with spatial correlation structure).

1.3 Data, software and flowcharts

Chapter 2 discusses data management and software. About 95% of the statistical analyses in this book were carried out in the low-budget software package Brodgar, which can be downloaded from www.brodgar.com. It is a user-friendly 'click-and-go' package that also has a link to the R software (www.r-project.org). However, most statistical analysis can be carried out in other excellent software packages as well, and we discuss our experience with some of them. We also provide flowcharts in this chapter to readers with an overview of the methods.

In Chapter 3, we discuss our experience with teaching the material described in this book and give general recommendations for instructors.

2 Data management and software

2.1 Introduction

This chapter reviews some statistical programmes with which we have experience and reinforces some ideas of good data management practice.

Although there is nothing more important than looking after your raw data, our experience shows this is often poorly implemented. Unfortunately, this criticism also often applies to the subsequent collation and management of data. We are sometimes given long-term datasets to analyse, where the data have been collected and stored in a different format for each year of the monitoring programme, different surveyors have collected data from a different range of variables at different levels of detail, some data 'have been lost' and some variables are labelled with cryptic codes whose meaning no one can remember. Although we are confident that 'our' readers will have an exemplary approach to data management, we also feel it is an important enough issue to use this chapter to (briefly) reinforce some key points of data management.

Also a word of warning about the reliability of raw data: The first author used to work at an institute that collected hydrological data used for a national monitoring network. The data were collected by farmers (among others) that provided the information for a small financial award. When employees of the institute asked one of the farmers to show them the location of the measurement apparatus, he could not find them. Hence, the farmer must have made up the data.

Choice of software, both for data management and analysis, is also something worth thinking about. Many large datasets are inappropriately managed in spreadsheet programmes when they should be managed in a database programme. A very wide range of statistical programmes are available, with many offering the same range of analytical techniques, and to a large extent the choice of programme is unimportant. However, it is important to be aware that different programmes often use slightly different algorithms and different default settings. It is therefore possible to get a different answer to the same analysis performed in two different programmes. This is one reason why the make of software as well as the procedures should stated whenever the results of an analysis are presented or published.

Excellent advice on statistical good practice is given in Stern et al. (2004), and for those who insist on using spreadsheets for data management, O'Beirne (2005) is essential reading.

2.2 Data management

Data management begins at the planning stage of a project. Even simple studies need established sampling protocols and standard recording methods. The latter take the form of a printed or electronic recording sheet. This goes some way to ensuring that each surveyor records the same variables to the same level of detail. Written field records should be transferred to a 'good copy' at the end of each day, and electronic records should be backed up to a CD and/or emailed to a second secure location.

Small datasets, where the number of observations is in the hundreds, can be stored in a spreadsheet. However, storing data in a spreadsheet is less secure than using a database programme, and even for small datasets, you should consider creating a proper database. Although increasing the time required to set up the project, your data will be safer and allow a more formal and consistent approach to data entry. Hernández (2003) and Whitehorn and Marklyn (2001) give good introductions to developing a relational database. Entering data into a spreadsheet or database is an obvious source of errors, but there are some tools available to reduce these errors. Some programmes allow a double entry system (e.g., GenStat and Epidata) where each entry is typed in twice, and the programme warns you if they do not match. The use of a drop-down 'pick list' is available in both spreadsheets and databases, which will keep entries consistent. You can also constrain the range of values that a spreadsheet cell or database field will accept, removing the risk of accidentally typing in a nonsensical value. For example if you are measuring pH, you can set up a constraint where the spreadsheet cell will only accept values between 1 and 14. Typing in data from paper records to a computer is also assisted if the format of the paper record mirrors the format of the electronic entry form. And one final comment on this, although it may seem to be an effective use of staff resources to use an admin member of staff to type in data, this can also result in serious problems. Whether this is a good idea or not will depend on how well you have set up your data management system. In one instance, bad handwriting from the ecologist led to records of several non-existent invertebrates appearing after the field notes were typed into the computer by an admin assistant.

Backing up is obviously a critical part of good data management and the low cost of CD or DVD burners, or even external hard drives, now makes this relatively easy and inexpensive. In our experience, the best back-up facility is online with software that makes a daily back up of the modified files. Obviously, this only works if you have a reasonably fast Internet connection. For example, this entire manuscript (with all its data files, graphs, word documents, and script codes) was backed up daily in this way; total size was about 2 Gigabyte but the incremental back up only took 5 minutes per day (as a background process). There is even software available that will automatically back up key directories at preset intervals or every time the computer is shut down (e.g., Second Copy, http://www.secondcopy.com). It is important to check the archival properties of your chosen back-up media and design a strategy to suit. In addition to daily back-up sets, a longer term approach to backing up and storing the project's raw data should be considered. For this set you should consider storing the data in ASCII and making sure a copy is stored in more than one location, preferably in a fire-proof safe. Although, for some, this may seem excessive, many datasets are the result of many years of study and irreplaceable. Even small datasets may play a critical role in future meta-analysis studies, and therefore worth looking after.

A critical aspect of backing up information is that occasionally you inspect the back-up files, and that you ensure that you know how to retrieve them.

2.3 Data preparation

Before any analysis can be done, the data have to be prepared in a database or spreadsheet programme. The data preparation is one of the most important steps in the analysis, and we mention a few key considerations below.

Data structure

Most statistics programmes assume data are in the form of a matrix with the measured values for each variable in their own column, and each row storing all the results from a single sample or trial. Even though some programmes will allow you to analyse data with a different structure, a standardised data structure can still save time and reduce risk of errors, particularly if you hand over data for someone else to help with the analysis.

Variable names

Variable names should be kept as short as possible. Some statistical programmes still require short variable names and, on importing, will truncate the name if it is longer than 10 (or 8) characters. Depending on your naming convention, this can result in several variables ending up with identical names. In graphs (particularly biplots and triplots) where you want to show the variable names on the plot, long variable names increase the chances of them overlapping and becoming unreadable. Shorter names also allow more variables to be seen and identified on one computer screen without scrolling through the columns. It is also important to check for any naming conventions required by the statistics programme you are going to use; for example, you cannot use an underscore as part of a variable name in either SPLUS or R. Developing a formal naming convention at the beginning of the project is well worth the time; for example, you can use a prefix to each variable name that indicates to which group the variable belongs. An example of this is where the same variables are measured from several different transects: 'a.phos', 'a.pot', 'a.ph' and 'b.phos', 'b.pot', 'b.ph' where the 'a' and 'b' identify the transect. A dot has been used in this instance as this is acceptable to both SPLUS and R as part of a variable name. However, it may cause problems in other software packages.

Missing and censored data

Inevitably some datasets have missing data and different programmes treat missing data differently. With SPLUS the letters NA are used to identify cells with missing data, but with other programmes, an asterisk may be used instead (GenStat) and others expect a completely blank cell (Minitab) as an indicator of missing data. If you use NA as your code for no data you will find that some statistics programmes will simply refuse to import the data, as they do not allow alphanumeric characters. Others will convert all the values for that particular variable into factors, even the numeric ones, because they recognise the NA as an alphanumeric character and assume the data refer to a nominal or categorical variable. Some programmes allow you to define the missing data code, and the use of '9', '99' or '999' is a common convention depending on the length of the variable (Newton and Rudestam 1999). If you already know the programme you will use for analysis, then the choice of code for missing data will be obvious, but whichever approach is adopted, it should be consistent and prominently documented.

Censored data are common when measuring the chemical content of an environmental variable. They arise because the concentration is below the detection limit of the technique being used to measure them. This gives results such as <0.001 ppm, and a decision needs to be made on how to deal with this type of data. Several approaches are possible, and an overview is given in Manly (2001). Different protocols seem to exist for different disciplines, and software help is available from www.vims.edu/env/research/software/vims_software.html. However, as with dealing with missing data, the key is to use a well-documented and consistent approach.

Nominal data

Nominal variables are variables of the form: yes/no; or yellow/green/blue; or transect 1, transect 2, transect3; or observer 1, observer 2, observer 3. Other examples of nominal variables are month, gender, location, etc. Some programmes can work with alphanumerical values such as 'yes' and 'no', and others cannot. For those that cannot work with alphanumeric variables, these variables have to be converted into numbers. For example, a 'yes' can be converted into a 1, and 'no' into a 0. If you do this process in a spreadsheet, you end up with an extra column containing only zeros and ones. The same principle holds if the variable has three classes; use a 1 for yellow, 2 for green and 3 for blue (or 1 for January and 12 for December). Table 2.1 shows an example for colour. The first column contains the sample number, and the second the colours. Table 2.2 shows the conversion to numerical values. Note that all data are numeric. Removing the original alphanumerical values and importing only the numerical data might still cause problems for some programmes. Table 2.3 shows how to prepare the data so that it can be used in specialised multivariate programmes that cannot convert a nominal variable with multiple classes in 0-1 dummy variables (Chapter 5).

Table 2.1. Artificial data to illustrate coding of nominal variables.

Sample	Colour	Other variables
1	Green	.....
2	Yellow	.....
3	Blue	.....
4	Blue	.....
5	Green	.....
6	Yellow	.....

Table 2.2. Artificial data to illustrate coding of nominal variables. The numbers 1, 2 and 3 represent yellow, green and blue, respectively.

Sample	Colour	Other variables
1	2	.....
2	1	.....
3	3	.....
4	3	.....
5	2	.....
6	1	.....

Table 2.3. Artificial data to illustrate coding of nominal variables.

Sample	Yellow	Green	Blue	Other variables
1	0	1	0	.....
2	1	0	1	.....
3	0	0	1	.....
4	0	0	1	.....
5	0	1	0	.....
6	1	0	0	.....

Coding the underlying question

In Chapter 28, we analyse a zoobenthic dataset measured in salt marshes in Argentina. The dataset contains four zoobenthic species and four explanatory variables measured at three transects (10 observations per transect) in two seasons. One underlying question is whether there is a transect effect and a season effect. To quantify this information, two new columns were made and labelled 'Season' and 'Transect'. Table 2.4 shows how we prepared the spreadsheet. Data of species and sediment variables are in columns. The first 30 rows are from Autumn and the next 30 rows from Spring (and there are 60 rows in total). The first 10 rows are from transect A in the Autumn, the second 10 from transect B in the Autumn, etc. The season (Autumn and Spring) was quantified using 0 and 1, and transect by 1, 2 and 3.

Table 2.4 Illustration of the preparation of a spreadsheet. The 30 observations in the three transect are denoted by A_1,..,A_{10}, B_1,..,B_{10}, C_1,..,C_{10}. We created the nominal variables Season (1 = Autumn, 2 = Spring) and Transect (1 = transect A, 2 = transect B, 3 = transect C).

Site	Season	Transect	*Laeonereis acuta*	*Heteromastus similis*	....	Mud
A_1	1	1	10	4	...	2
A_2	1	1	22	21	...	3
...	...	...	...	...	...	4
A_{10}	1	1	21	55	...	32
B_1	1	2	5	78	...	2
B_2	1	2	6	3	...	1
...	...	...	...	...	...	...
B_{10}	1	2	...	...	...	...
C_1	1	3	...	...	...	...
...	...	...	...	...	...	...
C_{10}	1	3	...	...	...	...
A_1	2	1	...	...	...	...
A_2	2	1	...	...	...	...
...	...	...	...	...	...	...
A_{10}	2	1	...	...	...	...
B_1	2	2	...	...	...	...
B_2	2	2	...	...	...	...
...	...	...	...	...	...	...
B_{10}	2	2	...	...	...	...
C_1	2	3	...	...	...	...
...	...	...	70	101	...	8
C_{10}	2	3	88	265	...	5

The underlying hypothesis for a study can sometimes be formulated using so-called dummy variables that consist of zeros and ones (or even more levels if required). For example, in Chapter 20, we analyse decapod data from two different areas and two different years. The question of whether there is an area effect can be investigated by introducing a new variable called 'Area' with values zero and one to represent the two different areas. It can then be included in the models as a main term or as an interaction term with other variables. Another example is presented in Chapter 23 in which the main underlying question is whether there is an effect of straw management on bird abundance in rice fields. To quantify the six different types of straw management, a new variable was introduced that had six different values (1 to 6), with each value representing a particular straw management regime. In Chapter 27, we use a nominal variable 'week' and 'exposure' to investigate whether there are differences in benthic communities between (i) the four weeks and (ii) beaches with different exposure types. For the data in Table 2.4, the variables Transect and Season can be used to answer the underlying questions.

2.4 Statistical software

In this section we discuss some statistics programmes with which we have experience. The absence of a programme from this chapter is simply because we have no experience using it, and is not a judgement on its quality or suitability.

Brodgar

Most of this book is based on statistics courses run by its authors. During these courses we use Brodgar (www.brodgar.com). Written by one of the books author's, Brodgar has been developed around the needs of the ecologist and environmental scientist. It is a low-cost, user-friendly programme with a graphical user interface (GUI). It allows users to apply data exploration tools, univariate, multivariate, time series and spatial methods with a few mouse clicks. About half of its methods call routines from R (www.r-project.org). This gives users a GUI alternative to the R command line interface, and with three exceptions allows all thc techniques used in the book to be run from a single programme. The exceptions are (i) the geometric morphometrics analysis of the landmark turtle data in Chapter 30, (ii) neural networks in Chapter 25, and (iii) kriging in Chapters 19 and 37. In these cases, we used specialised R routines.

SPLUS and R

Another programme we use extensively is SPLUS (http://www.insightful.com). SPLUS is a commercial implementation of the S language, which provides a user-friendly menu-driven interface. An alternative is the Open Source version of the S language called R (www.project.org). Its syntax is 95% identical to SPLUS, and an excellent range of books is available to help the beginning and the experienced user of R (and also SPLUS): for example, Chambers and Hastie (1992), Venables and Ripley (2002), Dalgaard (2002), Maindonald and Braun (2003), Crawley (2002, 2005) and Verzani (2005). If we could choose only one programme to use, then we would chose R as nearly every statistical technique available has been implemented in either the base programme or in one of its many add-in libraries. The only possible problem with R is the lack of a graphical user interface, and to make the most of R, the user needs to learn R programming. Some people would argue that this is good thing as it forces the user to know exactly what he or she is doing. However, for users concerned about this programming aspect, there are menu-based add-ins available, such as Rcmdr and Biodiversity.R. Rcmdr (http://socserv.mcmaster.ca/jfox/Misc/Rcmdr/) provides a menu system for general exploratory univariate and some multivariate statistics, whereas Biodiversity-R (www.worldagroforestry.org) provides a menu interface for a range of more specialised ecologically useful routines in R.

R is excellent for analysing data and, in our experience, can also be used for teaching basic statistics to undergraduate students. Things like means, medians, *p*-values, etc. are easily calculated, and most of the methods used in this book can be learned from a third-party book such as Dalgaard (2002) or Venables and Ripley (2002) in a few days of dedicated learning.

However, in our experience it is difficult to *teach* the statistical methods discussed in this book with R to a group of 10–80 undergraduate (or postgraduate) biology students who have had no previous experience of using R. A proportion of students simply refuse to learn a programming language when they know easy-to-use graphical user interface alternatives are available.

GenStat

In Chapter 5, we discuss linear regression, and extensions to linear regression such as generalised least squares and mixed modelling are explained in subsequent chapters. For these more advanced techniques we have found GenStat (www.vsn-intl.com/genstat/) to be one of the best programmes available, and for mixed modelling and related methods, it is probably even better than R (or SPLUS). GenStat also allows the user to apply generalised linear mixed modelling, a method not covered by this book. GenStat has roots in the applied biological sciences and has a user-friendly menu-driven interface making it a good teaching tool and a good choice for users looking for a high-quality menu-driven programme. As well as the friendly front end, GenStat has a powerful programming language comparable in power with SPLUS or R, which together with its capability with large datasets also makes it a serious tool for the advanced statistician. Instructors who are considering using GenStat for classroom teaching are advised to contact GenStat as temporary free classroom licences may be available.

Other programmes

We have also used other programmes such as Minitab, SYSTAT and SPSS for undergraduate statistics courses. Although these programmes can all apply basic statistical techniques, they have a limited range of exploratory graphics tools and are less suitable for specialised statistical techniques like generalised additive modelling, redundancy analysis, dynamic factor analysis, etc. (unless special add-on libraries are bought).

Specialised multivariate programmes

For multivariate analysis, three key programmes need to be mentioned: Canoco, PRIMER and PC-ORD. Canoco for Windows version 4.5 is a software tool for constrained and unconstrained ordination. The ordination methods are integrated with regression and permutation methodology, so as to allow sound statistical modelling of ecological data. Canoco contains both linear and unimodal methods, including DCA, PCA, CCA, RDA, db-RDA, or PCoA. Ordination diagrams can be displayed and exported in publication quality right after an analysis has been completed. Canoco is unique in its capability to account for background variation specified by covariables and in its extensive facilities for permutation tests, including tests of interaction effects. Canoco has been designed for ecologists, but it is also used in toxicology, soil science, geology, or public health research, to name a few.

PRIMER is popular in marine benthic fields, and its primary assets are measures of association combined with ANOSIM, BVSTEP and other methods that carry out permutation tests (Chapters 10 and 15). Primer has also been designed with ease of use in mind and only provides the more robust analytical tools, making it a good choice for the less experienced.

PC-ORD is a Windows programme that performs multivariate analysis of ecological data entered into spreadsheets. Its emphasis is on non-parametric tools, graphical representation, and randomization tests for analysis of community data. In addition to utilities for transforming data and managing files, PC-ORD offers many ordination and classification techniques not available in other major statistical packages. Very large datasets can be analyzed. Most operations accept a matrix up to 32,000 rows or 32,000 columns and up to 536,848,900 matrix elements, provided that you have adequate memory in your machine. The terminology is tailored for ecologists.

All three programmes are popular tools for multivariate analysis, but the user will need at least one other programme for data exploration and univariate methods.

Other multivariate programmes with a focus on ecological analysis are PATN, MVSP, CAP and ECOM. The last two are from Pisces Conservation Ltd. However, the authors have limited or no experience with these programmes.

Table 2.5 contains a list of all statistical methods discussed in this book and shows which programmes can be used for each method. It should be noted that several other programmes can do the majority of these statistical methods as well (e.g., SAS), but we are not familiar with them. The time series method, dynamic factor analysis (DFA) as in Zuur et al. (2003a) is, to the best of our knowledge, only available in Brodgar.

Table 2.5. Comparison of some statistical programmes. Note that several other excellent programmes are available (e.g., SAS), but they were not taken into account as the authors have not worked with them. The symbol 'X' is used to express that a method can easily be applied in the software package. In the column Brodgar, the symbol 'R' means that an interface to R is used, 'N' means native and 'NR' indicates both native and using R. The notation 'P_{simple}' means 'requires simple script programming' and 'P_{compl}' is ' requires complicated script programming'.

	Brodgar	Genstat	CANOCO	PRIMER	R	PC-Ord
Data exploration	NR	X			X	X
Linear regression	R	X	X		X	X
Partial linear regression	R	P_{simple}	X		P_{simple}	
GLM	R	X	X		X	
GAM	R	X	X		X	
Mixed modelling	R	X			X	
GLS	R	X			X	
Tree models	R	X			X	X
Neural networks	R	P			X	
Measures of association	NR	X	X	X	X	X
PCA	N	X	X	X	X	X
RDA	N		X		X	
Partial RDA	N		X		X	
CA	N	X	X	X	X	X
CCA	N		X		X	X
Partial CCA	N		X		X	
Discriminant analysis	N	X	X		X	X
NMDS	N	X	X	X	X	X
Geometric morphometric analysis		P_{compl}			P_{compl}	
De-seasonalising	N	X			X	
Repeated lowess smoothing	R	P_{compl}			P_{compl}	
MAFA	N	P_{compl}			P_{compl}	
DFA	N	P_{compl}			P_{compl}	
Chronological clustering	N	P_{compl}			P_{compl}	
Spatial statistics						
SAR	R				X	
SMA	R				X	
Variograms	R	X			X	
Surface variogram	R	X			X	
Kriging		X			X	

3 Advice for teachers

3.1 Introduction

In this chapter, we discuss our experience in teaching some of the material described in this book. Our first piece of advice is to avoid explaining too many statistical techniques in one course. When we started teaching statistics we tried to teach univariate, multivariate and time series methods in five days to between 8 and 100 biologists and environmental scientists. We did this in the form of in-house courses, open courses and university courses. The audiences in institutional in-house and open courses typically consisted of senior scientist, post-docs, PhD-students and a few brave MSc students. The university courses had between 50 and 100 PhD or MSc students. The courses covered modules from data exploration, regression, generalised linear modelling, generalised additive modelling, multivariate analysis (non-metric multidimensional scaling, principal component analysis, correspondence analysis, canonical correspondence analysis, redundancy analysis) and time series. Although these 'show-me-all' courses were popular, the actual amount of information that participants were able to fully understand was far less than we had hoped for. It was just too much information for five days (40 hours).

We now teach several modules across all levels of expertise, including large groups (up to 100) of undergraduate and first-year PhD-students. The teaching for the univariate modules is now broken down into the following components (1 day is 8 hours):

- Data exploration (1 day). The emphasis in this module is on outliers and data transformations. Most students will be obsessed by the idea of normality of data (Chapter 4). This idea is so strong that some will even make QQ-plots of explanatory variables. We suggest instructors emphasise that normality means normality at each X value (Chapter 4). As not all datasets have enough replicates to make histograms for the data at each X value, it may be better to convince the students to first apply linear regression and then test the model for normally distributed residuals. It may also be an option to expose students to the idea that normality is one underlying assumption of linear regression, but it is not the most essential one (Chapter 5). We suggest using no more than three datasets as students complain that it is difficult to remember all the details of the data. Any dataset used in the case study chapters would be suitable. The emphasis of this module should be outlier detection, transformations and collinearity between explanatory variables. Do not forget to explain the concepts (and differences) of interaction, collinearity and confounding!

- Regression (1 day). In every course we ask the participants the question: 'Do you understand linear regression?' Three quarters will respond positively, but most will fail to identify the four underlying assumptions of linear regression, and many will have never seen a graph of residuals plotted against explanatory variables. Because generalised linear modelling (GLM) and generalised additive modelling (GAM) are considerably easier to explain once the students understand the principle of Figures 5.5 and 5.6, we suggest you spend a lot of time establishing a sound understanding of regression. The challenge is to get the students to understand the numerical output and how to apply a proper model validation process. Homogeneity and residuals versus each explanatory variables are the key points here. The flowchart in Figure 3.1 is useful in this stage as it explains to the student how to decide among linear regression, GLM and GAM.
- Generalised linear modelling (1 day Poisson and half a day logistic regression). Many students find this subject difficult, especially with logistic regression. We suggest starting the students on an exercise with only one explanatory variable so that they can see the model fit in the form of a logistic curve.
- Generalised additive modelling (1 day). The second part of the GAM chapter is slightly more technical than the first part. If the audience consists of biologists, we suggest explaining only the underlying concepts of splines, degrees of freedom and cross-validation. We chose to present the more technical aspects of GAM in this book simply because it is difficult to find it in textbooks for biologists. Some PhD students that we have supervised were interrogated during their viva on GAM.
- Mixed modelling and generalised least squares (1 day). Having completed this book, we realised that the data in nearly every case study could have been analysed with mixed modelling or generalised least squares (GLS). In fact, most ecological datasets may require mixed modelling methods! Yet, it is highly complicated and difficult to explain to biologists.

The multivariate material also takes five days (40 hours) to explain. Relevant topics are measures of association (1 day), ANOSIM and the Mantel test (1 day), principal component analysis and redundancy analysis (1 day), correspondence analysis and canonical correspondence analysis (0.5 days) and half a day for methods like non-metric multidimensional scaling and discriminant analysis. We strongly advise combining the multivariate module with a day on data exploration. The important aspect of the multivariate module is to teach students when to choose a particular method, and this is mainly driven by the underlying question, together with the quality of the data (e.g., a species data matrix with many zero observations and patchy species means non-metric multidimensional scaling).

Figure 3.3 gives an overview on how most of the multivariate techniques discussed in this book fit together. Various case study chapters emphasise the options and choices that have to be made. Note that various other methods exist, e.g., distance based redundancy analysis (Legendre and Anderson 1999), but these are not discussed in this book.

Time series analysis is a difficult, but popular, subject. We tend to start with data exploration (1 day), followed by a brief summary of linear regression (0.5 day), and then we introduce the idea of having an auto-correlation structure on the data using generalised least squares (0.5 days). Once students are used to the concept of auto-correlation, topics such as auto-correlation, cross-correlation and auto-regressive moving average models can be introduced (0.75 day). Seasonality is another important subject and takes at least two hours to explain. More specialised methods like min/max auto-correlation factor analysis and dynamic factor analysis are popular, and it is important to emphasise the differences between them. It might also be an option to explain generalised additive modelling and regression methods and the possibility of adding an auto-correlation structure on the errors, leading to methods like generalised additive mixed modelling, generalised least squares and generalised linear mixed modelling (which is not discussed in this book).

As to the spatial modules, we tend to teach these methods in a follow-up course to students who are familiar with GAMs, regression, etc. Knowledge of time series, or at least the concept of auto-correlation, helps. Figure 3.3 shows a decision tree that can be used to explain how to decide upon the most appropriate method. The final decision on choice of method comes down to the small technical details and is part of the model validation process. This is expanded in the case study chapters.

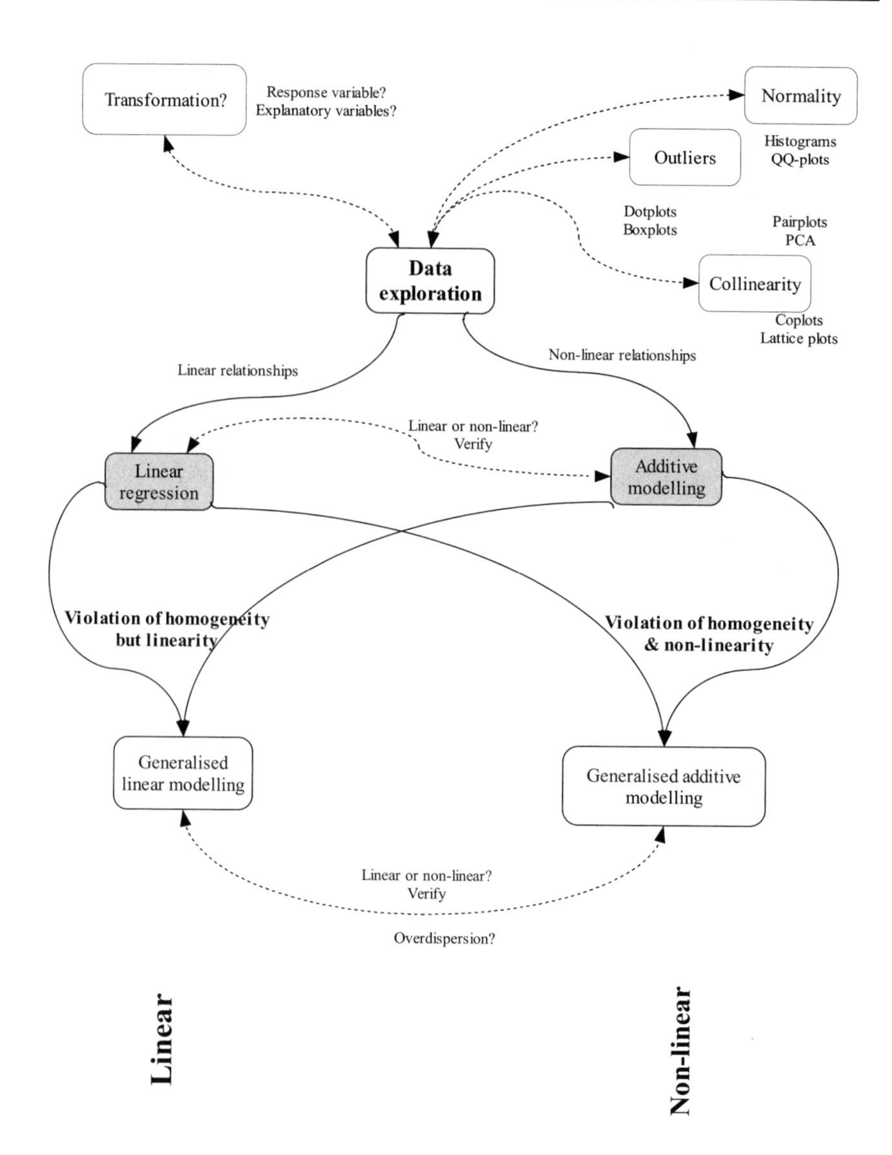

Figure 3.1. Flowchart showing how linear regression, additive modelling, generalised linear modelling (using the Poisson distribution and log-link function) and generalised additive modelling (using the Poisson distribution) are related to each other. In linear regression, violation of homogeneity means that the GLM with a Poisson distribution may be used. Normality but non-linear relationships (as detected for example by a graph of the residuals versus each explanatory variable) means that additive modelling can be applied. Non-linear relationships and violation of the normality assumption means a GAM with a Poisson distribution. The graph will change if another link function or distribution is used.

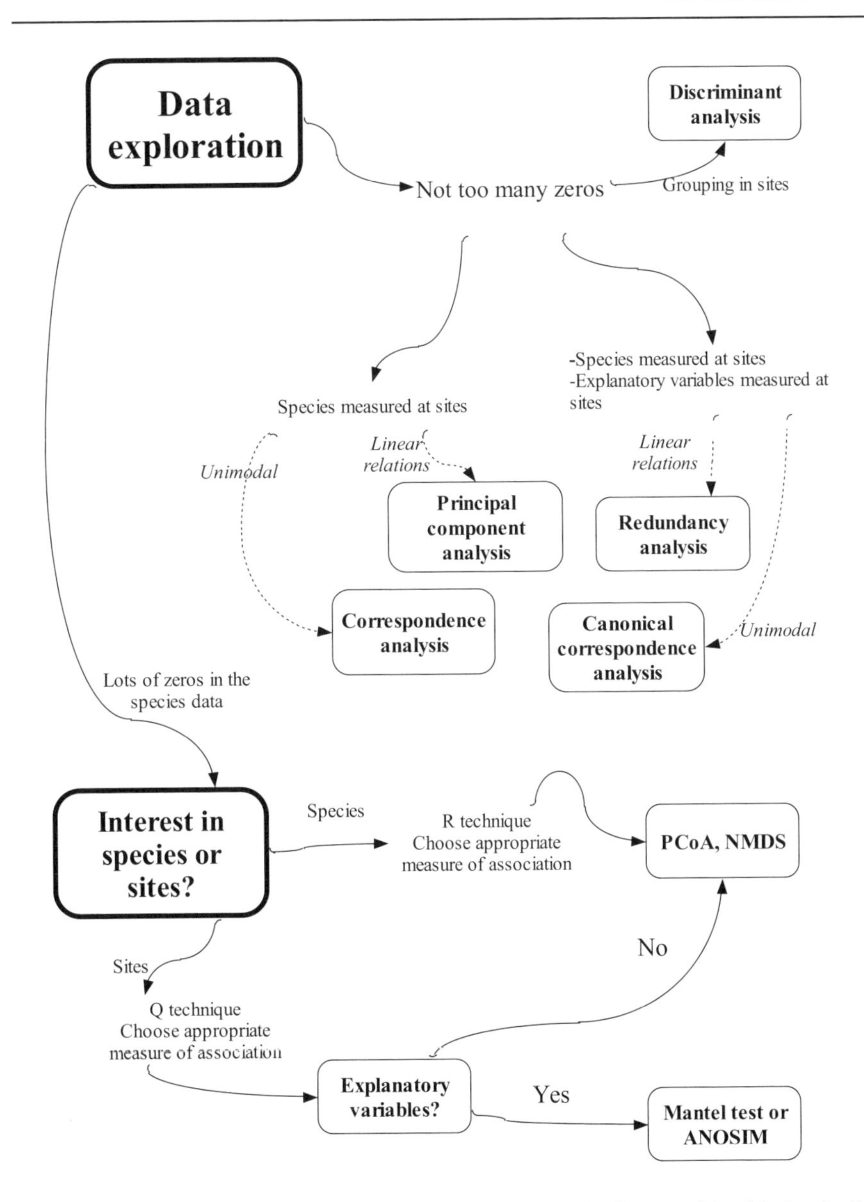

Figure 3.2. Overview of multivariate analysis methods discussed in this book. We assume that the data matrix consists of species measured at sites. If this is not the case (e.g., the data contain chemical variables), it becomes more tempting to apply principal component analysis, correspondence analysis, redundancy analysis or canonical correspondence analysis. If the species data matrix contains many zeros and double zeros, this needs to be taken into account, by choosing an appropriate association matrix and using principal co-ordinate analysis (PCoA), non-metric multidimensional scaling (NMDS), the Mantel test or ANOSIM.

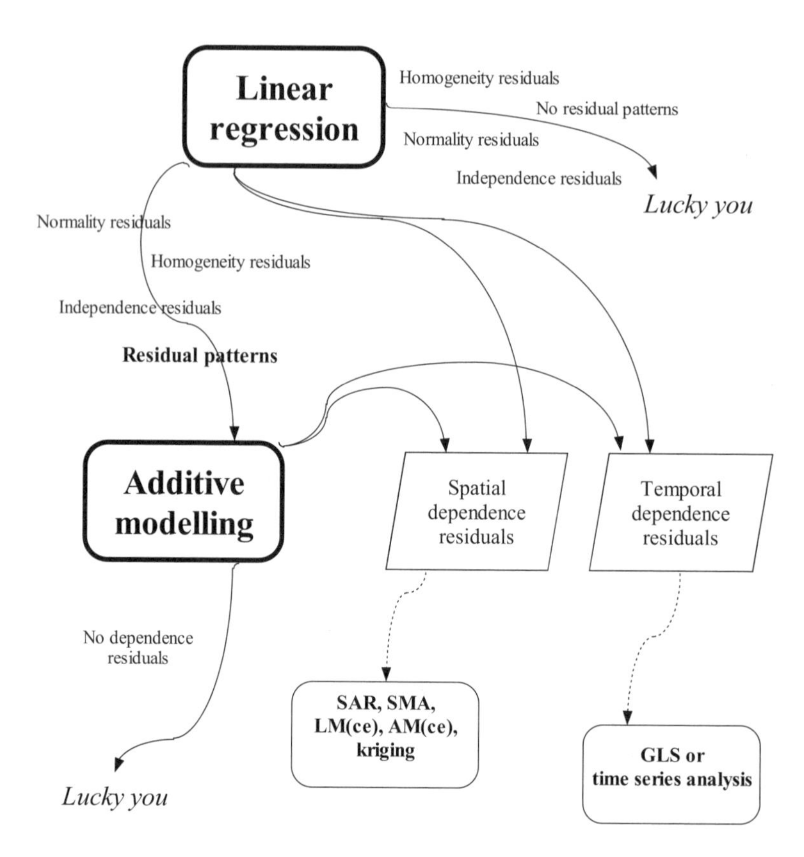

Figure 3.3. Decision tree to choose between linear regression models and additive models, and in case of temporal dependence GLS or time series methods can be applied. If there is spatial dependence in the additive or linear regression models, then SAR, SMA, LM(ce), AM(ce) or kriging techniques should be applied.

4 Exploration

The first step in analysing data is a graphical data exploration asking the following questions:

1. Where are the data centred? How are they spread? Are they symmetric, skewed, bimodal?
2. Are there outliers?
3. Are the variables normally distributed?
4. Are there any relationships between the variables? Are relationships between the variables linear? Which follow-up analysis should be applied?
5. Do we need a transformation?
6. Was the sampling effort approximately the same for each observation or variable?

We need to address all of these questions because the next step of the analysis needs the data to comply with several assumptions before any conclusions can be considered valid. For example, principal component analysis (PCA) depends on linear relationships between variables, and outlying values may cause non-significant regression parameters and mislead the analysis. Another example is large overdispersion in generalised linear modelling, which can also result in non-significant parameters. We therefore need a range of exploratory tools to address questions 1 to 6 with different tools aimed at answering different questions. For example, a scatterplot might suggest that a particular point is an outlier in the combined xy-space, but not identify it as an outlier within in the x-space or y-space if inspected in isolation. This chapter discusses a range of exploratory tools and suggests how they can be used to ensure the validity of any subsequent analysis. When looking at your data you should use all the techniques discussed and not rely on the results from a single technique to make decisions about outliers, normality or relationships.

Many books have chapters on data exploration techniques, and good sources are Montgomery and Peck (1992), Crawley (2002), Fox (2002a) and Quinn and Keough (2002). We have only presented the methods we find the most useful. Expect to spend at least 20% of your research time exploring your data. This makes the follow-up analysis easier and more efficient.

4.1 The first steps

Boxplots and conditional boxplots

A boxplot, or a box-and-whiskers plot (Figure 4.1), visualises the mean and spread for a univariate variable. Normally, the midpoint of a boxplot is the median, but it can also be the mean. The 25% and 75% quartiles (Q_{25} and Q_{75}) define the hinges (end of the boxes), and the difference between the hinges is called the spread. Lines (or whiskers) are drawn from each hinge to 1.5 times the spread or to the most extreme value of the spread, whichever is the smaller. Any points outside these values are normally identified as outliers. Some computer programmes draw the whiskers to the values covering most data points, such as 10% and 90% of the points, and show minimum and maximum values as separate points.

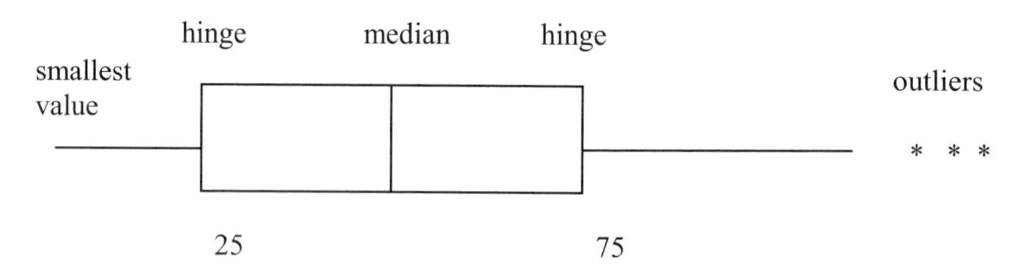

Figure 4.1. Boxplots show the middle of the sampled values, variability, shape of the distribution, outliers and extreme values.

The numbers below give the number of ragworms (*Laeoneris acuta*) recorded in an Argentinean salt marsh, and we use them to explain making a boxplot. The top row identifies the ranked sampling point, and the lower row gives the number of ragworm counted at that point.

					Q_{25}					M					Q_{75}					
1	2	3	4	5	**6**	7	8	9	10	**11**	12	13	14	15	**16**	17	18	19	20	21
0	0	0	1	2	**3**	6	7	9	11	**14**	14	14	16	19	**20**	21	24	27	35	121

The median value (denoted by **M**) for *L. acuta* is at observation number 11 (14 ragworms). The end points of the boxes in the boxplot are at Q_{25} and Q_{75}. Therefore, observation numbers 6 and 16 form the hinges. The spread for these data is 20 − 3 = 17, and 1.5 times the spread is 25.5. Adding 25.5 to the upper hinge of 20 (Q_{75}) allows the right line (or whisker) to be drawn up to 45.5. Observation number 21 (121 ragworms) would normally be considered as an extreme value or outlier. The resulting boxplot is given in Figure 4.2.

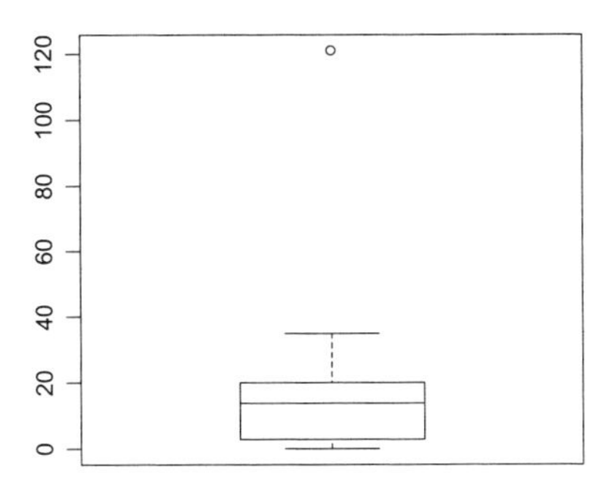

Figure 4.2. Boxplot for the ragworms (*L. acuta*). The upper hinge is calculated as having a value of 45.5, but as the most extreme value within this range is only 35, it is drawn at this latter point.

In Chapter 28, zoobenthic data from a salt marsh in Argentina are analysed. The data consist of measurements on four zoobenthic species in three transects. Each transect contained ten sites, and all sites were measured in Autumn and Spring, resulting in a 60-by-4 data matrix for the species data. Further details can be found in Chapter 28. Several boxplots for the species data are shown in Figure 4.3. Panel A in Figure 4.3 is a boxplot for the four zoobenthic species of the Argentinean zoobenthic dataset introduced in Chapter 2. It shows that some species have potential outliers, which prompted an inspection of the original data to check for errors in data entry. After checking, it was concluded that there were no data errors. However, the presence of outliers (or large observations) is the first sign that you may need to transform the data to reduce or down-weight its influence on the analysis. We decided to apply a square root transformation, and boxplots of the transformed data are shown in Figure 4.3-B. The reasons for choosing a square root transformation is discussed later. Note that the boxplots for the transformed data show that this has removed the outliers. The large number of dots outside the interval defined by 1.5 times the range might indicate a large number of zero observations for *Uca uruguayensis* and *Neanthes succinea*. This is called the double-zero problem, but how big a problem this is depends on the underlying ecological questions. If two variables have many data points with zero abundance, the correlation coefficient will be relatively large as both variables are below average at the same sites. This means that these two variables are identified as similar, only because they are absent at the same sites. It is like saying that butterflies and elephants are similar because they are both absent from the North Pole, the Antarctic and the moon. It sounds trivial, but the first few axes in a principal component analysis could be determined by such variables, and it is a common problem in ecological data of which we need to be aware.

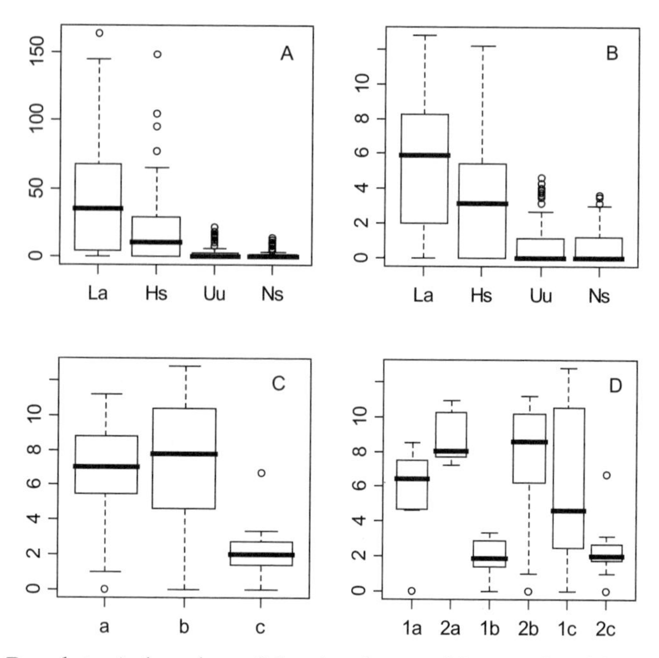

Figure 4.3. Boxplots. A: boxplots of the abundance of four zoobenthic species (La = *Laeonereis acuta*, Hs = *Heteromastus similis*, Uu = *Uca uruguayensis*, Ns=*Neanthes succinea*). B: boxplots of four square root transformed species. C: boxplot of square root transformed *L. acuta* conditional on the nominal variable transect with values a, b and c. D: Boxplot of square root transformed *L. acuta* conditional on season (1 = Autumn, 2 = Spring) and transect.

Boxplots are also useful to find relationships between variables. Panel C in Figure 4.3 shows the boxplot of square root transformed *L. acuta* abundance conditional on the nominal variable transect (a, b and c). It is readily seen that abundances are considerably lower in transect C. Panel D takes this one step further; the same species is now plotted conditional on season and transect. The first two boxplots from the left correspond to *L. acuta* from transect a in Autumn and Spring. Although this shows differences in abundances between the seasons, it also shows there appears to be no seasonal consistency between transects.

Depending on software, boxplots can be modified in various ways. For example, notches can be drawn at each side of the boxes. If the notches of two plots do not overlap, then the medians are significantly different at the 5% (Chambers et al. 1983). It is also possible to have boxplots with widths proportional to the square roots of the number of observations in the groups. Sometimes, it can be useful to plot the boxplot vertically instead of horizontally, where this might better visualise the characteristics of the original data.

Cleveland dotplot

Cleveland dotplots (Cleveland 1985) are useful to identify outliers and homogeneity. Homogeneity means that the variance in the data does not change along the gradients. Violation is called heterogeneity, and as we will see later, homogeneity is a crucial assumption for many statistical methods. Various software programmes use different terminology for dotplots. For example, with S-Plus and R, each observation is presented by a single dot. The value is presented along the horizontal axis, and the order of the dots (as arranged by the programme) is shown along the vertical axis. Cleveland dotplots for the abundance of four zoobenthic species of the Argentinean dataset are given in Figure 4.4. The 60 data points (30 sites in Spring and 30 sites in Autumn) are plotted along the vertical axes and the horizontal axes show the values for each site. Any isolated points on the right- or left-hand side indicate outliers, but in this dataset, none of the points are considered outliers. However, the dotplots also indicate a large number of zero observations, which can be a problem with some of the methods discussed in later chapters. Also note that the boxplots show that *L. acuta* and *H. similis* are considerably more abundant than the other two species. The dotplots were made using different symbols conditional on a nominal explanatory variable, which in this case is Transect. This means that data points from the same transect will have the same symbols. Note that *U. uruguayensis* has zero abundance along transect a in the Autumn (these are the bottom 10 data points along the *y*-axis); along transect c in the Autumn (these are the data points in the middle with a '+'); along transect a in the Spring (next 10 data points represented by 'o'); and along transect c in Spring (the upper 10 data points represented by '+'). Although we have not done it here, it would also be useful to make Cleveland dotplots for explanatory variables and diversity indeces.

You can also usefully compare boxplots with dotplots, as this can explain why the boxplot identified some points as 'outliers'. The boxplots and dotplots for the Argentinean zoobenthic data tell us that we have many zero observations, two species have larger abundances than the other species, there are no 'real' outliers, and there are differences in species abundances between transects and seasons.

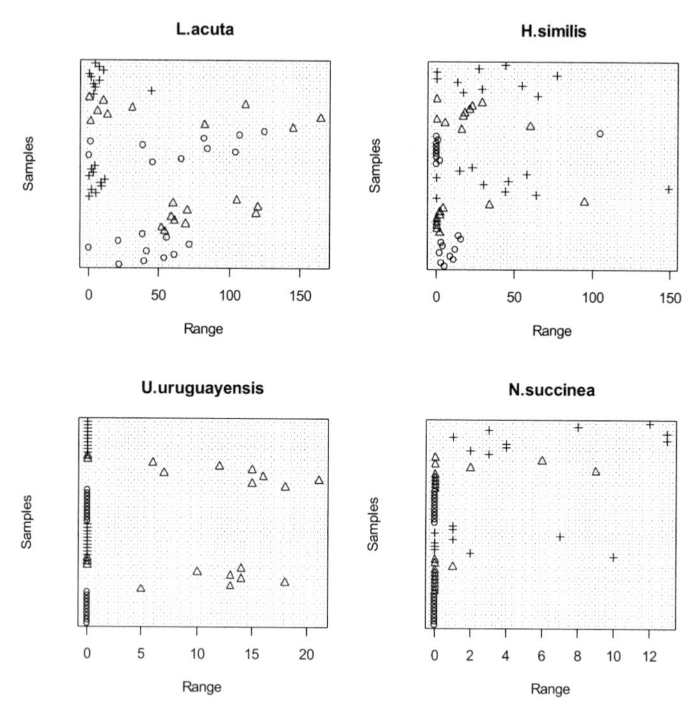

Figure 4.4. Dotplots for species of the Argentinean zoobenthic dataset. The horizontal axes show the value at each data point. The vertical axes represent the identity of the data points. The values at the top of the vertical axes are the data points at the end of the spreadsheet. It is also possible to group data points based on a nominal variable.

Histograms

A histogram shows the centre and distribution of the data and gives an indication of normality. However, applying a data transformation to make the data fit a normal distribution requires care. Panel A in Figure 4.5 shows the histogram for a set of data on the Gonadosomatic index (GSI, i.e., the weight of the gonads relative to total body weight) of squid (Graham Pierce, University of Aberdeen, UK, unpublished data). Measurements were taken from squid caught at various locations, months, and years in Scottish waters. The shape of the histogram shows bimodality, and one might be tempted to apply a transformation. However, a conditional histogram gives a rather different picture. In a conditional histogram the data are split up based on a nominal variable, and histograms of the subsets are plotted above each other. Panels B and C show the conditional histograms for the GSI index conditional on sex. Panel B shows the GSI index for female squid and Panel C for male squid. Notice that there is a clear difference in the shape and centre of the distribution. Hence, part of the first peak in panel A comprises mainly the male squid. This suggests the need to include a sex effect and interactions rather than transform the full dataset. We also suggest making conditional histograms on year, month and location for these data.

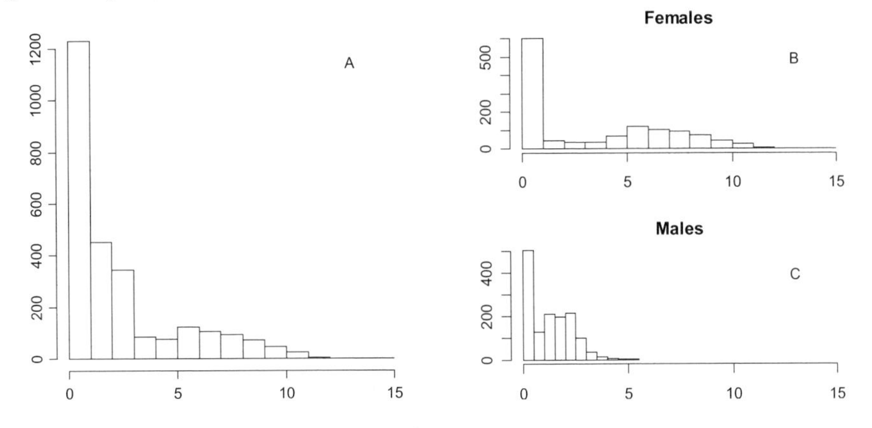

Figure 4.5. Histograms. A: histogram for GSI index of the squid data. B and C: conditional histograms for GSI index for the squid data. Panel B is for the female species and panel C for the male species.

QQ-plots

A Quantile-Quantile plot is a graphical tool used to determine whether the data follow a particular distribution. The QQ-plot for a normal distribution compares the distribution of a given variable to the Gaussian distribution. If the resulting points lie roughly on a straight line, then the distribution of the data is considered to be the same as a normally distributed variable. Let us discuss this in a bit more detail, as it can be quite confusing to understand what exactly it does. First, we need to revise some basic statistics. The pth quantile point q for a random variable y is given by $F(q) = P(y \leq q) = p$. If we want to know which q value belongs to the p, we write $q = F^{-1}(p)$. Suppose we have five observations Y_i with values 1, 2, 3, 4 and 5. We have sorted the observations from the smallest to the highest. By definition the first number is the 0% percentile, the middle is the 50% percentile and 5 is the 100% percentile. The difference between a quantile and percentile point is only a factor 100. QQ-plots are either based on these percentiles, or more typically they use the sample quantile points $(i - 0.5)/n$ where i is from 1 to 5 and $n = 5$ for this example. The sample quantile points for these data are 0.1, 0.3, 0.5, 0.7 and 0.9. These are the sample values for p. In the second step, we compare these sample quantile points with that of a normal distribution. This means that the density function used in $P(y \leq q)$ is now a normal density function and $F()$ is the corresponding normal cumulative distribution function. The QQ-plot is then a plot of the samples values Y_i versus q_i. Some software packages add a straight line to the plot, which is typically obtained by connecting the 25th and 75th quartile points.

It is useful to combine QQ-plots with a power transformation, which is given by

$$\frac{Y^p - 1}{p} \quad \text{if } p \text{ is not equal to 0} \qquad \text{and} \qquad \log(Y) \text{ if } p \text{ is 0} \tag{4.1}$$

Note that this p is not the p that we used for the quantiles. It is also useful to compare several QQ-plots for different values of p, and Figure 4.6 shows an example for the Argentinean data. In this example the square root transformation seems to perform the best, but this could be open to debate.

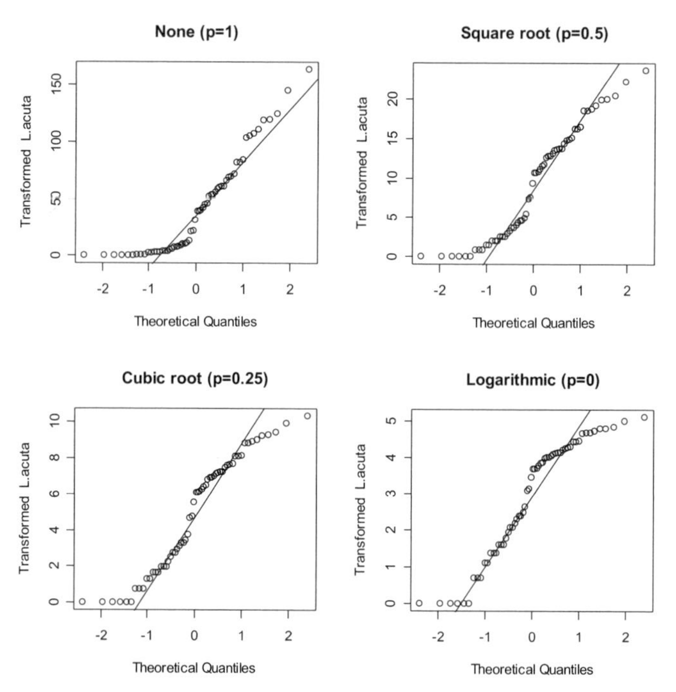

Figure 4.6. QQ-plots for the zoobenthic species *L. acuta* from the Argentinean zoobenthic dataset; for the untransformed data, square root transformed data, the cubic root transformed data, and $\log_{10}$ transformed data. In this example, the square root transformation seems to give the best results.

Scatterplot

So far, the main emphasis has been on detecting outliers, checking for normality, and exploring datasets associated with single nominal explanatory variables. The following techniques look at the relationships *between* variables.

A scatterplot is a tool to find a relationship between two variables. It plots one variable along the horizontal axis and a second variable along the vertical axis. To help visualise the relationship between the variables, a straight line or smoothing curve is often added to the plot. Figure 4.7 shows the pairplot for the variables biomass and length for the wedge clam *Donax hanleyanus*, measured on a beach in Buenos Aires province, Argentina (Ieno, unpublished data).

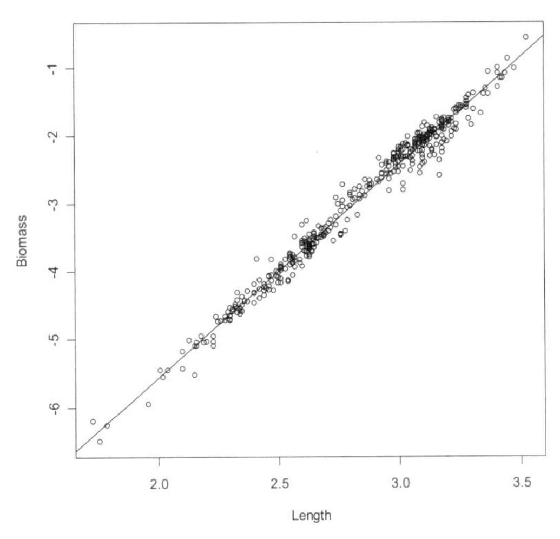

Figure 4.7. Scatterplot for biomass wedge clam dataset, using log transformed biomass, versus log transformed length.

Pairplot

If you have more than two variables, then a series of scatterplots can be produced: one for each pair of variables. However, the number of scatterplots required increases rapidly if you have more than three variables to explore. A better approach, for up to approximately 10 explanatory variables, is the pairplot, or scatterplot matrix (Figure 4.8). These show multiple pair-wise scatterplots in one graph and can be used to detect relationships between variables and to detect collinearity. The example in Figure 4.8 shows a pairplot for the response variable species richness and for four selected environmental variables. Species richness measures the different number of species per observation. The Decapoda zooplankton data form the basis for the case study in Chapter 20. Each panel is a scatterplot between two variables, with the labels for the variables printed in the panels running diagonally through the plot. A smoothing line has been added to help visualise the strength of the relationship. However, you can choose not to add a line, or you can add a regression line, whichever best suits the data. The pairplot in Figure 4.8 suggests a relationship between species richness (R) and temperature (T1m) and between species richness (R) and chlorophyll a (Ch). It also shows some collinearity between salinity at the surface (S1m) and at 35-45 meters (S45_35). Collinearity means that there is a high correlation between explanatory variables.

Figure 4.9 shows another pairplot for the same dataset where all the available explanatory variables have been plotted. The differences between this graph and the previous pairplot is that correlation coefficients between the variables are printed in the lower part of the graph. Note that there is strong collinearity between some of the variables, for example temperature at 1 m and temperature at 45 m.

Pairplots should be made for every analysis. These should include (i) a pairplot of all response variables (assuming that more than one response variable is available); (ii) a pairplot of all explanatory variables; and (iii) a pairplot of all response *and* explanatory variables. The first plot (i) gives information that will help choose the most appropriate multivariate techniques. It is hoped that the response variables will show strong linear relationships (some techniques such as PCA depend on linear relationships). However, if plot (ii) shows a clear linear relationship between the explanatory variables, indicating collinearity, then we know we have a major problem to deal with before further analysis. With plot (iii) we are judging whether the relationships between the response variables and the explanatory variables are linear. If this is not the case, then several options are available. The easiest option is to apply a transformation on response and/or explanatory variables to linearise the relationships. Other options are discussed later in this chapter.

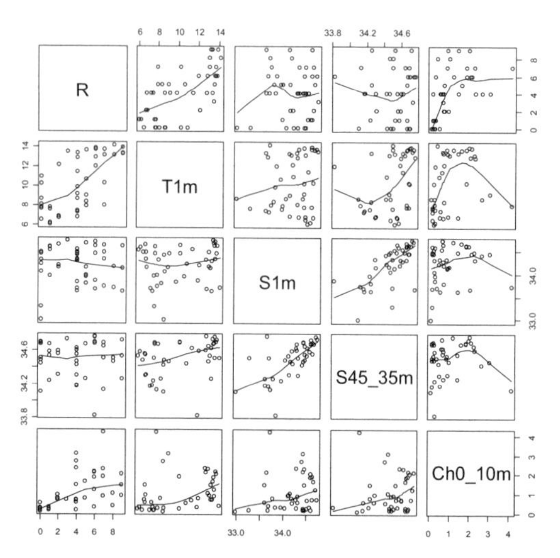

Figure 4.8. Pairplot for the response variable species richness and four selected environmental variables for the Decapoda zooplankton data. The pairplot indicates a linear relationship between richness and temperature. Each smoothing line is obtained by using one variable as the response variable and the other as an explanatory variable in the smoothing procedure. The difference between the smoothing lines in two corresponding graphs above and below the diagonal is due to what is used as the response and explanatory variable in the smoothing method, and therefore, the shape of the two matching smoothers might be different.

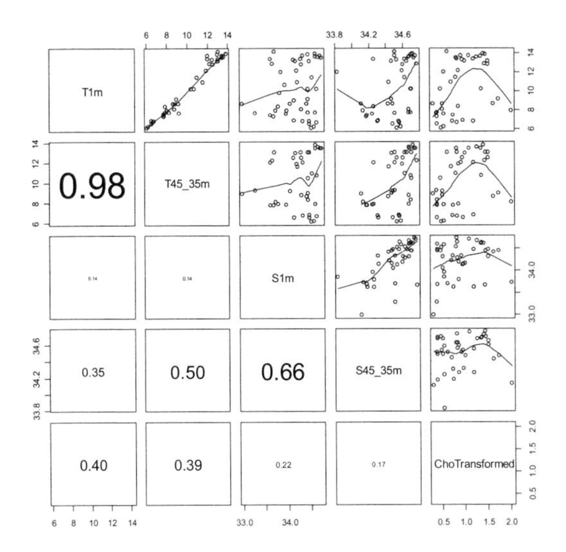

Figure 4.9. Pairplot for all environmental variables in the Decapoda zooplankton data. The lower diagonal part shows the (absolute) correlation coefficient and the upper diagonal part the scatterplots. The font size of the correlation is proportional to its size. There is strong collinearity between some of the variables, e.g., temperature at 1 m and temperature at 45 m.

Coplot

A coplot is a conditional scatterplot showing the relationship between y and x, for different values of a third variable z, or even a fourth variable w. The conditioning variables can be nominal or continuous. Figure 4.10 shows an example for the RIKZ data (Chapter 27). It is a coplot of the species richness versus NAP (which represents the average sea level height at each site), conditional on the nominal variable week. The panels are ordered from the lower left to the upper right. This order corresponds to increasing values of the conditioning explanatory variable. The lower left panel shows the relationship between NAP and the richness index for the samples measured in week 1, the lower right for the week 2 samples, the upper left panel for the week 3 samples, and the upper right panel for the week 4 samples. We did not add a regression line because in the fourth week, only 5 samples were taken. The richness values in week 1 are larger than in weeks 2 and 3, but the NAP range is smaller in week 1.

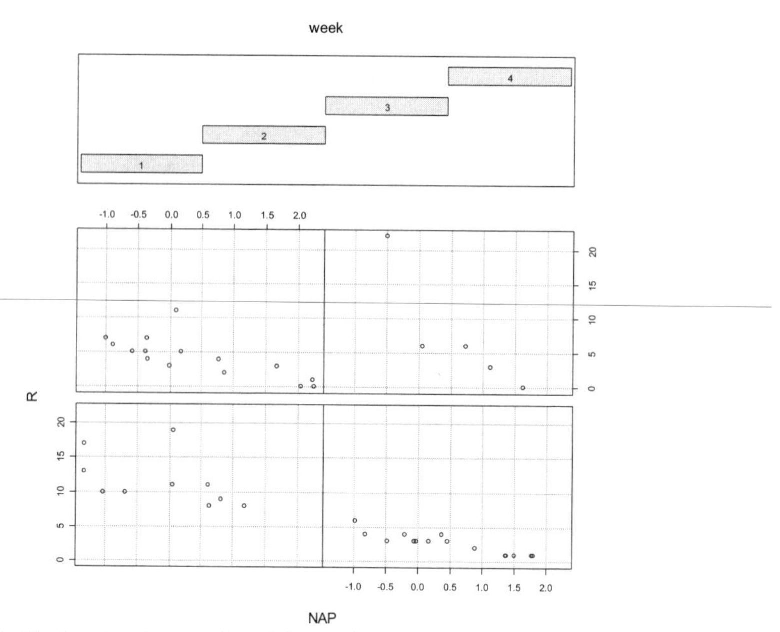

Figure 4.10. Coplot for the species richness index function of the RIKZ data versus NAP, conditional on week. The lower left panel corresponds to week 1, the lower right to week 2, the upper left to week 3, and the upper right for week 4. Note that the number of observations is different for each week.

For nominal variables, as shown in Figure 4.10, there is no overlap in ranges of the conditional variable. For continuous conditioning variables, we can allow for some overlap in the ranges of the conditioning variables, and the number of graphs, as well as the amount of overlap can be modified. This is illustrated in Figure 4.11, which shows another coplot for the RIKZ data. Each panel shows the relationship between the species richness index function and the explanatory variable NAP for a different temperature range. The lower left panel shows sites with temperatures between 15.5 and 17.5 degrees Celsius, and the upper right graph for temperature of 20 degrees and higher. The other panels show a range of different temperature bands between these two extremes. In this instance we have included smoothing curves (Chapter 7) and these highlight a negative relationship between species richness and NAP for all the measured temperature regimes. As this relationship between richness and NAP is common across all the temperature regimes, it suggests that it is not being influenced by temperature. Knowing that a specific variable is unrelated to the response variable is just as important as knowing that it is.

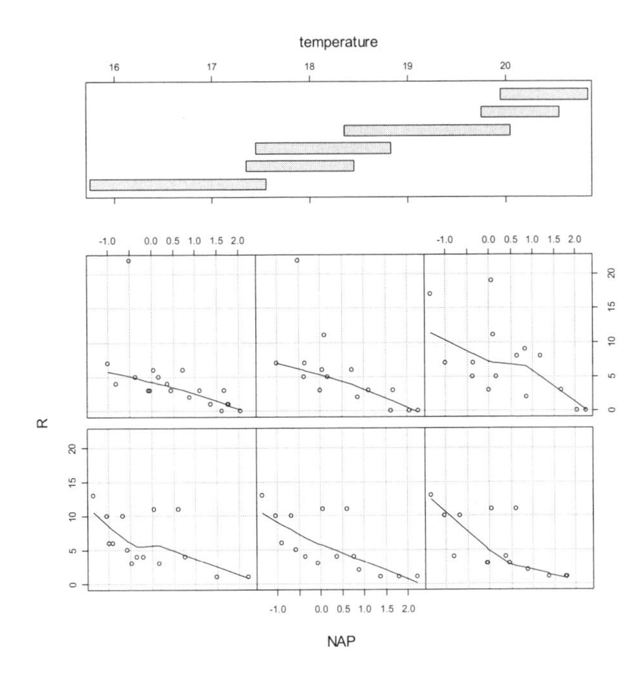

Figure 4.11. Coplot for RIKZ data where NAP is plotted against species richness for different temperature regimes.

Lattice graphs

Another useful tool are lattice graphs (called Trellis graphs in S-Plus). Like coplots these graphs show relationships between two variables, conditional on nominal variables. Lattice graphs have the advantage over coplots because they can work with larger numbers of panels. However, the conditional factor must be nominal. In coplots the conditional factor can be nominal or continuous. We use lattice graphs for time series data exploration and, to a lesser extent, to investigate sampling effort. Unless there are good reasons for deciding otherwise, you should normally use the same sample size and sampling effort across all the explanatory variables. Figure 4.12 shows a lattice graph for the squid data. Each panel shows the relationship between the GSI index and month. The conditional variable is area, and the plots clearly show an unbalance in the sampling effort. In some areas sampling largely took place in one month. Obviously, care is needed if these data were to be analysed in a regression model containing the nominal variables month and area.

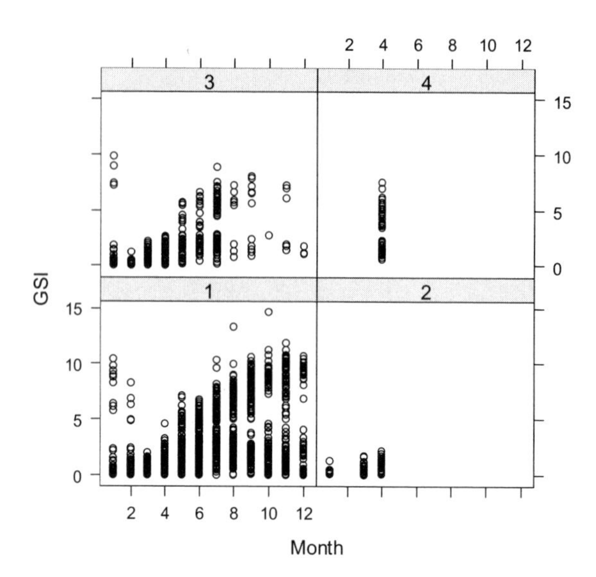

Figure 4.12. GSI index (vertical axis) versus month (horizontal axis) conditional on area (different panels) for the squid data. Note the unbalanced design of the data.

Design and interaction plots

Design and interaction plots are another valuable tool for exploring datasets with nominal variables and are particularly useful to use before applying regression, GLM, mixed modelling or ANOVA. They visualise (i) differences in mean values of the response variable for different levels of the nominal variables and (ii) interactions between explanatory variables. Figure 4.13 shows a design plot for the wedge clam data introduced earlier in this chapter. For these data there are three nominal variables: beach (3 beaches), intertidal or subtidal level on the beach (2 levels) and month (5 months). The design plot allows a direct comparison of the means (or medians) of all the nominal variables in a single graph. The graphs indicate that the mean value of the number of clams for beach 1 is around 0.26, with the mean values at the other two beaches considerably lower. It can also be seen that months 2 and 5 have relatively high mean values. However, the design plot shows little about the interaction *between* explanatory variables, and for this, we use an interaction plot (Figure 4.14). Panel A shows the interaction between month and beach. Mean values at beach 1 are rather different compared with beaches 2 and 3. It also shows that the interaction between season (month) and the mean clam numbers is similar for beaches 1 and 2, but very different for beach 3. Panel B shows the interaction between month and level, with mean values at level 1 in month 5 considerably larger than in the other levels.

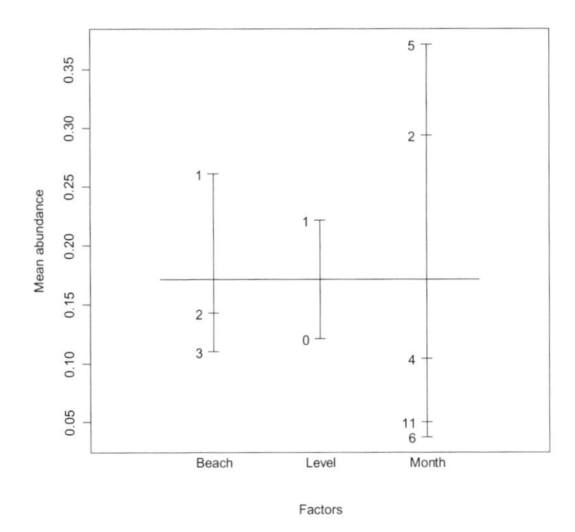

Figure 4.13. Design plot for the wedge clam data. The vertical axis shows the mean value per class for each nominal variable.

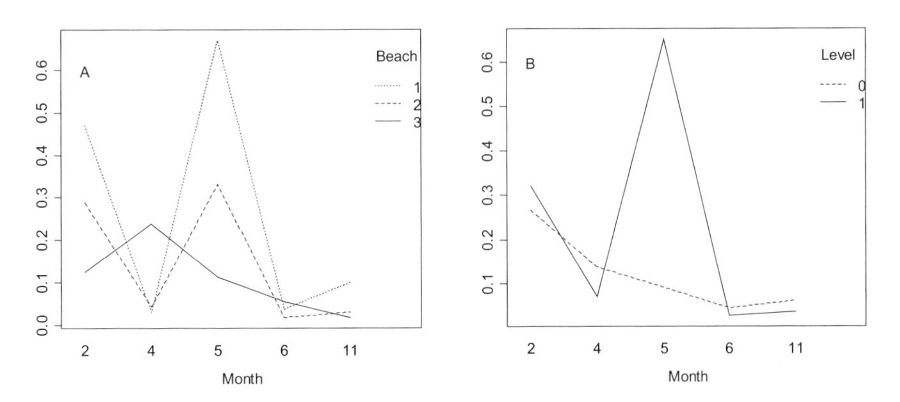

Figure 4.14. Design plot for the wedge clam data. The vertical axis shows the mean value and the horizontal axis the month. A: interaction between month and beach. B: interaction between month and level.

4.2 Outliers, transformations and standardisations

Outliers

An outlier is a data point that, because of its extreme value compared to the rest of the dataset, might incorrectly influence an analysis. So the first question is: 'how can we identify an outlier?' A simple approach might be to quantify everything as an outlier that is beyond a certain distance from the centre of the data. For example, the points outside the hinges of a boxplot could be considered as outliers. However, the dotplots and boxplots for the Argentinean data (Section 4.1) show that this is not always a good decision. Two-dimensional scatterplots can also highlight observations that may be potential outliers. For example, Figure 4.15 is a scatterplot for the variables NAP and species richness for the RIKZ data. These data are analysed in Chapter 27. The data consist of abundance of 75 zoobenthic species measured at 45 sites. NAP represents the height of a site compared with average sea level. The two observations with richness values larger than 19 species are not obviously outliers in the NAP (x) space. Although these sites have large richness values, they are not different enough from the other data points, to consider them extreme or isolated observations. However, as we will see in Chapter 5, these two observations cause serious problems in the linear regression for these data. So, although an observation is not considered an outlier in either the x-space or the y-space, it can still be an outlier in the xy-space. The situation that an observation is an outlier in the x-space, and also in the y-space, but not in the xy-space, is possible as well. A boxplot for the data presented in Figure 4.16 suggests that point A in the left panel would be an outlier in the y-space, but not in the x-space. However, fitting a linear regression line clearly identifies it as a highly influential observation. So, it is also an outlier in the xy-space. Point B is an outlier in the x-space *and* in the y-space, but not in the xy-space as it would not cause any major problems for a linear regression model. Point C is an outlier in the xy-space as it would strongly influence a regression. The right panel in Figure 4.16 shows a more serious problem: including point A in a linear regression or when calculating a correlation coefficient will show a strong positive relationship, whereas leaving out point A will give a strong negative relationship.

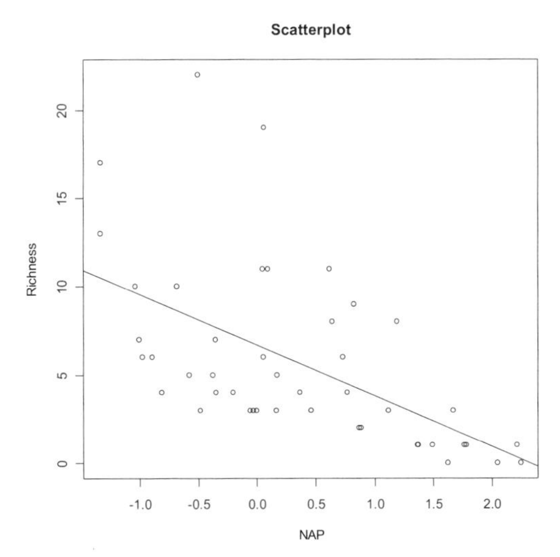

Figure 4.15. Scatterplot of species richness versus NAP for the RIKZ data. The two sites with high species richness are extreme with respect to the overall NAP-richness relationship.

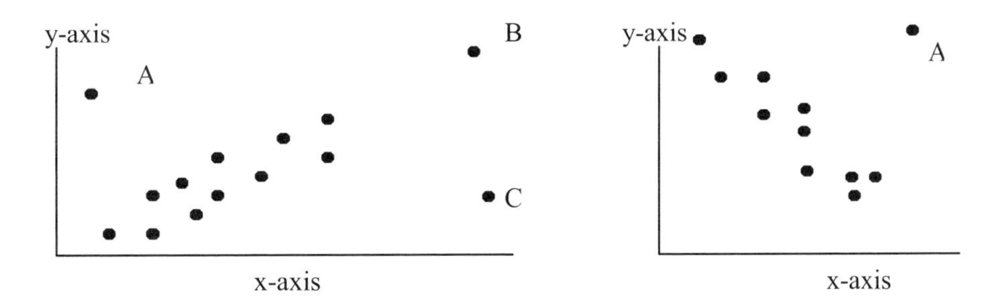

Figure 4.16. Left panel: scatterplot with two outliers. Right panel: scatterplot with 1 outlier.

Transformation

There are many reasons for transforming data, but it is normally because you have data with extreme outliers and non-normal distributions. Data transformation (on the response variables) will also be required when you plan to use discriminant analysis and there is clear evidence (e.g., by using a Cleveland dotplot) of heterogeneity.

Both the response variables and the explanatory variables can be transformed, and different types of transformations can be applied to different variables within the same dataset. Choosing the 'correct' transformation can be difficult and is usually, as least in part, based on experience. Additionally, the choice of transformation is influenced by the choice of follow-up analysis. For some techniques, such as classification or regression trees, the transformation of the explanatory variables makes no difference to the results. However, most techniques may require some transformation of the raw data before analysis.

The easiest problem to solve is where the extreme observations identified during the data exploration stage turn out to be typing errors. However, we will assume that this easy solution is not available and that we have a dataset with genuine extreme observations. If these extreme observations are in the explanatory variables, then a transformation of the (continuous) explanatory variables is definitely required, especially if regression, analysis of covariance, GLM, GAM or multivariate techniques like redundancy analysis and canonical correspondence analysis are applied. When the extreme observations are in the response variable, there is more than one approach available. You can either transform the data or you can apply a technique that is slightly better in dealing with extreme values, such as a GLM or a GAM with a Poisson distribution. The latter only works if there is an increase in spread of the observed data for larger values. Alternatively, quasi-Poisson models can be used if the data are overdispersed. Note, you should not apply a square root or log transformation on the response variable, and then continue with a Poisson GLM model, as this applies a correction twice. Yet, another option is to use dummy explanatory variables (Harvey 1989) to model the extreme observations. A more drastic solution for extreme observations is to simply omit them from the analysis. However, if you adopt this approach, you should always provide the results of the analysis with, and without, the extreme observations. If the large values are all from one area, or one month, or one sex, then it may be an option to use different variance components within the linear regression model, resulting in generalised least squares (GLS).

As an example, we will assume the aim is to carry out a linear regression. The Cleveland dotplot or boxplot indicate that there are no outliers of any concern, but the scatterplot of a response and explanatory variable shows a clear non-linear relationship. In this case, we should consider transforming one or both variables. But, which transformation should we use? The range of possible transformations for the response and explanatory variables can be selected from

$$..., y^{\frac{1}{4}}, y^{\frac{1}{3}}, y^{\frac{1}{2}}, y, \log(y), y^2, y^3, y^4, ...$$

These transformations can be written in one formula, namely the Box–Cox power transformation; see also equation (4.1). It is basically a family of transformations and they can only be applied if the data are non-negative, but a constant can be applied to avoid this problem. Alternative transformations are ranking and converting everything to 0–1 data. For example, if the original data have the values 2, 7, 4, 9, 22, and 40, the rank transformed data will be 1, 3, 2, 4, 5, 6. If the original data are 0, 1, 3, 0, 4, 0, and 100, then converting everything to 0–1 data gives 0 1 1 0 1 0 1. Converting to 0–1 data seems like a last resort, particularly if considerable expense and time has been spent collecting more detailed data. However, our experience shows this is often the only realistic option with difficult ecological datasets where the sample size or sampling quality is less than ideal. This is a common problem with zoobenthic species community and fisheries data and an example is given in the *Solea solea* case study in Chapter 21.

Several strategies are available for choosing the most appropriate transformation. The first approach is trial and error. Using the graphical data exploration techniques discussed earlier, especially the Cleveland dotplots, boxplots and pairplots, you can apply what appears to be the best transformation and see how well it corrects the identified issues. This is our preferred option, but it does require some existing expertise. It is also important that this trial-and-error approach is fully reported when presenting the results, including details of both the unsuccessful as well as the successful transformation approaches.

When the follow-up analysis is based on linear relationships, then a useful tool is the Mosteller and Tukey's bulging rule (Mosteller and Tukey 1977, Fox 2002a), which is from the Box–Cox family of transformations. This approach relies on identifying non-linear patterns in your data by inspecting a scatterplot. The required transformations for linearising the relationships can be inferred from the bulging rule illustrated in Figure 4.17. For example, if the scatterplot shows a pattern as in the upper right quadrant, then either the y's or the x's need to be increased to transform the data to linear.

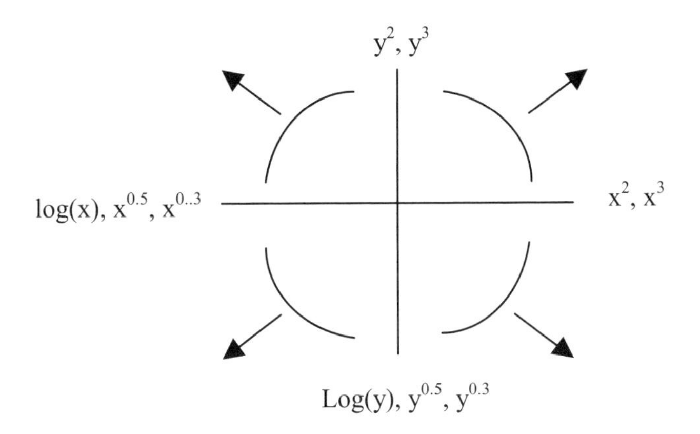

Figure 4.17. Mosteller and Turkey's bulging rule. When the arrow points downwards, y should be made smaller; if it points upwards, it should be increased. If the arrow points towards the left, x should be made larger, etc. See also Fox (2002a).

An example of the bulging rule is presented in Figure 4.18. Panel A shows a scatterplot of length versus biomass for the untransformed wedge clam data. This pattern matches with the lower right quadrant in Figure 4.17, and therefore the bulging rule suggests transforming either length to Length^2 (or higher powers) or taking the log or square root of biomass (panel B and C). Panel D suggests that transforming both length and biomass is the best option.

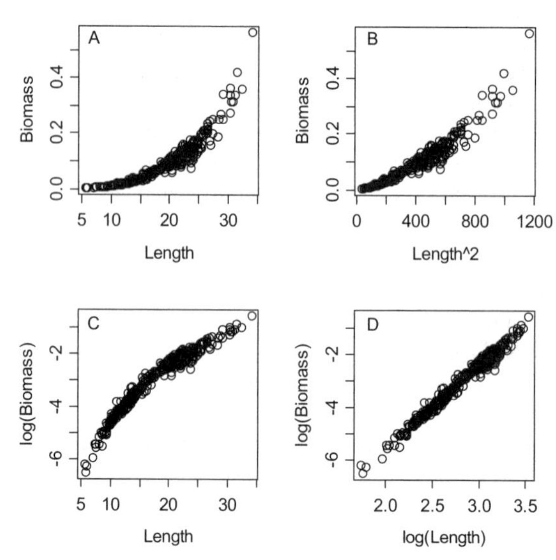

Figure 4.18. Scatterplot of (A) length versus biomass for the wedge clam data, (B) squared length versus biomass, (C) length versus log transformed biomass and (D) log length versus log biomass. The transformations shown in panels B and D follow those suggested by Mosteller and Tukey's bulging rule (Figure 4.17). Length-weight relationships typically require log-log transformations.

Automatic selection of transformations

Automatic transformation selection techniques are described by Montgomery and Peck (1992) and Fox (2002a), among others. Montgomery and Peck apply a series of power transformations, and for each power transformation, they calculate the residual sum of squares. These sums of squares cannot be compared directly as reducing the size of the data with a square root transformation, in most cases, also makes the residuals, and therefore residual sum of squares, smaller. Therefore, this power transformation contains a correction factor using the geometric mean that makes the residual sum of squares directly comparable. So, the challenge is to find the optimal value of p, where p defines the transformation in equation (4.1). Using a grid (range of values), and then increasingly finer grids, if required, the optimal value for p can be found as the one that has the smallest residual sum of squares. It is also possible to calculate a confidence interval for the power transformation parameter p. If this interval contains 1, then no transformation is required. This modified power transformation is

$$Y^p = \begin{cases} \dfrac{Y^p - 1}{p\dot{Y}^{p-1}} & \text{if } p \neq 0 \\ Y^p = \dot{Y}\ln(Y) & \text{if } p \neq 0 \end{cases} \tag{4.2}$$

$$\text{where } \dot{Y} = \exp(\frac{1}{n}\sum_{i=1}^{n} Y_i)$$

The difference between this formula and equation (4.1) is the correction factor $\dot{Y}$, also called the geometric mean. The confidence interval for p can be found by:

$$SS^* = SS_p(1 + \frac{t^2_{\alpha/2,v}}{v}) \tag{4.3}$$

where v is the residual degrees of freedom (Chapter 5) and SS_p is the lowest residual sum of squares. The confidence interval for p can be visualised by making a plot of various p values against SS_p, and using SS^* to read off the confidence bands. An example of using the biomass and log transformed length variables from the wedge clam dataset is given below. The log transformation for length was used as it contains various observations with rather large values. We are trying to find out which transformation for the biomass data is most optimal for fitting a linear regression model. Initially, the values of p were chosen to be between –3 and 3 with steps of 0.01. However, this was unsatisfactory and a finer grid of p was needed with values between –0.1 and 0.1. Figure 4.19 shows the sum of squares plotted against different values of p. The optimal value is $p = 0.025$ (lowest point on the curve), and SS^* is represented by the dotted line that allows us to read off the confidence intervals from where it intersects the curve. The 95% confidence band for p is therefore approximately between 0.005 and 0.035. Although 0, which is the log transformation by definition, is just outside this interval, in this instance for ease of interpretation a log transformation would probably be the best option.

Although we have only looked at transforming the response variable Montgomery and Peck (1992) also give a procedure for automatic selection of the transformation on the explanatory variable.

In conclusion, the main reasons for a data transformation are (in order of importance) as follows:

1. Reduce the effect of outliers.
2. Improve linearity between variables.
3. Make the data and error structure closer to the normal distribution.
4. Stabilise the relationship between the mean and the variance (this will be discussed further in Chapter 6).

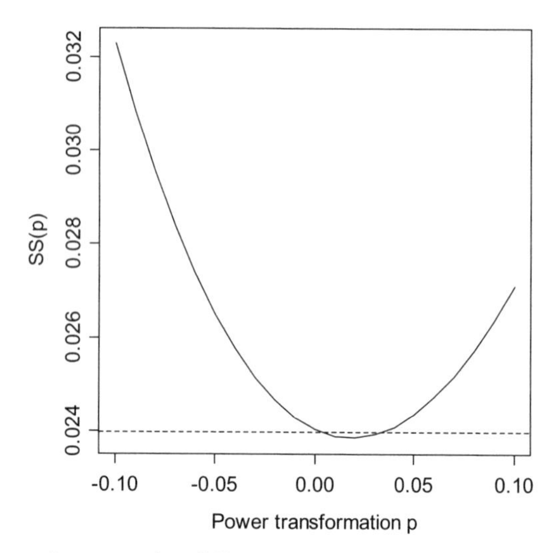

Figure 4.19. Sum of squares for different power transformations. The vertical axis shows the sum of squares for different power transformations p. The dotted line represents the 95% confidence band.

If the follow-up analysis is a generalised additive model with Poisson distribution, points 2 to 4 are irrelevant (Chapter 7). Although there are some rules of thumb for transformations, such as using a cubic or square root transformation for count data, and a log transformation where the relationships are multiplicative, it is still difficult to choose the best option. We suggest the following approach:

- Apply all data exploration techniques on the original data.
- If there are outliers in the explanatory variables, transform them.
- Apply linear regression or related techniques (e.g., GLM, analysis of variance), and judge whether the residuals show any patterns.
- If there is any residual information left, or if there are influential observations, then a data transformation might be an option.
- Choose the best transformation using trial and error, or use an automatic selection routine.

Unless replicates of the response variable are available, we believe it is unwise to apply a transformation purely on the argument that the 'response variable must be normally distributed'. The normality assumption is for the data at each X value (this will be discussed further in Section 5.1)! For example, a transformation on the GSI index for the squid data might remove the differences between male and female species (Figure 4.5). And normality of the explanatory variables is not assumed at all!

Other points to consider are whether to use the same transformation for (i) all response variables, (ii) all explanatory variables, or (iii) all response variables and all explanatory variables. In general, we recommend applying the same transformation to all response variables, and the same transformation to all explanatory variables. However, you can apply a different transformation to the response variables from the one applied to the explanatory variables. Sometimes, the explanatory variables represent different types of variables; e.g., if some are distance or size related, some are time related and some are nominal. In this case, there is nothing wrong in using a different transformation for each type of variable. Nominal explanatory variables should not be transformed, but distance and size-related variables tend to have a wider spread and might require a transformation. And the same approach should be adopted with response variables. For example, in an EU-funded project (WESTHER) on herring, biological, morphometric, chemical, genetic and parasite variables were measured, and were all considered as response variables. The parasite data were count data and required a square root transformation. The morphometric data were pre-filtered by length and did not require any further transformation, but the chemical data required a log transformation.

A final word on transformation is to be aware that, sometimes, the aim of the analysis is to investigate the outliers, e.g., the relationship between high water levels and the height of sea defences, or the analysis of scarce, and therefore only rarely recorded, species. In these cases, you cannot remove the extreme values, and the choice of analytical approach needs to take this into account. Useful sources on extreme values modelling are Coles (2004) and Thompson (2004), which both discuss sampling species that are rare and elusive.

Standardisations

If the variables being compared are from widely different scales, such as comparing the growth rates of small fish species against large fish species, then standardisation (converting all variables to the same scale) might be an option. However, this depends on which statistical technique is being used. For example, standardising the response variables would be sensible if you intend on using dynamic factor analysis (Chapter 17), but methods like canonical correspondence analysis and redundancy analysis (Chapters 12 and 13) apply their own standardisation before running the analysis. To make it more confusing, applying multidimensional scaling (Chapter 10) with the Euclidean distance function on standardised data is acceptable, but the same standardised data will give a problem if the Bray–Curtis distance function is used. There are several methods for converting data to the same scale, and one option is to centre all variables around zero by

$$Y_i^{new} = Y_i - \bar{Y}$$

where $\bar{Y}$ is the sample mean and Y_i the value of the ith observation. However, the most common used standardisation is given by:

$$Y_i^{new} = (Y_i - \bar{Y}) / s_y$$

where s_y is the sample standard deviation. The transformed values Y_i^{new} are now centred around zero, have a variance of one, and are unit-less. This transformation is also called normalisation. Other, less-used transformations are

$$Y_i^{new} = Y_i / Y_{max} \quad \text{and} \quad Y_i^{new} = (Y_i - Y_{min}) / (Y_{max} - Y_{min})$$

They rescale the data between zero and one. Centering or standardisation can be applied on response and/or explanatory variables. To illustrate the difference between no transformation, centring and normalisation, we use a North-American sea surface temperature (SST) time series. These data come from the COADS datasets (Slutz et al. 1985, Woodruff et al. 1987), and details on obtaining the mean monthly values used here can be found in Mendelssohn and Schwing (2002). The upper left panel in Figure 4.20 shows lattice graphs for four time series from this dataset. In the upper right panel, all series are centred around zero and are in their original units. This transformation takes away the differences in absolute value. However, there is still a difference in variation; the fluctuation in the upper left and lower right panels is considerably smaller. The standardisation (or: normalisation) removes these differences (lower left panel in Figure 4.20), and the amount of fluctuation becomes similar. This transformation is the most commonly used approach and rescales all time series around zero. The time series are now without units, and the normalisation makes all time series equally scaled, even if some time series had large fluctuations and others only small fluctuations. Centring only removes the differences in absolute values between series.

As with other transformations the decision to standardise your data depends on the statistical technique you plan to use. For example, if you want to compare regression parameters, you might consider it useful to standardise the explanatory variables before the analysis, especially if they are in different units or have different ranges. Some techniques such as principal component analysis automatically normalise or centre the variables.

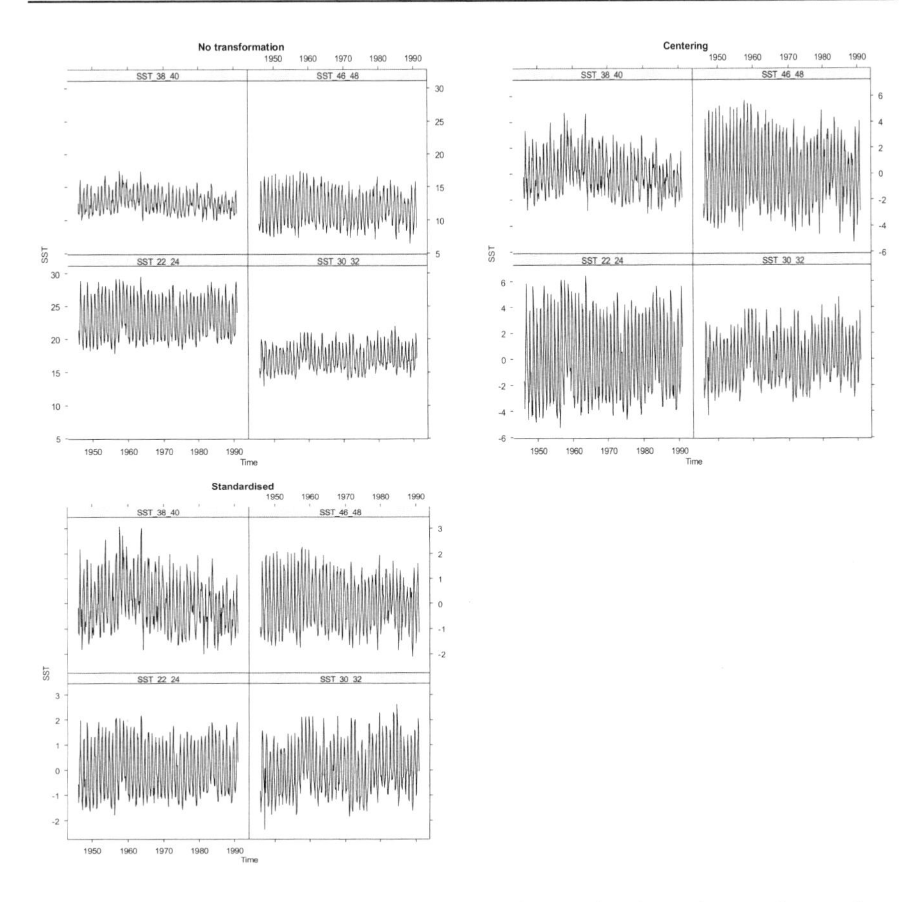

Figure 4.20. The upper left panel shows four time series from the North American SST dataset with no transformation applied. The upper right panel shows the same time series centred, and the lower left shows the same time series normalised.

4.3 A final thought on data exploration

Even if you don't see it, it might still be there

Even if the scatterplots suggest the absence of a relationship between Y and X, this does not necessarily mean one does not exist. A scatterplot only shows the relationship between two variables, and including a third, fourth or even fifth variable might force a different conclusion. To illustrate this, we have used the GSI index of the squid data again (Section 4.1). The left panel in Figure 4.21 shows the scatterplot of month against the GSI index. The most likely conclusion based on this graph is that that there is no strong seasonal effect in the GSI index. However, using a four-dimensional scatterplot, or coplot (right panel in Figure 4.21), a strong seasonal pattern is apparent for female squid in areas 1 and 3, and a weak seasonal pattern for males in area 3.

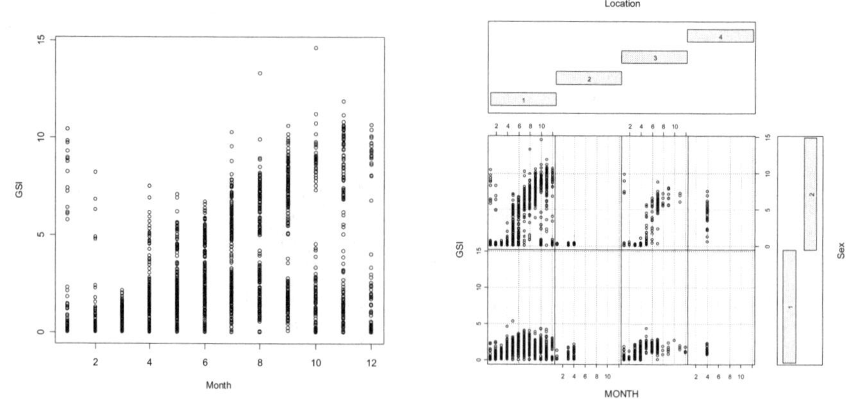

Figure 4.21. Left: scatterplot of GSI data. Right: coplot of GSI index for the squid data. The conditioning variables are Location (four areas) and Sex (1 is male, 2 is female).

This final thought here is to re-enforce the argument that a thorough data exploration stage is essential before moving on to the analysis stage of the data investigation.

What Next?

After completing the data exploration, the next step is to verify and investigate the patterns and relationships this step identified. Assuming the scatterplot indicates a linear relationship between the variables, then linear regression is the obvious next step. However, if the scatterplot suggests a clear non-linear pattern, then a different approach needs to be taken, which might include (i) using interactions and/or quadratic terms in the linear regression model, (ii) transforming the data, (iii) continuing with a non-linear regression model, (iii) using generalised linear modelling, (iv) applying generalised additive modelling techniques, or (v) applying (additive) mixed modelling techniques. All these approaches are investigated in later chapters. The first option means that you proceed with the linear regression model, but you need to ensure that all assumptions are met (e.g. no residual patterns).

To choose which approach is the most appropriate requires knowledge of the assumptions of the selected methods, and tools to detect violations (using residuals). These are all discussed in later chapters, but basically it all comes down to something very basic: learn from your errors.

5 Linear regression

In Chapter 4, we used various graphical tools (Cleveland dotplots, boxplots, histograms) to explore the shape of our data (normality), look for the presence of outliers, and assess the need for data transformations. We also discussed more complex methods (coplot, lattice graphs, scatterplot, pairplots, conditional boxplots, conditional histograms) that helped to see the relationships between a single response variable and more than one explanatory variable. This is the essential first step in any analysis that allows the researcher to get a feel for the data before moving on to formal statistical tools such as linear regression.

Not all datasets are suitable for linear regression. For data with counts or presence–absence data, generalised linear modelling (GLM) is more suitable. And where the parametric models used by linear regression and GLM give a poor fit, non-parametric techniques like additive modelling and generalised additive modelling (GAM) are likely to give better results. In this book we look at a range of tools suitable for analysing the univariate data commonly found in ecological or environmental studies, including linear regression, partial linear regression, GLM, additive modelling, GAM, regression and classification trees, generalised least squares (GLS) and mixed modelling. Techniques like GLM, GAM, GLS and mixed modelling are difficult to understand and even more difficult to explain. So we start by briefly summarising the underlying principles of linear regression, as this underpins most of the univariate techniques we are going to discuss. However, if you feel the need for more than a brief refresher then have a look at one of the many standard texts. Useful starting points are Fowler et al. (1998) and Quinn and Keough (2002), with more detailed discussions found in Montgommery and Peck (1992) and Draper and Smith (1998).

5.1 Bivariate linear regression

In Chapter 27 a detailed analysis of the RIKZ data is presented. Abundances of around 75 invertebrate species from 45 sites were measured on various beaches along the Dutch coast. In this study, the variable "NAP" measured the height of the sample site compared with average sea level, and indicated the time a site is under water. A site with a low NAP value will spend more time under water than a site with a high NAP value, and sites with high NAP values normally lie further up the beach. The tidal environment creates a harsh environment for the animals living there, and it is reasonable to assume that different species and species abundances will be found in beaches with different NAP values. A simple starting point is therefore to compare species diversity (species richness) with the NAP values from different areas of the beach. Although ecological knowledge suggests the relationship between species richness and NAP values is unlikely to be linear, we start with a bivariate linear regression model using only one explanatory variable. It is always better to start with a simple model first, only moving on to more advanced models when this approach proves inadequate.

The first step in a regression analysis is a scatterplot. Panel A in Figure 5.1 shows the scatterplot of species richness versus NAP. The scatter of points suggests that a straight line might provide a reasonable fit for these data. Panel B in Figure 5.1 shows the same scatterplot, but with a fitted regression line added. The slope and direction of the line suggests there is a negative linear relationship between richness and NAP. The slope is significantly different from 0 at the 5% level, and this suggests that the relationship between species richness and NAP values is significant. A histogram of the residuals suggests the residuals are approximately normally distributed, suggesting the data have a Gaussian or normal distribution, and the Cook distance function (which we explain later) shows there are no influential observations. Later in this chapter, we discuss the difference between influential and extreme observations. For some researchers, this evidence would be enough to decide there is a significant negative relationship between richness and NAP and allow them to consider the analysis complete. However, it is more complicated than this, and to understand why, we need to look at some linear regression basics.

Back to the basics

Figure 5.2-A shows the same scatterplot of richness versus NAP we used earlier, but this time, to keep the figure simple, we have only used 7 of the 45 available samples. The species richness values are essentially the result of a random process. It is random because we would expect to find a different range of values every time the field sampling was repeated, provided the environmental conditions were the same. To illustrate this randomness, Figure 5.2-B shows 30 simulated values of the same data. The large dots represent the observed values, and the small dots represent the simulated values (other realisations). However, in practise we do not have these extra observations, and therefore we need to make a series of assumptions before using a linear regression model.

At this point, we need to introduce some mathematical notation. Let Y_i be the value of the response variable (richness) at the ith site, and X_i the value of the explanatory variable (NAP) for the same site.

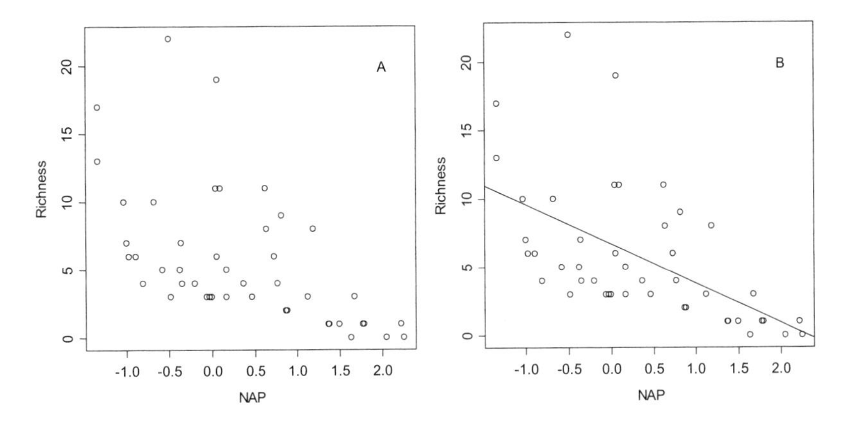

Figure 5.1. A: scatterplot of species richness versus NAP for the RIKZ data. B: scatterplot and regression line for RIKZ data.

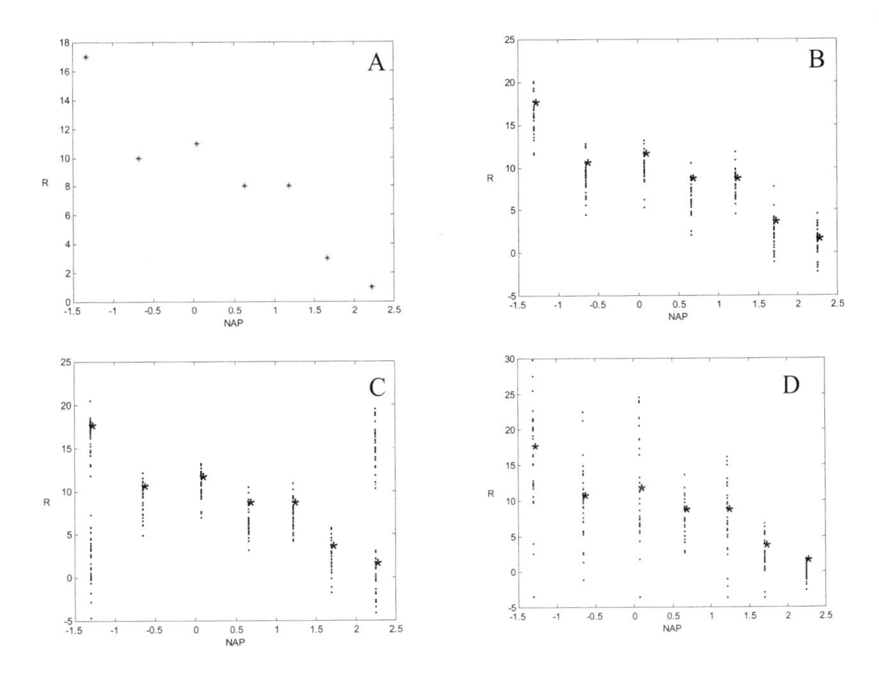

Figure 5.2. A: Scatter plot for NAP and species richness, B: Possible values of the same process. C: Violation of the normality and homogeneity assumptions. D: Violation of homogeneity assumption.

The bivariate (meaning two variables) linear regression model is given by

$$Y_i = \alpha + X_i \beta + \varepsilon_i \tag{5.1}$$

where α is the population intercept, β is the population slope, and ε_i is the residual, or the information that is not explained by the model. This model is based on the entire population, but as explained above, we only have a sample of the population, and somehow we need to use this sample data to estimate the values of α and β for the whole population. To do this, we need to make four assumptions about our data that will allow a mathematical procedure to produce estimated values for α and β. These estimators, called a and b, based on the sample data then act as estimators for their equivalent population parameters, α and β, respectively. The four assumptions that allow the sample data to be used to estimate the population data are (i) normality, (ii) homogeneity, (iii) independence and (iv) fixed X.

Normality

The normality assumption means that if we repeat the sampling many times under the same environmental conditions, the observations will be normally distributed for each value of X. We have illustrated this in the upper right panel of Figure 5.2 where observations at the same value of X are centred around a specific Y value. In a three-dimensional space you can imagine them as bell-shaped histograms. The observed value (indicated by a *) does not need to be exactly in the middle of the realisations at each X value. Figure 5.2-D shows another example, where the observations at each value of X are fairly evenly distributed around a middle value suggesting a normal distribution. Figure 5.2-C is a clear violation of the normality assumption as there are two distinct clouds of observations for some of the X values.

Up to now, we have discussed the normality assumption in terms of Y. However, because we have multiple Y observations for every X value, we also have multiple residuals for every value of X. As X is assumed to be fixed (see below), the assumption of normality implies that the errors ε are normally distributed (at each X value) as well. Using the normality assumption, the bivariate linear regression model is written as

$$Y_i = \alpha + X_i \beta + \varepsilon_i \quad \text{where } \varepsilon_i \sim N(0, \sigma_i^2) \tag{5.2}$$

The notation $N(0, \sigma_i^2)$ stands for a Normal distribution with expectation 0 and variance σ_i^2. If σ_i^2 is relatively large, then there is a large amount of unexplained variation in the response variable. Note that σ_i^2 has an index i.

Homogeneity

To avoid estimating a variance component for every X value, we assume the variance for all X values is the same. This is called the homogeneity assumption. This is important and implies that the spread of all possible values of the population is the same for every value of X. As a result, we have

$$\sigma_1^2 = \sigma_2^2 = \sigma_3^2 = \sigma_4^2 = \ldots = \sigma^2$$

And instead of having to estimate many variances, we only have to estimate one. Figure 5.2-B shows an example of this assumption where the spread of values for Y is the same at each value of X. Figure 5.2-C, however, shows a much wider variation at two values of X. In Figure 5.2-D, we simulated an example in which sites with low richness have a much smaller variation.

In ecological field studies, violation of the homogeneity assumption is common, as organisms often have a clumped distribution.

Independence

This assumption means the Y values for one observation (X_i) should not influence the Y values for other observations (X_j), for $i \neq j$. Expressed differently, the position of a realisation (or residual) at a particular X value should be independent of realisations at other X values. Clear violations of this assumption are time series data and spatial data. For example, when sampling vegetation over time, the vegetation present at the initial sampling period is likely to strongly influence the vegetation present at subsequent sampling periods. We will revisit independence later.

Fixed X

The term 'fixed' means that X is not random, and 'not random' implies that we know the exact values of X and there is no noise involved. In ecological field studies, X is hardly ever fixed, and it is normally assumed that any measurement error in X is small compared with the noise in Y. For example, using modern GPS, determining NAP can be reasonably accurate. An example of where X can be considered as fixed, is measurements of toxic concentrations in mesocosm studies in eco-toxicology. Other examples are nominal variables such as time and transects. Chapter 5 in Faraway (2004) contains a nice discussion on this topic.

The regression model

The four assumptions above give the linear regression model:

$$Y_i = \alpha + X_i \beta + \varepsilon_i \quad \text{where } \varepsilon_i \sim N(0, \sigma^2) \tag{5.3}$$

Note the variance term no longer contains an index i. The population parameters α, β and σ^2 (population variance) are estimated by a, b and s^2 (sample variance). The underlying mathematical tool for this is either ordinary least squares (OLS) or maximum likelihood estimation. OLS finds parameters a and b, based on minimising the residual sum of squares (RSS), where RSS is defined by $RSS = \Sigma_i e_i^2$ and $e_i = Y_i - a - X_i b$. We used an e instead of an ε to indicate that it is sample data, and not population data. Figure 5.3 shows the residuals, together with the observed and fitted values. The residual error (e_i) is the difference between the observed point Y_i and the fitted value. Note that β represents the change in the Y for a

1-unit change in the X, and α is the expected Y value if $X = 0$. Of main interest is the β as it measures the strength of the relationship between Y and X.

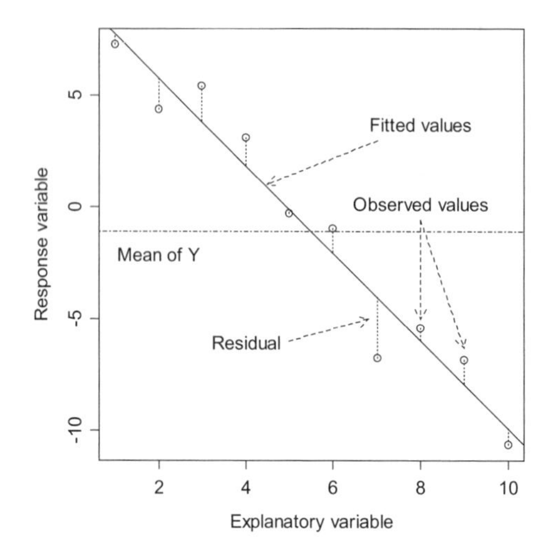

Figure 5.3. Illustration showing the fitted line and residuals. Observed values are the dots, and the straight line is the fitted regression curve. Residuals are obtained by the difference between these two sets of values. The mean of Y is the horizontal dotted line and is needed for the total sum of squares.

Now that we know what the underlying model and assumptions are, we show the results for the seven sampling points in Figure 5.2. The estimated regression parameters were $a = 10.37$ and $b = -3.90$, and the fitted values are obtained by

$$\hat{Y}_i = 10.37 - 3.90 NAP_i$$

The ^ on the Y is used to indicate that it is a fitted value. The equation produces a straight line shown in Figure 5.4 where the observed values are indicated by the squares. The implications of the assumptions are shown in Figure 5.5. The observed values Y_i are plotted as dots in the space defined by the R (richness) and NAP axes (which we will call the R-NAP space). The straight line in the same space defines the fitted values, and the fitted values at a particular value of X are represented by an 'x'. The Gaussian curves on top of the fitted values have their centre at the fitted values. The widths of these curves are all the same and are determined by the estimator of the standard deviation s, a function of the residuals and the number of observations calculated by

$$s = \sqrt{\sum_{i=1}^{n} \frac{e_i^2}{n-2}}$$

Each Gaussian curve shows the probability of other values for Y at any particular value of X. If these curves are wide, then other samples could have a range of very different Y values, compared with the observed value. From this figure it is easily seen why an extreme outlier can cause a large s.

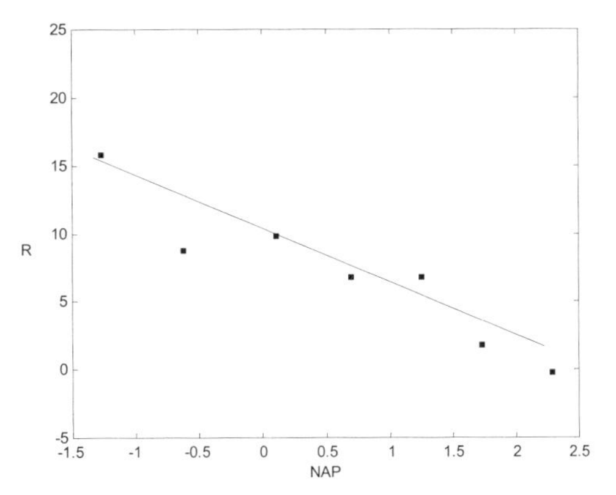

Figure 5.4. Fitted regression curve for seven points in the upper left panel in Figure 5.2.

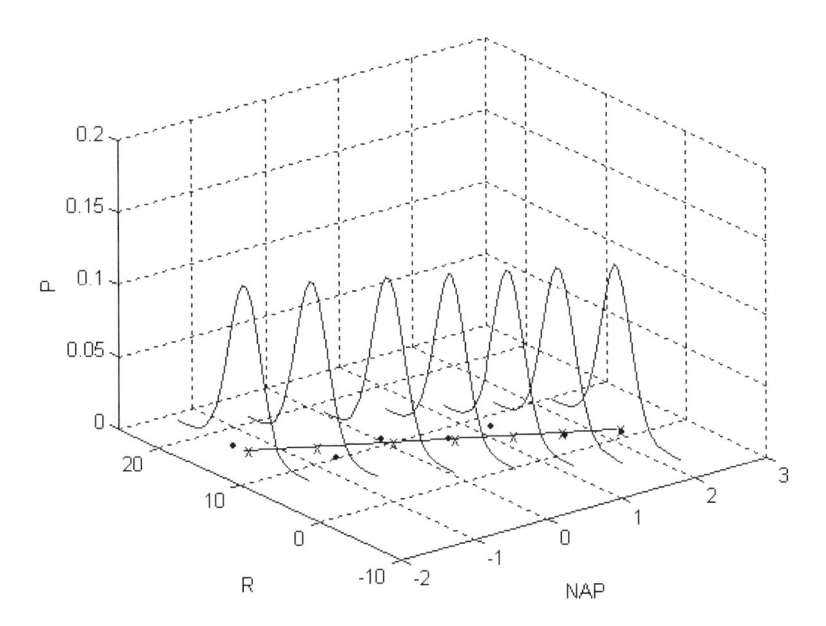

Figure 5.5. Implications of assumptions in linear regression. A dot represents the observed value, and the 'x' is the fitted value at a particular X value. The straight line contains all fitted values. Each Gaussian density curve is centred at a fitted value. The vertical axis shows the probability of finding a particular Y value.

In Figures 5.2, 5.4 and 5.5, we only used seven arbitrarily selected sample points from the 45 available. Using only seven points avoids cluttering the graphs with too much information. However, having gained some familiarity with the graphical interpretation of the Gaussian model, it is now valuable to look at the linear regression model using all 45 observations. Figurc 5.6 shows the same three-dimensional graph as in Figure 5.5, but using all 45 observations. The regression coefficients are slightly different in this example compared with the earlier one because they are based on a different sample size. Looking at these new figures highlights two concerns. The results show that the regression model now predicts negative values for species richness from sites with a NAP value of slightly larger than two metres. This is obviously impossible as the species richness can only ever be zero or greater than zero. Not only is the fitted value close to zero at two metres, but the Gaussian probability density function suggests there is a large probability of finding negative values for the response variable at these sites. Another concern becomes clear when viewing Figure 5.6 from a slightly different angle (Figure 5.7). At least three points in the tail of the distribution have high values for species richness. If there had been only one such observation, then it could probably be dismissed as an outlier, but with three points, we need to look at this in more detail. These three points are causing the wide shape of the Gaussian probability density functions (they have a high contribution to the standard deviation s). Imagine how much smaller the width of the curves would be if the three points were much closer to the fitted regression curve. They also cause large confidence intervals for the regression parameters. Figure 5.7 also shows a clear violation of the homogeneity assumption, and later in this section, we discuss some graphical techniques to detect this violation.

The final important assumption of regression analysis is independence. The independence assumption means that if an observed value is larger than the fitted value (positive residual) at a particular X value, then this should be independent of the Y value for neighbouring X values. We therefore do not want to see lots of positive (or negative) residuals next to one another, as this might indicate a violation of the independence assumption.

Explaining linear regression in this way simplifies introducing GLM and GAM, which can both be explained using a graph similar to Figure 5.6, except that the fitted curve has a slightly different form and the Gaussian density curve is replaced by a more appropriate distribution. GLM and GAM are discussed in a later chapter, but before moving on to these techniques we need to discuss a few linear regression concepts (ANOVA-tables, t-values, coefficient of determination and the AIC) that have equivalents in GLM and GAM. We also discuss the tools used to detect violation of the underlying linear regression assumptions described earlier.

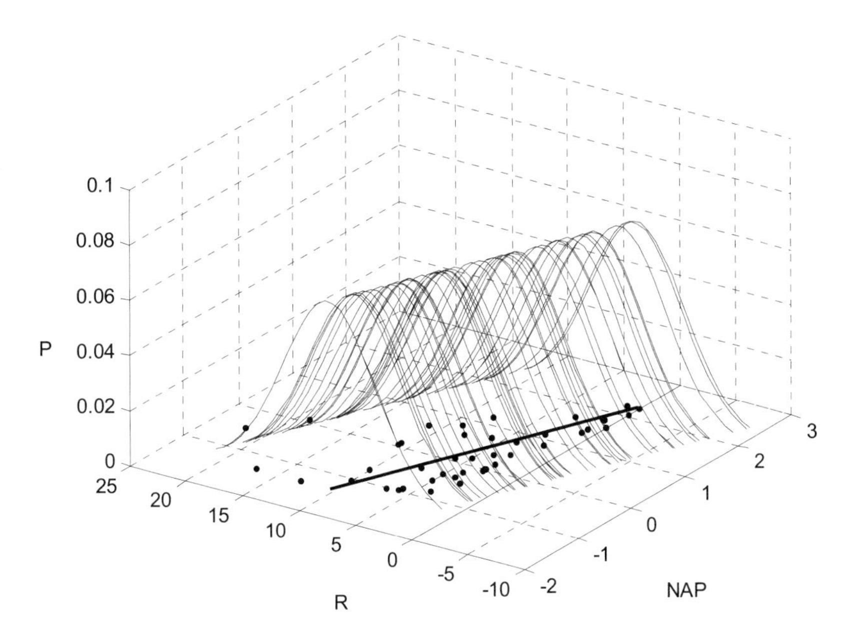

Figure 5.6. Regression curve for all 45 observations from the RIKZ data discussed in the text showing the underlying theory for linear regression.

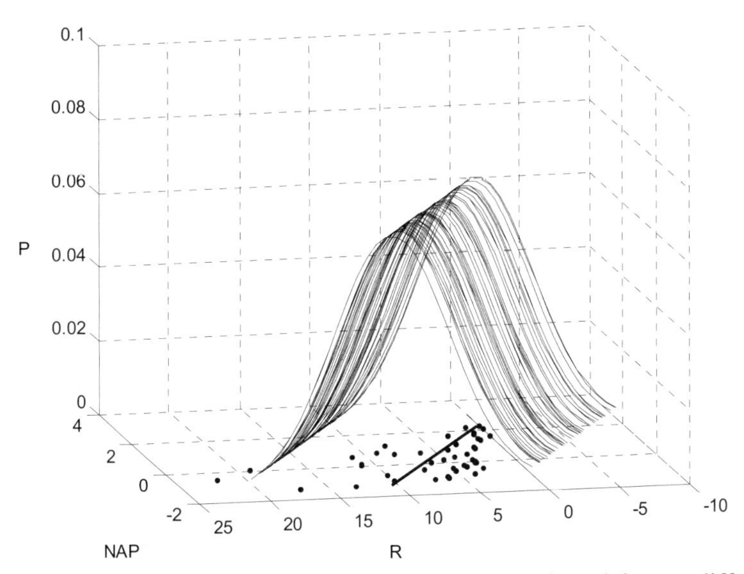

Figure 5.7. Regression curve for all 45 observations, viewed from a different angle.

Assessing the significance of regression parameters

An important aspect in linear regression is the partitioning of the total variability. The total variance in Y, denoted by SS_{total}, can be split up into the part explained by X ($SS_{regression}$) and the part not explained by X ($SS_{residual}$). $SS_{regression}$ measures how well the regression part (X) explains Y, and $SS_{residual}$ shows the amount of variability in the Y values that cannot be explained by the regression model. The values for these components are calculated using the formulae in Table 5.1, and Figure 5.3 shows a geometric interpretation. Most statistics programmes produce ANOVA tables in the style of Table 5.2. The sum of squares depends on the sample size n. If more observations are used, the sums of squares get larger. Therefore, the sums of squares are transformed to variance components by dividing them by the degrees of freedom[1]. The degrees of freedom for the regression sum of squares is the number of regression parameters minus 1. In this case: $2 - 1 = 1$. The degrees of freedom for the total sum of squares is $n - 1$. If there were no explanatory variables, the variance would be estimated from the ratio of the total sum of squares and $n - 1$. The degrees of freedom for the residual sum of squares is $n - 2$; two regression parameters were estimated to calculate this component: the intercept and the slope. The ratio of the two variance components is called the mean square (MS). The MSs are sample variances and, therefore, estimate parameters. $MS_{residual}$ estimates σ_ε^2 and $MS_{regression}$ estimates σ_ε^2 plus an extra term dependent on β and X. The fifth column in Table 5.2 shows what the MS components are estimating.

In bivariate regression, the ANOVA table is used to test the null-hypothesis that the slope of the regression line is equal to zero (H_0: $\beta = 0$). Under this null hypothesis, the expected MS for the regression component is equal to one. So, the ratio of the two variance components $MS_{regression}$ and $MS_{residual}$ is also 1. If for our sample data, the ratio is larger than one, then there is evidence that the null hypothesis is false. Assuming the four regression assumptions hold, then the ratio of $MS_{regression}$ and $MS_{residual}$ will follow an F-distribution with $df_{regression}$ and $df_{residual}$ degrees of freedom. The results for the RIKZ data are shown in Table 5.3.

---

[1] The degrees of freedom for a statistic is the number of observations that are free to vary. Suppose we want to calculate the standard deviation of the observations 1, 2, 3, 4 and 5. The formula for the standard deviation is given by

$$s = \sqrt{\sum_{i=1}^{n} \frac{(y_i - \bar{y})^2}{n-1}}$$

We first have to calculate the mean, which is three. In the next step, we have to calculate the sum of the squared deviations from the mean. Although there are five such squared deviations, only four of them are free to assume any value. The last one must contain the value of Y such that the mean is equal to three. For example, suppose we have the squared components 4 (=2(1 −3)), 1, 0 and 4. We now know that the last square component is calculated using $Y_i = 5$. For this reason, the standard deviation is said to have n-1 degrees of freedom.

Table 5.1. Three variance components.

Notation	Variance in	Sum of squared deviations of	Formula
SS_{total}	Y	Observed data from the mean	$\sum_{i=1}^{n}(Y_i - \bar{Y})^2$
$SS_{regression}$	Y explained by X	Fitted values from the mean value	$\sum_{i=1}^{n}(\hat{Y}_i - \bar{Y})^2$
$SS_{residual}$	Y not explained by X	Observed values from fitted values	$\sum_{i=1}^{n}(Y_i - \hat{Y}_i)^2$

Table 5.2. ANOVA table for simple regression model. df stands for degrees of freedom.

Source of variation	SS	df	MS	Expected MS
Regression	$\sum_{i=1}^{n}(\hat{Y}_i - \bar{Y})^2$	1	$\sum_{i=1}^{n}\frac{(\hat{Y}_i - \bar{Y})^2}{1}$	$\sigma_\varepsilon^2 + \beta^2\sum_{i=1}^{n}(X_i - \bar{X})^2$
Residual	$\sum_{i=1}^{n}(Y_i - \hat{Y}_i)^2$	$n - 2$	$\sum_{i=1}^{n}\frac{(Y_i - \hat{Y}_i)^2}{n-2}$	σ_ε^2
Total	$\sum_{i=1}^{n}(Y_i - \bar{Y})^2$	$n - 1$		

Table 5.3. ANOVA table for the RIKZ data.

	df	SS	MS	F-value	P(>F)
NAP	1	357.53	357.53	20.66	<0.001
residuals	43	744.12	17.31		

The ANOVA table (Table 5.3) shows that the ratio $MS_{regression}/MS_{residual}$ is 20.66. Under the null hypothesis that all slopes are equal to 0, the ratio of 20.66 is unlikely ($p < 0.001$), and H_0 can be rejected. The ANOVA table therefore provides a test to identify whether there is a linear relationship between Y and X.

For the bivariate regression model, a mathematically identical test to an ANOVA is the single parameter t-test. The null hypothesis is H_0: $\beta = 0$, and the test statistic is

$$t = \frac{b_1}{s_{b_1}}$$

The t-value can be compared with a t-distribution with $n - 2$ degrees of freedom. The estimated regression parameters, standard errors, t-values and p-values for the RIKZ data are shown in Table 5.4. Note that these values are slightly different then the earlier model where only seven observations were used, as we are now using all 45 observations. The t-statistic for the regression parameter β is

$$t = \frac{-2.87}{0.63} = -4.55$$

This statistic follows a t-distribution. The critical value for t is 2.02 (significance level is 0.05, two-sided, df = 43). So the null hypothesis can be rejected. Alternatively, the p-value can be used. So, assuming the four assumptions underlying the linear regression model are valid, we can conclude there is a significant negative relationship between species richness and NAP.

Table 5.4. Estimated regression parameters, standard errors, t-values and p-values for the RIKZ data using all 45 observations.

	Estimated Value	Std. Error	t-value	p-value
Intercept	6.69	0.66	10.16	<0.001
NAP	−2.87	0.63	−4.55	<0.001

Model validation in bivariate linear regression

Coefficient of determination

The proportion of total variance in Y explained by X can be measured by R^2, also called the coefficient of determination. It is defined by

$$R^2 = \frac{SS_{regression}}{SS_{total}} = 1 - \frac{SS_{residual}}{SS_{total}}$$

The higher this value, the more the model explains. For the RIKZ data, $R^2 = 0.32$, which means that NAP explains 32% of the variation in the species richness data. However, the value of R^2 should not be used to compare models with different data transformations. Nor should it be used for model selection, as a model with more explanatory variables will always have a higher R^2. R^2 can also often have high values for some non-linear models, even when the regression provides a poor fit with the data. This is shown in Figure 5.8 using data from Anscombe (1973). All four panels show data that share the same intercept, slope and confidence bands. Both the F-statistics and the t-values indicate that the regression parameter is significantly different from zero, and more worrying, all four R^2 values are equal to 0.67! Provided all assumptions hold and there are no patterns in the residuals, there is nothing wrong with an R^2 of 0.32 for the RIKZ data.

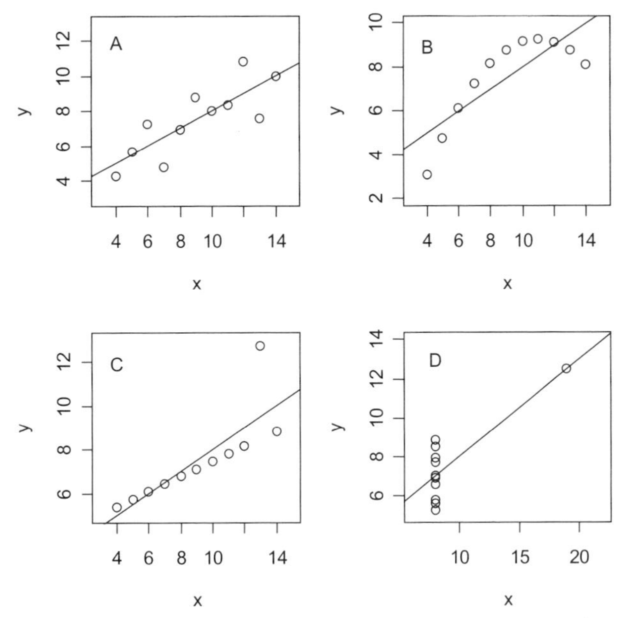

Figure 5.8. Famous Anscombe data. All regression parameters, t-values, F-values and R^2 values are the same.

Assessing the four assumptions

Linear regression is based on the four assumptions described above, and these need to be verified before placing any confidence in your regression model. Normality was the first assumption, and unless multiple observations at the same X value are available, normality cannot be confirmed. So, in practise, the normality assumption is checked using a histogram of the residuals. Quinn and Keough (2002) provide several references that consider normality to be an important but not a crucial assumption. More important is homogeneity, but without replicate observations, this assumption cannot be tested, either. However, homogeneity can be assessed by plotting the residuals against X to check for any increases (or decreases) in the spread of residuals along the x-axis. Alternatively, plotting fitted values against residuals can show increases in the spread for larger fitted values: a strong indicator of heterogeneity. So, we can pool residuals from different X values and use them to assess normality and homogeneity. As to independence, if the data are a time series, then residuals can be checked by the auto-correlation function found in most statistics programmes. This is further discussed in Chapter 16 and in various case study chapters. Spatial correlation is discussed in Chapters 18, 19 and 37.

As well as normality, homogeneity and independence, you should also check for residual patterns in the data. This assesses model misspecification and model fit. A useful tool is to plot residuals against each explanatory variable to check that no clear patterns are shown by the plotted residuals. Ideally the residuals should be scattered equally across the whole graph. To help with visual interpretation, a smoothing curve (Chapter 7) plus confidence bands can be added to this graph.

If the plotted residuals show an obvious non-random structure, several options are available:

1. Apply a transformation.
2. Add other explanatory variables.
3. Add interactions.
4. Add non-linear terms of the explanatory variables (e.g. quadratic terms).
5. Use smoothing techniques like additive modelling.
6. Allow for different spread using generalised least squares (GLS); see Chapters 8 and 23.
7. Apply mixed modelling (Pinheiro and Bates, 2000).

Some common problems and solutions are as follows:

1. There is a violation of homogeneity indicated by residuals versus fitted values. However, the residuals plotted against the explanatory variables do not show a clear pattern. Possible solutions are a transformation on the response variable, adding interactions or using generalised linear modelling with a Poisson distribution (if the data are counts).
2. There is a violation of homogeneity (as above), and the residuals plotted against the explanatory variables show a clear pattern. Possible solutions are as follows. Add interactions or non-linear terms of the explanatory variable (e.g., quadratic terms). Alternatively, consider generalised additive modelling.
3. There is no violation of homogeneity, but there are clear patterns in the residuals plotted against the explanatory variables. Possible solutions: Consider a transformation on the explanatory variables or apply additive modelling.

Instead of GLM with a Poisson distribution it is also possible to model the heterogeneity explicitly using generalised least squares (or mixed modelling). For example, instead of assuming that $\varepsilon_i \sim N(0,\sigma^2)$ in the linear regression model we can allow for heterogeneity using variance structures like (Pinheiro and Bates 2000):

- $\varepsilon_i \sim N(0, NAP_i \times \sigma^2)$
- $\varepsilon_i \sim N(0, |NAP_i|^{2\delta} \times \sigma^2)$
- $\varepsilon_i \sim N(0, \sigma^2 \times \exp(2\delta NAP_i))$
- $\varepsilon_i \sim N(0, \sigma_j^2)$

where δ is an unknown parameter. Further variance structures are described in Chapter 5 in Pinheiro and Bates (2000). The first three options allow for an increase (or decrease) in residual variance depending on the values of the *variance-covariate* NAP. The fourth option allows for different spread per level of a nominal variable. Only the fourth option is used in the case study chapters (23, 26, 36).

The Decapod case study chapter shows some of these approaches, but for this chapter we return to the RIKZ data. The linear regression for the RIKZ data is illustrated in Figure 5.9, where the assumptions discussed above can be checked. The upper left graph shows a scatterplot of the residuals versus the fitted values. This graph shows an increase in the spread of the residuals for the larger values of the fitted values, indicating a violation of the homogeneity assumption. The lower left graph shows a Scale-Location plot, which is a plot of square root transformed absolute standardised residuals versus fitted values. Standardised residuals are explained later in this section. Taking the square root of the absolute values reduces the skewness and makes non-constant variance more noticeable. This graph should show no pattern, but in this graph the values for Y increase as X increases, suggesting that variance is not constant. The upper right panel shows a QQ-plot of the residuals, which checks how closely the data follow a normal distribution. Normally distributed data points should lie approximately on a straight line, which is not the case here. Normality can also be checked using a histogram of the residuals (not shown) and, as with the QQ plot, suggests the data are not approximately normally distributed. The lower right panel in Figure 5.9 is discussed in the next section.

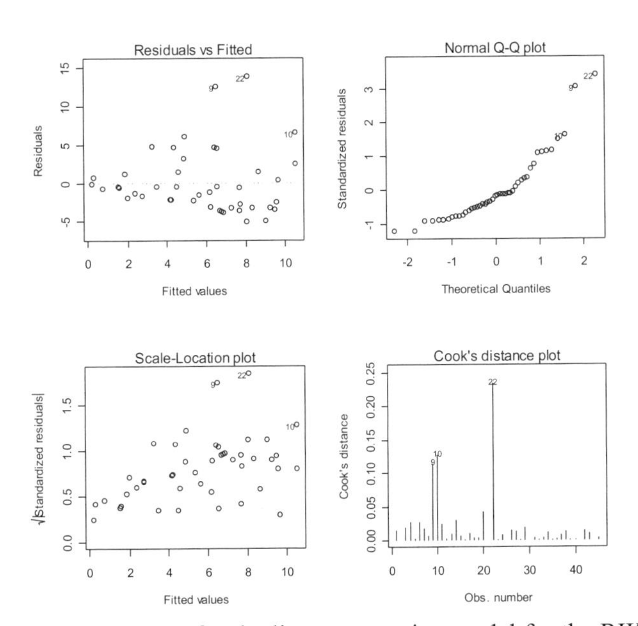

Figure 5.9. Graphical output for the linear regression model for the RIKZ data, allowing the assumptions of normality, homogeneity and independence to be checked. The Cook distance plot is used to check for influential observations.

Influential points

If a particular explanatory variable has one or more values that are much larger than the other observations, these observations could strongly influence the regression results. Leverage is a tool that identifies observations that have rather extreme values for the explanatory variables and may potentially bias the regression results. In Chapter 4, the left panel in Figure 4.16 showed three extreme observations A, B and C. Along the x-axis, A does not have any extremely large or small values, but B and C are both rather large and have a high leverage. However, applying a regression analysis with and without point B gives similar results, showing that B might have a high leverage, but it is not influential on the regression parameters. In contrast, C has high leverage and is also influential. A better measure for influential points is the Cook's distance statistic. This statistic identifies single observations that are influential on all regression parameters. It does this by calculating a test statistic D that measures the change in all regression parameters when omitting an observation. So, point B in the left panel in Figure 4.16 has a low Cook distance statistic, but A and C have high Cook distance statistics. The lower right panel in Figure 5.9 shows the Cook distance function for the RIKZ data. Each bar shows the value of the Cook distance statistic (which can be read from the vertical axis) for the corresponding observation along the x-axis. Relatively large bars indicate influential observations, and for the RIKZ data, observation 22 (as well as observations 9 and 10) looks influential. Fox (2002a) reports that the value $4/(n - k - 1)$ can be used as a rough cut off for noteworthy values of D_i where n is the number of observations, k is the number of regression slopes in the model, and D_i is the Cook value of observation i. Montgomery and Peck (1992) compare the Cook value with an F-value of approximately 1, and all D_i values larger than 1 could be influential. In this case, the Cook statistic at observation 22 is smaller than one and we can assume that this observation is not influential. We suggest using these graphs to inspect for points noticeably different from the majority. A mathematical formula for both the Cook distance statistic and the leverage can be found in Montgomery and Peck (1992) or Fox (2002a). A slightly modified Cook statistic is given in Garthwaite et al. (1995).

Summarising, leverage identifies observations with extreme explanatory variables and the Cook statistic detects points that are influential. It is easier to justify omitting influential points if they have extreme explanatory variables (these are points with a large Cook and a large leverage).

A related method is the Jackknife in which each observation i is omitted in turn, and regression parameters are estimated for the remaining $n - 1$ observations. Large changes in the regression parameters between iterations indicate influential points. Figure 5.10 shows the changes in the intercept and slope if the ith observation is omitted. Note the slope changes considerably if observation 22 is omitted, and leaving out this observation also reduces the slope (it becomes more negative). Similar changes are noted when observations 9 and 10 are omitted.

Note that these measures of influence (leverage, Cook distance, and change in parameters) only assess the effect of one observation at a time. If two observations have similar explanatory variables, and both are influential, these methods might not detect them.

We have already discussed using residuals to assess normality and homogeneity. As well as these ordinary residuals (calculated as observed values minus fitted values), alternative residuals can be defined as standardised residuals and Studentised residuals. These give a more detailed assessment on the likely influence of outlying values. The standardised residuals are defined as:

$$\frac{e_i}{\sqrt{MS_{residual}(1-h_i)}}$$

where e_i is the difference between the observed and fitted value and h_i is the leverage for observation i. Standardised residuals are assumed to be normally distributed with expectation 0 and variance 1; N(0,1). Consequently, large residual values (>2) indicate a poor fit of the regression model. Studentised residuals are a leave-one-out measure of influence. To obtain the ith Studentised residual, the regression model is applied on all data except for observation i, and the $MS_{residual}$ is based on the $n - 1$ points (but not the residual e_i or hat value h_i: These are based on the full dataset). If the ith Studentised residual is much larger than a standardised residual, then this is an influential observation because the variance without this point is smaller. Both types of residuals are shown in Figure 5.11. The Standardised residuals indicate that two observations have values larger than 2 and could be potential outliers.

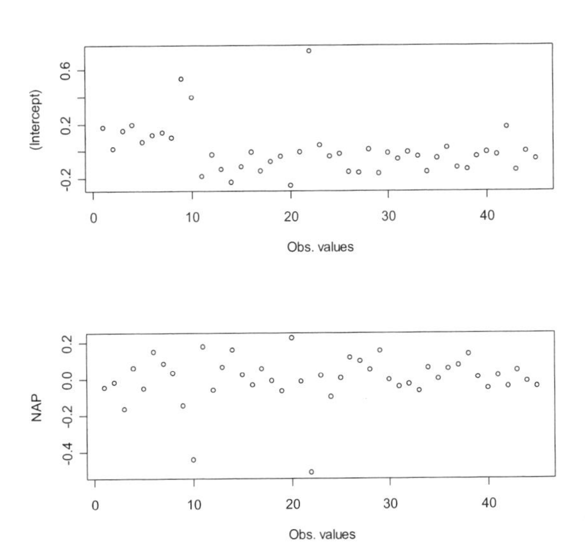

Figure 5.10. Changes in the intercept (upper panel) and slope for NAP (lower panel) if the ith observation is omitted. It can be seen that observations 9, 10 and 22 have a considerable influence on the value of the intercept regression parameter.

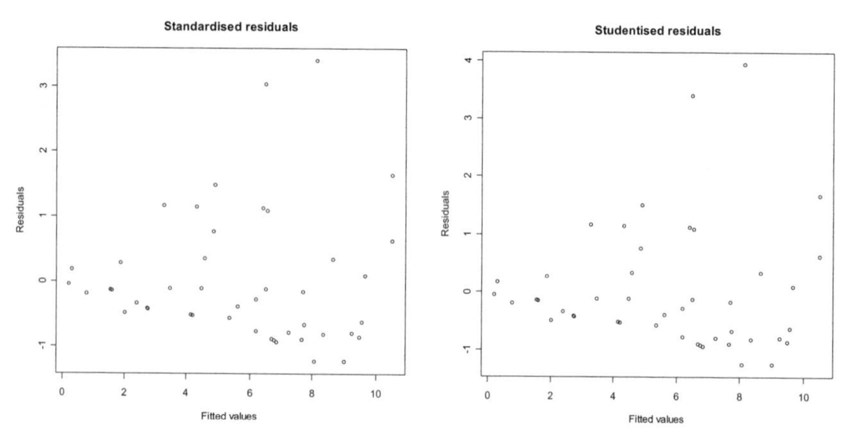

Figure 5.11. Standardised (left panel) and Studentised (right panel) residuals for the RIKZ data. The vertical axes show the values of the residuals and the horizontal axes the fitted values.

It is important to realise what the standardised residuals represent. Recall that Figure 5.5 showed three observations in the tails of the distributions. Suppose the NAP values of these three observations were very different in value from the other observations and located, say, on the far left-hand side of the NAP axis. In that case, you might conclude that the environmental conditions (in terms of NAP) are rather different from the other observations, and they could justifiably be excluded from the regression analysis. And this is exactly what leverage (or hat value) is measuring, the influence of a particular observation in the X space. An observation that has a relatively high leverage (compared with the other observations) is likely to have a relatively high standardised residual (because we are dividing by a small number). Figure 5.12 shows the leverage values, and there are no observations with much higher values than any other observations. As the observations with extreme NAP values do not have large standardised residuals, they cannot be left out of the analysis. This shows that comparing the leverage with the Cook distance can be useful.

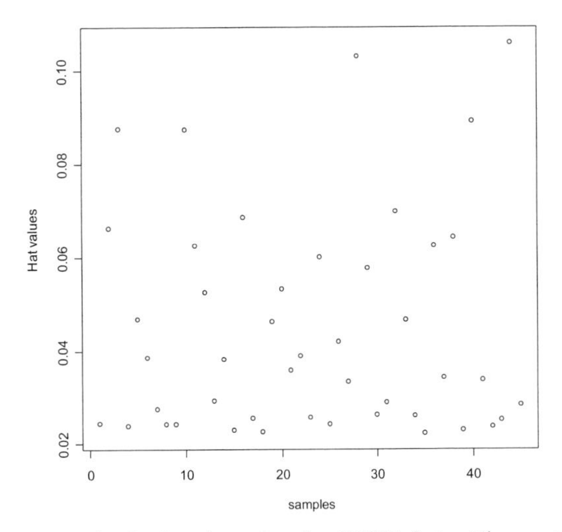

Figure 5.12. Leverage (or hat) values for the RIKZ data. The vertical axis shows the value of the leverage, and the horizontal axis indicates the identity of the observation (corresponding to the order of the data in the spreadsheet).

5.2 Multiple linear regression

In the previous section, we arbitrarily chose NAP as the explanatory variable in the bivariate regression model. However, this was only one of several explanatory variables available (e.g., grain size, humus, angle of the beach, exposure, and week, etc. were also measured; see Chapter 27). In this section, we discuss linear regression techniques that allow modelling a response variable (e.g., species richness) as a linear function of multiple explanatory variables, hence, the name multiple linear regression. This section expands our investigation into the RIKZ data using multiple regression; however, it does not seek to find the most optimal model as this is done later in the book. The general mathematical formula for a multiple regression model is

$$Y_i = \alpha + \beta_1 X_{1i} + \ldots + \beta_p X_{pi} + \varepsilon_i$$

For the species richness R in the RIKZ data this becomes:

$$\text{R}_i = \text{constant} + \beta_1\text{NAP}_i + \beta_2\text{Grainsize}_i + \beta_3\text{Humus}_i + \text{Week}_i + \beta_4\text{Angle}_i + \text{noise}_i$$

See Chapter 27 for an explanation of these explanatory variables. Week is fitted as a nominal variable. To reduce the numerical output, we only concentrate on these five explanatory variables. Selecting and assessing the best explanatory variables are discussed later in this section. On interpreting the regression parameters, β_1 shows the change in species richness for a one-unit change in NAP, while keeping all other variables constant. And β_2 represents the richness change for a one-unit change in grainsize, while keeping all other variables constant. These parameters are called partial regression slopes as they measure the change in Y for a particular value while keeping the remaining $p - 1$ values constant. The ANOVA table (Table 5.5) for a multivariate regression model is similar to the table produced for a bivariate regression model (Table 5.2).

Table 5.5. ANOVA table for multiple linear regression model. $\overline{Y}$ is the mean value, and $\hat{Y}$ is the fitted value.

Source of variation	SS	df	MS
Regression	$\sum_{i=1}^{n}(\hat{Y}_i - \overline{Y})^2$	p	$\frac{\sum_{i=1}^{n}(\hat{Y}_i - \overline{Y})^2}{p}$
Residual	$\sum_{i=1}^{n}(Y - \hat{Y}_i)^2$	n − p − 1	$\frac{\sum_{i=1}^{n}(Y - \hat{Y}_i)^2}{n-p-1}$
Total	$\sum_{i=1}^{n}(Y_i - \overline{Y})^2$	n − 1	

The null hypothesis tested by the ANOVA table is that all slope parameters are equal to 0. In formula: H_0: $\beta_1 = \beta_2 = .. = \beta_p = 0$. Just as in bivariate linear regression, the ratio of $MS_{regression}$ and $MS_{residual}$ follows an F-distribution and can be used to test the null hypothesis. If there is no evidence to reject the null hypothesis, then it can be concluded that none of the explanatory variables is related to the response variable. For the RIKZ model, the F-statistic is 11.18, which is highly significant ($p < 0\ .001$). This means the null hypothesis that all slope parameters are equal to 0 can be rejected. And consequently this means that at least one of the explanatory variables is significantly related to species richness. However, the F-statistic does not say which explanatory variables are significant. To identify the significant explanatory variables, the t-statistic introduced for the bivariate regression model can be used. The t-values of the regression parameters for NAP and Week (Table 5.6) indicate that both variables are significantly differently from zero at the 5% level. Occasionally a t-statistics will indicate non-significance even when you have a significant F-statistic. An explanation for this can be found in Montgommery and Peck (1992).

Table 5.6. Multiple linear regression results for the RIKZ data.

	Estimate	Std. Error	*t*-value	*p*-value
Intercept	9.30	7.97	1.17	0.25
angle2	0.02	0.04	0.39	0.70
NAP	−2.27	0.53	−4.30	<0.001
grainsize	0.00	0.02	0.11	0.92
humus	0.52	8.70	0.06	0.95
factor(week)2	−7.07	1.76	−4.01	<0.001
factor(week)3	−5.72	1.83	−3.13	<0.001
factor(week)4	−1.48	2.72	−0.55	0.58

Note that there are no entries for week 1. Most software will create dummy variables with values 0 and 1, indicating in which week an observation was taken. For week (with four levels), we get four of those dummy variables, which we could call W_1 to W_4. Using all of them as explanatory variables will result in an error or warning message as there is 100% collinearity; if an observation was not taken in week 1, 2 or 3, then it must be from week 4. A possible solution is to omit one of them, say W_1, and apply the GLM using W_2, W_3, W_4, and the other explanatory variables. The notation 'factor(week)2' just means W_2. Its estimated regression parameter is –7.07, which means that the species richness in week 2 is 7.07 lower than for week 1. The *p*-value suggests that it is significantly different from the baseline week. It is also possible to re-level the week and use a different baseline, for example week 2. Dalgaard (2002) showed how each level can be used as baseline in turn, and how to correct the *p*-values for multi-comparisons.

Instead of the *t*-statistic, you can compare two nested models. Two models are called nested if the explanatory variables in one model are a subset of those in the other model. For example, suppose we want to compare the following two models.

Model 1: $Y_i = \alpha + \varepsilon_i$
Model 2: $Y_i = \alpha + \beta \text{ Angle}_i + \varepsilon_i$

We have already introduced one form of *F*-statistic, but a more general form is

$$F = \frac{(RSS_1 - RSS_2)/(p-q)}{RSS_2/(n-p)}$$

RSS_1 and RSS_2 are the residual sum of squares of model 1 (nested model) and model 2 (the full model), respectively, and n is the number of observations. The number of parameters in models 2 and 1 are $p + 1$ and $q + 1$ respectively ($p > q$). The '+1' is because the intercept, and p and q, are the number of slopes in each model. Because model 1 contains fewer parameters ($q + 1$), the fit will always be equal or worse than model 2 and the same holds for the total sum of residuals. If both models give a similar fit, the *F*-statistic will be small. So, large values of the *F*-statistic indicate that the slope is not equal to zero (there is a relationship). Indeed, the null-hypothesis in this test is that the regression parameter for the extra term in model 2 is equal to zero (H_0: $\beta = 0$). Most statistics programmes produce analysis of variance tables in the following form

	df	Sum Sq	Mean Sq	F	Pr(>F)
angle	1	124.86	124.86	13.06	0.001
NAP	1	319.32	319.32	33.41	<0.001
grainsize	1	106.76	106.76	11.17	0.002
humus	1	19.53	19.53	2.04	0.161
factor(week)	3	177.51	59.17	6.19	0.003
Residuals	37	353.66	9.56		

The first line compares model 1 with model 2, and the F-value of 13.06 (p = 0.001) indicates that angle is significantly related to richness. The second line compares the following two models:

Model 2 $Y_i = \alpha + \beta_1 \text{ Angle}_i + \varepsilon_i$
Model 3 $Y_i = \alpha + \beta_1 \text{ Angle}_i + \beta_2 \text{ NAP}_i + \varepsilon_i$

The F-value is equal to 33.41 ($p < 0.001$) and indicates that adding NAP to a model that already contains angle gives a better model. Adding humus to a model that already contains angle, NAP and grain size does not result in a model improvement (the F-statistic is 2.04, which is not significant). Adding week to the model that contains all four explanatory variables still improves the model. The disadvantage of this table is that it depends on the order of the explanatory variables. Problems of collinearity may result in terms added at the end being not significant that would have been significant if added at the beginning.
This general F-statistic can also compare nested models in the following form:

Model 2 $Y_i = \alpha + \beta_1 \text{ Angle}_i + \varepsilon_i$
Model 4 $Y_i = \alpha + \beta_1 \text{ Angle}_i + \beta_2 \text{ NAP}_i + \beta_3 \text{ Grainsize}_i + \beta_4 \text{ Humus}_i + \varepsilon_i$

The underlying null hypothesis in the F-test is that the regression parameters for NAP, grain size and humus are equal to zero (H_0: $\beta_2 = \beta_3 = \beta_4 = 0$). In this case, both models give the same fit. The advantage of this general F-test is that it would give one p-value for a nominal variable that contains more than two classes.

Model selection — finding the best set of explanatory variables

One aim of regression modelling is to find the optimal model that identifies the parameters that best explain the collected data. In this instance these are the parameters that best explain why any specific species richness value occurs at any particular site on the beach. This means we are looking for the best subset of explanatory variables. The reasons for this are (i) a subset is easier to interpret, and (ii) precision of predicted intervals and confidence bands will be smaller (using fewer parameters). Defining 'optimal' is subjective, and to remove part of this subjectivity, statistical criteria are available, for example the AIC, the adjusted R^2 value, and the BIC. Here, we discuss the AIC (Akaike Information Criteria) and the adjusted R^2. They are defined by:

$$\text{AIC} = n \log(\text{SS}_{\text{residual}}) + 2(p + 1) - n\log(n)$$

$$\text{Adjusted } R^2 = 1 - \frac{SS_{residual}/(n-(p+1))}{SS_{total}/(n-1)}$$

The first part of the AIC definition is a measure of goodness of fit. The second part is a penalty for the number of parameters in the model. The AIC can be calculated for each possible combination of explanatory variables, and the model with the smallest AIC is chosen as the most optimal model. The disadvantage of R^2 is that the more explanatory variables are used, the higher the R^2. The adjusted R^2 accounts for different degrees of freedom and, hence, the extra regression parameters.

These criteria can be obtained for every possible combination of explanatory variables, and the model with the optimal values (lowest for AIC, highest for the adjusted R^2) can be selected as 'the' optimal model. Selecting the explanatory variables can be done manually, but if there are more than five explanatory variables, this can become tedious. Instead, automatic selection procedures can be used to apply a forward selection, backward selection, or a combination of forward and backward selection. The forward selection method first applies bivariate linear regression using each explanatory variable in turn. The variable with the lowest AIC is selected, and a multiple linear regression model is applied using the selected variable and each of the remaining variables in turn. The variable that gives the lowest AIC in combination with the first selected variable is then selected, and the process is repeated until the AIC starts to increase. Alternatively, a backward selection (starting with all variables and dropping one at a time) can be used, or even a combination of both. The output below shows the results for the RIKZ data. As most software is capable of applying a backward selection, we show the results for this approach.

$R_i = \text{constant} + \beta_1 \text{NAP}_i + \beta_2 \text{Grain size}_i + \beta_3 \text{Humus}_i + \text{Week}_i + \beta_4 \text{Angle}_i + \text{noise}_i$

Note that week is fitted as a nominal variable. The first iteration is presented:

```
Start: AIC= 108.78
 R ~ angle + NAP + grainsize + humus + factor(week)
```

	df	AIC
- humus	1	106.78
- grain size	1	106.79
- angle	1	106.96
<none>		108.78
- factor(week)	3	121.08
- NAP	1	124.98

This step shows that leaving out humus results in a drop in the AIC from 108.78 to 106.78, which means that humus is not important. Leaving out NAP or week gives a large increase in AIC (meaning these are important variables). Hence, the selection algorithm will continue without humus and produce the following output:

Step: AIC= 106.78. R ~ angle + NAP + grain size + as.factor(week)

	df	AIC
- grain size	1	104.80
- angle	1	104.98
<none>		106.78
- factor(week)	3	120.70
- NAP	1	123.32

This steps shows that grain size is the next variable that should be removed. We will omit the output from the next step and present the results of the final step:

Step: AIC= 103.2. R ~ NAP + as.factor(week)

	df	AIC
<none>		103.20
- NAP	1	122.04
- factor(week)	3	130.25

Further model simplifications lead to larger AIC values indicating a reduced fit. Hence, the optimal model contains both NAP and week. A selection method using both forward and backward selection gave the same results. Instead of a full selection procedure, you can drop one explanatory variable at a time and apply the general *F*-test. This was discussed earlier in this section, and the output is printed below. Note that one variable is dropped in turn.

	df	AIC	F-value	p-value
angle	1	106.96	0.15	0.70
NAP	1	124.98	18.45	0.00
Grain size	1	106.79	0.01	0.92
humus	1	106.78	0.00	0.95
factor(week)	3	121.08	6.19	0.00

Each row compares the full model versus the model with one variable dropped. For example, leaving out NAP resulted in an *F*-statistic of 18.45. Results indicate that angle, grain size and humus can be dropped from the model as their removal did not significantly affect the *F*-value. Note that they should not be dropped all at once!

The AIC should only be used as a general guide. Sometimes the AIC comes up with an 'optimal' model that has one or two non-significant regression parameters. In such cases, further model selection steps are required using, for example the *F*-test. On the other hand, things like model fit and residual patterns can also influence the choice of the final model. One can even argue that a model with a non-significant regression parameter, but with no clear residual pattern is better than a model where all parameters are significant but with clear residual patterns.

A problem with selection procedures is (i) multiple comparisons and (ii) collinearity. With multiple comparisons, every time we apply a regression model, there is a 5% chance of deciding a regression parameter is significantly differently from

0, even when it is not. This risk increases every time the regression is applied, and running a large number of forward and backward selections increases this chance. In ecological studies, there are three ways people deal with this problem; ignore it, avoid using selection methods, or apply a correction method such as the Bonferroni method where p-values (or significance levels) are adjusted for the number of tests carried out. When explanatory variables are highly correlated with each other, a forward selection and a backward selection might give different results due to collinearity. Assuming model simplification is the aim, it is better to avoid using explanatory variables that are likely to vary together (collinearity) because both variables may be reacting in the same way to changes in some other variable.

5.3 Partial linear regression

There are three reasons to discuss partial linear regression. The first reason is to answer why a particular explanatory variable is in the model. Is it significant because of a few outliers, because of collinearity, or is there a genuine relationship? The second reason is because it is used in some multivariate techniques (redundancy analysis and canonical correspondence analysis). The third reason is that variance partitioning (Chapter 12) is easier to explain now, using partial linear regression rather than trying to explain it when we get to the multivariate analysis chapters.

Based on the results in Chapter 4 for the Argentinean marine benthic data, we carried out a multiple regression analysis (results not shown). A backwards selection suggested the variables mud and transect were important in explaining biodiversity (Shannon–Weaver biodiversity index). The next question is how big a contribution does the mud variable make in explaining the different biodiversity indices recorded. Perhaps mud is only significant because it is collinear with transect; transects b and c might be muddier. Partial linear regression identifies the relationship between the biodiversity index and mud, while filtering out the effects of transect (or indeed any set of explanatory variables). In this section we are going to look at two slightly different approaches to partial linear regression. The first approach is discussed in Quinn and Keough (2002) and consists of three steps.

Step 1.

Assume there are one response variable Y and three explanatory variables X, W and Z. The basic linear regression model for these variables is given by

$$Y_i = \text{constant} + \beta_1 X_i + \beta_2 W_i + \beta_3 Z_i + \varepsilon_i$$

The residuals ε_i are estimated from the observed values minus the fitted values. Formulated differently, the residuals represent the information in Y that cannot be explained with X, W and Z. Suppose we fit the model

$$Y_i = \text{constant} + \beta_4 W_i + \beta_5 Z_i + \varepsilon_{1i}$$

The residuals ε_1 represent the information in Y that cannot be explained with W and Z. Hence, ε_1 can be seen as the information in Y after filtering out the effects of W and Z. Filtering out is also called 'partialling out'.

Step 2.

Suppose we also fit the following model:

$$X_i = \text{constant} + \beta_6 W_i + \beta_7 Z_i + \varepsilon_{2i}$$

The residuals ε_2 represent the information in X that cannot be explained with W and Z. Hence, ε_2 can be seen as the information in X after filtering out the effects of W and Z.

Step 3.

In step 1, we obtained residuals ε_{1i} that represent the information in Y after removing the linear effects of W and Z. In step 2, we obtained residuals ε_{2i} that represent the information in X after removing the linear effects of W and Z. In step 3, we regress ε_1 on ε_2 using the following model:

$$\varepsilon_{1i} = \beta\varepsilon_{2i} + \text{noise}_i$$

This model shows the relationship between Y and X, after partialling out the effect of W and Z. Hence, the regression model of ε_1 on ε_2 shows the pure X effect. A nearly horizontal fitted line (or a slope parameter that is not significant) means that X is not related to Y, once the information on the other variables has been considered. To illustrate the method, the Argentinean zoobenthic dataset (Chapters 4 and 28) was used, and the following three linear regression models were applied according to the partial linear regression steps (H is the Shannon–Weaver index):

Step 1: $H_i = \text{constant} + \text{factor}(\text{Transect}_i) + \varepsilon_{1i}$
Step 2: $\text{Mud}_i = \text{constant} + \text{factor}(\text{Transect}_i) + \varepsilon_{2i}$
Step 3: $\varepsilon_{1i} = \beta\varepsilon_{2i} + \text{noise}_i$

The results of the last regression model (ε_1 on ε_2) are given in Figure 5.13. The slope of the fitted linear regression line indicates that, after removing the effects of transect information on both the Shannon–Weaver index and the mud, there is still some information left unexplained in the data. The estimated regression slope b is 0.02 ($p = 0.03$), which is significant at the 5% level. Therefore, we conclude that the significance of mud in the original regression model is not because of collinearity with the transect variable. It would be interesting to omit the three observations with a low Shannon–Weaver value, reapply the model, and see whether mud is still significant. Further details on partial linear regression is in Fox (2002a) and Quinn and Keough (2002). In Fox (2002a), partial regression plots are called 'added variable' plots.

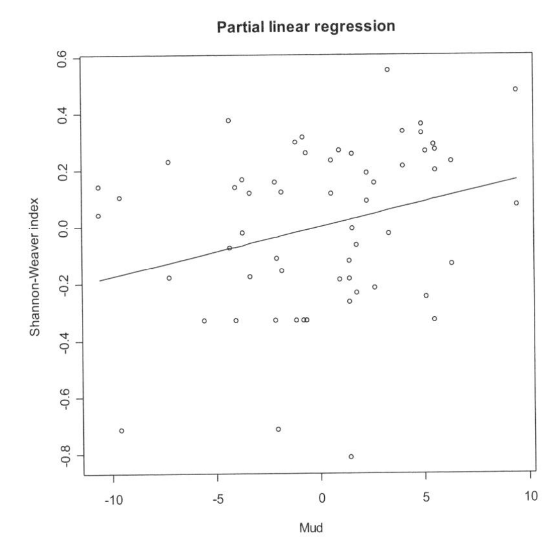

Figure 5.13. Results of partial linear regression. The pure Mud–Shannon–Weaver relationship is shown.

Legendre and Legendre (1998) take a similar approach, but their emphasis is on the decomposition of the variation. This approach forms the underlying mechanism for variance partitioning in redundancy analysis and canonical correspondence analysis (Chapters 12 and 13). Using a more general notation, the starting point for this approach is the following linear regression model:

$$y_i = X_i\beta + W_i v + \varepsilon_i$$

The matrices X and W contain p and m explanatory variables respectively. The parameters are β and v, and ε is noise. The aim is to find the relationship between y and X, while controlling for the variables in W, also called the covariables. For example, W can be spatial variables or observer effects that we want to partial out from the model. The process works as follows.

Step 1.

Regress y and each explanatory variable x_j against W. This means the following regression models are applied:

$$y_i = W_i v_0 + \varepsilon_{0i}$$
$$x_{1i} = W_i v_1 + \varepsilon_{1i}$$
$$\ldots$$
$$x_{pi} = W_i v_p + \varepsilon_{pi}$$

The residuals ε_j contain the information in y and X that cannot be explained with W. Denote these residuals by y_r and x_{jr}. Note this is a generalisation of steps 1 and 2 of the Quinn and Keough (2002) algorithm where only one explanatory variable X is used.

Step 2.

In the next step, a multiple linear regression is applied in which y_r is the response variable and the x_{jr}s are explanatory variables. It is then useful to compare the R^2 of the full model ($y = X\beta + Wv + \varepsilon$), partial model (step 2) and a model of the form $y = X\beta + \varepsilon$, and partial plots can also be usefully produced. Legendre and Legendre (1998) derived an algorithm to calculate the amount of variation purely related to X, purely related to W, and the information shared by both X and W, see Figure 5.14. The components [a] and [c] represent the 'pure X' effect and the 'pure W' effect, respectively, and are proportions. The unexplained variance [d], as a proportion, is equal to one minus the explained variation of a linear regression model in which both X and W are used as explanatory variables. Applying a linear regression model with only X as the explanatory variable, gives us [a + b]. The same holds for [b + c], and with a regression model with W as the explanatory variables.

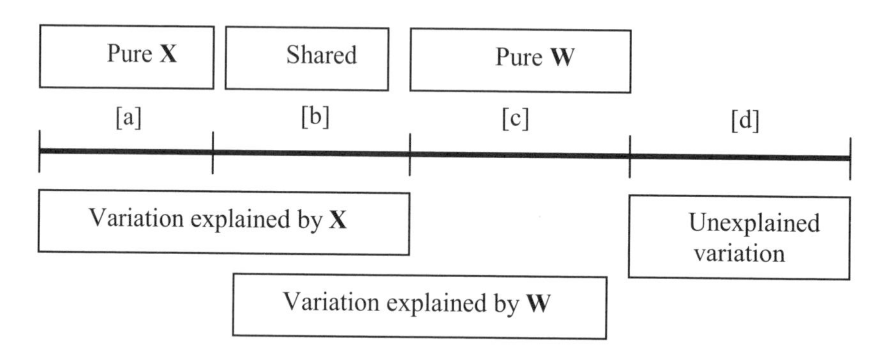

Figure 5.14. Partitioning of variance in 'pure X', 'pure W', shared and residuals components.

To obtain these variance components, Legendre and Legendre (1998) gave the following algorithm:

1. Apply the linear regression model $y_i = X_i\beta + W_iv + \varepsilon$ and obtain R^2. This is equal to [a + b + c], and [d] is equal to 1 – [a + b + c].
2. Apply the linear regression model $y_i = X_i\beta + \varepsilon_i$ and obtain R^2. This gives [a+ b].
3. Apply a linear regression model $y_i = W_iv + \varepsilon$ and obtain R^2. This gives [b + c].

The following formula gives [b] = [a + b] + [b + c] – [a + b + c].

Example: Variance partitioning for the RIKZ data

Again using the RIKZ data, we fitted the following linear regression model:

$$R_i = \text{constant} + \beta_1 \text{NAP}_i + \beta_2 \text{Angle}_i + \text{Exposure}_i + \text{noise}_i$$

The variable Angle represents the angle of the beach; a low value means that a large part of the beach will probably be flooded during high tide. Exposure is fitted as a nominal variable, and R_i is the species richness. All regression parameters are significantly differently from 0 at the 5% level. We would like to know the pure NAP effect. The R^2 for the full model is 0.637. Hence, [a + b + c]=0.637 and [d] = 1 –0.637 = 0.363, indicating that 36.3% of the variation in species richness is not explained by the model. The model with only NAP has R^2=0.325 (= [a + b]) and the model with only the variables Angle and Exposure as explanatory variables has R^2 = 0.347 (= [b + c]). Hence, [b] = 0.325 + 0.347 – 0.637 = 0.035, and therefore [a] = 0.29 and [c] = 0.312. The last result is based on [a + b] = [a] + [b], and [b + c] = [b] + [c].

These results indicate that the pure NAP effect is 29% (meaning that 29% of the variation in species richness is explained purely by NAP), and Angle and Exposure explain 31.2%. The shared amount of information between NAP and angle and exposure is 3.5%. It is possible for [b] to be negative, and a possible reason for this suggested by Legendre and Legendre (1998) is collinearity between [a] and [c].

6 Generalised linear modelling

6.1 Poisson regression

In the linear regression chapter, we analysed the RIKZ data and identified various problems:

- Some fitted values for the response variable were close to negative, but the response variable (species richness) can only take positive values. In this case, we were just lucky that they were positive.
- The Gaussian density curves suggest that, in theory, some values could be negative.
- There was a violation of the homogeneity assumption because the spread in the residuals increased for the larger fitted values indicating heterogeneity.

A data transformation might solve the heterogeneity problem, but this would not avoid the negative fitted values. We have even more problems if the data are of the form 0/1 (0 = absence and 1 = presence), proportions between zero and one, or percentages (between 0% and 100%). However, using Poisson or logistic regression can solve these problems, with Poisson regression used with count data and logistic regression used with presence–absence or proportional data.

Good starting points for generalised linear modelling are Chambers and Hastie (1992), Pampel (2000), Crawley (2002), Dobson (2002), Quinn and Keough (2002), Fitzmaurice et al. (2004), and especially for logistic regression Kleinbaum and Klein (2002). Note that some of these references do not use ecological data. At a more mathematical level we recommend McCullagh and Nelder (1989).

The Poisson regression provides the mathematical framework. First, it assumes that the values for the response variable Y_i are Poisson distributed with expectation μ_i. The notation for this is $Y_i \sim P(\mu_i)$, and as a direct consequence of this distributional assumption, the expectation of Y_i is equal to its variance: $E[Y_i] = \mu_i = \text{var}(Y_i)$. The probability density function is

$$P(x \text{ occurences}) = e^{-\mu} \times \frac{\mu^x}{x!}$$

Using this formula, it is easy to calculate P(x = 0), P(x = 1), P(x = 2), etc. for a given μ. Four different Poisson density curves are shown in Figure 6.1. Panels A and B show the Poisson density function for μ = 1 and μ = 5, where both curves are skewed to the right. Panels C and B show the probabilities for μ = 15 and μ = 25, where the skew is reduced. In panel C the density function is only slightly skewed to the right, and in Panel D, the second curve is approximately symmetrical. Indeed, for larger values of the expected value, the shape of density curve of a Poisson distribution becomes similar to that of the Gaussian density curve. The width of the curve in Panel D is the largest of the four curves, and this is inherent to the Poisson distribution; the larger the mean, the larger the variance. Note that the Poisson distribution is only used for discrete data (response variable). Rounding the data to the nearest integer will achieve this, but be careful if the data values of your response variable are all between a and a + 1; using the Poisson distribution is not sensible in this case.

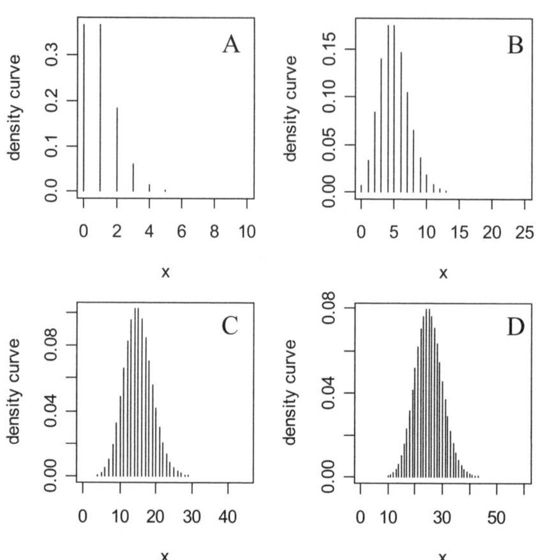

Figure 6.1. Poisson distributions for μ = 1 (A), μ = 5 (B), μ = 15 (C) and μ = 25 (D). In each panel the horizontal value shows possible values of x (the response variable) and the vertical axis gives the corresponding probability that this value is observed, given that the mean is μ.

In Poisson regression, Poisson density curves replace the Gaussian density curves used in linear regression. This step allows for some increase in the spread, and avoids density curves that suggest possible negative realisations. However, this does not stop the model from giving negative fitted values. It is now useful to introduce some mathematical notation. Let $g(x)$ be the so-called predictor function:

$$g(x_i) = \alpha + \beta_1 X_{1i} + \ldots + \beta_p X_{pi}$$

In the linear regression model, we have the distributional assumption $Y_i \sim N(\mu_i,\sigma^2)$, the model itself: $Y_i = g(x_i) + \varepsilon_i$, and as a consequence: $E[Y_i] = \mu_i = g(x_i)$. Negative fitted values occur if $g(x_i)$ becomes negative. In Poisson regression, we use a slightly different relationship between the expectation μ and linear predictor function $g(x_i)$:

$$\mu_i = e^{g(x_i)} \quad \text{or} \quad \log(\mu_i) = g(x_i) \tag{6.1}$$

Because the mean (or fitted value) is modelled as an exponential model, it is always positive. The Poisson regression model can be summarised as

$$R_i \sim P(\mu_i) \quad \text{and} \quad E[Y_i] = \mu_i = e^{g(x_i)} = e^{\alpha + \beta_1 X_{i1} + \ldots + \beta_p X_{ip}}$$

To illustrate this model, we applied a Poisson regression to the RIKZ data. For details on these data, see Chapter 31. The species richness at site i is denoted by R_i and is used as response variable and NAP as the explanatory variable. The model is

$$R_i \sim P(\mu_i) \qquad \text{and} \qquad E[R_i] = \mu_i = \exp(\alpha + \beta_I \text{NAP}_i)$$

The fit and the graphical interpretation of this model are shown in Figure 6.2. Dots represent the observed values in the R-NAP space, and R is the response variable. The fitted line, which is now an exponential curve, is plotted in the same space.

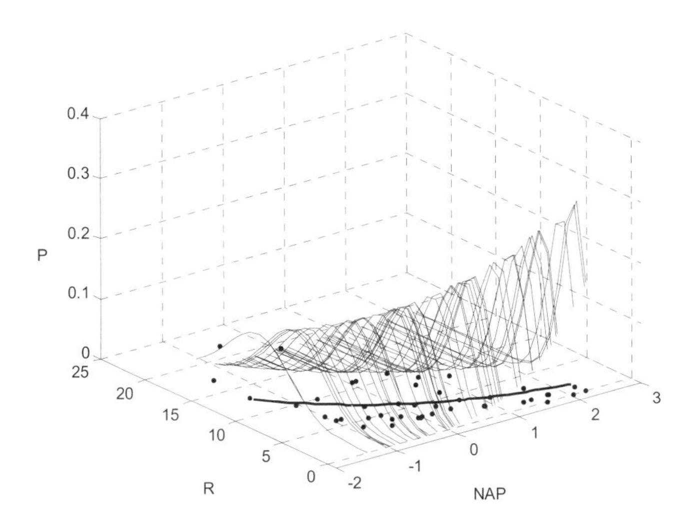

Figure 6.2. Fit of Poisson regression model for RIKZ data. R is the response variable (species richness), and NAP is the explanatory variable. The third axis identifies the (smoothed) probability of the density curves. Dots are the observed values, and the line in the R-NAP space is the fitted Poisson regression curve. The density curves show the probability of other realisations at the same NAP value.

On top of this curve, we plotted the Poisson density curves, and these define the most likely values of other potential values for R. Technically, these density curves are discrete, but for visualisation purposes, we drew a line. Note the fitted values are non-negative, Poisson density curves show that negative realisations are not possible, and only one observation is in the tail of the distribution.

Figure 6.3 shows the same graph, but from a different angle. It clearly shows the change in the Poisson density curves from small non-symmetric curves to wide and approximately symmetrically curves.

Dispersion

Sometimes the increasing spread in count data is even larger than can be modelled with the mean-variance relationship of the Poisson distribution. A possible solution is to introduce a dispersion parameter ρ such that $E[R_i] = \mu_i$, and the variance of R_i is modelled as $\rho\mu_i$, For $\rho > 1$, this allows for more spread than the standard Poisson mean-variance relationship and is called overdispersion. If $\rho < 1$, it is called underdispersion.

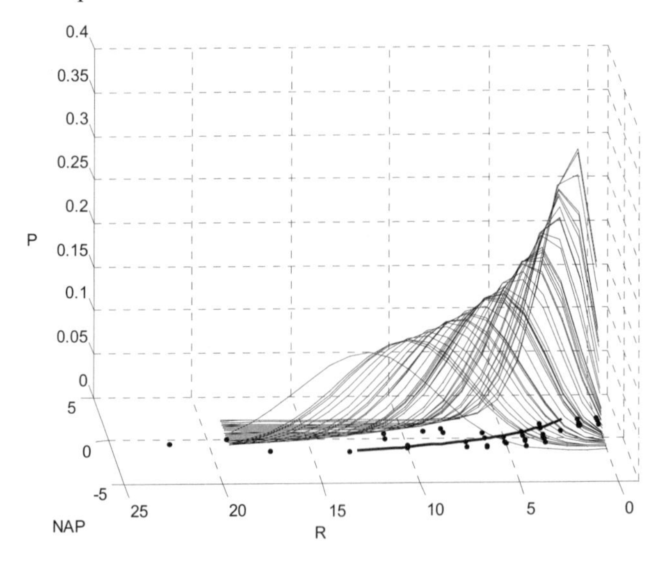

Figure 6.3. As Figure 6.2, but from a different angle.

Estimation of a Poisson regression model with a dispersion parameter is called quasi-likelihood (or quasi-Poisson), but technically it is no longer a Poisson model. A consequence of introducing an overdispersion parameter ρ is that all estimated standard errors are multiplied with the square root of ρ, and therefore ignoring overdispersion can lead to wrong conclusions. We demonstrate the solutions and effects of overdispersion using the RIKZ data. The Poisson regression model using richness as the response variable and NAP, as an explanatory variable, gave the following.

	Estimate	Std. Error	t-value	p-value
Intercept	1.79	0.06	28.30	<0.001
NAP	−0.56	0.07	−7.76	<0.001

The null deviance is 179.75 on 44 degrees of freedom, the residual deviance is 113.18 on 43 degrees of freedom and the AIC = 259.18. The null deviance is the equivalent of the total sum of squares in linear regression, whereas the residual deviance is the equivalent for the residual sum of squares. The estimated parameters are highly significant. The dispersion parameter can be estimated from the residuals of the Poisson regression model, or one can simply apply the quasi-Poisson model. The output after correcting for overdispersion (using the quasi-Poisson model) is:

	Estimate	Std. Error	t-value	p-value
Intercept	1.79	0.10	16.23	<0.001
NAP	−0.56	0.12	−4.45	<0.001

The overdispersion parameter was $\rho = 3.04$. We explain later how to estimate ρ. The standard errors are automatically corrected, and are now considerably larger, because they were multiplied by the square root of the dispersion parameter. This results in the estimated regression parameters being 'less' significant (the t-values are divided by the square root of the dispersion parameter). However, in this example, they are still significant at the 5% level. Note that for the Poisson regression model, there was just one observation in the tail (Figure 6.3), and this observation probably caused the overdispersion. Ignoring overdispersion can easily result in wrongly deciding a parameter is significant.

The Poisson regression model is also called a generalised linear model with Poisson distribution and log link function; the model $E[Y_i] = \exp(\alpha + \beta_1 \text{NAP}_i)$ can be written as $\log(E[Y_i]) = \alpha + \beta_1 \text{NAP}_i$, hence, the name log-link. Most model selection and validation techniques for Poisson regression are similar to those used in linear regression (ANOVA-tables, t-values, AIC, hat values, Cook's distances), but a few are different, and these are discussed next.

Deviance

The (residual) deviance D is the GLM equivalent of the residual sum of squares. Technically, it is the difference between the log-likelihood of the saturated model (using as many parameters as observations) and the log-likelihood for the fitted model. It is useful for model comparisons. A small deviance value indicates a good fit, and a large value a poor fit. To decide whether D is small or large, you can use a Chi-square distribution. The degrees of freedom for this distribution is $n - p - 1$, where n is the number of observations and p is the number of parameters (slopes) in the model. This test can only be used for reasonably large n, and many authors warn about its approximate nature (McCullagh and Nelder 1989; Hosmer and Lemeshow 2000). A safer use of the deviance is for model comparison of two nested models. Suppose we are fitting a Poisson model to the RIKZ data of the form:

$$R_i \sim P(\mu_i) \quad \text{and} \quad E[R_i] = \mu_i = \exp(\alpha + \beta_1 \text{NAP}_i + \text{Exposure}_i + \text{Week}_i)$$

where exposure (3 levels) and week (4 levels) are fitted as nominal variables. The question is whether a model with NAP, Exposure and Week is better than a model with NAP and Week. The deviance of the model containing all three variables is $D_1 = 47.80$, and the deviance for the model with only NAP and Week is $D_2 = 53.47$ (the smaller the deviance the better). The difference between D_1 and D_2, which is asymptotically Chi-square distributed with $p_1 - p_2$ degrees of freedom (p_1 and p_2 are the number of parameters in the two models), is 5.67 ($p = 0.02$). This test assumes there is no overdispersion. What we are testing here is whether the null hypothesis that all the regression parameters for Exposure (all levels) are equal to zero, and the results suggest there is evidence to reject this assumption. Another way to test whether Exposure can be dropped from the model is to look at the t-values:

	Estimate	Std. Error	t-value	p-value
Intercept	2.53	0.13	19.68	<0.001
NAP	–0.49	0.07	–6.57	<0.001
factor(week)2	–0.76	0.35	–2.16	0.03
factor(week)3	–0.51	0.21	–2.40	0.02
factor(week)4	0.12	0.23	0.55	0.58
factor(exposure)10	–0.43	0.19	–2.24	0.03
factor(exposure)11	–0.65	0.33	–1.96	0.05

Note that there are no entries for week 1 and exposure level 8 as these are nominal explanatory variables. This was discussed in Chapter 5.

The t-value is asymptotically normal distributed with expectation zero and variance one; hence, the 95% confidence interval for the NAP parameter is $-0.49 \pm 1.96 \times 0.07$. Since 0 is not in this interval the NAP parameter is significantly differently from 0 at the 5% level. One of the exposure levels is also significantly different from 0 at the 5% level. Instead of doing this process manually, several statistics programmes provide automatic 'drop 1 variable' tools. The output of this function is of the form:

	df	Deviance	AIC	LRT	p-value
<none>		47.80	203.80		
NAP	1	93.46	247.46	45.66	<0.001
factor(week)	3	58.37	208.37	10.57	0.01
factor(exposure)	2	53.47	205.46	5.67	0.06

If all explanatory variables are used, the deviance is 47.8. Each explanatory variable is deleted in turn. For example, if exposure is dropped the deviance increases to 53.47, which is an increase of 5.66. This difference follows a Chi-square distribution with 2 degree of freedom (the nested model has 2 parameters fewer than the full model). The associated p-value is 0.06, indicates that you cannot reject the null hypothesis at the 5% level, and that the regression parameters for exposure are zero. This, together with all other variables being significantly different from 0 at the 5% level, indicates that exposure can be dropped from the model. Clearly, NAP is the most important variable; leaving it out results in the highest deviance, indicating the poorest fit.

For small datasets the t-statistic may give a different message compared with the deviance test. In this case, you should use the deviance test, as it is more reliable that the t-test for small datasets. For large datasets, p-values obtained by the deviance test and t-test are likely to be similar (provided there is no strong collinearity). The deviance test can also be used to compare nested models where the difference is more than one explanatory variable. Instead of the deviance test, you can also use the AIC.

The deviance test described above assumes there is no overdispersion. However, as shown in the dispersion paragraph above, the RIKZ data might be overdispersed, and in this case, the difference between the deviances of two nested models will not be Chi-square distributed. If the dispersion parameter is estimated (as in Quasi-Poisson), the two models can be compared using:

$$\frac{D_2 - D_1}{\rho(p-q)} \sim F$$

where ρ is the overdispersion parameter, and p + 1 and q + 1 are the number of parameters in models 1 and 2, respectively. The '+1' is for the intercept. D_1 is the deviance of the full model and D_2 of the nested model. Under the null-hypothesis the regression parameters of the omitted explanatory variables are equal to zero, and the F-ratio follows an F-distribution with $p - q$ and $n - p$ degrees of freedom (n is the number of observations). For the model with all explanatory variables, we have $D_1 = 47.803$, and for the model without exposure $D_2 = 53.466$. The difference in number of explanatory variables is 2, and $n = 45$. The F-statistic is obtained from an analysis of deviance table:

Model	Resid. df	Resid. Dev	df	Deviance	F-statistic	p-value
1	38	47.80				
2	40	53.46	2	5.66	2.39	0.10

Model 1 contains the explanatory NAP, week (as a factor) and exposure (as a factor). Model 2 has NAP and week. This shows that by considering overdispersion, exposure is not significant at the 5% level. Repeating the analysis without exposure may change the estimated parameters, the overdispersion parameter, and even the order of importance of the explanatory variables.

Validation plots

Using GLM validation plots for the first time can be confusing because the fitted values can be plotted on either the predictor scale or on the original scale. In the first case, the function $g(x)$ is plotted, but in the second situation, we use $\exp(g(x))$. If observed values are close to zero, then $\exp(g(x))$ is hopefully close to zero as well, but this means that $g(x)$ has negative values, and these negative values can sometimes be rather large.

For the linear regression model, we used the ordinary, standardised and Studentised residuals. In non-normal GLM models (the Gaussian model is a GLM model as well) other types of residuals are defined. The two most important types of residuals are the deviance (E^D) and the Pearson (E^P) residuals. For the Poisson model they are defined by

$$E_i^D = sign(Y_i - \hat{\mu}_i)\sqrt{d_i}$$

$$E_i^P = \frac{Y_i - \hat{\mu}_i}{\sqrt{\hat{\mu}_i}}$$

where d_i is the contribution of the ith observation to the residual deviance D, and $\hat{\mu}_i$ the fitted value. The deviance is calculated from the sum of the squared deviance residuals. Hence, deviance residuals show which observations have a large contribution to the deviance. The underlying idea of Pearson residuals is as follows. The Poisson distribution allows for larger spread for larger fitted values and therefore it doesn't make sense to inspect observed values minus fitted values. Therefore, we scale these differences by the square root of the variance.

Both types of residuals are useful for detecting residuals with large influences. For non-Poisson and non-Binomial (see below) distributions, the dispersion parameter can be estimated from the sum of squared Pearson residuals divided by $n - q$ (n is the number of observations and q is the number of parameters in the model). An alternative estimator is $D/(n - q)$. For the normal distribution, the dispersion parameter is the variance.

Useful validation plots are as follows:

1. A scatterplot of the observed data versus the fitted values (Y_i versus μ_i). These should lie as much as possible on a straight line with slope 1.
2. Absolute deviance residuals versus the linear predictor $g(x)$. The discrete nature of the data might show a distinct pattern.
3. Individual explanatory variables versus residuals.
4. Hat (leverage) values and Cook's distances.
5. Influential observations (in terms of overall fitting criteria and estimated parameters).
6. Residuals versus each explanatory variable. If one can see a pattern in these graphs, generalised additive modelling is a possible follow-up analysis. Alternatives are including interaction terms or quadratic terms (Crawley 2005).

As we are now using techniques that do not need a normal distribution, a normal distribution of the residuals is no longer of concern. Therefore, histograms and QQ-plots of the residuals should be interpreted in terms of how well the model fits the data rather than the normality of the residuals. Several examples of these model validation techniques are shown in the case study chapters.

Overdispersion revisited

The overdispersion parameter in a Poisson GLM is estimated using Pearson residuals. The exact formula is

$$\rho = \frac{1}{n-p} \frac{\sum_{i=1}^{n} (Y_i - \hat{\mu}_i)^2}{\hat{\mu}_i}$$

There are various reasons why the model might give large Pearson residuals, for example:

1. There are observations with large values that are a poor fit with the model: outliers. Particularly observations with fitted values close to zero.
2. There is a model misspecification. An important explanatory variable is missing, interaction terms are not added, or there is a non-linear effect of an explanatory variable. In the latter case, the solution is to apply a transformation on the explanatory variable, add quadratic terms, or use smoothing components (e.g., GAM).
3. Violation of the independence assumption. There is correlation between the observations. In this case, random effect models (Chapter 8) or generalised linear mixed models (Fitzmaurice et al. 2004) can be used.
4. There is clustering of samples.
5. The wrong link function has been used.

The overdispersion correction gives the following model:

$$E[Y_i] = \mu_i \quad \text{and} \quad Var[Y_i] = \rho\mu_i$$

Although it is common in ecology to have datasets with a large overdispersion, it is unclear how large a value of ρ is acceptable. Some authors report that $\rho = 5$ or 10 is large, and other authors use overdispersion parameters of 50 or more. We suggest tackling overdispersion by trying to improve the systematic part of the model using GAM or by adding interactions. However, this is not always the best solution. A GLM modelling approach that allows for correlation is generalised estimation equations (GEE). These were introduced by Liang and Zeger (1986), and a useful introductory paper is Hardin and Hilbe (2003). GEE is similar to generalised least squares (GLS) and is an extension of regression models where a non-diagonal error covariance matrix is used. The underlying principle in GLS can be found in Greene (2000). An alternative to GEE is generalised linear mixed modelling (GLMM), which is a combination of GLM and mixed effects modelling. Currently, software for GLMM and its generalised additive modelling equivalent (GAMM) is being developed and improved by the scientific community (Wood 2006). It is likely that GLMM and GAMM will become a standard tool in the ecologist's statistical toolbox. Yet, another way to deal with overdispersed count data is Zero Inflated Poisson (ZIP) models (Tu 2002). The underlying model is

$$Y_i \sim \begin{cases} 0 & \text{with prob. } p_i \\ P(\mu_i) & \text{with prob. } 1 - p_i \end{cases}$$

Both components of the model are modelled in terms of the explanatory variables and are fitted simultaneously. Yet, another possible solution for overdispersion is to use the negative Binomial distribution (Lindsey 2004) instead of the Poisson distribution. It is also possible to use models of the form:

$$E[Y_i] = \mu_i \quad \text{and} \quad Var[Y_i] = \rho\mu_i^2$$

It can be used if the variance is considerably larger than the mean.

6.2 Logistic regression

In the previous section, we used the species richness as the response variable. This diversity index measures the number of different species per site giving a response variable with a vector of length 45 (number of observations) containing non-negative integers. Although Poisson regression is suitable for modelling count data, a different approach is needed for 0–1 and proportional data. An example of such data is where the response variable is a vector of ones and zeros representing presence and absence of a particular species at a site. For this type of data we need to consider logistic regression. The next few paragraphs are largely based on ideas from Pampel (2000).

Later in this book we look at a case study using the flatfish *Solea solea* measured in the Tagus estuary in Portugal. The data for this study include the abundances of *S. solea* and several explanatory variables, such as salinity and mud content measured at 61 sites in the estuary. The data were noisy and to compensate for the noise they were transformed to presence–absence data. Figure 6.4 shows a scatterplot of *S. solea* versus salinity. Define P_i as the probability that *S. solea* is present at site i, and $1 - P_i$ as the probability that it is not present at the site. A more formal notation is $P_i = P(Y_i = 1)$. We will model P_i as a function of the explanatory variables. However, P_i is always between 0 and 1 as it is a probability. The regression line in Figure 6.4 is a first attempt to estimate the probabilities P_i, but the line takes values larger than one for small salinity values. Additionally, the Gaussian density curves on top of the fitted values (not plotted here) suggest that realisations outside the 0–1 range are possible. Therefore, we need to apply a series of transformations on P_i such that the transformed values are not restricted to be in a certain interval. We can then back-transform so that the original probabilities are between 0 and 1. So, in logistic regression the transformed values (which can take any value) are modelled as a function of the explanatory variables.

The odds of an event occurring (i.e., the presence of *S. solea*) can be defined as

$$O_i = \frac{P_i}{1 - P_i}$$

Table 6.1 shows the value of the odds for various P_i values. If the probability that *S. solea* is measured at site i is equal to $P_i = 0.5$, then $1 - P_i$ is also 0.5 and the odds are 1. For larger values of P_i, the odds become larger as well, whereas for smaller values of P_i, the odds become small, but positive. An odds of nine means that it is nine times more likely that *S. solea* will be recorded than not recorded. Or stated slightly differently, at that level of salinity you would expect to have nine records of *S. solea* being present for every ten samples. To compare two odds with each other, their ratio can be used. This is the odds ratio. If the odds ratio of two samples is close to zero, then the odds of the second sample is much higher.

Note that the odds are always larger than zero. By taking the natural logarithm of the odds, also called the log odds, we can get values that are not restricted to lie between 0 and 1, and negative values are possible. The log odds are also symmetrically distributed around 0. Note that a small change in probabilities for P_i close to 0.5 has a different change on the log odds compared with the same change for P_i close to 1 or 0. We will visualise this later.

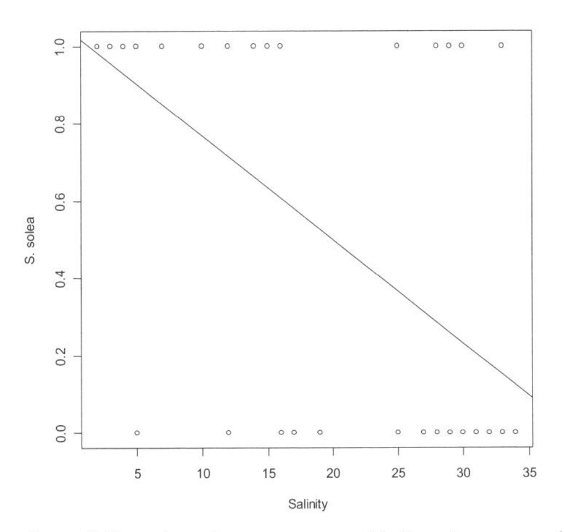

Figure 6.4. Scatterplot of *S. solea* data versus salinity. A regression line was added as a first attempt to estimate the probability of recording *S. solea*.

Table 6.1. Various probabilities, odds and log odds. The table shows how log odds are calculated from probabilities.

P_i	0.001	0.1	0.3	0.4	0.5	0.6	0.7	0.9	0.999
$1 - P_i$	0.999	0.9	0.7	0.6	0.5	0.4	0.3	0.1	0.001
O_i	0.001	0.11	0.43	0.67	1	1.5	2.33	9	999
$Ln(O_i)$	−6.91	−2.20	−0.85	−0.41	0	0.41	0.85	2.20	6.91

In logistic regression, the log odds are modelled as a linear function of the explanatory variables

$$\ln(O_i) = \ln(\frac{P_i}{1-P_i}) = g(x_i)$$

where $g(x_i) = \alpha + \beta_1 X_{1i} + \cdots + \beta_p X_{pi}$ is a linear function of the p explanatory variables. Using simple algebra, it follows that:

$$O_i = \frac{P_i}{1-P_i} = e^{g(x_i)} \Rightarrow P_i = \frac{e^{g(x_i)}}{1+e^{g(x_i)}} \quad (6.2)$$

It is easily verified that P_i is always between 0 and 1, whatever the value of the function $g(x_i)$. This provides a framework that gives fitted probabilities between 0 and 1. However, compared with linear regression and Poisson regression, we also need to replace the Gaussian (or Poisson) density curves by something more appropriate. More 'appropriate' means the realisations should be equal to 0 or 1, or between 0% and 100% for proportional data and the Bernoulli and Binomial distribution should be used. Choosing between the binomial or the Bernoulli distribution depends on the number of samples per X value. First, we assume there is only one observation per sample: $n_i = 1$. In this case the logistic regression model is of the form:

$$Y_i \sim B(1,P_i) \quad \text{and} \quad E[Y_i] = P_i = \mu_i = \frac{e^{g(x_i)}}{1+e^{g(x_i)}}$$

$B(1,P_i)$ is a Bernoulli distribution, and the variance of Y_i is given by $P_i(1-P_i)$. Before discussing the interpretation of the regression parameters, we look at an example of logistic regression using the *S. solea* data. The following model was used:

$$Y_i \sim B(1,P_i) \text{ and } E[Y_i] = P_i = \mu_i = \frac{e^{g(x_i)}}{1+e^{g(x_i)}} \text{ where } g(x) = \alpha + \beta \times \text{Salinity}_i.$$

Figure 6.5 shows the model fit of the logistic regression model and the numerical output is given below.

	Estimate	Std. Error	*t*-value	*p*-value
(Intercept)	2.66	0.90	2.95	0.003
sal	−0.12	0.03	−3.71	<0.001

Null deviance: 87.49 on 64 degrees of freedom
Residual deviance: 68.56 on 63 degrees of freedom. AIC: 72.56

The t-values (or p-values) indicate that Salinity is significantly different from 0 at the 5% level. The fitted curve in Figure 6.5 shows the typical S-shape of a logistic regression curve. Note that at P_i values around 0.5, the rate of change in probabilities is larger than the P_i values close to 0.9 and 0.2. Stated differently, for salinity values between 15 and 25, the rate of changes in the probability of observing

S. solea is larger compared with samples with smaller and larger salinity values. At extreme values of the salinity gradient, a change in salinity has less effect of the probability compared with average salinity values.

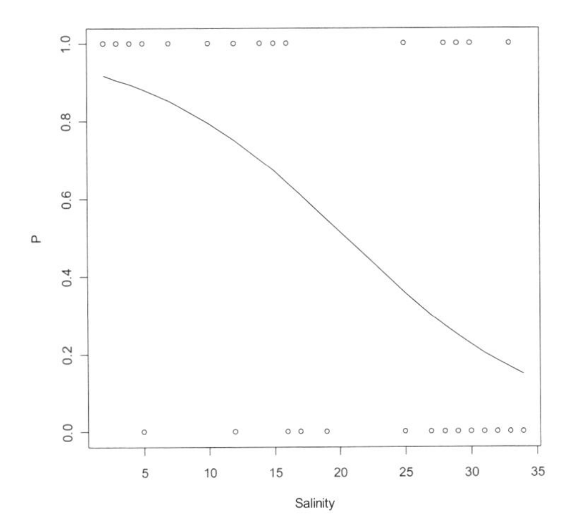

Figure 6.5. Observed and fitted values for *S. solea* data obtained by a logistic regression model.

In linear regression, interpreting a regression coefficient is relatively simple; it represents the one-unit change in Y for a one-unit change in X while keeping all other parameters constant. In logistic regression, this is slightly more complicated. The model we have applied to the *S. solea* data is of the form:

$$O_i = \frac{P_i}{1-P_i} = e^{\alpha+\beta*Salinity_i} = e^{2.66-0.13*Salinity_i} = e^{2.66} * e^{-0.13*Salinity_i}$$

So, the relationship between the odds and salinity is modelled in a non-linear way. If the regression parameter β is estimated as 0, then the exponentiated value is l, and has no effect on the odds. Therefore, a logistic regression parameter β that is zero has no effect on the odds. A positive regression parameter corresponds to an increase in the odds and a negative value to a decrease. For the *S. solea* data, a one-unit change in salinity matches a change in the odds of $e^{-0.13} = 0.88$. So, a one-unit change in salinity means the probability of recording *S. solea* at a site, divided by the probability of not recording it, will change by 0.88. As $e^0 = 1$ represents the effect of no change, the following formula gives the percentage increase or decrease in the odds due to a one-unit change in the explanatory variable: $(e^{\beta} - 1) \times 100$. In this case, for a one-unit change in salinity there is a 12% decrease in the odds of recording *S. solea*.

Validation graphs are shown in Figure 6.6. Panel A shows the residuals versus fitted values. Although this graph shows a distinct pattern, this is because of the presence–absence nature of the data and it does not indicate a lack of fit. So, one string of points corresponds to the samples with a 1, and the other to the samples with a 0. This graph is still useful as it allows you to check that there are no samples with a score of 1 in the string of points for the 0's, and vice versa. This is best done by using the values of the response variable as labels (0 or 1). The QQ-plot (often printed by default) can be ignored for presence/absence data as the residuals will never be normally distributed.

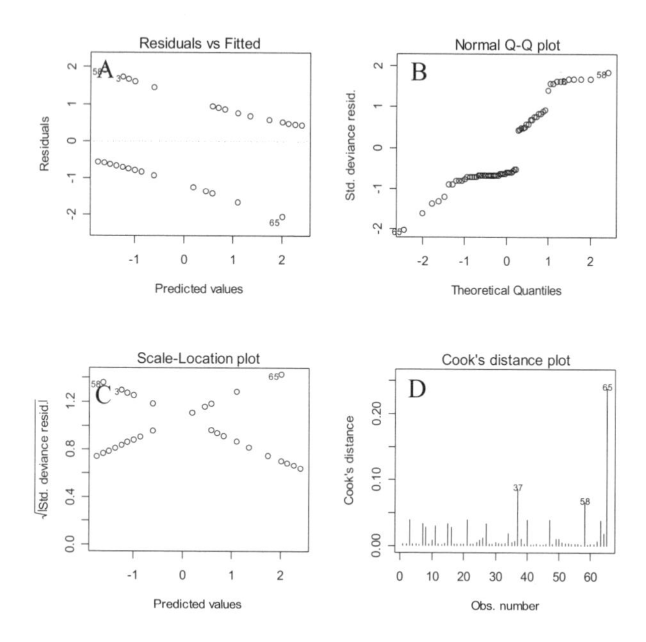

Figure 6.6. Validation graphs for *S. solea* data obtained by a logistic regression model.

Data on proportions can be expressed as Y_i successes out of n_i trials with probability P_i. To be more precise, if Z_j is Bernoulli distributed with $B(1,P_{ji})$, then the sum of Z_1, $Z_2 ... Z_{ni}$ is Binomial distributed: $B(n_i, P_i)$. The expectation and variance of Y_i is equal to

$$E[Y_i] = n_i P_i$$
$$Var[Y_i] = n_i P_i (1 - P_i)$$

This assumes that the n_i individuals are independent of each other, and that they all have the same probability of success, namely P_i. So, if we take 20 samples with a salinity of 15, and find 5 samples with *S. solea* present, this can be modelled as a binomial distribution with n_i = 20, a probability P_i (which is hopefully close to 5/20 = 0.25) and Y_i= 5. If similar data are available from stations with other salinity values, the probability P_i can be modelled in terms of the explanatory variable(s). The logistic regression model is now given by

$$Y_i \sim B(n_i, P_i) \quad \text{and} \quad E[Y_i] = P_i n_i = \mu_i = \frac{e^{g(x_i)}}{1 + e^{g(x_i)}} \quad \text{and} \quad \text{Var}(Y_i) = n_i P_i (1 - P_i)$$

where n_i is the number of samples at site i. Summarising, the Binomial distribution at a site i is specified by the probability P_i and number of observations n_i. The expected value of Y_i is given by $P_i n_i$, and its variance by $P_i n_i (1- P_i)$. The link between the expectation and the linear predictor function is called the logistic link function. Just as in Poisson regression, there can be over- and under-dispersion in logistic regression with the Binomial distribution ($n_i > 1$). One of the reasons for overdispersion is if the n_i individuals are not independent (if they are positively correlated). Overdispersion can be treated in the same way as in Poisson regression, by using the quasi-likelihood method and introducing an overdispersion parameter. However, for the logistic regression model with a Bernoulli distribution ($n_i = 1$), overdispersion does not exist, and therefore, one should not apply a correction for overdispersion.

Testing the significance of regression parameters

There are two options: (i) to use the ratio of the estimated parameter and its standard error or (ii) to use the maximum likelihood ratio test. For large sample size, the ratio of the estimated parameter and its standard error follows a z-distribution. Some programmes give the Wald statistic instead, which is the square of this ratio and it follows a Chi-square distribution. There is some criticism on the use of this test (McCullagh and Nelder 1989) and it should be used very carefully. Raftery (1995) used a BIC value to test the significance of regression parameters. For each regression parameter the following value is calculated:

$$\text{BIC} = z^2 - \log(n)$$

The null hypothesis is H_0: $\beta=0$, where z is the z-value (z^2 is the Wald statistic) and n the sample size. If the BIC is smaller than zero, there is no evidence to reject the null hypothesis. A BIC value between 0 and 2 indicates a weak relationship, values between 2 and 6 indicate a relationship, a strong relationship is indicated by values between 6 and 10, and a very strong relationship is indicated by values greater than 10.

Maximum likelihood

The regression parameters in ordinary regression models are estimated using the least squares method. In logistic regression, this is done with maximum likelihood, and a short discussion is presented next. Readers not interested in the statistical background can skip this section. Suppose we toss a coin 10 times and obtain 3 heads and 7 tails. Let P be the probability that a head is obtained, and $1 - P$ the probability for a tail. Using basic probability rules, the probability for 3 heads and 7 tails is

$$\text{P(3 heads and 7 tails)} = \frac{10!}{3!7!} P^3 (1-P)^7$$

If the coin is fair, you would expect P=0.5. However, suppose we do not know whether it is fair and want to know the most likely value for P. What value of P makes the probability of getting 3 heads and 7 tails as large as possible? Table 6.2 shows the value of the probability of 3 heads and 7 tails for various values of P. For P = 0.3 the probability has the highest value, hence P = 0.3 gives the highest probability for 3 heads and 7 tails. This is the underlying principle of maximum likelihood estimation.

Table 6.2. P(3 heads and 7 tails) for various values of P.

P	P(3 Heads and 7 Tails)	P	P(3 Heads and 7 Tails)
0.1	0.057	0.6	0.042
0.2	0.201	0.7	0.009
0.3	0.267	0.8	0.001
0.4	0.215	0.9	0.000
0.5	0.117	1.0	0.000

In logistic regression, we can also formulate a probability for finding the data:

$$L = \prod_{i=1}^{n} P_i^{Y_i} (1-P_i)^{1-Y_i}$$

where Y_i takes binary values translated to 0 or 1. In the coin tossing example, a head is Y_i = 1 and tail is Y_i = 0. Tossing the coin 10 times might give the Y sequence of 1 0 1 0 0 0 0 0 1 0 (3 heads and 7 tails). Except for the factorial values, this is the same probability as in the previous paragraph. Assuming the probability P(head) does not vary, the question is identical as above; what is the value of P such that the probability of 3 heads and 7 tails is the highest? Recall that in logistic regression P is a function of the regression parameters:

$$P_i = \frac{e^{g(x_i)}}{1+e^{g(x_i)}}$$

The new question is now: what are the values of the regression parameters such that the probability L of the observed data is the highest? Note that L is always between 0 and 1 because probabilities are multiplied with each other. To avoid problems with the 0–1 range (regression lines going outside this range), the natural log is taken of L, resulting in

$$L' = \log(L) = \sum_{i=1}^{n} Y_i \log(P_i) + (1 - Y_i)\log(1 - P_i)$$

Note that log(L) is now between minus infinity and 0, and the upper bound corresponds to a high probability for the data (which is what we want). To find the regression parameters that produce the highest value for the log likelihood, fast non-linear optimisation routines exist, called the iteratively weighted least squares (IWLS) algorithm. This algorithm is described in McCullagh and Nelder (1989), Garthwaite et al. (1995) or Chambers and Hastie (1992), and it is not discussed here. Some of these textbooks are mathematically oriented, but of the three, Chambers and Hastie is the least mathematically demanding.

To assess the significance of regression parameters, the log likelihood value L' itself is not useful. It is more useful to compare L' with the log likelihood value obtained by a reference model. Two candidates for this reference model are the null model and the saturated model. In the null model, we only use the intercept α in the linear predictor function $g(x_i)$. So, this is a poor model and the difference between both L' values only shows how much better our model performs compared with the worst-case scenario. The alternative is to compare L' with the log likelihood of a saturated model that produces an exact fit. The difference between the two log likelihood values show us how much worse our model is compared with the 'perfect' model. The deviance is defined as

$$D = 2 \times (L' \text{ saturated model} - L' \text{ model})$$

This value will always be positive, and the smaller the better. It can be shown that D is asymptotically Chi-square distributed with $n - p$ degrees of freedom (p is the number of parameters and n is the number of observations). Just as in Poisson regression, you can calculate the deviance of two nested models. The difference between the two deviances is again an asymptotically Chi-square distribution, but now with $p_1 - p_2$ degrees of freedom where p_1 and p_2 are the number of parameters in the two nested models. For small datasets ($n < 100$), the deviance test gives more reliable results compared with the z- or Wald test. Just as for Poisson regression, the Chi-square test needs to be replaced by an F-test in case of overdispersion.

Example

Returning to the *S. solea* data we can use the following logistic regression model:

$$Y_i \sim B(1, P_i) \quad \text{and} \quad E[Y_i] = P_i = \frac{e^{g(x_i)}}{1 + e^{g(x_i)}}$$

where $g(x) = \alpha + \beta_1 \times \text{Salinity}_i + \beta_2 \times \text{Temperature}_i$. This model states that the probability that *S. solea* is measured at a particular site i follows a binomial distribution with probability P_i. This probability is a function of salinity and temperature. The numerical output is given by

	Estimate	Std. Error	t-value	p-value
(Intercept)	5.21	3.52	1.48	0.13
temp	−0.10	0.13	−0.75	0.44
sal	−0.14	0.03	−3.60	<0.001
Null deviance: 87.49 on 64 degrees of freedom				
Residual deviance: 67.97 on 62 degrees of freedom. AIC: 73.97				

The deviance is $D = 67.97$ (called the residual deviance in the printout). The z-value indicates the regression parameter for temperature is not significantly different from 0. The BIC value is –3.58, which confirms the non-significance. For salinity, the BIC value of 8.87 suggests a strong relationship between salinity and *S. solea*. The deviance of the full model is 67.97, and the deviance of the model without temperature is 68.56 (see above). The difference is 0.58, and this is asymptotically Chi-square distributed with 1 degree of freedom. This is not significant at the 5% level (the p-value is 0.44[1]), indicating that temperature can be dropped from the model. Just as in Poisson regression, the AIC can be used and in this instance gives the same result for both of these models.

[1] The R command 1-pchisq(Statistic,df) can be used for this. For an F-distribution (in case of overdispersion), use 1-pf(Statistic,df$_1$,df$_2$).

7 Additive and generalised additive modelling

7.1 Introduction

When the data under investigation do not show a clear linear relationship, then additive modelling is a suitable alternative to linear regression. Figure 7.1 shows a scatterplot for two variables of the RIKZ data: species richness and grain size. See Chapter 27 for details on these data. A first look at the graph suggests there is a non-linear relationship between richness and grain size. Sites with large grain sizes seem to have a fairly constant but low species richness, with richness increasing as grain size decreases. Applying a linear regression model with richness as the response variable and grain size as the explanatory variable gives residuals showing a clear pattern, indicating a serious model misspecification.

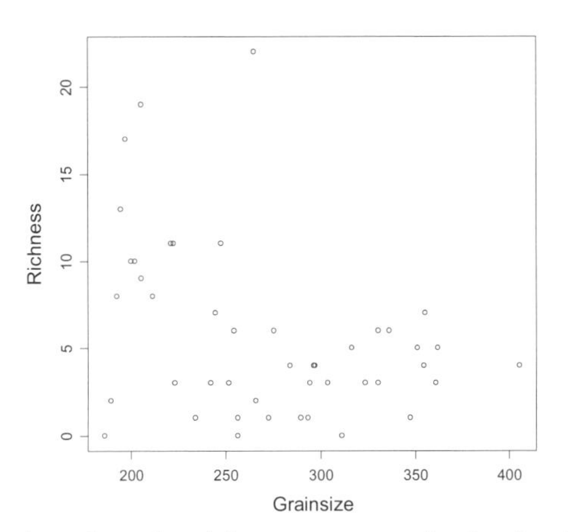

Figure 7.1. Scatterplot of species richness versus grain size for the RIKZ data.

If a data exploration suggests a clear non-linear relationship between the response variable and explanatory variable, then there are various options. First, you could apply a data transformation on the response variable or the explanatory variable (or both) to try to linearise the relationship. In this case, a square root or logarithmic transformation on grain size might produce a linear relationship between the two variables (see also the weight-length wedge clam example in Chapter 4) .

Another option is to model the non-linear relationship using interaction terms between explanatory variables. Suppose the samples with lower grain size were collected during the first few weeks of June 2002, and the other samples were collected towards the end of the month. Then adding a week-grain size interaction term will allow modelling of the non-linear relationship. The problem with interaction terms is deciding which interaction terms to use, especially if there are more than three explanatory variables.

Another alternative is to apply a smoothing method such as additive modelling or generalised additive modelling (GAM). These methods use smoothing curves to model the relationship between the response variable and the explanatory variables. Both allow for non-linear relationships, a common feature of many ecological datasets, and can be used to verify the results of linear regression or GLM models. This can provide confidence that applying a linear regression or GLM is the correct approach. In fact, you can consider linear regression and GLM as special cases of additive and generalised additive models. Indeed the smoothing equivalents of the bivariate and multiple linear regression models could be called bivariate additive models and multiple additive models. However, you are unlikely to see these names in the literature. Good GAM references are Hastie and Tibshirani (1990), Bowman and Azzalini (1997), Schimek (2000), Fox (2002a,b), Rupert et al. (2003), Faraway (2006) and Wood (2006). Although, all are fairly technical, Fox and Faraway are mathematically the simplest.

With linear regressions, we discussed several potential problems such as violation of homogeneity and negative fitted values for count data. As solutions we introduced the Poisson model for count data, and the logistic regression model for presence–absence data. This led to the general framework of generalised linear modelling. The same can be done for smoothing models. For count data, generalised additive models using the Poisson distribution with a log link function are used, and for presence–absence data, we use GAM with the binomial distribution and logistic link function. To understand the additive modelling text below, the reader needs to be familiar with linear regression as summarised in Chapter 5. And understanding GAM requires a detailed knowledge of additive modelling (discussed in this chapter) and GLM, which we discussed in Chapter 6.

Underlying principles of smoothing

The underlying principle of smoothing is relatively simple and is visualised in Figure 7.2 Figure 7.2-A shows a scatter plot of 100 artificial data points. Instead of having a line that goes through each point, we want to have a line that still represents the data reasonably well, but is reasonably smooth. The motivation for smoothing is simply because smooth lines are easier to interpret. A straight line is an extreme example of a smoothed line and although easy to interpret, it might not be the most representative fit of the data. The other extreme is a line that goes through every data point, which although providing a perfect fit may lack any smoothing and therefore be difficult to interpret. Before discussing how smooth we need the curve, we need to discuss the mathematical mechanism that provides the smoothing. In Figure 7.2-B a box is drawn around an arbitrary chosen target value of $X = 40$. At this value of X, we would like to obtain a smoothing value (a Y-value). The easiest option is to take the mean value of all samples in the box. Moving the box along the X gradient will give an estimated value of the smoother at each target value. The resulting smoother of this technique is called a running-*mean* smoother. Another option is to apply a linear regression using only the points in the box. The smoothed value is then the fitted value at the target value of $X = 40$. The fitted line within the box is shown in panel C. Moving the box along the X gradient gives the running-*line* smoother. As well as the running-mean and running-line smoothers, a range of other methods are available, for example: bin smoother, moving average smoother, moving median smoother, nearest neighbourhood smoother, locally weighted regression smoother (LOESS), Gaussian kernel smoother, regression splines, smoothing splines, among others. The mathematical background of most of these smoothers can be found in Hastie and Tibshirani (1990).

An important smoother is the LOESS smoother (Figure 7.2-D). This smoother works like the running-line smoother except weighting factors for the regression are used within the box. The further away a point is from the target value, the less weight it has. Most statistics programmes have routines for LOESS smoothing and smoothing splines, and we discuss these smoothers in detail later.

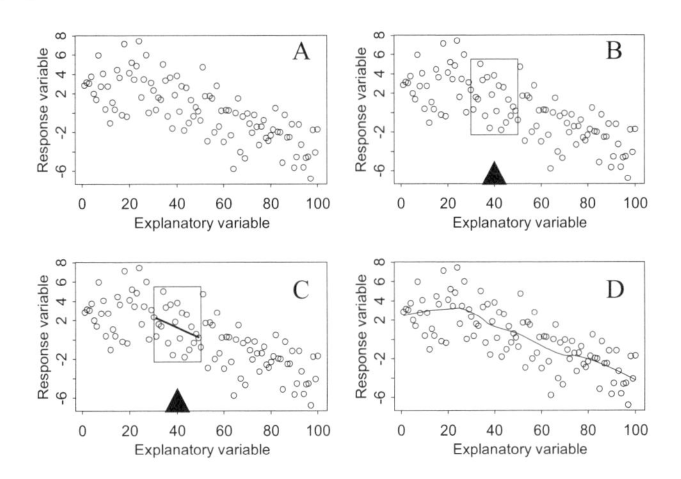

Figure 7.2. Underlying principle of smoothing. A: scatter plot of artificial data. B: A box around the target value of $X = 40$. The width of the box is 20 points (from 30 to 50). C: One option to get an estimated value at $X = 40$; the line within the box is obtained by linear regression. D: LOESS smoother obtained by applying weighted regression in the box and shifting the box along the gradient; the default span width of 0.5 was used (span width will be discussed later).

A crucial point we ignored in the above discussion is the size of the box. Regardless of the smoother used, the researcher still has to choose the width of the box. If the smoother does not use a box (like splines), it will have something equivalent. However, for the moment, we will explain this problem in terms of box width (also called span width), as it is conceptually easier to follow. To illustrate the effect of span width, a LOESS smoother with a span width of 0.7 is shown in panel A of Figure 7.3. A span of 0.7 means that at each target value, the width of the box is chosen such that it contains 70% of the data. If the points along the X gradient are unequally spaced, as with this example, then the width will change for different target values. Panel B is the LOESS smoother using a span of 0.3, which allows for more variation. Panel C shows the LOESS smoother with the default value of 0.5. Finally, panel D is a smoothing spline with the default amount of smoothing (4 df, which we explain later). Each panel gives different information; the smoother in the upper left panel indicates a slow decrease in species richness until a grain size of 275, and then it maintains an approximately constant value. The smoother in panel B shows two peaks in species richness and includes the two samples with low richness and grain size. The smoothing spline (panel D) gives a slightly smoother curve compared with LOESS with a span 0.5 (panel C).

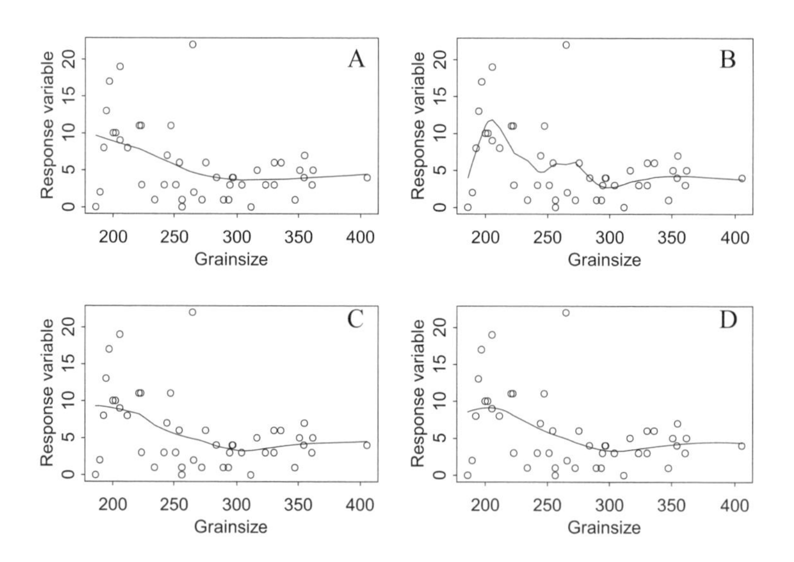

Figure 7.3. Four scatterplots and LOESS smoothers for species richness and grainsize for the RIKZ data. A: LOESS smoother with a span of 0.7. B: LOESS smoother with a span of 0.3. C: LOESS smoother with a span of 0.5, the default value in software packages like SPlus and R. D: Smoothing spline with 4 degrees of freedom (default value).

This arbitrary nature of span width might make smoothing methods confusing. However, if used with common sense, it is a useful method in the toolbox of any scientist. In the remaining part of this chapter we discuss the mathematical framework of additive modelling, the LOESS smoother and smoothing splines, the similarities with regression, model selection (including span width selection), model validation and extensions to the method for count data and presence–absence data.

7.2 The additive model

Before moving on to the additive model we begin by re-visiting the bivariate linear regression model given by

$$Y_i = \alpha + \beta X_i + \varepsilon_i \qquad \text{and} \qquad \varepsilon_i \sim N(0,\sigma^2)$$

where Y_i is the value of the response variable at sample i, X_i is the explanatory variable, and α and β are the population intercept and slope, respectively. The two most important assumptions are normality and homogeneity. The advantage of the linear regression model is that the relationship between Y and X is represented by the slope β. Hence, we only have to look at the estimated slope and its confidence interval to see whether there is a relationship between Y and X. The smoothing equivalent in an additive model is given by:

$$Y_i = \alpha + f(X_i) + \varepsilon_i \qquad \text{and} \qquad \varepsilon_i \sim N(0,\sigma^2)$$

The function $f()$ is the population smoothing function. This model is also called an additive model with one explanatory variable. So instead of summarising the relationship between Y and X using a slope parameter β, we use a smoothing function f. The disadvantage of this approach is that a graph is needed to visualise the function f(X). We will discuss later how to obtain estimates for the intercept and smoothing function.

The additive model $Y_i = \alpha + f(X_i) + \varepsilon_i$, is the equivalent of the bivariate linear regression model $Y_i = \alpha + \beta X_i + \varepsilon_i$. This equivalence allows us to present the underlying statistical principles of the additive model (Figure 7.4) just as we did for linear regression (Figure 5.5). In Figure 7.4, we have used species richness (R) and grain size from the RIKZ data to visualise the bivariate additive model. The observed richness values (points) and fitted values are plotted in the R-grain size space. Gaussian density curves are plotted on top of the fitted values indicating the range and likelihood of other realisations of richness against grain size. In the additive model we assume that R_i is normally distributed with expectation μ_i and variance σ^2, and $E[Ri] = \mu_i = \alpha + f(X_i)$. The only conceptual difference between Figure 7.4 here and Figure 5.5 from the regression chapter is the replacement of the regression curve with a smoothing curve. This similarity means that we can expect the additive model to share the same problems and solutions as we discussed for linear regression, and we discuss this later.

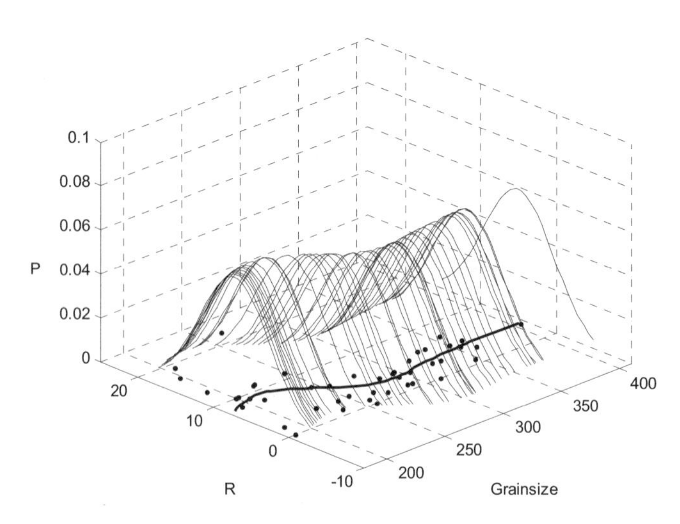

Figure 7.4. Visualisation of a bivariate additive model for the RIKZ data. The grain size richness relationship is modelled as a smoothing curve, and Gaussian density curves indicate the probability of other realisations.

7.3 Example of an additive model

To illustrate an additive model with only one explanatory variable, we use a Bahamas fisheries dataset (unpublished data from The Bahamas National Trust / Greenforce Andros Island Marine Study). The aim of this study was to find a relationship between reef fish and benthic habitat using data collected between February 2002 and September 2003. Here, the parrotfish density is used as the response variable and the explanatory variables are related to algae and coral cover. These are all nominal variables and include location, time (month), and survey method (method 1: point counts, method 2: transects). For this example we have only used the transect data (244 samples) with coral richness used as the explanatory variable. A scatterplot of the data is given in Figure 7.5-A. This shows that sample stations with the highest numbers of coral species also seem to have the highest densities of parrotfish. Dotplots and boxplots indicated there were no extreme observations. The additive model used is of the form:

$$\text{parrotfish}_i = \alpha + f(\text{coral richness}_i) + \varepsilon_i \quad \text{and} \quad \varepsilon_i \sim N(0,\sigma^2)$$

The smoothing function of coral richness is presented in panel B of Figure 7.5 and was estimated using a smoothing spline, which is explained later in this chapter. The numerical output (given below) shows that the intercept is equal to 6.45. This means that the fitted value at a particular station is obtained by

$$\text{parrotfish}_i = 6.45 + f(\text{coral richness}_i)$$

The smoothing function is the solid line in Figure 7.5-B. The confusing aspect of panel B is that both axes have labels that contain 'Coral richness'. The *y*-axis represents the value of the smoother, the software uses the notation s() instead of f(), and the *x*-axis contains the coral richness values. The shape of the smoothing curve indicates that for samples with a coral richness between 3 and 8, the expected parrotfish density is approximately 4.5 parrotfish (6.45 minus 2; the fitted value of the smoother minus the estimated intercept). The highest densities can be found at samples with a coral richness of about 12 (the density will be around 6.45 + 4 = 10.45), but there is a decline if the coral richness is higher.

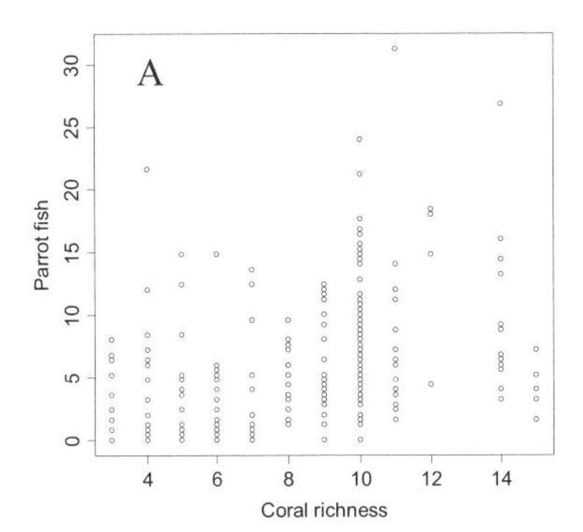

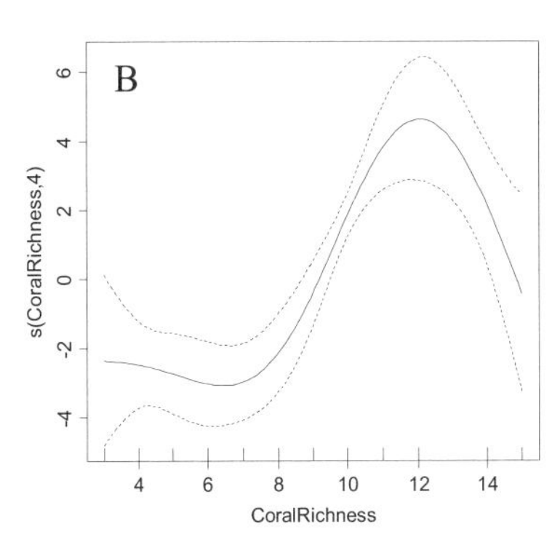

Figure 7.5. A: Scatterplot of coral richness against parrotfish density for the Bahama fisheries data. B: Smoothing function for the additive model applied on the Bahamas fisheries data. The solid line is the estimated smoother, and the dotted lines are point-wise 95% confidence bands.

The numerical output of the model is given by

Parametric coefficients:

	Estimate	std. err.	*t*-ratio	*p*-value
(Intercept)	6.44	0.311	20.74	<0.001

Approximate significance of smooth terms:

	edf	*F*	*p*-value
s(CoralRichness)	4	14.87	<0.001

R-sq.(adj) = 0.18 Deviance explained = 19.9%
Variance = 23.6 n = 244

This output indicates that the smoother is significant (an *F*-test is used); hence, there is a significant non-linear relationship between parrotfish and coral richness, provided a model validation would show that all assumptions are met. However, the explained variance (labelled as deviance by the software) is only 19.9%. This is the equivalent of the R^2 in linear regression. Adding more explanatory variables might improve the model. We will come back to model validation and model selection later in this chapter.

7.4 Estimate the smoother and amount of smoothing

The questions not yet addressed are (i) how do we estimate the smoother, and (ii) how to choose the optimal amount of smoothing. Starting with the first question, one of the simplest smoothers is the moving average smoother. Here is how it works. What is the mean value of the numbers 1, 2, 3, 4 and 5? The correct answer is 3, but to come to this you added them all up and divided by 5. Or in formula:

$$\text{mean} = \frac{1+2+3+4+5}{5} = \frac{1}{5}1 + \frac{1}{5}2 + \frac{1}{5}3 + \frac{1}{5}4 + \frac{1}{5}5$$

So, each value is multiplied with a weighting factor 1 over 5. In fact, we can easily extend the series with the values 0x6, 0x7, 0x8, etc. If we want to emphasise the importance of some values, then we can change the weighting factors. For the artificial data in Figure 7.2 we can use the following formula to estimate a smoothing value at the target value of $X = 40$:

$$\widehat{Y}_{40} = a_{30}Y_{30} + ... + a_{40}Y_{40} + ... + a_{50}Y_{50}$$

The box contains 21 neighbouring points around the target value (X=40). If we choose the weighting factors a_i as 1 over 21, we just get the mean value (as in the example above). But suppose we only want the five neighbouring observations to have an influence on the target value. A possible choice of coefficients is

$$\widehat{Y}_{40} = 0 + ... + 0 + 0.1 * Y_{38} + 0.2 * Y_{39} + 0.4 * Y_{40} + 0.2 * Y_{41} + 0.1 * Y_{42} + 0 + ... + 0$$

Hence, the fitted value at $X = 40$ is a weighted average of five neighbouring points. Note that this is the same principle as calculating the mean of the numbers 1 to 5. In matrix (or vector) notation, this can be written as

$$\widehat{Y}_{40} = \begin{pmatrix} 0 & \cdots & 0 & 0.1 & 0.2 & 0.4 & 0.2 & 0.1 & 0 & \cdots & 0 \end{pmatrix} \begin{pmatrix} Y_1 \\ \vdots \\ Y_{37} \\ Y_{38} \\ Y_{39} \\ Y_{40} \\ Y_{41} \\ Y_{42} \\ Y_{43} \\ \vdots \\ Y_{100} \end{pmatrix} = s \times Y$$

where $s = (0,\ldots,0, 0.1, 0.2, 0.4, 0.2, 0.1\ 0, \ldots, 0)$ and $Y = (Y_1,\ldots,Y_{100})'$. In fact, the fitted values at all target values can be obtained using a single matrix multiplication:

$$\hat{Y} = S \times Y$$

The matrix S contains all weighting factors and $\hat{Y}$ all target values. Filling in all elements of S is not an exercise to do by hand, but it is straightforward using a computer. The estimated smoother can also be written as (the difference between fitted values and smoother is only the intercept):

$$\hat{f}(X) = S \times Y$$

We use the hat notation to indicate that it is an estimated function. So, where does the underlying theory like p-values, F tests, t-values, etc., come from in additive modelling? In the next couple of paragraphs we will show how the expression for the smoother is related to the expression for the linear regression model. This means that we have to give some mathematical detail based on matrix algebra. You may skip these paragraphs, or only try to catch the concepts, if you are not familiar with matrix algebra.

Using simple algebra, the fitted values of a linear regression model $Y = \beta X + \varepsilon$ (for ease of notation the intercept was written within β by setting the first column of X equal to one) can be written as

$$\hat{Y} = X(X'X)^{-1}X'Y$$

This is a standard formula obtained by ordinary least squares and can be found in any good linear regression textbook (e.g., Montgomery and Peck 1992). A weighted linear regression model gives the following formula for the fitted values:

$$\hat{Y} = X(X'WX)^{-1}X'WY$$

The matrix W contains the weights. Again, this formula can be found in any textbook that covers weighted linear regression. It is easy to see that this formula can be written as

$$\hat{Y} = HY, \qquad \text{where} \qquad H = X(X'WX)^{-1}X'W$$

H is called the hat matrix and was used in Chapters 4 and 5 to identify possible influential observations (leverage). The LOESS smoother applies a weighted linear regression on the data within each box. The LOESS smoothing value at the target X is then the fitted Y value for this X value. The LOESS weights are obtained as follows: Points outside the box have a weight of 0 and points inside the box have weights following a unimodal pattern centred at the target value. The formula for the weights can be found in Chapter 2 of Hastie and Tibshirani (1990). However, you do not need to know exactly how the weights are calculated, and we do not discuss it further. The important thing to remember is that LOESS smoothing can be written as $\hat{f}(X) = S \times Y$; all that is required is to fill in the elements of S, which we can leave to the computer.

Splines

Only a short introduction into splines is given here (based on Fox 2000, 2002b), and the interested reader is referred to Hastie and Tibshirani (1990) or Wood (2006) for more details. A polynomial model for the Bahama's Parrotfish example has the following form:

$$\text{parrotfish}_i = \alpha + \beta_1 X_i + \beta_2 X_i^2 + \beta_3 X_i^3 + \beta_4 X_i^4 + \ldots + \beta_p X_i^p + \varepsilon_i$$

where $\varepsilon_i \sim N(0,\sigma^2)$ and X_i is the coral richness. A cubic polynomial uses terms up to, and including, X_i^3. The fit of the cubic polynomial model is given in Figure 7.6-A. The main problem with polynomial functions is that the fit is often rather poor, even though it seems to perform reasonably well in this example. To improve the fit, we can split the data into two parts based on the values of coral richness, and fit a separate cubic polynomial model on each dataset. We arbitrarily decided to split the data at a richness value of eight and the fitted values for both datasets are presented in the Figure 7.6-B. If required, we could continue splitting the data and repeat the exercise until a satisfactory fit is achieved.

The points along the x-axis where we split the data are called knots. The problem with this approach is seen in Figure 7.6-B, where the lines are not continuous at the point where they should meet. In some cases, they might not meet at all, and this defeats the main aim of obtaining a smooth curve.

A solution is to use a cubic regression spline, which is a third-order polynomial function with the following conditions:

- The curves must join at each knot.
- The first derivative must be continuous at each knot.
- The second derivate must be continuous at each knot.

A *natural* cubic regression spline is like a cubic regression spline, but with an additional constraint that forces a linear fitted line beyond the smallest and largest X values. This aims to avoid spurious behaviour at the ends of the gradient. But, how do you know how many parameters are used? It is important to know this as we may be over-fitting the data. Assume K knots are selected. This means there are $K + 1$ datasets, with a cubic polynomial applied with four parameters (three slopes, one for each term X_i, X_i^2 and X_i^3, and one intercept), to each dataset. However, at each knot we have three restrictions (the curves must join and first and second derivatives must be continuous), plus two at the edges of the gradient. This means that the total number of parameters is $4(K + 1) - 3K - 2 = K + 2$. The problem is knowing how many knots to choose: a similar problem as deciding on the span width in LOESS.

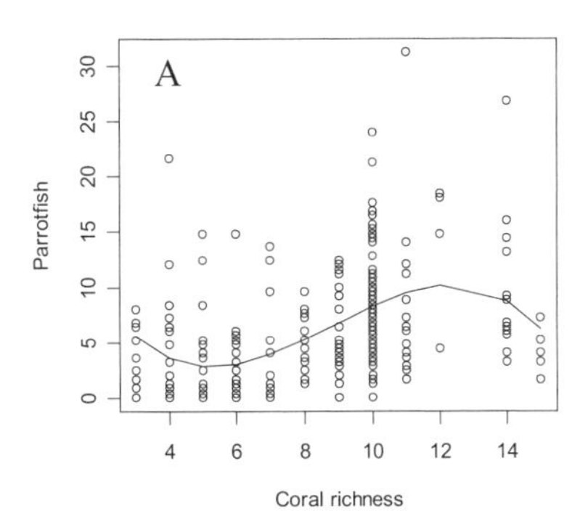

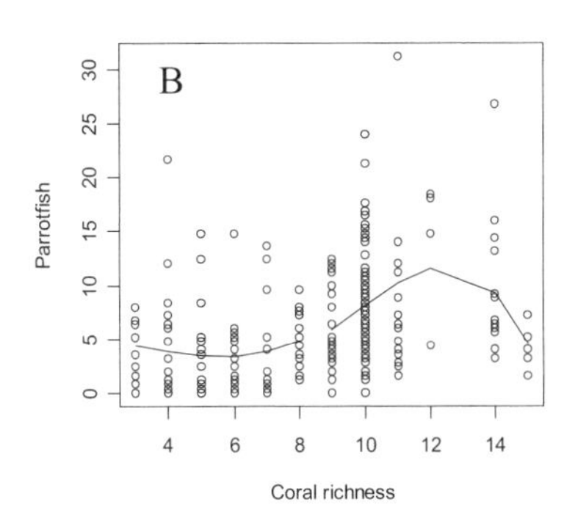

Figure 7.6. A: Fitted line obtained by a cubic polynomial model for the parrotfish of the Bahamas data. B: The data were split up in two parts (coral richness smaller and equal to eight, and larger than eight), and in each dataset a cubic polynomial model was applied.

Finally, we look at smoothing splines, starting with the formula called the penalised sum of squares:

$$SS(h) = \sum_{i=1}^{n} (Y_i - f(X))^2 + h \int_{x_{\min}}^{x_{\max}} f''(x)^2$$

This formula looks complex but its rationale is fairly simple (and is based on high school mathematics). The function *f*() is again the smoother. The first part is a sum of residual squares. The less smooth the function *f*(), the smaller the residual sum of squares. The integral measures the smoothness of the function *f*() and *h* is the penalty for non-smoothness. If *f*() is rough, its second derivative will be large, and therefore the integral will be large as well. The aim is to minimise SS(*h*). Choosing a small value of *h* will give a low penalty to roughness, whereas choosing a large value gives a high penalty. Extremely small values, such as $h = 0$, will give a line that goes through every point, and a very large value of *h* will give a linear regression line. The function *f*() that minimises SS(*h*) is a cubic smoothing spline, and the roughness penalty ensures that there are not too many parameters. Smoothing splines can also be written as $\hat{f} = SY$.

Degrees of freedom

We mentioned above that the fitted values in the linear regression model $Y = \beta X + \varepsilon$ are given by:

$$\hat{Y} = X(X'X)^{-1}X'Y$$

This is commonly written as:

$$\hat{Y} = HY, \qquad \text{where} \qquad H = X(X'X)^{-1}X'$$

The matrix H is called the hat matrix because it puts a hat on Y (it is estimated from the sample data) and, among other things, is used to calculate leverage. The degrees of freedom for the model is the number of parameters, and this is equal to the number of columns (variables) in X, assuming there are no problems of perfect collinearity that would require removing some collinear variables from the analysis. Another way to calculate the degrees of freedom is to use the rank or trace of the matrix H, where the trace is the sum of the diagonal elements. In linear regression, residuals can be calculated as observed values minus fitted values. In matrix notation, this is

$$e = Y - \hat{Y} = Y - HY = (I - H)Y$$

This shows the important role of the hat matrix in linear regression, with the degrees of freedom for the error given by $df_{res} = rank(I - H)$ or $n - trace(H)$.

The reason for introducing this equation is because in additive modelling degrees of freedom are defined in a similar way. Our starting point is:

$$\hat{f} = SY$$

And we only have to replace H by S to obtain the degrees of freedom. Although there are some theoretical problems with this method, software like R and SPlus use this analogous approach, which also gives confidence bands for the smoothers:

$$\text{Covariance } \hat{f} = \sigma^2 SS'$$

The same issues exist for hypothesis tests in additive modelling (testing whether a smoother is equal to zero), which also only mimic linear regression, with no formal justification. However, Hastie and Tibshirani (1990) mention that empirical studies have shown that hypothesis tests are reasonably robust.

7.5 Additive models with multiple explanatory variables

Recall that the multiple linear regression model is given by

$$y_i = \alpha + \beta_1 x_{i1} + \ldots + \beta_p x_{ip} + \varepsilon_i, \qquad \text{and } \varepsilon_i \sim N(0,\sigma^2)$$

leading to an additive model for p explanatory variables being defined by

$$y_i = \alpha + f_1(x_{i1}) + \ldots + f_p(x_{ip}) + \varepsilon_i, \qquad \text{and } \varepsilon_i \sim N(0,\sigma^2)$$

Each function $f_j()$ is a smoothing curve for the population, and can be estimated by using, for example, a running-mean smoother, a LOESS smoother or a smoothing spline.

As an example, we again use the RIKZ data to investigate whether there is a relationship between species richness and the explanatory variables: temperature, grain size and exposure. Exposure has only three values and is best modelled as a nominal variable with the results presented in a table rather than a graph. The additive model is of the form:

$$R_i = \alpha + \text{Exposure}_i + f_1(\text{Temperature}_i) + f_2(\text{Grain size}_i) + \varepsilon_i$$

where $\varepsilon_i \sim N(0, \sigma^2)$. The estimated smoothing curves for temperature and grain size, using the default amount of smoothing, are shown in Figure 7.7. The numerical information related to the intercept and exposure are shown in Table 7.1.

The fitted value for any particular sample receives a contribution from four terms: the intercept, exposure, temperature and grain size. The intercept is 17.68. Keep in mind that we are fitting species richness: the number of species recorded at a specific site. If a sample is from a site with exposure 3 (first class), a value of 0 is added. This is because the first exposure level is used as a baseline. So for sites with an exposure of 3, the exposure plus intercept is 17.68 + 0 (number of species). If a site has an exposure of 10, then 6.49 fewer species are counted. At sites with an exposure of 11, the richness is 19.46 lower, than the first class, with an exposure value of 3. The *p*-values suggests that exposure (at least for class 11) is highly significant at the 5% level. Later in this section, the *F*-test is used to obtain a single overall *p*-value for exposure. As well as exposure, the fitted values get some contribution from the smoothing curves for temperature and grain size. High temperature and low grain size is related to low richness, and high grain size and low temperature is related to higher richness (a larger number of species). Note that nothing stops the model from obtaining negative fitted values!

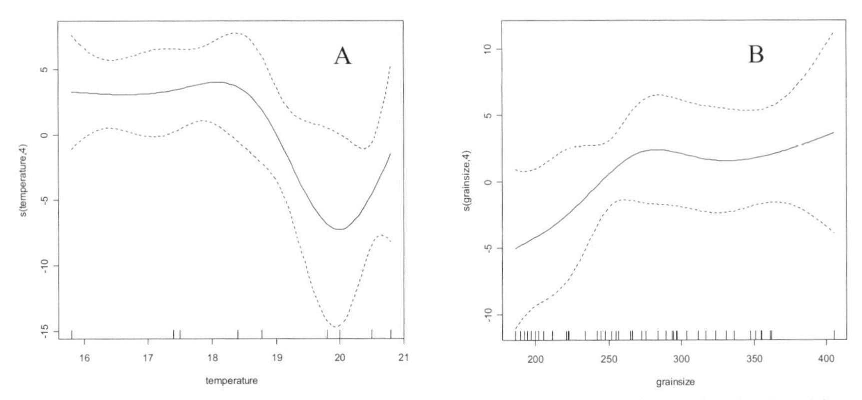

Figure 7.7. Smoothing curves for temperature (A) and grain size (B), obtained by additive modelling. The RIKZ dataset was used. The dotted lines are 95% confidence bands.

Table 7.1. Estimated parameters, standard errors, t-values and p-values of the parametric components in the additive model for the RIKZ data. Of the three nominal values for Exposure, the first exposure class is considered a baseline and adopts a value of zero. It is not reported in the table.

	Estimate	Std error	t-value	p-value
Intercept	17.32	3.40	5.09	<0.001
Exposure.10	−7.51	3.82	−1.96	0.06
Exposure.11	−18.67	4.81	−4.81	<0.001

Model selection

The question now is whether the smoothing terms are significantly differently from 0. The relevant numerical output produced is as follows.

```
Approximate significance of smooth terms:
                    edf    F      p-value
s(temperature)      4      2.45   0.06
s(grain size)       4      0.56   0.69

R-sq.(adj) = 0.359   Deviance explained = 50.5%
GCV score = 21.845   Scale est. = 16.05   n = 45
Dispersion parameter= 16.05,
Deviance= 545.85, df= 11
```

For the Gaussian additive model that we are looking at here, the dispersion parameter (also denoted by the scale estimator in the numerical output) is the variance σ^2 (16.05). The deviance is equivalent to the residual sum of squares and explains 50.5% of the null deviance, which is the equivalent of the total sum of squares in linear regression. GCV is the cross-validation score and is explained later. The p-value for grain size indicates that the smoothing component is not significant, and could probably be dropped from the model.

As with linear regression, we can omit an explanatory variable and then compare the fit of the two nested models with each other: with and without the dropped variable. This can be done with an F-test, or using the AIC. The F-test uses the residual sum of squares of the full model (RSS_2) and compares them with residual sum of squares of the nested model (RSS_1):

$$F = \frac{(RSS_1 - RSS_2)/(df_2 - df_1)}{RSS_2 / df_{res}}$$

where df_j is the degrees of freedom in model j, df_{res} is $n - df_2$, and n is the number of samples. The F-ratio can be compared with an F-distribution with $df_2 - df_1$ and df_{res} degrees of freedom. The models compared are:

$$R_i = \alpha + \text{factor}(\text{Exposure}_i) + f_1(\text{Temperature}_i) + f_2(\text{Grain size}_i) + \varepsilon_i \quad (1)$$
$$R_i = \alpha + \text{factor}(\text{Exposure}_i) + f_1(\text{Temperature}_i) + \varepsilon_i \quad (2)$$
$$R_i = \alpha + \text{factor}(\text{Exposure}_i) + f_2(\text{Grain size}_i) + \varepsilon_i \quad (3)$$
$$R_i = \alpha + f_1(\text{Temperature}_i) + f_2(\text{Grain size}_i) + \varepsilon_i \quad (4)$$

Comparing models 1 (the full model) and 2 gives an F-ratio of 0.5631 (p = 0.691) indicating that grain size is not important. Comparing models 1 and 3 (leaving out temperature) gives an F-ratio of 2.45 (p = 0.064) and finally leaving out exposure gives a ratio of 7.52 (p = 0.002). This indicates that grain size is not significant; there is a weak temperature effect and a strong exposure effect.

The alternative is to calculate an AIC for each model. The AIC for an additive model is defined by

$$\text{AIC} = -2\log(\text{Likelihood}) + 2df \tag{7.1}$$

where df takes the role of the total number of parameters in a regression model (Chapter 5). For this example, using all three explanatory variables gives an AIC of 264.01. The model with exposure and temperature has an AIC of 258.9, the model with exposure and grain size 267.43, and grain size and temperature 276.5. The lower the AIC, the better the model. Hence, we can drop grain size from the model, giving a new model with the form:

$$\text{R}_i = \alpha + \text{factor}(\text{Exposure}_i) + f_{1i}\,(\text{Temperature}) + \varepsilon_i$$

The same process can be applied to this new model to see whether exposure or temperature can be dropped. As well as applying the selection procedure on the explanatory variables, it should also be applied on the amount of smoothing for each explanatory variable. This means that the model selection procedure in additive modelling not only contains a selection of the optimal explanatory variables but also the optimal degrees of freedom per variables.

The back-fitting algorithm for LOESS

We haven't discussed yet how to obtain numerical estimated of intercept, regression parameters and smoothers. For splines, this is rather technical and involves integrals and second-order derivates, and the interested reader is referred to Wood (2006) for technical details, but be prepared for some complicated mathematics. For LOESS it is all much easier, and the principle is sketched below. We start with the additive model in which only one smoother $f()$ is used.

For the additive model we have to estimate the intercept α, the smoothing function $f()$ and the variance σ^2. In linear regression, ordinary least squares can be used to give simple formulae that estimate both the intercept and the slope. In additive modelling, we need to use a so-called back-fitting algorithm to estimate the intercept α and the smoothing curve $f()$. The algorithm follows these basic steps:

- Estimate the intercept α for a given smoothing function $f()$.
- Estimate the smoothing function $f()$ for a given intercept α.

These two steps are applied until convergence is reached, and in more detail look like this:

1. Estimate the intercept as the mean value of observed values Y_i, $i = 1,..,n$.
2. Subtract the intercept from the observed values: $Y_i - \hat{\alpha}$, and estimate the smoothing curve $f()$, using any of the smoothing methods discussed above.

3. To make the smoother unique, centre it around zero (calculate the overall mean of the smoother and subtract it from each value of the smoother).
4. Estimate the intercept as the mean value of $Y_i - \hat{f}(X_i)$. The mean is taken over all observations *i*.
5. Repeat steps 2 to 4 until convergence.

By convention a hat is used to indicate an estimator of *f* and α. Although this is an extremely *ad hoc* mathematical procedure, in practice it works well and convergence is nearly always obtained in a few cycles. If the additive model contains more than one smoother, a generalisation of the back-fitting algorithm is used to identify the different smoothers and the intercept. It has the following form:

1. Obtain initial estimates for all components in the model by using for example random numbers.
2. Estimate the intercept as the mean value of the observed values Y_i, $i = 1,..,n$.
3. Estimate $f_1()$ by smoothing on $Y_i - \hat{\alpha} - \hat{f}_2(X_{i2}) - ... - \hat{f}_p(X_{ip})$.
4. Estimate $f_2()$ by smoothing on $Y_i - \hat{\alpha} - \hat{f}_1(X_{i1}) - \hat{f}_3(X_{i3}) - ... - \hat{f}_p(X_{ip})$.
5. ....repeated until....
6. Estimate $f_p()$ by smoothing on $Y_i - \hat{\alpha} - \hat{f}_1(X_{i1}) - ... - \hat{f}_{p-1}(X_{ip-1})$.
7. Repeat steps 1-5 until nothing changes anymore (convergence).

In each step, the smoothing functions are made unique by mean deletion (centring around zero).

7.6 Choosing the amount of smoothing

Instead of specifically choosing the span width in LOESS smoothing, or the value for the penalty term *h* in smoothing splines, we normally only choose a degree of freedom for a smoothing term. This defines the amount of smoothing required, and the software then chooses the required span width or penalty term (h) used for the smoothing spline. The smoothers are calibrated so that a smoother with one degree of freedom gives an approximate straight line. The default value in programmes like SPlus and R is for four degrees of freedom, which approximately coincides with the smoothing of a third-order polynomial. But is the default always the best choice? This can be checked by:

1. Using trial and error and looking at the graphs.
2. Looking at residuals and changing the degrees of freedom to see whether there are any residual patterns.
3. Using a modified version of the AIC, see equation (7.1).
4. Using cross-validation to estimate the amount of smoothing automatically.

We discuss each of these points next. The lattice graph shown in Figure 7.8 shows a range of smoothing curves for the parrotfish data. The lower left panel is the smoothing curve for the (only) explanatory variable coral richness using one degree of freedom. The remaining panels show the smoothing curves with increasing degrees of freedom up to six. The shapes of the curves show there is little improvement after three or four degrees of freedom, and in this example, four degrees of freedom is the best choice. We can also look at the plot of residuals and if there is any discernable pattern choose to increase the degrees of freedom to allow for more variation in the smoother. This allows for stronger non-linear effects. When there are multiple explanatory variables, a plot of residuals versus each explanatory variable should be made. If any of these graphs shows a pattern, then the degrees of freedom of the smoother should be increased.

Earlier, we discussed using the AIC for selecting the best additive model and it can also be used to find the model with the optimal degrees of freedom. We can also use the F test to compare two nested models. A large *F*-ratio is evidence that the omitted explanatory variable is important. Note that a model with lower degrees of freedom can be seen as a nested model of a model with higher degrees of freedom. It is this principle that allows an additive model to be compared with a linear regression model (Fox 2000), because the linear regression can be seen as a nested model of the smoothing model.

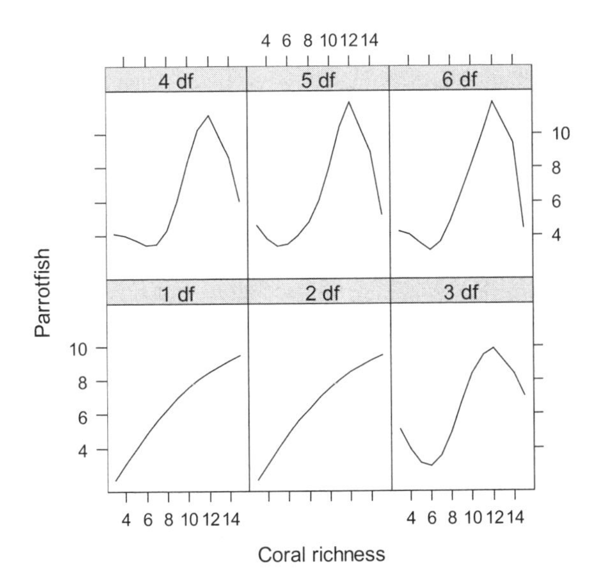

Figure 7.8. Smoothing curves for the explanatory variable coral richness in the Bahamas example. Each panel shows a smoothing curve using different degrees of freedom.

Staying with degrees of freedom, we now discuss cross-validation, which is a tool that automatically estimates the degrees of freedom for each smoother. In cross-validation, we leave out observation *i*, estimate a smoothing curve using the n - 1 remaining observations, predict the value at the omitted point X_i, and compare the predicted value at X_i with the real value Y_i. The difference between the original and predicted value at X_i is given by:

$$Y_i - \hat{f}_{\lambda}^{-i}(X_i)$$

The notation "–i " in the smoother is used to indicate that observation i was omitted. The parameter λ refers to the amount of smoothing. This process is repeated for each observation. Adding up all squared residuals gives the cross-validation error:

$$CV(\lambda) = \frac{1}{n}\sum_{1=1}^{n}(Y_i - \hat{f}_{\lambda}^{-i}(X_i))^2$$

The cross-validation error is calculated for various values of the smoothing parameter λ. The value of λ that gives the lowest value for CV is considered the closest to the optimal amount of smoothing. The generalised cross-validation (GCV) is a modified version of the CV. To justify the use of the cross-validation, we introduce two quantities called MSE(λ) and PSE(λ). Remember, the linear regression model $Y = \alpha + \beta X + \varepsilon$ is a model for the entire population, and the same holds for the additive model $Y = \alpha + f(X) + \varepsilon$. The smoothing function $f()$ is also for the entire population and estimated by $\hat{f}()$. The average mean squared error (MSE) measures the difference between the real smoother and the estimated smoother, and is defined by

$$\text{MSE}(\lambda) = \frac{1}{n}\sum_{i=1}^{n} E[f(X_i) - \hat{f}_{\lambda}(X_i)]^2$$

where $E[]$ stands for expectation. A related measure is the average predicted squared error (PSE):

$$\text{PSE}(\lambda) = \frac{1}{n}\sum_{i=1}^{n} E[Y_i^* - \hat{f}_{\lambda}(X_i)]^2$$

Y_i^* is a predicted value at X_i. It can be shown that PSE = MSE + σ^2. The theoretical justification for using cross-validation is that E[CV(λ)] is approximately equal to PSE (Hastie and Tibshirani 1990).

To illustrate cross-validation, the following model was applied on the Bahamas parrotfish dataset:

$$\text{parrotfish}_i = \alpha + f(\text{coral richness}_i) + \varepsilon_i \quad \text{and } \varepsilon_i \sim N(0,\sigma^2)$$

For the smoothing function, we first used two degrees of freedom and applied the cross-validation approach, and calculated CV(2). Then we used three degrees of freedom and obtained CV(3), continuing up to CV(10). A plot of CV versus degrees of freedom is given in Figure 7.9 and suggests that about four degrees of freedom is optimal. The R library mgcv allows for automatic application of the cross-validation method and gives a value of 4.37 degrees of freedom for coral richness.

You may expect that increasing the degrees of freedom always gives lower GCV values, but this is not the case as a high λ will ensure that observed values are fitted well, but it does not mean that omitted points are predicted well (Wood 2006).

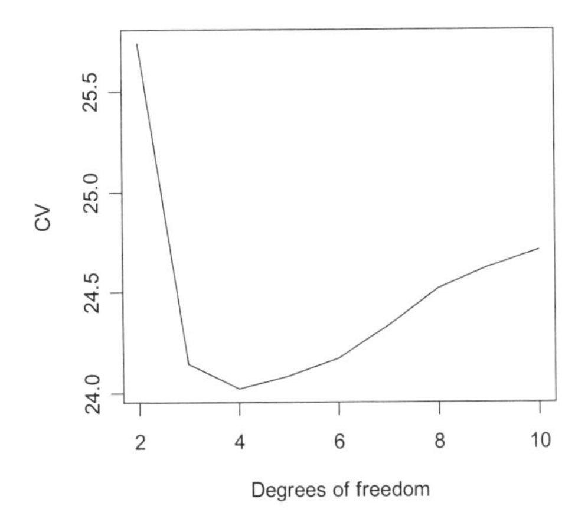

Figure 7.9. A plot of CV versus degrees of freedom for the Bahamas fisheries dataset.

7.7 Model selection and validation

This section works through the model selection and validation process using the squid data from Chapter 4. Recall from Section 4.1 that these data contain measurements for squid from various locations, months and years in Scottish waters. GSI is the Gonadosomatic index and is a standard index used by biologists.

Model selection

A data exploration using Cleveland dotplots and boxplots showed that there are no extreme observations. A coplot showed that month, year, sex and location effects might be expected. GSI is the response variable, with month, year, location and sex considered as explanatory variables. Of the explanatory variables only month has a reasonable number of different values (12). All other explanatory variables have less than five unique values, including year. In such situations, moving the box along the x-axis (Figure 7.2) will cause an error because there are not enough points in the box around a particular target value. If month is considered as a nominal variable, it will require 11 parameters because there are 12 months. It cannot be modelled as a parametric component because this would rank the dummy code 'twelve' for December higher than the 'one' for January, which would not make sense. An option is to model month as a smoother, and this reduces the number of parameters from 11 to 4, assuming the default amount of smoothing is used. The model we applied is in the form:

$$\mathrm{GSI}_i = \alpha + f(\mathrm{month}_i) + \mathrm{year}_i + \mathrm{location}_i + \mathrm{sex}_i + \varepsilon_i$$

Year, location and sex are modelled as nominal variables, and there are 2644 observations in the dataset. The function $f()$ was estimated with a smoothing spline and in the first instance, we used four degrees of freedom. The fitted smoothing function for month is presented in the Figure 7.10-A, and it shows a clear seasonal pattern. In panel B the smoothing curve obtained by cross-validation is shown, and is considerably less smooth. The estimated degrees of freedom is approximately nine, which is two degrees of freedom less compared with using month as a nominal variable.

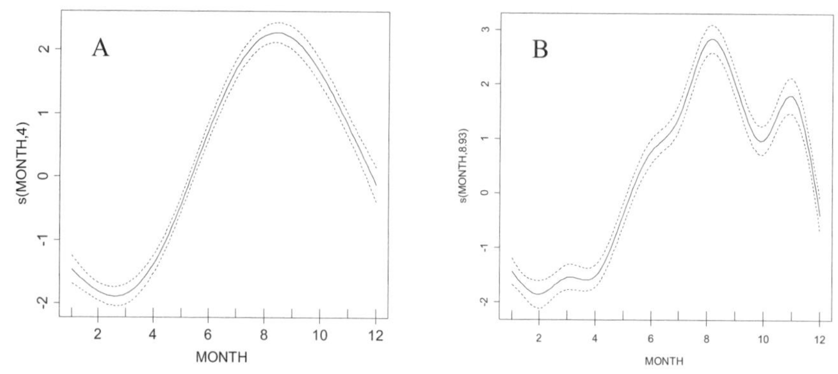

Figure 7.10. Smoothing curve for month for the squid data with 4 degrees of freedom (A) and 8.9 degrees of freedom (B).

Using the F-test to compare a full model with 8.9 degrees of freedom and a nested model with 4, 5, 6, 7 or 8 degrees of freedom confirms that 8.9 is indeed optimal. The numerical output of the model, with 8.9 degrees of freedom, is given below.

	Estimate	std. err.	t-ratio	p-value
Intercept	1.37	0.11	12.47	<0.001
factor(YEAR)2	0.14	0.12	1.22	0.22
factor(YEAR)3	−0.23	0.12	−1.88	0.05
factor(YEAR)4	−0.76	0.17	−4.27	<0.001
factor(Location)2	−0.33	0.18	−1.80	0.07
factor(Location)3	0.11	0.11	0.94	0.34
factor(Location)4	2.00	0.31	6.42	<0.001
factor(Sex)2	1.90	0.07	24.63	<0.001

For the smoothers we have the following numerical output

Approximate significance of smooth terms:

	edf	F-statistic	p-value
s(MONTH)	8.92	127.4	<0.001

R-sq.(adj) = 0.44. Deviance explained = 45%. GCV score = 3.93. Scale est. = 3.91. n = 2644. Dispersion parameter = 3.91. Deviance = 10275.3. df.residual (residual degrees of freedom) = 2627.07. df (n-df.residual) = 16.93. AIC according to formula: -2log(Likelihood) + 2df = 11128.29.

The explained deviance is 45%, which means that 45% of the total sum of squares is explained by the model. The estimated variance is $\sigma^2 = 3.91$ and the AIC is 11128.29. The smoothing term is significant at the 5% level and the p-values of the individual levels indicate the effects of location, year and sex are significant. The F-test can be used to obtain one overall p-value per nominal variable. Comparing the full model with a model without year gives an F-ratio of 14.34 ($p < 0.001$). Leaving out location gives $F = 16.79$ ($p < 0.001$). Both terms are highly significant in explaining GSI.

Model validation

As in linear regression, with additive models we need to verify the underlying assumptions of homogeneity and normality, and check for potential influential observations. Only if the model validation indicates that there are no problems, can we accept the smoothing curve in Figure 7.10-B, and the numerical output above. Figures 7.11 and 7.12 give a series of graphs, which can be used for model validation. Figure 7.11-A shows the fitted values against the observed values. Ideally, the points in Figure 7.11-A should lie on a straight line, but as the model only explains 45% of the variation in the data some discrepancies can be expected. Panel B illustrates the fitted values against the residuals, and these show a worrying violation of homogeneity. The concentration of points at the lower part of this panel are probably the samples with zero or low values. Panels C and D check on normality and suggest a lack of normality in our data. This violation of homogeneity and normality is enough justification to show the model is unsatisfactory. However, for the sake of this example, we will continue working through the model validation process. Panel E shows the leverage value for each sample, and this identifies a group of samples with considerably higher leverage than the rest. As none of the explanatory variables lend themselves to a transformation (there is no point in transforming a nominal variable and all months are sampled), there is little we can do about this particular problem. It might be caused by a set of samples only measured in one month in one location in one year. If this is indeed the case, it could be argued that they should be left out of the analysis, but a detailed data exploration using coplots and scatterplots is required before making this decision.

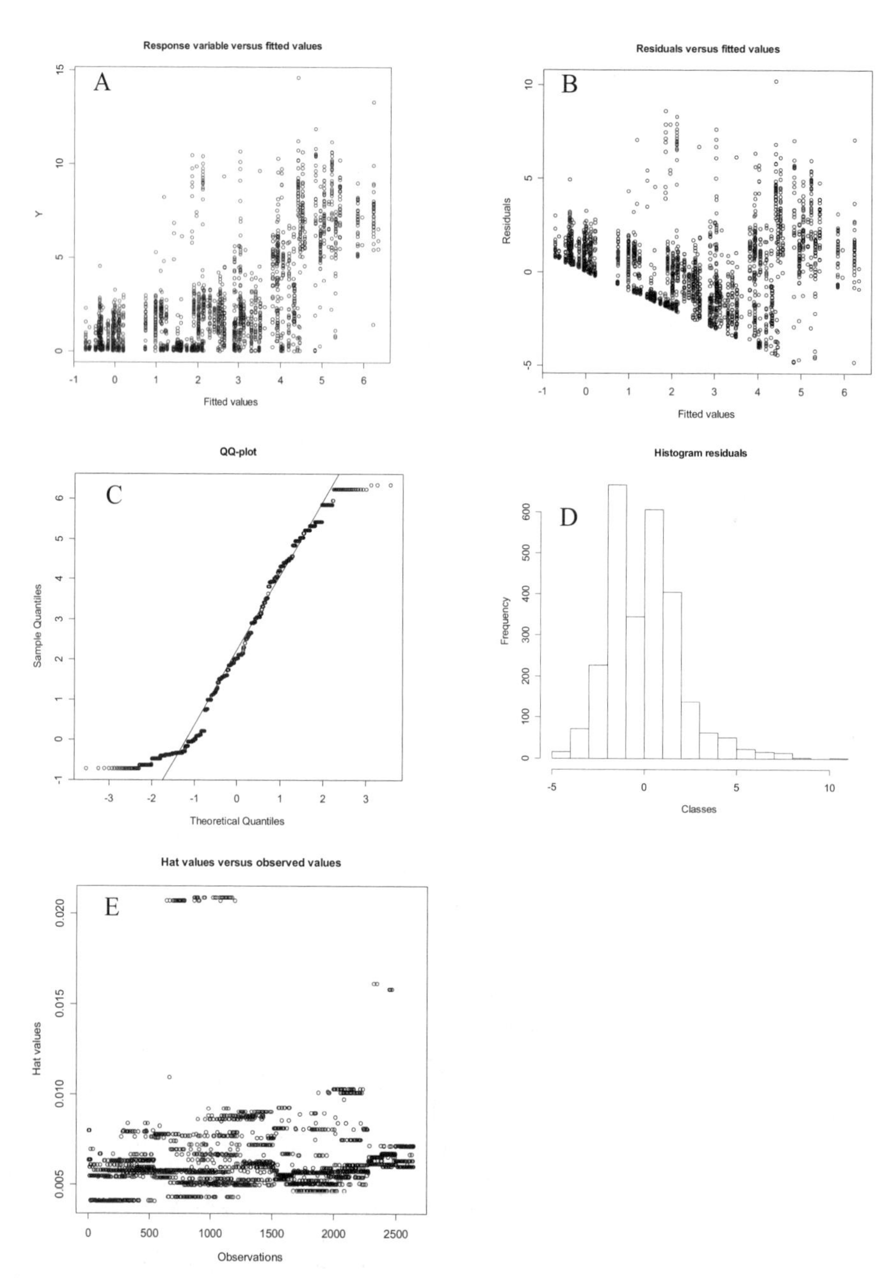

Figure 7.11. A: Observed values versus fitted values. B: Residuals versus fitted values. C: QQ-plot of residuals. D: Histogram of residuals. E: Leverage for each sample.

To detect a model misspecification, the residuals of the model can be plotted against the original explanatory variables (Figure 7.12). With nominal explanatory variables, this will be a boxplot, and for continuous variables, it will be a scatterplot. Under no circumstances should any of these plots show a pattern. One of the sex levels (Female) has considerably higher residuals (Figure 7.12A). The same holds for location one (Figure 7.12C). The scatterplot of residuals versus month shows that in months 2-6, the spread is smaller (Figure 7.12D). We cannot detect any residual-year relationship (Figure 7.12B). Although these graphs are trivial to make, the conclusions are critical; and they clearly show that the problem in the model fit is due to the female data in location 1 and that the heterogeneity is caused by a month (or more accurately: seasonal) effect. To show the importance of the data exploration, re-visit the coplot for these data in Figure 4.21. Most of the activity seems to occur in location one for females, and it might be advisable to analyse the female data from location one separately. An alternative approach is to use a *generalised* additive model using the Poisson distribution and log link function, and we look at this later.

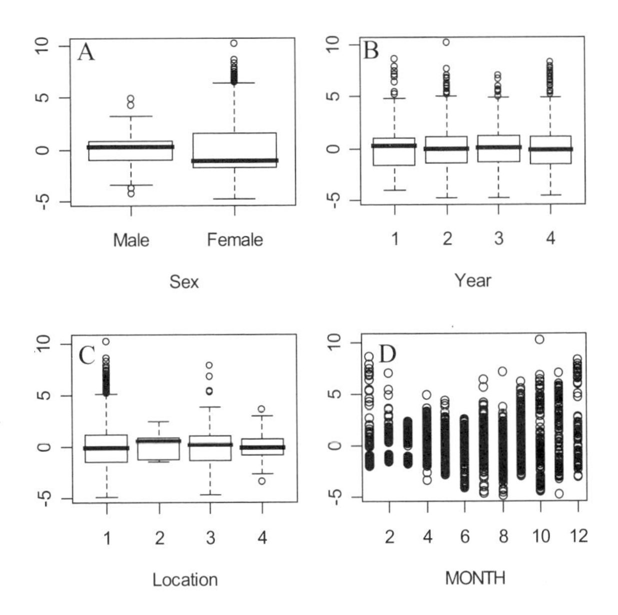

Figure 7.12. Residuals versus each explanatory variable. A: Residuals versus sex. B: Residuals versus year. C: Residuals versus location. D: Residuals versus month.

Concluding remarks on additive modelling

Although it lacks a theoretical justification, additive modelling is a useful data analysis tool. This is because it visualises the relationship between a response variable and multiple explanatory variables. And this ‘let the data speak’ approach can be used to obtain important insights into how to proceed with a parametric analysis. For example, the shape of the smoothing curve for temperature in the additive model for the RIKZ data suggests continuing with a simple (polynomial) regression for temperature, as it is capable of capturing the same pattern.

The smoothing curves in additive models are obtained rather arbitrarily, and the hypothesis tests rely on approximations and follow regression methods without any formal justification. We are not aware of any additive modelling simulation study in which the effects of non-homogeneity, non-normality, non-fixed X, and dependency have been studied. On the other hand, ecological data tend to be very noisy, and (generalised) additive modelling may be the only tool available that can give useful results. As to the criticism on the lack of underlying theory and foundation for the use of hypothesis tests (and p-values), one can always apply bootstrapping to get more reliable confidence bands. See Davison and Hinkley (1997) or Clarke et al. (2003) for a discussion and examples of bootstrapping for smoothing methods.

As discussed above, additive modelling has the same problems with heterogeneity, negative fitted values and negative realisation as linear regression, and it cannot be used to analyse 0–1 data. To deal with this, generalised additive modelling (GAM) can be used, and this can be considered as an extension to additive modelling, just as GLM can be considered as an extension of linear regression.

7.8 Generalised additive modelling

In Section 7.1, additive modelling was introduced and we used additive modelling to investigate the relationship between species richness and temperature and exposure for the RIKZ data. The final additive model was of the form:

$$R_i \sim N(\mu_i, \sigma^2), \text{ where } E[R_i] = \mu_i = g(x_i) = \alpha + \text{Exposure}_i + f(\text{Temperature}_i)$$

This additive model is a convenient tool to obtain some insights into the relationships between the response variable and the explanatory variables. For example, it might show that temperature has a non-linear relationship with species richness. However, nothing stops the model from obtaining negative fitted values, and the Gaussian density curves on top of the fitted values suggest that realisations with negative values are equally possible. To prevent this, and in the same way as linear regression was extended to Poisson regression, we can extend the additive model to a Poisson additive model. Better named: a generalised additive model with log link function. The mathematical formula for a GAM model of the RIKZ data (using richness as the explanatory variable, exposure as a nominal explanatory variable and temperature as a smoothing function) is given by:

$$R_i \sim P(\mu_i) \quad \text{and} \quad E[R_i] = \text{Var}(R_i) = \mu_i$$

$$\text{Log}(\mu_i) = g(x_i), \quad \text{where} \quad g(x_i) = \alpha + \text{Exposure}_i + \text{f}(\text{Temperature}_i)$$

The smoothing curve for temperature is presented in Figure 7.13, and the numerical output is given by

	Estimate	std. err.	t-ratio	p-value
(Intercept)	3.18	0.35	9.16	<0.001
factor(exposure)10	−1.25	0.41	−3.08	<0.001
factor(exposure)11	−2.36	0.45	−5.21	<0.001

All parameters are significantly different from zero at the 5% level, including the smoother ($p < 0.001$). Just as in GLM, there may be the possibility of overdispersion (Chapter 6). Because overdispersion was suspected (based on the results of GLM), the quasi-Poisson GAM was applied. The overdispersion parameter was 2.6, and therefore, all standard errors were corrected with the square root of 2.6. After this correction, all parameters were still significantly differently from 0 at the 5% level.

The general GAM model using the Poisson distribution and the log link function is similar to the GLM Poisson model, except that the predictor function g(x) is given by

$$g(x_i) = \alpha + f_1(x_{i1}) + \dots + f_p(x_{ip}).$$

The f_js are smoothing functions. It is also possible to have smoothing functions with parametric components, leading to semi-parametric models. In principle, GAM with a Poisson distribution follows the same requirements as GLM. We need to take into account overdispersion, and selecting the best model can be done by comparing deviances of nested models or by using the AIC. Just as additive modelling is mimicking linear regression, generalised additive modelling is mimicking generalised linear modelling. So, the tests for comparing models in GLM are also used in GAM: a Chi-square test if there is no overdispersion and an F-test if there is overdispersion.

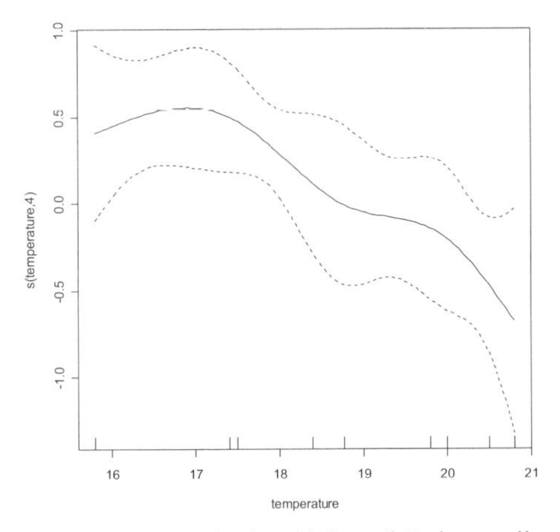

Figure 7.13. GAM for RIKZ data; the log link and Poisson distribution was used. The horizontal axis shows the temperature gradient and the vertical axis the contribution of the temperature smoother to the fitted values.

GAM for the squid data

The additive model applied on the squid data indicated violation of homogeneity, and therefore, a GAM model, using the Poisson distribution and log link function, was applied. Results indicated there was minor overdispersion (1.6), and a quasi-Poisson model was chosen. The smoothing curve is shown in Figure 7.14. Cross-validation was used to estimate the optimal degrees of freedom resulting in 8.8 df. All nominal variables were significantly differently from 0 at the 5% level and the model explained 50% of the deviance. As part of the model validation, deviance residuals were plotted against the explanatory variables (Figure 7.15). These residual indicate which samples contribute most to the deviance. As explained in the GLM section, the smaller the deviance the better. Observations from females in location one, in month one, have a relatively high contribution to the deviance, indicating a pattern in the residuals. This might be a reason not to accept the model as the 'best' model.

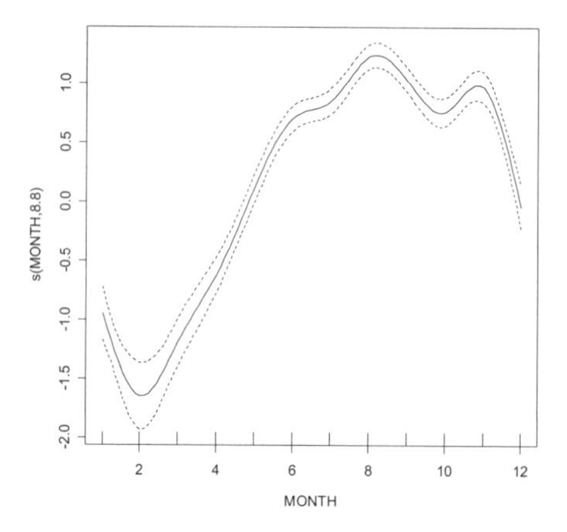

Figure 7.14. Smoothing curve for month in the GAM model with Poisson distribution and log-link.

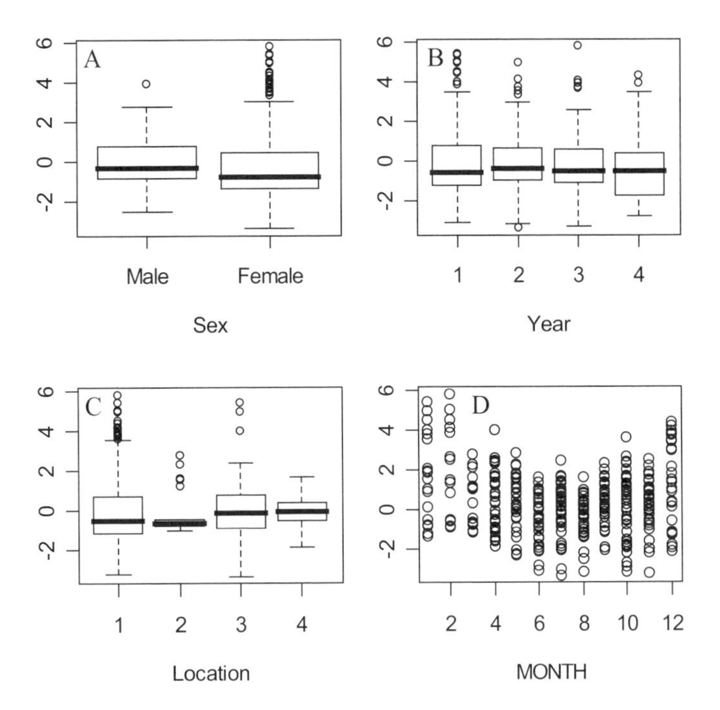

Figure 7.15. Deviance residuals versus explanatory variables. A: Residuals versus sex. B: Residuals versus year. C: Residuals versus location. D: Residuals versus month.

Presence–absence data

If the data are presence–absence data, the smoothing equivalent of a logistic regression model can be used. In this case there is only one observation per X value, and the formula becomes:

$$Y_i \sim \text{Bernoulli}(1, P_i) \quad \text{and} \quad \log\frac{P_i}{1-P_i} = \alpha + f_1(X_{1i}) + \ldots + f_p(X_{pi})$$

In case of data on proportions (multiple observations at the same X value), a binomial distribution should be used; see also the GLM section. Various examples of GAM applied to presence absence data and data on proportions are given in the case study chapters. It is also possible to have GAM models with interactions, and these are also discussed in the case study chapters.

7.9 Where to go from here

The additive modelling for the squid data showed that there was a different residual spread for females and for location 1. However, in the analysis we completely ignored the possible auto-correlation structure in the data. In the next chapter, we discuss using linear mixed modelling and additive mixed modelling that can deal with this problem.

Both modelling approaches allow for auto-correlation and multiple variances to be used (e.g., one for the male and one for the female data). The case study chapters contain several examples where we also show how to include interactions between smoothers and nominal explanatory variables.

These extensions can also be applied in GLM and GAM models and are called generalised linear mixed modelling (GLMM) and generalised additive mixed modelling (GAMM). These methods are not discussed in this book, but a good book for GLMM is Fitzmaurice et al. (2004), even though this has mainly medical examples, and for GAMM, Ruppert et al. (2003) or Wood (2006) are one of the few books available on this recently developed topic.

In the case studies we also present examples with interactions between smoothers and nominal variables, add auto-correlation, spatial correlation and model the heterogeneity. It is also possible to add random effects (Chapter 8).

8 Introduction to mixed modelling

8.1 Introduction

This chapter gives a non-technical introduction into mixed modelling. Mixed models are also known as mixed effects models or multilevel models and are used when the data have some sort of hierarchical form such as in longitudinal or panel data, repeated measures, time series and blocked experiments, which can have both fixed and random coefficients together with multiple error terms. It can be an extremely useful tool, but is potentially difficult to understand and to apply. In this chapter, we explain mixed models with random intercept and slope, different variances, and with an auto-correlation structure. In the case study chapters, several examples are presented (Chapters 22, 23, 26, 35 and 37). Some of these chapters are within a linear modelling context, whereas others use smoothing methods leading to additive mixed modelling. In the time series analysis and spatial statistics chapters (16, 17 and 19), temporal and spatial correlation structure is added to the linear regression and smoothing models. All of these techniques can be seen as extensions of mixed modelling.

Good references on mixed modelling and additive mixed modelling are Brown and Prescott (1999), Snijders and Bosker (1999), Pinheiro and Bates (2000), Diggle et al. (2002), Ruppert et al. (2003), Fitzmaurice et al. (2004), Faraway (2006) and Wood (2006). It should be noted that most of the literature on mixed modelling is technical. For non-mathematical text there is Chapter 35 in Crawley (2002) and Twisk (2006).

The easiest way to introduce mixed modelling is by using an example. In Chapters 4 to 7 and 27, we used a marine benthic dataset, referred to as the RIKZ data. Figure 8.1-A shows a scatterplot for these data with NAP plotted against species richness. Richness is the number of species found at each site, and NAP is the height of a site compared with average sea level. The data were sampled at nine beaches (five samples per beach), and the question being asked is whether there are any differences between the NAP-richness relationship at these nine beaches. As discussed in Chapters 5 and 6, species richness is a non-negative integer and we used it to explain both linear regression and generalised linear modelling with a Poisson distribution and log link. The same approach can be followed for mixed modelling. We will explain linear mixed modelling as an extension of linear regression (using the normal distribution) with species richness as the response variable. Generalised linear mixed modelling (GLMM) is the mixed modelling extension of GLM. Although GLMM may be more appropriate for species richness, it is outside the scope of this book. Panel A in Figure 8.1 contains a regression line using all 45 observations from all nine beaches. The regression line in Figure 8.1-A can be written as

Model 1 $$Y_i = \alpha + \beta \text{NAP}_i + \varepsilon_i \quad \text{and} \quad \varepsilon_i \sim N(0, \sigma^2)$$

This model contains three unknown parameters: the two regression parameters (one intercept and one slope) and the variance. The model assumes that the richness-NAP relationship is the same at all nine beaches. Figure 8.1-B contains the same data except that we have added a regression line for each beach. The directions and values of the nine regression lines indicate that the intercepts are different for each beach, and that two beaches have noticeably different slopes. This model can be written as

Model 2 $$Y_{ij} = \alpha_j + \beta_j \text{NAP}_{ij} + \varepsilon_{ij} \quad \text{and} \quad \varepsilon_{ij} \sim N(0, \sigma^2)$$

Where j = 1,...,9 and i = 1,..,5 (there are five observations per beach). The regression lines were obtained in one regression analysis using beach as a factor, NAP as a continuous explanatory variable and a beach-NAP interaction term. This can also be called an ANCOVA (Zar 1999). The total number of parameters adds up to 9 × 2 + 1 = 19 (a slope and intercept per beach plus one noise variance term). The total number of observations is only 45, so proportionally this model has a worryingly large number of parameters.

The model with only one regression line (Figure 8.1-A) is the most basic model, and the model with nine regression lines in which slope and intercept are allowed to differ per beach (Figure 8.1-B) is the most complex model we have considered here. Later in this chapter, we will discuss even more complex models by allowing for different variances per beach. There are two intermediate models: one where the intercept is allowed to differ between the beaches but with equal slopes, and one where the intercepts are kept same and the slopes are allowed to differ. This reduces the number of parameters. These models are given by

Model 3 $$Y_{ij} = \alpha_j + \beta \text{NAP}_{ij} + \varepsilon_{ij} \quad \text{and} \quad \varepsilon_{ij} \sim N(0, \sigma^2)$$

Model 4 $$Y_{ij} = \alpha + \beta_j \text{NAP}_{ij} + \varepsilon_{ij} \quad \text{and} \quad \varepsilon_{ij} \sim N(0, \sigma^2)$$

The fitted values of models 3 and 4 are given in panels C and D of Figure 8.1, respectively. The number of parameters in both models is 1 + 9 + 1 = 11. In Figure 8.1-D all lines intercept at NAP=0 because the intercepts are identical.

Using either an F-test or the AIC (see Chapter 5), it can be shown that of the four models, the one used in Figure 8.1-B is the best. This is the model using a different intercept and slope at each beach, giving a total of 19 parameters. The nine intercepts tell us which beaches have higher richness or lower richness at NAP = 0, and the nine slopes show the differences in the strength of the NAP-richness relationship.

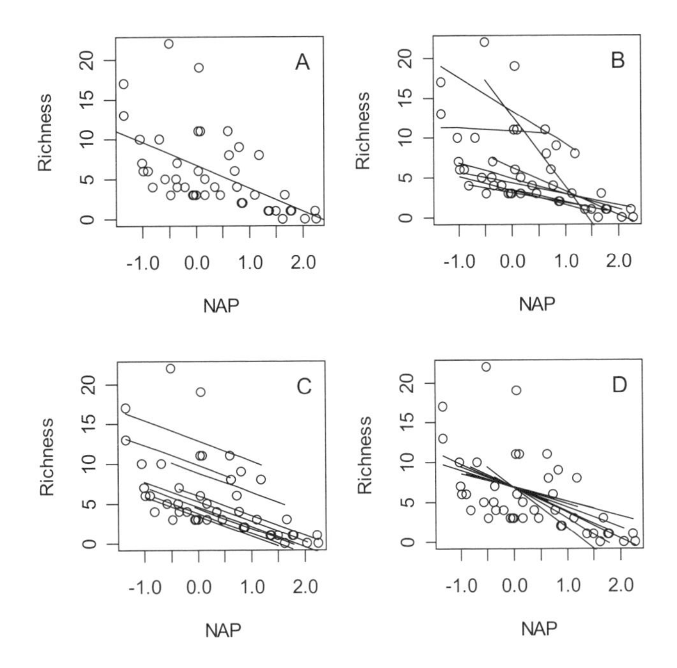

Figure 8.1. Scatterplots of NAP versus richness for the RIKZ data. A: One regression line was added using all data. B: One regression line for each beach was plotted. C: One regression line per beach but with all slopes equal. D: One regression line per beach but with intercepts equal.

If we are only interested in the general relationship between richness and NAP, and do not care about differences between beaches, then we could ignore the nominal variable beach. However, this means that the variance component might contain between-beach variation, and not taking this into account might affect standard errors and p-values of the fixed effects (e.g., suggesting a non-significant relationship between NAP and species richness even when the relationship is significant). But the price of 16 extra regression parameters can be rather large, namely in the loss of *precious degrees of freedom*! To avoid this, mixed modelling can be used. But there is another motivation for using mixed modelling with these data. If beach is used as a fixed term, we can only make a statement of richness-NAP relationships for these particular beaches, whereas if we use it as a random component, we can predict the richness-NAP relationship for all similar beaches.

8.2 The random intercept and slope model

For simplicity, we start with model 3 discussed earlier. In this model, each beach has a different intercept but the same slope. We extend it to:

Model 5
$$Y_{ij} = \alpha + \beta \mathrm{NAP}_{ij} + a_j + \varepsilon_{ij}$$
$$\text{where} \quad a_j \sim N(0, \sigma_a^2) \text{ and } \varepsilon_{ij} \sim N(0, \sigma^2)$$

The index j (representing beaches) takes values from 1 to 9, and i (representing samples on a beach) from 1 to 5. In model 3, we ended up with nine regression lines. The nine estimated slopes, intercepts and their standard errors tell us which beaches are different. In model 5, we assume there is only one overall regression line with a single intercept and a single slope. The single intercept α and single slope β are called the fixed parameters. Additionally, there is a random intercept a_j, which adds a certain amount of random variation to the intercept at each beach. The random intercept is assumed to follow a normal distribution with expectation 0 and variance σ_a^2. Hence, the unknown parameters in the model are α, β, the variance of the noise σ^2 and the variance of the random intercept σ_a^2, which adds up to only four parameters. So, basically, we have the same type of model fit as in panel C in Figure 8.1, but instead of nine estimated levels for each intercept, we now have nine realisations a_1, .., a_9 from which we assume that they follow a normal distribution. And we only need to estimate the variance of this distribution. The first part of the numerical output for this mixed model is given by: AIC = 247.580, BIC = 254.525 and logLik = −119.740. As with linear regression, GLM and GAM, we can measure how optimal the model is by using a selection criteria like the AIC. An alternative is the BIC, which is more conservative (the number of parameters have a higher penalty than in the AIC). Both terms use the log likelihood as a measure of fit. Further output gives:

```
Random effects:
            Intercept       Residual
StdDev:     2.944           3.060
```

The output for the random effects (above) shows that the variance of the noise is equal to $\sigma^2 = 3.06^2$ and the variance of the random intercept to $\sigma_a^2 = 2.944^2$.

```
Fixed effects:
             Value      Std.Error    df    t-value  p-value
Intercept    6.582      1.096        35     6.006   <0.001
NAP         −2.568      0.495        35    −5.191   <0.001
```

For the fixed effects part of the model (above), $\alpha + \beta \mathrm{NAP}_{ij}$, the intercept α is equal to 6.582 and the slope β is –2.568. Both parameters are significantly differently from 0 at the 5% level. The correlation between the estimated intercept and slope is small. The ANOVA table below also indicates the significance of the fixed slope β.

	numdf	dendf	*F*-value	*p*-value
Intercept	1	35	27.634	<0.001
NAP	1	35	26.952	<0.001

In summary, the model estimates a fixed component of the form: $6.582 - 2.568\text{NAP}_{ij}$. These estimated parameters are significantly differently from 0 at the 5% level. For each beach, the intercept is increased or decreased by a random value. This random value follows a normal distribution with expectation 0 and variance $\sigma_a^2 = 2.944^2$. The unexplained noise has a variance of $\sigma^2 = 3.06^2$. The type of model is called a mixed effects model with random intercept.

Extending the mixed model with a random slope

We will now extend model 5 to get the mixed modelling equivalent of model 2, which was illustrated in Figure 8.1-B. Model 5 was a mixed model with a random intercept that allowed the regression line to randomly shift up or down. In model 6, both the intercept and slope are allowed to randomly vary. The model is given by

Model 6
$$Y_{ij} = \alpha + a_j + \beta\text{NAP}_{ij} + b_j NAP_{ij} + \varepsilon_{ij}$$
$$\text{where} \quad \varepsilon_{ij} \sim N(0, \sigma^2) \text{ and } a_j \sim N(0, \sigma_a^2) \text{ and } b_j \sim N(0, \sigma_b^2)$$

This is the same model formulation as model 5, except for the term $b_j\text{NAP}_{ij}$. This new term allows for random variation of the slope at each beach. The model fit will look similar to the one in Figure 8.1-B, except that considerably fewer parameters are used: two for the fixed intercept α and slope β, and three random variances σ^2, σ_a^2 and σ_b^2. In general, one allows for correlation between the estimated variances for σ_a^2 and σ_b^2. The numerical output shows that the AIC is 244.397, which is slightly smaller than the AIC for model 5. As in regression, the same principle of 'the smaller the better' holds, so the lower AIC indicates that model 6 is more optimal than model 5. It should be noted that if the difference is smaller than two, the models are generally thought of as equivalent and the more simple one should be selected (Burnham and Anderson 2002).

Random effects:

	StdDev
Intercept	3.573
NAP	1.758
Residual	2.668

The printout (above) for the random effects component shows that $\sigma^2 = 2.668^2$, $\sigma_a^2 = 3.573^2$, and $\sigma_b^2 = 1.758^2$.

Fixed effects:

	Value	Std.Error	df	*t*-value	*p*-value
(Intercept)	6.612	1.271	35	5.203	<0.001
NAP	−2.829	0.732	35	−3.865	<0.001

The fixed component (above) is given by: 6.612 − 2.829NAP. Both intercept and slope are significant at the 5% level. This can also be inferred from the ANOVA table below.

ANOVA table

	Numdf	dendf	*F*-value	*p*-value
(Intercept)	1	35	12.427	0.001
NAP	1	35	14.939	0.001

8.3 Model selection and validation

We have now applied two mixed models: one where the intercept was allowed to vary randomly (model 5), and one where the intercept and slope were allowed to vary randomly (model 6). The question that arises is which model is better. It is important to realise that the only difference between model 5 and 6 is the random effects component; the fixed components are the same. To compare two models with the same fixed effect, but with different random components, a likelihood ratio test or the AIC can be used. In this case, we get:

Model	df	AIC	BIC	logLik	*L*-Ratio	*p*-value
6	6	244.40	254.96	−116.20		
5	4	247.48	254.53	−119.74	7.08	0.029

The AIC suggests selecting model 6, but the BIC picks model 5. The *p*-value indicates that the more complicated model (containing a random intercept and slope) is more optimal. However, there is one major problem with the likelihood ratio test. In Chapter 5, we used an *F*-test to compare two nested models and the residual sum of squares obtained by ordinary least squares of the two models were used to work out a test statistic. Estimation in mixed modelling is done with maximum likelihood (Chapter 7). The likelihood criteria of the full model L_0 and the nested model L_1 can be used for hypothesis testing. It is in fact the ratio L_0/L_1 that is used, hence the name likelihood ratio test. Taking the log or, more common, −2log gives a test statistic of the form: $L = -2(\log L_0 - \log L_1)$. It can be shown that under the null hypothesis, this test statistic follows *approximately* a Chi-square distribution with v degrees of freedom, where v is the difference in number of parameters in the two models. In linear regression, we can use this procedure to test H_0: $\beta_i = 0$ versus H_1: $\beta_i \neq 0$. So, what are we testing above? The test statistic is $L = 7.08$ with a *p*-value of 0.029. But what is the null hypothesis? The only difference between models 5 and 6 is the random component b_jNAP where $b_j \sim N(0, \sigma_b^2)$. By comparing models 5 and 6 using a likelihood ratio test, we are testing the null hypothesis H_0: $\sigma_b^2 = 0$ versus H_1: $\sigma_b^2 > 0$. The alternative hypothesis has to contain a '>' because variance components are supposed to be non-negative. This is called the boundary problem; we are testing whether the variance is equal to null, but null is on the boundary of all allowable values of the variance. The problem is that the theory underlying the likelihood ratio test, which gives us a *p*-value, assumes that we are not on the boundary of the parameter space. For this particular example, comparing a random intercept versus a random intercept plus slope model, it is still possible to derive a valid distribution for the test statistic, see p. 106 in Rupert et al. (2003). But as soon as we compare more complicated variance structures that are on the boundary, the underlying mathematics become rather complicated. Formulated differently, great care is needed with interpreting the *p*-value of the test statistic L if we are testing on the boundary. And in most applications, this is what we are interested in. Citing from Wood (2006): 'In practice, the most sensible approach is often to treat the *p*-values from the likelihood ratio test as "very approximate". If the *p*-value is very large, or very small, there is no practical difficulty about interpreting it, but on the borderline of significance, more care is needed'. Faraway (2006) mentions that the *p*-value tends to be too large if we are testing on the boundary. This means that a *p*-value of 0.029 means that we can reasonably trust that it is significant, but 0.07 would be a problem. Pinheiro and Bates (2000), Faraway (2006) and Wood (2006) all discuss bootstrapping as a way to obtain more trustable p-values if testing on the boundary.

Error terms

So far, we have *assumed* that the error term ε_{ij} has the same variance at each beach, and this is called homogeneity. If residuals plots indicate violation of homogeneity (see Chapter 5 for details), it might be an option to use different variance components per beach. A possible model could be as follows:

Model 7
$$Y_{ij} = \alpha + \beta \text{NAP}_{ij} + a_j + b_j NAP_{ij} + \varepsilon_{ij}$$
$$\text{where} \quad a_j \sim N(0,\sigma_a^2),\ b_j \sim N(0,\sigma_b^2) \text{ and } \varepsilon_{ij} \sim N(0,\sigma_j^2)$$

This model is nearly identical to model 6 except that the variance component of the noise now has an index j, where j = 1, .., 9. The output of model 7 shows that the AIC = 216.154 (BIC = 240.811), which is considerably smaller than model 6, and the variances of the random effects are also lower in model 7 than model 6 (below).

```
Random effects:
             StdDev
(Intercept)  2.940
NAP          0.011
Residual     1.472
```

In model 6 we had $\sigma^2 = 2.668^2$, $\sigma_a^2 = 3.573^2$, and $\sigma_b^2 = 1.758^2$. Both the residual variance and the variance for the random slope are considerably smaller in model

7 than model 6. An informal way to assess the importance of a variance component is its relative size. 0.011^2 is relatively small!

Variance function:
Parameter estimates:

1	2	3	4	5	6	7	8	9
1.0	3.387	0.578	0.269	4.891	0.426	0.450	0.745	2.33

We now have a residual variance component for each beach (above). The largest residual variance is at beaches two, five and nine.

Fixed effects:

	Value	Std.Error	df	*t*-value	*p*-value
(Intercept)	5.75	1.056	35	5.446	<0.001
NAP	−1.42	0.127	35	−11.188	<0.001

The output for the fixed effects part for model 7 (above) shows that it has the form: 5.75 −1.42NAP. The AIC or hypothesis testing with the likelihood ratio test can be used to judge which model is better.

Model	df	AIC	BIC	logLik	*L*-Ratio	*p*-value
6	4	247.480	254.525	−119.740		
7	12	212.154	233.289	−94.077	51.325	<0.001

The likelihood ratio test (above) indicates that model 7 is better than model 6. The effect of introducing nine variances is illustrated in Figure 8.2. Panels A and B show the standardised residuals versus fitted values and QQ-plot for model 6. Panels A and B show the same for model 7. Note that in model 6 there is clear evidence of heterogeneity (see panel A).

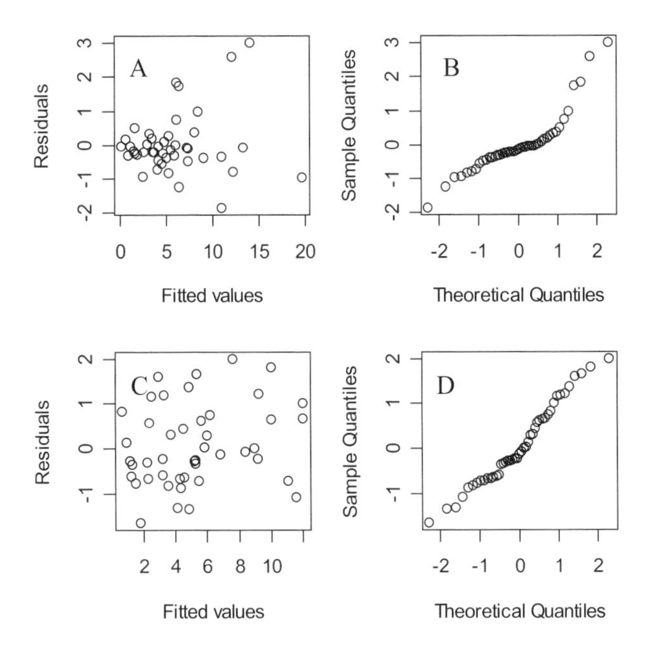

Figure 8.2. A: Standardised residuals versus fitted values for model 6. B: QQ-plot of residuals of model 6. C: Standardised residuals versus fitted values for model 7. D: QQ-plot of residuals of model 7.

Comparing models with different fixed effects

So far, we have only compared models with the same fixed components, but for the RIKZ data we also have exposure per beach. This is a nominal variable with three classes, and one of the prime interests of the project was to know the effects of exposure. As (i) exposure can only fall into one of three pre-defined categories and (ii) one of the prime underlying questions is whether there is an exposure effect, it is modelled as a fixed effect and not as random effect and we have extended model 7 to include this new effect:

Model 8

$$Y_{ij} = \alpha + \beta \text{NAP}_{ij} + \text{exposure}_{ij} + a_j + b_j NAP_{ij} + \varepsilon_{ij}$$
$$\text{where} \quad a_j \sim N(0,\sigma_a^2),\ \ b_j \sim N(0,\sigma_b^2) \ \text{ and } \ \varepsilon_{ij} \sim N(0,\sigma_j^2)$$

Exposure is modelled as a factor. There are different ways to include the levels of a nominal variable, and the default value in software packages like R and SPLUS is the treatment option. This means that the first level is set to 0 and is considered as the baseline level (Chapter 5). To assess whether adding exposure results in a better model, we can inspect the individual p-values, compare AIC values or apply a likelihood ratio test. The output is presented below and indicates that model 8 is better than model 7.

Model	df	AIC	BIC	logLik	L.Ratio	*p-value*
8	14	200.978	226.271	−86.488		
7	12	211.684	233.365	−93.842	14.707	<0.001

As we now compare two models with the same random structure but with different fixed terms, the maximum likelihood method is used instead of REML (see the next section).

Model selection strategy

We started this chapter with linear regression, then introduced the random intercept model, extended it to the random intercept and slope model, and finally added more explanatory variables. This order was chosen for reasons of clarity, but we now look at finding the optimal model for both the random and the fixed components. This discussion also applies to additive mixed modelling. Our starting point is either a multiple linear regression or an additive model:

$$Y_i = \alpha + \beta_1 X_{1i} + \cdots + \beta_p X_{pi} + \varepsilon_i \quad \text{or} \quad Y_i = \alpha + f_1(X_{1i}) + \cdots + f_p(X_{pi}) + \varepsilon_i$$

The part $\alpha + \beta_1 X_{1i} + \cdots + \beta_p X_{pi}$ or $\alpha + f_1(X_{1i}) + \cdots + f_p(X_{pi})$ is called the fixed component, and ε_i is the random component. We have already seen three forms of random components; random intercepts, random slopes and different variances. More structures will follow for time series and spatial data. The task of the researcher is to find the optimal fixed and random component structure. We assume that the prime interest of the analysis is the fixed component, so we need to get this one as good as possible. However, a poor fixed component structure may result in large residuals and therefore all types of random structures that could have been taken care of by the fixed component. For the RIKZ data, we started with a model using only NAP as an explanatory variable, from a full dataset of 10-15 explanatory variables. The different variances between beaches may be explained with a more appropriate fixed component, and we prefer to have this information as a fixed component, and not in the random structure, if possible. So, the consensus is to start with a model that has a fixed component structure that is as good as possible. Most textbooks advise including every possible explanatory variable and interaction, or a 'just beyond optimal' model. Using this 'just beyond optimal' model, we can select the most optimal random structure and ensure that the random structure does not contain any information that could have been modelled with fixed terms. As we will explain later in this chapter, this process requires one to use the restricted maximum likelihood (REML) estimation procedure. Once we have found the optimal random component structure, then we go to the third step and find the optimal fixed component structure. This requires a backwards selection approach where we drop the explanatory variables that are not significant using maximum likelihood (ML) estimation (which will be explained later). Summarising, the model selection strategy:

1. Start with a model that is close to optimal in terms of fixed components.

2. Search for the optimal random error structure, e.g., allowing for random intercepts and slopes, different variances, auto-correlation, spatial correlation or any of the many options described in Pinheiro and Bates (2000). Use REML.
3. Using the optimal random error structure from step 2, find the optimal fixed components using ML.
4. Present the estimated parameters and standard errors of the optimal model, but use REML estimation!

Needless to say, once the model selection has given us the optimal model, a model validation should be applied. This process is similar to linear regression, except that we now allow for different spread per strata, auto-correlation, etc.

Fixed or random

To find the differences between beaches for the RIKZ data, then clearly you should model these variables as fixed effects rather than as random effects. The advantage of treating the levels of a variable as random effects is that the model then holds for the entire population. Treating beach as a fixed effect in the RIKZ data means that the richness-NAP relationship found by the model only holds for those nine beaches. Treating it as a random component means that the species-richness relationship holds for all beaches, not just the nine beaches we sampled. This means that we need to assume that the nine sampled beaches represent the population of all beaches (with similar values of the other explanatory variables like grain size, etc.). If we cannot make this assumption, then treat them as fixed effects. Another important point is the number of levels. Are there enough levels (> 4, but preferable > 10) in the variable to treat it like a random effect? If there are only two or three levels, treat it as a fixed effect. But if there are more than 10 levels, then treat is as a random effect. In cases where an explanatory variable is a treatment effect, consider it as a fixed effect as in most situations the prime interest of the experiment is in the effects of the treatment (like toxic concentrations in mesocosm experiments).

8.4 A bit of theory

A linear regression model can be written as $Y_j = \beta X_j + \varepsilon_j$. The mathematical formulation of a mixed model is as follows:

$$Y_{ij} = \beta X_{ij} + b_j Z_{ij} + \varepsilon_{ij}$$
$$b_j \sim N(0, \Psi) \quad \text{and} \quad \varepsilon_{ij} \sim N(0, \sigma^2)$$

The component βX_{ij} contains the fixed effects and bZ_{ij} the random components. The random components b_j and ε_{ij} are assumed to be independent of each other. Covariance between the random effects b_j is allowed (e.g., between random intercept and slope) using the off-diagonal elements of Ψ. A two-stage algorithm, REML, is used to estimate the fixed regression coefficients and the variance components Ψ and σ. Full details of the REML estimation process can be found in Brown and Prescott (1999) or Pinheiro and Bates (2000). Most textbook discussions on REML and ML use complicated mathematics, but a reasonable non-technical explanation can be found in Fitzmaurice et al. (2004). In maximum likelihood estimation, the maximum likelihood function is specified, and to optimise it, derivates with respect to the regression parameters and variances are derived. The problem is that the estimates for the variance(s) are biased. In REML, a correction is applied so that less biased estimators are obtained. So, in general REML estimators are less biased than ML estimators. For large sample size (relative to the number of regression parameters), this issue is less important.

Covariances and Correlations

Revisiting some of the models we used for the RIKZ data we had:

Model 3
$$Y_{ij} = \alpha_j + \beta_j \text{NAP}_{ij} + \varepsilon_{ij} \quad \text{and} \quad \varepsilon_{ij} \sim N(0,\sigma^2)$$

Model 5
$$Y_{ij} = \alpha + \beta \text{NAP}_{ij} + a_j + \varepsilon_{ij}$$
$$\text{where} \quad a_j \sim N(0,\sigma_a^2) \text{ and } \varepsilon_{ij} \sim N(0,\sigma^2)$$

Model 6
$$Y_{ij} = \alpha + a_j + \beta \text{NAP}_{ij} + b_j NAP_{ij} + \varepsilon_{ij}$$
$$\varepsilon_{ij} \sim N(0,\sigma^2) \text{ and } a_j \sim N(0,\sigma_a^2) \text{ and } b_j \sim N(0,\sigma_b^2)$$

Model 3 was the ordinary regression model with interaction. Model 5 was called the random intercept model and model 6 the random intercept and slope model. We now focus on the covariance and correlation between two observations from the same beach j: samples Y_{ij} and Y_{kj}. Under model 3, this gives us:

$$Var(Y_{ij}) = Var(\varepsilon_{ij}) = \sigma^2$$
$$Cov(Y_{ij}, Y_{kj}) = \begin{cases} 0 & \text{if } i \neq k \\ \sigma^2 & \text{if } i = k \end{cases}$$

Hence, samples from the same beach are assumed to be independent. For the random intercept model 5, we have:

$$Var(Y_{ij}) = Var(a_j + \varepsilon_{ij}) = \sigma_a^2 + \sigma^2$$
$$Cov(Y_{ij}, Y_{kj}) = Cov(a_j + \varepsilon_{ij}, a_j + \varepsilon_{kj}) = \begin{cases} \sigma_a^2 & \text{if } i \neq k \\ \sigma_a^2 + \sigma^2 & \text{if } i = k \end{cases}$$

Hence, the variance is the same for all samples on a particular beach j. Furthermore, for two different samples i and k on the same beach j, the correlation is equal to (correlation is covariance divided by variance):

$$Cor(Y_{ij}, Y_{kj}) = \frac{\sigma_a^2}{\sigma_a^2 + \sigma^2} \quad (8.1)$$

So, whatever the value of the explanatory variables, the correlation between two different observations from the same beach is defined by the formula in equation (8.1). This type of correlation is also called a compound symmetry structure. Now, let us have a look at the random intercept and slope model 6.

$$Var(Y_{ij}) = Var(a_j + b_j NAP_{ij} + \varepsilon_{ij}) = \sigma_a^2 + 2\,\mathrm{cov}(a_j, b_j) NAP_{ij} + \sigma_b^2 NAP_{ij}^2 + \sigma^2$$

This formula looks a bit more intimidating, but it basically states that the variance of a particular observation depends on the explanatory variables. The same holds for the covariance between two different observations from the same beach. The formula is given by:

$$Cov(Y_{ij}, Y_{kj}) = \sigma_a^2 + \mathrm{cov}(a_j, b_j)(NAP_{ij} + NAP_{kj}) + \sigma_b^2 NAP_{ij} NAP_{kj}$$

This formula also shows why allowing for covariance between a_j and b_j can be useful.

8.5 Another mixed modelling example

The RIKZ example showed how the number of regression parameters in a model can be reduced by using random components instead of fixed components. We now show how mixed modelling can be used to analyse a short time series using some bee data as an example. A full analysis of these data is presented in Chapter 22, but for this exercise we only consider the data for honeybees (foraging on sunflower crops) at five different locations (transects). All locations were sampled over the same ten days.

Figure 8.3 shows a coplot of the bee data; numbers of bees are plotted versus time conditional on location. To aid visual interpretation, a regression line was added for each location, and there seems to be a general downwards trend over time.

Just as for the RIKZ data we apply a mixed model to replace the five regression curves by one curve, plus allowing for random variation in both the intercept and slope. This gives a model of the form:

Model 9

$$Bees_{ij} = \alpha + \beta \mathrm{Time}_{ij} + a_j + b_j \mathrm{Time}_{ij} + \varepsilon_{ij}$$
$$a_j \sim N(0, \sigma_a^2), \quad b_j \sim N(0, \sigma_b^2) \quad \text{and} \quad \varepsilon_{ij} \sim N(0, \sigma^2)$$

where j = 1, ..., 5 as there are five locations, and i = 1, ..,1 0 (10 days). The component $\alpha + \beta \mathrm{Time}_{ij}$ is the fixed component and $a_j + b_j \mathrm{Time}_{ij}$ is the random component ensuring variation around the intercept and slope.

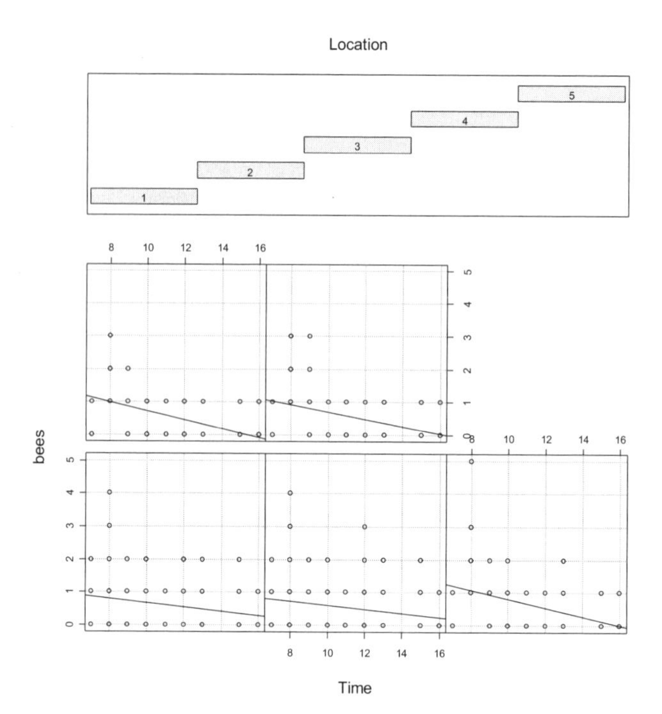

Figure 8.3. Coplot of bee numbers versus time conditional on location (transect). In each panel a linear regression line was added. The lower left panel shows the relationship between bees and time at location 1, and the upper right at location 5.

One of the assumptions in model 9 is that the errors ε_{ij} are independent. Hence, the residuals of two sequential days are not correlated to each other:

$$cor(\varepsilon_{ij}, \varepsilon_{i+1,j}) = 0$$

However, the observations are made sequentially over time and therefore we may be violating the independence assumption. One option to deal with this is by adding a correlation structure on the errors ε_{ij}:

$$cor(\varepsilon_{ij}, \varepsilon_{i+h,j}) = \rho_k$$

The problem is now how to model this correlation structure, and there are several ways of doing this. One option is to assume that the noise at day i is related to the noise at day $i - 1$, i - 2, etc. This is a so-called an auto-regressive model of order p:

$$\varepsilon_{ij} = \phi_1 \varepsilon_{i-1,j} + \phi_2 \varepsilon_{i-2,j} + ... + \phi_p \varepsilon_{i-p,j} + \eta_{ij}$$

The error term η_{ij} is independently normally distributed. The notation for this model is 'AR(p)' and the model allows for auto-correlation between the residuals. The coefficients $\phi_1,...,\phi_p$ are all smaller than 1 in the absolute sense. The unexplained information on day i is modelled as a function of the unexplained information on day $i-1$, $i-2$, etc. The AR(1) model is given by

$$\varepsilon_{ij} = \phi_1 \varepsilon_{i-1,j} + \eta_{ij}$$

Using some basic mathematics, it can be shown that the correlation between ε_{ij} and $\varepsilon_{i-k,j}$ is given by ϕ_1^k. This means that the further away two days are, the lower their correlation; the correlation between day i and $i-1$ is ϕ_1, between day i and $i-2$ is ϕ_1^2, etc. An auto-correlation function (Chapter 16) can be used to assess this assumption. For the bee data, an AR(1) structure seems to be a sensible choice, as what happens today is closely related to what happened yesterday, but much less to what happened two days ago. Another option is the so-called 'compound symmetry'. In this case the correlation between two error components is modelled as

$$cor(\varepsilon_{ij}, \varepsilon_{i-k,j}) = \rho$$

By definition if $k = 0$, the correlation is 1. This modelling approach assumes that the correlation between two days is always equal to ρ however far apart the days are. So, the correlation between the errors at time i and $i-k$ is the same whatever the value of k. Yet another option is to assume a general correlation structure. All days separated with k days have the same correlation ρ_k:

$$cor(\varepsilon_{ij}, \varepsilon_{i-k,j}) = \rho_k$$

All these options can be compared with each other using the AIC. We use the bee data to demonstrate mixed modelling and temporal auto-correlation. First, we fit the model with a random intercept and fixed slope but with no auto-correlation on the errors. The relevant output is given below.

AIC	BIC	logLik
2705.759	2726.308	-1348.880

Random effects:

	Intercept	Residual
StdDev:	0.013	0.702

Fixed effects:

	Value	Std.Error	df	*t*-value	*p*-value
Intercept	1.583	0.079	1254	19.934	<0.001
Time	-0.090	0.007	1254	-13.171	<0.001

The fixed effects are significantly differently from 0 at the 5% level. Refitting the model but specifying an AR(1) structure on the residuals gives:

AIC	BIC	logLik
2583.155	2608.841	-1286.577

Random effects:

	Intercept	Residual
StdDev:	0.004	0.703

Correlation Structure: AR(1)
Parameter estimate(s):
Phi
0.308

Fixed effects:

	Value	Std.Error	df	*t*-value	*p*-value
(Intercept)	1.577	0.108	1254	14.562	<0.001
Time	-0.090	0.009	1254	-9.585	<0.001

The coefficient ϕ_1 is estimated as 0.308, indicating that the correlation (in the error) between day i and $i-1$ is 0.308. And between days i and $i-1$ it is 0.308^2. To assess whether the AR(1) structure has improved the model, the likelihood ratio test can be used (note that this is not a boundary problem), or AIC values can be compared:

Model	df	AIC	BIC	logLik	*L*-Ratio	*p*-value
9	4	2705.76	2726.31	-1348.88		
9+AR(1)	5	2583.15	2608.84	-1286.57	124.604	<0.001

Both the AIC and the likelihood ratio test indicate that the second model (including the AR structure) is more optimal. Other options we tried were the compound symmetry correlation, but the AIC was 2707.758. In Chapter 22 additional explanatory variables and random components are used for the full analysis. Further discussion on auto-correlation and examples are given in Chapters 16, 26, 36 and 37.

8.6 Additive mixed modelling

Recall that in Section 8.3 we ended up with the following mixed model for the RIKZ data:

Model 7
$$Y_{ij} = \alpha + \beta \mathrm{NAP}_{ij} + a_j + b_j NAP_{ij} + \varepsilon_{ij}$$
$$\text{where} \quad a_j \sim N(0,\sigma_a^2),\ \ b_j \sim N(0,\sigma_b^2) \ \text{ and } \ \varepsilon_{ij} \sim N(0,\sigma_j^2)$$

It models the species richness as a linear function of NAP, with the intercept and slope varying randomly across the beaches. This model also dealt with the violation of homogeneity by allowing for different variation in the noise per beach. Although this is a complicated model, it is still fundamentally a linear relationship between richness and NAP. And, in the same way as the linear regression model was extended to additive models, we can extend mixed models into additive mixed models (Wood 2004, 2006; Ruppert et al. 2003). Two possible models are:

Model 10
$$Y_{ij} = \alpha + f(\text{NAP}_{ij}) + a_j + \varepsilon_{ij}$$
$$\text{where} \quad a_j \sim N(0, \sigma_a^2) \quad \text{and} \quad \varepsilon_{ij} \sim N(0, \sigma^2)$$

Model 11
$$Y_{ij} = \alpha + f(\text{NAP}_{ij}) + a_j + \varepsilon_{ij}$$
$$\text{where} \quad a_j \sim N(0, \sigma_a^2) \quad \text{and} \quad \varepsilon_{ij} \sim N(0, \sigma_j^2)$$

The linear component βNAP has been replaced by a smoothing function of NAP, denoted by f(NAP). The difference between models 10 and 11 is that the latter allows for different spread of the residuals per beach. So, the only difference between models 9 and 11 is the way the NAP effect is modelled**.** In model 9, we used a linear component plus random variation around the slope. In model 10, we have a smoothing function but we cannot have random variation around this smoother. The smoother itself can be estimated using, for example splines, and cross-validation can be used to estimate the optimal amount of smoothing. In model 11 we allow for different spread of the data per beach (heterogeneity). Both models 10 and 11 were applied on the RIKZ data and the AIC can be used to choose which one is best. In this example the AIC for model 11 was 211.71 and the AIC for model 10 was versus 249.52, suggesting model 11 to be the more optimal. The smoothing function for NAP is given in Figure 8.4.

In the same way as the linear regression model was extended to generalised linear modelling to analyse count data, presence–absence data or proportional data, so can the mixed model be extended to generalised linear mixed modelling. It is also possible to add auto-correlation structure to these methods. However, generalised linear mixed modelling and generalised additive mixed modelling are outside the scope of this book, and the interested reader is referred to Wood (2006) and Ruppert et al. (2003).

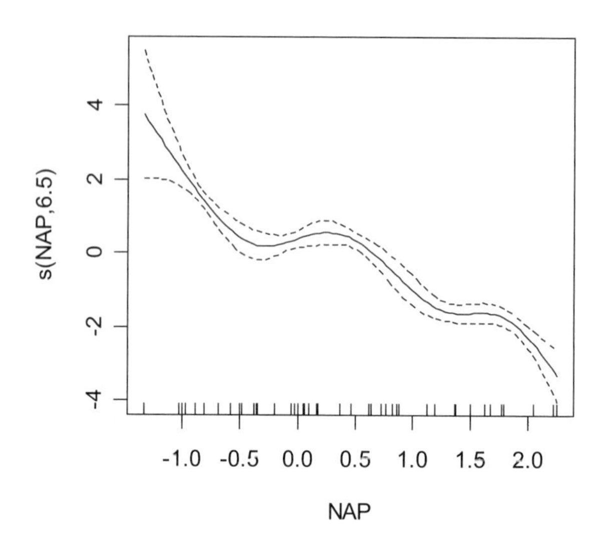

Figure 8.4. Smoothing function for NAP obtained by the additive mixed model 11. The amount of smoothing was estimated using cross-validation.

9 Univariate tree models

9.1 Introduction

A further tool to explore the relationship between a single response variable and multiple explanatory variables is a regression or classification tree (Chambers and Hastie 1992; De'Ath and Fabricus 2000; Fox 2000; Venables and Ripley 2002; Maindonald and Braun 2003). Classification trees are used for the analysis of a nominal response variable, and regression trees for a continuous response variable. Both types of tree models deal better with non-linearity and interaction between explanatory variables than regression, generalised linear models and generalised additive models, and can be used to find interactions missed by other methods. They also indicate the relative importance of different explanatory variables and are useful in analysing residuals from linear regression, GLM or GAM.

Tree models are relatively new in ecology, so we begin with a detailed, but non-technical, explanation of how they work using an artificial example inspired by Maindonald and Braun (2003). Suppose we count the numbers of bees found at seven sites on a particular day. Sampling took place in the morning and the afternoon, and the number of plant species found at each site were also recorded (Table 9.1). The response variable is the number of bees (abundance), and the explanatory variables are the number of plant species at a site and the time of sampling (morning or afternoon). Time of day is a nominal variable, and the number of plant species is considered a continuous variable. The questions of interest in this hypothetical example are whether bee abundance is related to time of day and/or number of different plant species, and which explanatory variable is the most important.

The overall mean of the data is $\bar{y} = (0 + 1 + 2 + 3 + 4 + 5 + 6)/7 = 3$ and we define the deviance D, or total sum of squares, as

$$D = \sum_{j=1}^{7} (y_j - \bar{y})^2$$

where y_j is the number of bees at site j. This gives a value for $D = 28$.

Table 9.1. Artificial bee data. The response variable is the bee abundance and the explanatory variables are number of plants and time of day (M = morning and A = afternoon).

Site	Bees	Number of Plant Species (P)	Time of Day
A	0	1	M
B	1	1	A
C	2	2	M
D	3	2	A
E	4	3	M
F	5	3	A
G	6	3	A

Now suppose that we want to split the seven observed bee data into *two* groups, based on the values of an explanatory variable. We have arbitrarily decided to start with time of day. As this explanatory variable has only two classes, we can readily split the bee data into two groups:

- Group 1: 0, 2 and 4 bees for the morning.
- Group 2: 1, 3, 5 and 6 bees for the afternoon.

Using basic algebra, it can be shown that the deviance can be rewritten as

$$\text{D} = \sum_{j=1}^{7}(y_j - \bar{y})^2$$

$$= \sum_{j\in\text{morning}}(y_j - \bar{y})^2 + \sum_{j\in\text{afternoon}}(y_j - \bar{y})^2 \qquad (9.1)$$

$$= \sum_{j\in\text{morning}}(y_j - \bar{y}_M)^2 + \sum_{j\in\text{afternoon}}(y_j - \bar{y}_A)^2 + n_M(\bar{y}_M - \bar{y})^2 + n_A(\bar{y}_A - \bar{y})^2 \qquad (9.2)$$

$$= 8 + 14.75 + 3 + 2.25 = 28$$

where $\bar{y}_M$ and $\bar{y}_A$ are the averages of the morning and afternoon data, and n_M and n_A are the number of observations in the morning and afternoon, respectively. Note that $\bar{y}_M = (0 + 2 + 4)/3 = 2$ and $\bar{y}_A = (1 + 3 + 5 + 6)/4 = 3.75$. Going from equation (9.1) to (9.2) requires basic algebra. The first component in equation (9.2) measures the variation within the morning observations. The second component determines the variation within the afternoon observations. So, the sum of the first two components represents the *within group variation*. The term $n_M(\bar{y}_M - \bar{y})^2$ measures the deviation of the morning average from the overall average, and the fourth term in equation (9.2) gives the deviation of the afternoon average from the overall average. The sum of the third and fourth components in equation (9.2) therefore represents the *between group variation*. The results are visualised in Figure 9.1, which can be interpreted as follows. If for a particular observation, the statement "Time = morning" is true, then follow the left branch, and if "Time = morning" is false, follow the right branch. The mean value for the number of bees from the morning observations is 2 (n = 3), and for the afternoon observations is 3.75 (n = 4).

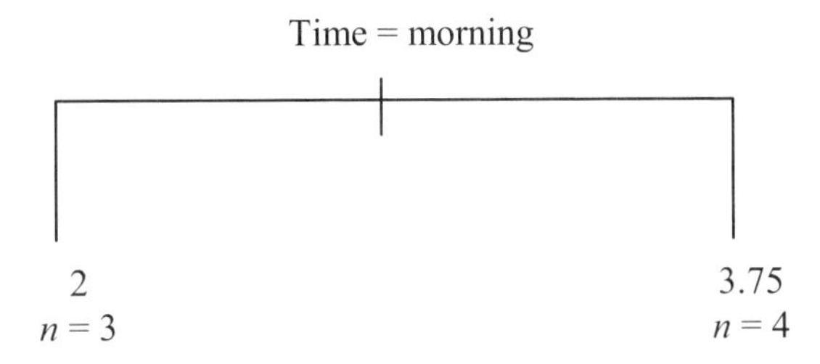

Figure 9.1. Graphical presentation for splitting up the bee data in morning and afternoon data. The mean value in the morning data is 2 (n = 3 observations), and in the afternoon 3.75 (n = 4 observations).

Instead of splitting the bee data into two groups using the time of day (morning and afternoon), we could equally have chosen the number of different plant species, and this is done next. This is a continuous explanatory variable. We now introduce an additional rule. The options for splitting the response variables are constrained by the need to keep any splits in the *continuous* explanatory variables in their original order. There are two possible options to split the bee data *into two groups* based on the continuous variable numbers of different plant species:

- Option 1: sites A and B versus C, D, E, F, G.
- Option 2: sites A, B, C, D versus E, F, G.

Note that groups C, D versus A, B, E, F, G is not an option as the splitting rule for a continuous variable is based on the order of the value of an explanatory variable. The first option results in:

$$
\begin{aligned}
\mathrm{D} &= \sum_{j=1}^{7}(y_j - \bar{y})^2 \\
&= \sum_{j=A,B}(y_j - \bar{y})^2 + \sum_{j=C,D,E,F,G}(y_j - \bar{y})^2 \\
&= \sum_{j=\mathrm{A,B}}(y_j - \bar{y}_{A,B})^2 + \sum_{j=C,\ldots,G}(y_j - \bar{y}_{C,\ldots,G})^2 + n_{A,B}(\bar{y}_{A,B} - \bar{y})^2 + n_{C,\ldots,G}(\bar{y}_{C,..G} - \bar{y})^2 \\
&= 0.5 + 10 + 12.5 + 5 = 28
\end{aligned}
$$

The mean value at the sites A and B is $\bar{y}_{A,B} = 0.5$ and at sites C,..,G it is $\bar{y}_{C,..,G} = 4$. Just as before, the results can be presented graphically (Figure 9.2). If an observation has only one species of plant or no plants present, then the mean value of bees is 0.5, and if there are two or more plant species present, then the mean value for number of bees is four.

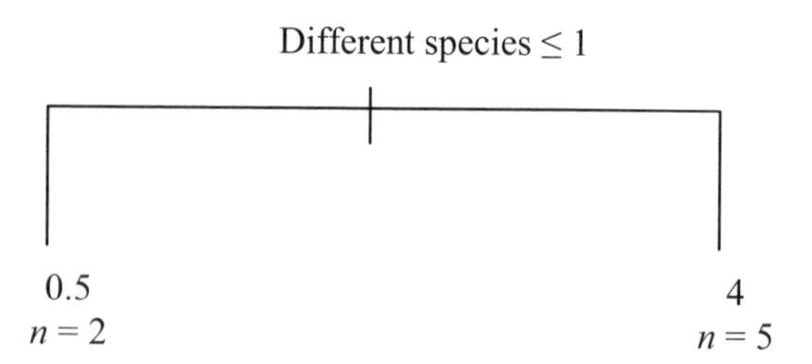

Figure 9.2. Graphical presentation for splitting the bee data based on the values of the number of plant species. The mean value for observations with 1 species of plant or no plants is 0.5 bees, and at observations with more than one plant species present, the mean number of bees is four.

Note that this division results in a within group variation of 10.5 and a between group variation of 17.5. We leave it as an exercise to the reader to check that splitting the data using the second option results in a within variation of 5+2=7 and between variation of 9 + 12 = 21. To summarise, using the explanatory variables we have divided the bee data three times into two groups. The within and between group variation for these splits are:

1. $D_{\text{time of day}} = D_{\text{within}} + D_{\text{between}} = 22.75 + 5.25$
2. $D_{\text{different species} \leq 1} = D_{\text{within}} + D_{\text{between}} = 10.5 + 17.5$
3. $D_{\text{different species} \leq 2} = D_{\text{within}} + D_{\text{between}} = 7 + 21$

The aim is to have groups that have a between variation as large as possible and a within variation as small as possible. Clearly dividing the data into two groups using the time of day is a poor choice as it gives the highest within group variation. In practice the statistical software automatically chooses the optimal grouping; and in this case it is group three, which has the smallest within group variation, and the largest between group variation. For each sub-group, the software will apply the same procedure on the remaining observations and continue until some stopping criteria are met. This process of repeatedly partitioning the observations into two homogenous groups based on the values (order) of an explanatory variable is called a regression (or classification) tree. Note that applying a transformation on the explanatory variables will not change the deviances. Hence tree models are not affected by transformations on the explanatory variables.

The terminal nodes are called leaves, and the regression tree gives a mean value for each leaf allowing the residuals for each observation to be calculated as observed values minus the average leaf value. The sum of squares of the residuals can then be calculated, and adding them together will give the residual sum of squares. Alternatively, the sum of the deviances of the leaves can be calculated (the sum of all D_{within} values).

Example: Bahamas fisheries data

Using the Bahamas fisheries dataset that we used earlier (see Section 7.3), we now look at a real example. In this example, parrotfish densities are used as response variable, with the explanatory variables: algae and coral cover, location, time (month), and fish survey method (method 1: point counts, method 2: transects). There were 402 observations measured at 10 sites, and the regression tree is shown in Figure 9.3. Looking at the regression tree you can see that the 402 parrotfish observations (the response variable) are repeatedly split into groups based on the values of the explanatory variables. In each step this division keeps each group as homogenous as possible. Splitting is conditional on the order of the values of the explanatory variable, and with nominal variables, the division is by (groups of) classes. For the parrotfish, the algorithm splits the data into two groups of 244 observations (left branch) and 158 observations (right branch) and is based on the explanatory variable 'fish survey method', which we called 'Method'. Hence, the most important variable in splitting the data is the survey method: point sampling versus transect sampling. The typical numerical output of SPlus and R is given below, and it shows that the overall deviance and mean are 50188.35 and 10.78, respectively.

```
node), split, n, deviance, yval      * denotes terminal node
 1) root   402   50188.35   10.78
   2) as.factor(Method)=2   244   7044.21   6.45
       4) CoralTotal< 4.955        87    1401.13    3.53 *
       5) CoralTotal>=4.955    157    4488.06    8.07 *
   3) as.factor(Method)=1    158  31494.20   17.47
       6) as.factor(Station)=3,4   25     781.74    3.157 *
       7) as.factor(Station)=1,2,5,6,7,8,9,10   133   24627.58   20.16
         14) as.factor(Month)=5,7,8,10   94   11587.55   17.09*
         15) as.factor(Month)=11   39   10023.52   27.56 *
```

This output looks a little confusing. We will explain the key aspects of interpretation here and look at it in more detail later in Chapter 23. The notation 'as.factor' is a computer notation for nominal variables. The first line shows that there are 402 observations, the total deviance is 50188.35 and the overall mean is 10.78. The lines labelled as 2) and 3) indicate the first split is based on the nominal explanatory variable 'Method'. The left branch is described by the lines labelled as 2), 4) and 5), and the right branch as 3) and all the lines below this one. The layout therefore mirrors the graphical tree representation of the data. Starting with the left branch, the information on the line labelled 2) shows that all the observations in 'Method=2' have a mean of 6.45, but it is not a final leaf as final leaves are labelled by a '*' at the end of the line. The 244 observations can be further split based on coral total and the cut off level is 4.955. All observations smaller than this threshold value for a group with a mean of 3.53, which is one of the smallest group means in the tree! The 157 observations with coral total larger than 4.955 have a mean value of 8.07. Now let us look at the main right-hand branch, starting at the line labelled 3). These observations are split further based on station and month, making these the second most important variables (for observations measured with method one).

So, what does this all mean? The group of observations with the highest mean value is obtained by method one, at stations 1, 2, 5, 6, 7, 8, 9 and 10, and is equal to 27.56 (39 observations). In fact, most groups in Method 1 have higher mean densities, which means that there is a structural difference between the observations due to sampling technique!

If a particularly explanatory variable appears repeatedly along the branches in a tree, this indicates a non-linear relationship between the response variable and the explanatory variable.

The results indicate that for the parrotfish, the type of sampling method is important and it might be useful to apply the same procedure on the other fish groups sampled. If similar results are found with other fish groups, then there is a strong argument for analysing the data from point and survey transects separately. Instead of applying a univariate tree model on each individual fish family, it is also possible to apply a multivariate regression tree (De'Ath 2002) on all the fish groups. But this method is not discussed in this book.

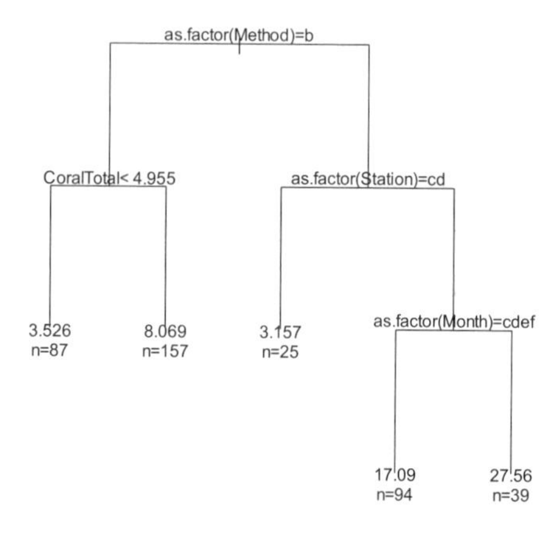

Figure 9.3. Regression tree for parrotfish of the Bahamas fisheries dataset. The observations are repeatedly split into two groups. The first and most important split is based on the nominal explanatory variable 'Method', where 'a' stands for method 1 and 'b' for method 2. If a statement is true, follow the left side of a branch, and if false follow the right side. Numbers at the bottom of a terminal leaf represent the mean value and the number of observations in that particular group.

9.2 Pruning the tree

A major task in linear regression is selecting the optimal subset of explanatory variables. Using too many explanatory variables results in a model that overfits the data and is difficult to interpret. Using only a few explanatory variables can lead to a poor model fit. Tools like the AIC (Chapter 5) can be used to judge which selection of explanatory variables is optimal in terms of model fit and model simplicity. In tree models, we have a similar problem deciding on the size of the tree: defined as the number of splits in the tree. In Figure 9.3 the size of the tree is five, because it has four splits. The number of splits is always equal to the number of leaves (terminal nodes at the bottom) minus 1. If a large tree size is used, we end up with very small groups but lots of terminal nodes, and therefore lots of information to interpret. On the other hand, using a small tree might result in a poor fit. The mean value of a group of observations is given at the end of a branch. The fitted value of a group is its mean value. Hence, if we have lots of terminal nodes with only a few values in it, it is likely that the mean value is close to the observations in the group. On the other hand, if there are only a few splits, then we have groups with lots of observations and the mean value may not represent all observations in that group (some points may have a large deviation from the group mean, causing large residuals).

Hence, the algorithm for repeatedly splitting up the data into two groups needs to stop at a certain stage. One option is to stop the algorithm if the 'to be split subset of data' contains less than a certain threshold value. This is called the minimum split value. However, this is like saying in linear regression: 'use only the best 5 explanatory variables' and a more sophisticated method is available. Define D_{cp} as

$$D_{cp} = D + cp \times \text{size-of-tree}$$

Recall that the deviance D measures the lack of fit, and cp is the so-called complexity parameter (always positive). Suppose that $cp = 0.001$. The criteria D_{cp} is similar to the AIC; if the size of the tree is small the term $0.001 \times$ size-of-tree is small, but D is relatively large. If the size of the tree is large, D is small. Hence, $D_{0.001}$ can be used to find the optimal size in a similar way as the AIC was used to find the optimal regression model. However, the choice $cp = 0.001$ is rather arbitrary, and any other choice will result in a different optimal tree size. To deal with this, the tree algorithm calculates the full tree, and then prunes the tree (pruning means cutting) back to the root. The graphical output of the pruning process is given in Figure 9.4 and is explained next. It is also useful to have the numerical output available, see Table 9.2. This numerical output was obtained by using the default cp value of 0.001. The column Rel-error gives the error of the tree as a fraction of the root node error (=deviance divided by the number of observations). For example, the tree of size 3 has a relative error of 0.59. This means that the sum of all leaf deviances is $0.59 \times 50188 = 29610.92$, and the error is $0.59 \times 125 = 73.75$. Recall that the total deviance was 50188.35. The more leaves used in a tree, the smaller the relative error. Note that there is a large decrease in the relative error for one, two and three splits, but thereafter differences become rather small. In this case choosing a tree with four splits seems to be a safe choice.

A better option for tree size selection is cross-validation. The tree algorithm applies a cross-validation, which means that the data are split into k (typically $k = 10$) subsets. Each of these k subsets is left out in turn, and a tree is calculated for the remaining 90% (if $k = 10$) of the data. Once the optimal tree size is calculated for a given *cp* value using the 90% subset, it is easy to determine in which leaves the observations of the remaining 10% belong by using the tree structure and the values of the explanatory variables. We already have the mean values per leaf so we can calculate a residual (observed value minus group mean) and prediction errors (sum of all squared difference between observed values and mean values) for each observation in the 10% group. This process is applied for each of the $k = 10$ cross-validations, giving 10 replicate values for the prediction error. Using those 10 error values, we can calculate an average and standard deviation. This entire process is then repeated for different tree sizes (and *cp* values) in a 'back-ward selection type' approach. This is illustrated in Figure 9.4. The average (dots) and the standard deviation (vertical bars) are plotted versus the complexity parameter *cp* and the tree size. Along the lower x-axis, the complexity parameter *cp* is printed, and the size of the tree runs along the upper x-axis. The y-axis is the relative error in the predictions, obtained by cross-validation. The vertical lines represent the variation within the cross-validations (standard deviation). This graph is used to select the closest to optimal *cp* value. A good choice of *cp* is the leftmost value for which the mean (dot) of the cross-validations lies below the horizontal line. This rule is called the one standard deviation rule (1-SE). The dotted line is obtained by the mean value of the errors (x-error) of the cross-validations plus the standard deviation (x-std) of the cross-validations upon convergence. Hence, this is $0.62 + 0.07 = 0.69$. The optimal tree is the first tree where the x-error is smaller than 0.69, which is either a tree of size 3 (Nsplit = 3) or 4 (Nsplit = 4) (Table 9.2). Figure 9.4 also suggests that a tree of size 3 or 4 is the optimal one. We decided to present the tree of size 4 (Figure 9.3), as the 1-SE rule should only be used as a *general* guidance. Figure 9.3 shows the tree with Nsplit = 4.

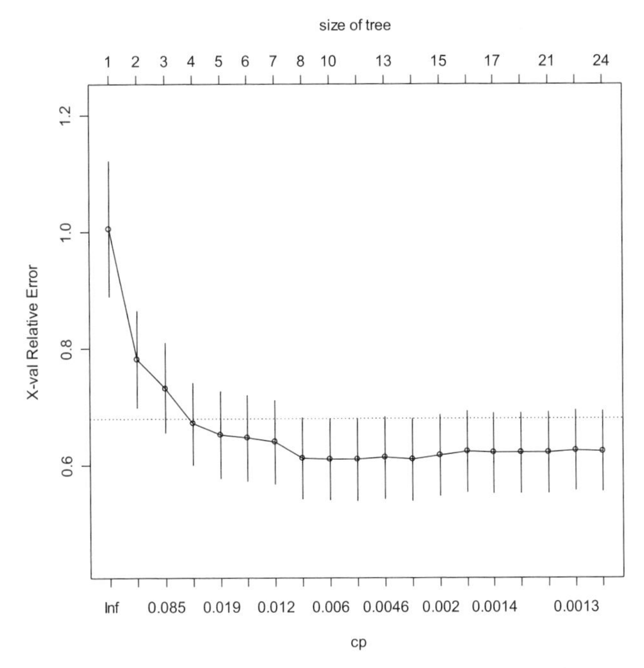

Figure 9.4. Pruning the tree. The lower horizontal axis shows the *cp* values and the upper horizontal axis the corresponding tree sizes. The vertical axis is the relative error in the predictions, obtained by cross-validation. The dots are the averages of the 10 cross-validations, and the vertical lines around the dots represent the variation within the cross-validations (standard deviation).

Table 9.2. Numerical output of the pruning process. The root node error is 50188/402 = 125, and there are 402 observations. X-error is the mean value of the *k* cross-validations (and is expressed as a percentage of the root node error), and x-std is the standard deviation. The column labelled "Rel-error" is the percentage of the root deviance explained by all terminal leaves of the tree.

	cp	Nsplit	Rel-error	x-error	x-std
1	0.2321	0	1.00	1.00	0.116
2	0.1212	1	0.77	0.78	0.083
3	0.0601	2	0.65	0.73	0.078
4	0.0230	3	0.59	0.67	0.071
5	0.0152	4	0.56	0.65	0.075
6	0.0132	5	0.55	0.65	0.074
7	0.0108	6	0.54	0.64	0.072
8	0.0062	7	0.52	0.61	0.070
..	...	..	...	...	...
19	0.0001	23	0.48	0.62	0.070

9.3 Classification trees

Classification trees work in the same way as regression trees except that (i) the response variable is a nominal variable, and (ii) the deviance is defined slightly differently. For 0–1 data, the deviance at a particularly leaf *j* is defined as

$$D_j = -2[n_{1j} \log \mu_j + n_{0j} \log(1 - \mu_j)]$$

where n_{1j} is the number of observation in leaf j for which $y = 1$ and n_{0j} is the number of observations for which $y = 0$. The fitted value at leaf *j*, μ_j, is the proportion $n_{1j}/(n_{1j} + n_{0j})$. The overall deviance of a tree is the sum of the deviances over all leaves. If the response variable has more than two classes, the deviance at leaf *j* is defined as

$$D_j = -2\sum_{i=1}^{n} n_{ij} \log \mu_{ij}$$

For example, if the response variable has the classes 1, 2 and 3, then the deviance at leaf *j* is defined as:

$$D_j = -2[n_{1j} \log \mu_{1j} + n_{2j} \log \mu_2 + n_{3j} \log \mu_{3j}]$$

where n_{1j} is the number of observations at leaf *j* for which $y = 1$, and $\mu_{1j} = n_{1j}/(n_{1j} + n_{2j} + n_{3j})$. The classification tree shows misclassification errors (errors/number of observations per leaf) instead of mean values. Further details and examples of classification trees are discussed in Chapter 24.

9.4 A detailed example: Ditch data

These data are from an annual ditch monitoring programme at a waste management and soil recycling facility in the UK. The soil recycling facility has been storing and blending soils on this site for over many years, but it has become more intensive during the last 8 to 10. Soils are bought into the stockpile area from a range of different sites locations, for example from derelict building sites, and are often saturated when they arrive. They are transferred from the stockpile area to nearby fields and spread out in layers approximately 300 mm deep. As the soils dry, stones are removed, and they are fertilised with farm manure and seeded with agricultural grasses. These processes recreate the original soil structure, and after about 18 months, the soil is stripped and stockpiled before being taken off-site to be sold as topsoil.

The main objective of the monitoring was to maintain a long-term surveillance of the surrounding ditch water quality and identify any changes in water quality that may be associated with the works and require remedial action. The ecological interest of the site relates mainly to the ditches: in particular the large diversity of aquatic invertebrates and plants, several of which are either nationally rare or scarce. Water quality data were collected four times a year in Spring, Summer, Autumn and Winter, and was analysed for the following parameters: pH, electrical conductivity (μS/cm), biochemical oxygen demand (mgl^{-1}), ammoniacal nitrogen (mgl^{-1}), total oxidised nitrogen, nitrate, nitrite (mgl^{-1}), total suspended solids (mgl^{-1}), chloride (mgl^{-1}), sulphate (mgl^{-1}), total calcium (mgl^{-1}), total zinc (μgl^{-1}), total cadmium (μgl^{-1}), dissolved lead (μgl^{-1}), dissolved nickel (μgl^{-1}), orthophosphate (mgl^{-1}) and total petroleum hydrocarbons (μgl^{-1}). In addition to water quality observations, ditch depth was measured during every visit. The data analysed here was collected from five sampling stations between 1997 and 2001. Vegetation and invertebrate data were also collected, but not used in this example.

The underlying question now is whether we can make a distinction between observations from different ditches based on the measured variables, and whether a classification tree can help with this.

A classification tree works in a similar way as a regression tree, except that the response variable is now a nominal variable with two or more classes. Here, the response variable is the ditch number with classes one, two, three, four and five. The explanatory variables are the chemical variables plus depth, month (nominal) and year (nominal). Tree models are not affected by a data transformation on the explanatory variables, and therefore we used the untransformed data. A detailed discussion of regression trees was given earlier in this chapter. Here, a short summary is given and we spend more time looking at the numerical output produced by software like Splus and R, as it can be rather cryptic.

As in linear regression (Chapter 5), we need to find the optimal regression or classification tree. A model with all variables is likely to overfit the data but using too few variables might give a poor fit. An AIC type criteria (Chapter 5) is used to determine how good or bad a particular tree is, and is of the form:

$$RSS_{cp} = RSS + cp * \text{size of tree} \qquad (9.3)$$

For regression trees, *RSS* stands for residual sum of squares and is a measure of the error. For 0–1 data, the *RSS*, or deviance at a particular leaf *j* was defined in Section 9.3. If the response variable has more than two classes, say five ditches, the deviance at leaf *j* is defined as

$$D_j = -2\sum_{i=1}^{5} n_{ij} \log \mu_{ij}$$

$$D_j = -2[n_{1j} \log \mu_{1j} + n_{2j} \log \mu_2 + n_{3j} \log \mu_{3j} + n_{4j} \log \mu_{4j} + n_{5j} \log \mu_{5j}]$$

where n_{1j} is the number of observations at leaf *j* for $y = 1$, and $\mu_{1j} = n_{1j}/(n_{1j} + n_{2j} + n_{3j} + n_{4j} + n_{5j})$. The parameter *cp* is a constant. For a given value, the optimal tree size can easily be determined in a similar way to choosing the optimal number of regression parameters in a regression model. Setting $cp = 0$ in equation (9.3) results in a very large tree as there is no penalty for its size and setting $cp = 1$ results in a tree with no leaves. To choose the optimal *cp* value, cross-validation can be applied. The underlying principle of this approach is simple: leave out a certain percentage of the data and calculate the tree. Once the tree is available, its structure is used to predict in which group the omitted data falls. As we know to which groups the omitted data belong, the actual and the predicted values can then be compared, and a measure of the error (the prediction error) can be calculated. This process is then repeated a couple of times, omitting different sets of observations. In more detail, the data are divided in 10 parts and 1 part is omitted. The tree is then estimated using 90% of the data. Once the tree has been estimated, the omitted 10% can be used to obtain a prediction error. This process is then repeated by leaving out each of the 10 datasets in turn. This gives 10 prediction errors. The mean values of these 10 cross-validation prediction errors are represented by dots in Figure 9.5. The vertical bars represent the variation in the 10 cross-validation errors. To choose the optimal *cp* value, the 1-SE rule can be used. This rules suggests choosing the *cp* value for the first mean value (dot) that falls below the dotted horizontal line (Figure 9.5). The dotted line is obtained from the average cross-validation mean multiplied with the standard deviation of the 10 mean values for the largest tree. In this case, the optimal tree size is four (Figure 9.5), and the corresponding *cp* value is slightly smaller than 0.1. Instead of the graph in Figure 9.5, the numerical output produced by most programmes can be used to choose the most optimal tree size (see below). The cross-validation mean value for the largest tree is 0.76 (this is a percentage of the root node error), and the standard deviation is 0.089. The sum of these two is 0.849. The smallest tree that has a smaller mean cross-validation error (x-error = 0.84) has three splits, and therefore has a tree size of four.

	cp	Nsplit	rel-error	x-error	x-std
1	0.237	0	1.00	1.18	0.044
2	0.184	1	0.76	1.16	0.050
3	0.132	2	0.58	1.00	0.074
4	0.079	3	0.45	0.84	0.086
5	0.053	4	0.37	0.82	0.087
6	0.026	7	0.21	0.76	0.089
7	0.001	8	0.18	0.76	0.089

Root node error: 38/48 = 0.79. $n = 48$

The root node error for a regression tree is the total sum of squares. For a classification tree it is the classification error. Because most observations in the dataset were from group one (10 observations), the algorithm classified the entire dataset as group one (in fact it was a tie because two other groups that also had 10 observations). Therefore, observations of all other groups, 38 in total, are wrongly classified, and the root node error is 38 out of 48 (=total number of observations). Using eight splits, which corresponds to a tree of size nine, gives an error of 18% (0.18) of the root node error. A tree of size four (Nsplit = 3) has an error of 45% of the root error. Further output produced by tree software is given below. An explanation is given in the next paragraph.

node) split	n	loss	yval	(yprob) *=terminal node
1) root	48	38	1	(0.21 0.19 0.21 0.21 0.19)
2) Total.Calcium>=118	25	16	5	(0.32 0.28 0 0.04 0.36)
4) Conductivity.< 1505	11	6	1	(0.45 0.45 0 0.091 0) *
5) Conductivity.>=1505	14	5	5	(0.21 0.14 0 0 0.64) *
3) Total.Calcium< 118	23	13	3	(0.087 0.087 0.43 0.39 0)
6) Depth>=0.505	8	0	3	(0 0 1 0 0) *
7) Depth< 0.505	15	6	4	(0.13 0.13 0.13 0.6 0) *

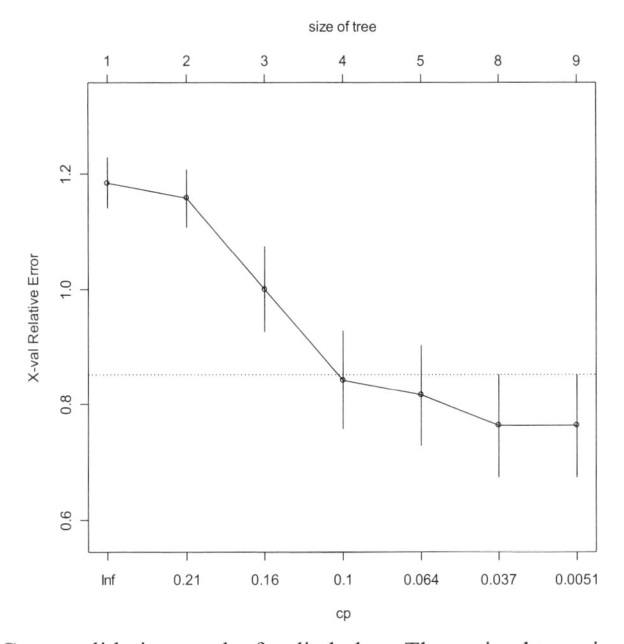

Figure 9.5. Cross-validation results for ditch data. The optimal tree size is four.

The optimal classification tree is presented in Figure 9.6. The most important variable to classify the 48 observations is total calcium. All observations with total calcium values, equal to, or larger than 118 are in the left branch. There are 25 such observations. The other 23 observations are in the right branch. These are not terminal nodes, but the relevant numerical output is:

node) split	n	loss	yval	(yprob)
2) Total.Calcium>=118	25	16	5	(0.32 0.28 0 0.04 0.36)
3) Total.Calcium< 118	23	13	3	(0.087 0.087 0.43 0.39 0)

The proportions per group (as a fraction of 25 observations) are 0.32/0.28/0/0.04/0.36. These proportions correspond to the ditches (groups) one, two, three, four and five. Hence from the 25 observations, 8 ($0.32 \times 25 = 8$) were from group 1. Expressed as real numbers per group, this is: 8/7/0/1/9. Most observations (nine) are from group five, and therefore, this group is classified as group five. However, it is not a terminal node, and therefore further splitting is applied. Sixteen observations are incorrectly classified (this is called the loss). Both splits can be further split. For example, the 25 observations in the left branch can be divided further on conductivity. Observations with conductivity values smaller than 1505 are classified as from ditch one, and those with larger conductivity as ditch five. These are terminal nodes and the relevant output is:

```
4) Conductivity < 1505   11  6  1  (0.45 0.45 0 0.091 0) *
5) Conductivity >=1505   14  5  5  (0.21 0.14 0 0 0.64) *
```

There are 11 observations in this group, and they were from the following ditches: 5/5/0/1/0. These can either be inferred from the numerical output or from the numbers at the end of each leaf. The number of wrongly classified observations in this branch is 6. Note that we actually have a tie, and the algorithm chooses for the first ditch. For observations with total calcium larger, or equal to 118 and conductivity larger than or equal to 1505, the predicted ditch is five. There are 14 such observations, and the observations per group are as follows: 3/2/0/0/9. Hence, this is clearly a ditch five group. The right branch makes a clear distinction between observations from ditch three and four, and involves depth.

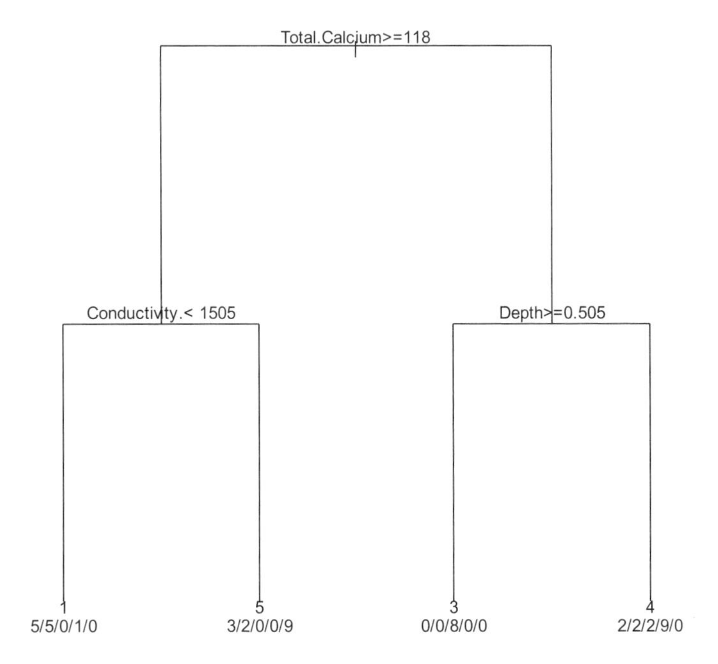

Figure 9.6. Optimal tree for ditch data. The numbers below terminal nodes represent groups (ditches), and the notation '5/5/0/1/0' in the far left terminal leaf means that in this group there were 5 observations from ditch one, 5 from ditch two, 0 from ditch three, 1 from ditch four and 0 from ditch five. The ditch with the highest number of observation wins and determines the name of the group. In this case it is a tie between ditch one and two and one is chosen arbitrarily. The tree shows that a classification can be made among ditches one, five, three and four based on total cadmium, conductivity and depth, but there is not enough information to make a distinction between ditch two and four.

The classification tree indicates that if observations have a total calcium smaller than 118 and depth values larger than 0.505, then they are likely to be from ditch three. If the depth is smaller than 0.505, then they are from ditch four. On the other hand, if total calcium is larger than 118, then conductivity can be used to classify observations in either group five or in groups one or two. Note that there is not enough information available in the explanatory variables to discriminate ditch two. It is the closest to group one (see the proportions in the leftmost terminal leaf). These results indicate that total calcium, depth and conductivity are the most important variables to discriminate among the observations from the five ditches.

To clarify the results obtained by the classification tree, a Cleveland dotplot of total calcium was made (Figure 9.7). The tree identified the value of 118. If one imagines a vertical line at 118, then most observations of ditches three and four have calcium values lower than 118 and most observations from ditches one, two and five have calcium values higher than 118. Similar Cleveland dotplots using depth and total conductivity can be made for the sub-groups.

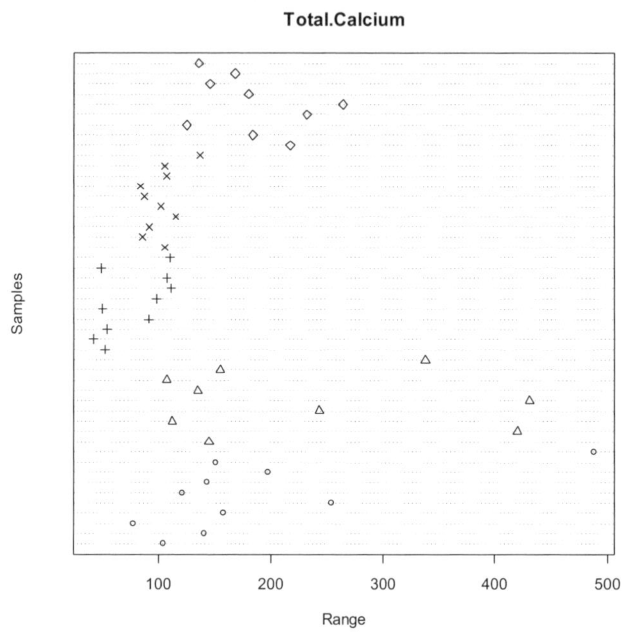

Figure 9.7. Cleveland dotplot for total calcium. Observations from the same ditch are represented by the same symbols. The horizontal axis shows the value of total calcium, and the vertical axis shows each observation in the same order as in the spreadsheet; the 10 observations at the top are from ditch five, and the first 10 at the bottom are from ditch one. Hence, the total calcium in the first observation from ditch one is slightly larger than 100.

Analysing the data in a different way: multinomial logistic regression

One of the confusing aspects of statistics is that the same data can be analysed using different techniques. For the ditch data, we could use classification trees, multinomial logistic regression and discriminant analysis (and neural networks). Discriminant analysis and neural networks will be discussed later, and the case study chapters give examples of using several statistical methods applied to the same data. However, none of the case study chapters use multinomial logistic regression, an extension of logistic regression. Although this technique should probably be explained in Chapter 6 after the logistic regression section, we decided to present it here. The reason for this is that data suitable for classification techniques, such as the ditch data, are also suitable for multinomial logistic regression.

We assume that the reader is familiar with logistic regression (Chapter 6). Suppose that the data consist of observations from two ditches (e.g., one and two). A possible logistic regression model is

$$\ln\left(\frac{P_i}{1-P_i}\right) = g(x_i)$$

P_i is the probability that an observation is from ditch one and $1 - P_i$ is the probability that it is not from ditch one. The function $g(x)$ can be of the form:

$$g(x) = \alpha + \beta_1 \text{ Total calcium} + \beta_2 \text{ Conductivity} + \beta_3 \text{ Depth}$$

where α is the population intercept and β_1, β_2 and β_3 the population slopes. Hence, in logistic regression the probability of an observation coming from ditch one divided by the probability that it is not from ditch one, is modelled as an exponential function of explanatory variables, such as total calcium, etc. In multinomial logistic regression, a similar model is used except that the response variable is allowed to have more than two classes. For the ditch data, we have five ditches so the response variable has five classes: 1, 2, 3, 4 and 5. In multinomial logistic regression, one of the classes is chosen as baseline, and by default most software chooses the first class as the baseline, which in this instance is ditch one. If there are only three explanatory variables, say (total) calcium, conductivity and depth, then the model is written as

$$\ln\left(\frac{P_{i2}}{P_{i1}}\right) = \alpha_2 + \beta_{12} Calcium_i + \beta_{22} Conductivity_i + \beta_{32} Depth_i$$
$$\ln\left(\frac{P_{i3}}{P_{i1}}\right) = \alpha_3 + \beta_{13} Calcium_i + \beta_{23} Conductivity_i + \beta_{33} Depth_i$$
$$\ln\left(\frac{P_{i4}}{P_{i1}}\right) = \alpha_4 + \beta_{14} Calcium_i + \beta_{24} Conductivity_i + \beta_{34} Depth_i$$
$$\ln\left(\frac{P_{i5}}{P_{i1}}\right) = \alpha_5 + \beta_{15} Calcium_i + \beta_{25} Conductivity_i + \beta_{35} Depth_i$$

The response variable has five classes, and therefore, the multinomial logistic regression model has four logistic regression equations. Each equation models the probability of an observation belonging to a particular ditch divided by the probability that it is from the baseline ditch (ditch one). This probability is modelled as an exponential function of the three explanatory variables. For each class, the regression parameters are estimated. This is done with maximum likelihood estimation, and all parameters are estimated simultaneously. To find the optimal model, identical tools as used in logistic regression are available, for example deviance testing, AIC values, backward selection, *t*-values, etc. A backward selection, starting with a model containing all explanatory variables, was used to find the most optimal model. It contained the variables depth, ammoniacal nitrogen, total oxidised nitrogen, total calcium and total zinc. Hence, the optimal model is:

$$\ln(\frac{P_{i2}}{P_{i1}}) = \alpha_2 + \beta_{12}Depth_i + \beta_{22}AN_i + \beta_{32}ON_i + \beta_{42}Calcium_i + \beta_{52}Zinc_i$$

$$\ln(\frac{P_{i3}}{P_{i1}}) = \alpha_3 + \beta_{13}Depth_i + \beta_{23}AN_i + \beta_{33}ON_i + \beta_{43}Calcium_i + \beta_{53}Zinc_i$$

$$\ln(\frac{P_{i4}}{P_{i1}}) = \alpha_4 + \beta_{14}Depth_i + \beta_{24}AN_i + \beta_{34}ON_i + \beta_{44}Calcium_i + \beta_{54}Zinc_i$$

$$\ln(\frac{P_{i5}}{P_{i1}}) = \alpha_5 + \beta_{15}Depth_i + \beta_{25}AN_i + \beta_{35}ON_i + \beta_{45}Calcium_i + \beta_{55}Zinc_i$$

where *AN* and *ON* stand for ammoniacal nitrogen and total oxidised nitrogen, respectively. The estimated regression parameters are:

Class	Intercept	Depth	AN	ON	Calcium	Zinc
2	16.93	−7.11	5.62	−3.25	−7.98	14.93
3	245.62	278.71	34.11	−66.85	−145.48	−40.18
4	212.82	4.92	37.26	−31.76	−102.88	−76.54
5	62.96	20.43	34.13	−11.49	−32.35	−163.51

These estimated regression parameters indicate that depth is important to discriminate between observations from ditches three and one. This regression parameter is also relatively large for ditch five. Ammoniacal nitrogen has relatively large values for ditches three, four and five, but it is small for ditch two. Hence, the probability that an observation is in ditch two, divided by the probability that it is in ditch one, is not influenced by ammoniacal nitrogen. The same holds for oxidised nitrogen. Calcium is important for ditches three and four, and zinc for ditches four and five (relative to ditch one).

The magnitude of most estimated regression parameters indicate that the multinomial logistic regression model cannot discriminate the observations from ditches one and two, but it is able to do this for ditch one versus three, four and five. The important variables for this are depth, ammoniacal nitrogen, total oxidised nitrogen, total calcium and total zinc.

The significance of the regression parameters can be determined by individual Wald statistics. These are obtained by dividing the estimated regression parameters by their standard errors, and can be compared with a *t*-distribution; Wald values larger than 1.96 in an absolute sense indicate non-significance at the 5% level. Often, an explanatory variable is significant for one specific (in this case) ditch, but not for the other ditches. However, such a variable must either be included or excluded in the model. Setting an individual regression parameter β_{ij} to null is not possible. Therefore, it is perhaps better to look at the overall significance of a particular explanatory variable. Just as in logistic regression, this can be done by comparing deviances of nested models with each other by using the Chi-square statistics. The AIC of the optimal model (presented above) is 77.16. Leaving out depth gives a deviance that is 42.15 larger compared with the optimal model. The difference of 42.14 is highly significant according to Chi-square test with 4 degrees of freedom. There are 4 degrees of freedom because 4 regression parameters are estimated for each explanatory variable. Results in Table 9.3 indicate that all explanatory variables are highly significant. Leaving out total calcium or depth causes the highest changes in deviance, and the highest AIC indicating that these are the two most important variables.

However, there are some concerns over this model and its conclusions. We started the analysis with 17 explanatory variables and used the AIC to decide on the optimal model presented above (the lower the AIC, the better the model). Month and year were the first variables to be removed. We then ended up where all the alternative models had nearly identical AICs: Differences were in the third digit. This indicates serious collinearity of the explanatory variables. To explain this, suppose we have a linear regression model with four explanatory variables x_1, x_2, x_3 and x_4. If x_4 is highly correlated with x_2 and x_3, then leaving out x_4 will give a model with a nearly identical model fit and AIC.

Only when working with 10 or fewer variables was this problem removed for the ditch data. To avoid these problems, a selection based on biological knowledge is required. If this is not done, one has to accept that total calcium might represent another gradient.

The results for the ditch data obtained by multinomial logistic regression (MLR) are similar to the classification trees. Yet, it provides a simpler alternative to discriminant analysis, classification trees or neural networks. MLR is a parametric approach (it is basically an extension of linear regression), and therefore more powerful. But if there are many explanatory variables, a pre-selection of the explanatory variables may be required to avoid problems with the numerical optimisation routines as many explanatory variables have to be estimated.

Table 9.3. Comparing deviances for the optimal multinomial logistic regression model. Log transformed data were used.

Leave out	df	AIC	LRT	Pr(Chi)
<none>		77.16		
Depth	4	111.31	42.14	<0.001
Ammoniacal Nitrogen	4	93.92	24.76	<0.001
Total Oxidised Nitrogen	4	92.76	23.60	<0.001
Total Calcium	4	125.58	56.42	<0.001
Total Zinc	4	89.06	19.90	<0.001

10 Measures of association

10.1 Introduction

In a multivariate dataset, more than one response variable can be analysed at the same time. In Chapter 4, we used Argentinean zoobenthic data where multiple species were measured at multiple sites with several explanatory variables measured at each sites. Possible underlying questions are as follows:

1. What are the relationships between the response variables (species)?
2. What are the relationships between response variables and explanatory variables?
3. What are the relationships between the explanatory variables?
4. What are the relationships between the observations (e.g., sites)? Are there differences between groups of observations (e.g., differences between Autumn and Spring data, or between the transects)?
5. Are there groups of species behaving similar?

Ordination and clustering are typically used to answer these questions. The reason we start by discussing measures of association is that both ordination and clustering techniques start by calculating a measure of association between the observations, or between the response variables. There is a wide range of choices of measures of association (see for example Legendre and Legendre 1998 or Jongman et al. 1995) and whichever measure of similarity is used will strongly affect the outcome of the analysis.

Once you have read this chapter, we strongly advise that you read Chapter 7 in Legendre and Legendre (1998), as we will closely follow it in Section 10.2. However, their chapter is more detailed (and technical) and has more measures of association. In Chapters 4 and 28 of this book, a zoobenthic dataset from Argentina is used. Here, we will use the same data to illustrate measures of association, but to keep the numerical output simple, we use totals per transect. The resulting data are given in Table 10.1. Three transects were sampled in Spring and Autumn giving six rows of data, but in this chapter we ignore the seasonal information. Formulated differently, we ignore the fact that sites 1 and 4, 2 and 5 and 3 and 6 are physically the same. See Chapter 28 for a detailed (and proper) statistical analysis of these data.

Table 10.1. Totals per transect for the Argentinean zoobenthic dataset. Transects were sampled in Autumn (labeled as 1, 2 and 3) and Spring (labeled as 4, 5 and 6).

Transect	Laeonereis acuta	Heteromastus similis	Uca uruguayensis	Neanthes succinea
1	407	79	0	0
2	769	139	87	1
3	44	429	0	22
4	654	108	0	0
5	563	189	110	17
6	84	327	0	63

Throughout this chapter, we discuss the statistical techniques from an ecological point of view. For example, we will talk about species measured at sites. However, all the methods can equally well be applied on other data, for example, financial or medical data. The best way to visualise the data is to imagine a spreadsheet where the rows correspond to observations (sites) and the columns to variables (species). In most of the ecological examples in this chapter, the species are the response variables and the sites are observations. In Table 10.1 we do not have sites but transects, but to avoid confusion we will just call them sites.

The first fundamental question that you have to address is whether you are interested in relationships between species or sites. All the measures of association that are to come can be divided into so-called Q and R analysis. Q analysis is used to define association between sites (objects, observations) and R analysis between species (descriptors, variables). This is a bit of an ecological thing; in some scientific fields, researchers may never have heard of it, or yet in other fields you may receive a lot of criticisms if you apply a Q analysis to define association between species, or worse, an R analysis to define association between sites. We start with Q analysis.

10.2 Association between sites: Q analysis

Once you have decided that interest is on similarity between sites, the next question is equally important: What about the zeros, and especially the double zeros? It is particularly important to know how a chosen technique treats double zeros and larger (or extreme) values. Double zeros refer to the situation in which there are many zeros in two rows. Suppose that observations on 10 species were made at two sites (S_1 and S_2). An artificial example is given below:

S_1: 0 0 0 0 0 0 0 2 2 1
S_2: 0 0 0 0 0 4 5 0 0 1

In some measures of association, the joint absence (double zeros) contributes to similarity. Other measures of association ignore the double zeros and instead focus on the joint presence. This is quite a crucial difference! A species not present at two sites may be due to environmental stress, and in this case, the sites should be similarly labelled. On the other hand, rare species or poor experimental design will result in lots of zeros and you do not want to say that two sites are similar because a rare species is not present in any of them. You need to decide (based upon your ecological knowledge) what to do with this. Once you have decided whether double zeros should have an influence, you need to choose a measure of association that indeed achieves this. This choice requires the knowledge of symmetrical and asymmetrical coefficients.

In a symmetrical coefficient, the zeros and non-zeros are treated in the same way. Three examples are the simple matching coefficient (S_1), the coefficient of Rogers and Tanimoto (S_2) and a coefficient that gives more weight to joint presence and joint absence (S_3). The original references can be found in Chapter 7 in Legendre and Legendre (1998). To avoid confusion we used the same notation.

Suppose we want to define the association between sites 1 and 2, sites 1 and 3 and sites 2 and 3 in Table 10.1. The starting point of these three coefficients is a simple 2-by-2 matrix showing the number of joint presence (a), observation unique to site 1 (b), unique to site 2 (c), and the joint absence (d) at both sites for each combination of sites. The values for a, b, c and d for various sites are given in Table 10.2.

The indices S_1, S_2 and S_3 between two sites are defined by:

$$S_1 = \frac{a+d}{a+b+c+d} \qquad S_2 = \frac{a+d}{a+2b+2c+d} \qquad S_3 = \frac{2a+2d}{2a+b+c+2d}$$

The simple matching coefficient S_1 between sites 1 and 2 is 2/(2 + 0 + 2 + 0) = 2/4. Hence, the simple matching coefficient considers the data as presence–absence, and takes into account joint absence (it is used in the formula via d). Computer software can be used to calculate an index for every possible combination of sites.

S_1, S_2 and S_3 are similarity coefficients; the larger the value the more similar. Software for dimension reduction techniques like non-metric multidimensional scaling (Chapter 15) is typically applied on dissimilarity coefficients. A conversion can be used to change a similarity coefficient into a dissimilarity coefficient, for example:

$$D = 1 - S \quad \text{or} \quad D = \sqrt{1-S}$$

We applied the first option. Table 10.3 shows the simple matching coefficient (S_1) and coefficient of Rogers and Tanimoto (S_2) for the Argentinean zoobenthic data. There is not much difference in the ecological conclusions obtained by these two coefficients (at least not for these data).

Table 10.2. Values unique to sites, and joint presence and absence for the Argentinean data.

		Site 2				Site 3				Site 3	
Site 1		present	absent	Site 1		present	absent	Site 2		present	absent
	present	2 (=a)	0 (=b)		present	2 (=a)	0 (=b)		present	3 (=a)	1 (=b)
	absent	2 (=c)	0 (=d)		Absent	1 (=c)	1 (=d)		absent	0 (=c)	0 (=d)

Table 10.3. Simple matching coefficient and coefficient of Rogers and Tanimoto. The smaller the value the more similar are the two sites.

	Simple Matching Coefficient						Coefficient of Rogers and Tanimoto					
	1	2	3	4	5	6	1	2	3	4	5	6
1	0	0.50	0.25	0.00	0.50	0.25	0	0.67	0.40	0.00	0.67	0.40
2		0	0.25	0.50	0.00	0.25		0	0.40	0.67	0.00	0.40
3			0	0.25	0.25	0.00			0	0.40	0.40	0.00
4				0	0.50	0.25				0	0.67	0.40
5					0	0.25					0	0.40
6						0						0

The crucial point with S_1, S_2 and S_3 is that double zeros contribute towards similarity. For example, both S_1 and S_2 indicate that sites 1 and 3 and 2 and 3 are equally similar; $S_1 = 0.25$ for both combinations and for S_2 we have 0.4. Yet, sites 2 and 3 have three species in common and sites 1 and 3 only two. It is the joint zero of *U. uruguayensis* that is causing this. If this makes ecological sense for your data, then you are OK.

We now discuss asymmetrical coefficients. In these coefficients double zeros do not contribute towards similarity. The Jaccard coefficient is defined by

$$S_7 = \frac{a}{a+b+c}$$

Again, we used the same notation as in Legendre and Legendre (1998). The Jaccard index is also called the coefficient of community. A slightly modified coefficient, the Sørensen coefficient (S_8) can be defined by giving more weight to joint presence by using the following formula:

$$S_8 = \frac{2a}{2a+b+c}$$

The Jaccard index and the Sørensen coefficient treat the data as presence–absence data. The Sørensen coefficient is also called the Dice index, critical success index and, meteorology, the threat score. Note that both S_7 and S_8 do not use the joint zeros (d), hence the name asymmetrical. Table 10.4 gives the Jaccard and Sørensen coefficients for the zoobenthic species. There are minor differences between them. However, S_7 (and S_8) for sites 1 and 3, and sites 2 and 3 are now different. The latter combination is more similar, which is what you would expect based on ecology (three species in common versus two).

Table 10.4. Jaccard coefficient and Sørensen coefficient. The smaller the value, the more similar are the two sites.

	Jaccard Coefficient						Sørensen Coefficient					
	1	2	3	4	5	6	1	2	3	4	5	6
1	0	0.50	0.33	0.00	0.50	0.33	0	0.33	0.20	0.00	0.33	0.20
2		0	0.25	0.50	0.00	0.25		0	0.14	0.33	0.00	0.14
3			0	0.33	0.25	0.00			0	0.20	0.14	0.00
4				0	0.50	0.33				0	0.33	0.20
5					0	0.25					0	0.14
6						0						0

Both the Jaccard and Sørensen coefficients treat the data as presence/absence data. We now discuss a series of asymmetrical coefficients that take into account the quantitative aspect of the data. The first one is the similarity ratio (SR). Its mathematical formulation between two species Y and X is given by

$$SR(X,Y) = \frac{\sum_k Y_k X_k}{\sum_k Y_k^2 + \sum_k X_k^2 - \sum_k Y_k X_k}$$

where the index k refers to species. Note that double zeros ($Y_k = X_k = 0$) do not contribute towards the coefficient (the product of two zeros is zero), hence why it is asymmetrical. For presence/absence data, the similarity ratio gives exactly the same results as the Jaccard coefficient. Yet another one is the percentage similarity index. For 0/1 data, the percentage similarity is identical to the Sørensen or Dice coefficient. For other types of data, it takes into account the quantitative aspect of data. Its mathematical formulation between two sites Y and X is given by

$$S_{17} = 2 \times \frac{\sum_k \min(Y_k, X_k)}{\sum_k Y_k + \sum_k X_k}$$

where the index k refers to species. Other names for this coefficient are the Bray–Curtis coefficient and the Czekanowski coefficient. Just like the similarity ratio, it ignores double zeros (hence it is asymmetrical). If two sites have no species in common, then the coefficient is equal to zero. Table 10.5 shows the similarity and the Bray–Curtis coefficients for the six sites of the Argentinean data. There are no spectacular differences between them, but there are small differences in ecological conclusions compared with the Jaccard and Sørensen coefficients.

Table 10.5. Similarity ratio and Bray–Curtis distance for the Argentinean data. The smaller the value the more similar are the two sites.

	Similarity Ratio						Bray–Curtis					
	1	2	3	4	5	6	1	2	3	4	5	6
1	0	0.30	0.83	0.18	0.17	0.74	0	0.34	0.75	0.22	0.29	0.66
2		0	0.87	0.04	0.09	0.82		0	0.75	0.13	0.16	0.70
3			0	0.86	0.76	0.09			0	0.76	0.64	0.19
4				0	0.07	0.81				0	0.18	0.69
5					0	0.70					0	0.57
6						0						0

We now move towards a more controversial measure of association, the Euclidean distances. Suppose we have counts for three species (A, B and C) from five sites (1-5); see Table 10.6. If we consider the species as axes, the five samples can be plotted in a three-dimensional space (Figure 10.1). The Euclidean distance between two sites *i* and *j* is calculated by

$$D_1 = \sqrt{\sum_{k=1}^{3}(X_k - Y_k)^2}$$

This is just Pythagoras. Y_k is the abundance of species *k* at site *Y*. It is easy to show that the Euclidean distance between sites one and two is the square root of 24, and between sites three and four it is the square root of 13. The smaller the D_1, the more similar the two sites. Hence, D_1 indicates that sites three and four are more similar than sites one and two, and this does make sense if you look at the three-dimensional graph. However, sites three and four do not have any species in common, whereas sites one and two have at least one species in common: species A. This shows that for certain types of data, the Euclidean distance function is not the best tool to use, unless you want to identify species or sites with large values or outliers as these will cause a large D_1.

The Euclidean distances between the six sites for the Argentinean data are given in Table 10.7. Again, we obtain a different ecological interpretation. The Euclidean distance between site 3 and all other sites (except for 6) are among the highest and this is because of the high values of *L. acuta*. Actually, all values in Table 10.8 are mainly driven by abundances of *L. acuta* and in lesser context *H. similis* due to the high values of these species, large variation and definition of D_1. The Euclidean distance is sensitive to large values and outliers.

Table 10.6. Artificial example of numbers of three species (A, B and C) measured at five sites (1-5).

	1	2	3	4	5
A	1	5	0	0	3
B	0	2	3	0	2
C	2	0	0	2	3

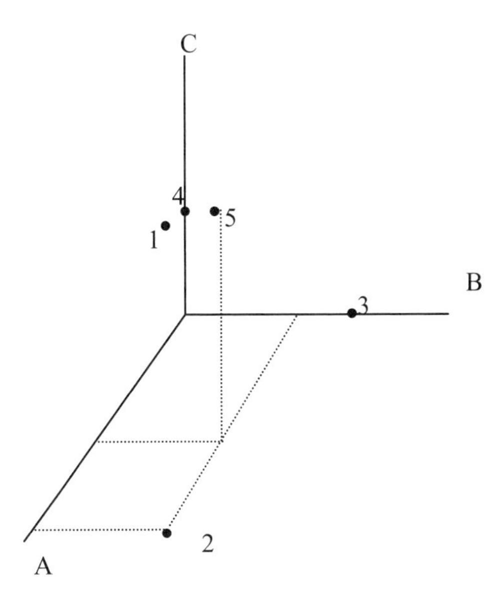

Figure 10.1. Three-dimensional graph of abundance at five sites of three species.

Table 10.7. Euclidean distances between the six sites for the Argentinean data. The smaller the value, the more similar are the two sites.

	1	2	3	4	5	6
1	0	377.11	504.73	248.70	220.96	412.07
2		0	785.96	147.50	213.83	718.32
3			0	689.66	582.31	116.98
4				0	165.02	613.87
5					0	512.54
6						0

As shown in the previous paragraph, absolute numbers influences the Euclidean distance function. To reduce this effect, the Orchiai coefficient (S_{14}) or Chord distance (D_3) can be used. To obtain the Orchiai coefficient, imagine a line from the origin to each site in Figure 10.1. The Orchiai coefficient between two sites is then the angle between the lines of the corresponding sites and can be calculated using simple geometry. The Chord distance between two sites is obtained by drawing a unit (= length 1) sphere around the origin in Figure 10.1, and calculating distances between the intersect points (these are the points where the sphere and the lines intercept). The Orchiai and Chord distances for the Argentinean data are presented in Table 10.8.

Table 10.8. Orchiai coefficient and Chord distance for the Argentinean data. The smaller the value the more similar are the two sites.

	Orchiai Coefficient						Chord Distance					
	1	2	3	4	5	6	1	2	3	4	5	6
1	0	0.29	0.18	0.00	0.29	0.18	0	0.11	1.19	0.03	0.23	1.08
2		0	0.13	0.29	0.00	0.13		0	1.20	0.11	0.16	1.09
3			0	0.18	0.13	0.00			0	1.22	1.09	0.20
4				0	0.29	0.18				0	0.24	1.10
5					0	0.13					0	0.97
6						0						0

We now look at two more measures of association that we use in later chapters: the Manhattan distance (also called taxicab or city-block distance function) and Whittaker's index of association. The Manhattan distance between two sites X and Y containing M species is defined by

$$D_7 = \sum_{k=1}^{M} \left| X_k - Y_k \right|$$

This is the sum of the absolute difference between X_k and Y_k, and it has the same problems as the Euclidean distance. The names Manhattan, taxicab and city-block indicate that this function has something to do with the real distance that a taxi would make if it drives around a city-block (distance in street 1 plus the distance in street 2, plus the distance in street 3, etc.). Indeed, this is how it calculates distance between two sites: the difference between species 1 and 2 at site A, plus the difference between species 1 and 2 at site B, etc.

The Whittaker index of association between two sites X and Y is given by

$$D_9 = \frac{1}{2} \sum_{k=1}^{M} \left| \frac{X_k}{\sum_k X_k} - \frac{Y_k}{\sum_k Y_k} \right|$$

With this index two sites are compared with each other, using the differences in proportions. The proportions are taken with respect of the total species at a site. To illustrate the mechanics of this index, suppose we have the abundances of five species at two sites:

X: 1 4 2 2 1
Y: 0 0 5 5 10

And we wish to calculate the Whittaker index of association between the two sites. The totals at the two sites are 10 and 20. The proportions and differences are:

X:	0.1	0.4	0. 2	0.2	0.1
Y:	0.0	0.0	0.25	0.25	0.5
|X − Y|:	0.1	0.4	0.05	0.05	0.4

Adding up the absolute differences, and multiplying by 0.5 gives $D_9 = 0.5$. It is well suited for species abundance (Legendre and Legendre 1998). The Manhattan and Whittaker index for the Argentinean data are given in Table 10.9. Note that the Manhattan distance is driven by abundant species.

Table 10.9. Manhattan distance and Whittaker index for the Argentinean data. The smaller the value, the more similar are the two sites.

	Manhattan Distance						Whittaker Index					
	1	2	3	4	5	6	1	2	3	4	5	6
1	0	510	735	276	393	634	0	0.09	0.75	0.02	0.20	0.66
2		0	1123	234	295	1022		0	0.77	0.09	0.13	0.68
3			0	953	874	183			0	0.77	0.68	0.18
4				0	299	852				0	0.22	0.68
5					0	773					0	0.59
6						0						0

10.3 Association among species: R analysis

We now discuss ways of defining association between two variables. Technically, we have to write that there are population variables y and x, and a sample of N (paired) observations $Y_1, X_1, Y_2, X_2, \ldots, Y_N, X_N$ is taken. We will define the association between y and x, and use the Y_i and X_i to calculate it. Again, there is the problem of double zeros.

Assume a sample of N observations with two variables y and x, for example the number of the zoobenthic species *L. acuta* (= Y) and mud content (= X). The structure of the sample data is as follows:

	Y	X
Observation 1	Value	Value
Observation 2	Value	Value
...	...	...
...	...	...
Observation N	Value	Value

The question we now wish to address is whether there is a *linear* relationship between y and x. Two obvious tools to analyse this are the covariance and correlation coefficients. Both determine how much the two variables covary (vary together): If the first variable increases/decreases, does the second variable increase/decrease as well? Mathematically, the (population) covariance between two random variables y and x is defined as

$$\text{cov}(y,x) = E[(y - E[y])(x - E[x])]$$

where $E[]$ stands for expectation. The (population) variance of y is defined as the covariance between y and y. If we take a sample of N observations $Y_1, X_1, \ldots, Y_N, X_N$, the sample covariance is calculated by:

$$cov(Y,X) = \frac{1}{N-1}\sum_{j=1}^{N}(Y_i - \bar{Y})(X_i - \bar{X})$$

The bars above Y and X indicate mean values. As an example, we have calculated the covariance among the four zoobenthic species from the Argentinean data (Table 10.1). The diagonal elements in the left side of Table 10.10 are the (sample) variances. Note that *L. acuta* has a rather large variance, which makes the comparison of covariance terms difficult. Although not relevant here, another problem with the covariance coefficient is that it depends on the original units. If for example, a weight variable is expressed in grams instead of kilos, one will find a larger covariance. The correlation coefficient standardises the data and takes values between –1 and 1, and is therefore a better option to use. The (population) correlation coefficient between two random variables y and x is defined by

$$cor(y,x) = \frac{\text{cov}(y,x)}{\sigma_y \sigma_x}$$

where σ_y and σ_x are the population standard deviations of y and x, respectively. If we have a sample of N observations $Y_1, X_1, \ldots, Y_N, X_N$, the sample correlation is calculated by

$$cor(Y,X) = \frac{1}{N-1}\sum_{i=1}^{N}\frac{(Y_i - \bar{Y})}{s_Y}\frac{(X_i - \bar{X})}{s_X} \quad (10.1)$$

s_y and s_x are the sample standard deviations of Y and X respectively. The correlation coefficients among the same four zoobenthic species are given in the right part of Table 10.10; *L. acuta* and *H. similis* have the highest (negative) correlation.

Table 10.10. Covariance and correlation coefficients among the four zoobenthic species from the Argentinean data. The abbreviations LA, HS, UU and NS refer to *L. acuta*, *H. similis*, *U. uruguayensis* and *N. succinea*. The higher the covariance and correlation coefficient, the more similar are the two species.

	Covariance Coefficients				Correlation Coefficients			
	LA	HS	UU	NS	LA	HS	UU	NS
LA	90289	–34321	9212	–5335	1	–0.83	0.6	–0.73
HS		18935	–1770	2314		1	–0.25	0.69
UU			2640	–285			1	–0.23
NS				595				1

In this dataset there are lots of zeros. For example, the percentage of observations with zero abundance for each of the four species in the original data is as follows: L. *acuta*: 10%, *H. similis*: 28%, *U. uruguayensis*: 75% and *N. succinea*: 67%. The problem with the correlation and covariance coefficients is that they may classify two species, which are absent at the same sites, as highly correlated.

The formula in equation (10.1) is the so-called Pearson (sample) correlation coefficient. In most textbooks, this is just called *the (sample) correlation coefficient.* Sometimes, the phrase product-moment is added because it is a product of two terms and involves the first and second moment (mean and variance). The Pearson correlation coefficient measures the strength of the linear relationship between two random variables y and x. To estimate the population Pearson correlation coefficient, observed data Y_1, X_1,..., Y_N, X_N are used, and therefore the estimator is called the *sample correlation coefficient*, or the Pearson sample correlation coefficient. It is a statistic and has a sample distribution. If you repeat the experiment m times, you end up with m estimations of the correlation coefficient. The most commonly used null hypothesis for the Pearson population correlation coefficient is: H_0: cor(y,x) = 0. If H_0 is true, there is no linear relationship between y and x. The correlation can be estimated from the data using equation (10.1), and a t-statistic or p-value can be used to test H_0. The underlying assumption for this test is that y and x are bivariate normally distributed. This means that both y and x need to be normally distributed. If either y or x is not normally distributed, the joint distribution will not be normally distributed neither. Graphical exploration tools can be used to investigate this. There are three options if non-normality is expected: Transform one or both variables, use a more robust measure of correlation, or do not use a hypothesis test.

More robust definitions for the correlation coefficient can be used if the data are non-normal, non-bivariate normal, a transformation does not help, or if there are non-linear relationships. One robust correlation function is the Spearman rank correlation coefficient, which is applied on rank transformed data. The process is explained in Table 10.11. The first two columns show data of two artificial variables Y and X. In the last two columns, each variable has been ranked. Originally the variable Y had the values 6 2 8 3 1. The first value, 6, is the fourth smallest value. Hence, it rank value is 4. Ranking all values results in 4 2 5 3 1. The same process is applied on X. Spearman's rank correlation coefficient is obtained by calculation the Pearson correlation coefficient between the ranked values of Y and X.

The correlation and covariance coefficients only detect monotonic relationships and not non-monotonic, non-linear relationships. It is therefore useful to inspect the correlation coefficients before and after a data transformation.

Table 10.11. Artificial data for Y and X. The first two columns show the original data and the last two columns the ranked data.

Original Data			Ranked Data		
Sample	Y	X	Sample	Y	X
1	6	10	1	4	4
2	2	7	2	2	3
3	8	15	3	5	5
4	3	3	4	3	5
5	1	−5	5	1	1

Chi-square distance

Suppose we are interested in the numbers of parasites on fish from four different locations in three different years (Table 10.12). The underlying questions are whether there is any association between location and years (e.g., are there particular areas and years where more parasites were measured), and whether a Chi-square test is the most appropriate test. The Chi-square statistic is calculated by:

$$X^2 = \sum \frac{(O-E)^2}{E}$$

where O represents the observed value and E the expected value. For the data in Table 10.12, the following steps are carried out to calculate the Chi-square statistic. First the null hypothesis is formulated:

H_0: there is no association between the rows and columns in Table 10.12.

If rows and columns are indeed independent, the expected number of parasites in area A in 1999 is 1254 × (272/1254) × (567/1254) = 122.909. Table 10.13 shows all the observed (in normal font) and expected values (in italic font). The $(O - E)^2/E$ values are also given in this table. These are the values in italic font and bold. The Chi-square statistic is equal to 20.77 (the sum of all values in bold in Table 10.13). The degrees of freedom for the statistic is equal to the number of rows minus one, multiplied with the number of columns minus 1. This information can be used to work out a p-value, which is $p = 0.002$. Hence, the null hypothesis is very unlikely. Based on the values in Table 10.13, the highest contribution to the Chi-square test was from area A in 2001, and area A in 1999. This information can be used to infer that parasites in area A were different in those two years. In later chapters, correspondence analysis is used to visualise this.

Table 10.12. Numbers of parasites in 4 areas and 3 years.

Area	1999	2000	2001	Total
A	147	55	70	272
B	98	50	107	255
C	183	75	157	415
D	139	50	123	312
Total	567	230	457	1254

Table 10.13. Observed (normal font), expected values (italic font) and contribution of each individual cell to the Chi-square statistic (italic font in bold) for the parasites in fish.

Area	1999			2000			2001		
A	147	*122.99*	***4.69***	55	*49.89*	***0.52***	70	*99.12*	***8.56***
B	98	*115.30*	***2.60***	50	*46.77*	***0.22***	107	*92.93*	***2.13***
C	183	*187.64*	***0.11***	75	*76.1 2*	***0.02***	157	*151.24*	***0.22***
D	139	*141.07*	***0.03***	50	*57.22*	***0.91***	123	*113.70*	***0.76***

The Chi-square statistic can be used to define similarity among variables (e.g., species). The Chi-square distance between two variables Z_1 and Z_2 is defined by

$$D(Z_1, Z_2) = \sqrt{Z_{++}} \sqrt{\sum_{j=1}^{M} \frac{1}{Z_{+j}} (\frac{Z_{1j}}{Z_{1+}} - \frac{Z_{2j}}{Z_{2+}})^2}$$

Z_{1j} is the abundance of the first species at site j, and a '+' refers to row or column totals. We have relaxed the mathematical notation with respect to populations and samples here. Let us have a look at what this formula is doing. Table 10.14 shows artificial data for two species measured at four sites. First of all, the two rows of data are changed into row-profiles by dividing each of them by row-totals Z_{1+} and Z_{2+}. This gives:

A: 0.5 0.1 0 0.4
B: 0.27 0.13 0.2 0.4

The profiles are subtracted from each other, and the differences are squared:

$(A - B)^2$: 0.05 0.0009 0.04 0

In the final step, the weighted average is calculated (weights are given by the site totals 18, 6, 6 and 20); take the square root and multiply with a constant:

$$D(Z_1, Z_2) = \sqrt{50} \sqrt{\frac{0.05}{18} + \frac{0.0009}{6} + \frac{0.04}{6} + \frac{0}{20}} = 0.69$$

The lower this value, the more similar the two species. Note that a site with a high total (e.g., 18) is likely to have less influence on the distance, but sites with low totals may have more influence. If the data contain more than two species, the Chi-square distance can be calculated for any combination of two species. The Chi-square *metric* is identical to the Chi-square *distance*, except for the multiplication of the square root of Z_{++} (total of Y and X)[1]. Table 10.15 shows the Chi-square

[1] A distance function is used to calculate association between two observations or two variables a and b. A metric has the following properties: (i) if $a = b$, then the distance between them is 0; (ii) if $a \neq b$, the distance between them is larger than 0; (iii) the distance between a and b is equal to the distance between b and a,

distances among the four zoobenthic species from the Argentinean data. Note that *U. uruguayensis* (UU) and *N. succinea* (NS) have the highest Chi-square distances. Hence, these species are the most dissimilar, as judged by the Chi-square distance function. The higher the Chi-square distance the greater the difference.

The disadvantage of the Chi-square distance function is that it may be sensitive to species that were measured at only a few sites (patchy behaviour) with low abundance. The underlying measure of association in correspondence analysis and canonical correspondence analysis is the Chi-square distance function. Patchy species (with low abundance) will almost certainly dominate the first few axes of the ordination diagram.

Table 10.14. Artificial data for two species (A and B) at 4 sites (1-4).

	Sites				
Species	1	2	3	4	Total
A	10	2	0	8	20
B	8	4	6	12	30
Total	18	6	6	20	50

Table 10.15. Chi-square distances among the four zoobenthic species from the Argentinean data. The abbreviations LA, HS, UU and NS refer to *L. acuta*, *H. similis*, *U. uruguayensis* and *N. succinea*. The smaller the values, the more similar are two species.

	LA	HS	UU	NS
LA	0	1.3	1.1	2
HS		0	1.7	1.2
UU			0	2.3
NS				0

10.4 Q and R analysis: Concluding remarks

As explained in Section 10.1, Legendre and Legendre (1998) grouped the measures of association into Q and R analyses. The Q analysis is for relationships among observations (e.g., sites) and the R analysis for relationships among variables (e.g., species). The correlation coefficient is an R analysis, and the Jaccard and related methods is a Q analysis. Legendre and Legendre (p289-290, 1998) gave five objections for not using the correlation as a tool to calculate association among observations. We discuss some of their objections using this example:

---

and (iv) the distance between *a* and *b* plus the distance between *b* and *c* is equal to or larger than the distance between *a* and *c* (Legendre and Legendre 1998).

	A	B	C	D	E	F
1:	2	4	2	3	3	99
2:	1	3	1	2	0	90
3:	0	2	3	3	1	95
4:	3	5	2	3	1	94

The four rows are sites and the six columns are species. Assume we want to calculate the correlation among the four sites. In this case, the correlation between any two rows will be close to 1 because F is causing a large contribution to the correlation (its value is far above the row average). This situation can arise if one species is considerably more abundant than the others, or if the variables are for example physical variables and one of them has much higher values (which could also be due to different units). One option to avoid this problem is to standardise each variable (column) before calculating the correlation between rows. However, this means that the correlation between rows one and two depend on the data in other rows as these are needed for the standardisation.

Another argument is that correlation between variables will standardise the data and therefore the data are without units. However, if the variables are in different units, and if the correlation is applied as a Q analysis in which the variables have different units, we get similar problems as in the artificial example above.

So, what about an R analysis to quantify relationships between species that contain many (double) zeros? Legendre and Legendre (p. 292, 1998) advise to (i) remove the species with lots of zeros, (ii) consider zeros as missing values, and (iii) eliminate double zeros from the computation of the calculations of the association matrix. Aggregation to a higher taxa or taking totals per transect/beach/area may help as well. This is what we did in Table 10.1, and the same principle (totals per pasture) is applied in Chapter 32 and Chapter 12 (totals per taxa). Note that the suggestions made by Legendre and Legendre (1998) are not a golden rule. Some people would argue that joint absence of species at sites is important information that should be taken into account. In this case, you do not want to aggregate, remove species with lots of zeros, etc. Whichever route you choose, you should be prepared to defend your choice. The suggestion from Legendre and Legendre (1998) that the Chi-square distance copes better with double zeros is valid, but then this one is sensitive to patchy species. The alternative is to use an asymmetrical coefficient in an R analysis (p. 294 Legendre and Legendre 1998) like the Jaccard index or the Sørensen coefficient, and focus on the question of which species co-occur. For example, the Jaccard indices among the four zoobenthic species are given in Table 10.16.

The message here is to think carefully about the underlying ecological question, and what the chosen measure of association is actually doing with the data.

Table 10.16. Jaccard index among the four zoobenthic species from the Argentinean data. The abbreviations LA, HS, UU and NS refer to *L. acuta*, *H. similis*, *U. uruguayensis* and *N. succinea*. The smaller the value, the more similar are the two species.

	LA	HS	UU	NS
LA	0	0	0.67	0.33
HS		0	0.67	0.33
UU			0	0.5
NS				0

Some measures of association cannot deal with negative numbers. The Jaccard index, for example, is typically designed for count data. Others have problems when the total of a row (and/or column) is equal to zero, for example the Chi-square distance. The correlation coefficient cannot deal with variables that have the same value at each site (standard deviation is null). So, again the message here is to know the technical aspects of the chosen measure of association.

It is also possible to transform (square root, log, etc.) the data prior to calculation of the measure of association. Some authors (e.g., Clarke and Warwick 1994) even advocate choosing only one measure of similarity, i.e., the Bray–Curtis coefficient, and then use different data transformations. This is tempting as it simplifies the number of steps to consider, but it is a lazy option. Bray–Curtis may well work fine for marine benthic data (the original focus of PRIMER), but for different application fields, other measures of association may be more appropriate. We recommend knowing and understanding the chosen measure of association, and trying different ones as necessary (depending on the data and questions). Remembering that it is possible to combine some measures of association with a data transformation. Although this approach sounds a bit more challenging, it forces the researcher to know and understand what he/she is doing, and why.

Which measure of association to chose?

Legendre and Legendre (1998) give around 30 other measures of similarity, and this obviously raises the question of when to use which measure of association. Unfortunately, there is no easy answer to this. A cliché answer is: 'it depends'. It depends on the underlying questions, the data itself, and the characteristics of the association function. The underlying question itself should guide you to a Q or R analysis. From there onwards it is the double zeros, outliers and your underlying question that determine which measure is appropriate. If you are interested in the outliers, the Euclidean distance function is a good tool. But for ordination and clustering purposes, the Sørensen or Chord distance functions seem to perform well in practice, and certainly better than the correlation coefficient and Chi-Square functions.

Jongman et al. (1995) carried out a simulation study in which they looked at the sensitivity (sample total, dominant species, species richness) of nine measures of association. The Jaccard index, coefficient of community and the Chord distance performed reasonably well. However, the (squared) Euclidean distance, similarity ratio and percentage similarity functions were sensitive to dominant species and sample totals.

Further guidance in the choice of a coefficient can be found in Section 7.6 in Legendre and Legendre (1998). The Argentinean data showed how important it is to choose the most appropriate measure of association. The Jaccard and Sørensen indices gave similar values, but other measures of association suggest different ecological interpretations. This in itself is useful, as long as the interpretation is done in the context of the characteristics of the different measure of association.

10.5 Hypothesis testing with measures of association

The measures of association discussed in the previous section result in an N-by-N matrix of similarities or dissimilarities, where N is the number of either observations or variables. We now discuss two techniques in where we want to link the (dis-)similarity matrix with external information. This will be done with ANOSIM (Clarke and Ainsworth 1993), the Mantel test (Sokal and Rohlf 1995) and the partial Mantel test (Legendre and Legendre 1998). In this section, we use real data.

The Data

The data are unpublished bird radar data (Central Science Laboratory, York, UK). During three days in October 2004, a new radar installation was used to measure birds. The radar gave X-band (vertical direction) and S-band data. The S-band data are used here. The radar turns 360 degrees in the horizontal plane and records latitude, longitude and time (among many other variables) for each bird. A certain amount of data preparation was carried out to avoid counting the same bird twice. Figure 10.2 shows the spatial distribution of the observations for the three sampling days. We use these data to illustrate ANOSIM and the Mantel test, but it should be stressed that the approach presented here is only one of several possible approaches. Later, we use specific spatial statistical methods on these data, and another valid approach would be to use a generalised additive (mixed) model.

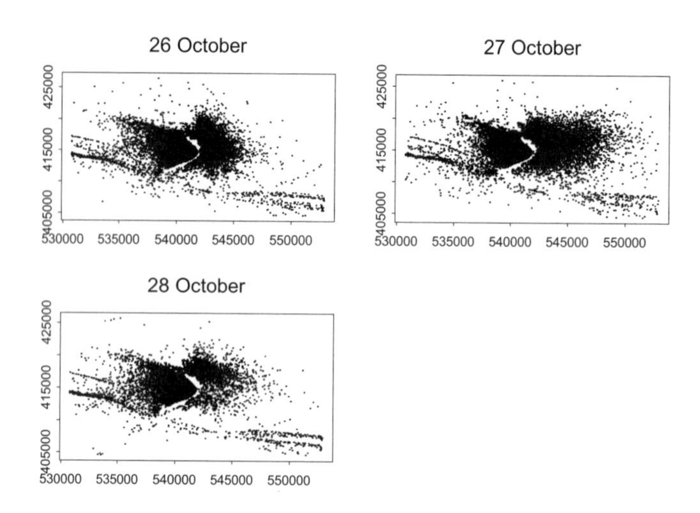

Figure 10.2. Observed values and spatial distribution for the three sampling days. Each dot represents an observation. The horizontal and vertical axes represent longitude and latitude, respectively (in metres).

To prepare the data for the ANOSIM and Mantel tests, totals per time unit per spatial unit were calculated. An arbitrary, but sensible choice for the time unit is one hour. Based on the spatial distribution, we decided to use data observed between a longitude of 535000 and 545000 and latitude between 410000 and 420000. This was to avoid too many cells with zero observations. The size of the spatial grid is again an arbitrary choice. The most convenient choice is to divide the axes in each panel in Figure 10.2 into M cells. We used $M = 10$, $M = 15$ and $M = 20$. The first option results in cells of 1000-by-1000 m, the second option in cells of 666-by-666 m, and the third value in cells of size 500-by-500 m. Figure 10.3 shows the spatial grid if $M = 15$ is used. Both the horizontal and the vertical axes are split up into 15 blocks, and as a result a total of 225 cells are used. $M =$ 10, gives us a 100 cells, and $M = 20$ gives us 200 cells. Total birds per cell per hour were calculated and used in the analyses. The first question to address is whether there is a difference in the spatial distribution and abundance of birds across the three time periods, and the Mantel test is used for this.

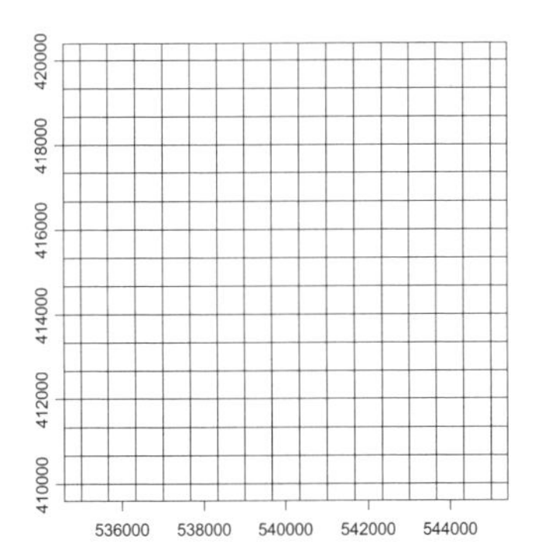

Figure 10.3. Example of a spatial grid using $M = 15$. All grids are of the same size (including grids on the boundary).

Spatial differences and the Mantel test

The data matrix is of the form:

$$\mathbf{D} = [\mathbf{D}_1 \quad \mathbf{D}_2 \quad \mathbf{D}_3] \qquad \text{and} \qquad \mathbf{S} = [\mathbf{X} \quad \mathbf{Y}]$$

If $M = 20$, then $\mathbf{D}_1$ is a 400-by-H_1 matrix, where H_1 is the number of hours in day one. We have $H_1 = 23$, $H_2 = 24$ and $H_3 = 23$ hours, which adds up to a total number of 70 hours. Hence, $\mathbf{D}$ is of dimension 400-by-70, and each row contains 70 bird observations in a particular cell. The vectors X and Y contain the longitude and latitude values for each grid, and $\mathbf{S}$ is of dimension 440-by-2.

The first ecological question we address is whether the relationships among cells (in terms of bird observations) is related to spatial distances; cells adjacent to each other might have a similar pattern over time. To test whether this is indeed the case, the Mantel test (Legendre and Legendre 1998; Sokal and Rohlf 1995) can be used. One way of applying this method is as follows:

- Calculate the similarity among the 400 cells in terms of observed number of birds. Call this matrix $\mathbf{F}_1$. The matrix $\mathbf{D}$ is used for this.
- Calculate the geographical similarity among the 400 grid cells. Use real distances for this. Call this matrix $\mathbf{F}_2$. The matrix $\mathbf{S}$ is used for this.
- Compared these two (dis-)similarity matrices $\mathbf{F}_1$ and $\mathbf{F}_2$ with each other using a correlation coefficient.
- Assess the statistical significant of the correlation coefficient.

The Mantel test calculates two distance matrices $\mathbf{F}_1$ and $\mathbf{F}_2$. Both matrices are of dimension 400-by-400. $\mathbf{F}_1$ represents the dissimilarity among the 400 grid points, and $\mathbf{F}_2$ the geographical (=Euclidean) distances among the cells. The Mantel test compares the two matrices $\mathbf{F}_1$ and $\mathbf{F}_2$ with each other. It does this by calculating a correlation between the (lower diagonal) elements of the two matrices. A permutation test is then used to assess the significance of the correlation. This test tells us whether there is a significant relation among bird numbers across the grids and the geographical distances of those grids.

A bit more detail on the Mantel test and the permutation

We now explain the Mantel test and permutation test in more detail. Recall that the 400-by-70 matrix of bird abundance was used to calculate $\mathbf{F}_1$, and the geographical coordinates for $\mathbf{F}_2$. A sensible measure of association for $\mathbf{F}_1$ is the Jaccard index, and Euclidean distances can be used to define the geographical distances in $\mathbf{F}_2$. Figure 10.4 shows a schematic outline of the Mantel test. Both matrices $\mathbf{F}_1$ and $\mathbf{F}_2$ are of dimension 400-by-400. These matrices are symmetric; the Jaccard index between cells 1 and 2 is the same as between 2 and 1. The diagonal elements are irrelevant as they represent the Jaccard index between cell 1 and cell 1. The same holds for $\mathbf{F}_2$; hence, we only have to concentrate on the part above (or below) the diagonal matrices.

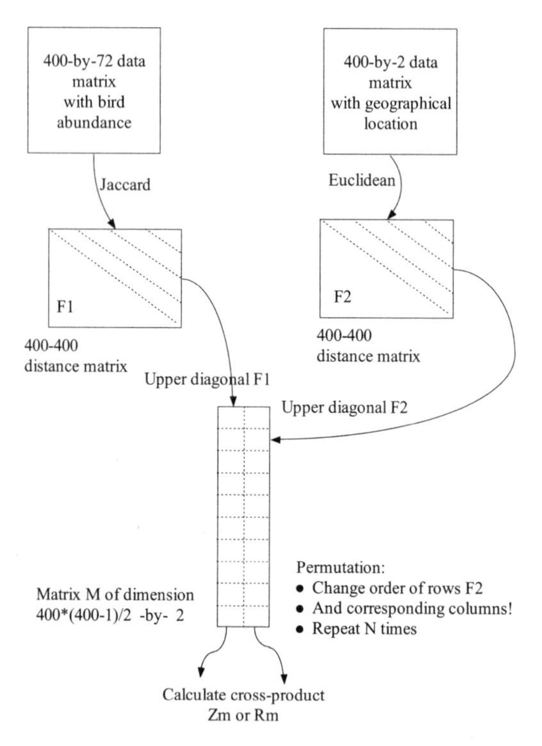

Figure 10.4. Schematic outline of the Mantel test. The data matrices are converted into distance matrices F_1 and F_2, and the upper diagonal matrices of these two distance matrices are compared using, for example, a correlation coefficient. A permutation test is used to assess its statistical significance.

To compare the upper diagonal elements in $\mathbf{F}_1$ and $\mathbf{F}_2$, they can be extracted and stored *side-by-side* in a new matrix **M** that has 400 × (400 − 1)/2 rows and 2 columns. With side-by-side we mean that the element $\mathbf{F}_{1,ij}$ is placed next to $\mathbf{F}_{2,ij}$ in **M** (see the lower part of Figure 10.4). To quantify how similar are the (dis-)similarity matrices $\mathbf{F}_1$ and $\mathbf{F}_2$, the cross-product of $M_{i,1}$ and $M_{i,2}$ can be calculated in two ways:

$$Z_m = \sum_{i=1}^{J} M_{i,1} M_{i,2}$$

$$R_m = \frac{1}{J-1} \sum_{i=1}^{J} \frac{(M_{i,1} - \overline{M}_1)}{s_{M_1}} \frac{(M_{i,2} - \overline{M}_2)}{s_{M_2}}$$

J is the number of rows in the matrix **M**, which is 400 × (400 − 1)/2 in this case. Z_m is the Mantel statistic, and R_m is the standardised Mantel statistic. In the latter, each column of **M** is mean deleted and divided by its standard deviation. The advantage of R_m is that it is rescaled to be between −1 and 1; it is a correlation coefficient.

Assume we find a certain value of R_m. The question now is what does it mean and is it significant? Recall that R_m measures the association among the elements of two (dis-)similarity matrices. So we cannot say anything in terms of the original variables, merely in terms of similarity among the matrices $\mathbf{F}_1$ and $\mathbf{F}_2$! In this case, the question is whether similarities between 400 sites in terms of the Jaccard index are related to the Euclidean distances (reflecting geographical distances in this case). The problem is that we cannot use conventional significance levels for the correlation coefficient R_m as the cells are not independent. Therefore, a permutation test is used. The hypotheses are as follows:

- H_0: The elements in the matrix $\mathbf{F}_1$ are not linearly correlated with the elements in the matrix $\mathbf{F}_2$. Or formulated in terms of the original data, the association among the 400 cells containing bird numbers is not (linearly) related to the spatial distances.
- H_1: There is a linear correlation between the *association* in the two original data matrices.

Under the null hypothesis, rows in one of the original data matrices can be permutated a large number of times, and each time the Mantel statistic, denoted by R_m^*, can be calculated. The number of times that R_m^* is larger than the original R_m, is used to calculate the p-value.

Results of the Mantel test

Several measures of similarity were used: the Jaccard index, the Chord distance, Whittaker index of association and Euclidean distances. The Chord distance function, the Jaccard index and Whittaker index are similar in the sense that the larger values are treated in the same way as the smaller values. The Jaccard index treats the data as presence–absence. The Whittaker index of association is using proportions. The results are presented in Table 10.17. The Jaccard index, the Chord distances and the Whittaker index indicate a significant relationship between bird abundance in grid cells and geographical distances. However, taking into account the absolute value of the data (Euclidean distance), there is no relationship between the bird numbers and geographical distances. This means that cells with large bird numbers are not necessarily close to each other. However, if we consider the data as presence–absence of birds in grid cells (Jaccard index) or "down-weight" the larger values (Chord), then there is a relationship.

Table 10.17. Results of the Mantel test. The number of permutations was 9999.

Measure of Association	Statistic	*p*-value
Chord	0.463	<0.001
Jaccard	0.415	<0.001
Euclidean	−0.161	1
Whittaker	0.405	<0.001

Extensions: The Partial Mantel test

The Mantel test identifies whether there is a correlation between the elements of two matrices. The way it was applied above was to compare a dissimilarity matrix for the bird counts in grid cells with a matrix representing geographical distances. Now suppose that the study area can be divided into two parts, let us call them area A and B. An artificial scenario is sketched in Figure 10.5. The question is whether there are any differences between the black cells (area A) and the white cells (area B). A possible scenario is that area A is close to a wind farm or airport. The question is now whether area A and B are similar in terms of bird observations, and the partial Mantel test (Legendre and Legendre 1998) can be used for this.

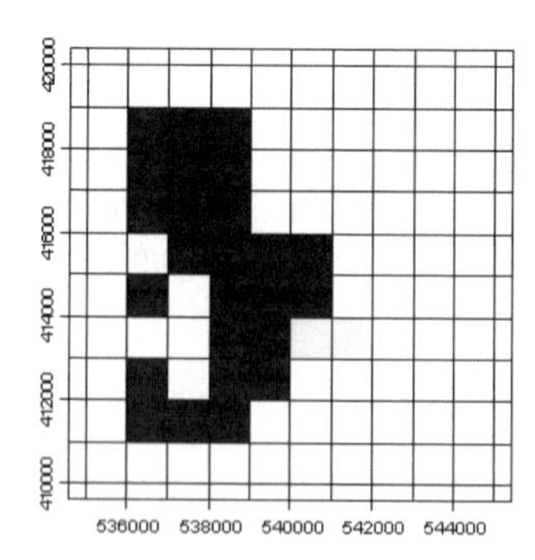

Figure 10.5. Scenario for partial Mantel test. Black cells form area A, and white cells area B.

The partial Mantel test uses three matrices. The first matrix is $\mathbf{F}_1$ and represents the dissimilarity among the 400 grid cells using the bird observations. The Jaccard index, or any other measure of association, can be used. The matrix $\mathbf{F}_2$ represents the difference between the two areas A and B. This sounds complicated, but all we need is a vector of length 400 with zeros (if a cell is from A) and ones (if a cell is from B). The Manhattan distance will ensure that the similarity among all cells from A are 0, and that the similarity among all cells from B are also 0. This is the within-area similarity. The between-area similarity contains only ones. The format of the resulting distance matrix $\mathbf{F}_2$ for a simple artificial example is sketched below:

$$\mathbf{F}_2 = \begin{array}{c} \\ A \\ \\ \\ B \\ \\ \end{array} \overset{\begin{array}{ccccccc} & A & & & B & & \end{array}}{\begin{bmatrix} 0 & & & & & & \\ 0 & 0 & & & & & \\ 0 & 0 & 0 & & & & \\ 1 & 1 & 1 & 0 & & & \\ 1 & 1 & 1 & 0 & 0 & & \\ 1 & 1 & 1 & 0 & 0 & 0 & \\ 1 & 1 & 1 & 0 & 0 & 0 & 0 \end{bmatrix}}$$

So $\mathbf{F}_2$ is only used to quantify whether cells are from the same area, yes (0) or no (1). The third matrix, $\mathbf{F}_3$, represents again the geographical distances among the 400 grid cells.

The partial Mantel test then compares $\mathbf{F}_1$ (similarity of bird abundance in the grids) with the design matrix $\mathbf{F}_2$, while partialling out similarities due to geographical distances. In other words, the association between bird abundance at the grid cells is compared with areas A and B in Figure 10.5, while taking into account (and removing) the effect of geographical distances. The idea of partialling out information is also discussed in Chapters 5 and 16. Further technical details can be found in Legendre and Legendre (1998).

Testing for differences in years using ANOSIM

The ANalyis Of SIMilarities (ANOSIM) method (Clarke 1993; Legendre and Legendre 1998) works in a similar way as the Mantel test. The starting point is the matrix:

$$\mathbf{D} = [\mathbf{D}_1 \quad \mathbf{D}_2 \quad \mathbf{D}_3]$$

If $M = 20$, the matrix $\mathbf{D}$ is of dimension 400-by-70. The matrix $\mathbf{D}_1$ contains the 400-cell-by-23 hour numbers. The question now is whether there is any difference in the association among the three days. Instead of comparing similarities among cells, we now calculate similarities among hours. This gives a 70-by-70 matrix $\mathbf{F}$

(sampling took place for 70 hours in total), using an appropriate measure of association. Each element in the matrix **F** represents the similarity between two hours. The 70-by-70 symmetric matrix **F** is of the form:

$$\mathbf{F} = \begin{pmatrix} \mathbf{F}_1 & & \\ \mathbf{F}_{12} & \mathbf{F}_2 & \\ \mathbf{F}_{13} & \mathbf{F}_{23} & \mathbf{F}_3 \end{pmatrix}$$

$\mathbf{F}_1$ represent the (dis-)similarities among the hours of day one, and the same holds for $\mathbf{F}_2$ (day two) and $\mathbf{F}_3$ (day three). The matrices $\mathbf{F}_{12}$, $\mathbf{F}_{13}$ and $\mathbf{F}_{23}$ represent the (dis-)similarities of hours from two different days.

Now let us try to understand how the ANOSIM method can help us, and which underlying questions it can address. Assume bird behaviour is the same on day one, day two and day three. If we then compare patterns between hour i and j on day 1, we would expect to find similar spatial patterns on day two, and also on day three. As described above, the matrix $\mathbf{F}_1$ contains a comparison among the (spatial) patterns in each hour. A high similarity between two hours means that the 400 spatial observations are similar (as measured by for example the Jaccard index). $\mathbf{F}_2$ and $\mathbf{F}_3$ represent the same information for days two and three, respectively. So what about $\mathbf{F}_{12}$? The information in this matrix tells us the relationship among the spatial observations in hour i on day one and hour j on day two. And a similar interpretation holds for day three. If spatial relationships are the same in all three days, we would expect to find similar dissimilarities in all the sub-matrices in **F**.

The ANOSIM method uses a test statistic that measures the difference between the within group variation (in $\mathbf{F}_1$, $\mathbf{F}_2$, and $\mathbf{F}_3$), and the between group variation (in $\mathbf{F}_{12}$, $\mathbf{F}_{13}$ and $\mathbf{F}_{23}$). In fact, the test statistic uses the ranks of all elements in **F**. If the difference between the *within* group similarity and *between* group similarity is large, then there is evidence that the association among the three time periods is different. The test statistic is given by

$$R = \frac{\bar{r}_b - \bar{r}_w}{n(n-1)/4}$$

where $\bar{r}_b$ and $\bar{r}_w$ are the between and within mean values of the ranked elements and n is the total number of observations. A permutation test similar to the one for the Mantel test can be used to assess its significance. The underlying null hypothesis is that there are no differences among the groups in terms of association. The approach summarised here is a so-called one-way ANOSIM test. The ANOSIM procedure can be extended to two-way ANOSIM. These methods are non-parametric multivariate extensions of ANOVA; see Clarke (1993) and Legendre and Legendre (1998) for further technical details.

Results ANOSIM

To define the matrix F, we used the Jaccard index, Chord distances, Euclidean distances and the Whittaker index of association, and the following results were obtained. For the Euclidean distances, we have $R = 0.192$ ($p < 0.001$). This means that there is a significant difference between the three time periods. The method also gives more detailed information like pair wise comparisons between days one and two ($R = 0.265$, $p < 0.001$), days one and three ($R = 0.346$, $p < 0.001$) and days two and three ($R = 0.016$, $p = 0.151$). Using the Jaccard index, the overall R is 0.234 ($p < 0.001$). The conclusions are the same as for the Euclidean distance. The same holds for the Chord distance ($R = 0.405$, $p < 0.001$) and the Whittaker index of association ($R = 0.431$, $p < 0.001$). With more groups, care is needed with the interpretation of p-values for the individual group-by-group comparisons; there is no such thing as a Bonferroni correction of p-values due to multiple comparisons in ANOSIM. However, a large p-value indicates we cannot reject the null-hypothesis that two time periods are the same, even if many multiple comparisons are made.

Instead of dividing the time into periods of three days, we could have chosen any other division, as long as there are not too many classes. The same principle as in ANOVA applies: If there are too many treatment levels (which are the three time periods here) with rather different effects, the null hypothesis will be rejected regardless of any real differences.

11 Ordination — First encounter

This chapter introduces readers with no experience of ordination methods before to its underlying concepts. If you are at all familiar with ordination, we suggest you may want to go straight to Chapter 12, as the method we use here, Bray–Curtis ordination, is rarely used. However, it is an excellent tool to explain the underlying idea of ordination.

In Chapter 10 several measures of association were applied on the Argentinean zoobenthic data. These gave a 6-by-6 matrix, and as long as this matrix is reasonably small, we can directly compare the numbers by looking at them to see which species are the most similar. But what do you do if the matrix has dimensions 10-by-10, or 500-by-500? Unless you have an excellent memory for numbers interpretation becomes difficult, if not impossible, by just looking at the matrix. In this chapter, and the next four chapters, we look at techniques that provide a graphical representation of either, the *N*-by-*N* association matrix, or the original data matrix. There are different ways of doing this, and the methods can be split into ordination or clustering. We only look at ordination. The aim of ordination is twofold. First, it tries to reduce a large number of variables into a smaller number of easier to interpret variables. And, secondly, it can be used to reveal patterns in multivariate data that would not be identified in univariate analyses.

Ordination methods give easy-to-read graphical outputs from relatively easy-to-use software, and this may partly explain their popularity.

We start by explaining the underlying idea of ordination. Probably, the best way to do this is to use one of the oldest techniques available: Bray–Curtis ordination. The mathematical calculations required for this method are so simple that they can be done with pen and paper.

11.1 Bray–Curtis ordination

Euclidean distances among the six sites for an Argentinean zoobenthic data set were presented in Table 10.7. We have reproduced these distances in Table 11.1, and these will be used to illustrate Bray–Curtis ordination. The choice for the Euclidean distance function is purely didactical; it provides higher contrast among sites. It is by no means the best choice for these data. Our goal is to visualise the distances along a single line. We start with a 6-by-6 matrix with dissimilarities, and then consider how we can graph them in a way that the positions of the sites on the map will say something about the relationships among the sites.

Table 11.1. Euclidean distances among the six sites for the Argentinean data used in Table 10.1. The larger the value, the more dissimilar are the two sites.

	1	2	3	4	5	6
1	0	377.11	504.73	248.70	220.96	412.07
2		0	785.96	147.50	213.83	718.32
3			0	689.66	582.31	116.98
4				0	165.02	613.87
5					0	512.54
6						0

The way Bray–Curtis ordination (or Polar ordination) does this, is as follows. Imagine sitting in a room with 10 other people. There is one person you like (person A), and one you don't like at all (person B). Place them at either side of the room, and draw a straight line between them. Now, place all other eight people along this line in such a way that distances between each person, and persons A and B, reflect how much you like, or dislike, them. So, all we do is identify the two extremes, and use these as a reference point for all other subjects. It sounds simple, and indeed it is.

For the zoobenthic data, Bray–Curtis ordination identifies the two sites that are the most dissimilar. These will have the highest value in Table 11.1, which is 785.96, and between sites 2 and 3. So, these are the two most dissimilar sites. The method places these two sites at the ends of a line (Figure 11.1). The length of this line is 1 unit. We then need to place the four remaining sites along this line. To determine the position of site along the line, say site 1, we have to estimate the distance between this site and each of the ones at the ends of the line. This can be done by drawing two circles. The first circle has its centre at the left end point, and the second circle has the right end point as its centre. The radius of the first circle is given by the Euclidean distance between sites 3 and 1, which is 504.73, but expressed as a fraction of 785.96 (the line has unit length, but represents the distance of 785.96). The radius of the second circle is determined by the Euclidean distance between sites 2 and 1, which is 377.11 (again expressed as a fraction of 785.96). The position of site 1 on the line is determined by vertical projection of the interception of the two circles on the line (Figure 11.1). So it represents a compromise between two distances to both sites at the end of the line. The same process is repeated for the other sites. Further axes can be calculated; see McCune and Grace (2002), and Beals (1984) for details.

Instead of the Euclidean distances, various other measures of similarity can be used (Chapter 10).

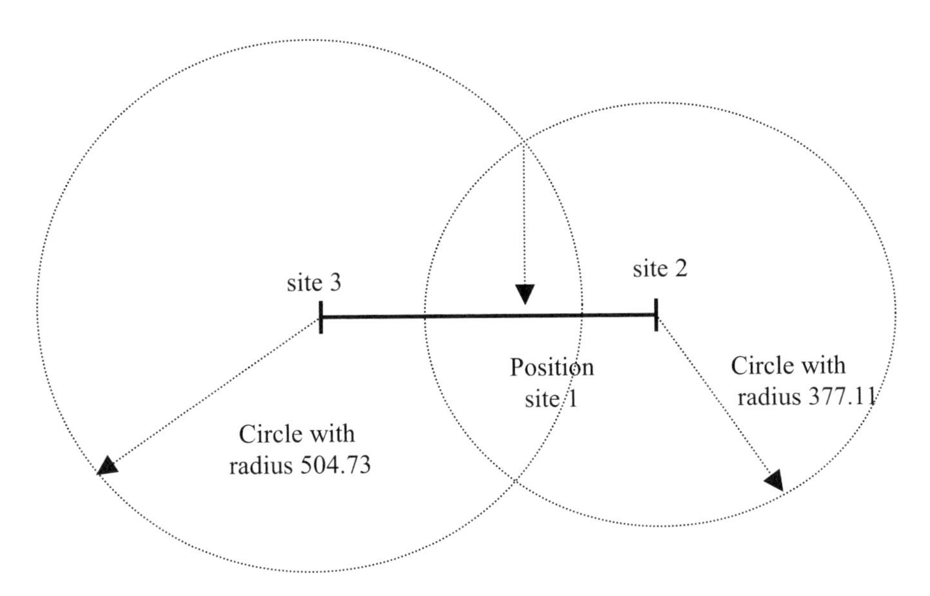

Figure 11.1. Underlying principle of Bray–Curtis ordination. The sites that are the most dissimilar are 3 and 2 and are plotted at both ends of a line with unit length. The other sites are placed on this line as compromise between distances to these two sites. An illustration for site 1 is presented.

There are different ways of selecting the end points of the line (or axis). The original method is shown in Figure 11.1, where the pair of variables that had the highest dissimilarity were selected as end points. However, this approach tends to give an axis that has one or two sites at one end of the axis, and all other sites at the other end of the axis, with isolated sites likely to be outliers. Indeed, Bray–Curtis ordination with the original method to select end points is a useful tool to identify outliers (provided these outliers have a low similarity with the other sites). An alternative is to select your own end points. This is called subjective end point selection. It allows you to test whether certain sites are outliers. The third option was developed by Beals (1984) and is called 'variance-regression'. The first end point calculated by this method is the point that has the highest variance of distances to all other points. An outlier will have large distances to most other points, and therefore the variation in these distances is small. Getting the second end point is slightly more complicated. Suppose that site 1 is the first end point. The algorithm will in turn consider each of the other sites as alternative end points. In the first step, it will take site 2 as the second end point. Let D_{1i} be the distance of site 1 to all other sites (except for sites 1 and 2). Furthermore, let D_{2i} be the distance of site 2 to all other sites. Then regress D_{1i} on D_{2i} and store the slope. Repeat this process for all other trial end points. The second end point will be the site that has the most negative slope. McCune and Grace (2002) justify this approach by saying that the second end point is at the edge of the main cloud of sites, opposite to end point 1.

The Bray–Curtis ordination using the zoobenthic data and the Euclidean distances in Table 11.1 is presented in Figure 11.2. The positions of the sites along the axis are represented by dots. The site names are plotted with an angle of 45 degrees above and below the points. Results indicate that sites 3 and 6 are similar, but dissimilar from the other 4 sites. The variance explained by the axis is 98%. Further technical details can be found in McCune and Grace (2002).

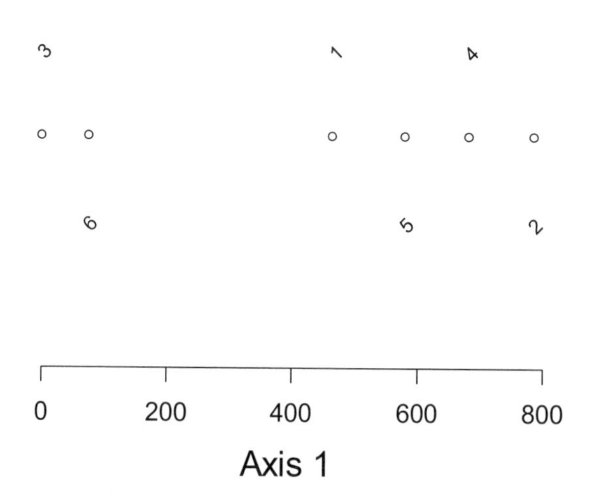

Figure 11.2. Bray–Curtis ordination using the Euclidean distance, applied to the zoobenthic data from Argentina. The original method to select end point was used, and 1 axis was calculated.

Bray–Curtis ordination is one of the oldest ordination methods in ecology and is a good didactical tool to introduce the concept of ordination. However, it might be difficult to get a scientific paper published that relies on a Bray–Curtis ordination. Since the 1950s, more complicated multivariate techniques have been developed and applied in ecology, e.g., principal component analysis, correspondence analysis, redundancy analysis (RDA), canonical correspondence analysis (CCA), canonical correlation analysis, partial RDA, partial CCA, variance partitioning or discriminant analysis, among many others. These techniques are discussed in the next four chapters.

12 Principal component analysis and redundancy analysis

In the previous chapter, Bray–Curtis ordination was explained, and more recently developed multivariate techniques were mentioned. Principal component analysis (PCA), correspondence analysis (CA), discriminant analysis (DA) and non-metric multidimensional scaling (NMDS) can be used to analyse data without explanatory variables, whereas canonical correspondence analysis (CCA) and redundancy analysis (RDA) use both response and explanatory variables. In this chapter, we present PCA and RDA, and in the next chapter CA and CCA are discussed.

The chapter is mainly based on Ter Braak and Prentice (1988), Ter Braak (1994), Legendre and Legendre (1998) and Jolliffe (2002). More easy readings are chapters in Kent and Coker (1992), McCune and Grace (2002), Quinn and Keough (2002), Everitt (2005) and especially Manly (2004)

12.1 The underlying principle of PCA

Let Y_{ij} be the value of variable j (j = 1, ..,N) for observation i (i = 1, ..,M). Most ordination techniques create linear combinations of the variables:

$$Z_{i1} = c_{11}Y_{i1} + c_{12}\,Y_{i2} + \ldots + c_{1N}\,Y_{iN} \tag{12.1}$$

Assume you have a spreadsheet where each column is a variable. The linear combination can be imagined as multiplying all elements in a column with a particular value, followed by a summation over the columns. The idea of calculating a linear combination of variables is perhaps difficult to grasp at first. However, think of a diversity index like the total abundance. This is the sum of all variables (all c_{ij}s are one), and summarise a large number of variables with a single diversity index.

The linear combination, $\mathbf{Z}_1 = (Z_{11},\ldots,Z_{M1})'$, is a vector of length M, and is called a principal component, gradient or axis. The underlying idea is that the most important features in the N variables are caught by the new variable $\mathbf{Z}_1$. Obviously one component cannot represent all features of the N variables and a second component may be extracted:

$$Z_{i2} = c_{21}Y_{i1} + c_{22}\,Y_{i2} + \ldots + c_{2N}\,Y_{iN} \tag{12.2}$$

Further axes can be extracted (in fact, there are as many axes as variables). Most ordination techniques are designed in such a way that the first axis is more important than the second, the second more important than the third, etc., and the axes represent different information. It should be noted that the variables in equations (12.1) and (12.2) are not the original variables. They are either centred (which means that the mean of each variable is subtracted) or normalised (centred and then divided by the standard deviation), and the implication of this is discussed later in this chapter.

The multiplication factors c_{ij} are called loadings. The difference among PCA, DA and CA is the way these loadings are calculated. In RDA and CCA, a second set of variables is taken into account in calculating the loadings. These latter variables are considered as *explanatory* variables, so we are modelling a cause-effect relationship.

12.2 PCA: Two easy explanations

PCA is one of the oldest and most commonly used ordination methods. The reason for its popularity is perhaps its simplicity. Before introducing PCA, we need to clear up some possible misconceptions. PCA cannot cope with missing values (but neither can most other statistical methods), it does not require normality, it is not a hypothesis test, there are no clear distinctions between response variables and explanatory variables, and it is not written as *principle* component analysis but as *principal* component analysis.

There are various ways to introduce PCA. Our first approach is based on Shaw (2003), who used the analogy with shadows to explain PCA. Imagine giving a presentation. If you put your hand in front of the overhead projector, your three-dimensional hand will be projected on a two-dimensional wall or screen. The challenge is to rotate your hand such that the projection on the screen resembles the original hand as much as possible. This idea of projection and rotation brings us to a more statistical approach of introducing PCA. Figure 12.1-A shows a scatterplot of the species richness and NAP variables of the RIKZ data (Chapter 27). There is a clear negative relationship between the two variables. Panel B shows the same scatterplot, except that both variables are now mean deleted and divided by the standard deviation (also called normalisation). We also used axes with the same range (from –3.5 to 3.5). Now suppose that we want to have two new axes such that the first axis represents most information, and the second axis the second most information. In PCA, 'most information' is defined as the largest variance. The diagonal line from the upper left to the lower right in Figure 12.1-B is this first new axis. Projecting the points on this new axis results in the axis with the largest variance; any line with another angle with the *x*-axis will have a smaller variance. The additional restriction we put on the new axes is that the axes should be perpendicular to each other. Hence, the other line (perpendicular to the first new axis) in the same panel is the second axis. Panel C shows the same graph as panel B except that the new axes are presented in a more natural direction. This graph is called a PCA ordination plot. In the same way as the three-dimensional hand was projected on a two-dimensional screen, we can project all the observations onto the first PCA axis, and omit the second axis (Figure 12.1-D). PCA output (which is discussed later) shows that the first new axis represents 76% of the information in the original scatterplot.

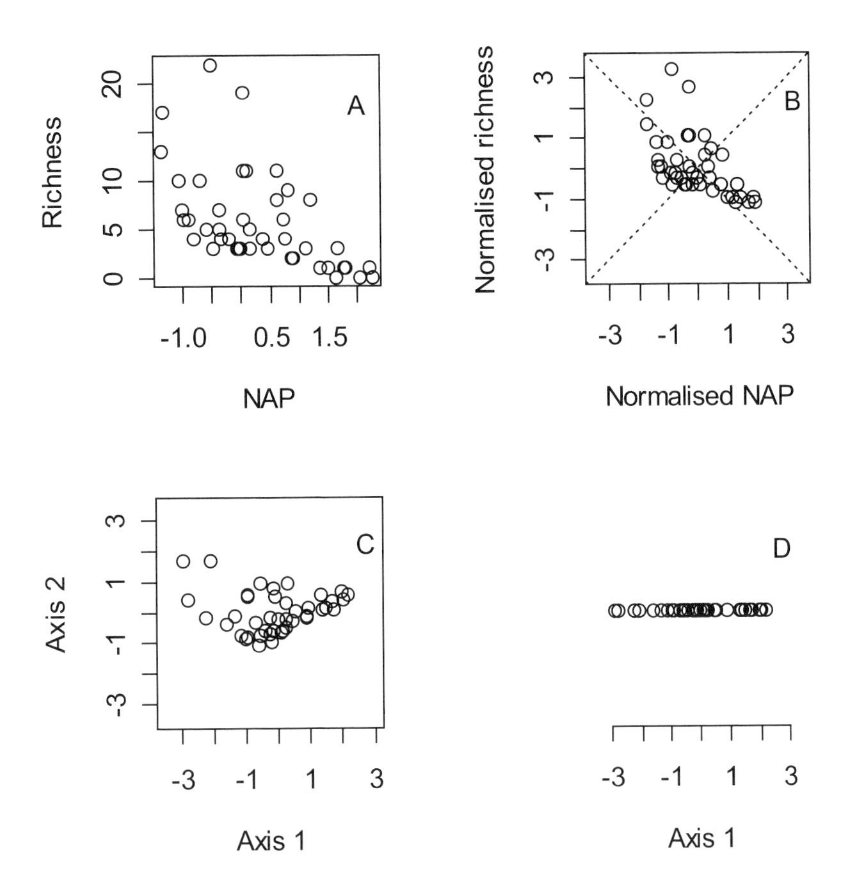

Figure 12.1. The underlying principle of PCA using the RIKZ data. A: Scatterplot of richness and NAP. B: Scatterplot of the same variables but now normalised. The two new axes have been added. C: The two PCA axes. D: Projection of all points on the first axis.

To go from Figure 12.1-A to Figure 12.1-D, we normalised and rotated the data. The two new axes are given by

$$Z_{i1} = -0.7 \times \mathrm{R}_i + 0.7 \times \mathrm{NAP}_i \quad \text{and} \quad Z_{i2} = 0.7 \times \mathrm{R}_i + 0.7 \times \mathrm{NAP}_i$$

In matrix notation, we have $\mathbf{Z} = \mathbf{XC}$, where $\mathbf{C}$ contains the multiplication factors, $\mathbf{X}$ the original variables and $\mathbf{Z}$ the principal components.

In applying PCA, one hopes that the variance of most components is negligible, and that the variation in the data can be described by a few (independent) principal components. Hence, instead of N original response variables, we end up with two or three principal components, which hopefully represent 70%-80% of the information in the data.

12.3 PCA: Two technical explanations

Having presented two relatively easy introductions to PCA, we feel it is necessary to present PCA in a mathematical context as well. Readers not familiar with matrix algebra may skip this section and continue with the illustrations of PCA in Section 12.4. We discuss two mathematical derivations of PCA.

PCA as an eigenvalue decomposition

More details on the technical aspects of PCA can be found in Jolliffe (2002), and only a short summary is presented below. The aim of PCA is to calculate an axis $\mathbf{Z}_1 = \mathbf{Y}\mathbf{c}_1$ that has maximum variance. Because the mean of each variable is equal to zero, the variance of the first axis $\mathbf{Z}_1$ is given by $\mathbf{Z}_1'\mathbf{Z}_1 = \mathbf{c}_1'\mathbf{Y}'\mathbf{Y}\mathbf{c}_1$. An aspect we have not mentioned so far is that the loadings are not unique. If a second axis $\mathbf{Z}_2 = \mathbf{Y}\mathbf{c}_2$ is calculated, then both $\mathbf{c}_1$ and $\mathbf{c}_2$ can be multiplied with 10, resulting in axes that are also uncorrelated. Therefore, a restriction on the loadings is needed to make the solution unique: $\mathbf{c}_i'\mathbf{c}_i = 1$ for all axes $i = 1,..,N$. Although it seems rather arbitrary to set the sum of the squared loadings to 1, it does not have any important consequences.

The question is now how do we get the $\mathbf{c}_i$'s? Once we have the $\mathbf{c}_i$'s, we can easily calculate the $\mathbf{Z}_i$'s. It is relatively easy to show that the loadings can be obtained by solving the following constrained optimisation problem. For the first axis, we have:

$$\text{Maximise var}(\mathbf{Z}_1) = \mathbf{Z}_1'\mathbf{Z}_1 = \mathbf{c}_1'\mathbf{Y}'\mathbf{Y}\mathbf{c}_1 \quad \text{subject to } \mathbf{c}_i'\mathbf{c}_i = 1.$$

The maximisation is with respect to the unknown parameters $\mathbf{c}_1$. Using matrix algebra, it is easy to show that the solution is given by $(\mathbf{S} - \mathbf{I}\lambda)\ \mathbf{c}_1 = 0$, see for example Jolliffe (2002). This may seem like magic to readers not used to matrix algebra, but this expression is fundamental in mathematics and is called the eigenvalue equation for $\mathbf{S}$, λ^* is the eigenvalue and $\mathbf{c}_1$ the eigenvector. $\mathbf{S}$ is either the covariance or correlation matrix, depending on whether the variables in $\mathbf{Y}$ are centred or normalised. This process can easily be extended to get the second or higher axes. In fact, the eigenvalue equation for all axes is given by $(\mathbf{S} - \mathbf{I}\mathbf{\Lambda})\mathbf{C} = \mathbf{0}$, where $\mathbf{C}$ contains the eigenvectors for all axes, $\mathbf{I}$ is the identity matrix and $\mathbf{\Lambda}$ the corresponding eigenvalues. Statistical software can be used to obtain **C** (which can be used to obtain the axes).

The motivation to present the eigenvalue equation for PCA is that it justifies the iterative algorithm presented in the next paragraph, and this algorithm is used to explain RDA in the next chapter.

PCA as an iterative algorithm

The last approach to explain PCA is by using an iterative algorithm, which was presented in Jongman et al. (1995) and Legendre and Legendre (1998), and references in there. The algorithm has the following steps.

1. Normalise (or centre) the variables in **Y**.
2. Obtain initial scores **z** (e.g., by a random number generator).
3. Calculate new loadings: **c** = **Y'z'**.
4. Calculate new scores: **z** = **Yc**.
5. For second and higher axes: Make **z** uncorrelated with previous axes using a regression analysis.
6. Scale **z** to unit variance: **z*** = **z**/λ, where λ is the standard deviation of the scores. Set **z** equal to **z***.
7. Repeat steps 2 to 6 until convergence. After convergence, divide λ by $M - 1$.

Once the algorithm is finished, the loadings **c** and principal components **z** (scores) are identical to those obtained from the PCA eigenvalue equation.

12.4 Example of PCA

To illustrate PCA, we use morphometric data from sparrows (Chris Elphick, University of Connecticut, USA), see also Chapter 14 for more details. This dataset consists of seven morphological variables taken from approximately 1000 sparrows. The morphological variables were wingcrd, flatwing, tarsus, head, culmen, nalopsi and weight (wt); see Section 14.3 for an explanation.

As the PCA will be based on the correlation matrix, it may be a good idea to inspect the correlation matrix before applying PCA. For the sparrow data, the correlation coefficients are relatively high (Table 12.1), and therefore we expect to find a PCA in which the first two axes (Z_1 and Z_2) explain a reasonable amount of information. The lengths of the wing are measured in two different ways, as the wing chord (wingcrd) and as the flattened wing (flatwing). Therefore the correlation between them is high, however the PCA can deal with this. Hence, for the moment we will use both variables.

Table 12.1. Correlations among the seven variables in the sparrow data.

	wingcrd	flatwing	tarsus	Head	culmen	nalopsi	Wt
wingcrd	1.00	0.99	0.55	0.53	0.42	0.38	0.66
flatwing		1.00	0.54	0.53	0.42	0.38	0.66
Tarsus			1.00	0.69	0.60	0.59	0.63
Head				1.00	0.72	0.73	0.62
culmen					1.00	0.70	0.49
nalopsi						1.00	0.50
Wt							1.00

The eigenvalue equation was solved and it gave the following values for the loadings of the first two axes (the variables were normalised):

$$Z_1 = -0.38 \text{ wingcrd} - 0.38 \text{ flatwing} - 0.39 \text{ tarsus} - 0.40 \text{ head} - 0.36 \text{ culmen} - 0.35 \text{ nalospi} - 0.38 \text{ wt}$$

$$Z_2 = 0.52 \text{ wingcrd} + 0.52 \text{ flatwing} - 0.12 \text{ tarsus} - 0.27 \text{ head} - 0.39 \text{ culmen} - 0.44 \text{ nalospi} + 0.16 \text{ wt}$$

The traditional way of presenting the graphical results of PCA is a scatterplot of $\mathbf{Z}_1$ and $\mathbf{Z}_2$; see Figure 12.2. However, the problem is the interpretation of this graph. The loadings for Z_1 and Z_2 indicate that the first axis is determined by all variables, and they all have approximately the same influence. This is typical for morphometric data; very often the first axis represents the overall shape of the animals. A more detailed discussion on morphometric data analysis is given in Chapter 30. The second axis seems to represent differences between wingcrd and flatwing versus culmen and nalospi.

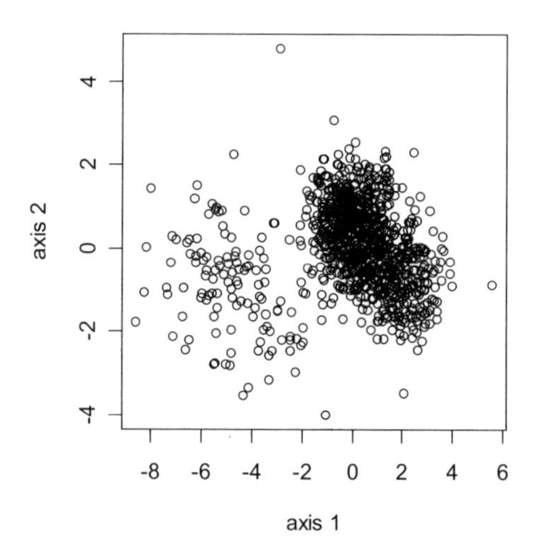

Figure 12.2. First two axes obtained by PCA for the sparrow data.

Eigenvalues in PCA represent the amount of variance explained by each axis. They can be expressed as numbers, percentage of the total variance, or as cumulative percentage of the total variance (Table 12.2). Some software packages present the eigenvalues that come out of the numerical routines, whereas others scale them so that the total sum is equal to one. The eigenvalues in the first column of Table 12.2 are scaled to have a sum of one. The first eigenvalue is 0.647, which means that it explains 64.7% of the variation in the data. The second eigenvalue is 0.163; hence the first two axes explain 81% of the variation.

Table 12.2. Eigenvalues and eigenvalues expressed as cumulative percentage. Some software packages rescale the eigenvalues so that the sum of all eigenvalues is equal to 1. These are given in the second column. Unscaled eigenvalues are in the third column. The sum of all eigenvalues is 7.

Axis	Eigenvalue (scaled)	Eigenvalue (unscaled)	Cumulative Eigenvalue as %
1	0.647	4.529	64.703
2	0.163	1.144	81.045
3	0.063	0.440	87.326
4	0.049	0.342	92.206

One problem with PCA is to decide how many components to present, and there are various rules of thumb. The first rule is the '80% rule'; present the first k axes that explain 80% (cumulative) of the total variation. In this case, two axes are sufficient. Another option is a scree plot of the eigenvalues (Figure 12.3). In such a graph, the eigenvalues are plotted as vertical lines next to each other. The aim is to detect an 'elbow-effect'. The justification is that the first k axes explain most information, whereas axes $k + 1$ and higher represent a considerably smaller amount of variation. A scree plot of the eigenvalues would then show which of the axes are important (long lines) and the change point (elbow) at which the axes become less important. In this case, one axis is sufficient. Some software packages use barplots or just lines in the scree plot.

Yet, another tool is the broken stick model approach (Jolliffe 2002; Legendre and Legendre 1998). If a stick of unit length is broken at random in p pieces, then the expected length of piece number j is given by

$$L_j = \frac{1}{p}\sum_{i=j}^{p}\frac{1}{i} \tag{12.3}$$

Comparing the eigenvalue of the jth axis with L_j, gives an idea of the importance of the eigenvalue; if the eigenvalue for axis j is larger than L_j, then it can be considered as important. In this case, the broken stick values of the first three axes are 0.37, 0.23 and 0.16. So based on the broken stick model, only the first axis is of interest. Other tools to select the number of axes (e.g., using cross-validation) are discussed in Jolliffe (2002). However, in many scientific publications the graphical output of PCA only consists of the first two axes, and occasionally the first four axes.

If the first few axes explain a low percentage, then it might be worthwhile to investigate whether there are outliers, or whether the relationships between variables are non-linear. If either occurs, then consider a transformation or accept the fact that ecological data are genuinely noisy.

The problem with the PCA ordination plot is that is does not tell us whether there are differences between groups of observations (e.g., two bird species), and neither does it give clear information on the influence of the original variables. Obviously, we could label species 1 as '1' and species 2 as '2' in the ordination diagram, and draw circles around groups of observations. Indeed, many authors do this. But with prior knowledge on a grouping structure in the observations available, this is not recommended as other statistical techniques can (and should) be applied for this purpose, for example discriminant analysis or classification models. In the next section, an extension of the ordination diagram is presented, which may give more visual information, namely the biplot.

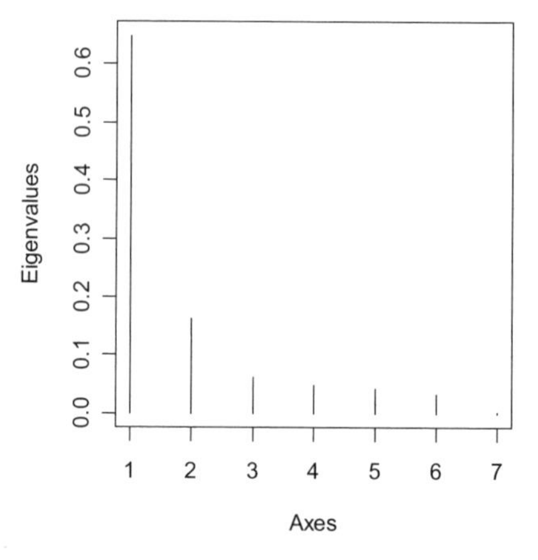

Figure 12.3. Eigenvalues for the sparrow data, obtained by PCA, are presented as vertical lines. The elbow rule suggests presenting only the first axis. Some programmes use barplots or a line plot.

12.5 The biplot

Figure 12.2 shows the general presentation of PCA as a plot of $\mathbf{Z}_1$ against $\mathbf{Z}_2$. Interpretation of the plot is difficult, and the biplot was developed to simplify this. We do not present the detailed mathematics behind the biplot, just the principle. However, if you are not familiar with matrix algebra, you may skip a few paragraphs. Useful references are Krzanowski (1988), Gabriel and Odoroff (1990),

Jongman et al. (1995), Gower and Hand (1996), Legendre and Legendre (1998) or Jolliffe (2002). Recall that the PCA loadings were obtained from an eigenvalue equation of the covariance or correlation matrix. Instead of this, we can also use the singular value decomposition (SVD): A mathematical technique closely related to the eigenvalue equation. It calculates matrices **U**, **L** and **V** such that **Y** = **U L V'**. The matrix **Y** contains all data in an observation-by-variable format. The variables are either centred or normalised. The matrices **U** and **V** are special in the sense that **U'U** = **I** and **V'V** = **I**, where **I** is an identity matrix. This property is also called orthonormal. The matrix **L** is diagonal and contains the square root of the eigenvalues. So, why do we need this? Well, if we take **YV**, we are back to our linear combination of the variables. So, as in the previous section we can write

$$\mathbf{Z} = \mathbf{YV} = \mathbf{UL}$$

The columns of **V** contain the loadings of each axis. Furthermore, each axis (which is a column in **Z**) is given by a column in **U** multiplied with a square root of an eigenvalue (the diagonal element on **L**). The reason we introduce the singular value decomposition is that it can be used for the following formula:

$$\mathbf{Y} = \mathbf{GH}' \quad \text{and } \mathbf{G} = \mathbf{UL}^{\alpha} \text{ and } \mathbf{H}' = \mathbf{L}^{1-\alpha}\mathbf{V}'$$

This formula is the basis for the biplot. The idea is that we are going to plot columns of **G** and columns of **H** in the same graph, and this will allow us to make statements on the variables and observations. The matrix **H** contains information on the loadings and **G** on scores. The problem is that the interpretation of the biplot is based on the choice of α. There are two conventional choices, $\alpha = 0$ and $\alpha=1$, and each of them is discussed below. This is also called the scaling of the biplot.

The biplot with α = 0 (correlation biplot)

When $\alpha = 0$, the biplot is also called a correlation biplot. In this case it is easy to show that: **Y'Y=HH'**. Except for a constant factor $N - 1$, the left part is the correlation or covariance matrix. This depends on whether Y was centred or normalised. Let us first assume that it is the covariance matrix. Denote the first two columns of **H** by $\mathbf{H}_2$. In the biplot, variable j is represented by $\mathbf{h}_j$, the jth row of $\mathbf{H}_2$. It is conventional to do this by drawing a line from the origin to a point with coordinates given by $\mathbf{h}_j$. Using basic algebra it can be shown that the angle between $\mathbf{h}_i$ and $\mathbf{h}_j$ represents the similarity between two variables. To be more precise, the cosine of the angle between two lines *approximates* (as it is a two-dimensional representation) the correlation. This means that lines pointing in the same direction have a high correlation. Variables with an angle of 90 degrees have a small correlation (close to zero), and variables pointing in the opposite direction have a large but negative correlation. The length of each line is proportional to the variance of a particular variable.

Denote the first two columns of **G** by $\mathbf{G}_2$. In the same graph, each observation can be plotted with coordinates $\mathbf{g}_i$ (the ith row of $\mathbf{G}_2$). This is typically done as a point (or label) and not as a line. Distances between two observations in the biplot (represented by $\mathbf{g}_i$ and $\mathbf{g}_j$) are an approximation of the Mahalanobis distance. This distance measure works in a similar way as the Euclidean distance except that variables with large variance and groups of highly correlated observations are down-weighted.

The last rule makes use of geometry. As the i,j^{th} element of $\mathbf{Y}$ is approximated by $\mathbf{g}_i{*}\mathbf{h}_j$, the observations (represented by points in the biplot) can be projected perpendicularly on the lines (variables). The position of the point along the line gives an indication of the value of this observation. The biplot only shows a line from the origin to a point with coordinates $\mathbf{h}_i$ for each observation. The origin represents the average, and the visible part of the line reflects above average values of the variable. One has to imagine a line pointing in the opposite direction representing the values below average.

In case the variables are normalised (centred and divided by the standard deviation), the length of the line shows how well the variable is represented in the two-dimensional approximation. Ideally, all lines should have length 1.

Figure 12.4 shows the PCA correlation biplot for the sparrow data. The variables were normalised. All lines are pointing to the left indicating that all variables are highly correlated with each other. In more detail, wingcrd and flatwing are highly correlated with each other. The same holds for culmen and nalospi. These results are in line with the correlation matrix in Table 12.1. Points (or dots) in the graph (=scores) can be projected on lines. The observations close to the origin have average values for all variables. The observations towards the left have above average values for all variables. However, we cannot easily interpret the Euclidean distances between the observations. The eigenvalues are not influenced by the scaling process.

The biplot with α = 1 (distance biplot)

Let us now turn to the situation in which $\alpha = 1$. This is also called the distance biplot. The reason for this is that Euclidean distances in the biplot are now a two-dimensional approximation of the Euclidean distances between observations. This makes it easier to compare the position of observations with each other. We can still project the points (observations) on the lines. However, angles between lines do not have a direct interpretation in terms of the correlation anymore. The principle of projecting points on lines applies just as before. This means that if two lines point in the same direction we can only say that they both have high or low values for the same observations. So, this is a slightly more descriptive interpretation compared with the interpretation based on angles in the correlation biplot.

Figure 12.5 shows the distance biplot for the same data. We cannot interpret the angles between lines directly, but we can still project observations on lines. The advantage of this approach is that distances between points can now be more easily interpreted: These are approximations of Euclidean distances between the observations. Most of the points on the left are actually from species 2. The fact that they are all separated from species 1 means that bird species 2 has higher values for all morphometric values.

Summary of the scaling process

A biplot is a visualisation tool to present results of PCA. There are various options for the PCA biplot and this is called the scaling process. There are two main options, the correlation biplot, and the distance biplot. In the correlation biplot, we can use the following rules:

- If the data are centred, then the length of a line is proportional to the variance of the corresponding variable. If the data are normalised, then the length of a line is an indication of how well it is represented by the two-dimensional approximation. Long lines are represented well, but short lines should be interpreted with care.
- Angles between lines approximate the correlation.
- Points (observations) can be projected perpendicular on lines, and this gives information of the observed values (abundances).
- Distances between observations are two-dimensional approximations of Mahalanobis distances, but these are difficult to interpret.

For the distance biplot we have:

- If the data are centred, then the length of a line is proportional to the variance of the corresponding variable. If the data are normalised, then the length of a line is an indication of how well it is represented by the two-dimensional approximation.
- Angles between lines are not directly interpretable in terms of correlations.
- Points (observations) can be projected perpendicular on lines, and this gives information of the observed values (abundances).
- Distances between observations are two-dimensional approximations of Euclidean distances.

The choice of scaling depends on the underlying question: Are we interested in making a statement on the variables or observations? If it is the first question, then use the correlation biplot, else the distance biplot. The biplot interpretation should be done with care if the two-dimensional approximation only explains a small percentage of variance.

In Chapter 10, we discussed the pros and cons of the correlation and covariance coefficients as a measure of association. One of the things we mentioned was that if the variation between the variables is important, then use the covariance as the PCA is dominated by such variables. If the variables are in different units, or if only relative changes are important, use the correlation coefficient. The same arguments hold in the choice between the covariance and the correlation matrix for the PCA. Do not confuse this with the correlation biplot as this is a scaling option.

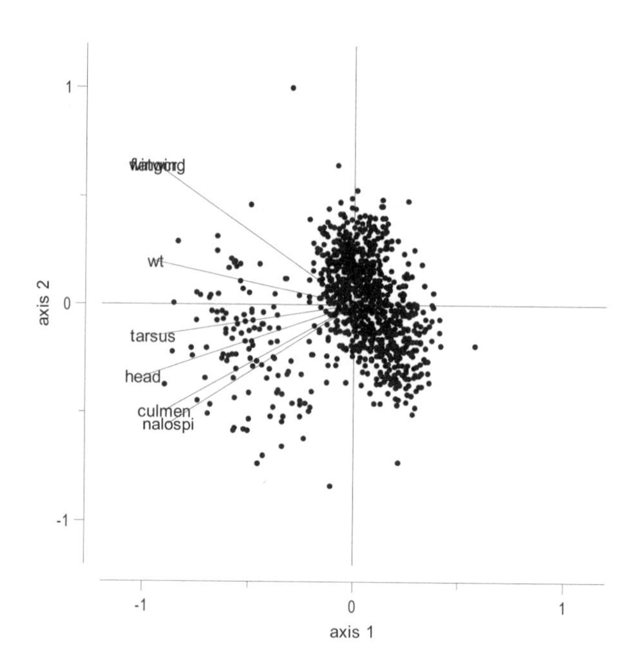

Figure 12.4. PCA biplot for sparrow data. The correlation biplot was used.

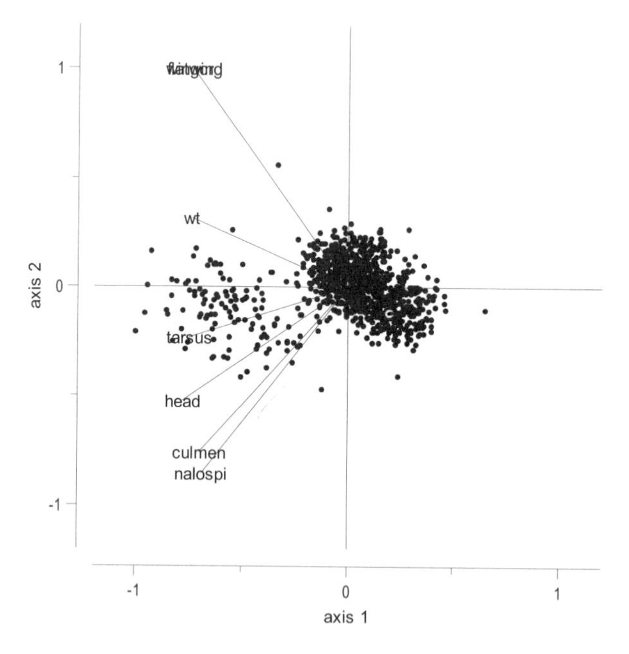

Figure 12.5. PCA distance biplot for the sparrow data.

12.6 General remarks

PCA, as with most other statistical methods, cannot cope with missing values. Missing values must be replaced by a sensible estimate (e.g., mean value of the response variable). Alternatively, consider omitting the variable or the entire observation from the data.

Jolliffe (2002) discusses how PCA can be used for outlier detection. He distinguishes between two situations: A variable with one (or more) large values that should have been detected using dotplots or boxplots, or observations that are outliers but cannot be detected with univariate plotting tools like the Cleveland dotplot (Chapter 4). We discussed this type of outliers in Chapter 4. Outliers of the first type tend to dominate one of the first few axes in PCA. The original sparrow data contained an observation with a large tarsus value, and this mainly determined the third PCA axis. According to Jolliffe (2002) the second type of outlier tends to dominate the last few axes.

Figure 12.6 shows the correlation biplot for the CSL bird radar data that we used in Chapter 10. The covariance matrix was used. The first two eigenvalues are 0.28 and 0.10, representing 38% of the variation. The original data matrix was of dimension 70-by-400. This means that there are fewer observations than variables and as a result there will be 330 axes with zero eigenvalues (these cannot be interpreted). In principle, one should have more observations than variables. However, this does not influence the first few axes and you can still use PCA, but avoid interpreting the higher axes (numbers 71 to 400).

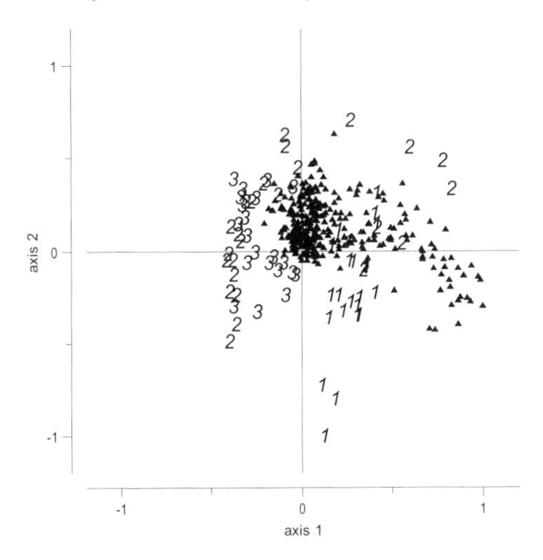

Figure 12.6. PCA correlation biplot for the CSL bird radar data. The triangles represent the variables (counts in grid cells), and the numbers are the observations. The labels 1, 2 and 3 indicate on which day an observation was taken. The covariance matrix was used. Grid cells with less than 10 birds were omitted from the analysis.

The biplot in Figure 12.6 is rather difficult to interpret. The 400 grid cells were used as variables. We have to omit the lines and labels to avoid the graph becoming cluttered with detail to the extent of becoming unreadable. Instead, we used triangles for the variables and the observations are labelled with values 1, 2 and 3, depending on which day the bird was measured. We decided to use the covariance matrix as some of the grid cells at the edge had considerably lower values than in the centre. The additional information we have are the spatial coordinates for each 'variable'. This allows us to make a contour plot of the loadings for an axis; see Figure 12.7. The contour plot for the loadings of the first axis shows that the grid cells with high loadings are all centred at a particular part of the study area, which is located on the right-hand side of the figure. Based on the position of the labels 1, 2 and 3 in Figure 12.6, the high loadings are mainly associated with days one and two. This means that 28% of the variation in the data is related to a geographical hotspot in the study area.

If the underlying question is whether there is a difference between the days, a distance biplot should be made. The resulting biplot is not shown here, but it is similar to Figure 12.6 except that the groups with the labels '1' are more isolated.

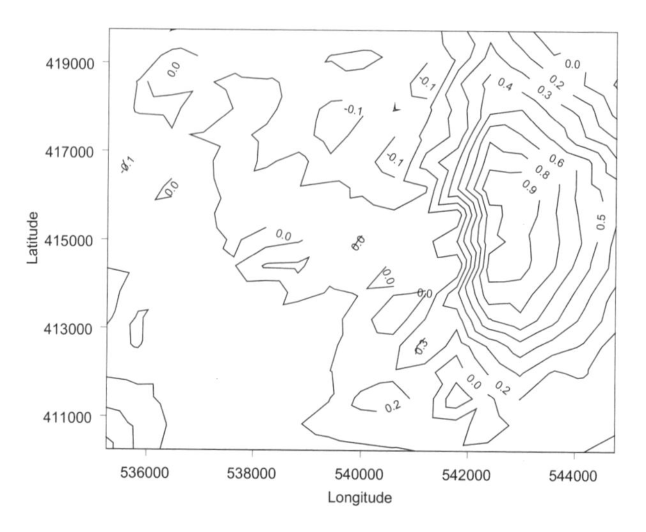

Figure 12.7. Contour graph of the loadings of the first PCA axis for the CSL radar data. Note the hotspot towards the midle right.

12.7 Chord and Hellinger transformations

Three potential problems with PCA are as follows: It measures linear relationships (because it is based on the correlation or covariance coefficient), the presence of double zeros (Chapter 10) and the arch effect. All three problems are mainly associated with ecological community data; counts of multiple species at multiple sites. For such data, relationships are typically non-linear, and there are many zeros in the data matrix. A data transformation might solve the first problem, but double zeros will stay a problem even after a transformation.

If two species contain many zeros, then both the correlation and the covariance coefficients will indicate that they are similar. Figure 12.8 shows species abundance along an artificial gradient. Species one and two are at different sides of the gradient; they were not sampled at the same sites. Therefore, these species should be labelled as dissimilar. A PCA should place such species and sites at different ends of the first axis. However, the correlation coefficient will indicate that species one and two are highly correlated because of the double zeros. As a result, the second PCA axis will bend the scores of both sides of the gradient slightly inwards, showing the so-called horseshoe or arch effect.

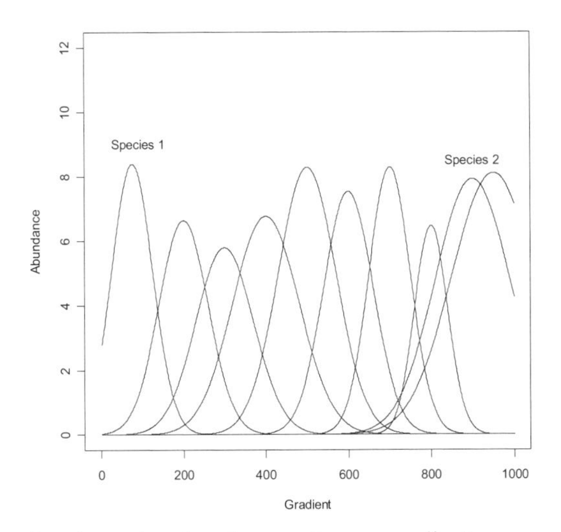

Figure 12.8. Simulated species abundances along a gradient.

Legendre and Gallagher (2001) showed that various other measures of association can be combined with PCA: The Chord distance, Hellinger distance, and two Chi-square related transformations. For the Chord distance, this process works as follows. Let y_{ij} be the abundance of species j at site i, and let y_{ij}' be the transformed abundance according to

$$y_{ij}^{'} = \frac{y_{ij}}{\sqrt{\sum_{j=1}^{p} y_{ij}^2}}$$

Then the Euclidean distance between two (transformed) observations is equal to the Chord distance between them, and the biplot shows a two-dimensional representation of these distances. The Hellinger and Chi-square distances can be obtained with similar data transformations but these are not given here. Legendre and Gallagher (2001) showed that the data transformation using the Chord and Hellinger distance function was less influenced by the arch effect than ordinary PCA (and correspondence analysis). The data transformation is applied before the PCA algorithm is started.

If a species is truly rare but locally abundant (patchy), and if it is felt that it should be included in the analysis, then one of these transformations can be applied. However, if such species are only rare because of the sampling design or if they are considered unimportant, then they should not be used.

The data used in Figure 12.6 contains a large percentage of zeros and a Chord transformation might be appropriate. The resulting biplot after the Chord transformation was applied is shown in Figure 12.9. Distances between observations in this graph are now two-dimensional approximations of the Chord distances.

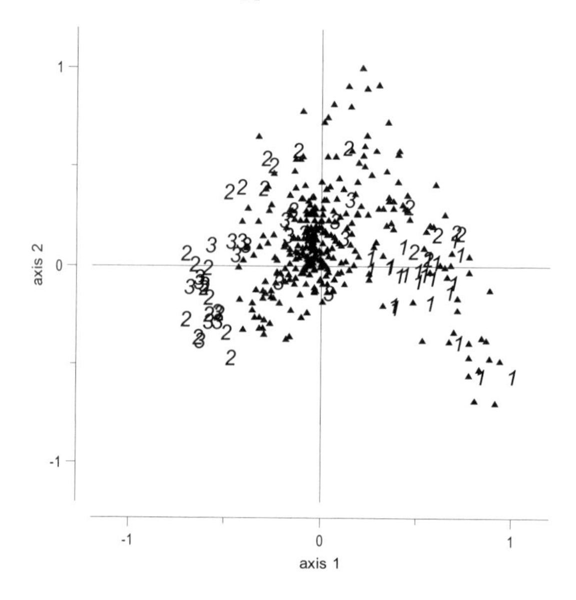

Figure 12.9. PCA distance biplot for the bird radar data. The covariance matrix and a Chord transformation was used.

12.8 Explanatory variables

In this paragraph, we set the scene for redundancy analysis. Figure 12.10-A shows the correlation biplot for the classes of the RIKZ data (Chapter 27) and indicates a clear zonation. Insecta are negatively correlated with Crustacea and Mollusca, and the later two are correlated with each other. The first two eigenvalues are 0.35 and 0.29, which means that they explain 64% of the variation.

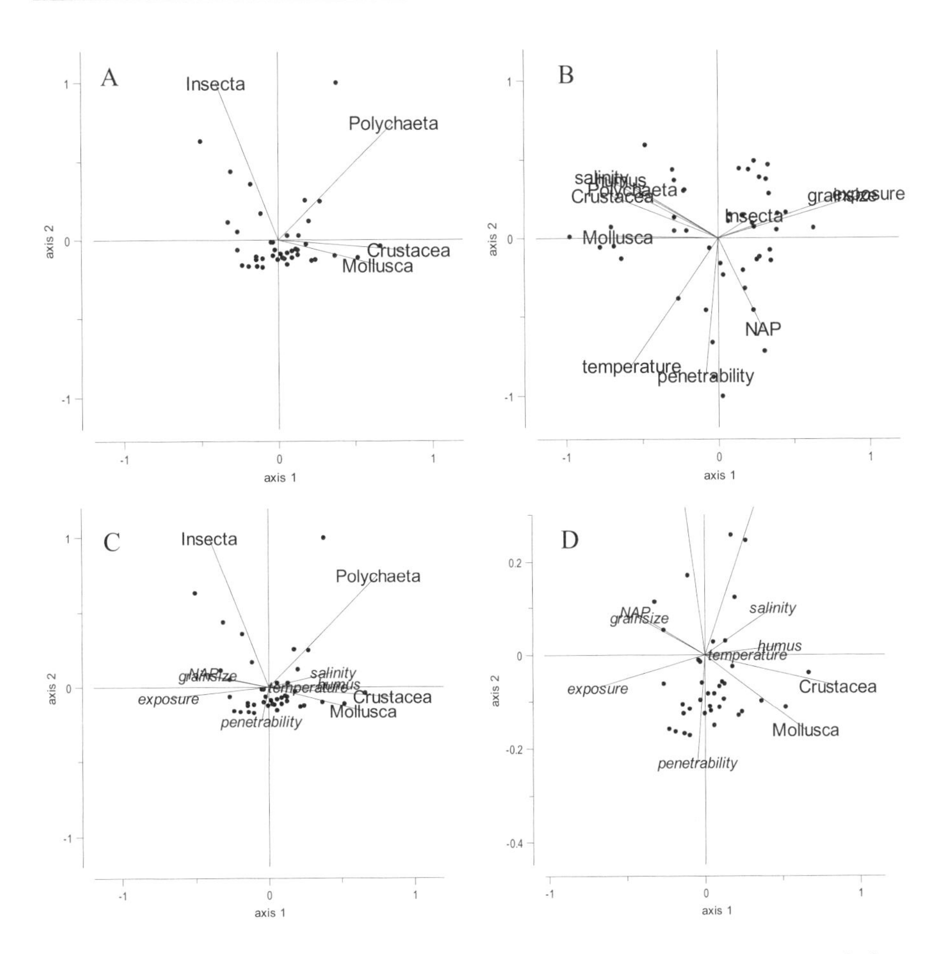

Figure 12.10. A: PCA correlation biplot for RIKZ class data. B: PCA correlation biplot for a combined (species classes and explanatory variables) data matrix. C: PCA correlation biplot applied on the species class data. Explanatory variables are superimposed using cross-correlations. D: As panel C but we zoomed in.

So far, the multivariate analysis has not taken into account any explanatory variables. The first option is to apply the PCA on a data matrix that contains both the class variables and the explanatory variables. The resulting correlation biplot is given in Figure 12.10-B. The first axis seems to be determined by grain size and exposure versus salinity, humus and three of the species classes. Insecta has a rather short line indicating that it is not represented well by the first two axes. The second axis is determined by the variables, NAP, penetrability and temperature. The first two eigenvalues are 0.31 and 0.17, corresponding to 49% of the variation. The problem with this approach is that the PCA will focus on the variables that have the highest correlations (or covariances). If the explanatory variables have higher correlations with each other, than with the response variables, the main gradients will only contain information on the explanatory variables. This situation is common in ecology. In this particular case, perhaps three species classes might be related to environmental variables, but there is no guarantee that this is the case with other datasets. Insecta is clearly not related to the main gradients in the combined data.

Another option is to apply PCA on the class data and calculate correlations between the PCA axes $\mathbf{Z}_1$ and $\mathbf{Z}_2$ and each of the explanatory variables. This information can then be superimposed on the biplot; see Figure 12.10-C. The line for an explanatory variable is obtained by drawing a line from the origin to a point (c_1,c_2), where c_1 is the correlation between the first axis and the explanatory variable, and c_2 with the second axes. A long line indicates that the explanatory variable is highly correlated with at least one of the axes. In this case, correlations between axes and explanatory variables are relatively low. We zoomed in to obtain a better view of the correlations (Figure 12.10-D). The first axis is mainly related to Crustacea and Mollusca. Using the biplot rule, observations that have positive $\mathbf{Z}_1$ values have above average values for these two species classes. These samples have below average values for exposure, grain size and NAP. These correlations are approximately 0.5. Correlations between the second axis and the explanatory variables are small. The problem with this approach is that the major gradients in the species data might not be related to the environmental variables. It might well be that the third and fourth PCA axes are related to the environmental variables. In such cases we would not detect it in Figure 12.10-C. In the next section, we will use a more appropriate technique to relate multiple species with multiple explanatory variables.

12.9 Redundancy analysis

In previous sections, it was shown that PCA is a useful tool to visualise correlations or covariances between variables, but the explanation of the results in terms of explanatory variables was cumbersome. Redundancy analysis (RDA) is an interesting extension of PCA that explicitly models response variables as a function of explanatory variables. We start by looking at RDA, by ignoring all the formulae and only discuss how to interpret the graphical and numerical output. We then look at the mathematics, which is an extension of the iterative algorithm discussed in the PCA section.

The graphical output of RDA consists of two biplots on top of each other, and is called a triplot. There are three components in such a graph:

1. The quantitative explanatory variables are represented by lines, and the qualitative (nominal) explanatory variables by squares (one for each level).
2. The response variables (often the species) by lines or labels. If there are many response variables, it may help visual interpretation to use labels.
3. The samples by points or labels. If there are many samples, using points instead of labels improves the visual quality of the graph.

The choices and interpretation are similar to the PCA biplot. First, we have to decide the scaling: The correlation ($\alpha = 0$) or distance triplot ($\alpha = 1$). Second, whether we want to use the covariance or correlation matrix for the response variables. In the triplot, the response variables (typically species) are represented by lines, and the observations by dots. The new thing is the lines for the quantitative explanatory variables and square for the levels of the nominal explanatory variables. The rules for the RDA *correlation triplot* interpretation are as follows (Ter Braak 1994; Legendre and Legendre 1998; Lepš and Šmilauer 2003):

1. Angles between species lines represent correlation between the species.
2. Points for observations can be projected perpendicularly on the species lines and indicate the values.
3. Points for observations cannot be compared directly with each other.
4. Angles between lines of quantitative explanatory variables represent correlations between them. But these are not as accurate as the ones obtained by a PCA applied on only the explanatory variables.
5. Angles between lines of species and explanatory variables represent correlations.
6. Observations can be projected perpendicular on the lines for explanatory variables indicating the values of the explanatory variables at those sites.
7. The nominal explanatory variables (coded as 0–1) can be represented as a square. Its position is determined by the centroid of the observations that have the value 1 (for this variable). Distances between centroids and between observations and centroids are *not* approximations of their Euclidean distances. A square can be projected perpendicular on a line for a species and represents the mean species abundances for that class. Squares cannot be compared with the qualitative explanatory variables

The first three rules are similar to the PCA correlation biplot. For the *distance triplot*, the following rules hold:

1. Angles between lines for species do not represent correlations.
2. Points for observations can be projected perpendicularly on the species lines and indicate the (fitted) values.
3. Distances between observations represent a two-dimensional approximation of their Euclidean distances.
4. Angles between lines of species and qualitative explanatory variables represent a two-dimensional approximation of correlations.
5. The nominal explanatory variables (coded as 0–1) can be represented as a square. Its position is determined by the centroid of the observations that have the value 1. Distances between centroids and between observations and centroids are approximations of their Euclidean distances. A square can be projected perpendicular on a line for a species and represents the mean species abundances for that class. Squares cannot be compared with the qualitative explanatory variables

The correlation and distance triplots are also called species-conditional triplot and the site conditional triplot, or scaling 2 and scaling 1. If the response variables are not species by sites, then perhaps the nomenclature *Y*-conditional and sample conditional scaling is more appropriate. Legendre and Legendre (1998) used the names correlation biplot and distance biplot. Mathematically, the difference between these two scalings is merely a few simple multiplications involving eigenvalues (Section 12.5, Legendre and Legendre 1998), but the ecological interpretation of the two types of triplots is different, as discussed above. If the interest is on observations, then the distance triplot should be used.

An example is presented next. We have used the same RIKZ data that we used earlier in Section 12.8. The response variables are the four class variables: Insecta, Polychaeta, Mollusca and Crustacea. There are 10 explanatory variables. One of them is nominal (week, with four levels) and one (exposure) could be considered as nominal or continuous. In first instance, we will omit week and consider exposure as continuous. The resulting triplot is presented in Figure 12.11. The rules to read this graph are similar to the PCA biplot. The triplot (Figure 12.11) is based on the correlation biplot, and because we want to make all 'species' equally important, the correlation matrix was used, and not the covariance matrix. The triplot for the RIKZ data shows that NAP and grain size are highly correlated, and exposure is negatively correlated to humus and temperature. Mollusca, and Crustacea are correlated with each other, and also with Polychaeta, but not with Insecta. In terms of species-environmental relations, there seems to be a negative effect of exposure, NAP and grain size on Crustacea, Mollusca and Polychaeta. Insecta seem to be positively related with chalk and sorting, and negatively to penetrability. We can also infer at which sites this is happening.

The numerical output of RDA for the RIKZ data is given in Table 12.3 and shows that all the explanatory variables explain 38% of the variation in the species data. This is the sum of all so-called canonical eigenvalues. From this 38%, the first two axes explain 87%. This is the column labelled 'eigenvalue as cumulative percentage of the sum of all eigenvalues'. Hence, the first two axes are a good representation of what can be explained with all the explanatory variables. This value tends to be high for most datasets due to high correlations between explanatory variables. The fourth column is more interesting. It shows that the first two axes explain 33% of the variation in the species data. This value is obtained by multiplying 0.38 (the sum of all canonical eigenvalues) with 0.87 (the variation explained by the first two axes). The RDA algorithm rescales the response variables to have a total sum of squares of 1.

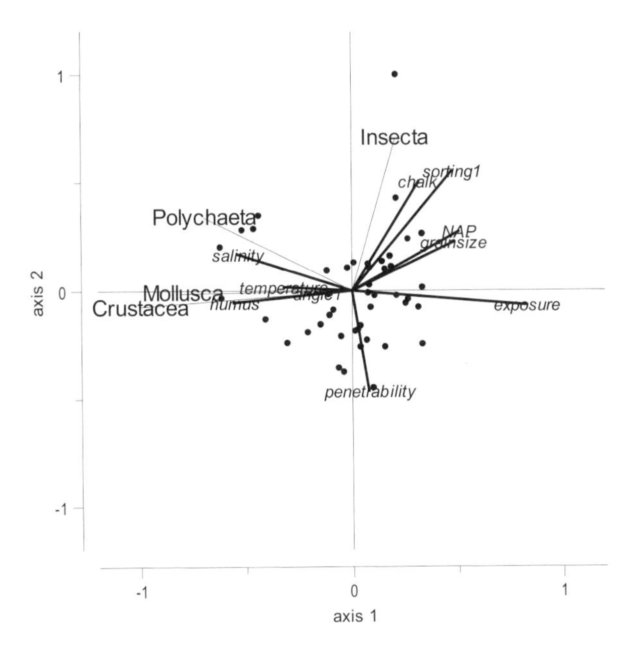

Figure 12.11. RDA correlation triplot for RIKZ class data.

Table 12.3.Numerical output of PCA applied to the RIKZ data. The sum of all canonical eigenvalues is 0.38, and the total variance is 1.

Axis	λ	λ as %	λ as cumulative %	λ as % of sum of all canonical eigenvalues	λ as cumulative % of sum of all canonical eigenvalues
1	0.26	26	26	68	68
2	0.07	7	33	19	87

A second approach to explain RDA is as follows. PCA calculates the first principal component as

$$Z_{i1} = c_{11}Y_{i1} + c_{12}\, Y_{i2} + \ldots + c_{1N}\, Y_{iN}$$

Redundancy analysis is a sort of PCA that requires that the components are linear functions of the explanatory variables:

$$Z_{i1} = a_{11}X_{i1} + a_{12}X_{i2} \ldots\ldots + a_{1q}\, X_{iQ}$$

Hence, the axes in RDA are not only a linear combination of the response variables, but also of the explanatory variables. It is as if you tell the computer to apply a PCA, but only show the information in the biplots, which can be (linearly) related to the explanatory variables. Further axes are obtained in the same way. RDA can be applied if there are N response variables measured at M sites and Q

explanatory variables measured at the same sites. Note that this technique requires an explicit division of the variables into response and explanatory variables. The maximum number of RDA axes is the minimum of N and M.

The more mathematical explanation of RDA requires the iterative algorithm for PCA, which was presented in the previous section. The algorithm has the following steps:

1. Normalise (or centre) the variables $\mathbf{Y}$, and normalise the explanatory variables $\mathbf{X}$.
2. Obtain initial scores $\mathbf{z}$ (e.g., by a random number generator).
3. Calculate new loadings: $\mathbf{c} = \mathbf{Y}'\mathbf{z}'$.
4. Calculate new scores: $\mathbf{z} = \mathbf{Yc}$.
5. For second and higher axes: Make $\mathbf{z}$ uncorrelated with previous axes using a regression analysis.
6. Apply a linear regression of $\mathbf{z}$ on $\mathbf{X}$, and set $\mathbf{z}$ equal to the fitted values.
7. Scale $\mathbf{z}$ to unit variance: $\mathbf{z}^* = \mathbf{z}/\lambda$, where λ is the standard deviation of the scores. Set $\mathbf{z}$ equal to $\mathbf{z}^*$.
8. Repeat steps 2 to 7 until convergence.
9. After convergence, divide λ by $M - 1$.

The only extra steps are the normalisation of $\mathbf{X}$ and the regression in step 6. The fitted values obtained by a linear regression model represent the information in the response variable that is linearly related with the explanatory variables. So the effect of this regression is that only the information in $\mathbf{z}$ (linear combination of $\mathbf{Y}$'s) that is related to the $\mathbf{X}$ is presented. This ensures that the scores $\mathbf{z}$ are a linear combination of the explanatory variables.

We also discussed the PCA eigenvalue equation. Indeed, RDA can also be solved with an eigenvalue equation and a couple of matrix multiplications, see Legendre and Legendre (1998) for details.

Because RDA is basically a series of multiple linear regression steps, we need more observations than explanatory variables.

Nominal variables

Different statistics programmes treat nominal variables slightly differently, and this is discussed next using an artificial example. Suppose that abundances of two species were measured at five sites and that sampling took place over a period of three months by two observers (Table12.4). Sites 1 and 4 were measured in April, site 2 in May and sites 3 and 5 in June. The first observer sampled sites 1 and 2, whereas the second observer sampled the remaining months. The explanatory variables Month and Observer are nominal variables. The last variable is easily dealt with. Define $\text{Observer}_i = 0$ if the ith site was sampled by observer A and 1 if observer B measured the data. Because Month has three classes (April, May and June), three new columns are defined: April, May and June. Where the value is 1 if sampling took place in the corresponding month and 0 elsewhere. However, there is one little problem: The variables April, May and June are linearly related and therefore one of the columns should be omitted. If this is not done, the software will give an error message. For the RDA and CCA, the nominal variables must be coded with values 0 and 1.

We re-applied the RDA model to the same RIKZ data, except that the explanatory variable 'week' was taken into account. This variable contains four classes; week 1, 2, 3 and 4, and four new variables W1, W2, W3 and W4 were created. If an observation was from week 1, W1 was set to 1, and W2, W3 and W4 to 0. The same was done for observations from other weeks. To avoid collinearity, the variable W4 was not used in the analysis. The RDA triplot is shown in Figure 12.12. Nominal variables are represented by squares. Adding the three extra variables did not change the main patterns in the triplot. The sum of all canonical eigenvalues is now 44%. Although difficult to see, the square for week 1 indicates that Polychatea, Mollusca, Crustacea and temperature were high in this week.

Table 12.4. Set up of an artificial dataset. Measurements were made in April, May and June (indicated by a '1') and by two observers.

	Species 1	Species 2	Temp	Wind	April	May	June	Observer
Site 1	...	...	...	...	1	0	0	0
Site 2	...	...	...	...	0	1	0	0
Site 3	...	...	...	...	0	0	1	1
Site 4	...	...	...	...	1	0	0	1
Site 5	...	...	...	...	0	0	1	1

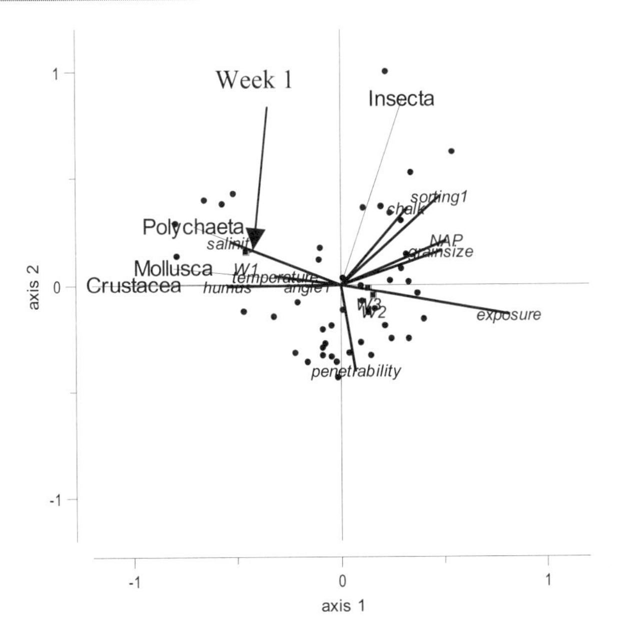

Figure 12.12. RDA triplot for the RIKZ data. The nominal variable week was added.

The order of importance

An interesting question is which of the explanatory variables is the most important, which are the least important, and which are irrelevant. Just as in linear regression, this question can be answered with a forward selection. However, in linear regression, we used the AIC to choose the optimal model. Here, we have a slightly different strategy. The sum of all canonical eigenvalues is used as a tool to assess how well a specific selection of explanatory variables explains the variance in the species data. But if there is only one explanatory variable, the total sum of all canonical eigenvalues is equal to the eigenvalue of the first and only axis. We call these the marginal effects. Table 12.5 shows the marginal effects for the same RIKZ data. It shows the eigenvalue and percentage of explained variance if only one explanatory variable is used in RDA. Results indicate that exposure is the single best explanatory variable, followed by week 1.

Conditional effects (Table 12.6) show the increase in the total sum of eigenvalues after including a new variable during the forward selection. The first variable is exposure, used because it was the best single explanatory variable (Table 12.5). To test the null hypothesis that the explained variation is larger than a random contribution, a Monte Carlo permutation test (see below) is applied. The *F*-statistic and *p*-value indicate that the null-hypothesis can be rejected. The second variable to enter the model is NAP, and the total sum of all eigenvalues increases with 0.06. The Monte-Carlo test indicates that it is significant. The next two variables to enter the model are sorting and week 1. After these four variables, the increase in the total sum of eigenvalues is only 0.02 and the Monte Carlo test shows that it is not significant.

Table 12.5. Marginal effects for the RIKZ. Group classes were square root transformed. The total sum of all eigenvalues is 0.44, and the total variance is 1. The second column shows the eigenvalue using only one explanatory variable, and the third column is the eigenvalue as percentage of the sum all eigenvalues (using all variables).

Explanatory Variable	Eigenvalue Using Only One Explanatory Variable	Eigenvalue as %
angle1	0.01	1.83
exposure	**0.18**	**40.52**
salinity	0.08	18.23
temperature	0.04	8.58
NAP	0.07	16.53
penetrability	0.02	4.10
Grain size	0.06	14.57
Humus	0.08	18.41
Chalk	0.05	11.14
sorting1	0.08	18.40
W1	0.17	39.64
W2	0.04	8.91
W3	0.03	6.02

Table 12.6. Conditional effects for the RIKZ data. The classes were square root transformed. The total sum of all eigenvalues is 0.44, and the total inertia is 1. The second column shows the increase in explained variation due to adding an extra explanatory variable. If one level of week is important, then all its levels should be selected in the final model (or none).

Order	Explanatory Variable	Increase Total Sum Eigenvalues After Including New Variable	F-statistic	p-value
1	**exposure**	**0.18**	**9.340**	**0.000**
2	**NAP**	**0.06**	**3.466**	**0.011**
3	**sorting1**	**0.04**	**2.532**	**0.068**
4	**W1**	**0.05**	**3.214**	**0.017**
5	salinity	0.02	1.099	0.352
6	W3	0.02	1.328	0.257
7	W2	0.02	0.951	0.450
8	chalk	0.01	0.835	0.456
9	penetrability	0.01	0.490	0.726
10	temperature	0.01	0.602	0.621
11	humus	0.01	0.389	0.780
12	angle1	0.01	0.210	0.878
13	grain size	0.00	0.228	0.906

A bit more info on the permutation tests

Details of the permutation test can be found in Legendre and Legendre (1998) or Lepš and Šmilauer (2003). Let us start simply. We have one explanatory variable and multiple response variables. The data are in the following format:

$$\begin{matrix} Y_{11} & \cdots & Y_{1N} & & X_1 \\ Y_{21} & \cdots & Y_{2N} & & X_2 \\ \vdots & & \vdots & & \vdots \\ Y_{M1} & \cdots & Y_{MN} & & X_M \end{matrix}$$

We have two data matrices, the N response variables Y and one explanatory variables $\mathbf{X}$, measured both at the same M sites. In linear regression, we used an F-test to compare nested models and this gave information on the importance of explanatory variables. Here, we do something similar. First, we apply the model with the explanatory variable and obtain an F-statistic, which is defined differently to the F-statistic in linear regression. Because it is for the original data, we call it F^*. Then we assume that there is no relationship between the rows in the $\mathbf{Y}$s and the $\mathbf{X}$. This is our null hypothesis. Under the null hypothesis, we can randomly change the order of the rows in the $\mathbf{X}$ (or indeed $\mathbf{Y}$, but let's do the $\mathbf{X}$). This is called a permutation. Each time we do a permutation, we can also calculate our F-statistic, which is a measure how much the Ys are related to the (permutated) $\mathbf{X}$. We could do this process a large number of times, say 9999 times. We will explain shortly why 9999 and not 10000. If the F^* from the original data has very similar values compared with permutated 9999 F values, then there is no reason to doubt the validity of the null hypothesis. However, if the F^* is considerably larger than the majority of the 9999 F values, then perhaps our assumption of no relationship between the rows of **Y** and **X** is incorrect. In fact, the number of times that F is larger than F^* is closely related to the p-value. To be more precise, the p-value is given by

$$p = \frac{\text{Number of times that } F \geq F^* + 1}{\text{Number of permutations} + 1}$$

The '+1' is used because F^* itself is also used as evidence against the null hypothesis. Hence, if we choose 9999 permutations in a software package, we basically have to write in the paper or report that 10000 permutations were carried out.

Instead of one explanatory variable, it is more likely that we have multiple (say Q) explanatory variables:

$$\begin{matrix} Y_{11} & \cdots & Y_{1N} \\ Y_{21} & \cdots & Y_{2N} \\ \vdots & & \vdots \\ Y_{M1} & \cdots & Y_{MN} \end{matrix} \qquad \begin{matrix} X_{11} & \cdots & X_{1Q} \\ X_{21} & \cdots & X_{2Q} \\ \vdots & & \vdots \\ X_{M1} & \cdots & X_{MQ} \end{matrix}$$

For multiple explanatory variables, exactly the same procedure can be carried out: The rows of **X** are permutated a large number of times, and each time an F-statistic is calculated. The next question is now, how to define the F-statistic. And here is where things get a bit complicated. There is an F-statistic for testing only the first axis, for the overall effect of the Q explanatory variables, for data involving covariables (these are discussed in the next section), and for doing a forward selection. They are all defined in terms of eigenvalues, information of explained variance and number of axes and explanatory variables. The interested reader is referred to Legendre and Legendre (1998) for more detail.

Just as in linear regression, the use of forward selection is criticised by some scientists (and routinely applied by others). If a large number of forward selection steps are made, it might be an option to apply a Bonferroni correction. In such a correction, the significance level α is divided by the maximum number of steps in the forward selection. Alternatively, the p-value of the F-statistic can be multiplied by the number of steps. The main message is probably, as always, be careful with explanatory variables that have a p-value close to significance.

The last point to discuss is the permutation process itself. We explained the permutation process as changing the order of the rows in the **X** matrix. This is called raw data permutation. Just as in bootstrapping techniques (Efron and Tibshirani 1993), you can also modify the order of the residuals. Within RDA, this is called permutation of residuals under a full model. In this case, the RDA is applied using the **Y**s and the **X**s, and residuals from the **Y**s are permuted. Legendre and Legendre (1998) mentioned that simulation studies have shown that both approaches are adequate. In case of permutation tests with covariables (next section), permutation of residuals is preferred if the covariables contain outliers. However, these should already have been identified during the data exploration.

When the rows in the data matrices correspond to a time series, a spatial gradient, or are blocks of samples, permuting the rows as discussed above might not be appropriate as it increases the type I error. For time series and spatial gradient data, permuting the rows with blocks might be an option. In this approach, blocks of samples are selected, and the blocks are permuted, but not the data within the blocks. If the time span of the data is not long enough to generate a sufficient number of blocks (> 20), it might be an option to consider the data as circular and connect the last observation (row) with the first one. However, this process does involve removing trends from data (Lepš and Šmilauer 2003), and this might be the main point of interest. If the data consist of different transects, and there is a large difference between transects, permuting the rows only within transects, and not between them, is better. This is also relevant if the data are obtained from an experiment that would normally (within a univariate context) be analysed with an ANOVA (different treatments).

Chord and Hellinger transformations

As with PCA, RDA is based on the correlation (or covariance) coefficient. It therefore measures linear relationships and is influenced by double zeros. The same special transformations as in PCA can be applied. This means that RDA can be used to visualise Chord or Hellinger distances. Examples can be found in Chapters 27 and 28.

12.10 Partial RDA and variance partitioning

If you are not interested in the influence of specific explanatory variables, it is possible to partial out their effect just as we did in partial linear regression (Chapter 5). For example, in Table 12.4 you might be interested in the relationship between species abundances and the two explanatory variables temperature and wind speed, with the effects of month and observer of less interest. Such variables are called covariables. Another example is the RIKZ data in Table 12.6 where you could consider the explanatory variables Weeks 1–3 as covariables and investigate the role of the remaining explanatory variables. Or possibly you are only interested in the pure exposure effect.

In partial RDA, the explanatory variables are divided into two groups by the researcher, denoted by **X** and **W**. The effects of **X** are analysed while removing the effects of **W**. This is done by regressing the covariables **W** on the explanatory variables **X** in step 1 of the algorithm for RDA, and continuing with the residuals as new explanatory variables. Additionally, the covariables are regressed on the canonical axes in step 6, and the algorithm continues with the residuals as new scores.

Suppose we want to know the relationships between all the variables after filtering out the effects of exposure. Simply leaving it out would not answer the question as various variables may be collinear with exposure, and would therefore take over its role. In the partial RDA, the effect of exposure is removed from the response variables, the explanatory variables and the axes. The resulting triplot is presented in Figure 12.13. Note that after filtering out the effects of exposure, Crustacea is positively related to humus and negatively to temperature and NAP. Insecta and Mollusca are not related to any of the remaining variables.

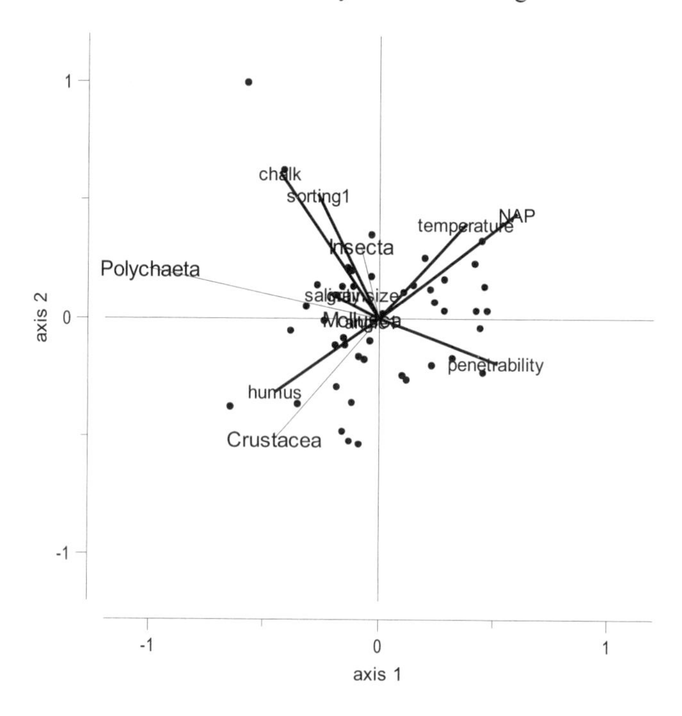

Figure 12.13. Triplot obtained by a partial RDA. The effect of exposure was removed. A correlation triplot using the covariance matrix is presented.

Variance partitioning for linear regression was explained in Chapter 5. Using the R^2 of each regression analysis, the pure **X** effect, the pure **W** effect, the shared effect and the amount of residual variation was determined. Borcard et al. (1992) applied a similar algorithm to multivariate data and used CCA instead of linear regression. Their approach results to the following sequence of steps for variance partitioning in RDA:

1. Apply a RDA on **Y** against **X** and **W** together.
2. Apply a RDA on **Y** against **X**.
3. Apply a RDA on **Y** against **W**.
4. Apply a RDA on **Y** against **X**, using **W** as covariates (partial RDA).

5. Apply a RDA on **Y** against **W**, using **X** as covariates (partial RDA).

Using the total sum of canonical eigenvalues of each RDA analysis (equivalent of R^2 in regression), the pure **X** effect, the pure **W** effect, the shared information and the residual variation can all be explained as a percentage of the total inertia (variation).

An example of variance partitioning for the RIKZ data is presented next. The question is: What is the pure exposure effect? As we suspected that week was strongly related to exposure, we excluded week from the analyses. The results from the five different RDA analyses are given in Table 12.7. Using this information, the pure exposure effect can easily be determined (Table 12.8) and is equal to 7% of the variation. The shared amount of variation is 10%. It is not possible to distinguish this information due to collinearity between exposure and some of the other variables.

Table 12.7. Results of various RDA and partial RDA analysis for the RIKZ data. Total variation is 1. Percentages are obtained by dividing the explained variance by total variance. The 'other' variables are the remaining nine explanatory variables.

Step	Explanatory variables	Explained Variance	%
1	Exposure and others	0.38	38
2	Exposure	0.18	18
3	Others	0.30	30
4	Exposure with others as covariable	0.07	7
5	Others with exposure as covariable	0.20	20

Table 12.8. Variance decomposition table showing the effects of exposure, and other variables for the RIKZ data. Components A and B are equal to the explained variances in steps 5 and 4, respectively. C is equal to variance step 3 minus variance step 5, and D is calculated as Total variance – the explained variance in step 1. In RDA, total variance is equal to 1. *: To avoid confusion, note that due to rounding error, the percentages do not add up to 100.

Component	Source	Calculation	Variance	%
A	Pure others		0.20	20
B	Pure exposure		0.07	7
C	Shared (3–5)	0.30 – 0.20	0.10	10
D	Residual	1.00 – 0.38	0.62	62
Total*				100

12.11 PCA regression to deal with collinearity

An interesting extension of PCA is PCA–regression. A detailed explanation can be found in Jolliffe (2002). In this method, the PCA components are extracted and used as explanatory variables in a multiple linear regression. Using estimated regression parameters and loadings, it can be inferred which of the original variables are important.

As an example, we use the same RIKZ data that we used in the previous section. The question we address is which of the explanatory variables is related to the class Crustacea using a linear regression model. In the previous section, we used a selection of the explanatory variables, but here we use all 10. Hence, the linear regression model is of the form:

$$\text{Crustacea} = \alpha + \beta_1 \text{Angle1} + \beta_2 \text{Exposure} + \ldots + \beta_{10} \text{Sorting} + \varepsilon$$

To find the optimal regression model, we need to apply a backward or forward selection to find the optimal selection of explanatory variables. However, due to high collinearity this may be a hazardous exercise. To visualise this potential problem, we applied a PCA on all explanatory variables and the resulting correlation biplot is presented in Figure 12.14. Note that several lines are pointing in the same direction, which is an indication of collinearity. However, before we can confirm this, we need to check the quality of the two-dimensional biplot. The first four eigenvalues are 0.37, 0.18, 0.12 and 0.11 and therefore all four axes explain 77% of the variation in all ten variables, and the first two axes explain 55%. This is enough to get nervous about (collinearity), so we need to do something. There seems to be three groups of variables. The quick solution is to select one explanatory variable from each group in the biplot in Figure 12.14, for example grain size, salinity and temperature, and use these in the linear regression model as explanatory variables. This is a valid approach, and is used in some of the case study chapters. The higher the explained amount of variation along the first two axes, the more confident you can be with the selected variables.

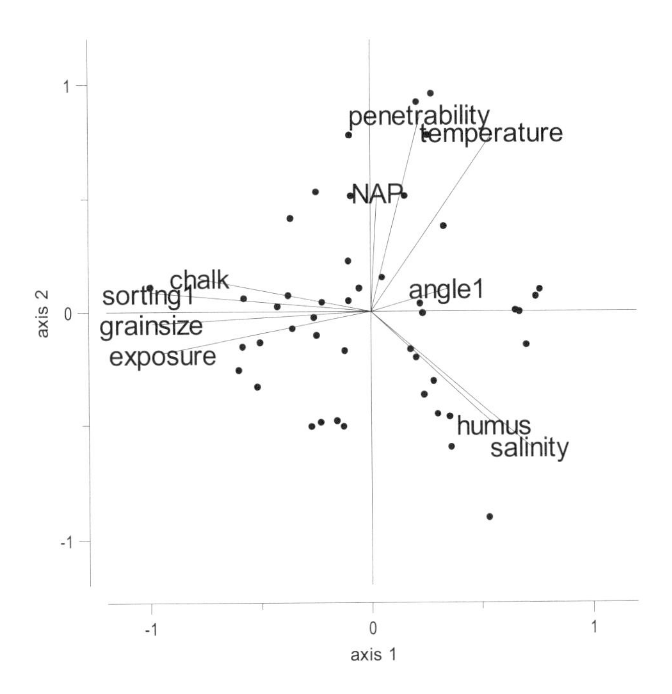

Figure 12.14. PCA biplot for the RIKZ data using all explanatory variables.

Alternative options to identify collinear explanatory variables are variance inflation factors (Chapter 26), pairplots or correlation coefficients. A more advanced method is PCA regression. All the PCA axes, which are uncorrelated, are used as explanatory variables in the linear regression model. The following linear regression model is applied:

$$\text{Crustacea} = \alpha + \beta_1 Z_1 + \beta_2 Z_2 + \beta_3 Z_3 + \beta_4 Z_4 + \ldots \beta_{10} Z_{10} + \varepsilon$$

Z_1 is the first axes (representing the major environmental gradient), Z_2 the second, etc. The estimated model was as follows:

$$\text{Crustacea} = 20.44 + 9.31\text{x } Z_1 + 3.30\, Z_2 - 6.07\, Z_3 + 0.97 \times Z_4 - 1.91\, Z_5 - 1.16 \times Z_6 + 12.08 \times Z_7 + 19.78 \times Z_8 - 32.51 \times Z_9 + 7.54 \times Z_{10} + \varepsilon$$

The *t*-values of the regression parameters (not presented here) indicate that only Z_1 and Z_9 are significantly related to Crustacea. A backward selection indicated the same. All axes explained 41% of the variation in Crustacea, Z_1 and Z_9 28%, Z_1 alone 20% and Z9 alone 8%. The optimal model can be written as

$$\text{Crustacea} = 20.44 + 9.31 \times Z_1 - 32.51 \times Z_9 + \varepsilon$$

However, the loadings of Z_1 and Z_9 tell us how the axes are composed in terms of the 10 original explanatory variables. So, we can multiply the loadings of each axis with the corresponding regression coefficient and then add up the loadings of the same explanatory variable (Table 12.9). The multiplication factors are the estimated regression coefficients 9.31 and –32.51. So, all the loadings for Z_1 are multiplied with 9.31, and those of Z_9 with –32,51. If we add up all these multiplied loadings (see the last column in Table 12.9), we obtain a regression model of the form

$$\text{Crustacea} = 20.44 + 4.17 \times \text{Angle} - \mathbf{21.82 \times Exposure} + 8.14 \times \text{Salinity} - \mathbf{15.89 \times Temperature} + 6.60 \times \text{NAP} + \mathbf{11.23 \times Penetrability} - 5.49 \times \text{Grain size} + 0.45 \times \text{Humus} + 6.95 \times \text{Chalk} - \mathbf{9.12 \times Sorting}$$

Because all explanatory variables were standardised, we can directly compare these loadings. The results suggest that high values of exposure, temperature and sorting are associated with low values of Crustacea, and high values of salinity and with high values of Crustacea. Other explanatory variables have less influence. Humus has a very small influence. The PCA regression approach allows you to assess which of the explanatory variables are important while avoiding problems with collinearity. It is also possible to obtain confidence bands and *p*-values (Jolliffe 2002, Chapter 8).

Jolliffe (2002) states that the low-variance axes can sometimes be better predictors than the high-variance ones. So you could also try applying the PCA regression method on the low-variance axes.

Table 12.9. Loadings for each axis and the total sum of loadings per variable.

Variable	Loadings Z_1	Loadings Z_9	Sum
angle1	0.16	–0.08	4.17
exposure	–0.43	0.55	-21.82
salinity	0.33	–0.16	8.14
temperature	0.25	0.56	–15.89
NAP	0.01	–0.20	6.60
penetrability	0.10	–0.32	11.23
grain size	–0.45	–0.30	5.49
humus	0.25	0.06	0.45
chalk	–0.36	–0.32	6.95
sorting1	–0.46	0.15	-9.12

13 Correspondence analysis and canonical correspondence analysis

In Chapter 12, we discussed PCA and RDA. Both techniques are based on the correlation or covariance coefficient. In this chapter, we introduce correspondence analysis (CA) and canonical correspondence analysis (CCA). We start by giving a historical insight into the techniques community ecologists have used most during the last two decades. This chapter is mainly based on Greenacre (1984), Ter Braak (1985, 1986), Ter Braak and Verdonschot (1995), Legendre and Legendre (1998) and Lepš and Šmilauer (2003).

13.1 Gaussian regression and extensions

Little is known about the relationships between abundances of marine ecological species and environmental variables. However, a feature that many species share is their change in abundances related to changes in environmental variables. Figure 13.1-A shows an artificial example of abundances of a particular species along the environmental variable temperature.

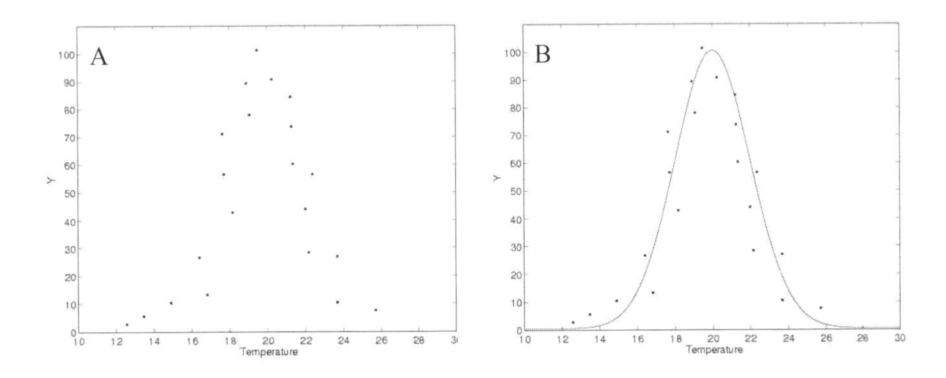

Figure 13.1. A: Observed abundance of a particular species along the environmental variable temperature. B: Fitted Gaussian response curve of a species along environmental variable X, with optimum $u = 20$, maximum value $c = 100$ and tolerance $t = 2$.

To model this behaviour, Whitaker (1978), Gauch (1982) and others used the so-called Gaussian response model. This is the simplest model to describe unimodal behaviour. For a particular species this Gaussian response model takes the form:

$$Y_i = ce^{-\frac{(X_i - u)^2}{2t^2}} \tag{13.1}$$

Where $i = 1,..,N$, Y_i is the abundance of the species at site i, N is the number of sites, c is the maximum abundance of the species at the optimum u, and t is its tolerance (measure of spread). Finally, X_i is the value of environmental variable X at site i. In Figure 13.1-B the Gaussian response curve is plotted. Note that equation (13.1) is a *response* function and not a probability density function.

The Gaussian response model is a very simple model, and real life processes in ecology are much more complex. Alternative models are available when many sites (e.g., 100 or more) are monitored. For example, Austin et al. (1994) used so-called β-functions, which allow for a wide range of asymmetric shaped curves. Because the (original) underlying model for CA and CCA is the Gaussian response model, we restrict ourselves to this model.

Several methods exist to estimate the three parameters c, t and u of the Gaussian response model. The easiest option is to use generalised linear modelling (GLM). In order to do so, we need to rewrite equation (13.1) as

$$Y_i = \exp(\ln c - \frac{u^2}{2t^2} + \frac{u}{t^2}x_i - \frac{1}{2t^2}x_i^2) = \exp(b_1 + b_2 x_i + b_3 x_i^2) \tag{13.2}$$

where $t = 1/\sqrt{-2b_3}$, $u = -b_2/2b_3$ and $c = \exp(b_1 - b_2^2/4b_3)$. If Y_i represents count data, it is common to assume that the Y_i are independent Poisson distributed. Now GLM (Chapter 6) can be applied to the right most part of equation (13.2). This gives estimates of the parameters b_1, b_2 and b_3. From these estimates, the parameters c, t and u can be derived (Ter Braak and Prentice 1988).

Multiple Gaussian regression

Assume that two environmental variables, say temperature and salinity, are measured at each of the N sites. The Gaussian response model can now be written as

$$Y_i = \exp(b_1 + b_2 x_{i1} + b_3 x_{i1}^2 + b_4 x_{i2} + b_5 x_{i2}^2) \tag{13.3}$$

where x_{i1} denotes temperature at site i and x_{i2} the salinity. This model contains five parameters. It is assumed that x_1 and x_2 do not interact. The bivariate Gaussian response curve is plotted in Figure 13.2. These can be constructed for every species of interest.

If M species and Q environmental variables are observed, and interactions are ignored, $(1 + 2Q)M$ parameters have to be estimated. If for example 10 species and 5 environmental variables are used, you need to estimate 110 parameters.

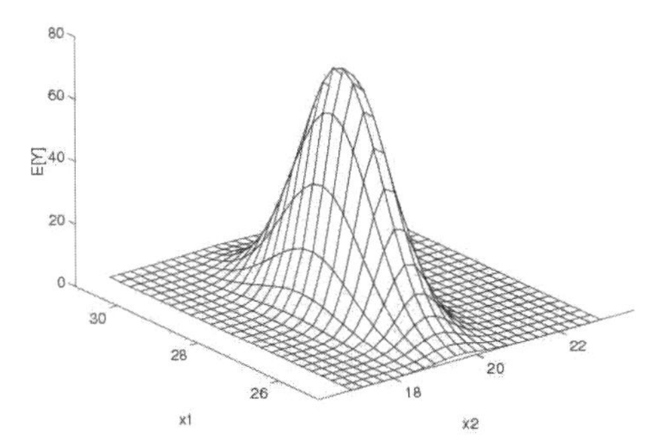

Figure 13.2. Bivariate Gaussian response curve of one species. The x_1 and x_2 axes are explanatory variables and the y-axis shows the expected counts.

Restricted Gaussian regression

Let x_{ip} be the value of environmental variable p at site i, where $p = 1,..,Q$. Instead of using all environmental variables as covariates, we now use a linear combination of them as a single covariate in the Gaussian response model. The model becomes

$$
\begin{aligned}
Y_i &= ce^{-\frac{(z_i-u)^2}{2t^2}} = \exp(b_1 + b_2 z_i + b_3 z_i^2) \\
z_i &= \sum_{p=1}^{Q} \alpha_p x_{ip}
\end{aligned}
\tag{13.4}
$$

The model is called restricted Gaussian regression (RGR), and it tries to detect the major environmental gradients underlying the data. The parameters α_p, denoted as *canonical coefficients*, are unknown. To interpret the gradient z_i, the coefficients α_p can be compared with each other. For this reason, environmental variables are standardised prior to the analysis. Just as in PCA, more gradients (or axes) can be used. This gradient is orthogonal with previous axes. Up to Q gradients can be extracted. The formulae of the RGR model with two or more axes are given in Zuur (1999).

Geometric interpretation of RGR

The geometric interpretation of restricted Gaussian regression is as follows. In Figure 13.3-A, abundances of three species are plotted against two covariates x_1 and x_2. A large point indicates a high abundance and a small point low abundance. We now seek a line z that gives the best fit to these points. A potential candidate for this line is drawn in Figure 13.3-A. If abundances of each species are projected perpendicular on z we obtain Figure 13.3-B. Now, the Gaussian response model in equation (13.4) can be fitted for all species, resulting in an overall measure of fit (typically the maximum likelihood). So, now we only need to find the combination of α_p's that gives the best overall measure of fit. Formulated differently, we need to find a gradient z along which projected species abundances are fitted as well as possible by the Gaussian response model in equation (13.4). The mathematical procedure for this is given in Zuur (1999). The number of parameters to be estimated for the model in equation (13.4) is $3M + Q - 1$, where M is the number of species and Q is the number of environmental variables. If for example 10 species and 5 environmental variables are used, one has to estimate 34 parameters for the first gradient. This is considerably less than in Gaussian regression. In general, if s gradients are used, then the Gaussian response model has $M(2s + 1) + Qs - \sum_{j=1}^{s} j$ parameters.

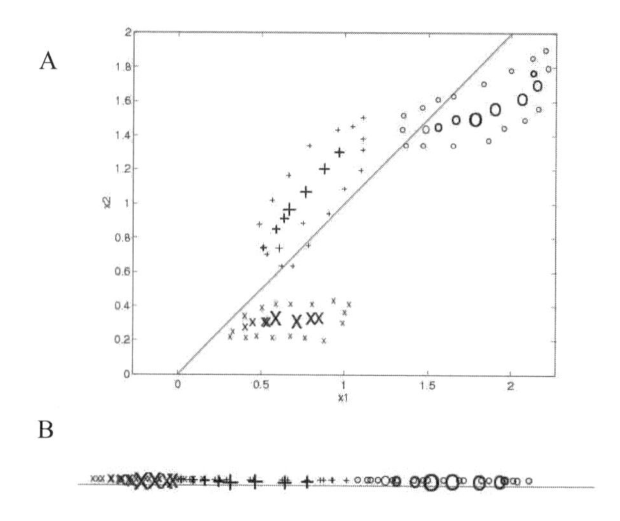

Figure 13.3. A: abundances of three species plotted in the (x_1, x_2) space. The species are denoted by o, x and +, respectively. A thick point indicates a high abundance. The straight line is the gradient z. B: Abundances projected on z.

Gaussian ordination

In Gaussian ordination, we do not use measured environmental variables. Instead, we try to estimate a hypothetical gradient. Other names for this hypothetical gradient are latent variable, synthetic variable, or factor variable. This hypothetical gradient is estimated in such a way, that if abundances of species are projected on the gradient, then this gives the best possible fit (measured by the maximum likelihood) by the Gaussian response model. The Gaussian response model now takes the form

$$Y_{ik} = c_k e^{-\frac{(l_i - u_k)^2}{2t_k^2}} \tag{13.5}$$

where l_i is the value of the latent variable at site i, $i = 1,..,N$, and the index k refers to species. Hence in Gaussian ordination we estimate c_k, u_k, t_k and l_i from the observed abundances Y_{ik}. So we have to estimate $N + 3M$ parameters for the first gradient. If 30 sites and 10 species are used, the Gaussian response model contains 60 parameters. Numerical problems arise if more than one latent variable is used (Kooijman 1977).

Heuristic Solutions

In Ter Braak (1986), the following four assumptions are made:

1. Tolerances of all species along an environmental variable are equal: $t_k = t$ for all k.
2. Maximum values of all species along an environmental variable are equal: $c_k = c$ for all k.
3. The optimum values u_k are equally spaced along the environmental variable, which is long compared with the species tolerance t_k.
4. The sites (samples) cover the whole range of occurrence of species along the environmental variable and are equally spaced.

Using these assumptions, restricted Gaussian regression reduces to a method that is computationally fast and produces easily interpretable information on the parameters u_k of the Gaussian response model. This method is called canonical correspondence analysis (CCA). As a result of these assumptions the Gaussian response curves of the species along an environmental variable simplify considerably; see Figure 13.4. This is the so-called *species packing* model. Due to these assumptions, CCA gives only information on the optimum values of species and on the canonical coefficients.

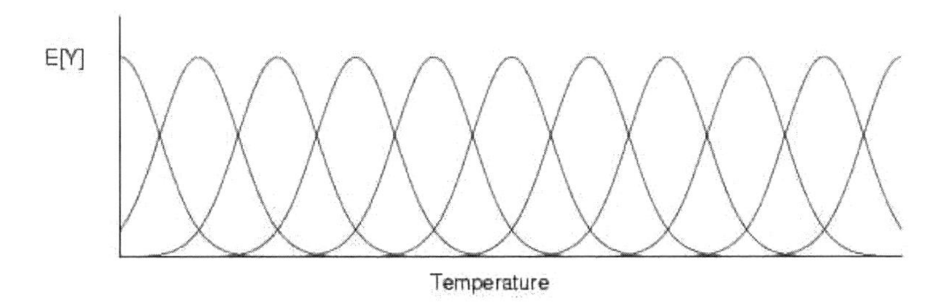

Figure 13.4. Gaussian response curves in the species packing model.

Obviously, the assumptions 1–4 do not hold in practice and they have lead to criticism (Austin and Gaywood 1994). Palmer (1993) showed that CCA is robust against violations of the assumptions. Unfortunately, the simulation studies carried out by Palmer (1993) only concentrate on the estimation of values of z_i. As CCA estimates the optimum values and canonical coefficients of the RGR model, one would expect a simulation study that compares these estimated parameters. Zuur (1999) carried out such a simulation study. He looked at what happened if (i) species scores were not evenly spaced along the gradients, and if (ii) species optima and tolerances were not equal for all species. Results indicated that CCA is robust against violations of the assumptions as long as the fourth assumption of Ter Braak (samples cover the whole range of occurrence of species along the environmental variable and are equally spaced) holds.

Using the assumptions 1–4, Ter Braak (1985) showed that Gaussian ordination reduces to a simple iterative algorithm, which gives the same results as correspondence analysis (Greenacre 1984). Based on simulation studies, Ter Braak and Looman (1986) showed that this heuristic solution is robust against violations of the assumptions.

Historical developments

We started this introduction with the Gaussian response model. Estimating its parameters is basically a regression problem, and this was called (multiple) Gaussian regression. To reduce the number of parameters, we introduced restricted Gaussian regression, which is basically a regression problem with constraints. If no environmental variables have been monitored, Gaussian ordination can be used. This is an *ordination* method. It creates its own latent variables. Finally, we introduced the ordination methods CCA and CA as heuristic solutions for restricted Gaussian regression and Gaussian ordination, respectively. The reason for explaining these techniques in this order (Gaussian regression, restricted regression, Gaussian ordination, CA and CCA) is mainly a logical (in terms of mathematical complexity) one.

Surprisingly, the historical development of these techniques went the other way around. The Gaussian response model has been used by ecologists for many decades. Correspondence analysis was introduced to ecologists by Hill (1973). The method became popular when the software package DECORANA (Hill 1979) was released. CA can probably be considered as the state-of-the-art technique of the 1980s in community ecology. Independently of this, various attempts were made to estimate the parameters of the latent variable model (Equation 13.5) of Gaussian ordination (Kooijman 1977). In 1985, Ter Braak showed that correspondence analysis provides a heuristic approximation of Gaussian ordination if assumptions 1–4 hold. This gave correspondence analysis an ecological rationale. Ter Braak (1986) introduced a restricted form of correspondence analysis, which was called canonical correspondence analysis. CCA is a restricted version of CA in the sense that the axes in CCA are restricted to be linear combinations of environmental variables.

So, the historical development of these techniques went via Gaussian regression, Gaussian ordination, correspondence analysis to canonical correspondence analysis. Ter Braak (1986) argued that CCA is a heuristic approximation of canonical Gaussian ordination, the technique that Zuur (1999) called restricted Gaussian regression.

13.2 Three rationales for correspondence analysis

In this section, correspondence analysis (CA) is explained. The reason for this is that CA can be seen as *the* state-of-the-art technique in community ecology in the 1980s, and it forms the basis of canonical correspondence analysis (CCA). We start by presenting three rationales for correspondence analysis, namely a heuristic approximation of Gaussian ordination, reciprocal averaging and weighted principal component analysis. These approaches are computationally equivalent, but they differ in their initial assumptions and interpretations.

Rationale 1: Heuristic approximation of Gaussian ordination

Recall from Section 13.1 that in Gaussian ordination we use the following model

$$E[Y_{ik}] = \mu_{ik} = c_k e^{-\frac{(l_i - u_k)^2}{2t_k^2}}$$

Y_{ik} is the number of species k at site i (i = 1,..,N and k = 1,..,M), l_i is the value of the latent variable l at site i, with N, the total number of sites and M the total number of species. If we assume that Y_{ik} is Poisson distributed with expectation μ_{ik}, the log likelihood function F is given by

$$F(c_1,..,c_M,t_1,...,t_M,u_1,...,u_M,l_1,...,l_N) = \sum_i \sum_k (Y_{ik} \log(\mu_{ik}) - \mu_{ik})$$

Note that c_k, t_k, u_k and l_i are all estimated from the data. So, the total number of parameters in this model is $3M + N$. We now want to know the values of u_k and l_i that maximise the likelihood. Basic (high school) mathematics dictates calculating the partial derivates of F with respect to u_k and l_i, setting them to zero, and solving them. This gives an expression for u_k and l_i that look rather intimidating and we do not show it here. Using the same four assumptions as in Section 13.1, Ter Braak (1985) showed with the help of a simulation study that the partial derivates can be approximated (and simplified) by

$$u_k = \sum_i \frac{Y_{ik}}{Y_{+k}} l_i \qquad \text{and} \qquad l_i = \sum_k \frac{Y_{ik}}{Y_{i+}} u_k$$

where Y_{i+} and Y_{+k} are sums over all species and sites respectively. In matrix notation this becomes

$$\mathbf{u} = \mathbf{D}_c^{-1}\mathbf{Y}'\mathbf{l} \qquad \text{and} \qquad \mathbf{l} = \mathbf{D}_r^{-1}\mathbf{Y}\mathbf{u}$$

where $\mathbf{u}$ = $(u_1, \ldots ,u_M)'$, $\mathbf{l}$ =$(l_1, \ldots ,l_N)'$ and $\mathbf{Y}$ is a N-by-M matrix containing the data. Furthermore, $\mathbf{D}_c$ is an M-by-M diagonal matrix with Y_{+k} as k,k^{th} element, and $\mathbf{D}_r$ is a N-by-N diagonal matrix with Y_{i+} as i,i^{th} element. The vectors $\mathbf{u}$ and $\mathbf{l}$ are also referred to as species scores, respectively, site scores. In the next two paragraphs, we show that the scores **u** and **l** are the same scores as those obtained by reciprocal averaging, another name for correspondence analysis.

Rationale 2: Reciprocal averaging

Reciprocal averaging (RA) was introduced to ecologists in Hill (1973) and Hill (1974). The aim of RA is to obtain species scores that are weighted averages of site scores and, reciprocally, site scores that are weighted averages of species scores. The algorithm for RA is simple. It starts with arbitrary site scores, and species scores are calculated as the weighted average of the site scores. Then new site scores are calculated as weighted averages of the species scores. This process of reciprocally calculating weighted averages continues until the site scores stabilise. As the range of the weighted averages is smaller than the range of the scores that are used to calculate them, a scaling of site scores is used in each iteration. This prevents the algorithm from drifting into a small range of scores.

Before we present the algorithm for RA, we first need to introduce some more mathematical notation. Let **r** be a N-by-1 vector containing the site (row) proportions of **Y**. So, the ith element of **r** is equal to Y_{i+}/Y_{++}. Similarly, let **c** be an M-by-1 vector containing species (column) proportions. Finally, let $\mathbf{x} = (x_1,\ldots,x_N)'$ and $\mathbf{u} = (u_1,\ldots,u_M)'$. The algorithm for RA has the following form:

1. Start with arbitrary site scores x_i.
2. Calculate species scores by $u_k = \sum_i Y_{ik} x_i / Y_{+k}$.
3. Calculate site scores by $x_i = \sum_k Y_{ik} u_k / Y_{i+}$.
4. Standardise the site scores x_i using a weighted mean and standard deviation. The weights are given by site totals.
5. Stop on convergence; else go to step 2.

To obtain further axes, an orthogonalisation procedure can be used between steps 3 and 4 in each iteration (Ter Braak and Prentice 1988). In such a procedure, the site scores are kept uncorrelated with previous axes by a weighted multiple regression of x_i on previous axes. The calculations of the algorithm in the last iteration were

$$\mathbf{u} = \mathbf{D}_c^{-1}\mathbf{Y}'\mathbf{x} \quad \text{and} \quad \mathbf{x} = s^{-1}\mathbf{D}_r^{-1}\mathbf{Y}\mathbf{u} \tag{13.6}$$

The s comes from the standardisation step. So species scores **u** are weighted averages of the site scores **x**, and site scores are *proportional* (due to the s) to the weighted averages of species scores. Now suppose that the roles of **u** and **x** interchange (this is called the dual problem). We start with arbitrary species scores, calculate site scores as a weighted average of species scores, and calculate species scores as a weighted average of site scores. The results of the last iteration of the algorithm would be

$$\mathbf{u} = s^{-1}\mathbf{D}_c^{-1}\mathbf{Y}'\mathbf{x} \quad \text{and} \quad \mathbf{x} = \mathbf{D}_r^{-1}\mathbf{Y}\mathbf{u} \tag{13.7}$$

Thus species scores are proportional to weighted averages of site scores and site scores are weighted averages of species scores. For $\alpha = 0$ respectively $\alpha = 1$, equations (13.6) and (13.7) can be summarised by

$$\mathbf{u} = s^{-\alpha}\mathbf{D}_c^{-1}\mathbf{Y}'\mathbf{x} \quad \text{and} \quad \mathbf{x} = s^{1-\alpha}\mathbf{D}_r^{-1}\mathbf{Y}\mathbf{u}$$

Rationale 3: Weighted principal component analysis

Calculations in principal component analysis (PCA) are made in a Euclidean metric. In the early 1970s, the Frenchman Benzecri and co-workers developed a similar method using weights in a Chi-square metric. This method was called 'analyses des correspondence', translated by Hill (1974) into correspondence analysis. A good overview, with applications, of this weighted principal component analysis is given in Greenacre (1984). The biplot can be used in combination with CA (Krzanowski 1988; Gabriel and Odoroff 1990; Greenacre 1993; Gabriel 1995; Ter Braak and Verdonschot 1995; Jongman et al. 1995; Gower and Hand 1996; Legendre and Legendre 1998; Jolliffe 2002).

The starting point in CA is a contingency table. This is a table that gives the counts in the dataset for all combinations of categories of each variable (Krzanowski and Marriott 1994). Most of the theory of CA is based on two-way contingency tables, but extensions to higher dimensions are popular in fields like psychology. Analysing a two-way contingency table, the first question that arises is whether there is a relation between the two variables in the contingency table. This can be investigated by performing a Chi-square test; see also Chapter 10 (Table 10.4) for a worked example. The null hypothesis is the independence of the two variables, and the test statistic (Pearson Chi-square) is

$$\mathrm{X}^2 = \sum\sum \frac{(\text{Observed} - \text{Expected})^2}{\text{Expected}} = \sum_i \sum_k \frac{(Y_{ik} - Y_{i+}Y_{+k}/n)^2}{Y_{i+}Y_{+k}/n}$$

The notation Y_{i+}, Y_{+k} and n stand for row total, column total and overall total ($n = Y_{++}$). Using this statistic, you can easily test the null hypothesis. If this hypothesis is rejected, it is valuable to analyse why it is rejected and to see which cells account for the relations between the two variables. In Chapter 10, we calculated the Chi-square statistic for an artificial dataset and determined the contribution of each cell to the test statistic. Define q_{ik} for species k at site i as

$$q_{ik} = \frac{p_{ik} - p_{i+}p_{+k}}{\sqrt{p_{i+}p_{+k}}}$$

where $p_{ik} = Y_{ij}/n$, n is the sum of all species at all sites, p_{i+} is the total abundance at site i, and p_{+k} the total abundance for species k. Using basic algebra it can be shown that $n \times q_{ik}^2$ is the contribution of cell i,k to the Chi-square statistic (these are the values in bold font in Table 10.4). The matrix $\mathbf{Q}$, containing the elements q_{ik} for all sites and species, is the starting point in correspondence analysis. Just as in PCA (Chapter 12), the singular value decomposition is used to decompose **Q** into three special matrices:

$$\mathbf{Q} = \mathbf{U}\ \mathbf{L}\ \mathbf{V}' \tag{13.8}$$

The matrices **U** and **V** are orthonormal and **L** contains the square roots of the eigenvalues. Equation (13.1) can be used to provide a low-dimensional approximation of (i) the relationships between sites, (ii) the relationships between species, and (iii) the relationships between sites and species. Just as in PCA, we can choose which aspect to focus on by converting the right part in equation (13.8) into 'something' related to the sites and 'something' to the species. This requires pre-multiplying **U** and **V** with diagonal matrices containing sites totals and species totals, and we also need to decide where to put the **L**. Technical details can be found in Legendre and Legendre (1998). Just as in PCA, the first few, say two, rows and columns of these matrixes can be plotted in one graph, and these provide a low-dimensional approximation of the information in **Q**.

This process is called scaling, and there are three main choices. We will discuss and illustrate three scaling choices using coverage indices from lowland plant species from 20 sites in Mexico. A full analysis of these data is given in Chapter 32. Here, we only use the 15 most frequently measured families that give us a data matrix of dimension 20-by-15.

In CA, the position of a species represents the optimum value in terms of the Gaussian response model (niche) along the first and second axes. For this reason, most ecological software packages present a species scores as a point or label, and not by a line.

Scaling 1 is appropriate if one is interested in sites, because distances between sites in the ordination diagram are two-dimensional approximations of their Chi-square distances. The sites are at the centroid of the species. This means that the sites are scattered near the species that occur at those sites. An example is given in Figure 13.5. Site 3 is rather different from the other sites (it has a relatively large Chi-square distance to the other sites). The centroid rule dictates that this site has relatively large values for ci, vo and com.

Scaling type 2 (or species conditional scaling) is appropriate if one is interested in species because distances between species are two-dimensional approximations of their Chi-square distances. The species are at the centroid of the sites. This means that the species points are close to the sites where they occur. An example is given in Figure 13.6. Grcyn is rather different from the other species as it has a large Chi-square distance to all other species. The centroid rule indicates that grcyn has high values at all sites in the lower left quadrant.

Scaling type 3 results in a graph in which distances between sites are two-dimensional approximations of their Chi-square distances, and the same holds for the species. But the species and the sites cannot be compared with each other. An example of this scaling is presented in Figure 13.7. The family grcyn is rather different (in terms of the Chi-square distance) from all other families. Site 3 is rather different from the other sites. Distances between sites and species cannot be interpreted. The joint plot of species and sites under this scaling has caused a lot of confusion in the literature. Greenacre (1984) warns not to interpret the joint plot, because it has no formal justification.

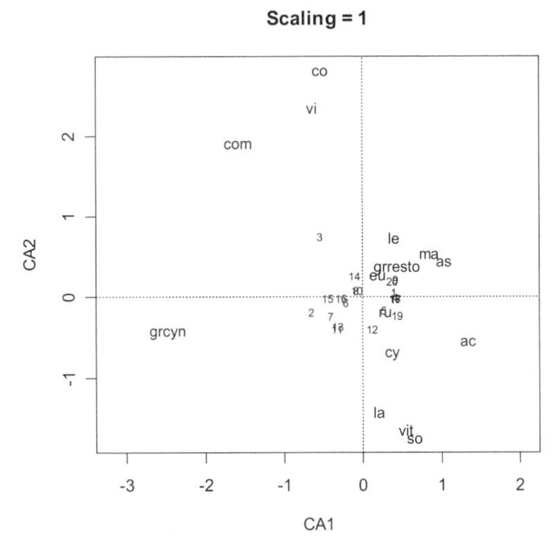

Figure 13.5. Site conditional biplot. Distances between sites are two-dimensional approximations of their Chi-square distances. The first two eigenvalues are 0.13 and 0.06, and the total variation (inertia) is 0.34.

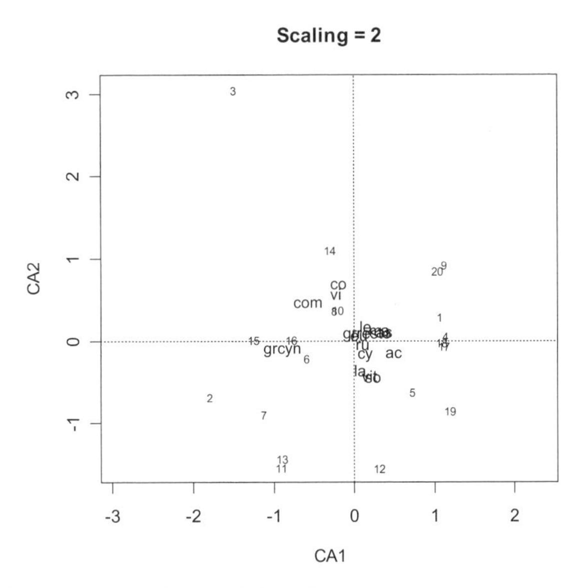

Figure 13.6. Species conditional biplot. The first two eigenvalues are 0.13 and 0.06, and the total variation (inertia) is 0.34.

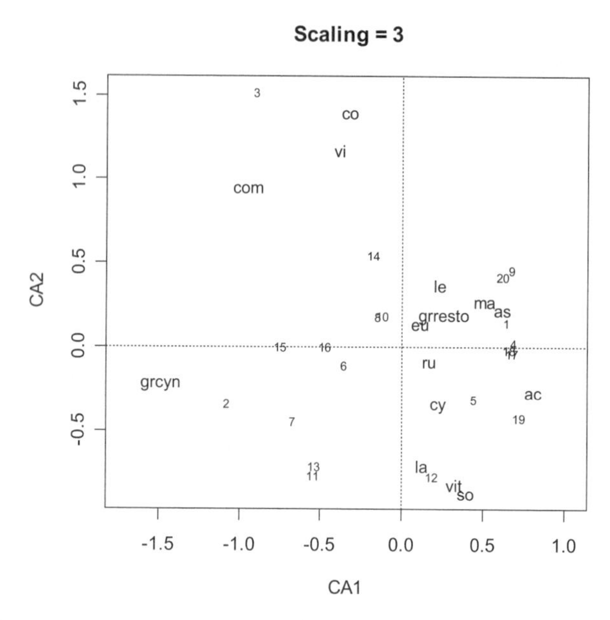

Figure 13.7. Joint plot of species and site scores. Distances between species are two-dimensional approximations of their Chi-square distances. Distances between sites (the numbers) are also two-dimensional approximations of Chi-square distances. Distances between species and sites cannot be interpreted. The first two axes explain 57% of the variation.

The total inertia (or total variance) in CA is defined as the Chi-square statistic of the site-by-species table divided by the total number of observations. Points far away from the origin in each diagram are the most interesting, because these points make a relatively higher contributions to the Chi square statistic than points nearer the origin. So the further away from the origin that a site is plotted, the more different it is from the average site. A numerical example may help. Suppose we have the following (artificial) data:

	Species 1	Species 2	Species 3	Total
Site 1	1	2	3	6
Site 2	0	1	2	3
Site 3	2	1	0	3
Site 4	4	3	3	10
Total	7	7	8	22

In scaling 2, we look at species profiles; the abundance of each species is divided by the species total, and the profiles are then compared with the average profile:

	Species 1	Species 2	Species 3	**Average**
Site 1	1/7	2/7	3/7	**6/22**
Site 2	0	1/7	2/7	**3/22**
Site 3	2/7	1/7	0	**3/22**
Site 4	4/7	3/7	3/7	**10/22**

In scaling 1, we do it the other way around. Row (site) profiles are calculated, and these are compared with the average profile:

	Species 1	Species 2	Species 3
Site 1	1/6	2/6	3/6
Site 2	0/6	1/3	2/3
Site 3	2/6	1/3	0
Site 4	4/10	3/10	3/10
Average	**7/22**	**7/22**	**8/22**

Section 10.1 contains an example of how these profiles are compared with each other (Chi-square distance). Just as in PCA, eigenvalues can be used to assess how much variation is explained by each axis. Instead of total variance, the total variation is called inertia.

Heuristic approximation, CA and RA

Because RA and CA share the same eigenvalue problems, the estimated scores are identical. Furthermore, the species scores **u** and site scores **x** obtained by RA are similar to the scores obtained by the heuristic approximation (rationale 1). This means that we now have three different approaches: Reciprocal averaging, weighted PCA, and the heuristic approximation of Gaussian ordination, which all give the same estimated species and site scores. As well as these three approaches, several other approaches exist, for example, dual scaling (Greenacre 1984).

The difference between RA, CA and the heuristic approximation of Gaussian ordination, besides the estimation procedure, concerns the type of data on which they can be used. RA can analyse any data, as long as the data are non-negative and have the same units. The heuristic approximation of Gaussian ordination assumes that data are Poisson distributed. Correspondence analysis was presented as a method that analyses contingency tables. However, in many textbooks CA is applied to other kinds of datasets; see for example Gauch (1982), Greenacre (1984), Jambu (1991) or Jongman et al. (1995). The use of CA on such datasets can be justified by considering the table as a distribution of a certain amount of *mass* (e.g., weight, length, volume) over the cells. Although it is still interesting to look at row-column interactions, the Chi-square statistic cannot now be used to *test* row-column independence. It merely serves as a measure of association.

In the rest of this chapter, the name *correspondence analysis* will be used for reciprocal averaging, weighted PCA and the heuristic approximation of Gaussian ordination. In the next section, the biplot interpretation is introduced in the context of weighted PCA. Because of the similarities among these approaches, the biplot can be used with all three methods.

13.3 From RGR to CCA

Recall from Section 13.1 that the restricted Gaussian response model has the following form:

$$Y_i = ce^{-\frac{(z_i-u)^2}{2t^2}} = \exp(b_1 + b_2 z_i + b_3 z_i^2) \qquad (13.9)$$

$$z_i = \sum_{p=1}^{Q} \alpha_p x_{ip}$$

Y_{ik} is the abundance (counts) of species k at site i, x_{ip} is the value of the pth environmental variable at site i, z_i is the value of the gradient at site i, and α_p, c_k, t_k and u_k are unknown parameters. Because Y_{ik} are counts, it is common to assume that Y_{ik} is Poisson distributed with expectation μ_{ik}. The log likelihood function is given by

$$F(\boldsymbol{\alpha},\mathbf{c},\mathbf{t},\mathbf{u}) = \sum_i \sum_k (Y_{ik} \log(\mu_{ik}) - \mu_{ik})$$

where $\boldsymbol{\alpha} = (\alpha_1,\ldots,\alpha_Q)$, $\mathbf{c} = (c_1,\ldots,c_M)$, $\mathbf{t} = (t_1,\ldots,t_M)$ and $\mathbf{u} = (u_1,\ldots,u_M)$. An option to obtain parameter estimates for $\boldsymbol{\alpha}$, $\mathbf{c}$, $\mathbf{t}$ and $\mathbf{u}$ is to formulate the partial differential equations of F with respect to the parameters, set these to zero, and use numerical optimisation routines to solve them. Instead of doing this, Ter Braak (1986) derived partial differential equations of F with respect to u_k, c_k, t_k and α_p. Just as for the Gaussian response model, these partial derivatives do not look friendly. In Section 13.2, four assumptions were used to simplify the partial differential equations resulting in CA. Ter Braak (1986) used the same four assumptions and obtained a considerably easier set of equations for RGR. The resulting technique is called CCA. As a result of assumptions 1–4, we obtain the following set of equations:

$$\mathbf{u} = \mathbf{D}_c \mathbf{Y}'\mathbf{Z} \quad \text{and} \quad \mathbf{Z}_{wa} = \mathbf{D}_r^{-1}\mathbf{Y}\mathbf{u} \quad \text{and} \quad \alpha = (\mathbf{X}'\mathbf{D}_r\mathbf{X})^{-1}\mathbf{X}'\mathbf{D}_r\mathbf{Z}_{wa}$$

Note that the parameter $\boldsymbol{\alpha}$ is the weighted least-squares solution of the regression of $\mathbf{Z}_{wa}$ on $\mathbf{X}$. The scores $\mathbf{u}$ will be referred to as species scores, $\mathbf{Z}_{wa}$ as site scores that are weighted averages of species scores (weights are given by site totals), and $\mathbf{Z}$ as site scores that are a linear combination of environmental variables. The latter scores are also denoted by $\mathbf{Z}_{env}$. To calculate the species and site scores, the following algorithm can be used:

1. Start with arbitrary site scores $\mathbf{Z}$.
2. Calculate species scores $\mathbf{u}$, which are weighted averages of site scores, by $\mathbf{u} = \mathbf{D}_c^{-1}\mathbf{Y}'\mathbf{Z}$.
3. Calculate site scores $\mathbf{Z}_{wa}$, which are weighted averages of species scores, by $\mathbf{Z}_{wa} = \mathbf{D}_r^{-1}\mathbf{Y}\mathbf{u}$.
4. Use weighted linear regression of the site scores $\mathbf{Z}_{wa}$ on environmental variables $\mathbf{X}$, and obtain the regression coefficients by $\boldsymbol{\alpha} = (\mathbf{X}'\mathbf{D}_r\mathbf{X})^{-1}\mathbf{X}'\mathbf{D}_r\mathbf{Z}_{wa}$.

5. Obtain new, estimated site scores $\mathbf{Z}$, which are a linear combination of environmental variables, by $\mathbf{Z}_{env} = \mathbf{X}\,\boldsymbol{\alpha}$ and set $\mathbf{Z}$ equal to these.
6. Standardise the estimated site scores $\mathbf{Z}$.
7. Stop on convergence; else go to step 2.

As the range of the weighted averages is smaller than the range of the scores that are used to calculate them, a scaling of scores $\mathbf{Z}_{env}$ (= $\mathbf{Z}$) is used in each iteration (step 6). This prevents the algorithm from drifting into a small range of scores and avoids a trivial solution. To obtain a second axis (or further axes), a weighted regression can be carried out in each iteration. In such a procedure, we regress $\mathbf{Z}_{env}$ on the first axis (or previous axes) and continue to work with the residuals of this regression. See Ter Braak and Prentice (1988) for more details. In a similar way, the linear effects of particular environmental variables can be filtered out. This technique is called partial CCA. It can be useful if one is not interested in the effects of these particular environmental variables.

Note that the algorithm for CCA is similar to the algorithm for reciprocal averaging (alias correspondence analysis). From a mathematical point of view, CCA can be seen as correspondence analysis in which the axes are restricted to be linear combinations of environmental variables. Put simply, CCA is a CA in which the axes are restricted to be linear combinations of explanatory variables.

Inertia

The inertia (or total variance) in CCA is identified the same way as in CA. The eigenvalue of an axis is given by the weighted standard deviation s, which is calculated in step 6 of the CCA algorithm. This can be seen by making similar substitutions as in CA. These eigenvalues are also called canonical eigenvalues. The amount of variation that can be explained by all the environmental variables is equal to the sum of all canonical eigenvalues. The amount of variation explained by the first two axes can be expressed as a percentage of the total inertia, and as a percentage of the variance that can be explained by the environmental variables.

Canonical coefficients and intraset correlations

So, how do we know which explanatory variables are important? There are two tools for this: The final regression coefficients $\boldsymbol{\alpha}$, also called canonical coefficients, and the intraset correlations defined as the correlations between environmental variables and axes $\mathbf{Z}_{env}$. The intraset correlations are also called environmental scores. The environmental variables are standardised, which makes a comparison of the canonical coefficients possible. How to interpret the species scores $\mathbf{u}$, site scores $\mathbf{Z}_{wa}$ and $\mathbf{Z}_{env}$, intraset correlations and canonical coefficients is explained in the next section.

13.4 Understanding the CCA triplot

The species scores **u**, site scores $\mathbf{Z}_{env}$ and intraset correlations obtained by CCA are plotted in a figure called a triplot. For the same reason as in CA, species are represented by labels, sites by points or labels, and the explanatory variables by lines. Species points represent the optimum parameter of the Gaussian response model (niche) and can be projected on the axes but also on the explanatory variables showing the optimum value along each of them. Nominal explanatory variables are dealt with in the same way as in RDA. An example of a triplot is presented in Figure 13.8. This triplot was obtained by applying CCA on the Mexican plant data. The same families as in the previous section were used, and we only used four explanatory variables. Families are represented by their name (abbreviated, see Chapter 32), sites by numbers 1–20 and intraset correlations by lines starting at the origin to the point with coordinates given by the intraset correlations of the two axes.

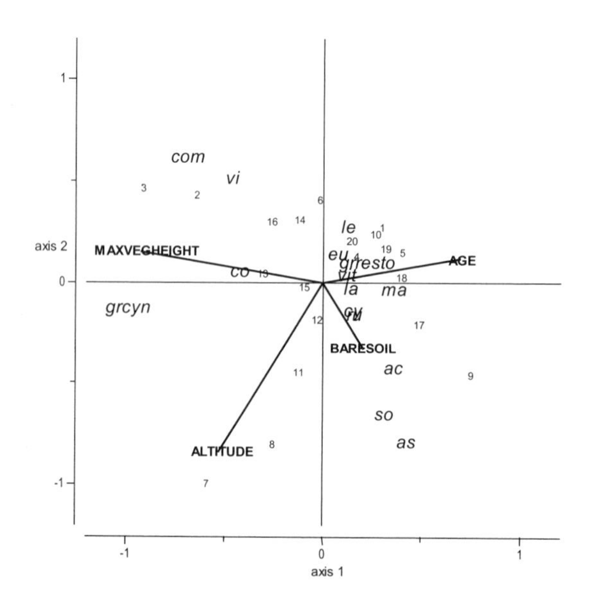

Figure 13.8. Triplot for the Mexican plant data, obtained by canonical correspondence analysis. Species conditional scaling was used.

In a triplot, species scores, site scores and environmental scores are plotted in the same graph and form a series of biplots. These biplots are based on the species scores and site scores, the species scores and intraset correlations (representing the explanatory variables), and the site scores and intraset correlations. CCA produces two sets of site scores; $\mathbf{Z}_{wa}$ and $\mathbf{Z}_{env}$. The standard choice is to use the scores $\mathbf{Z}_{env}$

in the triplot. The motivation for this is that $\mathbf{Z}_{env}$ can be used for two biplots, whereas this is not the case for $\mathbf{Z}_{wa}$.

As with CA, you can choose from various scaling options, such as the species conditional scaling (called scaling 2 in CA) or the site conditional scaling (called scaling 1 in CA). If the interest is on species, then the most sensible choice is species conditional scaling. The same holds for the sites and scaling 1.

Concentrating on the species scores and site scores in the species conditional triplot in Figure 13.8, the interpretation is identical to that in CA. So comparing species (families in this example) scores give information about which species behave similarly. Species close to each other are similar in terms of Chi-square distances, and species relatively far away from the origin, contribute more to the inertia than species close to the origin. The species scores are at the centroids of the sites scores, which allows us to infer relative abundances (just as in CA). Using this interpretation, one can infer from Figure 13.8 that the families so, as, and ac deviate from the average profile at sites 7, 8 and 9. These families are positively related to each other, but negatively to vi and com.

The species scores and intraset correlations (the explanatory variables) can also be compared. Species can be projected perpendicular on the lines showing the species optima. If sites scores are projected perpendicular on the lines, we can infer the values of the environmental variables at those sites. Results indicate that the family grcyn occurs at high values of maximum vegetation height, and low values of age. At sites 2 and 3, maximum vegetation height is large. In fact, we could draw the species packing model (Figure 13.4) along each line.

Recall that the intraset correlations are the correlations between the axes and the original explanatory variables, and the canonical coefficients are the $\boldsymbol{\alpha}$s defining the linear combination of explanatory variables constituting the gradient.

As to the intraset correlations, the tip of a line (representing intraset correlations) can be projected perpendicularly on another line, and the weighted correlation between them is inferred. The tip of a line can also be projected on the axes, and the correlation between the corresponding environmental variable and the axes is inferred. The lines in the triplot indicate that the environmental variables age and maximum vegetation height are negatively correlated. Projecting the lines on the axes shows that the first axis is highly correlated with maximum vegetation height. The second axis is correlated with altitude.

With the canonical coefficients, recall the gradients are linear combinations of environmental variables. The exact form of this linear combination is determined by the canonical coefficients. The canonical coefficients for the first axis are 0.18 (altitude), −0.28 (age), −0.23 (bare soil) and 0.32 (maximum vegetation height). The first axis is mainly determined by maximum vegetation height versus age.

Finally, we discuss the eigenvalues. The total inertia (variation) is 0.33, and the sum of all canonical eigenvalues is 0.15. Hence, all four explanatory variables explain 45% (= 100 × 0.15/0.33) of the variation in the data. The first two eigenvalues are 0.10 and 0.02. Making 82% of the total inertia explained by the first two axes. And 37% (= 82% of 45%) of the variation that can be explained with the environmental variables is explained by the first two axes. Both percentages are relatively high for ecological datasets.

In the distance biplot (scaling 1), distances between sites represent (approximate) Chi-square distances, but distances between species cannot be interpreted.

13.5 When to use PCA, CA, RDA or CCA

At this point we introduce two measures of diversity: Alpha and beta diversity. Alpha diversity is the diversity of a site and beta diversity measures the change in species composition from place to place, or along environmental gradients. Examples of these diversity measures are given in Figure 13.9. The total beta diversity is the 'gradient length'. A short gradient has low beta diversity. As explained above, Ter Braak (1986) showed that CA is an approximation of Gaussian ordination and CCA is an approximation of restricted Gaussian regression. This is the ecological rationale of CA and CCA. PCA and RDA analyse linear responses along the gradient, and CA and CCA look at unimodal responses along the gradient. This is summarised in Table 13.1 and described in more detail below:

1. PCA should be used to analyse species data if the relations along the gradients are linear.
2. RDA should be used to analyse linear relationships between species and environmental variables.
3. CA analyses species data and unimodal relations along the gradients.
4. CCA can be used to analyse unimodal relationships between species and environmental variables.
5. PCA or RDA should be used if the beta diversity is small, or if the range of the samples covers only a small part of the gradient.
6. A long gradient has high beta diversity, and this indicates that CA or CCA should be used.

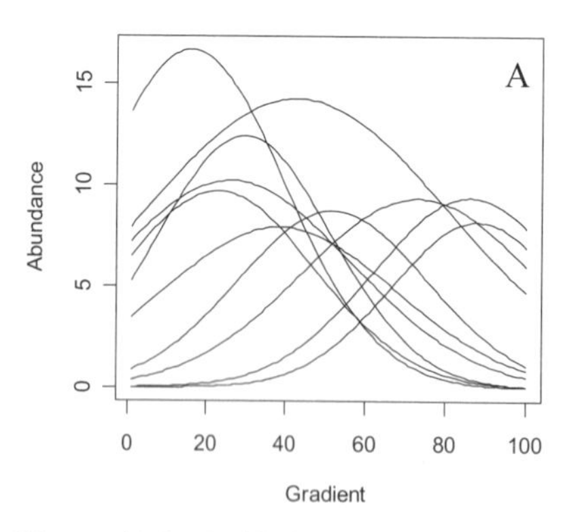

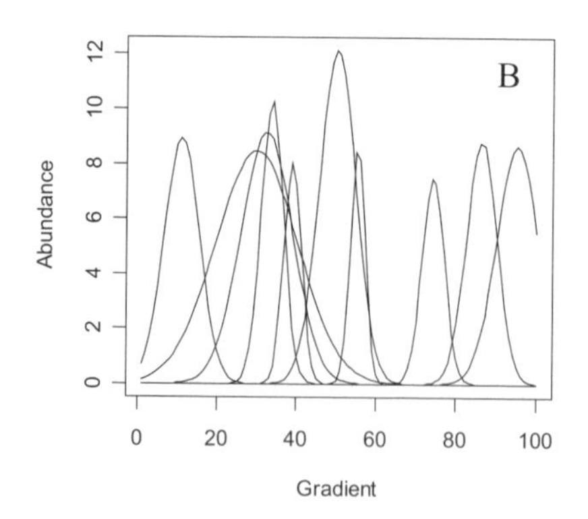

Figure 13.9. Artificial response curves showing high alpha and low beta diversity (A) and low alpha and high beta diversity (B). PCA and RDA should be applied on data in panel A and CA and CCA in panel B.

Table 13.1. Summary of methods. Relationships in PCA and RDA are linear. In RDA and CCA two sets of variables are used, and a cause-effect relationship is assumed.

	Indirect Gradient Analysis	Direct Gradient Analysis
Linear model	PCA	RDA
Unimodal model	CA	CCA

13.6 Problems with CA and CCA

CA and CCA are useful techniques as long as the data matrix does not contain too many zeros. Figure 13.10 shows what happens if correspondence analysis is applied on the RIKZ data used in Chapter 27. The data contain many observations equal to zero; there are a few species measured at only one site and a few sites where only one species was observed. These more extreme observations for species and sites dominate the first few axes! We also had to remove two sites because no species were observed. Obviously, we can remove such species and sites, but the question is how much data can you afford to remove.

Another potential problem is the arch effect (Chapter 12), which again is due to the many observations equal to zero. Figure 13.11 shows a CCA triplot (species conditional) for the full Mexican plant data. Instead of using averages per pasture, we use all 200 observations. Note that the shape of the site scores may indicate the presence of an arch effect. If this is the case, then there are three options: (i) Argue that the arch shape is a real pattern in the site scores caused by the explanatory variables (risking the referee rejecting the paper because they think that detrended canonical correspondence analysis should have been used), (ii) apply detrended canonical correspondence analysis to bring down both ends of the arch (risking the referee rejecting the paper because they do not agree that detrended correspondence analysis is an appropriate technique), or (iii) applying a special (Chord or Hellinger) transformation followed by an RDA and visualise Chord distances. Detrended correspondence analysis is an artificial way to remove the arch effect by splitting up the axis in segments and detrending the scores in each segment. Obviously, any real pattern will also be removed. Our choice is option (iii); the other two each have 50% chance of getting past a referee. Some books will condemn detrended canonical correspondence analysis (and detrended correspondence analysis), with some software programmers even refusing to add it to their software; yet other books are positive about it. McCune and Grace (2002) say: 'Detrended CA unnecessarily imposes assumptions about the distribution of samples and species in environmental spaces. .... There is no need to use detrended CA'. And Legendre and Legendre (1998) write: 'Present evidence indicates that detrending should be avoided, ...'. Both books advise to use non-metric multidimensional scaling if there is an arch effect.

The case study chapters contain various examples of PCA, RDA, CCA and variance partitioning and expand on the ideas presented in this chapter.

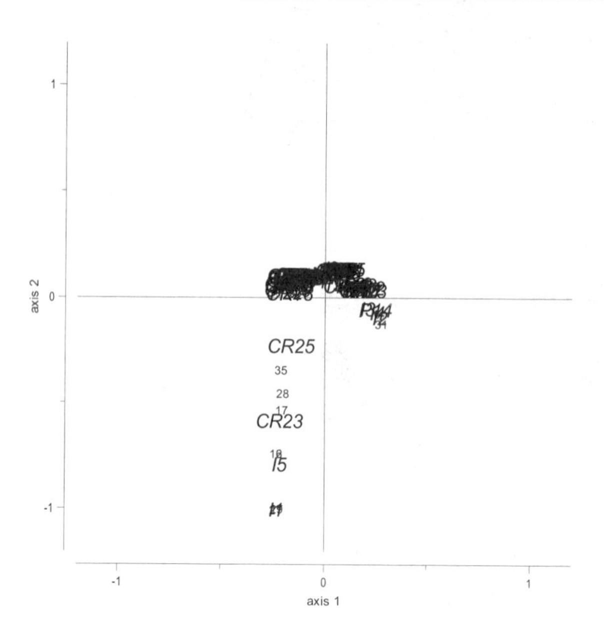

Figure 13.10. Correspondence analysis on the RIKZ data. The species conditional scaling was used. A few species were only measured at one site, and a few sites only had one species (with low values). As a result, the first few axes are dominated by these species and sites.

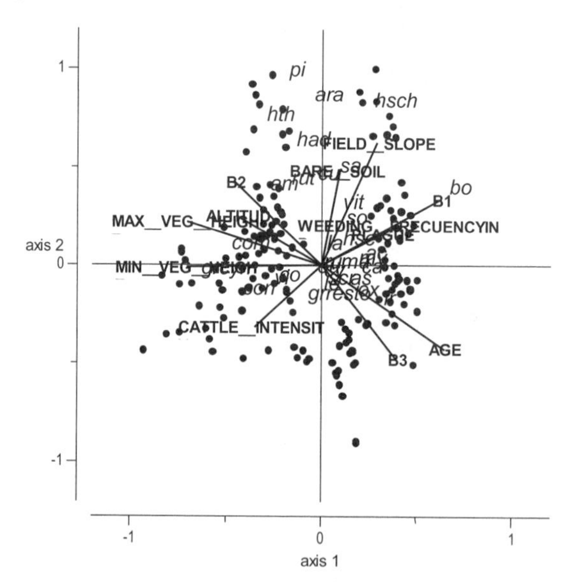

Figure 13.11. CCA applied on the full Mexican plant data. The sites scores (dots) may exhibit the arch effect as they show a U-shape.

14 Introduction to discriminant analysis

14.1 Introduction

In Chapter 12, principal component analysis (PCA) was introduced, which can be applied when you have M observations on N variables, denoted by Y_1 to Y_N. Recall that the aim of PCA is to create linear combinations of the N variables (principal components or axes), such that the first principal component (PC) has maximum variance, the second PC, the second largest variance, etc. The first PC, denoted by Z_1, is given by

$$Z_{i1} = c_{11}Y_{i1} + c_{12}\,Y_{i2} + \ldots + c_{1N}\,Y_{iN} \tag{14.1}$$

The index i refers to the observations, and all we need to find are the multiplication factors c_{ij}. However, if we already know that there is an *a priori* grouping structure in the M observations, then we can use discriminant analysis (DA) and take advantage of this additional information. As an example we will use a sparrow dataset (unpublished data, Chis Elphick, University of Connecticut). This dataset has seven body measurements from approximately 1100 saltmarsh sharp-tailed sparrows (*Ammodramus caudacutus*), e.g., size of the head, size of the wings, tarsus length, weight, etc. Over the course of the study, 10 different observers were involved in taking measurements. Some measured over 300 birds, whereas others measured much less. This gave a dataset with M = 1100 rows of measurements for N = 7 variables, plus an extra variable identifying the observer. We could apply a PCA on these data and use the identity of the observers as labels in the biplot, but this does not take advantage of the extra grouping information given by knowing who was the observer. Nor does it help answer the possible underlying question on whether the observers have influenced the results. PCA cannot answer this question, but discriminant analysis (DA) can.

Before applying DA on these data, we will look at some examples from the case study chapters. In Chapter 28, a zoobenthic dataset from an Argentinean salt marsh area is used. There are four species (variables), and measurements were taken across the seasons from three transects with 10 samples per transect (per season). The data matrix (per season) for the species data gives a 30-by-4 matrix plus an extra column with value 1, 2 or 3 to identify the transect where the sample was collected. And we can make the situation more complicated by combining the data from two seasons resulting in a 60-by-4 data matrix. Assuming we are interested in how the relationships between the four species differ among the transect, we can use DA to investigate these relationships by discriminating between season, between transect, or between season and transect to see which are the most important in understanding the species relationships.

In Chapter 29, fatty acid concentrations in blubber of stranded dolphins are analysed. There are 31 fatty acids (variables), and 89 dolphins (observations) were measured. Although the chapter uses PCA to analyse these data (focussing on the relationship between different fatty acids), you can also ask the question whether the fatty acid values can be used to discriminate between male and female species, type of death and area of stranding. The data matrix is of dimension 89-by-31 and has three extra columns identifying sex, type of death and area. We could apply three different discriminant analyses: one for each question.

In Chapter 24, classification trees are applied on bird observations obtained by radar. The radar measures a large number of variables per bird, for example velocity, size of the target, etc. About 650 observations are used in the chapter. As well as the observations from the radar, field observations from the ground were available and these allowed the observations to be grouped by species, clutter, etc. The question is then whether we can only discriminate between birds and clutter, or whether we can also discriminate between species.

The common feature shared by all these datasets is that the dataset is of dimension *M*-by-*N*, and there is an extra column identifying groups of observations. The structure of the data is visualised in Table 14.1. In all the datasets the question is: 'Do the variables differ per group of observations, and if they do, which variables?' Stated differently, can we discriminate between *a priori* defined groups of observations using the variables? And which variables are the best at discriminating?

Table 14.1. Structure of the data for discriminant analysis. The variables Y_1 to Y_N are for each observation and 'Group' identifies either the observations made by different observers, male and female (g = 2), different transects, areas, etc.

Observation	Y_1	Y_2	...	Y_N	Group
1	10	16	...	21	1
2	21	18	...	52	1
3	31	41	...	2	1
4	12	15	...	34	1
5	1	10	...	1	2
6	12	20	...	2	2
...	...	...	...	...	...
...	...	...	...	...	...
N	15	21	...	6	G

To answer this question, we could apply one-way ANOVA on each of the *N* variables Y_j. The explanatory variable would be 'Group' and the response variable Y_j. The problem with this approach is that it is rather time consuming to apply *N* one-way ANOVAs, and we are not taking advantage of the multivariate nature of the data. Discriminant analysis (DA), also called canonical variate analysis, although similar to PCA, uses all variables Y_1 to Y_N and extracts a linear combination of them. This linear combination is of the form:

$$Z_{i1} = \text{constant}_1 + w_{11} Y_{i1} + w_{12} Y_{i2} + \ldots + w_{1N} Y_{iN} \quad (14.2)$$

The unknown parameters are the constant and the multiplication factors (or weighting factors) w_{ij}. In PCA, the multiplication factors are chosen such that Z_1 has maximum variance. In DA the aim is different as we are interested in discrimination between the *a priori* defined groups of observations. To illustrate the underlying principle of DA, we use the sparrow data described earlier. Nine observers were involved in the sampling process and the observations consist of seven body measurements on each sparrow. A scatterplot for two variables is given in Figure 14.1. To keep the graph simple, we only used observations made by two observers, and they are identified by the symbols '+' and 'o'. If we apply PCA on these data, the first axis would probably go from somewhere in the lower left corner to the upper right corner as this would give a line that when all points are projected on it, has maximum variance. In DA, the objective is different; we look for a line that, when all points are projected on it, the observations of the same group are close to each other and observations from different groups are far away from each other. The dotted line in Figure 14.1 is a potential candidate. If we project all points on this line, the points with a '+' will be mostly on the left side and the observations with 'o' are mainly right of it.

So, how do we get this line? Just as in PCA it is a matter of finding the optimal rotation. We could define a criteria that measures the discrimination and try every possible rotation, but just as in PCA, it turns out that the solution can be obtained with an eigenvalue equation. Further axes can be calculated, and these are uncorrelated with each other. In this chapter, we will not present too much mathematical detail and refer you to Legendre and Legendre (1998), Huberty (1994) or Klecka (1980). Material presented here is mainly based on these three references.

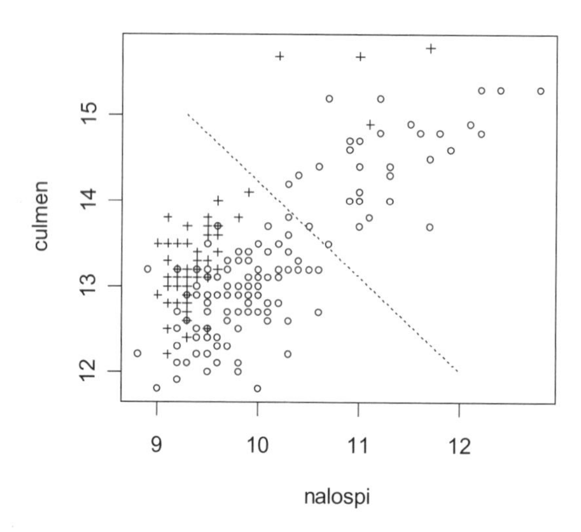

Figure 14.1. Example of the underlying principle of discriminant analysis. Measurements made by two observers for a sparrow dataset are used. The two observers are plotted using 'o' and '+'. The variables nalospi and culmen are variables measured by the observers. The dotted line shows a possible axis that gives maximum separation of the two groups, if all points are projected on it.

14.2 Assumptions

In PCA there were no underlying assumptions except that the relationships between the variables should be linear. In DA, there is a whole series of assumptions, and their validity dictates how useful the method is for your own data. These assumptions are as follows:

- The observations can be divided *a priori* into at least two groups. Each observation can only be in one group. This means that one should not apply clustering on the data to obtain groups, and then apply DA on the same data using the grouping structure obtained by clustering. A valid approach is to split up the data into two groups, apply clustering on one dataset, and use the results in the DA for the second dataset (Legendre and Legendre 1998).
- There are at least two observations per group. But a common recommendation is to have at least four or five times as many observations per group as the number of variables.
- The number of observations in the smallest group is larger than the number of variables. This assumption is stated in most books on DA, but interestingly half of the authors then give examples of where this assumption is violated.

- The variables are on a continuous scale. Because covariance matrices are calculated, you cannot use nominal variables in the dataset. You can convert them to dummy variables with zeros and ones, but this then violates other assumptions.
- Relationships between the variables are linear. This is because covariance matrices are used.
- There is no 100% collinearity between the variables. Always inspect the correlation matrix before applying DA. If variables have a correlation of 0.9 or higher, remove one of them.
- Within group variances are approximately similar between the groups. This is also called the homogeneity assumption. It means that the spread of a particular variable must be (approximately) the same for each group of observations. But different variables are allowed to have a different spreads. Cleveland dotplots or boxplots conditional on the grouping of observations can be used to assess this assumption (Chapter 4).
- During the calculations a covariance matrix for each group will be calculated and these will be pooled. Therefore, we also assume that the covariance matrices for the g groups are similar. This means that relationships between variables must be similar for different groups. It also means that a variable within a group is not allowed to have the same value (e.g., zero) for each observation as this prevents calculating the covariance matrix. Some programmes will still produce sensible results even if this assumption is not met.
- The hypothesis tests assume multivariate normality of the N variables within each group. This means that each variable must be (approximately) normally distributed within each group. Conditional histograms (Chapter 4) can be used to assess this assumption.
- The observations are independent. This means that time series, spatial data and before-after comparisons cannot be used.

If the normality and homogeneity assumptions do not hold, a logarithmic or square root transformation might help. Normality is required for the hypothesis tests, but not for the method itself (Hair et al. 1998). Violation of homogeneity in combination with small group sizes is seen as a serious problem. In such cases (multinomial) logistic regression might be a better alternative; the group variable is used as a response variable and the rest as explanatory variables and no conditions are imposed on the explanatory variables. Quadratic discriminant analysis is an alternative option if there is violation of homogeneity.

Equal group size is not required, but as with all statistical methods, common sense dictates that they should be similar. A ratio of largest group size versus the smallest group size of 9:1 has been suggested by some authors as the maximum value before one should decide not to apply DA.

Due to this long list of assumptions, some authors (e.g., McCune and Grace 2002) have suggested that DA has 'limited application in community ecology'. We do not agree with this, as DA can be used in about one quarter of the case study chapters in this book. Obviously, it is less useful if the dataset consists of a large number of plant species sampled at 200 sites, and 95% of the observations is equal to zero as we are violating various assumptions. The answer to the question of whether DA is useful for your own data is simple: 'It all depends'. The example presented in the next section should help you decide.

14.3 Example

Recall that the sparrow data consist of approximately 1000 birds measured by 10 observers. The measured variables are the lengths of the wing (measured in two different ways, as the wing chord and as the flattened wing), leg (a standard measure of the tarsus), head (from the bill tip to the back of the skull), culmen (the top of the bill from the tip to where the feathering starts), nalospi (the distance from the bill tip to the nostril) and weight. The question we want to look at is whether the observers are producing similar measurements or, stated slightly differently, is there an observer effect? Discriminant analysis is one of the most appropriate methods to answer this (alternative methods are redundancy analysis and multivariate regression trees), but before we can apply DA, we need to verify the assumptions.

There are 10 observers, and the number of observations per observer was between 9 and 332. This means that we have a serious problem with unequal group sizes. If we use the (arbitrary) 9:1 rule, we need to drop the two observers that only made between 9 and 30 observations, from the analyses. Even without this rule, it is common sense not to compare results of an observer with only 9 observations with one with 332 observations.

The homogeneity assumption was checked using Cleveland dotplots or boxplots conditional on observer, and one such graph for flatwing is shown in Figure 14.2. It shows that the spread is approximately the same in each group, indicating homogeneity. The assumption also holds for the other variables (results are not shown here). To have faith in the hypotheses tests (which will be discussed later), the normality of each variable in each group is required. A conditional histogram for each variable (not shown here) indicates that this is a valid assumption.

Another assumption is that the variables are not collinear. The correlation coefficient between each pair of variables was calculated, and all were smaller than 0.75, except wingcrd and flatwing; their correlation coefficient was 0.99. This high correlation was entirely expected as the two variables represent the same thing. Hence, one of these variables should be dropped and we decided to use flatwing in the analysis.

A pairplot (not shown here) showed that relationships between all variables are approximately linear. The last assumption we need to check is the independence of the observations. Sampling took place in different months, and the variable weight shows a strong seasonal pattern, which means violation of the independence assumption. We de-seasonalised the weight variable by subtracting the monthly average (Chapter 16). The resulting dataset now complies with all assumptions, and we can apply linear discriminant analysis.

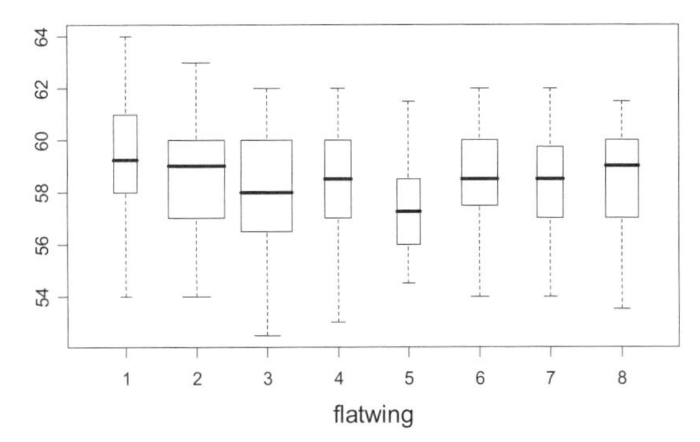

Figure 14.2. Boxplot of flatwing conditional on observer. Each number along the horizontal axis is an observer. The prime question is whether the spread is the same in each group. The width of a box is proportional to the number of observations made by an observer, but all observers (presented here) measured more than 55 birds.

The first discriminant function is given by

$$Z_{i1} = \text{constant}_1 - 0.04 \times \text{flatwing}_i + 0.25 \times \text{tarsus}_i - 0.68 \times \text{head}_i - 1.62 \times \text{culmen}_i + 2.91 \times \text{nalospi}_i + 0.05 \times \text{wt}_i \quad (14.3)$$

The multiplication factors 0.04, .., 0.05 are called the unstandardised discrimination coefficients, and their interpretation is difficult as the original variables were not normalised (centred and divided by the standard deviation). If we normalise the variables prior to the analysis, it is easier to compare them with each other and the intercept also vanishes:

$$Z_{i1} = -0.08 \times \text{flatwing}_i + 0.17 \times \text{tarsus}_i - 0.47 \times \text{head}_i - 1.00 \times \text{culmen}_i + 1.55 \times \text{nalospi}_i + 0.07 \times \text{wt}_i \quad (14.4)$$

The multiplication factors obtained for normalised variables are called the standardised discrimination coefficients, and these can be used to assess which variables are important for the discrimination along the first axis. For example, the variable nalospi has a large positive multiplication factor and head and culmen have large negative factors. These three variables play an important role for discrimination along the first axis. All three variables indicate something about the size of the bird's head.

The standardised discrimination coefficients can either be obtained by standardising the variables prior to the analysis or by using a short-cut formula (Klecka 1980). Some authors mention that the (standardised) discrimination coefficients can be instable and instead advise using the correlation coefficients between the discriminant functions and each original variable. These are called canonical correlations, but note that terminology differs between software and authors.

Further axes are obtained, and the traditional graphical presentation of DA is to plot two axes against each other, in most cases Z_1 versus Z_2 as these explain most of the separation; see Figure 14.3. Each observation is plotted as a group number. If there is an observer effect, you would expect to see observations from the same group close to each other with a clear separation between groups. One way to enhance visual detection of group effects is to calculate the average group scores per axis and to plot these for example as large triangles (Figure 14.3). If the triangles are clearly separated, then there is a visual indication of a group effect. In this case, some triangles are separated, but we have seen examples with more separation. Instead of the scatterplot of Z_1 and Z_2, Krzanowski (1988) used tolerance intervals. These are presented in Figure 14.4 and show the group means again, but with these now identified by a number representing the group (observer). The circles around the group means represent the 90% tolerance regions where 90% of the whole population in a group is expected to lie (Krzanowksi 1988, pp. 374-375). These graphs are easier to interpret, as there is less clutter compared with plotting all the scores, but they are based on the normality assumption. The canonical correlation coefficients (these are the correlations between the original variables and the DA axes) are plotted in Figure 14.5 and indicate that the first axis is positively correlated with nalospi.

The DA graphs indicate that there is some marginal evidence of discrimination, possibly related to nalopsi. As this is probably the hardest variable to measure, it does not come as a surprise that it contributes the most to differences among individuals. The question is now whether the discrimination is significant, and this is discussed in Section 14.5. However, the fact that the groups are not clearly separated in the ordination diagrams indicates that even if the statistical tests indicate that there is an observer effect, it is not particularly strong.

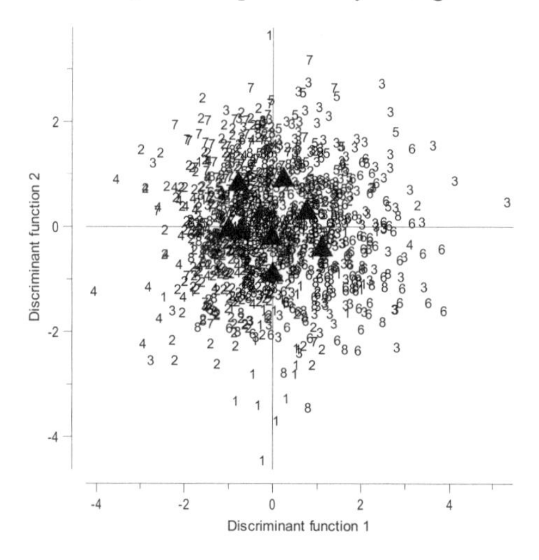

Figure 14.3. Scatterplot of the first discriminant function versus the second. Observations are represented by their group (=observer) number. The triangles represent the group averages.

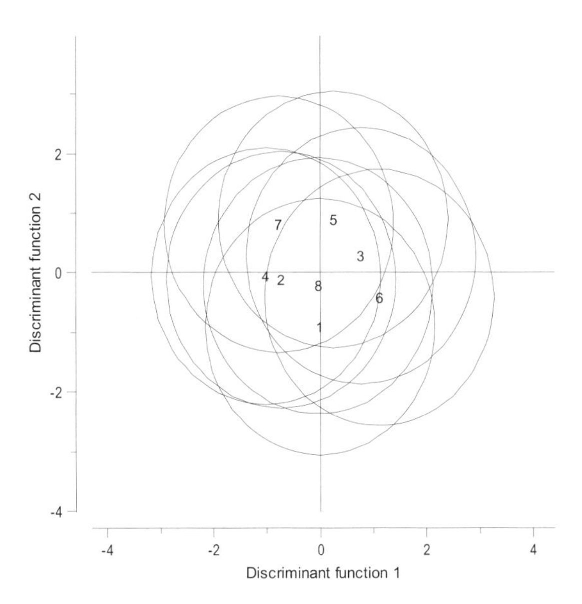

Figure 14.4. 90% tolerance intervals. The numbers refer to the observers, and the circles are the 90% tolerance intervals; they define the range in which 90% of the population values are found.

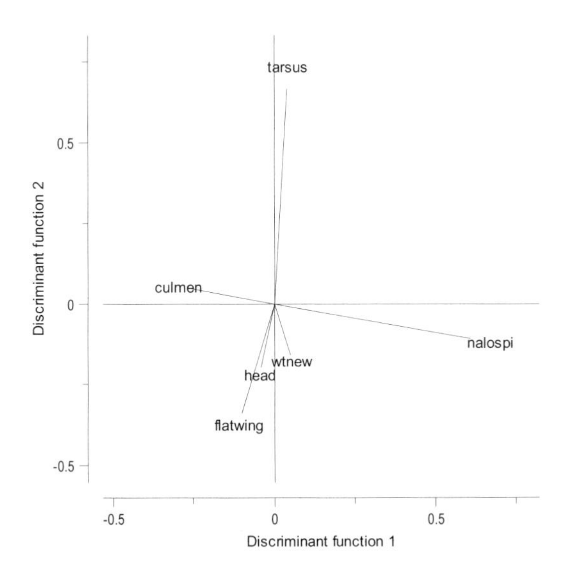

Figure 14.5. Correlation coefficients between the first two axes and each of the original variables.

14.4 The mathematics

The reader not interested in the principles of the underlying mathematics for linear discriminant analysis may skip this section. We follow the same notation and outline as Legendre and Legendre (1998). DA is based on three important matrices, and these are extensions of linear regression where we decomposed the total variation (SS_{total}) in $SS_{regression}$ and $SS_{residual}$, and used these in an F-test (Chapter 5). Here, we use multivariate extensions of these terms.

Define **X** as the matrix containing all the variables. The rows contain the M observations and the columns the N variables giving a matrix of dimension N-by-M. The N-by-N matrix measuring the overall variation in the data is given by

$$\mathbf{T} = (\mathbf{X} - \bar{\mathbf{X}})'(\mathbf{X} - \bar{\mathbf{X}})$$

It can easily be converted to the covariance (or correlation) matrix; simply divide **T** by $N - 1$:

$$\mathbf{S} = (\mathbf{X} - \bar{\mathbf{X}})'(\mathbf{X} - \bar{\mathbf{X}})/(N - 1)$$

The matrix **S** is the covariance (or correlation) matrix, and it represents the total variation in the data. Define $\mathbf{W}_j$ as the sum of squares for group j. This is calculated in the same way as **T**, but we only use data of group j. This matrix can be calculated for each group, giving $\mathbf{W}_1,..,\mathbf{W}_g$. Because we assumed homogeneity, we can pool the g within-group matrices:

$$\mathbf{W} = \mathbf{W}_1 + ... + \mathbf{W}_g$$

The matrix **W** can be converted into a covariance matrix by dividing it by N–g:

$$\mathbf{V} = \mathbf{W}/(N - g)$$

V represents the within group covariance. Having a total variation (**S**), within group variation (**V**), there is one last term we need, namely the between group variation. It is obtained by subtracting **W** from **T**:

$$\mathbf{B} = \mathbf{T} - \mathbf{W}.$$

It can be converted into a covariance matrix by dividing **B** by $g - 1$:

$$\mathbf{A} = \mathbf{B}/(g - 1)$$

The matrices **A** and **V** form the basis for the eigenvalue equation for linear discriminant analysis, or to be more precise: $\mathbf{V}^{-1}\mathbf{A}$. Because this matrix is not symmetrical the following eigenvalue equation is solved (Legendre and Legendre 1998, p. 621):

$$(\mathbf{A} - \lambda_k \mathbf{V})\mathbf{u}_k = 0$$

To make the axes orthogonal, the eigenvectors are rescaled: $\mathbf{C} = \mathbf{U}(\mathbf{U}'\mathbf{V}\mathbf{U})^{-0.5}$, where **U** contains the eigenvectors and **C** the scaled eigenvectors.

14.5 The numerical output for the sparrow data

Discriminant analysis comes with a whole suite of hypotheses, and full details on these tests can be found in Huberty (1994). These tests are based on the normality assumption, but this is a valid assumption for the sparrow data. Instead of giving a detailed explanation for each of these hypotheses and tests, we present the results of some of them within the context of the sparrow data. It is not only the statistical tests that make the numerical output of DA intimidating, there is also a lot of other information presented. We have grouped this information into three types: the importance of the axes, the importance of the variables and finally, classification.

Importance of the axes

It does not really matter which software is used as in essence they all produce similar output (although not necessarily identical). The first part of the output is as follows:

```
The number of variables is              6
The number of observations is           1102
The number of groups is                 8
The number of discriminant functions is       6
The number of rows (observations) containing missing values is    0
```

This information shows that there are 1102 observations and 8 groups (observers). The number of discriminant functions that one can calculate is the minimum of $g - 1$ and N (g is the number of groups and N the number of variables). In this case it is 6. As with most other multivariate statistical methods, DA cannot cope well with missing values. What it does with missing values depends on the software: It might delete the entire row or fill in an average value. For these data, there are no missing values. Further output is as follows:

```
Linear discriminant analysis is used. Observations per group:
  1   56
  2  332
  3  271
  4   73
  5   54
  6  135
  7   67
  8  114
```

So far we have only discussed linear discriminant analysis (see first line of the output). There are also extensions that can cope better with violation of homogeneity (quadratic discriminant analysis), but these are outside the scope of this book. The additional output (above) shows the number of observations per group. The problem of observers with small numbers of observations was solved by omitting the two with the lowest values from the analysis. We now discuss how many axes to present. As in PCA, the eigenvalues indicate the importance of each axis (discriminant function). The information on eigenvalues is as follows:

Eigenvalues (=lambda)

axis	lambda	lambda as %	lambda cumulative %
1	0.572	67.792	67.792
2	0.171	20.320	88.112
3	0.078	9.249	97.361
4	0.013	1.575	98.936
5	0.007	0.867	99.804
6	0.002	0.196	100.000

It shows that the first two axes represent 88% of the variation, which is more than sufficient. Hence, there is no point in inspecting higher axes. Tests relevant for the number of axes give:

Dimensionality tests for group separation
H_0: No separation on any dimension
B0= 774.142
B0 is a chi-squared statistic with degrees of freedom: 42.000
Probability that a Chi-squared with larger value is found: 0.000

The statistical background for this test is described in Huberty (1994), and it shows that we can reject the null hypothesis that there is no separation on any dimension. This means that we need at least one discriminant function to describe group differences. Further tests (results are not shown here) all indicate that the separation along the first and second axes is significant.

To test the hypothesis that there is no overall group effects, three test statistics are available: the Wilks lambda statistic, the Barlett–Pillai statistic and the Hotelling–Lawley statistic. The first is the most popular. For the sparrow data, all three statistics indicate that there is a significant group effect. Again, in our experience these tests tend to reject the null hypothesis of no group separation even if the ordination diagrams do not show a clear separation.

Statistic	Value	F	Num df	Den df	*p*-value
Wilks lambda	0.493	19.817	42	5111	<0.001
Barlett–Pillai	0.604	17.509	42	6564	<0.001
Hotelling–Lawley	0.843	21.834	42	6524	<0.001

Importance of the variables

If the data contain a large number of variables, it may be interesting to identify which of the variables are responsible for the discrimination of the groups and which ones can be omitted. The discrimination between all the groups is measured by the total sum of Mahalanobis distances. It uses the distances between all the group means. In the backwards selection approach, one can leave out one variable in turn and the total sum of Mahalanobis distances between group means is calculated again. An important variable, with respect to discrimination between groups, will give a large change in the total sum of Mahalanobis distances, compared with a less important variable. The information below shows the results of a backwards selection procedure.

Variables	Total Mah. distance	Dropped variable
6	61.9532	none
5	60.9435	wt
4	53.5066	flatwing
3	42.0006	head

Variables that were not dropped are tarsus, culmen and nalospi (as the DA was run with at least three variables). To decide how many variables to drop, a stop criterion needs to be used. One option is to make a so-called scree-plot, just as we did for PCA (not shown here); draw the total sum of Mahalanobis distances versus the number of variables and try to detect a cut-off point. This is similar to PCA where eigenvalues can be plotted versus the number of axes.

Classification

We can also apply classification in DA. The classification process is actually very simple. Once the discriminant functions are calculated, you can easily determine to which group average a particular observation is the closest. If an observation from group 1 is the closest to the group average of group 1, then it is classified correctly as group 1 (in this case 22 observations). But if it is closer to the group average of group 2, we classify it as group 2 (in this case three observations).

The classification table below indicates that from the 56 observations made by observer 1 (see group totals above), DA classified 22 of those to group 1 and 3 to group 2, 1 to group 3, etc. So, 39.25% of the observations of group 1 were classified correctly (=22/56).

	1	2	3	4	5	6	7	8
1	22	3	1	8	2	9	4	7
2	56	61	9	75	16	7	61	47
3	24	12	83	2	44	57	15	34
4	4	6	1	34	4	3	9	12
5	2	2	8	6	19	2	8	7
6	17	5	19	4	10	66	3	11
7	3	6	3	13	9	1	30	2
8	11	10	15	22	5	12	11	28

The percentages of correctly classified samples per group are as follows:

1	39.29
2	18.37
3	30.63
4	46.58
5	35.19

```
6   48.89
7   44.78
8   24.56
```

The problem with these numbers is that they were obtained using the same data that were used to create the classification rules. Tools exist to obtain more objective classification scores, and one option is a cross-validation process in which the data are split into two parts. The first dataset is used to derive classification rules and the second dataset for testing and obtaining classification scores (Tabachnick and Fidell 2001). Another option is the leave-one-out classification. An observation is left out, the classification rules are determined, and the left out observation is classified. This process is then applied on each observation in turn; see Huberty (1994) and references therein.

Several statistical programmes have routines for DA. Results obtained from these programmes can vary considerably due to different choices for scaling, standardisation, the centring of discriminant functions and the estimation method.

15 Principal coordinate analysis and non-metric multidimensional scaling

15.1 Principal coordinate analysis

In Chapter 12, principal component analysis (PCA) was introduced. The visual presentation of the PCA results is by plotting the axes (scores) in a graph. Some books use the phrase 'scores are plotted in a Euclidian space'. What this means is that the scores can be plotted in a Cartesian axes system, another notation is $\mathbb{R}^2$, and the Pythagoras theorem can be used to calculate distances between scores. The problem is that PCA is based on the correlation or covariance coefficient, and this may not always be the most appropriate measure of association. Principal coordinate analysis (PCoA) is a method that, just like PCA, is based on an eigenvalue equation, but it can use any measure of association (Chapter 10). Just like PCA, the axes are plotted against each other in a Euclidean space, but the PCoA does not produce a biplot (a joint plot of the variables and observations).

The aim of PCoA is to calculate a distance matrix and produce a graphical configuration in a low-dimensional (typically two or three) Euclidean space, such that the distances (as measured by the Pythagoras theorem) between the points in the configuration reflect the original distances as good as possible. The PCoA can be applied either on the variables or on the observations. Other names for PCoA are metric multidimensional scaling and classical scaling.

How does PCoA work?

To illustrate how PCoA works, we will use data from a plant study in Mexico. A detailed analysis of these data is presented in Chapter 32. The data consist of 200 observations (percentage cover) on a large number of plant families (32). In Chapter 32, totals per pastures (20 in total) are used, but here we will use the original 200 observations. Further details are given in Chapter 32. There are many observations equal to zero, and various species have a patchy distribution, which makes the correlation, covariance and Chi-square functions less appropriate tools to define association. Hence, we have a data matrix of dimension 200-by-32 with many zeros, and the question we focus on here is which families co-occur. The motivation for this question is that one of the plant families was introduced, and it may cause stress for the indigenous families. So, we need to calculate a distance matrix of dimension 32-by-32, which will tell us how dissimilar the families are, and we want a graphical representation of this to help with the interpretation. Because PCA and correspondence analysis are not suitable for these data (they are based on covariance, correlation or the Chi-square function), we use PCoA.

The first choice we have to make is how to define association between the families. In Chapter 10, and in Legendre and Legendre (1998, p. 294), it is argued that the Jaccard index can be used to measure 'co-occurrence of species'. We used the Jaccard index to measure the association between the families. There are 32 families and 200 sites; hence the similarity matrix **S** is of dimension 32-by-32. It is converted into a distance matrix **D** by $\mathbf{D} = 1 - \mathbf{S}$ (Thorrington-Smith 1971). In the next step of the algorithm, the ij^{th} element of **D** is transformed as follows:

$$A_{ij} = -\frac{1}{2} D_{ij}^2$$

This gives a matrix **A** where the ij^{th} element is calculated as above, and it has the same dimension as **D**. A second transformation is then applied on the elements of **A**:

$$E_{ij} = A_{ij} - \overline{A}_i - \overline{A}_j + \overline{A}$$

The notations $\overline{A}_i$, $\overline{A}_j$ and $\overline{A}$ stand for row, column and overall average. It can be shown that this transformation preserves the distances relationship between family i and j (Legendre and Legendre 1998, pg 430). In the last step, an eigenvalue equation is solved for **E,** and after applying the appropriate scaling on the eigenvectors, the eigenvector (or axes) can be plotted against each other in a two- or three-dimensional graph, just like in PCA. Distances between the points on this graph approximate the distances in **D**. An example is given in Figure 15.1-A. The first two axes explain 30% of the variation in the distance matrix (just as in PCA, the eigenvalues are used for this). Families close to each other co-occur at the same sites. Our special interest is on the family grcyn, and it co-occurs with ac, co and the group of families at the right side of the graph (ma, ru, eu, la, le, cy and grresto).

To assess how good the PCoA approximates the original distances in **D**, these distances can be plotted versus the ones obtained by PCoA. Such a graph is called a Shepard plot, and ideally, the points should lie on a straight line. If that is indeed the case, the PCoA distances are identical to the original ones. The Shepard plot in Figure 15.1-B shows there is a considerable mismatch for the larger distances. Increasing the number of axes (e.g., three) seems to be an appropriate step. Indeed this gives better results, but the axes are not presented here.

Occasionally, the PCoA may produce negative eigenvalues and solutions exist to solve this problem (Legendre and Legendre 1998). In most occasions this problem will not affect the first few axes used for plotting.

The results from the PCoA in Figure 15.1 indicate that more axes are needed; alternatively the method discussed in the next section can be applied, as it is more capable in reducing a high-dimensional space into a small number of axes.

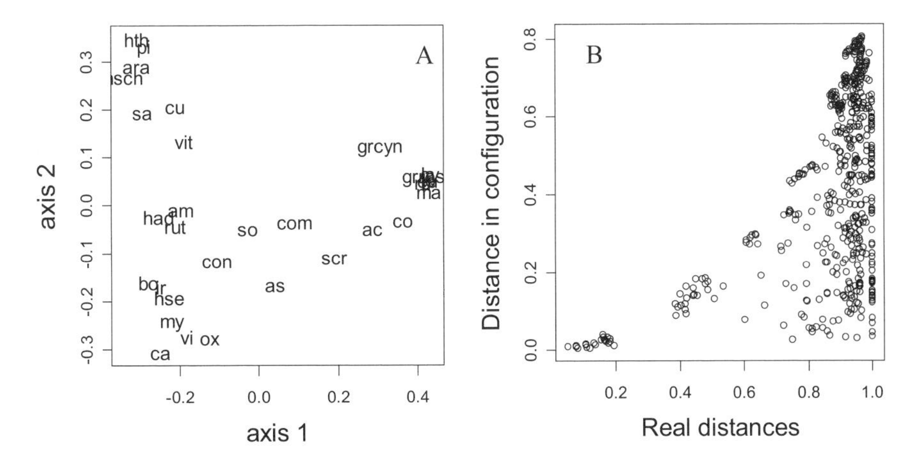

Figure 15.1. A: Results obtained by PCoA. Two axes were estimated. The first two eigenvalues are 2.68 and 0.97, which corresponds to 22% and 8% of the variation in the distance matrix, respectively. B: Shepard plot for the PCoA. The horizontal axis represents the original Jaccard distances between the families and the vertical axis contains the distances in the two-dimensional PCoA configuration. Ideally, the points should be on a straight line.

15.2 Non-metric multidimensional scaling

PCA, CA and PCoA are all methods that solve an eigenvalue equation. The advantage of PCoA above PCA is that any measure of association can be used; PCA is limited to the correlation and covariance coefficients. We will now discuss a method that has the same aim as PCoA, it can also use any measure of association, but it is better in preserving the high-dimensional structure with a few axes. It is called non-metric multidimensional scaling (NMDS). Its disadvantage is that it is not based on an eigenvalue solution but on numerical optimisation methods and for larger datasets the calculations tend to become time consuming, even on fast computers.

The aim of NMDS is to calculate a distance matrix **D** and visualise this matrix in a low (typically two- or three-) dimensional configuration. The difference between PCoA and NMDS is that in PCoA the distances in the configuration should match the original distances as closely as possible. In NMDS it is the order, or ranking, of the distances in **D** that we try to represent as closely as possible. As an example, Jaccard indices are given for four families in Table 15.1. The indices were converted to dissimilarities. PCoA uses this type of matrix to obtain the configuration such that Euclidian distances match the numbers in the table as closely as possible. In Table 15.2 we have converted the distances into ranks. As and Ac had the lowest dissimilarity (they were the most similar) and therefore have rank

1, etc. NMDS will produce a configuration that matches the data in Table 15.2 as closely as possible. Hence, points close to each other in the NMDS ordination diagram represent families that are more similar than others.

Table 15.1. Matrix with Jaccard dissimilarities among four families.

	Ac	Aam	Ara	As
Ac				
Am	0.949			
Ara	0.963	0.929		
As	0.642	0.952	0.962	

Table 15.2. Matrix with the dissimilarities from Table 15.1 transformed to ranks.

	Ac	am	ara	as
Ac				
Am	3			
Ara	6	2		
As	1	4	5	

Most books, including the text above, tend to say that 'NMDS is better in preserving relationships in a low dimensional space compared to PCoA'. In fact this is not a fair comment as we are comparing two different things: absolute distances and ranks. It is just that in most ecological studies, one is still content with the information that families A and B are more similar than C and D (NMDS), as to knowing that A and B are five times more similar than C and D (PCoA).
The NMDS ordination diagram for the Mexican plant species is given in Figure 15.2-A and the Shepard diagram in Figure 15.2-B. We used only two axes. The interpretation of the ordination diagram is simple; families close to each other are more similar than points far away from each other, but we do not know by how much. The only problem is then, what does similar mean? As we used the Jaccard index, it means that families close to each other in the graphical configuration coexist at the same sites. As the method is not based on an eigenvalue decomposition, the ordination diagram can be rotated and scaled. This does not affect its interpretation.

To give an impression of how NMDS works, the underlying mathematical algorithm is summarised next. Readers not interested in the underlying maths may skip this paragraph.

1. Choose a measure of association and calculate the distance matrix **D**.
2. Specify m, the number of axes.
3. Construct a starting configuration **E**. This can be done with PCoA.
4. Regress the configuration on **D**: $D_{ij} = \alpha + \beta E_{ij} + \varepsilon_{ij}$.
5. Measure the relationship between the m dimensional configuration and the real distances by fitting a non-parametric (monotonic) regression curve in

the Shepard diagram. A monotonic regression is constrained to increase. If a parametric regression line is used, we obtain PCoA.

6. The discrepancy from the fitted curve is called STRESS.
7. Using non-linear optimisation routines, obtain a new estimation of **E** and go to step 4 until convergence.

There are different ways to quantify STRESS in step 6, but all use quadratic sums of the original distances and those in the reduced space; see Legendre and Legendre (1998) for the exact formulations.

The NMDS algorithm is iterative, and for large datasets, different starting values might give different results. It may be an option to run the algorithm a couple of times with different starting values and see how the STRESS is changing.

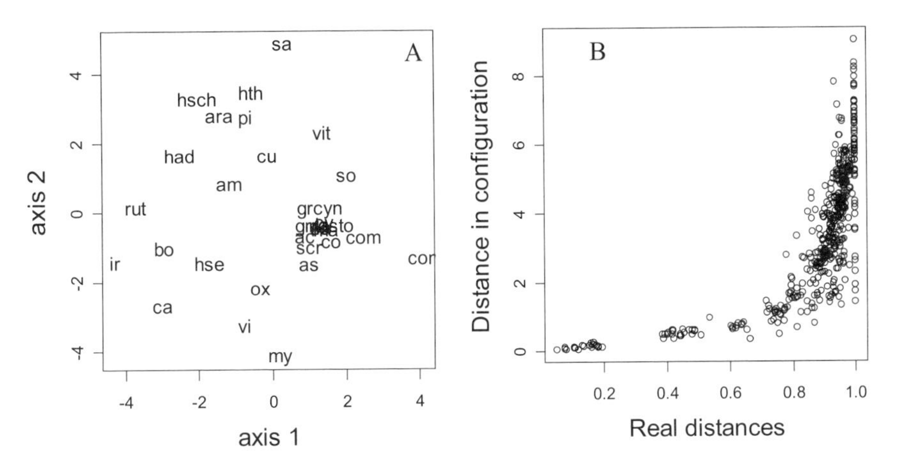

Figure 15.2. A: Results obtained by NMDS for the Mexican plant data. The STRESS is equal to 0.2336, and two axes were extracted. B: Shepard plot for the NMDS. The horizontal axis represents the Jaccard real distances between the families, and the vertical axis contains the distances obtained by the two NMDS configurations. The STRESS is 0.2336. If three axes are estimated, the STRESS is 0.1432.

Assessing the quality of the display

In PCA, CA and PCoA we have eigenvalues that can be used to assess the quality of the display. In NMDS we do not have eigenvalues and instead the STRESS is used to judge how good is the m-dimensional configuration. There are different ways to do this. One option is to calculate STRESS for different values of m (number of axes) and make a scree diagram, just as we did for the eigenvalues in PCA (Chapter 12). Along the x-axis we plot the number of axis m, and along the y-axis the STRESS. A clear change in stress (elbow effect) would indicate the optimal value of m. An alternative approach is to use the following rule of thumb (PRIMER manual):

- STRESS smaller than 0.05. The configuration is excellent and allows for a detailed inspection.
- STRESS between 0.05 and 0.1. Good configuration and no need to increase m.
- STRESS between 0.1 and 0.2. Be careful with the interpretation.
- STRESS between 0.2 and 0.3. Problems start, especially in the upper range of this interval.
- STRESS larger than 0.3. Poor presentation and consider increasing m.

Example

In Chapter 29, PCA is applied on fatty acid data measured in stranded dolphins in Scotland. There are 31 fatty acids (variables) and 89 dolphins (observations). Here, PCoA and NMDS are applied on the same data. In the first step of both methods, we calculate an appropriate distance matrix between the 89 dolphins. Both methods give a low-dimensional configuration of this matrix. Because the variation between the fatty acids differed (Chapter 29) we decided to normalise the fatty acid variables and use Euclidian distances between the dolphins. A PCoA gave exactly the same ordination diagram (not presented here) as the PCA (Chapter 29). Indeed, PCoA with the Euclidean distance function gives the same results as PCA. The NMDS ordination is given in Figure 15.3. The graph is similar (though not identical) to the PCoA and PCA ordination plots. The STRESS is 0.129, which means that the two-dimensional configuration is reasonable. Note the four observations in the upper left corner. An explanation of their separation is given in Chapter 29.

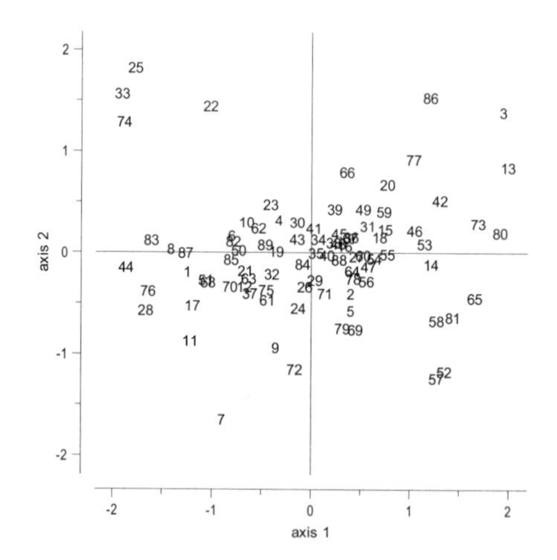

Figure 15.3. NMDS ordination diagram for the fatty acid dolphin data obtained with $k = 2$. The STRESS is 0.129. The numbers identify individual dolphins.

16 Time series analysis — Introduction

What makes a time series a time series? The answer to this question is simple, if a particular variable is measured repeatedly over time, we have a time series. It is a misconception to believe that most of the statistical methods discussed earlier in this book cannot be applied on time series. Provided the appropriate steps are made, one can easily apply linear regression or additive modelling on time series. The same holds for principal component analysis or redundancy analysis. The real problem is obtaining correct standard errors, t-values, p-values and F-statistics in linear regression (and related methods), and applying the appropriate permutation methods in RDA to obtain p-values. In this chapter, we show how to use some of the methods discussed earlier in this book. For example, generalised least squares (GLS) applied on time series data works like linear regression except that it takes into account auto-correlation structures in the data. We also discuss a standard time series method, namely auto-regressive integrated moving average models with exogenous variables (ARIMAX). In Chapter 17, more specialised methods to estimate common trends are introduced.

Questions in many time series analysis studies are as follows: what is going on, is there a trend, are there common trends, what is the influence of explanatory variables, do the time series interact, is there a sudden change in the time series, can we predict future values and are there cyclic patterns in the data? A series of different techniques are presented in this (and the next) chapter to address these questions.

16.1 Using what we have already seen before

Most of the material that has been presented in previous chapters can easily be adapted or modified so that it can be applied on time series. In Chapter 4, it was shown how time (season) can be used as a conditional variable in boxplots for the Argentinean zoobenthic data. In Chapter 4, we used the Gonadosomatic index (GSI) for a squid dataset. The boxplot in Figure 16.1 shows that there is a clear monthly pattern in the GSI index. Figure 16.2 shows how lattice graphs can be used to visualise a North American sea surface temperature (SST) time series. Mendelssohn and Schwing (2002) used the Comprehensive Ocean–Atmosphere Dataset (COADS) to generate monthly sea surface temperature time series from various grid points. COADS is a database containing data of oceanographic parameters taken mainly by 'ships-of-opportunity'. Mendelssohn and Schwing

(2002) defined 15 two-degree latitude by two-degree longitude regions 'based on a combination of ecological and oceanographic features, and data density'. The regions cover a large part of the American west coast. For each region, monthly SST time series were extracted from COADS covering the time period 1946 to 1990. The SST time series used in this book are the same as in Mendelssohn and Schwing (2002), and the interested reader is referred to their paper for further details. Figure 16.2 suggests the presence of a monthly effect, trends and differences in absolute values. Another way of visually inspecting the time series is plotting them all in one graph. Standardising the variables to ensure they all have the same range and mean value may help visual interpretation.

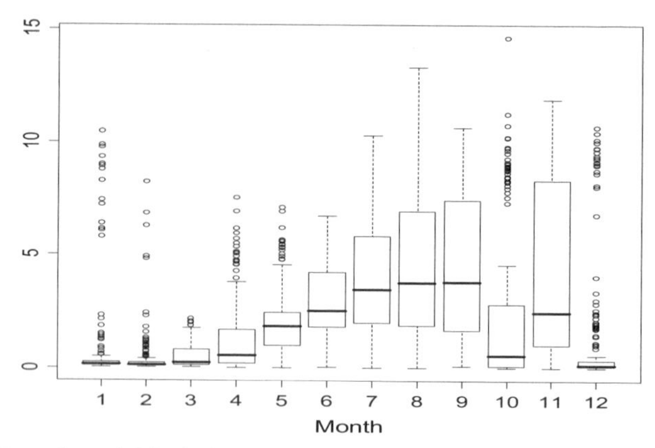

Figure 16.1. Boxplot of GSI index for squid data conditional on time. Months are represented by the numbers 1 (January) to 12 (December).

It is also interesting to enumerate all correlations or cross-correlations larger (in absolute sense) than a certain threshold value. Cross-correlations are correlations (Chapter 10) between two time series with a certain time lag k. How to calculate it, is discussed later in this section. Some statistical methods may produce errors if collinear variables are used. For the North American SST data, various correlations are higher than 0.9; see Table 16.1. The problem is that the high correlations are mainly due to the month effects in the time series. It may be useful to remove the month effect and look at correlations between the trends, or between deseasonalised time series (the difference between these two will be explained later in this chapter).

Figure 16.3 shows the annual time series of Nephrops catch per unit effort (CPUE) measured in 11 areas South of Iceland in the Atlantic Ocean between 1960 and 2002 (Eiríksson 1999). Most of the time series follow a similar pattern over time. The second most important step in a time series analysis is to look at auto-and cross-correlations. Carrying out this step, together with plotting the series, should give a first impression of what the data are showing.

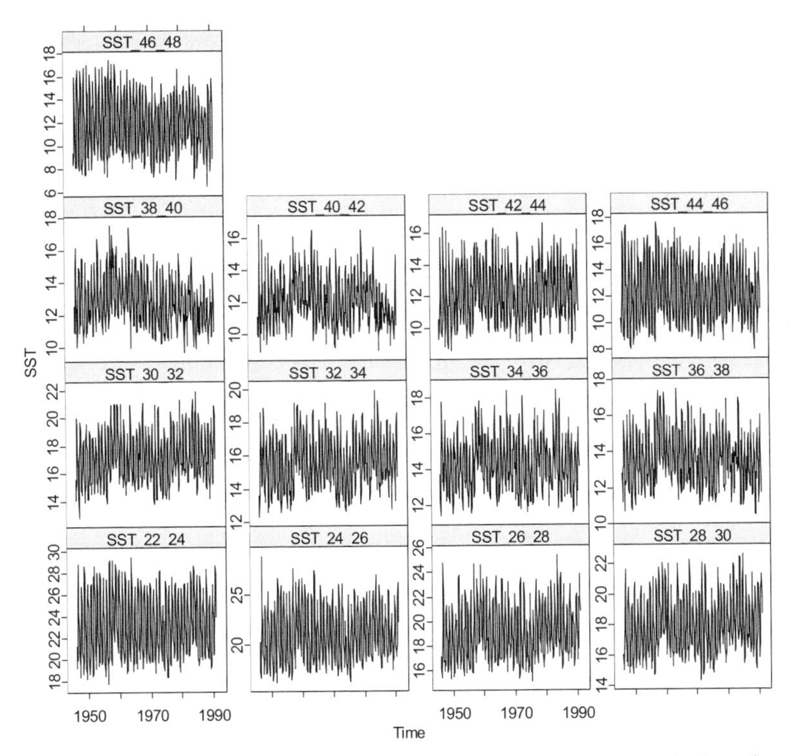

Figure 16.2. North American SST time series. The labels refer to the location.

Auto-correlation and cross-correlation

Define Y_t as the CPUE at a particular station, say station 1, in year t. The question we now address is whether there are any temporal relationships at this station. The auto-correlation function gives an indication of the amount of association between the variables Y_t and $Y_{t+k,}$ where the time lag k takes the values 1, 2, 3, etc. Hence, for k = 1, the auto-correlation shows the relationship between Nephrops in year t and year t + 1 at station 1. Data of all years are used to calculate this relationship. Formulated differently, the auto-correlation with a time lag of k years represents the overall association between time points that are k years separated.

The Pearson correlation coefficient (Chapter 10) is used to quantify this association. It is always between –1 and 1, and it is calculated (hence it is the sample auto-correlation) by (Chatfield 2003)

$$\hat{\rho}(k) = cor(Y_t, Y_{t+k}) = \frac{1}{N}\frac{\sum_{t=1}^{N-k}(Y_t - \bar{Y})(Y_{t+k} - \bar{Y})}{s_Y^2}$$

Table 16.1. Correlations larger (in absolute sense) than 0.9 for the North American SST data. The numbers 1 to 13 refer to SST-24-26, SST-26-28, SST-28-30, SST-30-32, SST-32-34, SST-34-36, SST-36-38, SST-38-40, SST-40-42, SST-42-44, SST-44-46 and SST-46-48, respectively. The numbers in the table are the variables and the associated correlation.

Variable 1			
	1	2	0.980
	1	3	0.946
	1	4	0.905
Variable 2			
	2	1	0.980
	2	3	0.973
	2	4	0.921
Variable 3			
	3	1	0.946
	3	2	0.973
	3	4	0.961
	3	5	0.915
Variable 4			
	4	1	0.905
	4	2	0.921
	4	3	0.961
	4	5	0.969
	4	6	0.932
Variable 5			
	5	3	0.915
	5	4	0.969
	5	6	0.941
Variable 6			
	6	4	0.932
	6	5	0.941
	6	7	0.931
Variable 7			
	7	6	0.931
	7	8	0.932
Variable 8			
	8	7	0.932
	8	9	0.930
Variable 9			
	9	8	0.930
Variable 11			
	11	12	0.941
	11	13	0.920
Variable 12			
	12	11	0.941
	12	13	0.967
Variable 13			
	13	11	0.920
	13	12	0.967

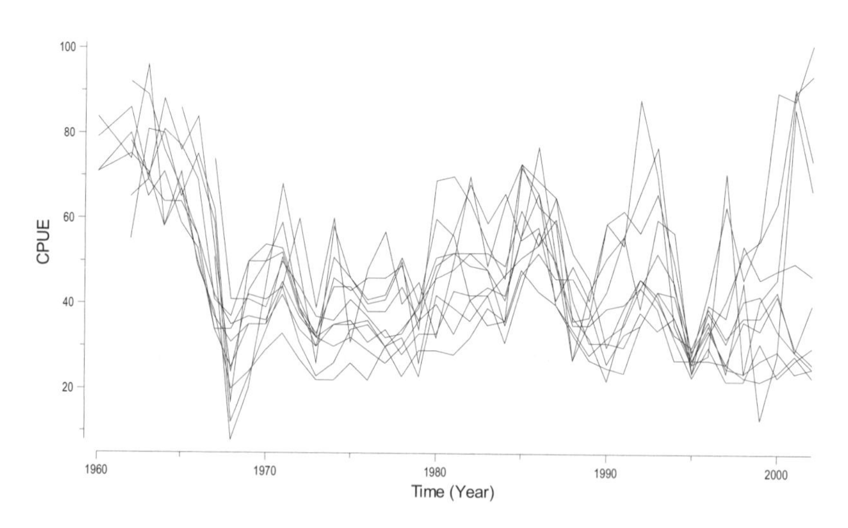

Figure 16.3. Nephrops CPUE time series.

The auto-correlation function is simply the Pearson correlation of a time series with itself, after applying a shift (lag) of k years. The term s_Y is the sample standard deviation of the time series Y_t. The notation $\hat{\rho}$ is used to indicate that we are working with the *sample* auto-correlation. One problem with the auto-correlation is that for larger time lags k, fewer points are used to calculate it. Therefore it is better to limit interpretation of the auto-correlation to the first few time lags only. But how many time lags is 'a few'? Most software packages produce time lags up to around 40% of the length of the time series ($0.4 \times N$). However, for some datasets this may already be too large. We advise to think first about the question 'what time lags would be sensible for the data?' before calculating it. Figure 16.4 shows the auto-correlation for the Nephrops CPUE time series at station 1. The horizontal axis shows time lags, and the corresponding correlation can be read off from the y-axis. The dotted lines can be used to test the null hypothesis that the correlation is equal to 0 for a certain time lag. These dotted lines are obtained from $\pm 1.96/\sqrt{N-k}$, where N is the length of a time series. Note that by chance alone, 1 out of 20 auto-correlations may be significant, whereas in reality they are not (type I error). For station 1 there is a significant auto-correlation with time lags of 1, 2 and 3 years. This means that if the CPUE is high in year t, it is also high in year $t-1$, $t-2$ and $t-3$.

The auto-correlation function is a useful tool to infer the pattern in the time series for which it was calculated. If a time series shows a strong seasonal pattern with a period of 12 months, then its values are likely to be high at time t, low at time $t+6$ and high again at time $t+12$. The same holds for $t+18$ and $t+24$, $t+30$, $t+48$, etc. The auto-correlation for such a time series shows large negative correlations for $k = 6$, $k = 18$, $k = 30$, etc., and large positive correlations for $k = 12$, $k = 24$, etc. Hence, a monthly pattern results in an oscillating auto-correlation function. The same holds for cyclic patterns. The distinction between a seasonal and cyclic component is that in the former we know the length of the cyclic period, whereas in the latter it is unknown. The auto-correlation function can then be used to get some idea on the length of the cycle. Another scenario is that the time series shows a slow-moving pattern. In this case, if Y_t is above the average, then so is Y_{t+1}. Hence, the auto-correlation with a time lag of $k = 1$ is large. The same holds for other small values of k. For larger values of k, one would expect that if Y_t is above average, then Y_{t+k} would be below average (or vice versa); hence we would obtain a negative correlation. Therefore, an auto-correlation that shows a slow decrease may indicate a trend. For example, Figure 16.4 is the auto-correlation function for the CPUE Nephrops time series at station 1. The shape of the function indicates the presence of a trend.

Now let us assume that Y_t and X_{t-k} are the CPUE values at station 1 in year t and at station 2 in year $t - k$ respectively. The question we address is whether there is any association between CPUE values at these two stations. Do the two stations have high CPUE at the same time, or has station 1 high CPUE values in year t, and station 2 in year $t - k$? Or is it just the other way around: Station 1 has high CPUE values in year t, and station 2 low values in year $t - k$? In this case, it makes sense to compare cross-correlations between Y_t and X_{t-k} for positive and negative values of k. But what if X_{t-k} is the temperature in year $t - k$? Obviously, there is not much point in comparing the CPUE in year t with temperature in year $t + 1$.

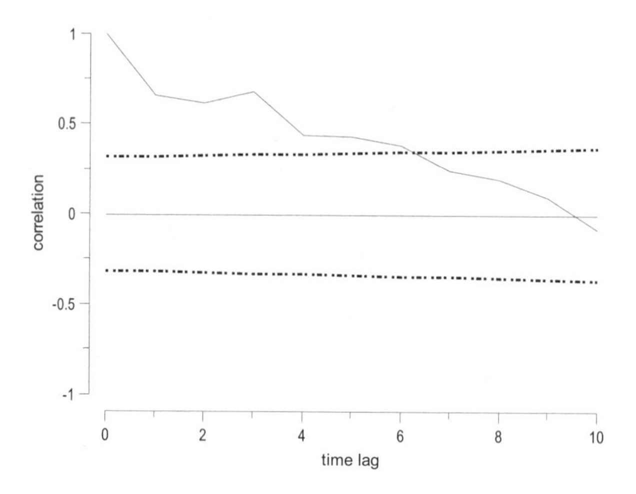

Figure 16.4. Auto-correlation for the CPUE Nephrops time series at station 1.

The cross-correlation function quantifies the association between two variables with a time lag of k years. It is again based on the Pearson correlation function, except that the second variable is shifted in time with a lag of k years. The (sample) cross-correlation is calculated by (Chatfield 2003; Diggle 1990)

$$\hat{\rho}_{YX}(k) = \begin{cases} \dfrac{1}{N} \dfrac{\sum_{t=1}^{N-k} (Y_t - \bar{Y})(X_{t+k} - \bar{X})}{s_Y s_X} & \text{if } k \geq 0 \\ \dfrac{1}{N} \dfrac{\sum_{t=1-k}^{N} (Y_t - \bar{Y})(X_{t+k} - \bar{X})}{s_Y s_X} & \text{if } k < 0 \end{cases}$$

This is the same mathematical formula as the auto-correlation except that the second bit of the formula now contains the variable X. The terms s_Y and s_X are sample standard deviations of the time series Y_t and X_t respectively. The cross-correlation can be calculated for various time lags, and the results can be plotted in a graph in which various time lags (positive and negative) are plotted along the horizontal axis and the correlations along the vertical axis. The cross-correlations for stations 1 and 2 are presented in Figure 16.5. The dotted lines can be used to test the null hypothesis that the correlation between Y_t and X_{t-k} is equal to 0. Points outside this interval are significant (at the 5% level) cross-correlations with time lag k. The same warning as for the auto-correlation holds: 1 out of 20 cross-correlations may be significant by chance only. There is a significant cross-correlation between the CPUE at station 1 and 2 at the time lags of −3, −2, .., 3.

This may indicate the presence of a slow-moving trend in both series (the same interpretation as for the auto-correlation function applies here). Table 16.2 shows the cross-correlations between CPUE time series of all 11 stations. Estimated cross-correlations in bold typeface are significantly different from 0 at the 5% significance level (Diggle 1990). Note that most correlations are significant!

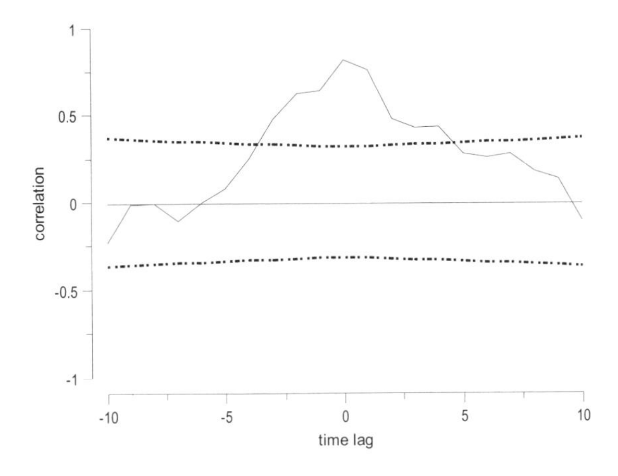

Figure 16.5. Cross-correlations for stations 1 and 2 plotted versus time lags for the CPUE Nephrops time series. Dotted lines represent the 95% upper and lower confidence bands.

Table 16.2. Correlations among 11 Nephrops time series. Values in bold typeface are significantly different from 0 at the 5% level. The numbers 1 to 11 correspond to the stations.

	2	3	4	5	6	7	8	9	10	11
1	**0.82**	**0.87**	0.29	**0.84**	**0.79**	**0.68**	**0.36**	0.28	0.08	0.27
2	1	**0.85**	**0.36**	**0.80**	**0.73**	**0.78**	**0.37**	**0.31**	0.04	0.22
3		1	**0.36**	**0.88**	**0.80**	**0.80**	**0.43**	**0.38**	0.12	0.26
4			1.00	**0.35**	**0.37**	0.22	0.09	0.14	0.11	0.24
5				1.00	**0.88**	**0.77**	**0.37**	**0.34**	0.04	0.24
6					1.00	**0.78**	**0.50**	**0.45**	0.12	**0.37**
7						1.00	**0.71**	**0.65**	0.22	**0.42**
8							1.00	**0.83**	**0.50**	**0.54**
9								1.00	**0.69**	**0.49**
10									1.00	**0.55**
11										1.00

It is also interesting to know at which time lag the maximum cross-correlation between two time series was obtained. This information is given in Table 16.3. The upper-diagonal shows the maximum cross-correlations between each of the two combinations of CPUE time series, and the lower diagonal elements show the corresponding time lag k. For example, the maximum cross-correlation between stations 1 and 2 was obtained at a time lag of 0 years. And the correlation between station 4 and station 7 had the highest value (in the absolute sense) for a time lag of $k = 2$. One can now look at the numbers in Table 16.2 (which are correlations) or Table 16.3 (which are cross-correlations) and try to find a pattern in them. Alternatively, one can visualise these association matrices using tools we have already discussed, for example multidimensional scaling (MDS). As we saw in Chapter 15, MDS can be used to graphically represent a matrix of dissimilarities. All we have to do is convert the measures of association (correlation) into a measure of dissimilarity; see Chapters 10 and 15. We leave it as an exercise for the reader to carry out this analysis.

Table 16.3. Maximum cross-correlations over time lags for the Neprophs time series data. The lower triangular part shows the time lags for which the maximum value was obtained. The numbers 1 to 11 correspond to the 11 stations. The range of the time lags was set to 25% of the length of the series.

	1	2	3	4	5	6	7	8	9	10	11
1		0.82	0.87	0.34	0.84	0.79	0.68	0.40	0.29	−0.29	0.27
2	0		0.85	0.42	0.80	0.73	0.78	0.37	0.31	−0.27	0.25
3	0	0		0.39	0.88	0.80	0.80	0.43	0.38	−0.30	0.32
4	4	−4	−1		0.41	0.44	0.45	0.31	0.36	0.25	0.56
5	0	0	0	1		0.88	0.77	0.39	0.34	−0.17	0.24
6	0	0	0	2	0		0.78	0.51	0.45	0.25	0.37
7	0	0	0	2	0	0		0.71	0.65	−0.25	0.42
8	−1	0	0	2	−1	−1	0		0.83	0.50	0.54
9	1	0	0	2	0	0	0	0		0.69	0.52
10	−3	−5	−4	1	−4	1	−4	0	0		0.55
11	0	−2	1	2	0	0	0	0	−1	0	

The portmanteau test

Instead of looking at individual time lags of the auto-correlation, it is also possible to apply a test that combines a number of time lags. The portmanteau test aggregates the first K time lags of the auto-correlation:

$$Q = N\sum_{j=1}^{K}\hat{\rho}(j)^2$$

where N is the length of the time series. If the first K (sample) auto-correlations are relatively small, Q will be small as well. Q is also called the Box–Pierce statistic (Box and Pierce 1970), and it follows a Chi-square distribution with K degrees of freedom. If the statistic is applied on residuals obtained by a model with p regression parameters, the degrees of freedom is $K - p$. As to the value of K, common values are $K = 15$, $K = 20$ and some packages give Q for a range of different K values.

A slightly 'better' statistic is the Ljung–Box (Ljung and Box 1978) statistic:

$$Q^* = N(N+2)\sum_{j=1}^{K} \frac{1}{N-j} \hat{\rho}(j)^2$$

Hence, this test takes the first K auto-correlations of the residuals, squares them, and adds them all up using weighting factors $T-j$. The statistic is typically presented for various values of K, and for $K > 15$ it can be shown that Q^* is Chi-square distributed with K degrees of freedom.

Large values of Q or Q^* are an indication that the time series (or residuals) are not white noise. White noise is defined as a variable that is normally distributed with mean 0 and variance 1. If Q is larger than the critical value, one rejects the null hypothesis that the time series (or residuals) are normally distributed. Alternatively, the p-value can be used.

The partial auto-correlation function

Suppose we have three variables Y, X and Z. In partial linear regression we removed the information of Z from both the Y and X variables and compared the residuals with each other (Chapter 5). With the auto-correlation we can do something similar. Suppose that Y_t and Y_{t-1} are highly correlated. As a result Y_{t-1} and Y_{t-2} are highly correlated as well. Because both Y_t and Y_{t-2} are highly correlated with Y_{t-1}, it is likely that Y_t and Y_{t-2} are correlated with each other! Would it not be nicer if we can calculate the correlation between Y_t and Y_{t-2} after removing the effect of Y_{t-1}? This is what the partial auto-correlation does (Makridakis et al. 1998). Technically, the partial auto-correlation α_k is obtained by applying the following linear regression and setting α_k equal to the estimated value of β_k.

$$Y_t = \text{intercept} + \beta_1 Y_{t-1} + \beta_2 Y_{t-2} + \ldots + \beta_k Y_{t-k}$$

To obtain the partial auto-correlation for another time lag, k is increased in the regression model. The first partial auto-correlation is always equal to the first ordinary auto-correlation, and the same critical values can be used to assess its significance. The partial auto-correlation turns out to be particular useful in some of the techniques that are discussed later in this chapter.

Another example

An SST time series was taken from the oceanographical database COADS, available on the Internet. In this database, one can select a grid point and obtain monthly SST data. It should be noted that these data should be interpreted with care as a certain amount of smoothing and interpolation is applied. We selected the grid point with coordinates 2.5E and 57.5N (as it is close to the home of the first author), and extracted monthly time series from January 1945 until December 1992. This grid point is located east of Scotland. Various studies have compared SST with the North Atlantic Oscillation (NAO) index, which can be seen as an environmental index function. Here, we will compare the SST with the monthly

NAO index. To visualise the two time series, we used a lattice plot (Figure 16.6) and a coplot (Figure 16.7). The lattice plot shows the general increase in the SST series since the late 1980s, and the coplot indicates that there is a strong relationship between the NAO index and SST during the months January, February and March.

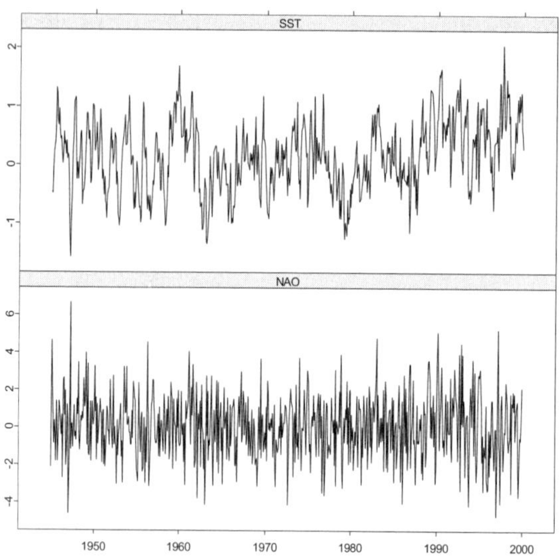

Figure 16.6. NAO and SST time series.

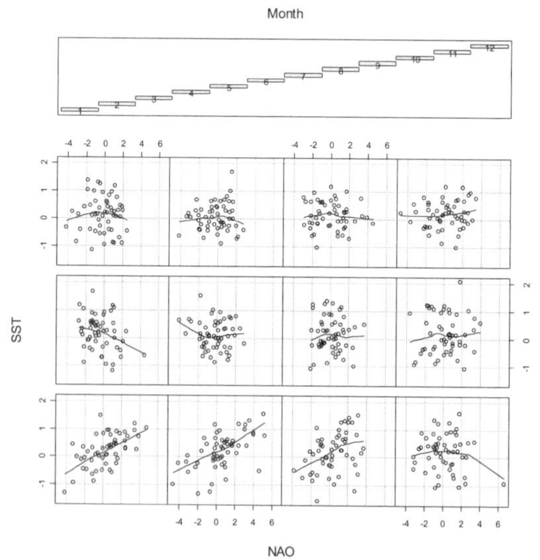

Figure 16.7. Coplot of SST and NAO index conditional on month. The lower left panel shows the relationship between the NAO and SST for all January data and

the upper right panel for the December data. A LOESS smoother was added in each panel.

The auto-correlation plot (not shown here) for the SST series shows a clear seasonal pattern. For the NAO index it is considerably weaker. However, we decided to remove the seasonal pattern from both series in order to avoid saying that they are highly related to each other just because of the seasonality. The process of removing the seasonal component is discussed later in this chapter. The deseasonalised SST and NAO series are plotted in Figure 16.8, and their cross-correlations are given in Figure 16.9. The latter graph indicates that the deseasonalised SST and NAO have similar patterns over time.

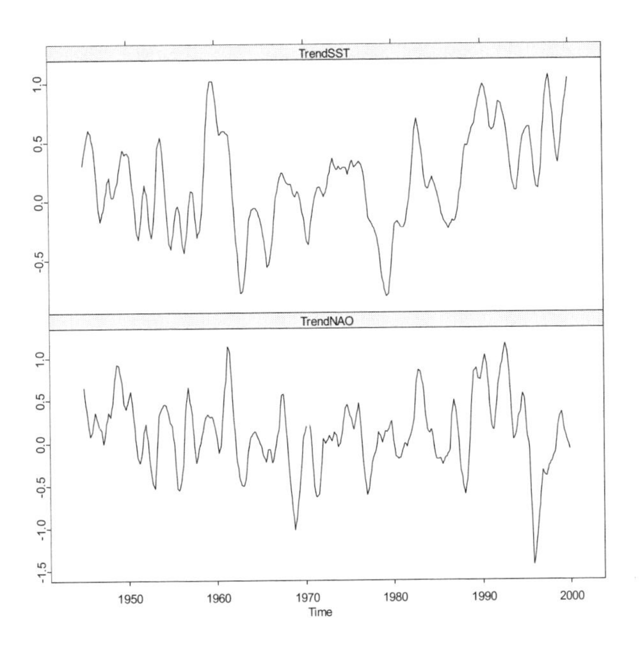

Figure 16.8. Deseasonalised SST and the NAO index.

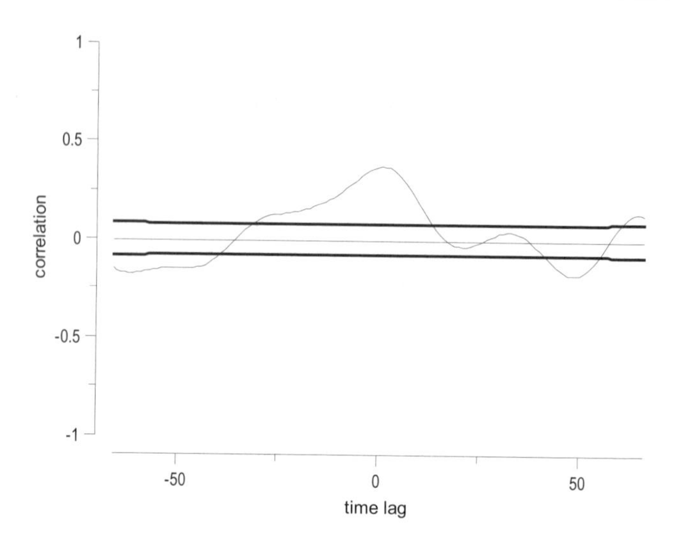

Figure 16.9. Cross-correlations between deseasonalised SST and NAO index.

Multivariate techniques

Figure 16.10 shows a time series plot of annual abundance indices derived from counts on six native duck species wintering on Scottish wetland sites (Musgrove et al. 2002). The data are expressed as a percentage of the abundance in the first year (1996 =100); see http://www.scotland.gov.uk/stats/envonline for further details and the data (source: The Wetland Bird Survey (WeBS)). Suppose that one is interested in determining which of the series are related to each other. One option is to calculate the cross-correlation matrix and inspect the numbers. However, it is also possible to apply a principal component analysis on these time series. Figure 16.11 shows the PCA correlation biplot (Chapter 12) obtained for the six time series. To simplify the interpretation, we used labels 6, 7, 8 and 9 for observations from the 1960s, 1970s, 1980s and 1990s, respectively, and 0 refers to 2000. The biplot explains 72% of the variation and shows that Pochards were abundant in the 1970s, Mallards in the 1960s, and Goosanders, Goldeneyes and Gadalls in the 1990s. All these results are in line with the time series plot (Figure 16.10). If one also has multiple explanatory variables, then redundancy analysis could be applied. To account for the time series aspect in the data we need only modify the permutation methods that we use to obtain the p-values that determine the significance of explanatory variables. Permutation techniques in RDA (or CCA) could make use of block permutation provided the series are long enough (Efron and Tibshiran 1993; Davison and Hinkley 1997; Lepš and Šmilauer 2003).

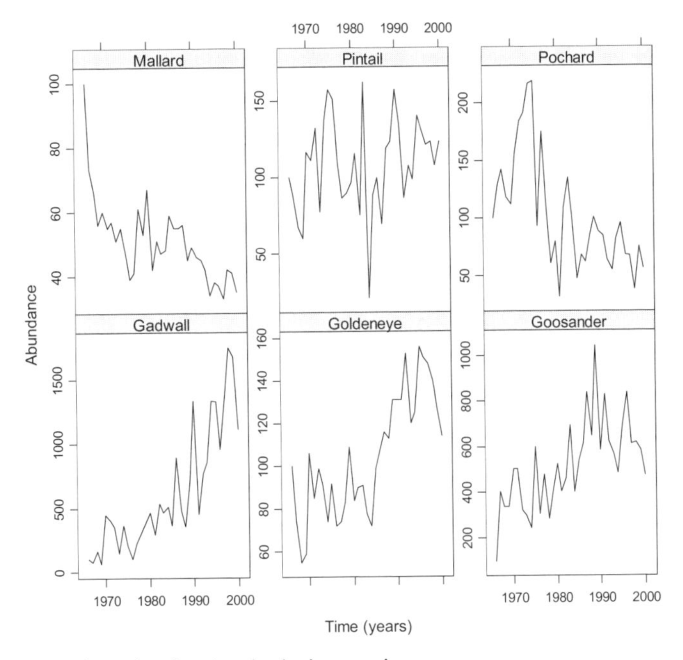

Figure 16.10. Lattice plot for the duck time series.

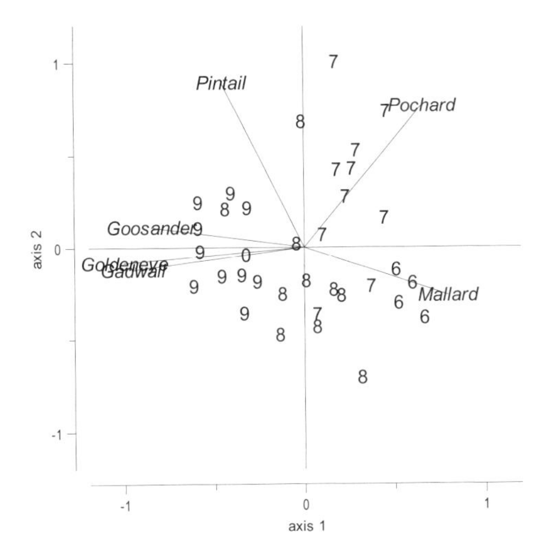

Figure 16.11. PCA biplot on six duck time series. The first two eigenvalues are 0.52 and 0.20, which means that the two axes explain 72% of the variation in the data (the eigenvalues are scaled to have sum 1). The labels 6, 7, 8 and 9 refer to observations from the 1960s, 1970s, 1980s and 1990s respectively, and 0 to 2000.

Generalised least squares

So far, we have not done anything new but merely applied tools we have discussed before. We now present the first extension. Recall that one of the assumptions in the linear regression model was independence of the data. This works it way through as independence of the residuals. For time series data, this assumption is clearly violated. It is relatively easy to show that if there is a temporal structure in the errors, then the linear regression model can seriously underestimate the standard errors of the slopes (Ostrom 1990) and this can lead to all kinds of trouble, including type I errors. So, how do we avoid this? We discuss two options. The first one is simple, add covariates such that there is no auto-correlation between residuals. For example, month can be used as a nominal explanatory variable in linear regression or as a smoother in an additive model. However, during the model validation process one must ensure that the residuals do not exhibit any temporal structure or else the model is incorrect. Chapter 20 shows an example of this approach. The second option is to extend the model in such a way that auto-correlation between residuals is allowed (Chapter 8). Recall that the underlying model in linear regression is (in matrix notation)

$$\mathbf{Y} = \mathbf{X}\boldsymbol{\beta} + \boldsymbol{\varepsilon}, \qquad \text{where} \quad \boldsymbol{\varepsilon} \sim N(0,\sigma^2\mathbf{I}) \quad \text{or more generally} \quad \boldsymbol{\varepsilon} \sim N(0,\sigma^2\mathbf{V})$$

where $\mathbf{V}$ is a diagonal matrix with only ones. We formulated the underlying assumptions in terms of the underlying response variable Y, but as one of the assumptions states that the explanatory variables are fixed, the assumptions can also be expressed in terms of the noise. The independence assumption means that the noise at one time should not be related to noise at any other time.

The matrix $\mathbf{V}$ is the key to extending the linear regression (or additive) model so that it allows an auto-correlation structure between the residuals. Let us have a more detailed look at the structure of $\mathbf{V}$. In linear regression it assumed to be the identity matrix:

$$\mathbf{V} = \begin{pmatrix} 1 & 0 & \cdots & 0 \\ 0 & 1 & \vdots & \vdots \\ \vdots & \vdots & \ddots & \vdots \\ 0 & \cdots & \cdots & 1 \end{pmatrix}$$

The observed data are stored in a vector $\mathbf{Y}$ of the form $\mathbf{Y} = (Y_1,..,Y_N)'$ and the same holds for the residuals: $\boldsymbol{\varepsilon}=(\varepsilon_1,\ldots,\varepsilon_N)'$. As the variance of $\boldsymbol{\varepsilon}$ is equal to $\sigma^2\mathbf{V}$, we have:

$$\text{cov}(\varepsilon_i,\varepsilon_j) = \begin{cases} 0 \text{ if i} \neq \text{j} \\ \sigma^2 \text{ if i} = \text{j} \end{cases}$$

Hence, by using the identity matrix $\mathbf{V}$ we are forcing an independence structure on the residuals. This is fine as long as the residuals are indeed uncorrelated. Ostrom (1990) gives a simple example in which linear regression is applied on a US defence expense time series. Comparing the linear regression model with a model that takes into account auto-correlation between residuals, he showed that the *t*-values obtained by linear regression were inflated by approximately 400%! Therefore, we better make sure that we do something appropriate with the time dependence structure if it is indeed present! The matrix **V** can be used for this. Although we will use it for time series applications, we might as well discuss it in a wider context. For example, one option is allowing groups of residuals to have a different variance. This is one way to tackle violation of homogeneity in linear regression. Suppose that we are trying to model (using linear regression) the relationship between weight and length of any species with a remarkable sexual dimorphism. A model validation may indicate that residuals of the males have a larger spread of residuals compared with the females (or vice versa), which is a violation of the homogeneity assumption. Let us assume that the first *k* observations in **Y** are of males and the remaining observations are of females. Instead of applying a data transformation to stabilise the variance, we could introduce two different variance components, one for the males and one for the females. Technically, this is done by using a diagonal matrix **V** with different values on the diagonal:

$$\mathbf{V} = \begin{pmatrix} v_1 & 0 & \cdots & 0 \\ 0 & v_2 & \vdots & \vdots \\ \vdots & \vdots & \ddots & \vdots \\ 0 & \cdots & \cdots & v_N \end{pmatrix} \tag{16.1}$$

In the example of the male and female we would have: $v_{male} = v_1 = v_2 = v_3 = \ldots = v_k$ and $v_{female} = v_{k+1} = v_{k+2} = \ldots = v_N$. As a result we have:

$$\text{cov}(\varepsilon_i, \varepsilon_j) = \begin{cases} 0 & \text{if } i \neq j \\ v_{male}\sigma^2 & \text{if } i = j \quad \text{and both are males} \\ v_{female}\sigma^2 & \text{if } i = j \quad \text{and both are females} \end{cases}$$

Hence, there is no covariance between the different sexes, but males and females are each allowed to have a different variance. Specialised computer software can be used to estimate the variance components. Now suppose that sampling effort was different for the *N* observations. A possible scenario is if monthly averages are used but the number of observations per month differs due to a lower sampling effort during certain months (e.g., high sampling effort during spring and low during the winter). In this case, we know the sampling effort per observation and we can give more weight to observations with higher sampling effort by using the sampling effort as a weighting factor. Technically, the matrix **V** in equation (16.1) is used and the *v*'s are set to the (known) weighting factors.

Now let us assume that the vector Y contains observations made repeatedly over time. In none of these extensions have we allowed for covariance between two observations. In the context of a time series, this means that $cov(Y_t, Y_{t+k}) = 0$ for $k \neq 0$. Now suppose that Y_t and Y_{t+1} are related to each other. We assume that the relationship between two observations that are one time unit apart is the same;

Y_1 and Y_2, Y_2 and Y_3, Y_3 and Y_4, etc. The same holds for two observations that are two units apart: Y_1 and Y_3, and Y_2 and Y_4, etc. This dependence structure implies that the covariance between ε_t and ε_{t+k} is given by

$$\text{cov}(\varepsilon_t, \varepsilon_{t+k}) = \begin{cases} \sigma^2 & \text{if } k = 0 \\ v_k \sigma^2 & \text{if } k \neq 0 \end{cases}$$

For example, the covariance for $k = 1$ is $cov(\varepsilon_t,\varepsilon_{t+1}) = v_1\sigma^2$, for $k = 2$ we have $cov(\varepsilon_t,\varepsilon_{t+2}) = v_2\sigma^2$, etc. So, all what we need is an estimate for v_1, v_2 and σ^2. It is straightforward to implement such an error covariance structure using a (positive definite) matrix $\mathbf{V}$ of the form

$$\mathbf{V} = \begin{pmatrix} 1 & v_1 & v_2 & \cdots & \cdots & v_{N-1} \\ v_1 & 1 & v_1 & \cdots & \cdots & v_{N-2} \\ v_2 & v_1 & 1 & & & v_{N-3} \\ \vdots & v_2 & v_1 & \ddots & & \vdots \\ \vdots & \vdots & \vdots & & \ddots & v_1 \\ v_{N-1} & v_{N-2} & v_{N-3} & \cdots & \cdots & 1 \end{pmatrix}$$

The problem with this approach is that there are many elements in $\mathbf{V}$ to estimate. A common approach is to assume that the covariance between observations Y_t and Y_{t+k} only depends on the time lag between them; observations made close after each other have a much higher covariance than points separated more in time. This can be modelled as

$$\text{cov}(\varepsilon_t, \varepsilon_{t+k}) = v^{|k|} \sigma^2$$

Where v is between 0 and 1. The larger the time lag k, the smaller the covariance. As an example, assume that $v = 0.5$:

$$\begin{aligned} \text{cov}(\varepsilon_t, \varepsilon_{t+0}) &= v^0 \sigma^2 = \sigma^2 \\ \text{cov}(\varepsilon_t, \varepsilon_{t+1}) &= v^1 \sigma^2 = 0.5\sigma^2 \\ \text{cov}(\varepsilon_t, \varepsilon_{t+2}) &= v^2 \sigma^2 = 0.25\sigma^2 \end{aligned}$$

Points close to each other have a much higher covariance than points with a larger time lag. This covariance structure can be modelled using a matrix $\mathbf{V}$ of the form

$$\mathbf{V} = \begin{pmatrix} 1 & v & v^2 & \cdots & \cdots & v^{N-1} \\ v & 1 & v & \cdots & \cdots & v^{N-2} \\ v^2 & v & 1 & & & v^{N-3} \\ \vdots & v^2 & v & \ddots & & \vdots \\ \vdots & \vdots & \vdots & & \ddots & v \\ v^{N-1} & v^{N-2} & v^{N-3} & \cdots & \cdots & 1 \end{pmatrix}$$

All we need is some clever software that estimates the value of v, together with σ, α and β. Hence, if we apply linear regression and use this matrix $\mathbf{V}$, we are allowing for auto-correlation in the time series. Other correlation structures are possible (Pinheiro and Bates 2000). The auto-correlation structure can also be combined with random intercept and slope models within a mixed modelling context (Chapter 8). Generalised least squares examples can be found in Chapters 18, 19, 23, 26, 35 and 37.

16.2 Auto-regressive integrated moving average models with exogenous variables

Now that we have applied data exploration techniques, and auto-and cross-correlation functions, it is time for more advanced time series techniques. One of these methods is the auto-regressive integrated moving average model with exogenous variables, abbreviated as ARIMAX. Useful reading sources are Ljung (1987), Diggle (1990), Chatfield (2003), Brockwell and Davis (2002), among others. The text and ideas presented here owe much to Makridakis et al. (1998). The ARIMAX framework is based on the assumption that the time series is stationary, and therefore we first need to discuss this very important issue. ARIMAX consists of various building blocks, and it is perhaps easier to break them down in more simple models, namely the auto-regressive (AR) model, followed by the moving average models before dealing with the ARIMAX model itself.

Stationarity

So, what is stationarity? Well, roughly speaking it means that the time series Y_t does not contain a trend and the variation is approximately the same during the entire time span. We never said that ARIMAX models were useful for estimating trends! In fact, they cannot be used for this purpose! Stationarity is best assessed by making a time plot of the time series; it should fluctuate around a constant value, and the spread should be the same everywhere. None of the SST time series in Figure 16.2 are stationary as each series contains a seasonal component and some exhibit a long-term trend. None of the CPUE series in Figure 16.3 are stationary as they all contain a trend and variation during the 1970s is much lower for some time series compared with other periods. Neither the SST series nor the

NAO index in Figure 16.6 is stationary; the first series shows a clear trend and seasonal effect, and the NAO has changes in spread. None of the duck time series in Figure 16.10 are stationary as all contain either a trend or variation in the spread over time.

The auto-correlation can also be used to assess whether a time series is stationary. The auto-correlation function of a stationary series should drop to small values reasonably quickly. Slowly decaying auto-correlation functions (Figure 16.4) are an indication of non-stationarity, and the same holds for oscillating ones (seasonality). The partial auto-correlation function of non-stationary data tends to show spiky behaviour.

So, before moving on to ARIMAX models, we need to discuss how to get rid of non-stationarity. One option is to remove the trend and/or seasonal components and this will be discussed in Chapter 17. Within the world of ARIMAX models one often applies a different approach, namely differencing the time series. The first order difference is defined by

$$Y_t' = Y_t - Y_{t-1}$$

Note that because Y_0 does not exist, the length of the differenced series is one unit less than the length of the original series. Figure 16.12-A shows the CPUE Nephrops time series at station 3; the series is clearly non-stationary as it contains a trend and changes in spread. The time series containing the first differences is presented in Figure 16.12-B and is much closer to being stationary.

If the time series with the first differences is not stationary, the time series can be differenced a second time.

$$Y_t'' = (Y_t - Y_{t-1})' = Y_t' - Y_{t-1}' = Y_t - 2Y_{t-1} + Y_{t-2}$$

In practise, one rarely applies higher order differences as the interpretation becomes rather difficult.

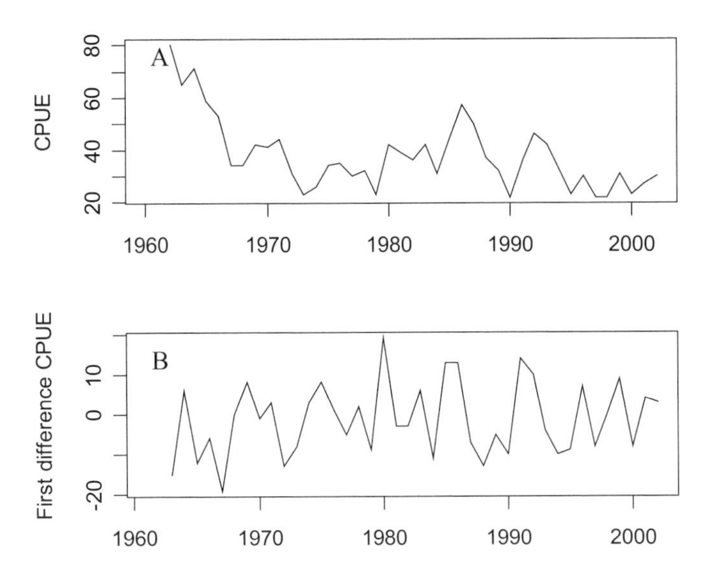

Figure 16.12. A: Non-stationary CPUE Nephrops time series at station 3 south of Iceland. B: First difference of CPUE series at station 3.

So, what do we do about monthly data? We can either analyse the first-order difference or look at seasonal differences:

$$Y_t^{'} = Y_t - Y_{t-12}$$

This series shows the change between the same months in consecutive years. The same can be done for quarterly or even weekly data. As before, if the seasonal differenced series is not stationary, the time series can be differenced again to give

$$Y_t^{''} = (Y_t - Y_{t-1})' = Y_t^{'} - Y_{t-1}^{'} = Y_t - 2Y_{t-1}^{'} + Y_{t-2}$$

It is also possible to difference the other way around (giving $Y_t - 2Y_{t-12} + Y_{t-24}$), but this makes the interpretation rather difficult. Figure 16.13-A and Figure 16.13-B show the time series of the Scottish SST and the auto-correlation, respectively. Both indicate clear non-stationarity. Panels C and D show the same graphs but now for seasonal differenced data. There is still strong evidence for non-stationarity. Panels E and F contain the first differences of the seasonal differenced series, and although there is still some auto-correlation with time lag 12, it is much closer to stationarity now.

One may also consider applying a transformation on the time series to stabilise the variance, if this is causing the non-stationarity.

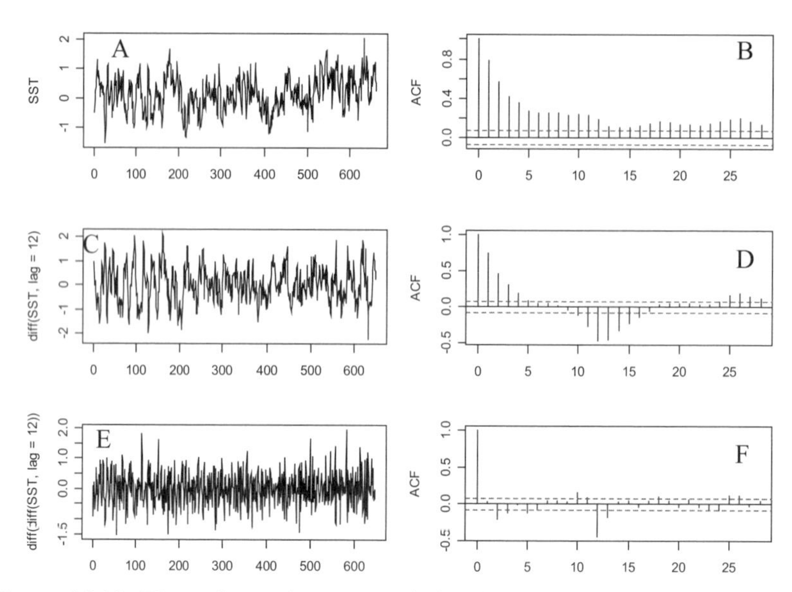

Figure 16.13. Time plot and auto-correlation for the Scottish SST series for the original data (A and B), seasonal differenced series (C and D) and first differences of the seasonal differenced series (E an F).

The AR model

In this class of models, the *stationary* time series Y_t is modelled as a function of past observations Y_{t-1}, Y_{t-2},...,Y_{t-p}, and a noise components e_t. Stationarity is rather important, and, techniques discussed in the previous paragraph should be applied if the series is non-stationary. Let us formulate an AR model for the SST time series northeast of Scotland:

$$Y_t = \alpha + \beta_1 Y_{t-1} + \beta_2 Y_{t-2} + \beta_3 Y_{t-3} + ... + \beta_p Y_{t-p} + \varepsilon_t$$

To ensure stationarity of the time series Y_t, we took the first-order differenced seasonal differences (see previous paragraph and Figure 16.13). The term ε_t is normally distributed noise with mean 0 and variance σ^2, and α is the intercept. The model is called an AR model of order p, also written as AR(p). Just as in linear regression we have an intercept and explanatory variables with regression slopes β_j, but the explanatory variables are now lagged response variables! The question that arises is why we are allowed to do this, and how can we obtain 'good' standard errors, t-values and p-values for the estimated regression parameters? What is the mechanism for this? Before addressing these issues, have a look at the structure of the model first. If we know the values of the intercept and all p slopes, we can easily predict the value for Y_t, Y_{t+1},Y_{t+2}, etc. Hence, for prediction purposes, this is a fantastic tool. But it is perhaps less suitable for the 'what is going on' question.

So, why are we allowed to include lagged response variables and obtain valid standard errors and *p*-values? The answer is stationarity and *asymptotic normality*. The last point is easy; just make sure you have at least 25 to 30 observations in time and then the central limit theory ensures that the normal distribution can be used to say that, in 95% of the cases, the population parameter β_j is within the interval given by the estimated slope $b_j \pm 1.96$ times its standard error. If the dataset is smaller, then do not apply ARIMAX or any of its subset models. Stationarity was discussed earlier in this section.

Let us work out the example for the SST series. In fact, this is not an AR(p) model, but an ARI(*p*) model as the series are integrated. The first question we have to address is how many lagged explanatory variables to take, or formulated differently, what is the order of the AR model? There are a couple tools we can use. First of all, based on theory, the partial auto-correlation of an AR(*p*) model will show spikes up from time lag 1 to *p*, and then drops to 0. So, we could calculate the partial auto-correlation function for the SST series and see when it drops to 0. Another tool is our best friend from Chapters 5 to 8, the AIC. It is defined in a similar way as in the linear regression model (a function of the maximum likelihood and the number of parameters). It can be calculated for different values of *p*, and the model with the smallest AIC can be selected as the optimal model. Just as in linear regression, the model selection procedure needs to be followed up by a model validation. In this process, the residuals need to be inspected and one should not be able to detect any patterns in it.

Table 16.4 shows the AIC values obtained by applying an AR(*p*) model on the differenced seasonal differenced Scottish SST data for various values of *p*. As can be seen, the larger the number of auto-regressive terms, the lower the AIC. In this case, one should use at least 14 AR terms. We did not try models with higher values of *p* as computing time becomes an issue. Instead of fully exploring the AR(14) model, we will extend the AR to ARMA models in the next two paragraphs.

Table 16.4. AIC for various values of *p* in an AR model for the differenced seasonal differenced Scottish SST time series.

p	AIC	p	AIC
1	1044.27	8	993.05
2	1016.95	9	995.03
3	1011.87	10	978.03
4	1012.62	11	972.78
5	994.86	12	851.29
6	989.93	13	844.88
7	991.56	14	834.39

The MA model

In an moving average model the *stationary* time series Y_t is modelled as a function of present and past error terms ε_t, ε_{t-1}, $\varepsilon_{t\ t-2}, \ldots, \varepsilon_{t\ t-q}$. A possible MA model for the SST time series northeast of Scotland is:

$$Y_t = \alpha + \varepsilon_t + \phi_1 \varepsilon_{t-1} + \phi_2 \varepsilon_{t-2} + \phi_3 \varepsilon_{t-3} + \ldots + \phi_q \varepsilon_{t-q}$$

Just as in the AR(p) model we need to estimate the optimal number of MA parameters. The auto- and partial auto-correlation functions give some clues: The auto-correlation of an MA(q) model has spikes at lags up to q and then goes to zero, whereas the partial auto-correlation may show exponential decay or damped sine wave patterns (Makridakis et al. 1998). We could produce a similar table as for the AR(p) model in Table 16.4 but leave this as an exercise for the interested reader.

The ARIMA model

The more useful approach is the combination of the AR(p) and MA(q) model. It is of the form:

$$Y_t = \alpha + \beta_1 Y_{t-1} + \ldots + \beta_p Y_{t-p} + \varepsilon_t + \phi_1 \varepsilon_{t-1} + \ldots + \phi_q \varepsilon_{t-q}$$

This model is called an ARMA(p,q) model, and if the series are differenced, we call it an ARIMA(p,q) model and the challenge is to find the optimal values of p and q. Table 16.5 shows the AIC for various values of p and q for the Scottish SST series. We took more values of p than of q as the ecological interpretation of lagged error terms is rather difficult. The ARIMA(14,3) model has the lowest AIC. The numerical output for this model is given in Table 16.6. Note that the first auto-regressive parameter is relatively large, which may indicate that the integrated seasonal differenced series is still non-stationary. The problem with most software routines is that if one specifies an ARIMA(14,3) model, the software will estimate all time lags from 1 to 14. Based on the standard errors in Table 16.6 it may be better to omit some of the AR components as they are not significant at the 5% level (the 95% confidence band for each parameter is given by the estimated value ± 1.96 times the standard error).

Table 16.5. AIC for various values of p and q in an ARIMA(p,q) model for the differenced seasonal differenced Scottish SST time series. The five models with the lowest AIC are in bold face.

	q		
p	1	2	3
1	1041.008	940.919	942.919
2	943.294	942.919	943.635
3	942.523	944.463	894.725
4	944.315	945.794	918.098
5	990.200	943.168	894.037
6	991.451	951.403	888.152
7	993.443	924.387	936.914
8	995.047	919.169	887.076
9	995.120	932.378	882.649
10	978.820	910.351	911.431
11	932.888	907.887	897.664
12	814.875	799.124	787.017
13	**781.672**	**783.672**	**782.750**
14	**783.659**	785.665	**772.192**

Table 16.6. Estimated parameters and standard errors for the ARIMA(14,3) model for the Scottish SST series.

AR	Estimate	S.E.	AR	Estimate	S.E.	MA	Estimate	S.E.
1	−0.973	0.055	8	0.032	0.051	1	0.890	0.050
2	0.039	0.077	9	−0.009	0.048	2	−0.459	0.085
3	0.326	0.058	10	0.006	0.048	3	−0.842	0.049
4	−0.078	0.050	11	0.076	0.047			
5	0.007	0.050	12	−0.386	0.046			
6	− 0.133	0.049	13	−0.632	0.056			
7	− 0.100	0.052	4	−0.264	0.044			

Extending the ARIMA to ARIMAX models

So far, we have ignored the explanatory variables. It is relatively simple to extend the ARIMA with explanatory variables. For Scottish SST data we also have the NAO index, which could be a driving factor for sea surface temperature. A possible ARMAX model is of the form:

$$Y_t = \alpha + \beta_1 Y_{t-1} + \ldots + \beta_p Y_{t-p} + \varepsilon_t + \phi_1 \varepsilon_{t-1} + \ldots + \phi_q \varepsilon_{t-q} + \gamma \text{NAO}_t$$

where Y_t is the SST in month t. The underlying assumption is again that the series Y_t is stationary. If this is not the case, there are two options. The first option is to take the differences of both the Y_t and the explanatory variable NAO_t until the Y_t is stationary. For this specific example, we would take the integrated seasonal differences for both series. The second option is to consider the model as

$$Y_t = \alpha + \beta_1 Y_{t-1} + \ldots + \beta_p Y_{t-p} + \gamma \text{NAO}_t + N_t$$
$$N_t = \varepsilon_t + \phi_1 \varepsilon_{t-1} + \ldots + \phi_q \varepsilon_{t-q}$$

And assume that the noise series ε_t is stationary. This brings us back to the world of generalised least squares. There is no differencing of the Y_t or NAO_t series involved. In the Scottish SST data, it may be an option to add a nominal variable month to the model:

$$Y_t = \alpha + \beta_1 Y_{t-1} + \ldots + \beta_p Y_{t-p} + \gamma \text{NAO}_t + \text{factor(Month)} + N_t$$
$$N_t = \varepsilon_t + \phi_1 \varepsilon_{t-1} + \ldots + \phi_q \varepsilon_{t-q}$$

The component 'factor(Month)' creates 11 dummy variables with zeros and ones identifying in which month a measurement was taken. This process is identical to linear regression (Chapter 5). Things can be made even more complex if we have models with lagged explanatory variables. An example is given by

$$Y_t = \alpha + \beta_1 Y_{t-1} + \ldots + \beta_p Y_{t-p} + \gamma_1 \text{NAO}_t + \gamma_2 \text{NAO}_{t-1} + \gamma_3 \text{NAO}_{t-2} + N_t$$
$$N_t = \varepsilon_t + \phi_1 \varepsilon_{t-1} + \ldots + \phi_q \varepsilon_{t-q}$$

The SST in year t is modelled as a function of SST in the past (the AR components), the NAO, the NAO in the past (lagged terms), noise, and noise from the past. Indeed, a complicated model, but we never said that it would be easy. This set of models is called a dynamic regression model, and a more detailed discussion can be found in Chapter 8 of Makridakis et al. (1998). It is also possible to use models in which the regression parameters are allowed to change over time, and these will be discussed in Chapter 17.

ARIMAX models are useful for prediction but not for understanding what goes on. This statement holds especially if one starts to take differences. If ARIMAX models are applied on non-stationary data, standard errors of estimated parameters should be interpreted with great care. ARIMAX models are typically applied on univariate response variables. Recently, multivariate extensions of these models have been developed and are called vector AR models (Shumway and Stoffer 2000; Lütkepohl 1991). Specialised software and considerable statistical knowledge are required to apply these multivariate AR models, and they are not discussed here.

17 Common trends and sudden changes

In this chapter, we start discussing various methods to estimate long-term patterns in time series. We call these patterns 'trends', but it should be noted that they are not restricted to being straight lines. Some of the methods can be applied on univariate time series and others require multiple time series. If the data are available on a monthly basis, one should make a distinction between seasonal variation and long-term patterns. In order to do this, the seasonal component needs to be determined and dealt with in some way. We will discuss three methods, of increasing mathematical complexity, for estimating common patterns in time series. In the last section, we discuss a technique that can be used to identify sudden changes.

17.1 Repeated LOESS smoothing

Figure 16.3 showed the annual time series of CPUE of the lobster species Nephrops at 11 stations in the Atlantic Ocean south of Iceland, between 1960 and 1999. We noted that most time series follow a similar pattern over time. To focus on the question of whether there are any common trends in the data, various methods can be used. The conceptually easiest approach is based on repeated LOESS smoothing. More difficult methods are discussed in the following two sections.

To visualise the general pattern in the CPUE time series, we will fit a LOESS smoother (Chapter 7) through each time series, and plot these smoothers in one graph. This process was applied on each time series and the results are plotted in Figure 17.1. The thick line shows the mean value of the smoothing curves. It was calculated by averaging the smoothing values in each year. It may be an option to standardise or mean delete the series before doing this. In our case, the mean curve seems to capture the patterns of most smoothing curves reasonably well. We used a span width of 0.5 for each LOESS smoother. This means that 0.5×40 years = 20 years are used in the windows around each target value. A small span width results in a curve that fits the data reasonably well, but it may not be smooth. On the other hand, a large span gives a smoother curve, but it may not follow the patterns of the data. The decision of which span width to use is discussed later.

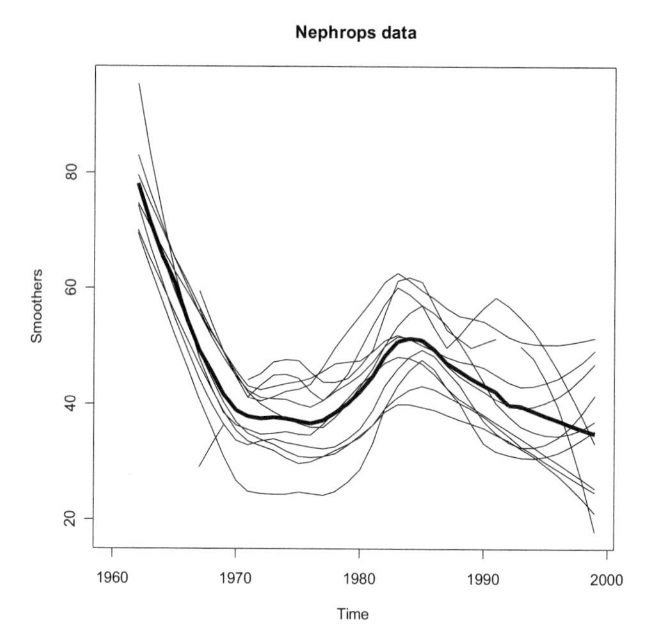

Figure 17.1. Eleven smoothing curves for the Nephrops time series data. The thick line represents the mean smoothing curve.

Mathematically, each smoothing curve in Figure 17.1 was estimated using the following model:

$$Y_t = f(\text{Time}_t, \lambda_1) + e_t$$

where $f(.)$ is a function obtained by the LOESS smoothing method; see Cleveland (1993) for details. The amount of smoothing is also called the degrees of freedom λ_1. Once a smoothing curve has been estimated for a univariate time series, the residuals can be calculated using the equation:

$$e_t = Y_t - f(\text{Time}_t, \lambda_1)$$

The function $f(.)$ can also be called the trend because it captures the long-term pattern in the time series. To capture the shorter-term variation, LOESS smoothing can be applied again, but now on the residuals, alias detrended series. Figure 17.2 shows the 11 LOESS curves capturing the shorter-term variation in the residuals. A smaller span of 0.2 was used. All the LOESS curves show a cyclic pattern, which is also captured by the mean value of the smoothing curves.

This repeated application of LOESS smoothing is discussed in Section 3.11 of Cleveland (1993). He suggested iterating the repeated LOESS smoothing process a couple of times because the smoothing curves may compete for the same variation. In the first step of the iteration process, LOESS smoothing is applied to the original time series with a span width of λ_1, followed by LOESS smoothing on the residuals with a span width of λ_2. In the second step of the iteration, the short-term smoother is subtracted from the data, and the long-term smoother is estimated again. This process is repeated a couple of times, and in practise convergence is obtained quickly. Convergence itself can be assessed by comparing the changes in smoothing curves of two sequential iterations. It is important that λ_1 is relatively larger than λ_2, and that the time series are of approximately the same length.

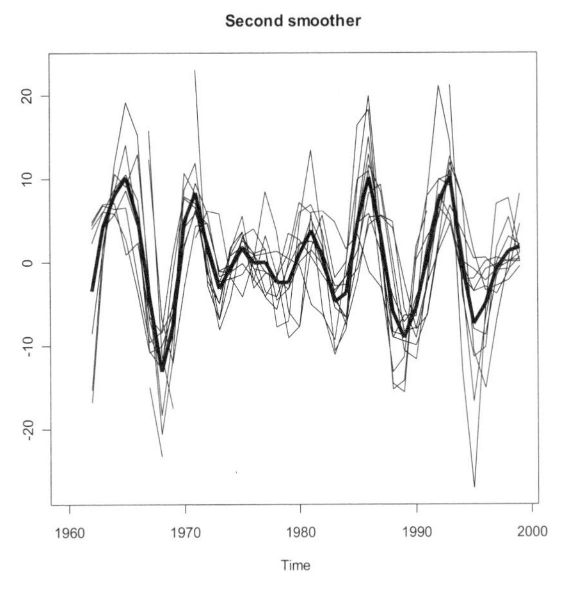

Figure 17.2. Eleven smoothing curves for the Nephrops time series data. These curves represent the short-term variation. The thick line is the mean smoothing curve.

Repeated LOESS smoothing applied on the SST and NAO index

As a second example, repeated LOESS smoothing is applied on the Scottish SST time series and the NAO index. (Chapter 16). Because of the clear patterns in the CPUE *Nephrops* time series and their relative short length, it was relatively easy to find a span width for the first and second smoothers. If the series are longer and have a less clear pattern, then this can be more difficult. Figure 17.3 shows the smoothing curve for the SST time series using different span widths. There are minor differences in the curves with span widths between 0.5 and 0.9. The curve with a span of 0.5 shows more small-scale variation. The question is what we consider as the long-term trend. The curves with the large span widths clearly indicate a rise in SST since the early 1980s. The curves with intermediate span width show three bumps and the curve with span = 0.1 shows cyclic behaviour. We decided to choose $\lambda_1 = 0.6$. The same process was applied on the detrended series, and we chose $\lambda_2 = 0.3$. The resulting long-term and short-term smoothers are presented in Figure 17.4. The long-term smoothers for the SST and

NAO series both show a general decrease followed by an increase, but the NAO trend decreases again in the late 1980s. The shorter term trends (right panel) show similar patterns over time. As a follow-up analysis, the shorter term trends of the SST and NAO can be compared with each other using auto-correlation functions, ARIMAX models, MAFA or dynamic factor analysis.

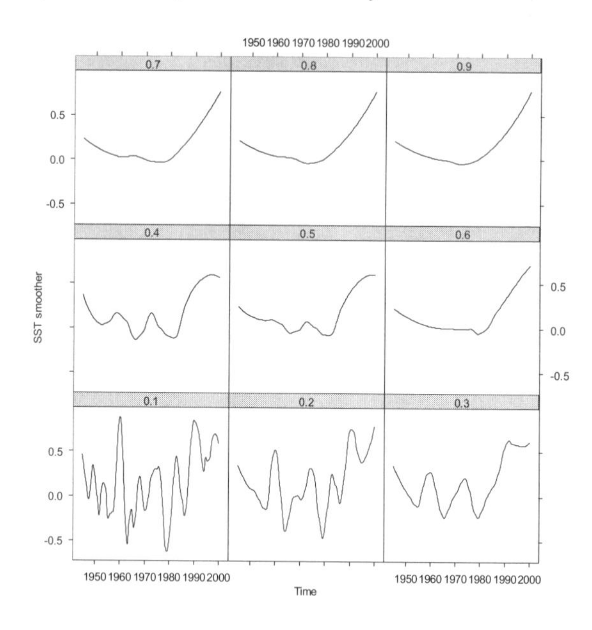

Figure 17.3. Smoothing curves for the SST time series using different span widths. The lower left panel shows the smoothing curve using a span of 0.1, and the upper right of 0.9.

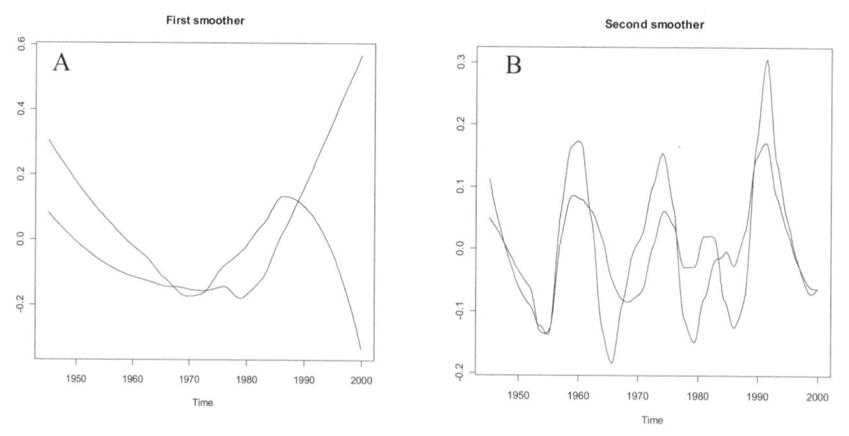

Figure 17.4. Smoothing curves for SST and NAO series obtained by repeated LOESS smoothing. A: Long-term smoothers obtained with $\lambda_1 = 0.6$. B: Short-term smoothers obtained with $\lambda_2 = 0.3$.

Neat et al. (2006) used repeated LOESS smoothing on a group of tagged cod time series in the coastal areas of the Shetland Isles in the northern part of the North Sea. Time series on depth and temperature on approximately 20 cod were available on a 10-minute basis throughout the year. Long-term, intermediate and short-term patterns were extracted using repeated LOESS smoothing for the depth time series and compared with each other. Various fish showed similar depth patterns for some of these components. The cyclic patterns may have been linked to seasonal and diet variation in prey resources and local variation in seabed substrate and bottom-depth.

The repeated LOESS smoothing method, as discussed here, can be extended in various ways. Two relatively easy to implement extensions are (i) automatic selection of the amount of smoothing per component (e.g., using cross-validation; see also Chapter 7), and (ii) applying bootstrapping methods to obtain confidence bands around the estimated components (Davison and Hinkley 1997).

17.2 Identifying the seasonal component

If one is looking for common trends in multiple time series that were measured monthly, then the main part of the variation may be related to seasonal fluctuation. It may be an option to remove the seasonal pattern in each time series and focus on the remaining information instead. Suppose we have a univariate time series Y_t, where measurements were taken monthly (or quarterly or indeed anything from which we know the length of the period). Y_t can be modelled as

$$Y_t = \text{Trend}_t + \text{Seasonal}_t + \text{remainder}_t \tag{17.1}$$

Hence, the univariate time series is decomposed into a trend, a seasonal component and residual information. The method is described in Cleveland (1993) and is sketched here. Repeated LOESS smoothing (Section 17.2) is used to estimate these components. Suppose that the monthly time series Y_t was measured from January 1980 to December 2000. Instead of storing it as one long vector of length 252 (= 12 months × 21 years), we can also present it as in Table 17.1. To identify the three components in equation (17.1), a two-step algorithm is applied. In the first step of the algorithm, the mean value per month is calculated. This gives a January average, a February average, etc. Once the 12 monthly averages are estimated, they are concatenated to form a univariate time series of length 252. In the second step of the algorithm, the seasonal time series is subtracted from the original time series Y_t. To identify the 'Trend' a LOESS smoother with a large span size is applied on the 'deseasonalised' data. This gives the long-term trend. However, there is one problem, the seasonal component may contain part of the long-term trend signal. We simulated a small dataset to visualise this; see Figure 17.5. The simulated data contains a strong linear trend and a seasonal component. For these data, the algorithm will first calculate a January average, then a February average, etc. However, the January data during the first few years have much lower values than in later years. And therefore, the mean value of January will contain some information of the upward trend and the same holds for the other months! Starting the other way around, estimating first a trend and then a seasonal component will give a general upward trend with some seasonal information in it. The solution to this problem is as in repeated LOESS smoothing. The algorithm (i) estimates the seasonal component, (ii) subtracts the seasonal component from the data, (iii) estimates the long-term pattern, (iv) subtracts the long-term pattern from the data, and (v) estimates a seasonal component. The process of estimating 12 monthly averages and a long-term trend is iterated a couple of times until convergence is reached and provides the terms Trendt, Seasonalt and Remainder$_t$. We can either decide to apply subsequent analyses on only the trend or on the Trend+Remainder component.

Table 17.1. Monthly time series measured in 21 year.

	Jan	Feb	Mar	Apr	May	Jun	Jul	Aug	Sep	Oct	Nov	Dec
1980	…	…	…	…	…	…	…	…	…	…	…	…
1981	…	…	…	…	…	…	…	…	…	…	…	…
…	…	…	…	…	…	…	…	…	…	…	…	…
…	…	…	…	…	…	…	…	…	…	…	…	…
1999	…	…	…	…	…	…	…	…	…	…	…	…
2000	…	…	…	…	…	…	…	…	…	…	…	…

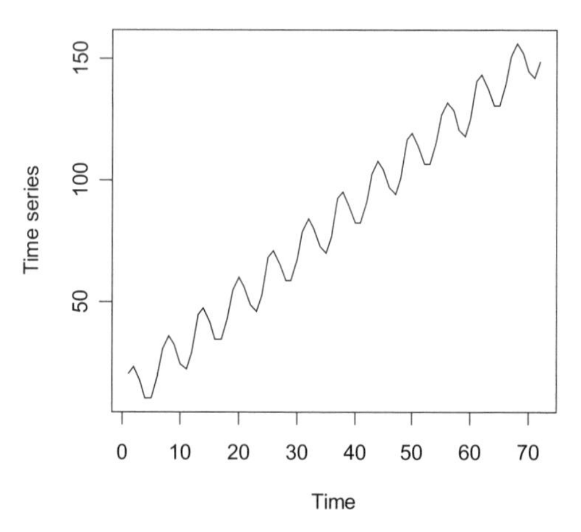

Figure 17.5. Simulated monthly data. The response variable was obtained by $Y = 10 + 2X + 10\sin(X)$, where X takes values between 1 and 72 (six years).

The problem with this approach is that the seasonal components for each month are always the same. This is fine if the seasonal pattern does not change over time. However, it is rather easy to extend the above algorithm to allow for changes in seasonal patterns. Instead of calculating a mean value for all January data, LOESS smoothing can be applied on the January time series. The same can be done for the other 11 monthly time series. As a result, the seasonal components are allowed to change slowly over time. For multiple time series, this process can be applied on each univariate time series and corresponding components can be plotted in the same graph. It is also possible to predict missing values.

Other methods to remove (or better: identify) the seasonal component can be found in Makridakis et al. (1998). Instead of the decomposition in equation (17.1), it is also possible to use a decomposition of the form

$$Y_t = \text{Trend}_t \times \text{Seasonal}_t \times \text{Remainder}_t \tag{17.2}$$

This decomposition is useful if the variation in the seasonal component varies with the trend. The trend and seasonal components are multiplied allowing for larger seasonal fluctuation in Y_t if the trend is high and small seasonal variation if the trend is close to zero. To estimate the individual components, a similar algorithm as for the repeated LOESS smoothing is used except that instead of subtracting a term from Y_t, we now divide by it: $Y_t/(\text{Seasonal}_t \times \text{Remainder}_t) = \text{Trend}_t$. This gives immediately a problem if one tries to apply the decomposition in equation (17.2) on for example monthly zooplankton data. During most months measured values tend to be low, if not zero (which causes the problem as you cannot divide by something that is zero) and the peak is in spring. An interesting solution may be the pseudo-additive decomposition (Baxter 1994):

$$\begin{aligned} Y_t &= \text{Trend}_t \times (\text{Seasonal}_t + \text{Remainder}_t - 1) \\ &= \text{Trend}_t \times \text{Remainder}_t + \text{Trend}_t\,(\text{Seasonal}_t - 1) \end{aligned} \tag{17.3}$$

Further details and a justification of this approach can be found in Findley et al. (1997). There is actually a whole range of methods available (enough to write a book about) that estimate the trend, seasonal and remainder components, and a few names we want to mention are the X-11, X-11 ARIMA and X-12 ARIMA. They are all variations of each other, and a more detailed description can be found in Findley et al. (1997), Makridakis et al. (1998), among many others.

An example

The seasonal decomposition is illustrated for the North American SST data. Figure 17.6 shows a lattice plot containing auto-correlation functions for all SST time series and the NAO index. The shape of the auto-correlation functions indicates strong seasonal patterns for all SST time series. Figure 17.7 and Figure 17.8 show the estimated trend and seasonal components for the time series at the coordinates 22–24. The difference between the two figures is that in the first one, the mean value per month is used and in the second graph the LOESS smoother is applied on monthly time series to allow for changes over time in the seasonal components. The seasonal component in Figure 17.8 had a span width of 11 months. Figure 17.9 shows the trends obtained by this method for all time series. These trends can be further analysed by the multivariate time series techniques to estimate common trends, which are discussed later in this chapter.

All approaches discussed so far in this section are based on the assumption that the signal measured in January in year 1 is the same signal as in year 2, 3, etc. Hence, extracting all the January data and obtaining a mean value or smoother makes sense. But what if the data are monthly zooplankton values, and the big peak is in April in year 1, March in year 2 and May in year 3? The answer is simple: The method falls down. Analysing anomalies obtained from fitting a moving average may be an option.

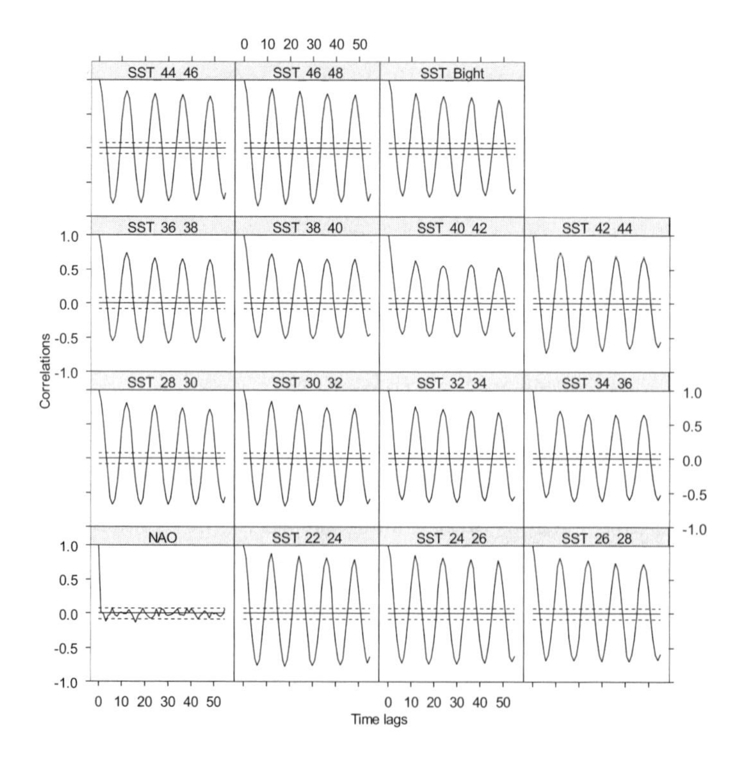

Figure 17.6. Auto-correlation functions for the North American SST. The x-axis shows time lags (months) and the y-axis the auto-correlations.

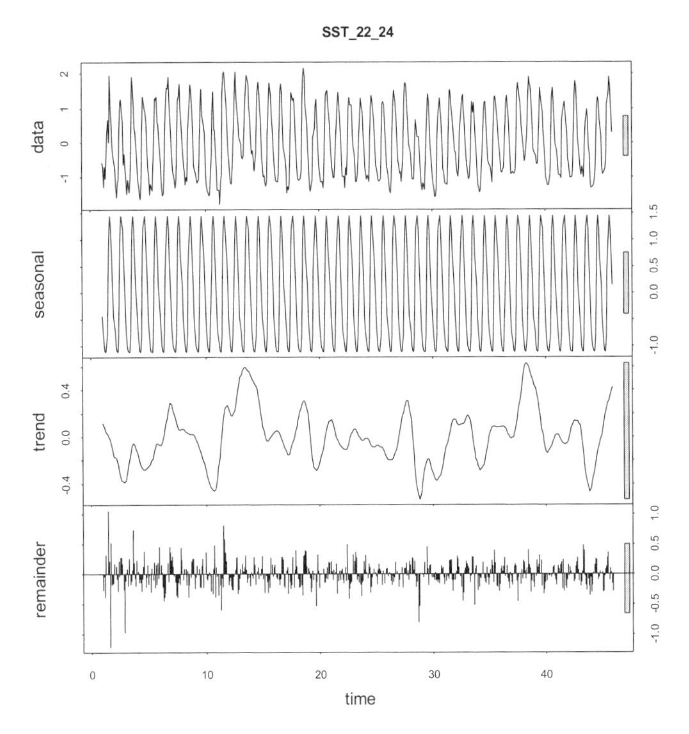

Figure 17.7. Decomposition of a SST time series into seasonal, trend and remainder components. The seasonal component was obtained by taking the average per month. The original time series were normalised.

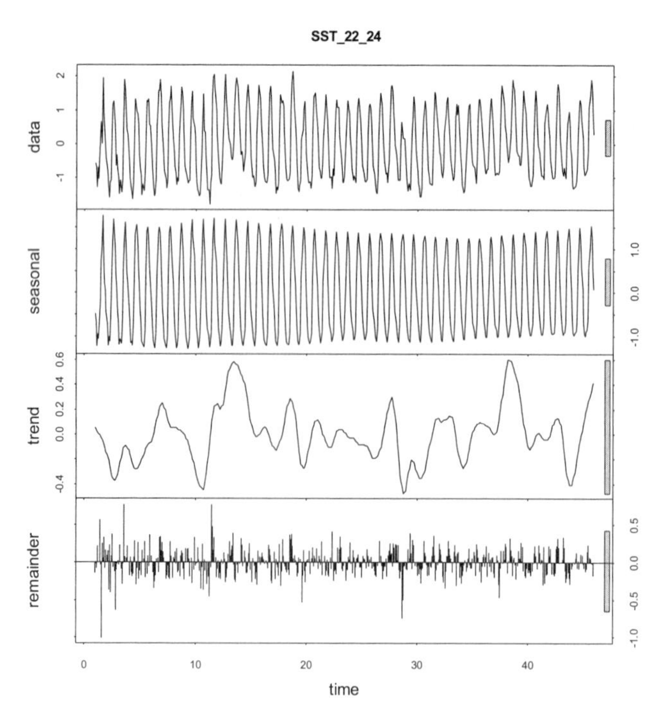

Figure 17.8. Decomposition of an SST time series into seasonal, trend and remainder components. The seasonal component was estimated by a LOESS smoother using a span width of 11 points. The time series were normalised.

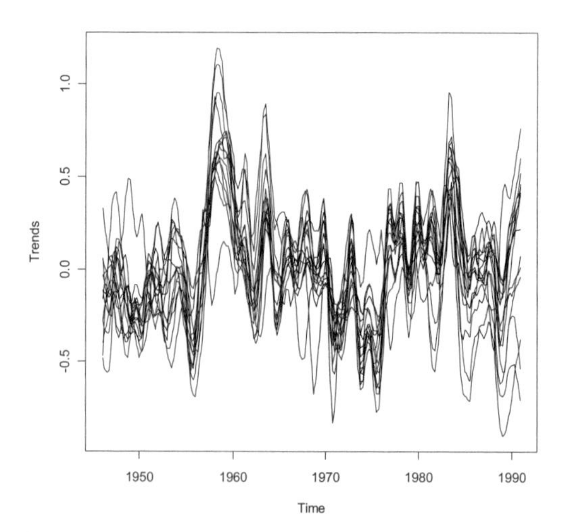

Figure 17.9. Trends for all time series. The seasonal components had a span width of 11. The original time series were normalised.

17.3 Common trends: MAFA

If the data contains a seasonal pattern, methods discussed in the previous section can be applied to remove it. If this is not done, methods to estimate common patterns will mainly pick up the seasonal patterns. We now discuss the first of two more formal methods (compared to repeated LOESS smoothing) to estimate common trends in a multivariate time series dataset.

What does it do?

MAFA (Solow 1994) stands for min/max auto-correlation factor analysis. It can be described in various ways, e.g., a type of principal component analysis for time series, a method for extracting trends from multiple time series, a method for estimating index functions from time series or a smoothing method. Let us recall the underlying formula for principal component analysis (Chapter 12):

$$Z_1 = c_{11} Y_{i1} + c_{12} Y_{i2} + \ldots + c_{1N} Y_{iN} \tag{17.1}$$

Y_{i1}, Y_{i2}, .., Y_{iN} contain the values of the N variables at site i and Z_1 is the principal component. The factors c_{ij} are calculated in such a way that the variance of Z_1 is maximal. In discriminant analysis we have a similar underlying formula, but it uses a different optimisation criteria namely that observations from the same group were as close to each other as possible and maximum separation of group averages. In MAFA we use a different optimisation criteria namely Z_1 should have maximum auto-correlation with time lag 1. Hence, in MAFA, the first axis has the highest auto-correlation with lag 1. The second axis has the second highest auto-correlation with time lag 1. The underlying idea is that a trend is associated with high auto-correlation with time lag 1. Therefore, the first MAFA axis represents the trend, or the main underlying pattern in the data. This axis can also be seen as an index function or smoothing curve that summarises the original N time series in the best possible way. The second MAFA is the second most important trend. Just as in PCA, the axes (or trends) are uncorrelated and estimating a second or third trend does not change the shape of the earlier estimated trends. One of the main differences is that we do not get eigenvalues that tell us how important is a trend. Instead, Solow (1994) presented a permutation test that determines whether the auto-correlation of the axes is significantly different from zero.

An example

We will use a zoobenthic dataset from an intertidal in the Dutch part of the Wadden Sea, namely the Balgzand data (Beukema 1974, 1979, 1992; Zuur et al. 2003a). The data used here consist of 15 zoobenthic species. The original data were measured at 15 locations on the Balgzand but for illustration purposes, total abundances over all sites were taken. The species used in this example are given in Table 17.2, and Figure 17.10 shows a plot of all 15 standardised time series.
Table 17.2. List of zoobenthic species used in the MAFA example.

No	Species	No	Species
1	*Arenicola marina*	9	*Mya arenaria*
2	*Cerastoderma edule*	10	*Mytilus edulis*
3	*Corophium spec.*	11	*Nephtys hombergii*
4	*Eteone longa*	12	*Nereis spec.*
5	*Heteromastus filiformis*	13	*Phyllodoce sp.*
6	*Hydrobia ulvae*	14	*Scoloplos armiger*
7	*Lanice conchilega*	15	*Tellina tenuis*
8	*Macoma balthica*		

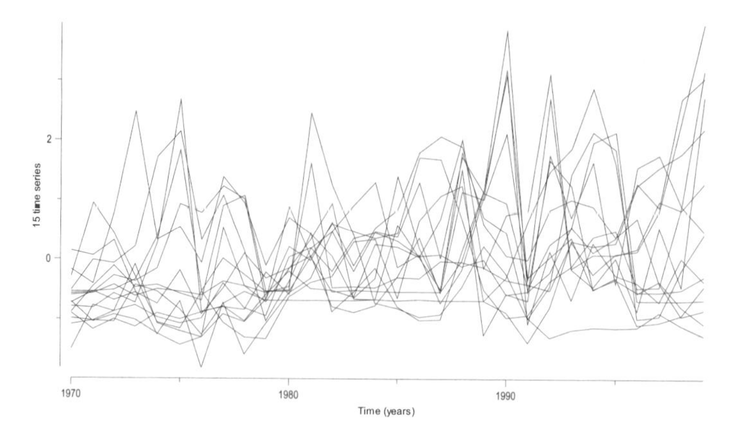

Figure 17.10. Time series plot of all 15 standardised zoobenthic species. The x-axis shows the year and the y-axis the value of the standardised series.

MAFA was applied on the 15 time series, and Figure 17.11 shows the first two MAFA axes. The first axis shows an increase between 1973–1982 and 1992–2000. Note that this is the main trend underlying the time series. The second axis shows an increase from 1970 until 1986 and a decrease there after (except for the period 1994–1995).

Just as in PCA, loadings are estimated. These can be used to infer which species are related to a particular MAFA axis. Another option, used here, is to calculate the cross-correlation between the MAFA axes and each of the original species time series. We called these canonical correlations (Figure 17.12). Results indicate that the first MAFA axis is important for *A. marine*, *H. ulvae*, *C. edule*, *E. directus*, *M. balthica*, *H. filiformis*, *M. edulis*, *P.* species, *M. arenaria* and *S. armiger*. Hence, all these species are characterised by a general increase (or decrease if the correlation was negative) in abundance. In Figure 17.13, the original time series of some of these species are highlighted, and one can (vaguely) see a general increase. A similar graph can be made for some of the species related to the second MAFA axis (not shown here).

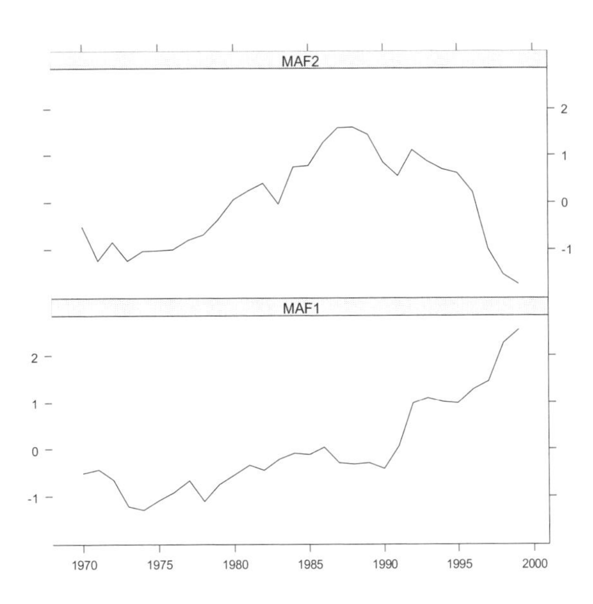

Figure 17.11. First and second MAFA trends. The x-axis shows time in years and the y-axis the value of the MAFA axis (which is unitless).

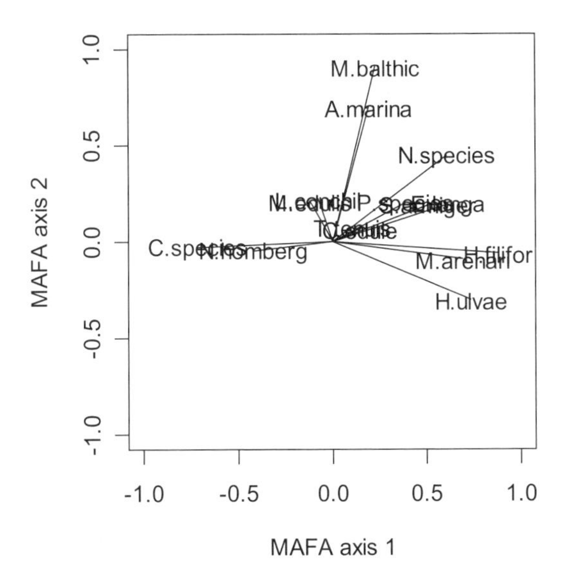

Figure 17.12. Canonical correlations for MAFA axis 1 and 2.

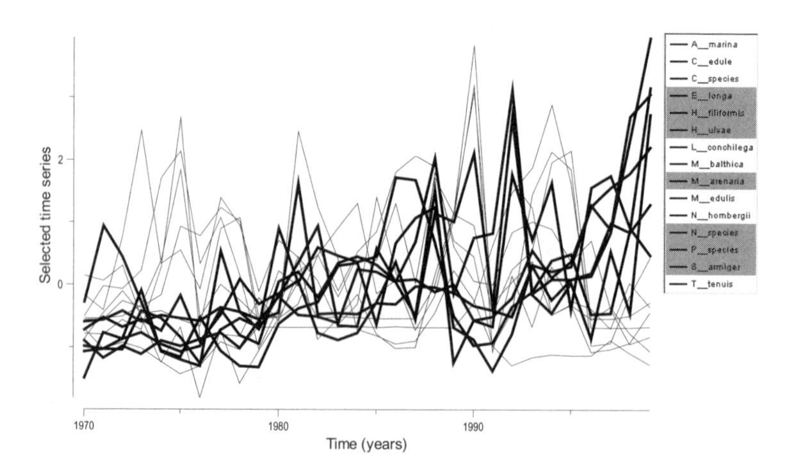

Figure 17.13. Highlighted time series are important for the first MAFA axis. The x-axis shows time in years and the y-axis the value of the standardized series (which are unitless).

If explanatory variables are available, one can estimate the correlations between the MAFA axes and all explanatory variables. For this dataset water temperature was measured also. The correlation between the temperature and the first MAFA axis is 0.095, and it is not significant at the 5% level. The correlation between temperature and the second MAFA axis is –0.062, and for the third axis it is –0.206. Summarising, temperature is not significantly related to any of the first three MAFA axes.

Solow (1994) described a randomisation process to obtain p-values for the MAFA axes. These can be used to decide how many axes to present. Results for the zoobenthic dataset are as follows:

Axis	Auto-correlation lag 1	p-value
1	0.953	<0.001
2	0.906	<0.001
3	0.879	0.006
4	0.727	0.047

These results indicate that the first three MAFA are important.

How does it work?

The reader not interested in the underlying mathematical details may skip this part. The mathematics underlying MAFA are described in two technical reports (Shapiro and Switzer 1989; Switzer and Green 1984) and Solow (1994). Here, a short sketch of the mathematics is presented. The underlying formula is similar to that in PCA:

$$\mathbf{z}_t = \mathbf{a'y}_t$$

where $\mathbf{y}_t$ contains the N variables measured at time t, **a** is a vector of dimension N-by-1 containing the loadings, and $\mathbf{z}_t$ represents the first MAFA axis. In PCA, the loadings in **a** are chosen such that $\mathbf{z}_t$ has maximum variance. In MAFA, we chose them such that $\mathbf{z}_t$ has maximum auto-correlation with lag 1. So, how does it do this? It takes a couple of pages of formulae and a few hours, but at the end of the day it is plain high school algebra to show that the following relationship holds:

$$\frac{\mathbf{a'Va}}{\mathbf{a'Ca}} = 2(1 - r_1)$$

C is the covariance matrix of the original time series $\mathbf{Y}_t$, **V** is the covariance matrix of the first-order difference $\mathbf{Y}_t' = \mathbf{Y}_t - \mathbf{Y}_{t-1}$, and r_1 is the auto-correlation coefficient of the first MAFA axis $\mathbf{z}_t$ at time lag 1. The aim is now to maximise r_1 as a function of the loadings **a**. Just as in PCA (Chapter 12), we can take derivatives, set them to zero, solve the equations, etc. Using first year university algebra it can be shown that the solution is based on an eigenvalue equation of $\mathbf{C}^{-1}\mathbf{V}$. In fact, **a** is proportional to the eigenvector corresponding to the smallest eigenvalue of $\mathbf{C}^{-1}\mathbf{V}$. Just as in PCA, we have a certain flexibility in making the solution unique. Recall that in PCA the sum of squared loadings was set to 1. In MAFA, they are made unique by setting (i) the variance of $\mathbf{z}_1$ to 1, and (ii) the weights on **Y** to be positive. Just as in PCA we can derive a second MAFA axis. This is the second smoothest curve, and it is uncorrelated with the first MAFA axis. In fact, we can rewrite the MAFA formula as $\mathbf{Z}_t = \mathbf{Ay}_t$, where $\mathbf{Z}_t$ contains all MAFA axes and **A** the corresponding loadings.

All we need is an eigenvalue equation of the matrix $\mathbf{C}^{-1}\mathbf{V}$, but this is a bit tricky, as the matrix is not symmetric. For this reason, Shapiro and Switzer (1989) used an indirect method for extracting the eigenvectors of $\mathbf{C}^{-1}\mathbf{V}$. It involves PCA on centred data **Y**, followed by a first-differencing on the principal components, and a second PCA on these differenced components to give the matrix **A**. As a result, the MAFA axes are mutually uncorrelated with unit variance, and the axes have decreasing auto-correlation with time lag 1. A requirement for MAFA is that there are more time points than variables.

17.4 Common trends: Dynamic factor analysis

Dynamic factor analysis (DFA) is a method to estimate common trends, effects of explanatory variables and interactions in a multivariate time series dataset. DFA has been applied in many different fields, and all have their different flavours and modifications. Zuur et al. (2003a) presented a detailed mathematical derivation in a pilot paper with a zoobenthic example. This paper was followed by two applied papers (Zuur et al. 2003b; Zuur and Pierce 2004) in which fisheries data were used. Erzini (2005) and Erzini et al. (2005) applied both DFA and MAFA as complimentary methods to examine fisheries data. Other fisheries applications can be found in Chen et al. (2006), among others. Hydrological applications can be found in Muñoz-Carpena et al. (2005) and Ritter and Muñoz-Carpena (2006). The underlying DFA method in all these papers was discussed in Harvey (1989), Shumway and Stoffer (1982) and Shumway and Stoffer (2000). Mendelssohn and Schwing (1997, 2002) used a slightly modified version of DFA to estimate common trends in large oceanographical time series datasets albeit their approach cannot easily deal with explanatory variables. DFA has also been used extensively in psychological related fields (Molenaar 1985; Molenaar et al. 1992, among others).

What does it do?

In Section 17.1, we used 11 Icelandic Nephrops CPUE annual time series. Let the vector $\mathbf{Y}_t=(Y_1,..,Y_{11})'$ contain the values at year t. The symbol ' is mathematical notation for the transpose, hence $\mathbf{Y}_t$ is of dimension 11-by-1. The main aim of DFA is to estimate underlying common trends. The simplest DFA model contains only one common trend and is given by

$$\mathbf{Y}_t = \mathbf{A}z_t + \boldsymbol{\varepsilon}_t \tag{17.2}$$

where $\mathbf{A}$ is a vector of dimension 11-by-1 with *unknown* loadings, z_t is the trend and $\boldsymbol{\varepsilon}_t$ is normally distributed noise (of dimension 11-by-1) with expectation $\mathbf{0}$ and variance $\sigma^2\mathbf{V}$ and where $\mathbf{V}$ is a positive definite diagonal matrix. The unknown parameters are the multiplication factors $\mathbf{A}$, the trend z_t and the variances. Later in this section we show that a special construction for z_t is used to ensure that it captures the long-term pattern. It is perhaps easier to present this model in full mathematical notation:

$$\begin{pmatrix} Y_{1,t} \\ Y_{2,t} \\ \vdots \\ Y_{11,t} \end{pmatrix} = \begin{pmatrix} a_1 \\ a_2 \\ \vdots \\ a_{11} \end{pmatrix} z_t + \begin{pmatrix} \varepsilon_{1,t} \\ \varepsilon_{2,t} \\ \vdots \\ \varepsilon_{11,t} \end{pmatrix}$$

The model with one common trend assumes that all the eleven time series follow the same pattern, namely that of z_t. To obtain the fitted value for each time series, we multiply the trend z_t by a loading. If the loading is relatively large and positive, we know that the corresponding time series follows the pattern of the trend. If the loading is close to zero, we know it does not follow this pattern. A loading that is relatively large and negative indicates that the time series follows the opposite pattern of the trend. These statements assume that the spread in the 11 time series is the same. Otherwise, it is more difficult to compare the factor loadings with each other! One way to ensure this is normalisation of the time series prior to the analysis. It is also an option to include an intercept:

$$\mathbf{Y}_t = \mathbf{c} + \mathbf{A}z_t + \boldsymbol{\varepsilon}_t \tag{17.3}$$

The intercept $\mathbf{c}$ is a vector of dimension 11-by-1 containing an intercept for each time series. Hence, we are saying that all time series follow the underlying pattern (the trend), the factor loadings are used to determine the strength of this relationship, and each time series is shifted up or down by an intercept. If the time series are normalised or centred, the intercepts will all be close to zero. The model in equation (17.3) can be extended in various ways. The easiest option is to add explanatory variables, for example the NAO index:

$$\mathbf{Y}_t = \mathbf{c} + \mathbf{A}z_t + \boldsymbol{\beta}\text{NAO}_t + \boldsymbol{\varepsilon}_t \tag{17.4}$$

The term '$\boldsymbol{\beta}\text{NAO}_t$' models the effect of the NAO index on the 11 time series. It is always useful to do a dimension check after a model specification. The NAO index is a global variable, and in this case, we have only one value per time unit (year). So, the term NAO_t is of dimension 1-by-1. The regression parameter $\boldsymbol{\beta}$ is of dimension 11-by-1; the NAO is allowed to have a different effect on each CPUE time series. Just as in linear regression, standard errors and t-values are obtained for the estimated values of $\boldsymbol{\beta}$ and these can be used to assess whether it is significantly different from 0. Another extension is modifying the error structure. In equation (17.4) we assumed that

$$\boldsymbol{\varepsilon}_t \sim N(\mathbf{0}, \sigma^2\mathbf{V}), \qquad \text{and } \mathbf{V} \text{ is diagonal and positive definite} \tag{17.5}$$

So each error component has a different variance. In the same way as the linear regression model was extended to generalised least squares (GLS) in Chapter 16, the error components of different time series of the DFA model can be allowed to covary. This can be done by using an unstructured (positive definite) error covariance matrix $\mathbf{V}$ of the form

$$\mathbf{V} = \begin{pmatrix} 1 & v_{1,2} & v_{1,3} & \cdots & \cdots & v_{1,N-1} \\ v_{2,1} & 1 & v_{2,3} & \cdots & \cdots & v_{2,N-2} \\ v_{3,1} & v_{3,2} & 1 & & & v_{3,N-3} \\ \vdots & v_{4,2} & v_{4,3} & \ddots & & \vdots \\ \vdots & \vdots & \vdots & & \ddots & v_{N-1,N} \\ v_{N-1,1} & v_{N-2,2} & v_{N-3,3} & \cdots & \cdots & 1 \end{pmatrix}$$

Let us just recap what we have so far. We have 11 CPUE time series, and the NAO may have a different effect on each of them. This is done in a regression style ($\mathbf{c}$ + $\boldsymbol{\beta}\text{NAO}_t$). A trend is used to capture the remaining common long-term variation. Factor loadings determine how important this trend is for each of the time series. On top of this, we allow for residual interactions between the time series via a non-diagonal error covariance matrix. Obviously, we have to estimate a lot of parameters, especially if a non-diagonal error covariance matrix $\mathbf{V}$ is used, and the technical aspects of this are discussed in Zuur et al. (2003a).

Hopefully, the explanatory variables explain most variation in the data as it makes interpretation of the model much easier. If this is not the case, the researcher has the task of explaining the meaning of the common trend. If the factor loadings show a clear grouping, then this can be easy as it indicates a grouping in the time series.

So far, we have discussed a DFA model with one common trend. But what happens if groups of time series are behaving differently over time? If only two common patterns behave oppositely, we can still model it with a model containing only one common trend; the model will use positive and negative loadings. But this does not work if the patterns are not opposite to each other, or if there are more than two common patterns. The solution is to extend the DFA model in equation (17.2) with an extra common trend:

$$\mathbf{Y}_t = \mathbf{A}\mathbf{Z}_t + \boldsymbol{\varepsilon}_t \tag{17.6}$$

The model specification looks the same as before, but the difference is the dimension of $\mathbf{A}$ and $\mathbf{Z}_t$. Suppose we have two common trends. The factor loading matrix $\mathbf{A}$ is now of dimension 11-by-2 and $\mathbf{Z}_t$ is of dimension 2-by-1. Writing out the full formula may clarify the model:

$$\begin{pmatrix} Y_{1,t} \\ Y_{2,t} \\ \vdots \\ Y_{11,t} \end{pmatrix} = \begin{pmatrix} a_{1,1} & a_{2,1} \\ a_{1,2} & a_{2,2} \\ \vdots & \\ a_{1,11} & a_{2,11} \end{pmatrix} \begin{pmatrix} z_{1,t} \\ z_{2,t} \end{pmatrix} + \begin{pmatrix} \varepsilon_{1,t} \\ \varepsilon_{2,t} \\ \vdots \\ \varepsilon_{11,t} \end{pmatrix}$$

There are now two common trends, and each one has factor loadings associated with it. Hence, each time series is modelled as the sum of (i) a factor loading multiplied with the first common trend plus another factor loading multiplied with the second common trend plus noise. It is now interesting to compare the factor loadings with each other. The signs and magnitudes will tell us which CPUE time series are driven by the first, the second, and by both trends. The model with two common trends can easily be extended to include explanatory variables, an intercept and a non-diagonal error covariance matrix $\mathbf{V}$:

$$\mathbf{Y}_t = \mathbf{c} + \mathbf{A}\mathbf{Z}_t + \boldsymbol{\beta}\mathrm{NAO}_t + \boldsymbol{\varepsilon}_t \qquad \boldsymbol{\varepsilon}_t \sim N(\mathbf{0}, \sigma^2\mathbf{V}) \tag{17.7}$$

One can add even more trends, but just as in PCA the interpretation of three or more axes (trends) becomes difficult. Further explanatory variables can also be used. Zuur et al. (2003a) used the AIC to find the optimal model in terms of (i) the number of common trends, (ii) which explanatory variables to select and (iii) what type of error covariance matrix $\mathbf{V}$ to use. Another aspect one should keep in mind is that in MAFA and PCA, the first axis does not change if two or three axes are calculated. In DFA, all trends are estimated simultaneously using the maximum likelihood method. This means that if two trends are used, the first trend may not be the same as the trend in a model with only one common trend. To get some idea of the importance of the trends, we advise the following: (i) Apply the model with one common trend, (ii) apply a model with two common trends, and (iii) compare trends of both models. In most cases, one of the trends in the model with two common trends will look similar (although not identical) to the trend obtained in the first model. This is the dominant pattern.

An example

To illustrate dynamic factor analysis, we use the CPUE Nephrops data. A time series plot was given in Figure 16.3 and showed that various time series followed a similar pattern over time. The DFA model was applied on the standardised CPUE series, and various combinations of number of trends, explanatory variable and error covariance matrix were investigated. Each time we obtained the AIC, and these are given in Table 17.3. The model containing three common trends, no explanatory variables and a diagonal error covariance matrix was the most optimal, as judged by the AIC. The problem with this model is that the third common trend was mainly related to one time series, and its contribution to the model was small (as could be inferred from the small factor loading). This is typically an indication that too many trends are used. The second most optimal model contained three common trends and the NAO index. It had the same problem with the third trend and *t*-values for the NAO were either not significant or only slightly larger than two in the absolute sense, which indicates only borderline significance. For these reasons, we decided to present results of the model with two common trends, no explanatory variables and a diagonal error covariance matrix.

Table 17.3.Values of AIC using a dynamic factor model with 1 to 4 common trends. The model in bold typeface is selected.

V Diagonal			**V** Non-diagonal		
Number of trends	Explanatory variables	AIC	Number of trends	Explanatory variables	AIC
1	-	850.951	1	-	778.537
2	-	**755.427**	2	-	769.933
3	-	749.387	3	-	764.764
4	-	757.197	4	-	773.499
1	NAO	858.911	1	NAO	778.139
2	NAO	757.998	2	NAO	770.916
3	NAO	750.483	3	NAO	764.483
4	NAO	757.458	4	NAO	770.519

The estimated two common trends are presented in Figure 17.14 and corresponding factor loadings in Figure 17.15. The factor loadings indicate that station S4 is driven mainly by the first common trend, stations S8, S9 and S10 are related to the second common trend, S7 and S11 are driven mainly by the second trend but with a contribution from the first, and stations S1, S2, S3, S5, S6 and S7 by the first and also a bit by the second. This grouping corresponds with the location of the stations. Eiríksson (1999) found a similar pattern in these data and argued that the distinction between the stations may be due to differences in temperature and sediment type. To illustrate the differences between the time series of the groups, fitted values for all stations are presented in Figure 17.16. Thick lines in this figure correspond to stations S1, S2, S3, S5, S6 and S7. These stations tend to have lower CPUE values in most years. Also note the differences between the two groups of time series from 1995 onwards. The fitted curve of station 4 is the curve that had the highest fitted values from 1965 until 1995.

Just as in linear regression, one has to apply a model validation. Useful tools are graphs in which fitted lines and the observed data are plotted (Figure 17.17). They identify the extent to which the model can capture the patterns in the time series. Other tools are residuals plotted versus time (Figure 17.18), residuals versus fitted values and QQ-plots and histograms of residuals. In our experience, these graphs will always show some patterns as we are summarising N time series with only 2 or 3 common trends. The more patterns we see, the more we need to improve the model.

The dynamic factor analysis applied on the 11 Nephrops time series indicated that there are two underlying common trends. The factor loadings indicated that the distinction between the series is probably based on geographical differences. The NAO index was not significantly related to the time series. To understand which biological mechanisms are driving the two common trends, further information is needed (e.g., sediment type).

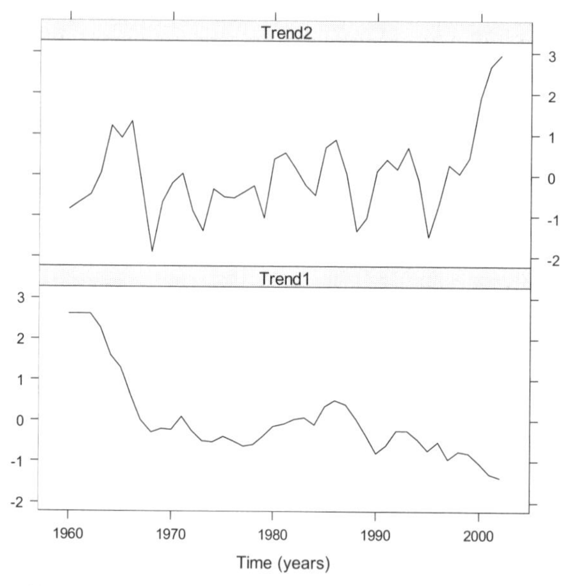

Figure 17.14. Estimated common trends obtained by the DFA model. The x-axis shows time in years and the y-axis the values of the trends (which are unitless).

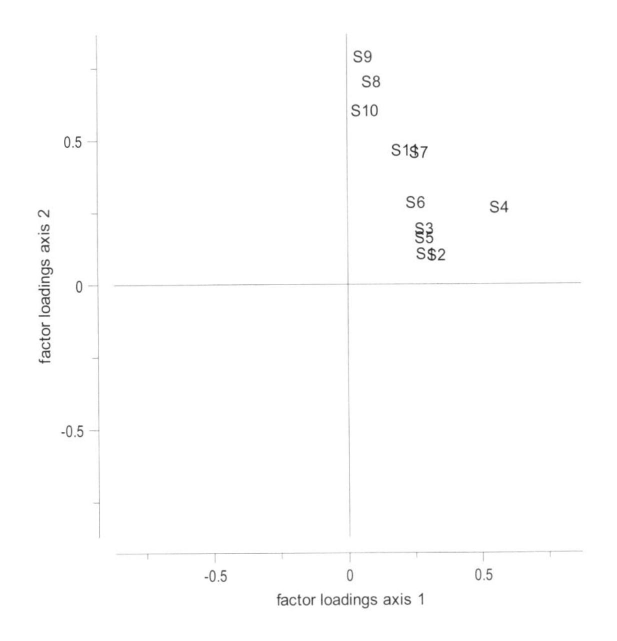

Figure 17.15. Factor loadings corresponding to the first two common trends.

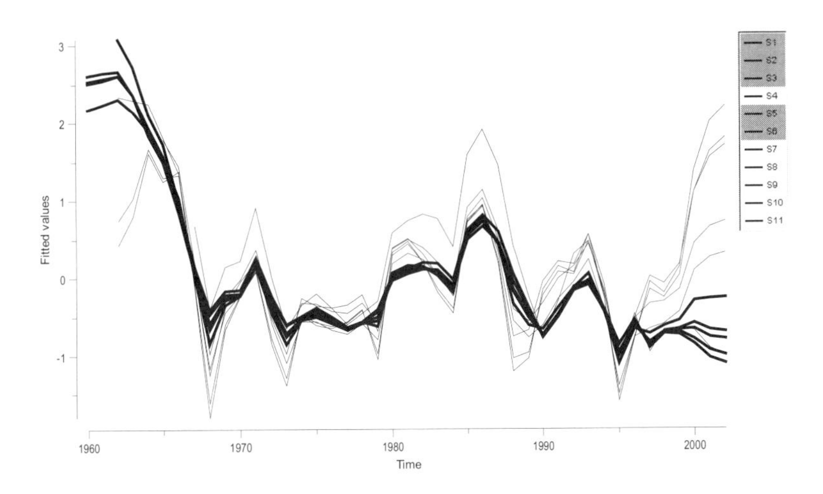

Figure 17.16. Fitted values obtained by the DFA model with two common trends. The thick lines correspond to stations S1, S2, S3, S5, S6 and S7. The x-axis shows time in years and the y-axis the values of the normalised series (which are unitless).

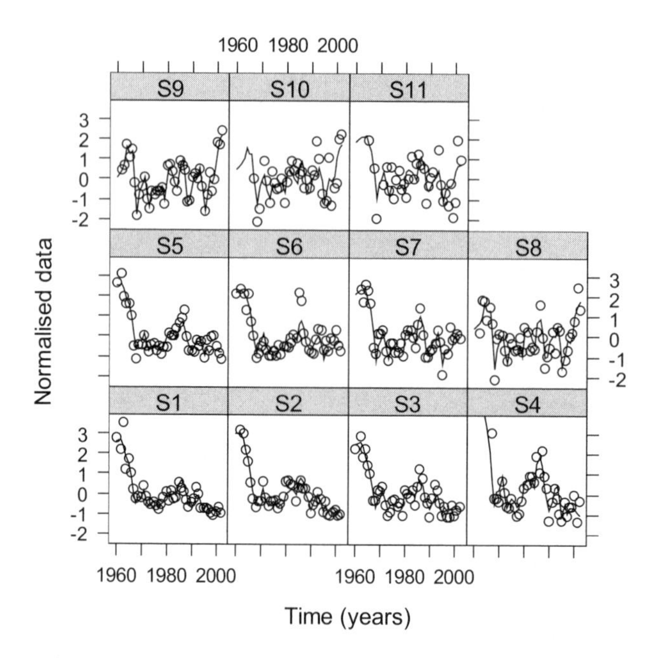

Figure 17.17. Observed data and fitted values obtained by a DFA model with two common trends. The fitted curves follow the patterns of most time series reasonably well.

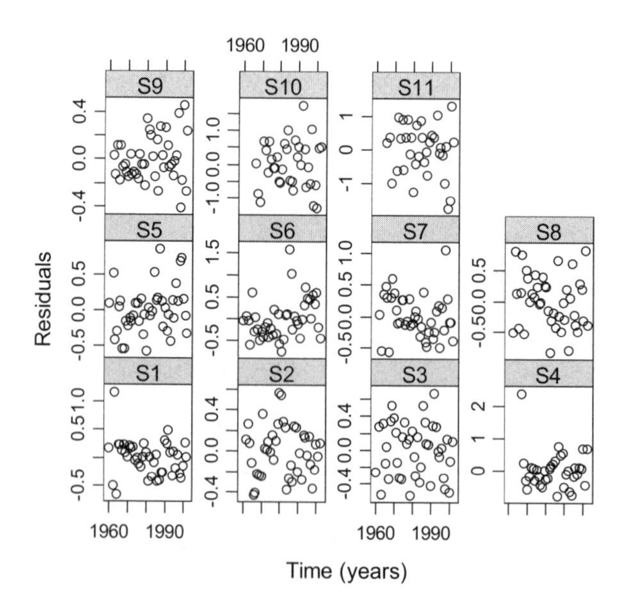

Figure 17.18. Residuals versus time for the DFA model with two common trends.

How does it work?

The reader not interested in the mathematical details underling DFA can skip the next two pages. DFA is based on multivariate structural time series models. For simplicity, we start with univariate structural time series models. The underlying principle in univariate structural time series models is that the time series are modelled in terms of a trend, seasonal component, cycle, explanatory variables and noise. Each of these components is allowed to be stochastic. This means that a trend is not restricted to being a straight line, and the cyclic and seasonal components are not necessarily smooth-looking cosine functions, but their shape can change over time. Obviously, if the time series are measured on an annual basis for a period of less than 50 years, the seasonal component can be dropped. Furthermore, using a cyclic component for such short time series is less suitable. For such data, the following univariate time series model can be used:

$$\text{1 time series} = \text{trend} + \text{explanatory variables} + \text{noise} \qquad (17.8)$$

The 'trend' component in (17.8) is based on a so-called random walk, which is mathematically defined by

$$z_t = z_{t-1} + \eta_t \quad \text{and} \quad \eta_t \sim \text{N}(0, \sigma_\eta^2) \qquad (17.9)$$

The trend at time t is given by z_t. This type of model is popular in econometrical fields (Harvey 1989; Durbin and Koopman 2001). They typically have the form of a slow-moving pattern with occasionally a sharp drop or sudden increase. It is also possible to add explanatory variables:

$$\begin{aligned} y_t &= z_t + \boldsymbol{\beta}\mathbf{X}_t + \varepsilon_t \\ z_t &= z_{t-1} + \eta_t \\ \varepsilon_t &\sim N(0,\sigma^2) \quad \eta_t \sim N(0,\sigma_\eta^2) \quad z_0 \sim N(a_0,\sigma_z^2) \end{aligned} \qquad (17.10)$$

where the vector $\mathbf{X}_t$ contains the values of the explanatory variables at time t, and $\boldsymbol{\beta}$ the corresponding slopes. The error terms are assumed to be independent of each other. The unknown parameters are the slopes, the three variance components and the value of the trend at $t = 0$. However, the trend z_t is unknown as well! So, how does one estimate all these parameters, variances and the trend? The answer is a combination of (i) the state-space formulation, (ii) the Kalman filter and smoother and (iii) the maximum likelihood. The state-space is a special mathematical formulation that takes the form:

$$\begin{aligned} \mathbf{y}_t &= \mathbf{A}\mathbf{z}_t + \mathbf{B}\boldsymbol{\varepsilon}_t \\ \mathbf{z}_t &= \mathbf{C}\mathbf{z}_{t-1} + \mathbf{D}\boldsymbol{\eta}_t \end{aligned} \qquad (17.11)$$

It turns out that a lot of time series models, including ARMAX, structural time series and DFA can be reformulated in this format. It is just a matter of choosing the right format for the matrices **A**, **B**, **C** and **D**. The Kalman filter and smoother is a mathematical procedure that estimates the so-called state vector $\mathbf{z}_t$ and its standard errors for given values of **A**, **B**, **C** and **D** and the variances of the noise components. The EM method (Zuur et al. 2003a) is used to estimate all the unknown parameters. For example, let the response variable be the Nephrops CPUE at station 1 and let us include a trend and the NAO as an explanatory variable to give:

$$\begin{aligned} & y_t = z_t + \beta NAO_t + \varepsilon_t \\ & z_t = z_{t-1} + \eta_t \\ & \varepsilon_t \sim N(0, \sigma^2) \quad \eta_t \sim N(0, \sigma_\eta^2) \quad z_0 \sim N(a_0, \sigma_z^2) \end{aligned} \tag{17.12}$$

The first step is to rewrite this model in state-space format:

$$\begin{aligned} & y_t = \begin{pmatrix} 1 & NAO_t \end{pmatrix} \begin{pmatrix} z_t \\ \beta \end{pmatrix} + \varepsilon_t \\ & \begin{pmatrix} z_t \\ \beta \end{pmatrix} = \begin{pmatrix} 1 & 0 \\ 0 & 1 \end{pmatrix} \begin{pmatrix} z_{t-1} \\ \beta \end{pmatrix} + \begin{pmatrix} 1 & 0 \\ 0 & 0 \end{pmatrix} \begin{pmatrix} \eta_{1t} \\ \eta_{2t} \end{pmatrix} \end{aligned} \tag{17.13}$$

η_{2t} is a new error term but it has no influence. The unknown parameters are the state vector and the variances. For given values of the variances, the Kalman smoother estimates the state vector and its standard errors. The construction of the design matrix **D** is such that β is constant over time. Using maximum likelihood, the unknown variance components can be estimated. The trend z_t is given in Figure 17.19 and $\beta = -0.21$, and its t-value is –3.55, indicating a significant negative effect of the NAO at the 5% level.

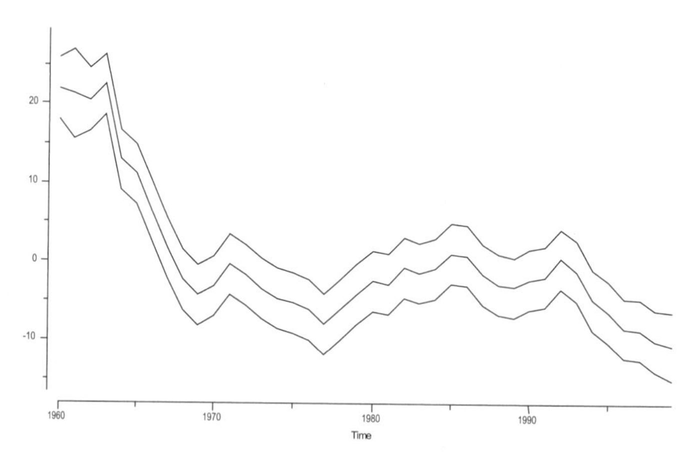

Figure 17.19. Estimated trend obtained by Kalman smoothing and 95% confidence bands for Nephrops CPUE at station 1. The x-axis shows time (in years) and the y-axis the values of the trend.

The framework in equation (17.11) allows for all kinds of exotic models, for example seemingly unrelated time series, dynamic regression models and also the DFA. Let us start with the first. Suppose we have 11 CPUE time series, which we assume are unrelated so that each has its own trend and dependency on the explanatory variables but where the error components are interacting:

$$\begin{aligned} \mathbf{y}_t &= \mathbf{z}_t + \mathbf{B}\mathbf{X}_t + \boldsymbol{\varepsilon}_t \\ \mathbf{z}_t &= \mathbf{z}_{t-1} + \boldsymbol{\eta}_t \\ \boldsymbol{\varepsilon}_t &\sim N(0, \sigma^2 \mathbf{V}) \end{aligned} \tag{17.14}$$

So, we have 11 trends ($\mathbf{z}_t$ is of dimension 11-by-1), and effects of explanatory variables, but the noise components are related to each other using a non-diagonal error covariance matrix $\mathbf{V}$. Indeed, we have seen this before within the context of generalised least squares!

Another interesting extension is to allow the regression parameters for the explanatory variables to change over time. For example, the NAO may only have an effect on some years. For a univariate structural time series, this can be modelled as

$$\begin{aligned} y_t &= \begin{pmatrix} 1 & NAO_t \end{pmatrix} \begin{pmatrix} z_t \\ \beta_t \end{pmatrix} + \varepsilon_t \\ \begin{pmatrix} z_t \\ \beta_t \end{pmatrix} &= \begin{pmatrix} 1 & 0 \\ 0 & 1 \end{pmatrix} \begin{pmatrix} z_{t-1} \\ \beta_{t-1} \end{pmatrix} + \begin{pmatrix} 1 & 0 \\ 0 & 1 \end{pmatrix} \begin{pmatrix} \eta_{1t} \\ \eta_{2t} \end{pmatrix} \end{aligned} \tag{17.15}$$

Note the difference with equation (17.13). In this model the slope β is allowed to change over time. The estimated smoother and the slope are presented in Figure 17.20. Note that the NAO is only significant during the first 8 years!

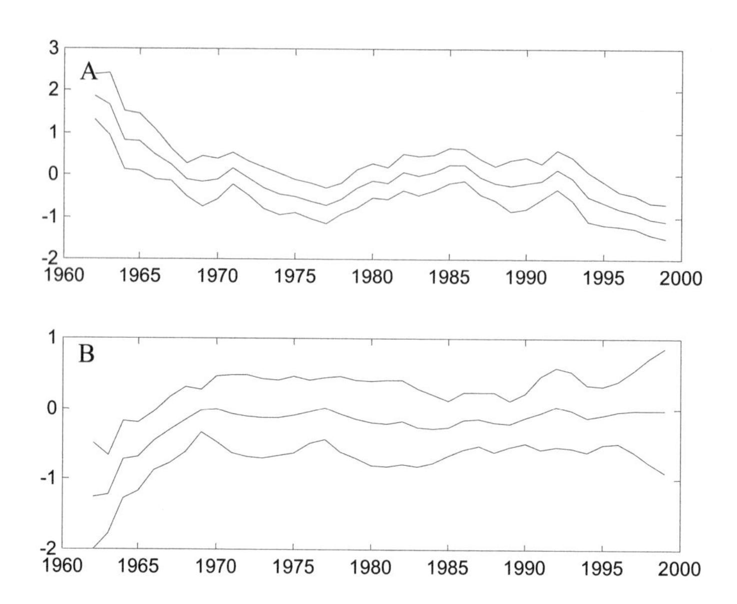

Figure 17.20. Smoothing trend (A) and effect of NAO (B) obtained by dynamic regression.

The model can easily be extended to have multiple response variables. The first part of the state-space formulation becomes:

$$\begin{pmatrix} y_{1,t} \\ y_{2,t} \\ y_{3,t} \\ \vdots \\ y_{11,t} \end{pmatrix} = \begin{pmatrix} 1 & 1 & 0 & \cdots & 0 \\ 1 & 0 & 1 & & \vdots \\ \vdots & \vdots & & \ddots & 0 \\ 1 & 0 & \vdots & 0 & 1 \end{pmatrix} \begin{pmatrix} z_t \\ \beta_1 \\ \beta_2 \\ \vdots \\ \beta_{11} \end{pmatrix} + \varepsilon_t \tag{17.16}$$

The mathematical formulation of the dynamic factor analysis is actually much easier. Full details can be found in Zuur et al. (2003a), Harvey (1989) and Lütkepohl (1991). For the 11 Nephrops series we used a model with two common trends. The full underlying model is:

$$\begin{pmatrix} Y_{1,t} \\ Y_{2,t} \\ \vdots \\ Y_{11,t} \end{pmatrix} = \begin{pmatrix} a_{1,1} & a_{2,1} \\ a_{1,2} & a_{2,2} \\ \vdots & \\ a_{1,11} & a_{2,11} \end{pmatrix} \begin{pmatrix} z_{1,t} \\ z_{2,t} \end{pmatrix} + \begin{pmatrix} \varepsilon_{1,t} \\ \varepsilon_{2,t} \\ \vdots \\ \varepsilon_{11,t} \end{pmatrix}$$

$$\begin{pmatrix} z_{1,t} \\ z_{2,t} \end{pmatrix} = \begin{pmatrix} 1 & 0 \\ 0 & 1 \end{pmatrix} \begin{pmatrix} z_{1,t-1} \\ z_{2,t-1} \end{pmatrix} + \begin{pmatrix} \eta_{1,t} \\ \eta_{2},t \end{pmatrix}$$

The technical aspects of estimating the parameters, of how to deal with missing values, including explanatory variables, etc. are rather complex and are outside the scope of this book. The interested reader is referred to Zuur et al. (2003a) for these details but be prepared for some complex mathematics. There are three case study chapters that make use of DFA.

17.5 Sudden changes: Chronological clustering

MAFA and dynamic factor analysis are techniques, that can be used to estimate trends in multivariate time series. Application of these techniques on biological data assumes that the underlying ecosystem is gradually changing over time. The reason for this is that the mathematical mechanism is based on smoothing techniques, and these are less suitable to capture fast changing patterns. So, for ecosystems that change rapidly from one state into another, common long-term trend estimation is less suitable.

If an ecosystem is under study that changes rapidly from one stage to the next, then identifying years (assuming that year is the time unit) in which the ecosystem is the same, and when it is changing, becomes one of the aims of the analysis.

Standard multivariate methods like principal component analysis, correspondence analysis, etc. produce continuous gradient, and are also less suitable (i) to extract discontinuous gradients and (ii) identify breakpoints.

Cluster analysis

Cluster analysis is a multivariate method that can be used if one knows *a priori* that the observations (e.g., sites or time units) form groups. Cluster analysis can also be applied on time series, leading to groups of years in which the variables are similar. Figure 17.21 shows a so-called dendrogram for the Nephrops time series dataset. Recall that the dataset consists of Nephrops abundances measured at 11 sites south of Iceland since the early 1960s. A dendrogram shows the arrangement of years in clusters, as obtained by the cluster algorithm, and we will explain shortly how it is created. The main problem with clustering applied on time series is to explain groupings of non-sequential years. For example, in Figure 17.21 we have 1964, 1965, 1966, 1985 and 1986 in one group; hence these years are similar, but the ecological reason for this is unclear.

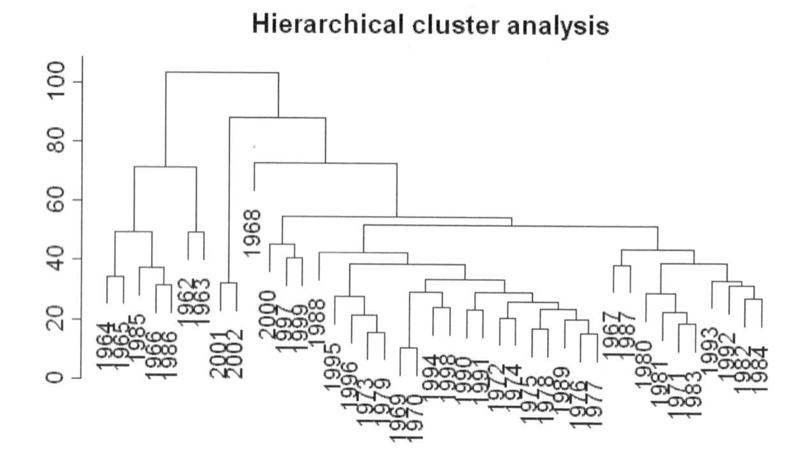

Figure 17.21. Results of hierarchical cluster analysis presented as a dendrogram. The time series were normalised prior to the cluster analysis, and hierarchical clustering using the Euclidean distance function was used with the average linkage. The vertical axis shows at which value of the Euclidean distance groups were fused.

Chronological clustering, as the name already suggests, is specially designed for clustering of time series. The method is fully described in Legendre et al. (1985), Bell and Legendre (1987), and Legendre and Legendre (1998), and a short description is given here. Before we can explain chronological clustering, we need to explain clustering, as this method has not been discussed in this book yet. A short explanation of the basic underlying principle of clustering is given here. A full discussion can be found in, for example, Legendre and Legendre (1998), Krzanowsky (1988), among many other books. The subject deserves more space than the two pages we dedicate to it. Clustering is probably the most misused technique in multivariate analysis. First of all, one should only apply it if there is prior knowledge that the observations (or variables) form groups. Formulated slightly stronger, a justification for why clustering is applied should be given! Possible justifications are that sampling took place in different areas or countries, or in different habitats.

Cluster analysis will always produce a grouping structure. The main problem with clustering methods is that one not only has to choose a measure of association, but also the grouping rule, and the type of clustering method itself. It is possible to produce a large number (>50) of different clustering structures by changing the settings of the software and choose the one that suits the report; it has a high cheating potential. Basically, if you apply clustering methods, you need to know exactly what all options in the software mean.

Standard cluster analysis methods usually start with a similarity or dissimilarity matrix. Table 17.4 shows an artificial dissimilarity matrix for four sites, and we will use these data to explain the principle of agglomorative clustering. A high value in the dissimilarity matrix means that two sites are not similar. Cluster analysis applied on this matrix forms groups of similar sites. In the first step, it will compare all dissimilarities and it chooses the two sites that are the most similar. These are sites B and C, and they are fused in one group. So, we now have group 1 (formed by sites B and C), site A and site D. The next step is visualised in Figure 17.22. We need to know the dissimilarity between group 1 and A, and also between group 1 and D. Once we know all three dissimilarities, we can choose the smallest value and fuse the corresponding sites/group. This process can then be repeated until all sites/groups are fused into one single group. The problem is now, how do we get the dissimilarity between a group and a site? And a relevant question (for the next stage) is how to calculate the dissimilarity between two groups.

Table 17.4. An artificial dissimilarity matrix for four sites. The numbers in the cells are dissimilarity values among four sites, labeled A to D. The smaller a value is, the more similar are the two sites.

	A	B	C	D
A	–	0.4	0.7	0.9
B		–	0.3	0.5
C			–	0.6
D				–

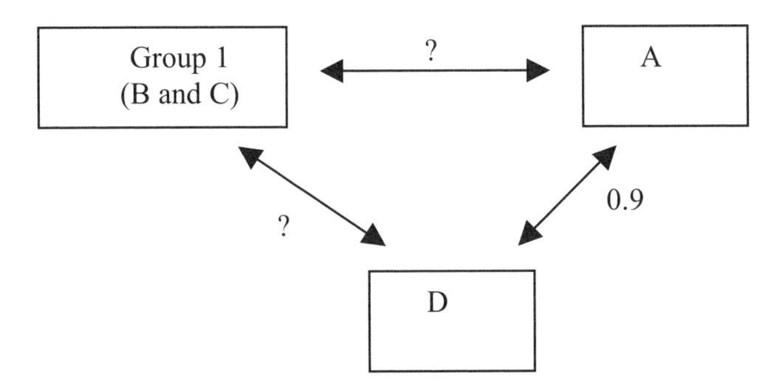

Figure 17.22. Visualisation of the second step in the clustering process. In the first step, sites B and C were fused. In the next step, we need to define the dissimilarity between group 1 and A, and between group 1 and D, and fuse the groups/sites with the smallest value.

There are different ways of quantifying this dissimilarity. The easiest (but not best) way is called single linkage. For group 1 and site A it works as follows. Within group 1, we have sites B and C. The dissimilarity between B and A was 0.4, and between C and A it was 0.7. Single linkage will choose the smallest of these, 0.4. Hence, the question mark between group 1 and A will be replaced by 0.4. The same process will give a dissimilarity of 0.5 between group 1 and D (minimum of 0.5 and 0.6). Therefore, the algorithm will fuse group 1 and A. Instead of taking the minimum value of the dissimilarities between two groups (or one group and a site), we could use the average (average linkage) or the maximum (maximum linkage), among many other options. The choice for linkage will greatly determine the outcome! The average linkage is the default value of most software packages.

The results of the clustering process (as described above) can visually be presented in a dendrogram, and we have already seen an example (Figure 17.21). The first two sites (B and C) were fused at the value 0.3. Site C was fused into this group at the value 0.4. So, we can visualise this in a vertical tree structure in which B and C are group first, then B and C with A, and then B, C, A with D. A vertical axis can be added that shows at which value groups fuse.

Chronological clustering

Chronological clustering works in a similar way as ordinary clustering. In the first step of the algorithm, a dissimilarity matrix between the time units is calculated. For simplicity, let us assume that the time unit is years. The only difference with ordinary clustering is that (i) the fusing rule is slightly different, and (ii) candidate groups (or years) for fusing are restricted to be sequential in time. Table 17.5 shows the same artificial data as in Table 17.4, except that we have replaced the site labels A–D by the years 2002–2003. Due to the restriction of sequential years, we can only fuse 2000–2001, 2001–2002 or 2002–2003. In this case, we would fuse 2001 and 2002 in one group as it has the lowest dissimilarities. In the next step, we only consider fusing 2000 with group 1 (=2001 and 2002) and group 1 with 2003. Fusing 2000 with 2003 is not an option as these years are not sequential. The fusing process itself is based on a permutation test similar to the Mantel test (Chapter 10), and further details can be found in Legendre and Legendre (1998). An example is given next.

Table 17.5. An artificial dissimilarity matrix for four years.

	2000	2001	2002	2003
2000	–	0.4	0.7	0.9
2001		–	0.3	0.5
2002			–	0.6
2003				–

Example chronological clustering

Explaining chronological clustering is best done with an example. Hare and Mantua (2000) used 100 biological and physical time series from the North Pacific Ocean. These were variables like atmospheric indices (teleconnection index, North Pacific index, Southern oscillation index, Arctic oscillation, etc.), terrestrial indices, oceanic indices and biological indices (zooplankton biomass, Coastal Washington oyster condition index, etc.). Most of the time series were available annually from 1965 onwards. Hare and Mantua (2000) concluded that there were two major shifts in these time series, namely in 1977 and 1989.

Here, we show that chronological clustering identifies the same regimes. Chronological clustering requires two parameters to be set, namely the connectedness and the fusion level alpha, which is a clustering sensitivity parameter. Legendre et al. (1985) suggested to use different values of alpha and to keep the connectedness constant. The effect of alpha is as follows. Small values (0.05, 0.01, 0.1) provide a birds-eye overview, and the most important breaks in the time series are visualised. Higher values of alpha (0.2, 0.3, 0.4) give more detailed information and therefore show more breaks in the time series.

The results for chronological clustering are given in Figure 17.23. Vertical lines identify the breaks, and the numbers along the lines represent the groups. Figure 17.23 shows that at the birds-eye view, there are two major breaks, namely in 1977 (a bar represents the first year of a new group) and 1989. For larger values of alpha, we can see that the 1980s were reasonably stable, and that more (short term) variation occurred during the 1970s.

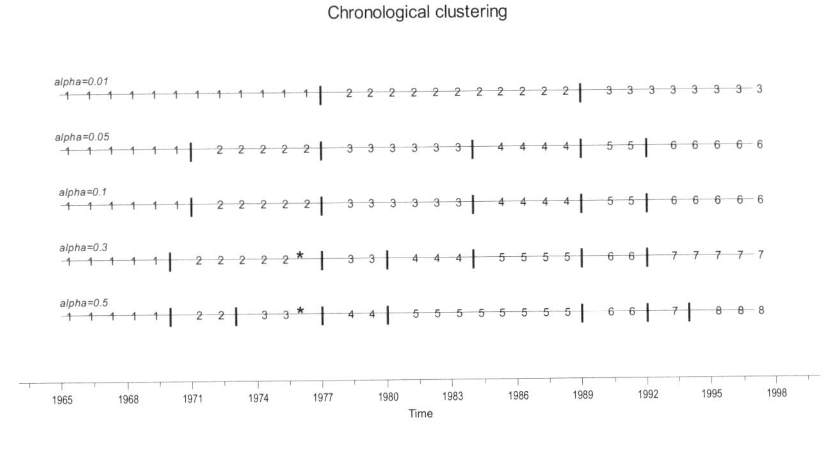

Figure 17.23. Results of chronological clustering applied to the 100 time series analysed in Hare and Mantua (2000). A vertical line corresponds to the start of a new group. Numbers refer to groups. The smallest alpha value (0.01) gives the most important breakpoints.

Legendre et al. (1985) developed a posterior test. It can be used to test if, for example of groups 1 and 3 (for alpha=0.01) belong to the same group. Formulated differently, one might ask the question whether the ecosystem changes back to its original state. If this is indeed the case, the posterior test would indicate that groups 1 and 3 were similar. The stars in the figure are so-called singletons. This is a point that does not belong to the group immediately before and after it. See Legendre et al. (1985) for a detailed interpretation of singletons.

Just as in PCoA and NMDS (Chapter 15), there is one other important point we need to discuss, namely the measure of similarity between time points (years in this case). Suppose that the time-by-variable matrix is in the following format:

	$\mathbf{Y_1}$	$\mathbf{Y_2}$	$\mathbf{Y_3}$	$\mathbf{Y_4}$
$\mathbf{T_1}$	Y_{11}	Y_{12}	Y_{13}	Y_{14}
$\mathbf{T_2}$	Y_{21}	Y_{22}	Y_{23}	Y_{24}
$\mathbf{T_3}$	Y_{31}	Y_{32}	Y_{33}	Y_{34}
$\mathbf{T_4}$	Y_{41}	Y_{42}	Y_{43}	Y_{44}
$\mathbf{T_5}$	Y_{51}	Y_{52}	Y_{53}	Y_{54}
$\mathbf{T_6}$	Y_{61}	Y_{62}	Y_{63}	Y_{64}
$\mathbf{T_7}$	Y_{71}	Y_{72}	Y_{73}	Y_{74}
$\mathbf{T_8}$	Y_{81}	Y_{82}	Y_{83}	Y_{84}
$\mathbf{T_9}$	Y_{91}	Y_{92}	Y_{93}	Y_{94}

Chronological clustering calculates the association between the rows T_1 and T_2, between the rows T_2 and T_3, etc. Legendre et al. (1985), Bell and Legendre (1987) and Legendre and Legendre (1998) used Whittaker's index of association in combination with chronological clustering. Its mathematical formulation is given by:

$$D(T_1,T_2) = 0.5 \times \sum_{j=1}^{p} \left| \frac{Y_{1j}}{Y_{1+}} - \frac{Y_{2j}}{Y_{2+}} \right|$$

The '+' stands for row totals and p is the number of variables (species). This index transfers a row in the table into a row of fractions (of the row total), and then it compares two rows by taking the sum of the absolute differences of the fractions. The index should only be used if the values are non-negative. If the variables are on a different scale (e.g., Y_1 goes from 0 to 1 and Y_2 from 1 to 1000), or if they are different types of variables (e.g., Y_1 is species abundance, Y_2 is the NAO index, Y_3 is temperature and Y_4 is wind speed), a sensible approach is to standardise the variables (Y_j's) first, and then use the Euclidean or absolute difference metric. Figure 17.23 was obtained by standardising the 100 time series and using the Euclidean distance function.

18 Analysis and modelling of lattice data

Saveliev, A.A., Mukharamova, S.S. and Zuur, A.F.

18.1 Lattice data

In this chapter we consider statistical techniques for analysing spatial units arranged in a lattice pattern. A lattice structure is created when a landscape or region is divided into sub-areas (Cressie 1993). The sub-areas can also be called cells, units or locations. None of the sub-areas can intersect each other, but each shares a boundary edge with one or more of the other sub-areas. An example of a lattice is shown in Figure 18.1. A *regular* lattice is formed if all of the cells have the same form and size. Regular lattices are usually obtained if a region is divided into cells based on a regular grid (e.g., Figure 10.3 for the bird radar data). If a region is divided into cells based on the outlines of natural objects, such as river basins, national boundaries, counties, or postal codes, an *irregular* lattice results. The lattice shown in Figure 18.1 is an example of an irregular lattice.

Figure 18.1. Hypothetical irregular lattice. The dots represent the arbitrarily chosen centre of each unit (also called a cell) and the numbers identify each unit.

A typical example of spatial data arranged in a regular lattice pattern is remotely sensed satellite imagery. Satellite data provide spatially distributed information on soil, topography, vegetation, surface temperature, and much more.

These data can also be collected at a variety of spatial resolutions. Image areas can be divided into cells based on a regular grid, and data values for each variable on the grid can be analysed. A wide variety of statistical techniques are used to process satellite imagery. Statistical models based on regular lattices are among the most powerful.

In many biological studies, field data are collected at sites or stations that occur in more irregular configurations than satellite data. The data from these sites are assumed to be representative of the characteristics at the sample unit's position within the lattice. Each unit is usually compared for differences with other sampling sites. Some ecosystem parameters, like regional species richness, however, can only be estimated if the data are aggregated. Aggregation can be done in two ways, namely by (i) using natural spatial units like landscape patches or (ii) using an arbitrary spatial unit like a county or forest inventory stand. Aggregating areas, although sometimes necessary, can cause problems if the phenomenon under study extends beyond the boundaries of the study area. If arbitrary units are used for the analysis, an additional problem called the *modifiable areal unit problem*, can result, in which the size of the spatial unit on which aggregation is applied influences the correlation between the variables. When the size of the cells match the phenomenon under investigation, however, aggregating cells can be a powerful analysis tool, e.g., Openshaw and Taylor (1979), Fotheringham and Wong (1991) or Cressie (1996).

Spatial patterning within the lattice structure of a ecological habitat depends on the interaction of several different forces acting at different spatial scales. These forces range from global climatic factors to local microclimate variations caused by differences in relief and soil characteristics that affect moisture and nutrition availability.

Most of the statistical techniques applied on data that are arranged like a lattice structure take spatial patterning of neighboring cells into account. The purpose of the statistical analysis of lattice data is (i) the detection of spatial patterns in the values over the lattice (rainfall in connected units may be similar) and (ii) an explanation of these patterns in term of covariates measured at the same cells. The dependence (auto-correlation) of neighbouring values requires that special statistical techniques be used to avoid the bias of statistical inference results because most parametric statistical analyses are based on assumptions of independence among samples (Legendre 1993). The analysis of spatial dependence, also referred to as spatial auto-correlation or spatial association, originates from Tobler's (1979) First Law of Geography: 'everything is related to everything else, but near things are more related than distant things'.

Notation

To study spatial associations in data, we assume that the data can be thought of as a random process **Y** on a lattice D. The lattice is then a fixed (regular or irregular) collection of spatial objects, and each object has a distinct neighbourhood structure (see Figure 18.1). Mathematically, S denotes the region under study. A_i

denotes the spatial units within S, and S is part of a two-dimensional space $\mathbf{R}^2$. These notations have the following relationships:

$$D = \{A_1, A_2, ..., A_n\} \quad A_i \subset S \subset \mathbf{R}^2 \quad A_1 \cup A_2 \cup ... \cup A_n = S,$$

The first formula defines a lattice consisting of n units (or cells or objects with aerial extent). Within the lattice in Figure 18.1, there are n individual units (A_1 to A_n). The second formula tells us that each unit is contained within the region of interest and the region is part of a two-dimensional space. The third formula designates that the n units together cover the entire study area S. Finally, we need to verify that the lattice consists of units that do not overlap, or that:

$$A_i \cap A_j = \varnothing \text{ for } i \neq j$$

A crucial point, also emphasised in Schabenberger and Pierce (2002), is that all possible units can be enumerated. Hence, n is finite. In Figure 18.1 we can see that none of the units overlap. The random variable at location A_i is defined as $Y_i = Y(A_i)$ and $\mathbf{Y}$ contains the random variables at all n units: $\mathbf{Y} = (Y_1,..,Y_n)$. We denote the observed value for object A_i as y_i and the vector of all n observed values as $\boldsymbol{y} = (y_1,..,y_n)$. So, if we were interested in daily rainfall in Figure 18.1, we have a vector $\boldsymbol{y} = (y_1,..,y_{10})$, and each element represents the measured rainfall in a particular unit. We will use the usual statistical concepts of first- and second-order moments of the variable of interest. These correspond to modelling first-order variation in the mean value $\mu_i = E[Y_i]$ simultaneously with second order variation or spatial dependence between Y_i and Y_j, that is, the covariance between values at two objects (Bailey and Gatrell 1995).

In the analysis of lattice-arranged data, we have to take into account spatial correlation between values at adjacent units. Units with high values (or: above average) are likely to be located near other units with high values; and units with low values should be adjacent to other units with low values. Alternation of neighbouring areal units with high and low values gives negative spatial correlation. The term 'auto-correlation' is used to denote the spatial dependence of the same variable, but the usage of the term 'correlation' instead of the 'auto-correlation' is a common practice that will also be followed in this chapter.

18.2 Numerical representation of the lattice structure

We now discuss how spatial relationships can be quantified using a spatial weight matrix $\mathbf{W}$ in which the elements represent the strength of the spatial structure between the units (Cliff and Ord 1973; Anselin et al. 2004). This spatial weight matrix will be used to calculate the spatial auto-correlation. There are various ways of defining $\mathbf{W}$, and the choice of which one to choose is subjective. All of them are based on the concept of the neighborhood of a unit (A_i). The easiest option is described in Cliff and Ord (1981) and Upton and Fingleton (1985). Construct a binary contiguity matrix (only zeros and ones) by specifying the units that are adjacent (one), and those that are not (zero). As an example, consider the irregular lattice in Figure 18.1. **W** is of dimension 10-by-10, and its ij^{th} element w_{ij} is given by

$$w_{ij} = \begin{cases} 1 & \text{if the } A_i \text{ and } A_j \text{ have a common border} \\ 0 & \text{otherwise} \end{cases}$$

By definition we have $w_{ii} = 0$. The resulting matrix **W** is in Table 18.1. One of the problems with this definition of the w_{ij}'s is that the common borders between units vary in length. Some areas are only connected by a short border; see for example the pairs (4,8) or (5,6) in Figure 18.1. Cells with such a connection are coloured in grey in Table 18.1. Other units share a longer expanse of border.

Schabenberger and Pierce (2002) used the movements of the chess pieces rook, bishop and queen ('king' would have been more appropriate) to define **W**. Figure 18.2 shows the movements of these pieces. Using the rook movement to define the neighborhood, if A_i is the black unit in the middle, then there are only four other units that have a common border with A_i. The queen movement produces six units with a common border. The matrix **W** in Table 18.1 can be seen as some sort of queen movement for an irregular lattice.

Other options, especially for an irregular lattice, to define **W** are given in Haining (1990) or Schabenberger and Pierce (2002). Using the notation of the latter:

- $w_{ij} = \|A_i - A_j\|^{-\gamma} \qquad \gamma \geq 0$
- $w_{ij} = \exp(\|A_i - A_j\|^{-\gamma})$
- $w_{ij} = (l_{ij} / l_i)^{\gamma}$
- $w_{ij} = (l_{ij} / l_i) / \|A_i - A_j\|^{-\gamma}$

The underlying principle of these definitions is simple; $\|A_i - A_j\|$ defines the spatial separation between units. The closer in space two units A_i and A_j are, the larger the weighting factor w_{ij}. In the last two definitions, l_{ij} is the length of the common border between units A_i and A_j, and l_i is the perimeter of unit A_i. Basically we are defining association between spatial units as a function of physical distance between each unit. Because the distance between regions cannot be uniquely defined, the centre or any other meaningful point can be used. Euclidean distances (Chapter 10) between the ith and j^{th} unit centers are given in Table 18.2.

Table 18.1. Matrix **W** for the objects. An '1' indicates that two areas have a joint border. Empty cells represent zeros (no border in common). Grey cells indicate that the two areas have a short border.

	1	2	3	4	5	6	7	8	9	10
1		1		1						
2	1		1	1						
3		1		1						
4	1	1	1		1	1		1		
5				1		1	1	1	1	
6				1	1			1		
7					1			1	1	
8				1	1	1	1		1	1
9					1		1	1		1
10								1	1	

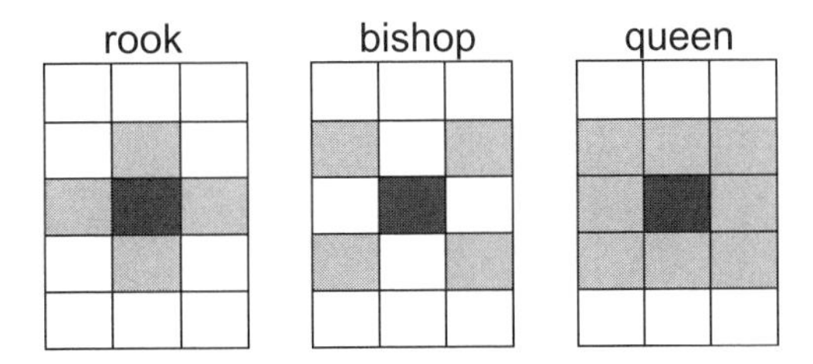

Figure 18.2. Movements of the chess pieces rook, bishop and queen (or actually the king) defining units with a common border for a regular grid.

Table 18.2. Centroid based distance matrix (km). Each value represents the Euclidean distance between the centres of two units. The lower diagonal elements are equal to the upper diagonal elements and were omitted.

	2	3	4	5	6	7	8	9	10
1	5.8	7.2	6.4	7.3	12.9	7.9	13.7	12.6	15.5
2		3.6	6.4	9.4	11.8	11.4	14	15.2	16.5
3			4.1	7.5	8.3	10	10.9	12.9	13.6
4				3.4	6.5	6	7.8	8.9	10.2
5					7	2.8	6.7	5.8	8.1
6						9.3	3.6	8.6	6.5
7							8.1	4.8	8.6
8								5.7	3
9									4.6

After a distance matrix is created using one of the methods above, it is converted into a contiguity matrix. A contiguity matrix is a binary matrix (only zeros and ones) specifying the units that are adjacent (one), and those that are not (zero). In order to do this for Table 18.2, we first need to define a distance cut off point.

Two objects are then considered to be contiguous if their centroids are less than the specified 'cut distance' apart. Figure 18.3 shows the distances from unit 8. The smallest inner circle in Figure 18.3 represents the area that is within 3 km from the centroid of area 8. If we use a 'cut distance' of 3 km, then only unit 10 falls into the neighbourhood of object 8. Hence, for unit 8, we would designate the relationship of the each unit to unit 8 with the following row:

	1	2	3	4	5	6	7	8	9	10
8	0	0	0	0	0	0	0	0	0	1

where '1' designates the only site (10) that is within a radius of 3 kilometres of point 8 and the zeros indicate the units that are outside this interval.

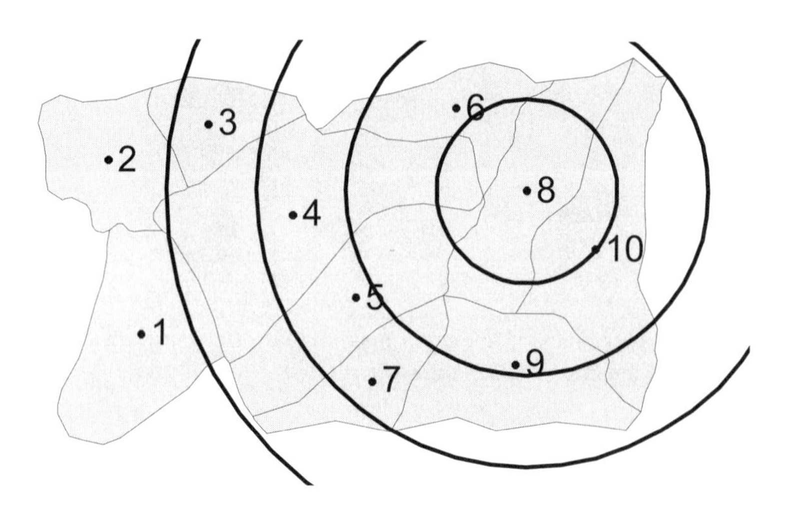

Figure 18.3. Increasing circular neighbourhoods around the object 8. The smallest inner circle has a radius of 3 km and shows that only the centroid of unit 10 is within 3 km of the centroid of unit 8.

Increasing the cut distance to 6, 9 and 12 km, gives the neighbourhoods {6, 9, 10}, {4, 5, 6, 7, 9, 10} and {3, 4, 5, 6, 7, 9, 10}, respectively. So, for the last example, we have:

	1	2	3	4	5	6	7	8	9	10
8	0	0	1	1	1	1	1	0	1	1

This means that only objects 1 and 2 are not within a radius of 12 km of object 8. Hence, we now have a mechanism to convert the distances in Table 18.2 into a contiguity matrix containing only zeros and ones based on the value of a cut distance. Mathematically, the indictor function can be used for this process. An indictor function is a useful mathematical tool of the form:

$$I(x) = \begin{cases} 1 & \text{if } x \text{ is true} \\ 0 & \text{is } x \text{ is not true} \end{cases}$$

This function can be used as follows. Define $w_{ij}^{(k)}$ as

$$w_{ij}^{(k)} = w_{ij}(d_k, d_{k+1}) = \mathrm{I}(d_k < d_{ij} \le d_{k+1})$$

where the d_k are the cut distances and the indices i and j refer to centroids. Suppose we use the following cut-distance sequence {0, 3, 6, 9, 12, 15}. For unit $i = 8$, $w_{ij}^{(0)}$ is defined by

$$w_{8j}^{(0)} = w_{8j}(d_0, d_1) = \mathrm{I}(d_0 < d_{8j} \le d_1) = I(0 < d_{8j} \le 3)$$

And this gives the same row of zeros and ones as on the previous page:

	1	2	3	4	5	6	7	8	9	10
8	0	0	0	0	0	0	0	0	0	1

The information in the form of zeros and ones tells us which objects are within a 3 km radius of object 8. We can do the same for

$$w_{8,j}(3,6),\ w_{8,j}(6,9),\ w_{8,j}(9,12), w_{8,j}(12,15)$$

For example, $w_{ij}^{(1)}$ gives $I(3 < d_{8j} < 6)$, which are all objects that have a distance between 3 and 6 kilometres from object 8. All these terms provide a row of zeros and ones. For $w_{ij}^{(1)}$ we have:

	1	2	3	4	5	6	7	8	9	10
8	0	0	0	0	0	1	0	0	1	0

18.3 Spatial correlation

In previous chapters on time series analysis, we introduced terms like auto- and cross-correlation to quantify the relation within and between time series. In spatial statistics, we use similar tools that allow us to analyse the second-order properties, i.e., the dependence of the variable of interest (e.g., rainfall) between the spatially separated locations. The general formulation is based on a cross-product of the following form (Cliff and Ord 1973; Hubert et al. 1981; Getis 1991):

$$\sum_i \sum_j w_{ij} U_{ij}$$

where U_{ij} is a measure of dissimilarity between the measured variable of interest (e.g., rainfall) at the units i and j and w_{ij} are spatial weights as defined in Section 18.2. We will discuss one such dissimilarity measure in this section, namely Moran's I (Moran 1950) .

The Morans I coefficient

The Moran I coefficient is used to quantify the degree of spatial correlation between neighbouring units. It is defined by

$$I = \frac{n}{w_{++}} \frac{\sum_{i=1}^{n}\sum_{j=1}^{n} w_{ij}(y_i - \bar{y})(y_j - \bar{y})}{\sum_{i=1}^{n}(y_i - \bar{y})^2} = \frac{n}{\mathbf{1'W1}} \frac{\mathbf{u'Wu}}{\mathbf{u'u}} \tag{18.1}$$

In this equation, n is the number of units in the lattice, $w_{++,}$ is the sum of all weights w_{ij}, y_i is the value of the variable of interest (e.g., rainfall) in object i, and $\bar{y}$ is the mean value of the variable for the whole region. The weights w_{ij} are obtained by any of the methods discussed in Section 18.2. The rightmost part of the equation is just matrix notation for the same thing; **1** is a vector of ones, and **u** contains the centred elements y_i.

The term $w_{ij}(y_i - \bar{y})(y_j - \bar{y})$ is used as a dissimilarity measure (see also the definition of the covariance and correlation coefficients in Chapter 10). In fact, the interpretation is similar to the correlation coefficient. If the values at two adjacent objects are both above average, or below average, then this suggests that there is a positive spatial correlation. If one value is above and the other is below average, a negative correlation is suggested.

Under the null hypothesis of no spatial correlation, the expected value of I is $E[I] = -1/(n - 1)$, which is close to 0 for large n. A p-value can be obtained by either assuming asymptotic normality or using a permutation test; see Cliff and Ord (1981), Anselin et al. (2004) or Schabenberger and Pierce (2002) for details. An I value larger than $-1/(n - 1)$ means that similar values of the variable of interest, either high or low, are spatially clustered. Negative spatial auto-correlation is harder to interpret.

Example of the Moran I index for tree height data

Figures 18.4A-B show the locations of trees in two 20-by-20 m square plots situated in the Raifa section of the Valga-Kama State reserve (Tatarstan, Russia; see also Chapter 37). Data on the spatial distribution, height, and diameter at breast height (dbh) for five tree species, including *Betula pendula Roth.*, *Acer platanoides L.*, *Tilia cordata Mill.*, *Pinus sylvestris L.* and *Picea* × *fennica (Regel) Kom.*, were collected by Rogova and co-workers at the Faculty of Ecology, Kazan State University. To create the lattice the Voronoi tessellation[1] (Moller 1994) was used; see Figures 18.4C-D.

[1] The word 'tessellation' means that a particular shape is repeated a large number of times, covering a plane without gaps or overlaps (http://mathforum.org/). Another word for tessellation is tiling, and it is derived from the Greek word *tesseres* which means four. Just think of the square tiles in the bathroom. But just as in the bathroom, the tiles do not have

The Moran I coefficient is used to investigate whether there is any spatial autocorrelation in tree height. The null hypothesis is that there is no spatial correlation. The Moran's I statistic for height of trees for the plot-1-6 data is 0.246. A permutation test on the tree height data indicates that the null hypothesis can be rejected ($p < 0.001$). For plot plot-22-7, the I value is 0.021 ($p = 0.31$) indicating that there is no evidence to reject the null hypothesis. These results suggest, therefore, that plot-1-6 shows spatial correlation of tree heights, but plot-22-7 does not.

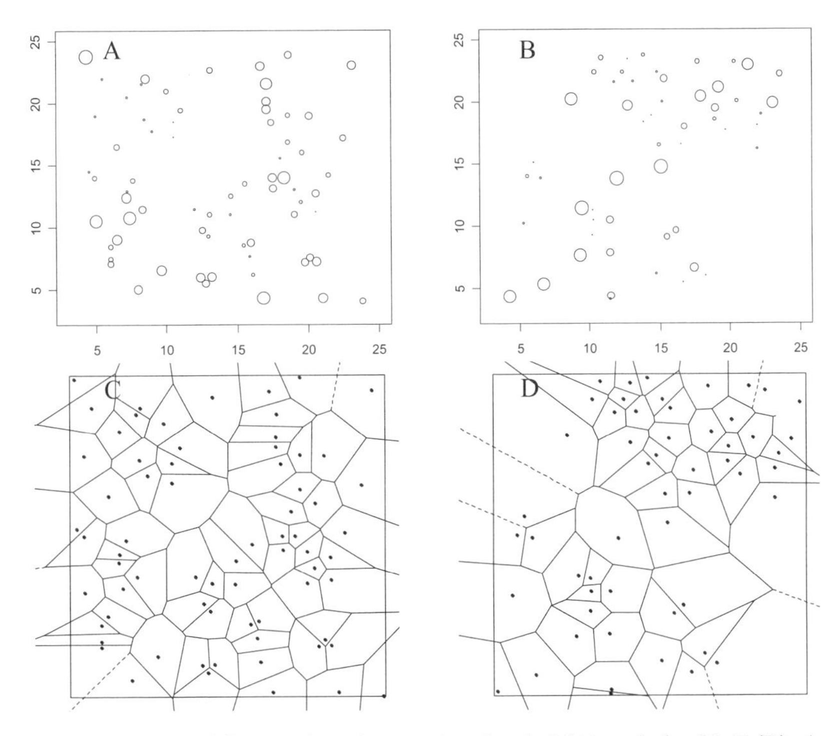

Figure 18.4. Map of the tree locations at the plot-1-6 (A) and plot-22-7 (B). A circle represents the spatial position of a tree, and the diameter of the circles is proportional to tree height. Tree diameter will be used later as an explanatory variable of tree height. C and D: The tree locations were converted into a lattice using so-called Voronoi tessellation. In this process cell border is created; see also Moller (1994).

to be squares. Roman mosaic, for example, can have a huge variety in tessellation patterns. Voronoi diagrams are a special form of tessellation patterns and have been used in many applications. Data other than lattice, e.g., points pattern, can be converted to the lattice using Voronoi tessellation (Moller, 1994). We applied this nearest-neighbourhood tessellation, and as a result the lattice borders are drawn at half the distance between the points.

The spatial correlogram

The Moran I index depends on the choice of how the w_{ij}'s are defined. The w_{ij}'s are zero if the distance is larger than a threshold value or they are equal to one if they are smaller than the threshold. So, choosing the proper threshold value to calculate this index is crucial. To expand the information that we can get from a variable of interest, a range of threshold values can be used. The weights w_{ij} depend on *cutting values*, as defined by the spatial weights $w_{ij}^{(k)}$ (see Section 18.3). We only have to choose a series of cutting distances d_k, d_{k+1}, etc. For each value of k we calculate the Moran statistic. This shows us how the spatial correlation is changing for different distances. The resulting function is called a *spatial correlogram* (Cliff and Ord 1981; Upton and Fingleton 1985). The Moran spatial correlogram is defined by

$$I^{(k)} = \frac{n}{w_{++}^{(k)}} \frac{\sum_{i=1}^{n}\sum_{j=1}^{n} w_{ij}^{(k)}(y_i - \bar{y})(y_j - \bar{y})}{\sum_{i=1}^{n}(y_i - \bar{y})^2} \tag{18.2}$$

Note that the only difference between equations (18.1) and (18.2) is the use of the weighting factors. The index $I^{(k)}$ is calculated for different values of k (0, 1, 2,…), and the graphical presentation is a plot of the index k versus $I^{(k)}$. The graph shows how the strength of the dependence changes with distance between units. A permutation test or assymptotic distribution can be used to obtain critical values and p-values (Cliff and Ord 1981; Anselin et al. 2004).

Example of the Moran spatial correlogram for the tree height data

The Moran spatial correlograms for tree height in plot-1-6 and plot-22-7 are shown in Figure 18.5. The cut distance sequences {2.5, 5, 7.5, 10, 12.5, 15} were used. The Moran index is calculated first using only points that are separated by 2.5 m and then it is calculated with points between 2.5 and 5 m, etc. For plot-1-6, , the Moran's $I^{(k)}$ is significantly different from 0 ($p < 0.001$) for the distance band 7.5-10 m; so there is evidence that trees that are separated by 7.5 to 10 m are dependent. For all other classes in plot-1-6 and for plot-22-7, there is no evidence to reject the null hypothesis of no spatial correlation.

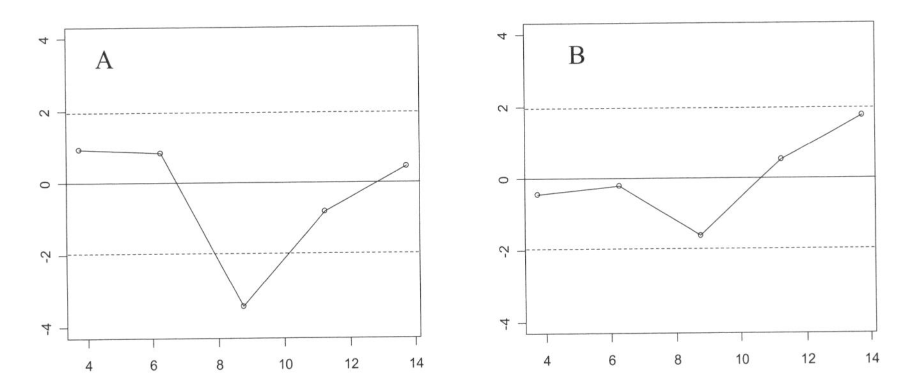

Figure 18.5. Moran's $I^{(k)}$ calculated for trees height at five distance bands (k = 2.5, 5, 7.5, 10, 12.5, 15). A: Plot-1-6. B: Plot-22-7. Expectation (solid line) and two-sided 95% asymptotic critical values (dashed lines) are also given. The y-axis is expressed as a z-score (the difference between $I^{(k)}$ and its mean (over all k) was divided by the standard devition of the $I^{(k)}$s).

18.4 Modelling lattice data

In this section, we will combine regression (and smoothing) models with spatial auto-correlation. Therefore it is essential that you are familiar with the material in Chapters 5 and 7, and knowledge of the time series methods (ARMAX, GLS) in Chapter 16 is also beneficial.

In the previous section, we introduced tools to test whether there is spatial correlation between values at units or cells of lattice-structured data. If the tests indicate that there is indeed spatial auto-correlation, we need to incorporate that correlation within the models that will be applied in the next step of the analysis. Ignoring the spatial auto-correlation structure may cause type I errors and may thus lead to neglecting important exploratory variables and inadequate model selection. The general approach to incorporating auto-correlation into a linear regression model (or any of the extensions discussed in Chapters 6–8) is to model the variable of interest as a function of (i) a systematic part and (ii) a residual component with a covariance structure reflecting fine-scale spatial variation. The systematic part is a phrase used by geostatisticians to express $\boldsymbol{X\beta}$, the effect of explanatory variables. Non-linear effects can easily be incorporated using quadratic terms (Legendre and Legendre 1998) or smoothers (Chapter 7). It is also called the 'spatial trend'. The relevant models are of the form:

Response variable = F(explanatory variables) + spatial correlated noise

Modelling the systematic spatial trend, i.e. the function F of the explanatory variables, is done using linear regression, generalised linear modelling or generalised additive modelling (Chapters 5 to 7).

In previous chapters, noise within the mathematical model was handled in different ways. In Chapters 5 to 7, noise was assumed to be uncorrelated. In the time series chapters (e.g., chapters 16, 23, 26, and 34), noise was allowed to correlate in time, resulting in generalised least squares (GLS) estimation. To incorporate residual spatial structure within the mathematical models, it is common to assume that neighbouring units have similar values. Some random spatial process is causing the residuals to be spatially correlated.

For lattice data, two popular approaches incorporate residual spatial correlation. These include the *conditional auto-regressive model* (CAR) and the *simultaneously auto-regressive* (SAR) *model*. SAR models are widely used and easy to understand. They are most appropriate for inference studies. CAR models are useful for prediction and spatial interpolation. In this chapter, we will only discuss SAR and their related types. See Cressie (1993) for references and a detailed discussion on both CAR and SAR models. Before we present a discussion of the SAR family of models, however, we need to present the results of a linear regression analysis on the tree height data so that we can refer to these results when presenting the SAR models.

Example of linear regression applied on the tree height data

The following linear regression model was applied on the data from plot-1-6:

$$\text{Tree height}_i = \alpha + \beta \text{ Tree diameter}_i + \varepsilon_i$$

The numerical output for the linear regression model is as follows:

Variable	Estimate	Std.Err	*t*-value	*p*-value
Intercept	5.820	0.441	13.19	<0.001
Diameter	0.311	0.021	14.45	<0.001

Residual standard error: 2.622 on 68 degrees of freedom
Multiple R-Squared: 0.754, AIC = 337.547
F-statistic: 208.9 on 1 and 68 df, p-value: < 0.001

The interpretation of this type of numerical output was discussed in Chapter 5. The model shows that there is a positive and significant relationship between tree height and diameter. However, the model also assumes that the residuals are independently distributed. To verify this assumption, we applied the Moran's I test on the residuals. We can do this because we have access to the spatial coordinates of the residuals (Figure 18.4). The result was $I = 0.115$ ($p = 0.026$), which means that there is evidence of spatial auto-correlation in the residuals. The Moran's I coefficient for this specific situation is given by

$$I = (n / \mathbf{1}'\mathbf{W}\mathbf{1}) \times (\mathbf{e}'\mathbf{W}\mathbf{e} / (\mathbf{e}'\mathbf{e})) \tag{18.3}$$

This is a similar matrix notation as in equation (18.1). The residuals $\mathbf{e}$ are from the linear regression of tree height on the explanatory variable tree diameter. The matrix **W** contains the weights (Section 18.2), and details on the distribution of the statistic can be found in Cliff and Ord (1981).

If the test indicates that the spatial error auto-correlation is significant, we violate the underlying assumption of linear regression that requires independent residuals. This means that we cannot trust the *p*-values for the *t*- and *F*-statistics in the above regression analysis. One option to lessen spatial auto-correlation is to try to include more explanatory variables in the model. The residual pattern may be present because an important explanatory variable is missing from the analysis. Alternatively, the SAR model can be applied, which is described next.

Simultaneous auto-regressive model

At this point, you may want to read the time series chapter (Chapter 16), which contains relevant key terminology and notation applicable to the SAR models. In time series, the auto-regressive time series model (AR) of order p is given by

$$Y_t = \alpha + \beta_1 Y_{t-1} + \beta_2 Y_{t-2} + \ldots + \beta_p Y_{t-p} + \varepsilon_t$$

The α and β_js are unknown regression parameters, and the errors ε_i are independent and normally distributed with the zero mean and variance σ^2. So, the explanatory variables in an AR time series model are lagged response variables. For spatial data, we consider the spatial weights of units in the lattice to construct regression models that take into account spatial auto-correlation. The SAR model is defined similarly (Ord 1975):

$$Y_i = \mu_i + \rho \sum\nolimits_j w_{ij}(Y_j - \mu_j) + \varepsilon_i \tag{18.4}$$

The w_{ij} are spatial weights (Section 18.2), Y_i is a random variable corresponding to the ith unit, $E[Y_i] = \mu_i$, ρ is a model parameter, and the errors ε_i are assumed to be independent and normally distributed with the zero mean and variance σ^2. Note the similarity between the AR time series model and the SAR model. Both contain lagged response variables as explanatory variables, and both have the same assumptions on the error term. This explains why we call the model in equation (18.4) a simultaneous auto-regressive model. The parameter ρ measures the strength of the spatial correlation. It is common to assume that $\mu_i = \alpha + \mathbf{X}_i\boldsymbol{\beta}$, and it represents the *F*(explanatory variables) component mentioned above. The intercept can be included in **X** and $\boldsymbol{\beta}$ by using a column with only ones in **X**. As a result we have $E[Y_i] = \mathbf{X}_i\boldsymbol{\beta}$. In matrix notation, the SAR model in equation (18.4) can be written as

$$\mathbf{Y} = \mathbf{X}\boldsymbol{\beta} + \rho\mathbf{W}(\mathbf{Y} - \mathbf{X}\boldsymbol{\beta}) + \boldsymbol{\varepsilon} \quad \text{and} \quad \boldsymbol{\varepsilon} \sim N(\mathbf{0}, \sigma^2\mathbf{I}) \tag{18.5}$$

W contains the weights and **Y**, ε are n-by-1 vectors, and **I** is the identity matrix. **X** is the n-by-m matrix containing explanatory variables and $\boldsymbol{\beta}$ is an m-by-1 vector of the regression coefficients. A common approach is to use the spatial coordinates (e.g., latitude and longitude) as explanatory variables in **X**. Details for parameter estimations can be found in Ord (1975) and Anselin (1988). Standard statistics (pseudo-R^2 and AIC, BIC) for the model, goodness-of-fit and confidence intervals for parameters can be calculated, and parameter significance estimation is based on the asymptotic normality (Ord 1975).

Example of a SAR model for the tree height data

A SAR model was applied on the plot-1-6 tree height data with the same explanatory variables that were used to create the linear model. The following model was applied:

$$Height_i = \alpha + \beta Diameter_i + \rho \sum_j w_{ij}(Height_j - \beta Diameter_i) + \varepsilon_i$$

The numerical output from the fitted SAR model is given below:

Variable	Estimate	Std.Error	*z*-value	*p*-value
Intercept	5.144	1.107	4.645	<0.001
Diameter	0.304	0.023	13.022	<0.001
rho	0.074	0.113	0.660	0.509

Residual standard error: 2.574
AIC: 339.06
LM test for residual auto-correlation test value: 1.553 (p=0.212)

Although adding the auto-regression parameter ρ to the model removes the residual auto-correlation (Moran's I coefficient was calculated for the residuals), ρ is not significantly different from 0 (p = 0.509). Also note that the residual standard error is only slightly lower than the residual of the linear regression model (2.574 against 2.622), but the AIC for the SAR model is slightly higher than the linear regression model (339.06 against 337.55). Just as in linear regression, the model with the lowest AIC is preferred.

18.5 More exotic models

We now discuss a series of models that are closely related to SAR models. They differ from the SAR model discussed above mainly in how the residual error is incorporated into the model.

Spatial moving average model

When we introduced time series techniques in Chapter 16, we first introduced AR models, then moving average (MA) models, and then combined the AR and MA models into ARMAX models. Recall that the MA time series model was given by

$$Y_t = \alpha + \varepsilon_t + \gamma_1 \varepsilon_{t-1} + \gamma_2 \varepsilon_{t-2} + \ldots + \gamma_p \varepsilon_{t-p}$$

The variable Y_t is modelled in terms of current and past error terms with unknown regression coefficients γ_i. We can do something similar for the spatial models. The variable of interest is modelled as a function of explanatory variables and residual patterns from different units, and not surprisingly the resulting model is called a spatial moving average model (SMA). The model is given by (Ord 1975):

$$\begin{aligned} Y_i &= \alpha + \beta_1 X_{1i} + \ldots + \beta_m X_{mi} + u_i \\ u_i &= \lambda \sum\nolimits_j w_{ij} u_j + \varepsilon_i \end{aligned} \tag{18.6}$$

The regression parameters β_i model the effect of the explanatory variables on the response variable Y_i, u_i is spatially correlated noise and λ is the error auto-regression coefficient. If λ is zero, there is no spatial correlation and u_i becomes independently distributed noise. In matrix notation, equation (18.6) becomes

$$\mathbf{Y} = \mathbf{X\beta} + \mathbf{U} \qquad \mathbf{U} = \lambda\mathbf{WU} + \boldsymbol{\varepsilon} \qquad \boldsymbol{\varepsilon} \sim N(\mathbf{0}, \sigma^2\mathbf{I})$$

Y contains the data, and **U** is a vector of the spatially correlated errors.

Example of a SMA model for the tree height data

The following SMA model was applied on the plot-1-6 tree height data:

$$\begin{aligned} Height_i &= \alpha + \beta Diameter_i + u_i \\ u_i &= \lambda \sum\nolimits_j w_{ij} u_j + \varepsilon_i \end{aligned} \tag{18.7}$$

The numerical output of the SMA model is given below:

Variable	Estimate	Std.Error	*z*-value	*p*-value
Intercept	5.806	0.520	11.147	<0.001
Diameter	0.310	0.022	14.006	<0.001
Lambda	0.265	0.175	1.521	0.128

Residual standard error: 2.523
AIC: 337.34

Although adding the nuisance parameter λ to the model gives minor improvement in the standard error (2.523 against the 2.622 for the linear regression model) and AIC (337.34 against 337.55 for the linear regression), it is not significant (*p*-value 0.128). The AICs for the linear regression, SAR and SMA model are, respectively, 337.55, 339.06 and 337.34, which indicates that the SAR model is less adequate in describing tree height than the SMA model, although the significance of the nuisance parameter λ in the SMA model is low ($p = 0.128$).

As mentioned, spatial auto-correlation in the models may be caused by not including an important explanatory variable in the model. For the tree data, we have ignored the species information of the tree (there are four different species). Including 'species' as a nominal explanatory variable to the linear regression model gives:

$$Height_i = \alpha + \beta Diameter_i + factor(Species_i) + \varepsilon_i \tag{18.8}$$

The numerical output shows that adding the nominal variable gives a better overall fit of the model to the data:

Variable	Estimate	Std.Error	*t*-value	*p*-value
Intercept	3.058	1.673	1.828	0.072
Diameter	0.320	0.030	10.423	<0.001
Species2	2.262	1.605	1.409	0.163
Species3	3.910	1.311	2.982	0.004
Species4	0.790	1.792	0.441	0.660

Residual standard error: 2.354 on 65 degrees of freedom
Multiple *R*-Squared: 0.8107
F-statistic: 69.61 on 4 and 65 df, *p*-value: < 2.2e-16
AIC = 325.317

Note that the AIC is now 325.317, which is considerably lower compared with the model without the nominal variable species. Again using the Moran's I test for spatial correlation of the residuals gave $I = 0.041$ ($p = 0.18$) and there was no evidence to reject the null hypothesis of no spatial auto-correlation. At this point we could stop and consider the model in equation (18.8) as the most optimal model. But purely for curiosity, we also applied the SMA equivalent of the model in equation (18.8):

$$\begin{aligned} Height_i &= \alpha + \beta Diameter_i + factor(Species_i) + u_i \\ u_i &= \lambda \sum\nolimits_j w_{ij} u_j + \varepsilon_i \end{aligned} \tag{18.9}$$

The results are as follows:

Variable	Estimate	Std.Error	*z*-value	*p*-value
Intercept	2.827	1.611	1.754	0.079
Diameter	0.325	0.029	10.862	<0.001
Species2	2.441	1.531	1.594	0.110
Species3	3.987	1.261	3.159	0.001
Species4	1.185	1.735	0.683	0.494
Lambda	0.156	0.186	0.841	0.401

Residual standard error: 2.253
AIC: 326.67

The nuisance parameter λ is not significant ($p = 0.401$), and although the residuals standard error has slightly decreased, the AIC favours the linear regression model with species included as a nominal explanatory variable.

Locally linear spatial models

Recall that in Chapter 17 we defined a regression model in which the regression parameters were allowed to change over time (dynamic regression models). We can formulate a similar model for spatial data. In the linear regression model $\mathbf{Y} = \boldsymbol{\alpha} + \mathbf{X}\boldsymbol{\beta} + \boldsymbol{\varepsilon}$, let the regression coefficients $\boldsymbol{\beta}$ be a function of the spatial location. One example of such an approach is when the regression coefficients are linear functions of the additional variables, including the spatial coordinates (Casetti and Emilio 1972):

$$\beta_j = \gamma_{j,0} + \gamma_{j,1} z_1 + \ldots + \gamma_{j,k} z_k$$

where $j = 1,..,m$ and m is the number of exploratory variables. For example, if z_1 and z_2 are longitude and latitude coordinates, the regression coefficients can be linear functions of the form

$$\beta_j = \gamma_{j,0} + \gamma_{j,1}\text{Longtitude} + \gamma_{j,2}\text{Latitude}$$

Examples of such models can be found in Jones and Casetti (1992), and details of the parameter estimation method are in Casetti (1982). Such a model for the tree example would be of the form (results are not presented here):

$$\text{Height}_i = \alpha + \beta\ \text{Diameter}_i + \varepsilon_i$$

$$\beta = \gamma_1\ \text{Latitude} + \gamma_2\ \text{Longitude}$$

Linear regression model with correlated errors — LM(ce)

The last model we want to introduce is the linear regression model with correlated errors, denoted by LM(e). As with the spatial moving average, we can link the LM(ce) process to the time series models. Recall that for the dynamic factor analysis, we used a symmetric non-diagonal error covariance matrix. The same is done here. The LM(ce) model can be written as

$$\mathbf{Y} = \mathbf{X}\boldsymbol{\beta} + \boldsymbol{\varepsilon}, \qquad \text{and} \quad \boldsymbol{\varepsilon} \sim N(\mathbf{0}, \sigma^2\boldsymbol{\Lambda}),$$

Technically, the error covariance matrix is modelled as symmetric and positive-definite. The unknown parameters are the regression parameters $\boldsymbol{\beta}$ and the elements in $\boldsymbol{\Lambda}$. Using matrix algebra, it is easy to rewrite the LM(ce) model to

$$\mathbf{Y}^* = \mathbf{X}^*\boldsymbol{\beta} + \boldsymbol{\varepsilon}^*, \ \text{where} \quad \boldsymbol{\varepsilon}^* \sim N(\mathbf{0}, \sigma^2\mathbf{I})$$

and $Y^*=\mathbf{\Lambda}^{-0.5}\mathbf{Y}$, $\boldsymbol{\varepsilon}^*=\mathbf{\Lambda}^{-0.5}\boldsymbol{\varepsilon}$ and $\mathbf{X}=\mathbf{\Lambda}^{-0.5}\mathbf{X}$. As this is again the linear regression model with uncorrelated residuals (note the identity matrix **I** in the normal distribution), the ordinary least squares (Chapter 5) can be used to estimate the parameters. However, alternative estimation routines are also available (Amemiya 1985; Greene 2000). Just as in Chapter 16, one has to choose the structure of **Λ**. This modelling approach is further discussed in Chapter 37 and presented with a detailed example of the modelling approach..

18.6 Summary

The process of analysing spatial data can follow many pathways. Figure 18.6 shows the decision processes to analyse spatial data that have either a regular or irregular lattice structure. In its most simple form, spatial analysis is modelled using linear regression. If a linear regression model is applied, the residuals are assumed to be independent. The Morgan *I* index is used to test whether the spatial residuals from the linear regression are really independent. With luck, the test give no reason to reject the null hypothesis that there is no residual spatial auto-correlation. If there is spatial residual correlation, several options exist. More covariates or interaction terms can be added to the regression model. Smoothing methods can be applied. If simple procedures do not create an appropriate model, then alternative models exist that are specially designed to take spatial dependence into account. Choosing the appropriate spatial model, whether it is a SAR, SMA, local linear spatial model, or LM(ce), will depend on the data structure, the underlying questions, the residual patterns, and the AIC results computed for each model. An example of these methods is given in Chapter 37.

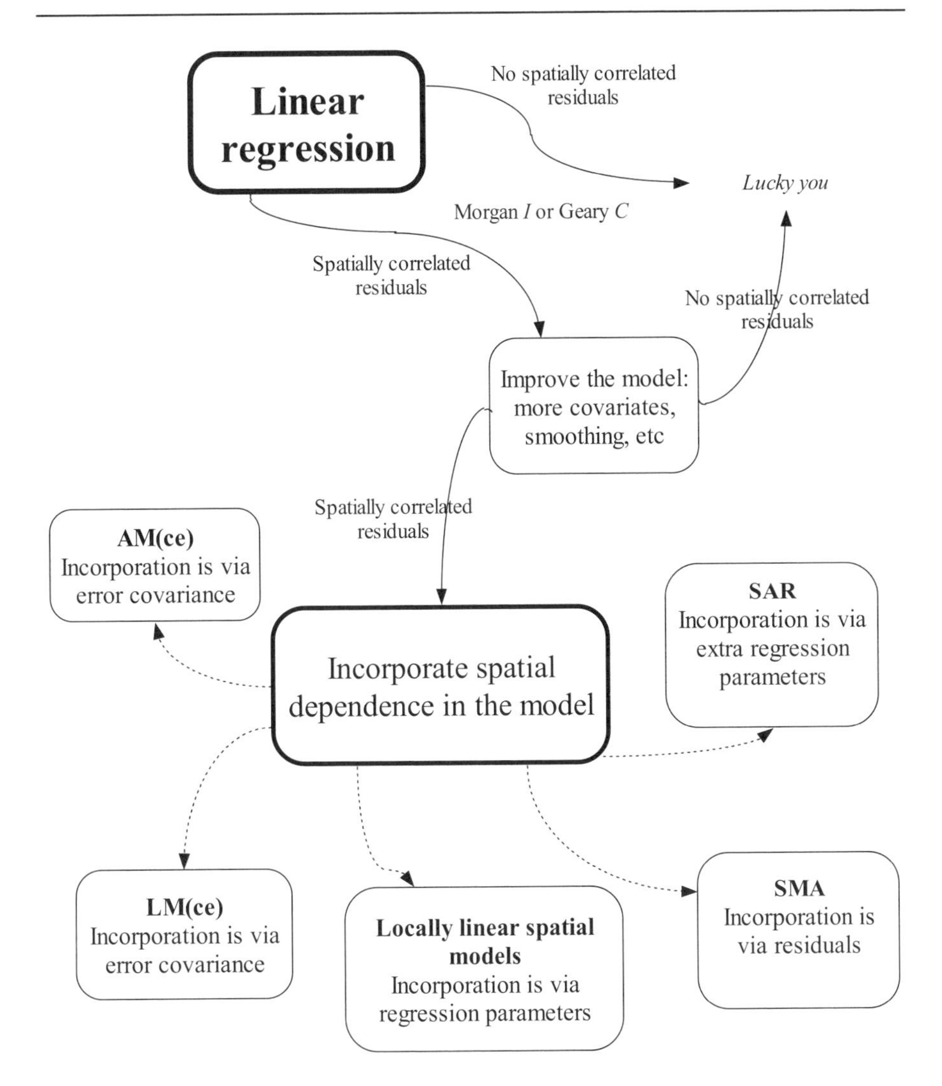

Figure 18.6. Flowchart showing the decision process for the modelling of data measured on a lattice. If a linear regression model is applied, the residuals are assumed to be independent. The Morgan *I* index can be used to test for spatial independence in the residuals. If you are lucky, there is no evidence to reject the null hypothesis of no spatial correlation. If there is spatial correlation, it is time for action. Either add more covariates or interaction terms, using smoothing methods, and if this does not help, consider one of the many alternative models that take into account spatial dependence. The choice, which model to choose, depends on the data, underlying questions, and residual patterns, AIC, etc. of each model. AM(ce) represents the smoothing equivalen of the LM(ce) model. 'Extra regression parameters' in SAR refers to using lagged response variables.

19 Spatially continuous data analysis and modelling

Saveliev, A.A., Mukharamova, S.S., Chizhikova, N.A., Budgey, R. and Zuur, A.F.

19.1 Spatially continuous data

In the previous chapter, we explored techniques to analyse data collected on a lattice. In this chapter, we will consider techniques to model continuous spatial data. The term *continuous* does not mean that the variable of interest is continuous, but merely that the variable can be measured in any location in the study area. Such continuously distributed variables are widely used in ecology and geoscience. Examples are relief elevation and bathymetry, temperature, moisture, soil nutrients, and subsurface geology. Spatially continuous data are often referred to as *geostatistical data* (Bailey and Gatrell 1995). The set of statistical techniques that can be used for analysing and modelling this type of data is called *geostatistics*.

Mathematical tools used to analyse and model geostatistical data are similar to those applied in time series analysis. Methods from both fields consider discrete observations, e.g., time points or spatial locations, and use correlation structure to describe the data dependence with tools like the auto-correlation function for time series and the variogram for geostatistical data. However, there are two important differences. Data in time series have only one direction, which is from past to more recent. However, there are multiple directions in space. Additionally, the main aim in time series analysis is to determine what has happened and occasionally to predict future values. In general, there is no need for data interpolation at intermediate time points. In geostatistics, the prediction of values at new locations is one of the prime aims of the analysis.

Geostatistical analysis usually consists of the same steps of the traditional statistical analysis workflow, and the following steps will be discussed in this chapter:

- Exploratory data analysis.
- Analysis of spatial correlation. This step is usually called the structural analysis or variogram modelling.

- Prediction of the data values at unsampled locations and estimation of the prediction uncertainty using *kriging*. Note that the goal is now shifting from parameter estimation to prediction of data.

To illustrate some of the geostatistical methods in subsequent sections of this chapter, we will refer to a simple relief elevation data from an area in the Volga River Valley in Russia. The data are part of the SRTM data set; see Rabus et al. (2003). We randomly selected a subset of 172 locations from our main data set for this area to illustrate analysis methods. The subset of points represents about 20% of the total relief data (26 × 33 grid) for this river valley. The location of samples within this relief surface are shown in Figure 19.1. A more challenging data set (bird radar data; see also Chapter 10) is analysed at the end of this chapter, and Chapter 37 shows a case study.

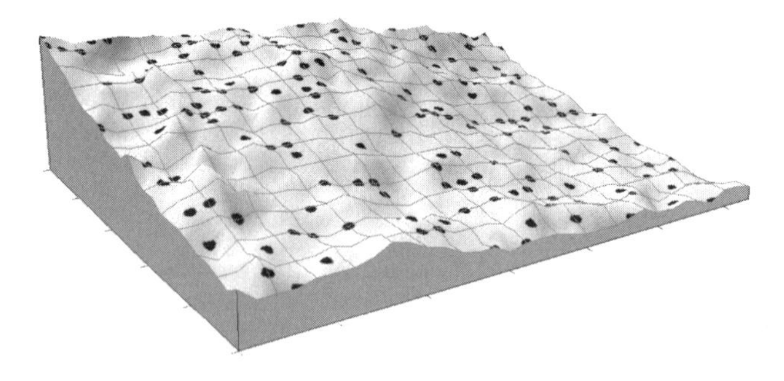

Figure 19.1. Relief surface and sampled locations (dots). The area is 2600 by 3300 meters.

19.2 Geostatistical functions and assumptions

In geostatistics, the term *regional variable* refers to some phenomenon that takes place in geographical space, i.e., on a plane (Isaaks and Srivastava 1989; Pannatier 1996). The phenomenon under study is modelled as a *random function* $Z(\boldsymbol{x})$ that depends on the spatial coordinates $\boldsymbol{x}$ on a plane and contains information about (i) some statistical regularity, which is also called the spatial trend, and (ii) local randomness. The random variables $Z(\boldsymbol{x})$ and $Z(\boldsymbol{x} + \boldsymbol{h})$, which are obtained at the points $\boldsymbol{x}$ and $\boldsymbol{x} + \boldsymbol{h}$, are separated by a vector $\boldsymbol{h}$. Figure 19.2 shows an example of the vector $\boldsymbol{h}$. It places the point $\boldsymbol{x}_1$ with coordinates (1,2) at the point (8,4). For neighbouring sites (i.e., those with small values of $\boldsymbol{h}$), $Z(\boldsymbol{x})$ and $Z(\boldsymbol{x} + \boldsymbol{h})$ may be statistically dependent (spatially correlated). Finding the spatial dependence of these variables is the aim of geostatistical analysis.

In geostatistics, we have a model for the population and we need to estimate its parameters using sample data just like we do in linear regression (Chapter 5). In spatial analysis, $z(\boldsymbol{x})$ is considered a realisation of the random variable $Z(\boldsymbol{x})$, so we ultimately want to characterise the population $Z(\boldsymbol{x})$ using the observed value $z(\boldsymbol{x})$.

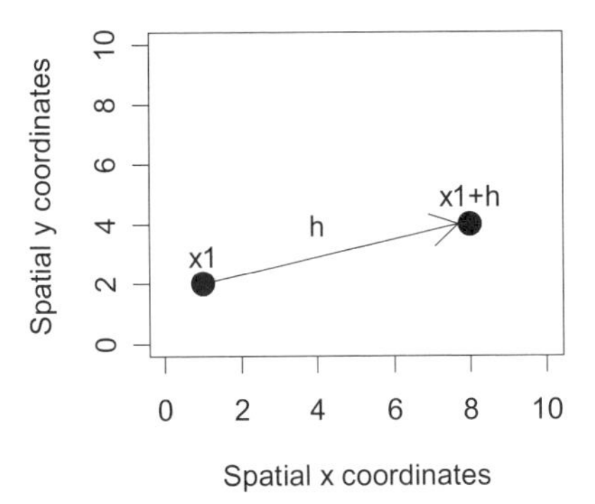

Figure 19.2. Illustration of the separation vector ***h*** that places the point x_1 with coordinates (1,2) at (8,4). The length of ***h*** is 7.3 (using the Pythagoras theorem), and ***h*** adds 7 and 2 units, respectively, to the x and y coordinate of x_1.

A random variable like $Z(\boldsymbol{x})$ is typically described by a density function. However, we have multiple spatial locations $\boldsymbol{x}_1,\ldots,\boldsymbol{x}_k$. and at each location we have a population $Z(\boldsymbol{x}_i)$ and an observed value $z(\boldsymbol{x}_i)$. In our illustration in Figure 19.1, we have a similar situation. Sample position on the landscape is designated by $\boldsymbol{x}_i$ (and determined by longitude and latitude). The population value of whatever we measure (relief, rainfall, or number of birds) at the site is designated $Z(\boldsymbol{x}_i)$, and the observed value of the measured variable is $z(\boldsymbol{x}_i)$. The cumulative distribution function for $Z(\mathbf{x}_1),\ldots,Z(\boldsymbol{x}_k)$ is given by[1]

$$F_{\mathbf{x}_1,\ldots,\mathbf{x}_k}(z_1,\ldots z_k) = P(Z(\mathbf{x}_1) < z_1,\ldots,Z(\mathbf{x}_k) < z_k) \tag{19.1}$$

[1] A refresher in density curves. We have explored density functions already when we examined the normal, Poisson, and binomial distributions in Chapters 5 and 6. Remember the simple exercises from a basic course in statistics that ask you to determine the probability of getting a particular number of items given a specified distribution function. For example, if the number of birds is Poisson-distributed with a mean of 5, what is the probability that we count not more than 10 birds in a particular location $\boldsymbol{x}$? The question is formulated mathematically as find $P(Z(\boldsymbol{x}) < 10)$, if $Z(\boldsymbol{x})$ follows a Poisson distribution with mean 5. The probability $P(Z(\boldsymbol{x}) < 10)$ is also called the cumulative distribution function $F(\boldsymbol{x})$; and, for this specific example, the notion for the distribution function is $F(10)= P(Z(\boldsymbol{x}) < 10)$. We have a similar notation in spatial statistics; see equation (19.1). Suppose that k represents the total number of sample locations in Figure 19.1. If we count birds at all of these locations and assume that bird abundances in the k locations are independent, then the cumulative density function in equation (19.1) can be simplified using the probability rule $P(A,B) = P(A) \times P(B)$. However, if the k locations are spatially dependent, then this is incorrect and, instead, we would use the cumulative distribution function in equation (19.1).

This equation is also called the *spatial distribution* of the random function $Z(\boldsymbol{x})$ (Chiles and Delfiner 1999). In a normal distribution, the density curve is fully specified by the mean (first moment) and the variance (second moment). Linear geostatistics also uses the first two moments of the random function. The first moment is the expectation or mean. The second moment is represented by the covariance or variogram. The mathematical notations for the mean and covariance are as follows:

$$m(\mathbf{x}) = E[Z(\mathbf{x})]$$

$$Cov(\mathbf{x}_1, \mathbf{x}_2) = E[(Z(\mathbf{x}_1) - m(\mathbf{x}_1))(Z(\mathbf{x}_2) - m(\mathbf{x}_2))]$$

The definition of the covariance is similar to a time series. However, instead of finding the association or relationship between two points in time, we want to find the relationship between two points in space. The association is quantified by coherence of the $Z(\boldsymbol{x}_1)$ and $Z(\boldsymbol{x}_2)$ deviate from their respective means.

On a real spatial landscape with continuous variables, we cannot sample such that we have values for every point on the landscape. Sometimes we need to estimate values for a regional variable $z(\boldsymbol{x})$ at a point where it is not measured. To estimate values, we use a geostatistical interpolation process called *kriging* (Chiles and Delfiner 1999). To interpolate values by kriging, however, we need to find an unbiased estimator for $Z(\boldsymbol{x})$ that has minimal uncertainty. This requires estimating parameters, and structural analysis or variography is used for this. Its main tool is the variogram. The variogram is a commonly used tool in geostatistics that measures how much of the variation in a measured variable is due to the spatial location of the sample itself. The variogram is defined as follows:

$$\gamma(\mathbf{x}_1, \mathbf{x}_2) = \frac{1}{2} E[(Z(\mathbf{x}_1) - Z(\mathbf{x}_2))^2]$$

The variogram or $\gamma(\mathbf{x}_1,\mathbf{x}_2)$ measures spatial dependence. Usually, if sample sites are located close to each other in space, the variable of interest will have similar values at each site and the difference calculated for the variogram value will be small. A low value of $\gamma(\mathbf{x}_1,\mathbf{x}_2)$ may indicate dependence between $Z(\mathbf{x}_1)$ and $Z(\mathbf{x}_2)$, whereas a large value may indicate independence.

Structural analysis is used to create a quantitative model for spatial correlation. In the geostatistical estimation process, certain conditions are imposed on the random function and its spatial correlation. These assumptions allow us to apply spatial analysis tools like variography and kriging. The most important geostatistical assumptions require that the random function $Z(\boldsymbol{x})$ is *stationary* and *ergodic,* and that it follows a multivariate *normal* distribution. The distribution assumption means that for any set of k variables, $Z(\mathbf{x}_1),\ldots,Z(\boldsymbol{x}_k)$ must be multivariate normally distributed.

The ergodicity assumption requires that the average over all possible realisations is equal to the average over the single realisation. For example, whether you examine 10 maps to get information about an area or just examine 1, the resulting impressions should be similar if map information follows the ergodicity assumption. The ergodicity assumption allows parameter estimations based on only one realisation of the random variables.

As to stationarity, this is a slightly more complex issue. The random function *Z(x)* (the variable of interest) is stationary when its spatial cumulative distribution function does not vary when the points are translated by the vector **h**:

$$F_{\mathbf{x}_1,\ldots,\mathbf{x}_k}(z_1,\ldots z_k) = F_{\mathbf{x}_1+\mathbf{h},\ldots,\mathbf{x}_k+\mathbf{h}}(z_1,\ldots z_k)$$

The stationarity assumption would hold, for example, if we count birds in k locations, and the probability that we count certain values at these k locations is the same at k other locations, translated by **h**, elsewhere in the study area. In practice, this form of stationarity is unrealistic. This definition of stationarity is also called strong stationarity, and as the name suggests we can also define weak stationarity or second-order stationarity. There is even a third definition of stationarity, namely intrinsic stationarity.

Second-order stationarity means that the mean does not depend on location and the covariance is only a function of the separation vector ***h***. In mathematics:

$$E[Z(\mathbf{x})] \equiv m \quad \text{and} \quad Cov[Z(\mathbf{x}), Z(\mathbf{x}+\mathbf{h})] = C^*(\mathbf{h})$$

Strong stationarity implies weak stationarity but not the other way around. It is important to realise that so far, we have not said anything about the direction of ***h***. Indeed, the ***h*** in C^* is causing a translation of points in the same direction. This is visualised in Figure 19.3A, in which we connected points that are on a distance $h = 2$, but all connecting lines have the same direction. The covariance C^* at $h = 2$ can be estimated using only the points that are connected in Figure 19.3-A. *Isotropy* (discussed in more detail below) means that the direction is not important, only the distance between the points.

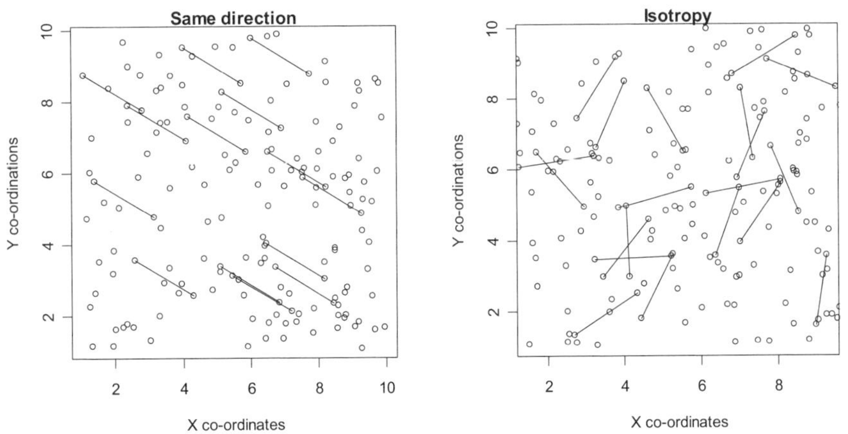

Figure 19.3. A: Points that are on a distance h with the same direction are connected. B: Points that are on a distance h in *any* direction are connected. In both graphs $h = 2$ was used.

Second-order stationarity and isotropy means that the covariance is calculated by

$$Cov[Z(\mathbf{x}), Z(\mathbf{x}+\mathbf{h})] = C(\|\mathbf{h}\|)$$

The notation $\|\mathbf{h}\|$ stands for Euclidean distances (Chapter 10). We omitted the * above the C to emphasise the effect of isotropy. In the spatial statistical literature one often uses $C()$ for both $C^*()$ and $C()$. The assumption of isotropy implies which one it is. Under the second-order stationarity and isotropy assumptions, the variance of $Z(\mathbf{x})$ is constant and location independent:

$$C(\mathbf{0}) = Cov[Z(\mathbf{x}), Z(\mathbf{x}+\mathbf{0})] = \text{var}(Z(\mathbf{x}))$$

If a time series is not stationary, one can look at differences between time points (first differences, see Chapter 16). In spatial statistic, we can do something similar, we can look at the difference $Z(\mathbf{x} + \mathbf{h}) - Z(\mathbf{x})$, and assume that this is second-order stationary. This is called intrinsic stationary. This is a weak form of stationarity, which assumes that the mean can have a local linear drift and the variogram only depends on the separation vector between points. These assumptions are represented mathematically as follows:

$$E\left[Z(\mathbf{x}+\mathbf{h}) - Z(\mathbf{x})\right] \equiv m \quad \text{and} \quad Var[Z(\mathbf{x}+\mathbf{h}) - Z(\mathbf{x})] = 2\gamma(\mathbf{h}).$$

Note that intrinsic stationarity only uses the first two moments. Again, we can assume that $\gamma(\mathbf{h})$ does not depend on the direction (isotropy). Strong stationarity, as the name suggests, is much stronger because it makes a statement on the spatial cumulative distribution functions.

Before some spatial analyses like variography and kriging can be applied on a data set, the three assumptions given above must be verified. Multivariate normality can be verified by making QQ-plots or histograms of individual variables. The intrinsic hypothesis can be verified by checking for a spatial trend. Ergodicity cannot be verified, unless samples are taken repeatedly in time and space.

19.3 Exploratory variography analysis

The aim of variography (Isaaks and Srivastava 1989; Pannatier 1996; Chiles and Delfiner 1999) is to estimate the strength and direction of spatial data dependency. As stated, the basic tool of variography analysis is the variogram or semi-variogram (these terms are interchangeable). A variogram measures the spatial dependence between the values of the variable of interest in the points (locations) $\boldsymbol{x}$ and $\boldsymbol{x} + \boldsymbol{h}$, where $\boldsymbol{h}$ is the separating vector or lag (Figure 19.2).

Experimental variogram

In Section 19.2, we defined a variogram for the population, but how do we calculate such a population variogram using sample data? First, we need to calculate a *variogram estimator* using sample values just like we use samples to estimate population parameters in linear regression. This estimator is called an *experimental* variogram and it is calculated by

$$\hat{\gamma}(\mathbf{h}) = \frac{1}{2\mathrm{N}(\mathbf{h})} \sum_{i=1}^{N(\mathbf{h})} [z(\mathbf{x}_i + \mathbf{h}) - z(\mathbf{x}_i)]^2$$

We used the notation ^ to emphasise that it is an estimator based on sample data, not the population variogram. All the pairs of the sampled points that are separated by a specific *lag* $\boldsymbol{h}$ are used to calculate $\hat{\gamma}(\mathbf{h})$. If there is spatial dependence, then points close to each other will tend to have similar values and $\hat{\gamma}(\mathbf{h})$ will be small. If $\hat{\gamma}(\mathbf{h})$ is large, it indicates spatial independence.

$\boldsymbol{h}$ can be referred to as either a vector or as a number denoting the vector's length in this analysis. The context should be clear enough in the following discussions to indicate whether we mean the vector itself or its length. Both the length of $\boldsymbol{h}$ and its direction are important in this variogram.

The number of points that are separated from each other by a distance $\boldsymbol{h}$ is denoted by $N(\boldsymbol{h})$. In general, the number of points that are exactly separated by a distance of $\boldsymbol{h}$ will be very small so a small tolerance around the value of $\boldsymbol{h}$ is usually used instead of exact values. To illustrate this characteristic, we simulated the spatial distribution of 100 points (the coordinates were drawn from a univariate distribution), and we plotted the results in Figure 19.4. The figure contains four panels, each with a precise $\boldsymbol{h}$ value. In each panel, we connected the points that were *exactly* separated by a distance $\boldsymbol{h}$, where $h = 1$ (upper left panel), 2, 3 and 4 (lower right panel). As you can see, there are only two points that have exactly a distance of 1. There are no other connected points in any of the other panels. Obviously, we cannot calculate the variogram using only two points. Therefore, we need a small tolerance adjustment around the lag distance. We arbitrarily selected the interval $h - 0.05$ to $h + 0.05$, where h is the lag. The effect of introducing a lag tolerance is shown in Figure 19.5. Note that there are now several connected points in $\mathrm{N}(\boldsymbol{h})$ (which is therefore larger).

The variogram $\hat{\gamma}(\mathbf{h})$ is calculated by subtracting the observed values $z(\boldsymbol{x}_i)$ at both ends of a line from each other and squaring the difference. This is then repeated for all points that are connected by the given lag. If there are not enough connecting lines, the tolerance interval can be increased.

If we calculate an experimental variogram using all the points that are connected by the lines in Figure 19.4, we assume that the spatial relationships are the same in all directions, which is referred to as *isotropy*. To assume that the spatial dependence is the same in every direction, or *omnidirectional*, is a reasonable starting point for any spatial analysis. The resulting experimental variogram is considered as an average variogram of all spatial directions.

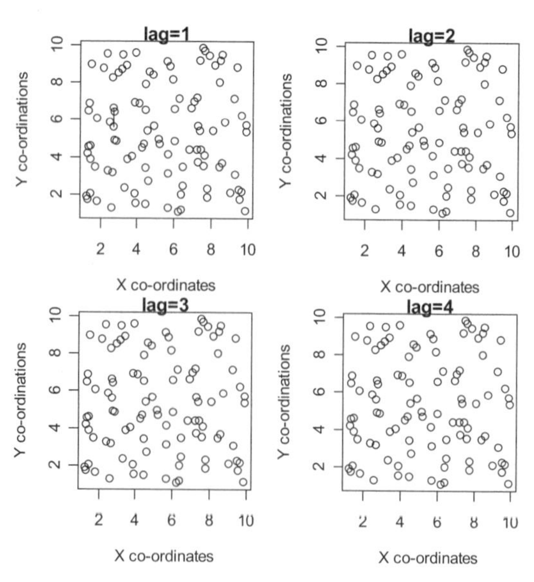

Figure 19.4. 100 Randomly selected sites. Points that are exactly 1 (upper left), 2 (upper right), 3 (lower left) or 4 (lower right) units separated are connected by a line. The numbers in the title are the lags. As can be seen (with some effort) from the upper left panel, there is only one combination of points that are exactly separated by the lag = 1.

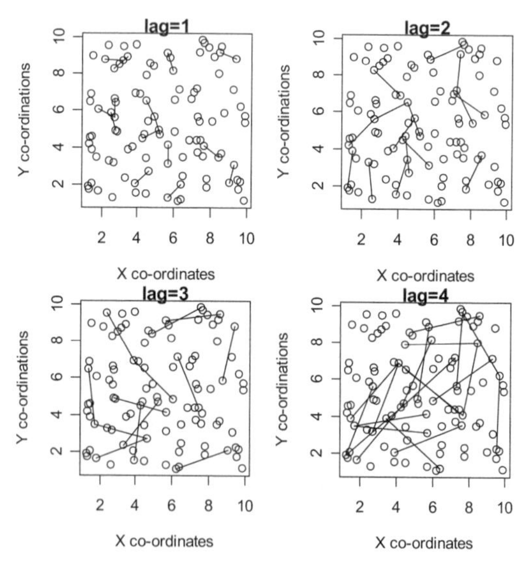

Figure 19.5. The same 100 randomly selected sites as in Figure 19.3. Points that are separated by a distance between $h - 0.05$ and $h + 0.05$ are connected by a line. The lag represents the value of $\boldsymbol{h}$.

How do you verify, however, that the variogram has the same spatial relationships in each direction? The easiest method is to calculate *directional* experimental variograms, each focusing on a specific direction, and then compare the variograms with each other to see whether the shape changes. This can be done by calculating $\hat{\gamma}(\mathbf{h})$ for only those lines that point in the same direction, say from north to south or from east to west. During these tests, however, we may end up with the same problem as we encountered when using lag values without tolerances (Figure 19.4), namely that there are only a few lines pointing exactly in the right direction. The solution is to again use a tolerance range, but this time to apply it on the direction. Usually, optimum values of the tolerance around distances and angles are estimated iteratively and details can be found in Cressie (1993). Another way to verify isotropy is using a variogram surface, which is discussed next.

Variogram surface

Recall from Figure 19.2 that the vector $\boldsymbol{h}$ adds a horizontal and vertical replacement to the x-y coordinates and has a directional angle for the change. Instead of using a specific angle and length of $\boldsymbol{h}$, however, we can express the same concept as a replacement in terms of West-East and North-South map units. For example, Figure 19.6 shows a replacement of two (with some tolerance) units in the West-East direction and two units in the South-North direction. For this replacement, we can calculate the experimental variogram as described in the previous paragraph but allowing for a small tolerance around the length and direction of $\boldsymbol{h}$, defined on a grid. This exercise is done for each possible grid cell and results in a experimental variogram with values in tabular (grid surface) form; e.g., see Table 7.5 in Isaaks and Srivastava (1989).

Instead of trying to analyse all of the numbers in such a table/figure for spatial relationships, we make a contour plot of the experimental variogram values called an *experimental variogram surface* (Isaaks and Srivastava 1989). This graph is used to detect violations in isotropy or *anisotropy*.

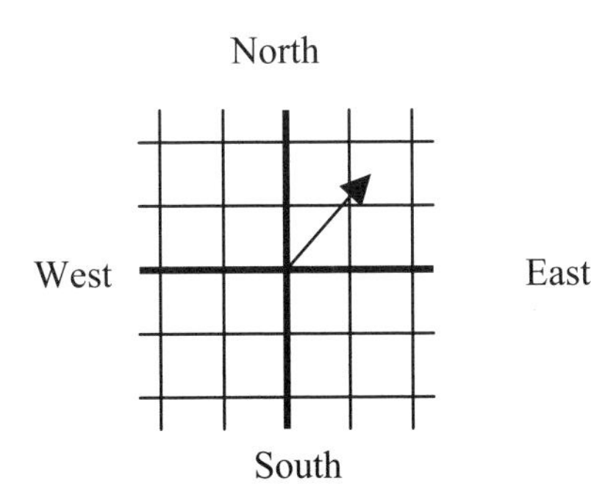

Figure 19.6. Example of replacement along the West-East and North-South axes.

Anisotropy indicates dependence in the data that changes in different directions. Recall that in Chapter 10 we used bird abundance data measured by a radar. If the birds use the wind direction for their flight path, then the dependencies are anisotropic. Plant abundances measured along an altitude gradient are also anisotropic because their spatial dependencies will be different based on map directions. When we calculate experimental variograms for different directions, they should be referred to as *experimental directional variograms*. In practice, we will use covariates to model the trends in the data and hope that the residuals are isotropic.

h-Scatterplots

An *h*-scatterplot can be used to visualise violation of the stationary and multivariate normality assumptions. Instead of plotting two different variables against each other, we take a particular site *i* and determine which other sites are on a distance *h* (scalar value), again allowing for a small tolerance around h. We then plot our observed value of $z(\boldsymbol{x}_i)$ against the values of $z(\boldsymbol{x}_i + \boldsymbol{h})$, for all the stations that are on a distance *h*. Let us explain this in a bit more detail using our lag values from Figure 19.5. First, we take a particular point $\boldsymbol{x}_i$ and make a list of sites $\boldsymbol{x}_j$, $\boldsymbol{x}_{j+1}$, $\boldsymbol{x}_{j+2}$, etc., that are all at the same distance *h* from $\boldsymbol{x}_i$. In the *h*-scatterplot, we plot $z(\boldsymbol{x}_i)$ versus $z(\boldsymbol{x}_j)$, $z(\boldsymbol{x}_i)$ versus $z(\boldsymbol{x}_{j+1})$, $z(\boldsymbol{x}_i)$ versus $z(\boldsymbol{x}_{j+2})$, etc. This gives a vertical band of points because we are using the same coordinate $z(\mathbf{x}_i)$ along the *x*-axis. We repeat this process for all sites, which gives us layers of vertical dots. Obviously, the distribution of points in this graph depends on the value of h. It is common to make a couple of *h*-scatterplots for different values of h and plot them next to each other to see how *h* values affect the distribution.

If the values of $z(\boldsymbol{x}_i)$ and $z(\boldsymbol{x}_{j+k})$ are similar for all *i* and *k*, (implying spatial dependence), then most points should lie close to a 45-degree line halfway between the horizontal and vertical axes and starting at the origin. For small *h*, we expect to see a narrow band along this 45-degree line. When *h* is larger, the spatial dependence decreases, and the cloud of points is likely to get 'fatter' (Figure 19.7-A). If the univariate distribution of $Z(\boldsymbol{x})$ is highly skewed to the right, then small values of $z(\boldsymbol{x}_i)$ correspond to large values of $z(\mathbf{x}_i+\boldsymbol{h})$ and vice versa. These relationships form lines of dots adjacent to the axes of the *h*-scatterplot and are referred to as a 'butterfly wing' pattern (Figure 19.7-B). Note that a histogram ignores the spatial dependences and is therefore less suitable to verify stationarity. A single cloud located above or below the diagonal line indicates a systematic drift in the direction of the vector ***h***. It means that the value of $z(\boldsymbol{x}_i+\boldsymbol{h})$ has systematic bias from $z(\boldsymbol{x}_i)$. If the cloud is located above the diagonal, values of $z(\boldsymbol{x}_i)$ systematically increase in the direction of ***h*** (Figure 19.7C). Clouds of points far from the diagonal line in the scatter plot are usually formed by pairs belonging to different statistical populations. For example, if we have two populations with different means (as in Figure 19.7-D), the *h*-scatterplot has four individual clouds: The two clouds on the diagonal correspond to values of $z(\boldsymbol{x}_i)$ and $z(\boldsymbol{x}_i+\boldsymbol{h})$ from the same populations; the cloud in the right bottom corner corresponds to values of $z(\boldsymbol{x}_i+\boldsymbol{h})$ from the smaller mean population and $z(\boldsymbol{x}_i)$ from the larger mean population; and the cloud in the left top corner corresponds to values of $z(\boldsymbol{x}_i+\boldsymbol{h})$ from the larger mean population and $z(\mathbf{x}_i)$ from the smaller mean population. In this case, $Z(\mathbf{x})$ cannot be considered stationary.

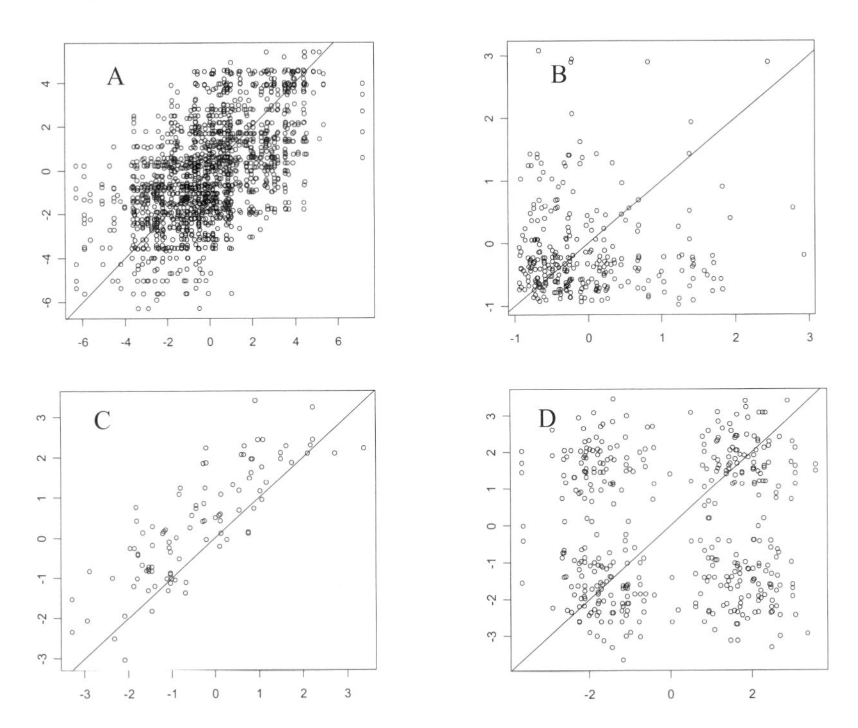

Figure 19.7. *h*-Scatterplots for various different simulated data sets. A: *h*-Scatterplot for the data drawn from the multivariate Gaussian distribution. B: Data violating the multivariate Gaussian distribution properties: the 'butterfly wing' for the highly skewed distribution. C: Non-stationarity caused by the mean drift. D: Two populations.

Illustration: Volga River relief data — detecting trends in spatial data

The shape of the relief data in Figure 19.1 suggests a spatial trend. To eliminate this trend, a linear regression model using the longitude (X) and latitude (Y) coordinates as covariates was fitted. The model gave the following results:

	Estimate	Std.Error	*t*-value	*p*-value
Intercept	86.970	0.420	209.54	<0.001
X	-0.005	0.0002	-32.11	<0.001
Y	0.008	0.0002	39.34	<0.001

Residual standard error: 1.97 on 169 degrees of freedom
Multiple R-Squared: 0.941
F-statistic: 1348 on 2 and 169 df, *p*-value: <0.001

The residual standard deviation is about 2 m. The trend surface model was subtracted from the observed values, and the residuals were normally distributed (as indicated by histograms and QQ-plots). The residuals were plotted against their locations in Figure 19.8. In this example we have access to a much larger part of the data, but we only used 20% in the trend fitting. This allows us to calculate the trend residuals from the true elevation values at the 'un-sampled' (the other 80% of the data) locations. Although the trend was fitted at discrete sample points, we present all the trend residuals because it is visually easier to interpret. White or dark grey areas indicate large positive or large negative residuals. The distribution of the colours indicate areas of spatial dependence. If there are clear directional patterns formed by the shades of colour, we have evidence of anisotropy.

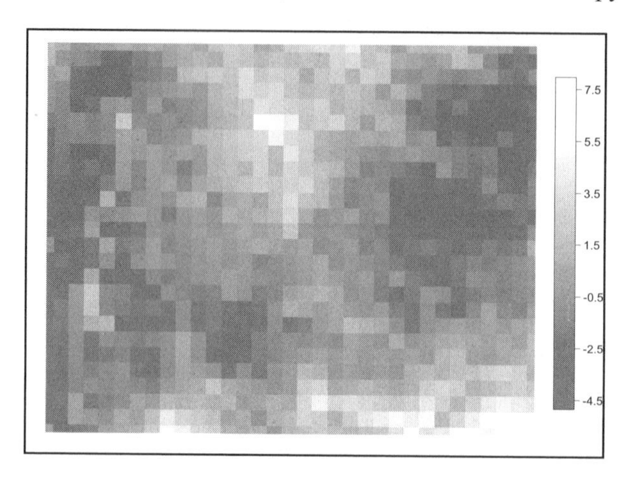

Figure 19.8. Residual pattern for the linear trend model (residuals for all locations are shown). The horizontal and vertical axes represent spatial coordinates, and the grey-scaling indicates the relative value of the residuals. Large white or dark grey areas, or directional patterns, indicate violation of stationarity and isotropy.

To assess whether the residuals are independent (i.e., whether we violate the assumption of independence) in the linear regression model, the experimental variogram for the residuals was calculated for 20 lags and with a lag distance of 200 m. The resulting variogram is shown in Figure 19.9-B. Because no anisotropy was detected on the variogram surface (Figure 19.9-A), the omnidirectional variogram was used. Under second-order stationarity, it can be shown (Schabenberger and Pierce 2002) that $\gamma(\mathbf{h}) = C(\mathbf{0}) - C(\mathbf{h})$; hence, the variogram is always smaller than the variance. In this case, the residual experimental variogram value exceeds the theoretical limit (residual variance) for lags above 1000 m (Figure 19.9-B). The reason for this is the deviation from the Gaussian model for larger lags as can be seen in Figure 19.9-D. At lags above 1000 m, the h-scatterplot reveals the 'butterfly wing' pattern with large values of $z(\boldsymbol{x})$ corresponding to small values of $z(\boldsymbol{x} + \boldsymbol{h})$ and vice versa. This means that the values are negatively correlated at this distance and the average value of $[z(\boldsymbol{x}_i + \boldsymbol{h}) - z(\boldsymbol{x}_i)]^2$ becomes larger than the variance. This may suggest some spatial structure like a hill, which has a typical size in the area of about 1500 m (implying violation of the second-order stationarity assumption).

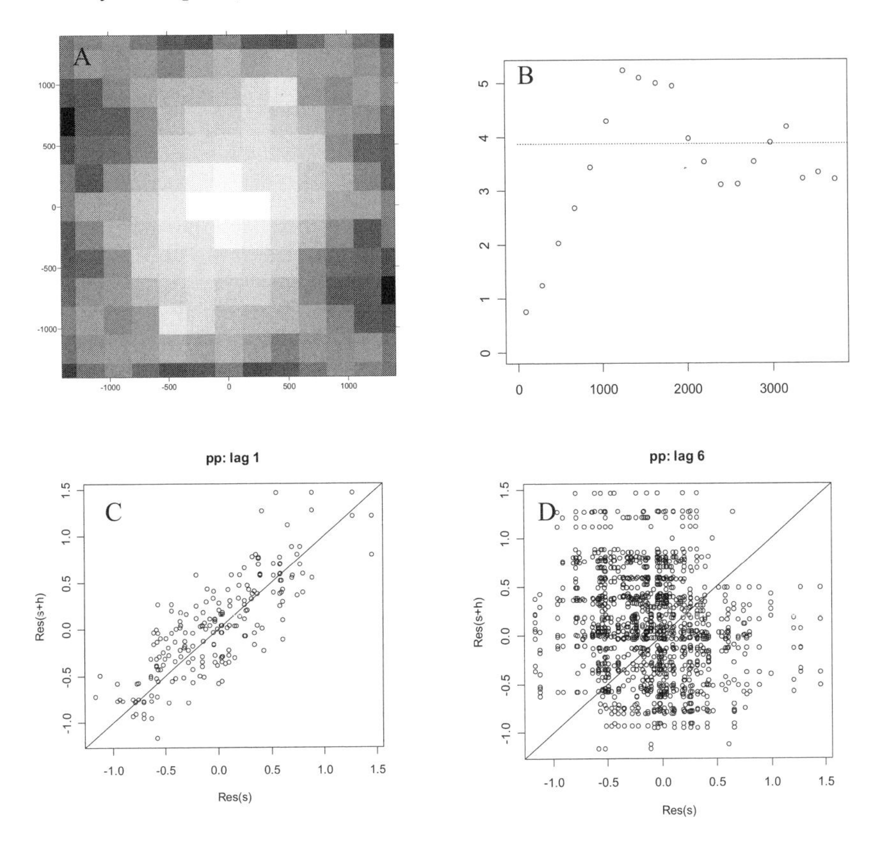

Figure 19.9. A: Experimental variogram surface for residuals from the linear trend. The variogram values for various West-East and North-South displacements were calculated, and a grey-level was used to visualise the values. B: Experimental variogram with the residuals variance drawn. h-Scatterplots for lag = 200 m (C) and for lag = 1200 m (D).

We will now work towards models for the variogram. Note that the experimental variogram in Figure 19.9-B only contains dots, and these are based on distances between the sampled locations. To analyse the entire area, we need to know the value of the variogram at intermediate distance values. In order to obtain these values, we need to fit a model on the experimental variogram and obtain a fitted line through the points in the experimental variogram. This model has parameters that need to be estimated. Once we know the parameters, the model can also be used in kriging or regression type models (Chapter 37). We first discuss possible line shapes that fit the points and then discuss their associated models.

A typical variogram has low values for small distances $\boldsymbol{h}$ (points close to each other tend to be similar) and increasing values for larger $\boldsymbol{h}$ that eventually reach an asymptotic value beyond which sites are independent (Figure 19.10). The asymptotic value itself is called *sill*, and the distance $\boldsymbol{h}$ at which it is reached is the *range*. If $Z(\mathbf{x})$ exhibits stationarity, then the sill is equal to the variance. See Figure 19.10 for an illustration of each of these variogram parts.

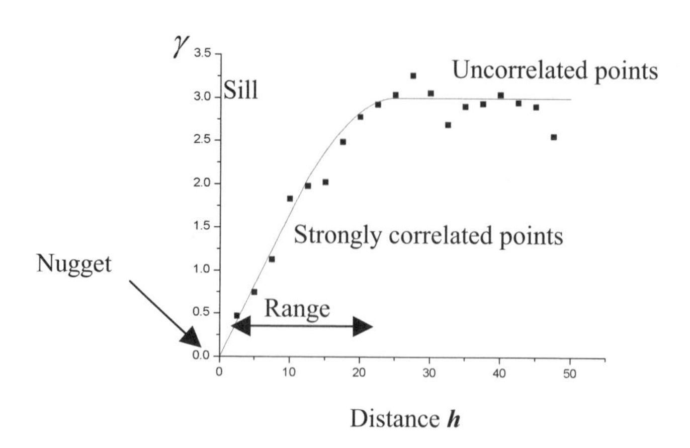

Figure 19.10. Variogram with fitted line. The sill is the asymptotic value, and the range is the distance at which this is reached. Points on a distance larger than the range are uncorrelated. The nugget effect occurs if $\hat{\gamma}(\mathbf{h})$ is not 0 for small $\boldsymbol{h}$.

What kinds of patterns are usually depicted in the experimental variogram? Typically, the variogram takes on one of four shapes. The first shape looks like part of a *parabolic* function (Figure 19.11-A). It shows a smooth pattern for small $\boldsymbol{h}$. The second shape looks *linear* for small $\boldsymbol{h}$ values (Figure 19.11-B). This is most common in areas that are small in comparison with the range of the phenomenon. The experimental variogram may not reach the asymptotic value and has only increasing values of $\boldsymbol{h}$. Note that both the parabolic and the linear patterns start at the origin. The third shape for the variogram may have distinct patterns that start above the origin (0,0). This result occurs when $\hat{\gamma}(\mathbf{h})$ is not 0 for small $\boldsymbol{h}$, and it is also known as the *nugget effect* (Figure 19.11-C). It represents the discontinuity of the regional variable caused by spatial structure at distances less than the minimum lag (also called the microstructure) or measurement error. The fourth shape of the variogram may resemble a *flat* line (Figure 19.11-D). It occurs when there is no correlation between the values taken at two locations at any distance. This is an extreme case of an absence of spatial structure.

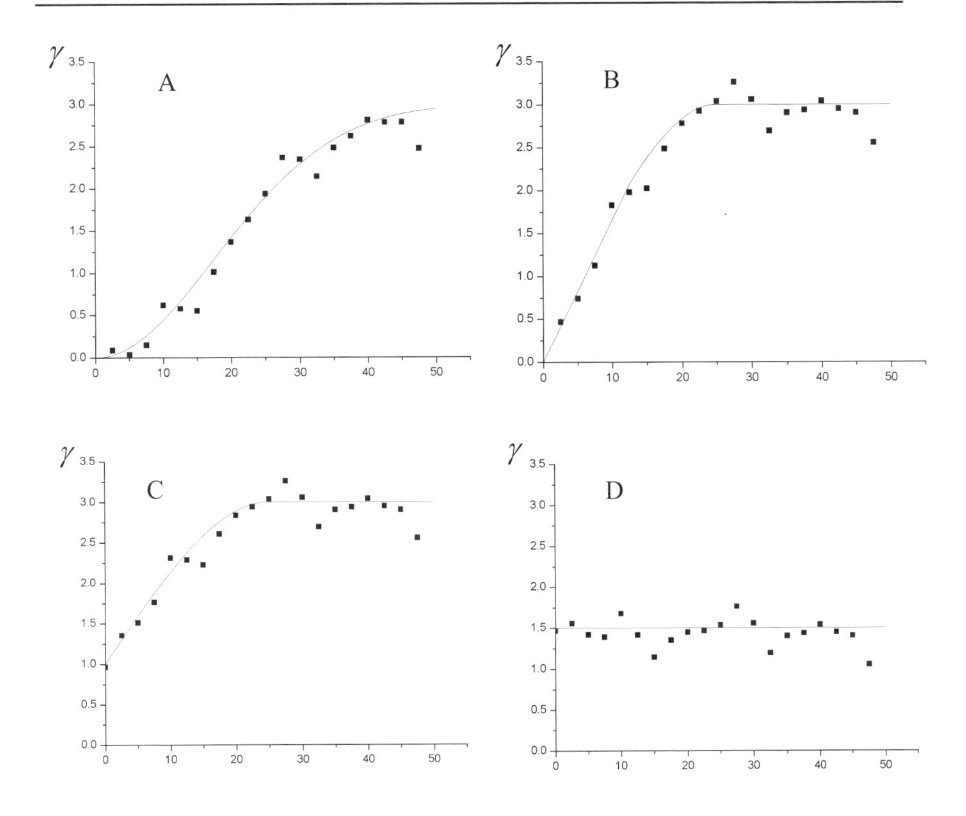

Figure 19.11. Variogram behaviour at short distances examples: parabolic (A), linear (B), nugget (C) and pure nugget (D). The horizontal axis in each graph contains values of ***h***.

Variogram models

The shape of the experimental variogram allows one to define and fit a variogram model or a *theoretical variogram*. Only a certain number of basic models meet the special requirements of the variogram modelling (Chiles and Delfiner 1999). The most commonly used are the spherical, power and wave models; Gaussian models sometimes better fit the experimental variogram but are hard to implement in kriging. The spherical model $\gamma_{sph}(h,r,\sigma)$ and the Gaussian model $\gamma_{sph}(h,r,\sigma)$ are defined by:

$$\gamma_{sph}(h,r,\sigma^2)=\begin{cases}\sigma^2(1.5\dfrac{h}{r}-0.5\dfrac{h^3}{r^3}) & \text{if } h<r\\ \sigma^2 & \text{otherwise}\end{cases}$$

$$\gamma_G(h,r,\sigma^2)=\sigma^2(1-\mathrm{e}^{-(h/r)^2})$$

The wave model $\gamma_w(h,\phi,\sigma^2)$ is defined by:

$$\gamma_w(h,\varphi,\sigma^2)=\sigma^2(1-(\varphi/h)\sin(h/\varphi)),$$

The nugget model is given by

$$\gamma_{ne}=\begin{cases}0 & \text{if } h=0\\ c & \text{otherwise}\end{cases}$$

Within each of these models, r is the range, σ^2 is the variance or sill value, and φ is the period parameter. The variable $\boldsymbol{h}$ defines the distance between the sites and is the separating lag. The unknown parameters in each of these models are the range and sill (variance). The nugget effect is modelled by adding a known constant. If the area under investigation is small in comparison with the correlation range, the basic model for the variogram without asymptotic behaviour can be used, such as a power function with $0 < a < 2$:

$$\gamma_{pow}(h,a,\sigma^2)=\sigma^2h^a$$

More examples of variogram models are presented in Chiles and Delfiner (1999) and Cressie (1993). Figure 19.12 shows some of the patterns that can be obtained for these models. To decide which model is most appropriate for your data set, you need to consider both the shape of the experimental variogram and the spatial characteristics of the phenomenon under study. Improper variogram selection may result in unstable kriging results. Cross-validation (see also Chapters 7 and 9) may be used to test whether the selected variogram model is appropriate.

For the variogram models that we have discussed so far, we assumed isotropy. If the data show anisotropy, there are two possible types. The first is directional (or geometrical) anisotropy. In this case, the sill is the same for all $\boldsymbol{h}$ directions, but the ranges may vary. In that case we need to calculate directional variograms for several directions instead of the omnidirectional variogram. Usually the anisotropy direction (direction of higher correlation), direction perpendicular to it and one or two intermediate directions are used. Then we need to allow for different ranges within the model and this can be done by a scaling of $\boldsymbol{h}$ according to the directions (Cressie 1993; Chiles and Delfiner 1999). The extreme case of the directional anisotropy is *zonal anisotropy* in which the sill varies with the $\boldsymbol{h}$ direction (Chiles and Delfiner 1999).

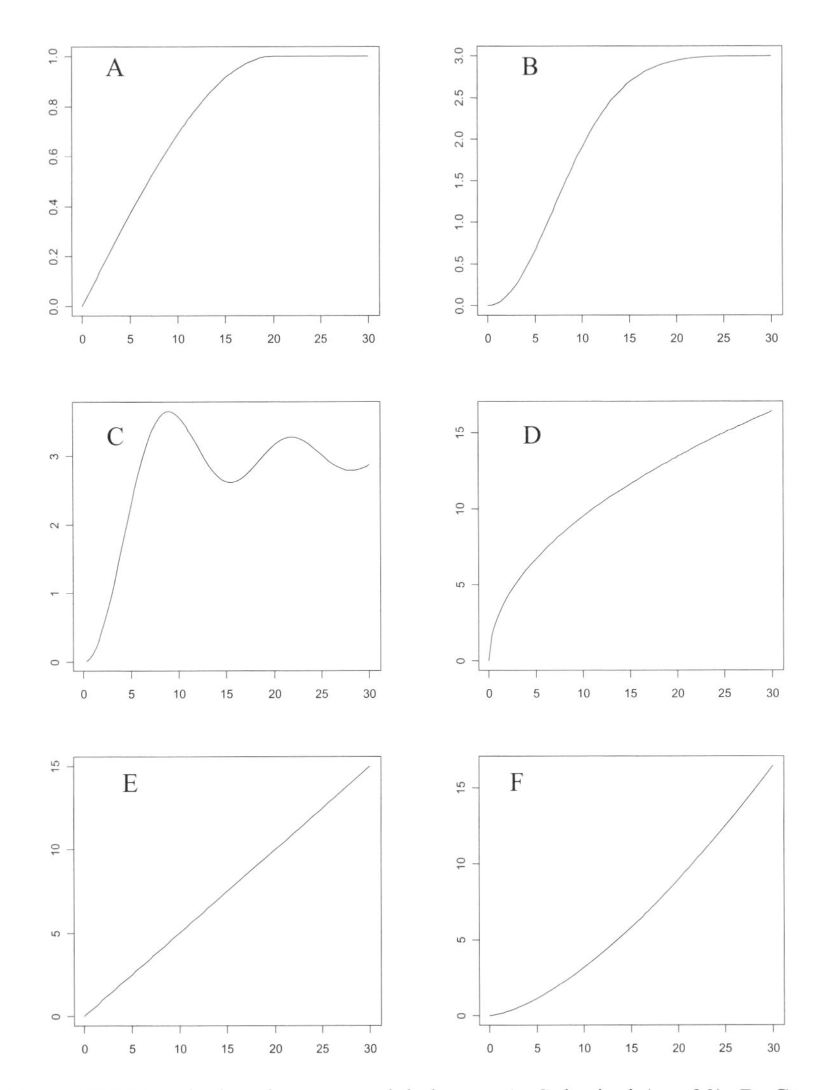

Figure 19.12 Typical variogram model shapes. A: Spherical (r = 20), B: Gaussian (r = 10), C: Wave (φ =2), D: Power (a = 0.5), E: Linear (power with a = 1), F: Power (a = 1.5) models.

Illustration: Volga River relief data — fitting a variogram model

For our illustration data, we fitted two theoretical variogram models to the experimental variogram for the relief data in Figure 19.1. We chose to test the spherical and wave models. For the spherical model, we obtained the following variogram-model parameters: range = 1300, nugget = 0.17 and sill = 3.7. For the wave model, the parameters were period = 310, nugget = 0.6 and sill = 3. The fitted variogram models are shown in Figure 19.13. The spherical variogram incorporates the medium range behaviour and does not include the large-scale residuals pattern; the wave variogram models take into account the large-scale residuals pattern, but it gives a larger nugget effect.

To estimate the variogram model parameters, we used a direct minimization of the sum of squared difference between the experimental variogram $\hat{\gamma}(\mathbf{h})$ and its theoretical model $\hat{\gamma}(\mathbf{h},\theta)$ as follows:

$$\sum_i (\hat{\gamma}(\mathbf{h}_i) - \hat{\gamma}(\mathbf{h}_i,\theta))^2$$

θ is the vector of the parameters of variogram model. Calculations were performed over all the lags $\mathbf{h}_i$. Other robust methods for variogram model fitting can be found in Cressie (1993). The best method to test whether either of these models fit the relief data is to substitute the models into kriging equations and to compare the kriging results, which we will do in the next section. Alternatively, we could use independent sample tests or modified cross-validation tests to select the model that best fits the phenomenon at both sampled and unsampled locations.

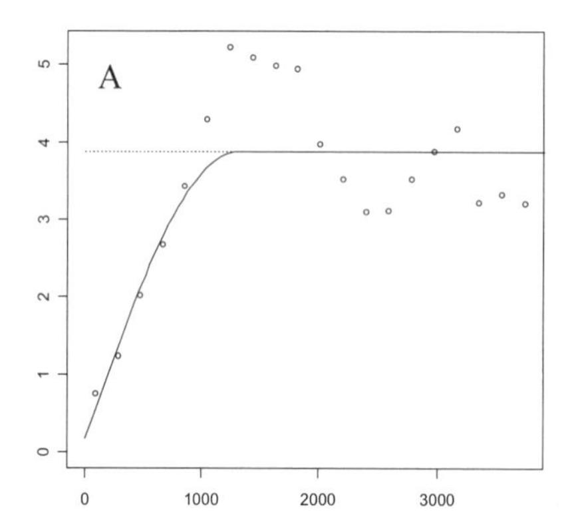

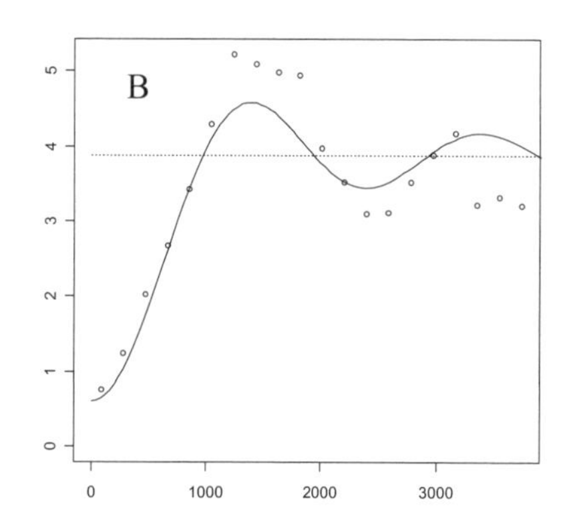

Figure 19.13. Fitted variogram models for Volga River relief data. A: Fitted spherical variogram model with range = 1300, nugget = 0.17, and sill = 3.7. B: Wave variogram model with period = 310, nugget = 0.6 and sill = 3.8. The residuals variance is drawn as a dotted line.

19.4 Geostatistical modelling: Kriging

Kriging is one of those magical terms used in geostatistics that sounds exotic but is simply a term for the process of interpolating data values from areas that have been sampled on a landscape to areas that have not. In Figure 19.14, x_0 is surrounded by locations that have values for some variable of interest but x_0 itself has not been sampled. In nine sites, the variable of interested was measured (e.g., rainfall, relief, number of birds, and tree height) and we would like to predict the value at the site $\boldsymbol{x}_0$. However, assigning a value to $\boldsymbol{x}_0$ cannot be done with a simple interpolation from the nearest site. It involves considering the data's gradient in all directions around each sample point near $\boldsymbol{x}_0$. We may even want to predict the variable of interest between each sampled site in the study area and create a contour plot of all possible values between them. For this interpolation process, kriging employs a weighted-average approach to assign values to unsampled areas based on their proximity to each other.

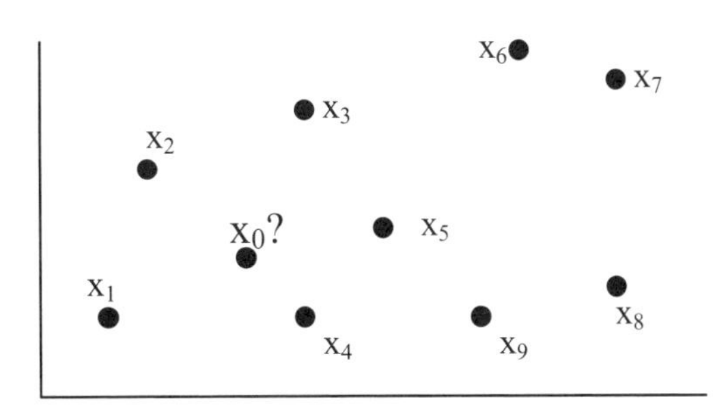

Figure 19.14. Nine sites labelled as $\boldsymbol{x}_1$ to $\boldsymbol{x}_9$, and at each site the variable of interest was measured: $z(\boldsymbol{x}_1)$ to $z(\boldsymbol{x}_9)$. The site labelled $\boldsymbol{x}_0$ was not sampled and we want to predict its z value.

Statistically, if we have sampled N locations $\boldsymbol{x}_1, \boldsymbol{x}_2, \ldots, \boldsymbol{x}_N$, at each location $\boldsymbol{x}_i$, we have realisations $z(\boldsymbol{x}_i)$ of the variables $Z(\boldsymbol{x}_i)$. Our question then is, what is value of Z at an unsampled location $\boldsymbol{x}_0$? To estimate the value $z(\mathbf{x}_0)$, it seems sensible to use information from neighbouring sites. Due to spatial dependence it is likely that the values in the points $\boldsymbol{x}_1$ to $\boldsymbol{x}_5$ contain more information on the variable $Z(\boldsymbol{x}_0)$ than the sites $\boldsymbol{x}_6$ to $\boldsymbol{x}_9$, simply because the first five locations are closer. So, it makes sense to use some sort of weighted average of information from nearby locations to obtain the estimator for $z(\boldsymbol{x}_0)$. The weighted average is denoted by $z^*(\boldsymbol{x}_0)$ and can be estimated by

$$z^*(x_0) = \mu(x_0) + \sum_i \lambda_i (z(x_i) - \mu(x_i)) \tag{19.2}$$

where $\boldsymbol{x}_0$ is the unsampled site, $\mu(\mathbf{x}_i)$ is the value of the mean in $\boldsymbol{x}_i$, and λ_i are weighting functions that determine the importance of neighbouring sites. Sites that are far away will have a λ_i close to zero and for sites close to $\boldsymbol{x}_0$, λ_i is relative large. As defined in Section 19.2, $Z(\boldsymbol{x}_i)$ is the random variable and $z(\mathbf{x}_i)$ is the observed value. $z^*(\boldsymbol{x}_0)$ is the predicted value of $Z(\boldsymbol{x}_0)$ at the unsampled location. Equation (19.2) tells us that the value $z^*(\boldsymbol{x}_0)$ is estimated as the trend plus a weighted sum of deviations from the trend at nearby sites.

Equation (19.2) looks similar to the linear regression model, where $z^*(\boldsymbol{x}_0)$ is the Y, $\mu(\boldsymbol{x}_0)$ the intercept, $z(\boldsymbol{x}_i) - \mu(\boldsymbol{x}_0)$ the X, and the λ_is the slopes. However, this expression differs from a linear regression equation in that the coefficients λ_i are not fixed constants. Each are allowed to differ for each location $\mathbf{x}_0$.

There are three questions that we now need to address. What are the values of λ_i? Do we actually know the trend values $\mu(\boldsymbol{x}_i)$? Can we ensure that $z^*(\boldsymbol{x}_0)$ is precise? The last question implies that we need to find an expression for the variance of $z^*(\boldsymbol{x}_0) - Z(\boldsymbol{x}_0)$ that is as small as possible. You might wonder how the variogram is used in this analysis. The variogram is used to calculate the λ_is and in the expression of the variance of $z^*(\boldsymbol{x}_0) - Z(\boldsymbol{x}_0)$. We also need to determine the mean or trend value $\mu(\boldsymbol{x}_i)$ in equation (19.1). There are three basic forms of kriging (Goovaerts 1997; Chiles and Delfiner 1999), namely:

1. Simple kriging (SK). Here we assume that the mean $\mu(\boldsymbol{x})$ is known.
2. Ordinary kriging (OK) in which we assume that the mean $\mu(\boldsymbol{x}) = \mu$ is unknown but constant.
3. Universal kriging (UK); $\mu(\boldsymbol{x})$ is unknown and varies over the area.

The exact definitions of the estimator in equation (19.2) and the variance $z^*(\boldsymbol{x}_0) - Z(\boldsymbol{x}_0)$ depend on whether we know the mean $\mu(\boldsymbol{x})$ and if not whether it is constant or varies over the area. Once we know which type of kriging we have, finding the unknown weights λ_i follows the standard least squares approach. It is a matter of taking derivatives of the variance $z^*(\boldsymbol{x}_0) - Z(\boldsymbol{x}_0)$, setting them to zero, and solving linear equations. In terms of matrix algebra, the solution for the known-mean case (SK) looks simple:

$$\boldsymbol{\lambda} = \mathbf{K}^{-1}\mathbf{k}$$

where $\mathbf{K}$ and $\mathbf{k}$ are a matrix and vector of covariance terms (which are closely related to the variogram model fitted to the experimental variograms), respectively, and $\boldsymbol{\lambda}$ contains all the weighting factors. Although the mathematical expressions used to calculate for the variance of $z^*(\boldsymbol{x}_0) - Z(\boldsymbol{x}_0)$ are slightly more complex, calculations for simple kriging use the following relationships (supposing a linear dependence on covariates):

$$\begin{aligned} z_0^* &= \beta_0 + \beta_1 X_1 + \ldots + \beta_m X_m + u_0 \\ u_0 &= \sum\nolimits_j \lambda_j u_j + \varepsilon \\ \varepsilon &\sim N(0, \sigma_{err}^2) \end{aligned}$$

This technique is very similar to the spatial moving average (SMA) model discussed in Chapter 18. The X_i are covariates (e.g., spatial coordinates) giving the trend model mean $\mu(\boldsymbol{x})$, and u_0 is the residual from the trend model. The regression parameter β_i models the effect of the explanatory variables on the $Z(\mathbf{x}_0)$, u_i is spatially correlated noise and λ_i are the kriging (averaging) coefficients for it. SK differs from an SMA model, however, because the averaging coefficients λ_i are not derived from the fixed neighbourhood matrix $\boldsymbol{W}$ but are newly calculated for each location $\mathbf{x}_0$ from the variogram model.

Illustration: Volga River relief data — selecting the variogram model

In Section 19.3, we identified two potential variograms that might model the residuals of the Volga River data well. In this section, we illustrate how to choose between the two variogram models using kriging. Recall that the following linear regression model was applied prior to our variogram model calculations in Section 19.3:

$$\text{Elevation}(\boldsymbol{x}) = \alpha + \beta_1 \text{ Latitude}(\boldsymbol{x}) + \beta_2 \text{ Longitude}(\boldsymbol{x}) + Z(\boldsymbol{x}) \quad (19.3)$$

Another way to express this relationship is:

$$H(\boldsymbol{x}) = H_{LM}(\boldsymbol{x}) + Z(\boldsymbol{x}),$$

where $H(\boldsymbol{x})$ is elevation at location $\boldsymbol{x}$ and $H_{LM}(\boldsymbol{x})$ is the spatial trend as modelled by the covariates in the linear regression model. The mean of the residuals of a linear regression model is always zero. However, $z(\boldsymbol{x})$ may be locally unequal to zero. We will apply two types of kriging on the relief data to estimate the trend residuals $z(\boldsymbol{x})$ at each location, namely: the wave variogram tested with ordinary kriging and the spherical variogram tested with universal kriging.

Wave variogram tested with ordinary kriging.

To apply this method, we assume that the expectation of the residuals $z(\mathbf{x})$ is an unknown constant, denoted by μ. We assume that large-scale smooth spatial changes in Figure 19.7 are caused by a stochastic process, and the changes are modelled using the wave variogram. In this case, equation (19.2) becomes

$$z^*(\mathbf{x}_0) = \mu + \sum_i \lambda_i (z(\mathbf{x}_i) - \mu)$$

Note that the spatial trend in the residuals is a constant μ. In this case we use ordinary kriging equations with the wave variogram fitted to calculate both μ and λ_i, and this allows us to estimate $z^*(\boldsymbol{x}_0)$. It is common to denote the estimator by $z^*_{OK}(\mathbf{x})$.

Spherical variogram tested with universal kriging.

In this approach, we assume that large-scale smooth spatial changes in Figure 19.7 are the result of the residual mean varying over the territory (non-stationarity). Therefore, equation (19.2) becomes

$$z^*(\mathbf{x}_0) = \mu_{\mathbf{x}}(\mathbf{x}_0) + \sum_i \lambda_i (z(\mathbf{x}_i) - \mu_{\mathbf{x}}(\mathbf{x}_i))$$

where $\mu_x()$ means that the mean values in $Z(\mathbf{x})$ differ for each location $\mathbf{x}$ and needs to be estimated simultaneously with the λ_i in the kriging system. It is common to model $\mu_x(\boldsymbol{x})$ as a polynomial function of spatial coordinates x and y (e.g., latitude and longitude):

$$\mu_{\mathbf{x}}(\mathbf{x}) = \beta_{\mathbf{x},0} + \beta_{\mathbf{x},1}x + \beta_{\mathbf{x},2}y + \beta_{\mathbf{x},3}xy + \beta_{\mathbf{x},4}x^2 + \beta_{\mathbf{x},5}y^2$$

Universal kriging (UK) with the local mean estimation and fitted spherical variogram can be used to estimate both the local trend and the λ_is, and it is denoted by $z^*_{UK}(\mathbf{x})$.

Figure 19.15-A shows the same relief elevation graph as in Figure 19.1, except that we now used a contour graph to plot all available data. Recall that in the examples in this chapter we used only 20% of the data to fit models. The fit of the linear regression model in equation (19.2) is shown in Figure 19.15-B. The results of the two kriging methods applied on the residuals, as described above are presented in Figure 19.15-C and Figure 19.15-D. So, panels B–D were obtained with 20% of the data, and panel A shows the true picture. Note that the kriged relief is smoother than the original source data (panel A). This is a generic aspect of the kriging estimations.

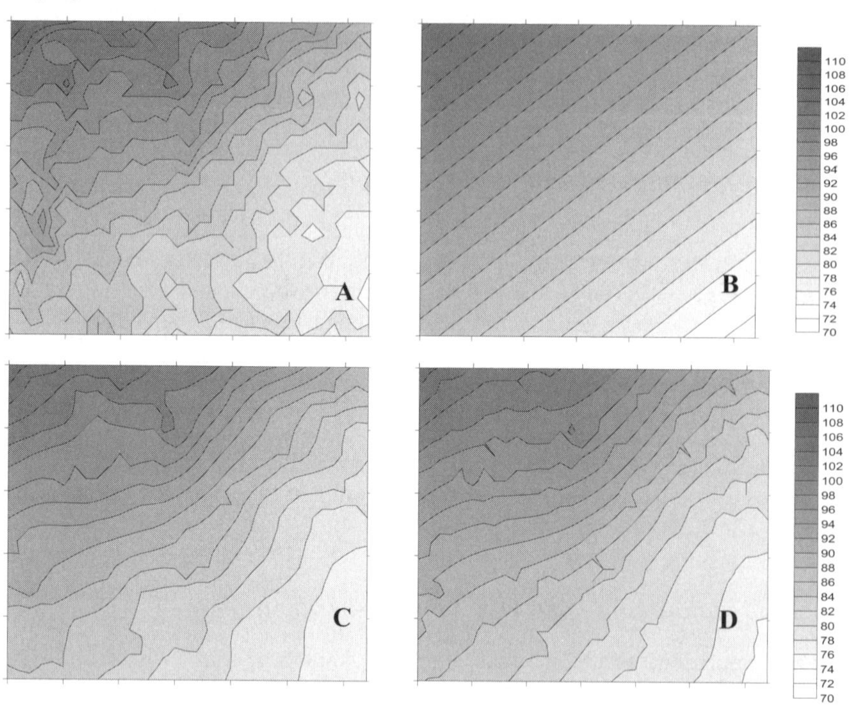

Figure 19.15. A: Contour plot for the original elevation relief data. B: Linear trend model. C: OK with spherical variogram. D: UK with wave variogram.

Because we have access to the detailed relief data for the area, we can compare the original relief structure and the kriged relief structure. The linear trend was used to calculate residuals from the true elevation $z(\boldsymbol{x})$. Differences between $z(\boldsymbol{x})$

values and the residuals estimations $z^*(\boldsymbol{x})$ obtained from kriging are calculated using $z(\boldsymbol{x}) - z^*(\boldsymbol{x})$. For our first approach of testing the wave variogram with ordinary kriging, we obtained a difference mean of 0.011 and a variance of 1.45. In the second approach using universal kriging with the spherical variogram model, the difference mean was −0.001 and the variance 1.17. Both types of kriging incorporated most of the spatial variation of the residuals and gave similar results. However, the spherical model showed slightly better prediction in the error mean and variance.

19.5 A full spatial analysis of the bird radar data

In Chapter 10, multivariate techniques were applied on a bird radar data set. A spatial grid was used to calculate numbers of birds per hour for 3 days. In this section, we apply the geostatistical analysis process that was introduced in this chapter and Chapter 18 to model the spatial relationships in the bird radar data. For this analysis, we will use totals per grid for only the first day. Recall from Section 19.1 that a full geostatistical analysis follows three steps: data exploration, analysis of spatial correlation, and kriging.

Data exploration

Moran's I index is one test for auto-correlation in spatial data. The Moran's I statistic for the bird radar data is 0.842, which indicates that there is positive correlation in bird abundance at neighbouring sites (Chapter 10). A permutation test also indicates that the spatial correlation of bird counts is significant ($p < 0.001$). So the closer the sites are located to each other, the more similar are their bird counts.

Moran spatial correlogram explores the correlation of bird counts at sites against the distances that separate each site. For the bird radar data, the correlogram in Figure 19.16 shows that bird counts at sites that are spaced 3000 m or less apart are significantly positively correlated. If bird counts are separated by 4000 m or more, they are significantly negatively correlated.

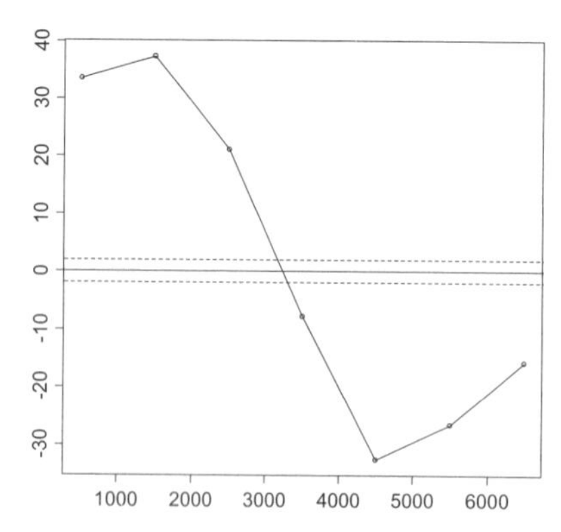

Figure 19.16. Moran's $I^{(k)}$ calculated for birds count at 7 distance bands (k = 0, 1000, 2000, 3000, 4000, 5000, 6000, 7000). Expectation (solid line) and critical values (dashed lines) are also given.

Sometimes, techniques are applied on spatial data that ignore the spatial correlations. However, ignoring spatial auto-correlation structure may cause important exploratory variables to be missed or an inadequate model to be selected to describe the data. As an example of how neglecting spatial correlations affects a linear model, we use the bird radar data in the following linear regression:

$$Count_i = \alpha + \beta_1 \ Latitude_i + \beta_2 \ Longitude_i + \varepsilon_i$$

where the response variable *Count* of birds at a site i is a function of *latitude* and *longitude* (of the site) plus an error component ε_i that is assumed to be *independently* normally distributed with a mean 0 and a variance σ^2. A log-transformation of counts was used to ensure homogeneity. Also latitude and longitude were centred by subtracting the mean latitude and longitude, respectively. This reduces collinearity. The full output of this linear regression model is not given here. Both latitude ($p < 0.001$) and longitude ($p = 0.029$) were significantly different from 0 at the 5% level. The R^2 and AIC were 14% and 1072.56, respectively. Moran's I test applied to the residuals showed that there was significant spatial auto-correlation. The test statistics was $I = 0.834$ ($p < 0.0001$), so the p-values for the t- and F-statistics produced by this linear regression model were unreliable.

We also applied an additive model on the bird radar data in which longitude and latitude were used as smoothers (Chapter 7). The R^2 for this additive model was 0.73, and the AIC was 724.35. The additive model is preferred over the linear regression model because its AIC is lower. ANOVA tests indicated that both smoothers were highly significant (for both: $p < 0.0001$). The Moran's I test statistic for residuals is 0.544 ($p < 0.0001$); so, in this case, we cannot trust the p-values for the t- and F-statistics either.

The linear regression and additive models indicate violation of the independence assumptions. One possible approach to incorporate the spatial dependence within the model is to use the simultaneous autoregressive model (SAR) or the spatial moving average model (SMA) that are described in Chapter 18. For the bird count data, the SAR model specification is:

$$Counts_i = \alpha + \rho \sum_j w_{ij} Counts_j + \beta_1 Latitude_i + \beta_2 Longitude_i + \varepsilon_i$$

The SAR models bird counts at site i as a function of latitude, longitude and bird counts at nearby sites. The output of the SAR model (not shown here) showed that latitude and longitude were not significant at the 5% significance level, and ρ was 0.94 ($p < 0.001$). The large value of ρ confirms the spatial dependence. The AIC was 535.22. The SMA model is given by

$$Counts_i = \alpha + \beta_1 Latitude_i + \beta_2 Longitude_i + u_i$$
$$u_i = \lambda \sum_j w_{ij} u_j + \varepsilon_i$$

In the SMA, bird counts are modelled as a function of latitude, longitude, and error components of nearby sites. In the SMA model, latitude was significantly different from zero ($p = 0.026$), λ was 0.95 ($p < 0.001$) and the AIC was 530.8. The AICs for the linear regression, additive model, SAR, and SMA model are, respectively, 1072.56, 724.35, 535.22 and 530.87. Therefore, the SMA model fits the data best.

Analysis of spatial correlation

Two assumptions made for most geostatistical modelling tools are normality and stationarity. Before we proceed with the geostatistical analysis of bird counts, these assumptions must be verified using data for this area. The spatial trend (if present) in the count data can be subtracted from the data in order to remove any dependence of site locations or other covariates. This means that the following model (in words) is applied:

$$\text{Counts}_i = F(\text{Latitude}_i, \text{Longitude}_i, \text{Other Covariates}_i) + \text{residuals}_i \qquad (19.4)$$

F stands for 'Function of' and can be modelled as in linear regression or additive modelling, among others. For the bird radar data, we only have latitude and longitude and there are no other covariates. In geostatistics, F is confusingly called the trend. The entire model in equation (19.4) is sometimes called a trend model. Earlier in this section, we applied two trend models to relate the spatial coordinates of sites with bird counts. These were the linear regression (LM) and additive model (AM). We will now use the residuals of these models as a variable in the geostatistical analysis and modelling. Hence, the normality and stationarity assumptions are for the residuals and not the raw data. QQ-plot, histograms, and graphs with fitted values versus residuals were made to verify normality and homogeneity for the residuals of the LM and AM models. These plots (not shown here) show a clear violation of normality and homogeneity for the regression residuals and to a lesser extend also for the AM residuals.

Possible ways to overcome violation of normality and homogeneity were discussed in Chapters 4 and 5 and are (i) to apply a transformation on the bird count data, (ii) to model birds counts using a Poisson distribution, (iii) add more covariates to the models (or interactions), and (iv) apply smoothing methods, among others. In this example, we applied a log-transformation on the count data. We chose a log transformation and not a square root because of the results of the log transformation produced better results in the QQ-plots. Histograms and QQ-plots of residuals after the log-transformation (not shown here) of the bird count data have a 'more' normal distribution than they had without the transformation.

The stationarity assumption for the residuals is tested with *h*-scatterplots. Residuals of the linear model show strong departure from stationarity; i.e., a swarm of points is above the diagonal line (Figure 19.17). The figure shows a strong drift in the data and indicates that we have underestimated the trend. So, we need to eliminate the trend from the data. One option to do this is to use the additive model, which is more flexible than the linear model.

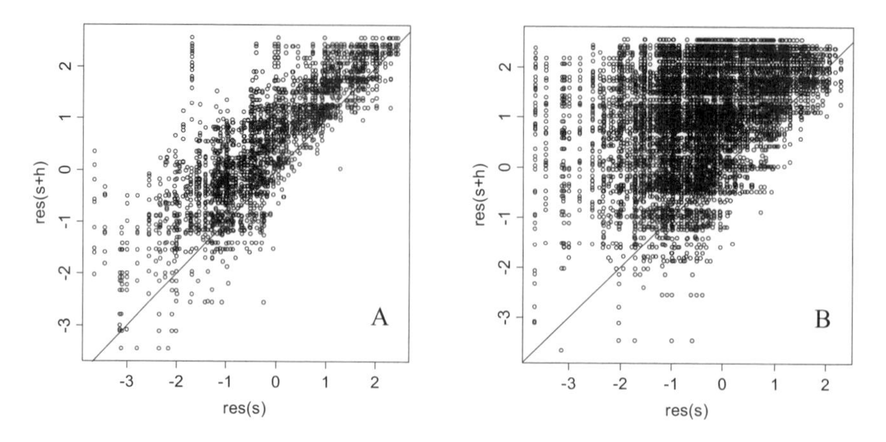

Figure 19.17. Omnidirectional *h*-scatterplots (all directions of lag ***h*** are used). A: Plot for LM residuals for lag h = 1300 m. B: Plot for LM residuals for lag h = 3900 m. The value of the residual at **s** is along the horizontal axis, and residuals of all points that are ***h*** units away from **s** are plotted along the vertical axis. This is done for each site in the study area. The bird data were log-transformed.

h-Scatterplots for AM residuals (Figure 19.18) look better than for the LM residuals. There is no strong butterfly effect. Although we have points that are far from the diagonal line, there are only a few of them. A drift is present, but it is not as strong as for the LM residuals. *h*-Scatterplots for different directions (Figure 19.19) confirm that the drift is weak. Therefore, we proceed by assuming that the residuals are stationary and normally distributed.

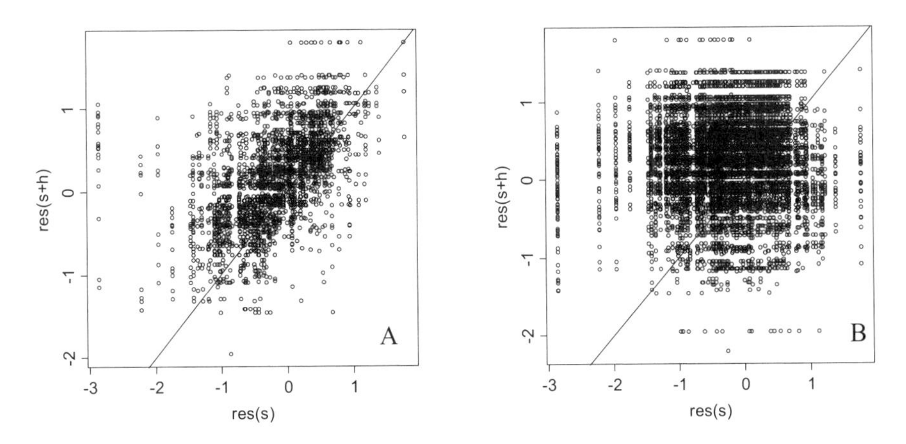

Figure 19.18. Omnidirectional h-scatterplots (All directions of lag $\boldsymbol{h}$ are used). A: Plot for AM residuals for lag h = 1300 m. B: Plot for AM residuals for lag h = 3900 m. The bird data were log-transformed.

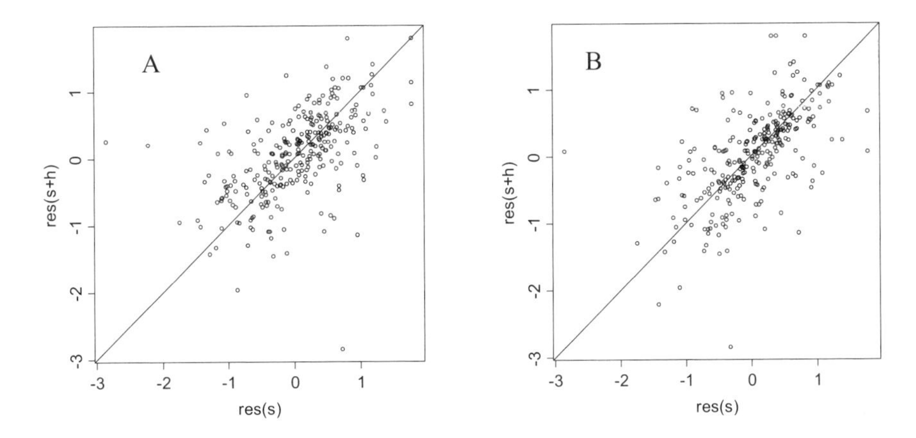

Figure 19.19. Directional h-scatterplots (lag $\boldsymbol{h}$ is taken only along selected directions). A: Plot for AM residuals for lag h= 1300 m at 90° direction. B: Plot for AM residuals for lag h = 1300 m at 0° direction. The value of the residual at **s** is along the horizontal axis, and residuals of all points that are $\boldsymbol{h}$ units away from **s** are plotted along the vertical axis. The bird data were log-transformed.

In the previous paragraph the following model (in words) was fitted on the log-transformed bird data (the value of 1 was added to each observation to avoid problems with logs of zero):

$$\text{Birds} = \text{intercept} + f(\text{Latitude}) + f(\text{Longitude}) + \text{noise}$$

where $f()$ was either a smoothing function (additive modelling) or a parametric term (linear regression). The noise component is assumed to be independently normally distributed. If $f()$ is a smoothing function, the model above is called the AM trend model. We can refine the AM trend model by taking into account the spatial auto-correlation directly within the error structure of the noise as this produces better smoothing curves.

In Chapter 18 we introduced linear regression models with correlated errors, LM(ce), to do this, but the same can be done for additive models. We call it an additive model with correlated error AM(ce). The estimation process makes use of generalised least squares (GLS), and details can be found in Pinheiro and Bates (2000). Within the GLS, we need to specify the model for the correlation structure and we tried two different correlation structures. The first model with a spherical residual auto-correlation structure gave an AIC of 553, and the second model with an exponential correlation structure has AIC = 556. Therefore, we decided to use the spherical correlation structure because it had a lower AIC.

Recall that the spherical correlation structure is fully specified once we know the range and sill. Numerical optimisation gave a range of 3764 m and nugget effect (nugget/sill) ≈ 0.08.

Due to software implementations, AM(ce) models can only be used on spatial data with isotropic auto-correlation processes, so we need to explore the residuals of the trend model and verify that we indeed have isotropy. We use a variogram surface to determine the presence and directions of anisotropy; see Section 18.3 for an explanation. A variogram surface for the AM trend residuals is plotted in Figure 19.20. Anisotropy is present, but it is not strongly pronounced. Until the ranges of 4000–5000 m the direction with the largest auto-correlation observed is approximately 135°.

We can also explore anisotropy using a directional variogram. It shows variogram values for lags taken only in specified directions. Directional variograms are presented for residuals of the AM trend model in Figure 19.20. Usually four directions are tested: the first along the direction with maximum spatial auto-correlation detected by the variogram surface (in this example, at 135°); the second, perpendicular to the maximum (at 45°); and the third and fourth at intermediate directions (at 0° and 90° in this example). Isotropy would be confirmed if all directional variograms are the same. If they are not, our models possess anisotropy; and we need to look at the values of the nugget, sill, and range of the auto-correlation process.

The directional variograms in Figure 19.21 are not the same, and show strong evidence for zonal anisotropy. Zonal anisotropy means that the sill varies with direction (recall that in geometrical anisotropy sills of all variograms are the same but are reached at different ranges). It is near 0.9 in the 45° direction and near 0.7 in the 135° direction. This may also be caused by the weak drift indicated in the *h*-scatterplots above.

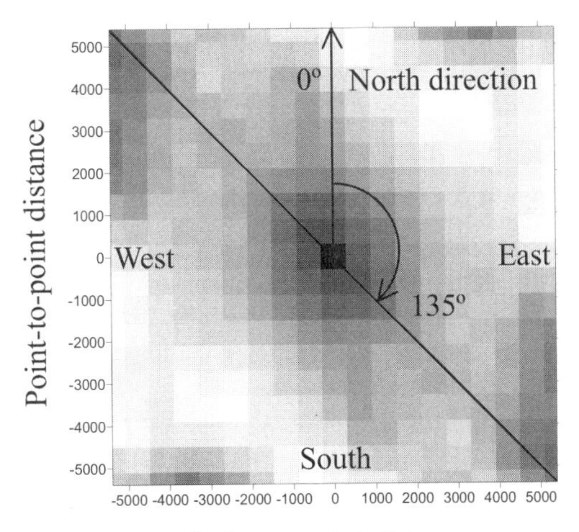

Figure 19.20. Variogram surface. Darker cells indicate lower variogram values. The direction with the largest auto-correlation observed is approximately 135°.

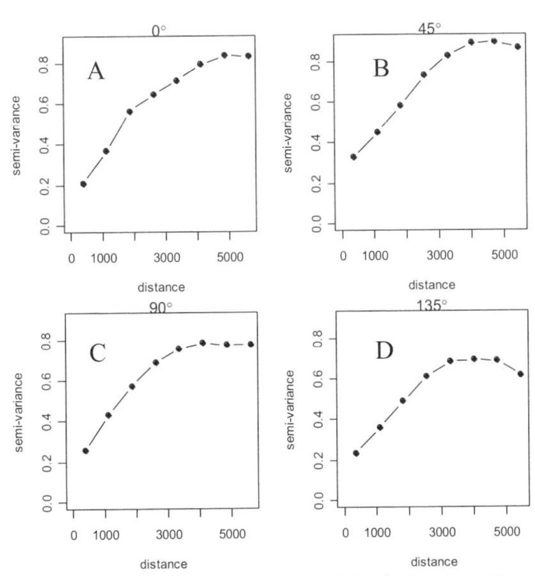

Figure 19.21. Directional variograms for AM residuals for lag distances at four directions selected. A: 0°, B: 45°, C: 90°, D: 135°. Differences may be caused due to different flight paths of the birds.

Figure 19.21 shows two problems; the range differs per direction (shorter range for the 90° and longer for the 0° direction), and the sill differs per direction (approximately 0.2 higher for the 45° direction than for the 135° direction). The good news is that the nugget is approximately the same in each direction. We need to take the two problems in account, but how? The solution is to use the following mechanism:

$$\text{Anisotropy variogram} = \text{omnidirectional variogram} + \text{deviation in direction } i$$

The variogram is presented as the sum of two components: An omnidirectional variogram and a correction allowing for anisotropy. For the first component we will just pretend that there is isotropy and calculate an omnidirectional variogram. This can be seen as an average variogram for all directions. Because we will add a second component, the resulting variogram values can only increase. The omnidirectional variogram is not the variogram that we would obtain by clumping all the directions. Instead, we will use the shortest range and lowest sill of the directional variograms to guarantee a positive difference between the experimental and the omnidirectional variograms. Once we have the experimental directional variograms, we can easily calculate the difference between the experimental variogram in direction i, and fit an anisotropic variogram on the difference (component 2). It is the deviation in direction i. The process is illustrated for the bird radar data in Figure 19.22. Hollow dots represent experimental variograms in each of the four directions. The dotted line in each panel is the omnidirectional variogram model. Note that it has the same shape for all directions. The heavy dashed line is the anisotropic component, and it models the difference between these two. The solid line is a sum of two components, and it fits the experimental variogram reasonably well in each of the four directions. A variogram was applied on the difference. Full technical details can be found in Chiles and Delfiner (1999) or Cressie (1993).

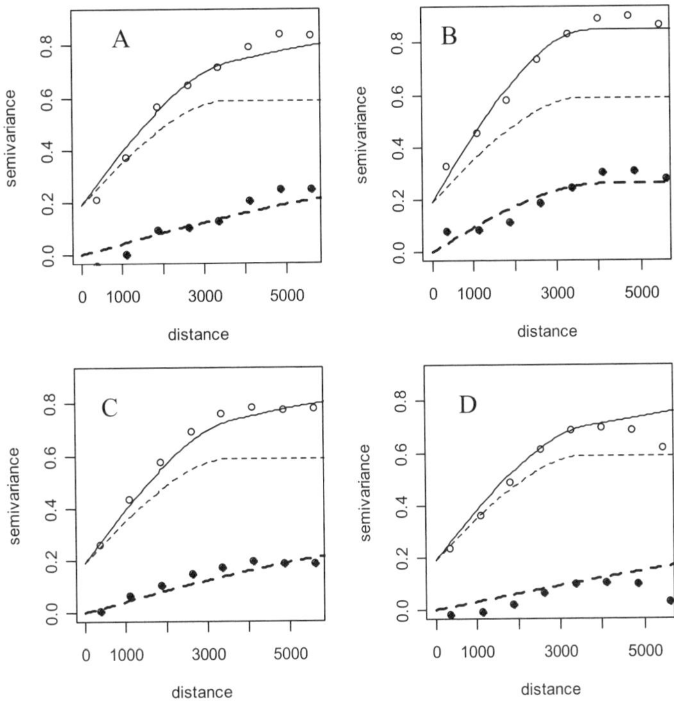

Figure 19.22. Hollow dots represent empirical variograms at the four directions selected. A: 0°, B: 45°, C: 90°, D: 135°. Dotted lines represent the behaviour of the isotropic (general) variogram component. Filled dots represent the difference between empirical variogram and isotropic general variogram component. Dashed lines represent the behaviour of anisotropic variogram component. Solid lines represent the behaviour of the final variogram model that is the sum of isotropic and anisotropic components.

Kriging

We will now incorporate the resulting variogram model into a kriging analysis that interpolates the sampled bird counts into unsampled areas and allows us to predict how birds are distributed over this area.

Kriging combines information from the trend models and the variogram model of trend residuals to predict the spatial distribution of a variable of interest. In this case, we want to predict how birds are spatially distributed over the entire area when only a minimal number of actual bird counts have been done. Figure 19.22 shows how the bird-count information differs using just the AM trend model (Figure 19.23-A), the predicted trend residuals from the AM model after kriging (Figure 19.23-B), and the combined results from the trend and residuals predictions (Figure 19.23-C). Because we used the log-transformed bird counts, we also included a prediction based on the back-transformed values on the original count scale (Figure 19.23-D). Each map shows different characteristics of the bird count data, and each analysis refines our predictions about how birds may be distributed in this area.

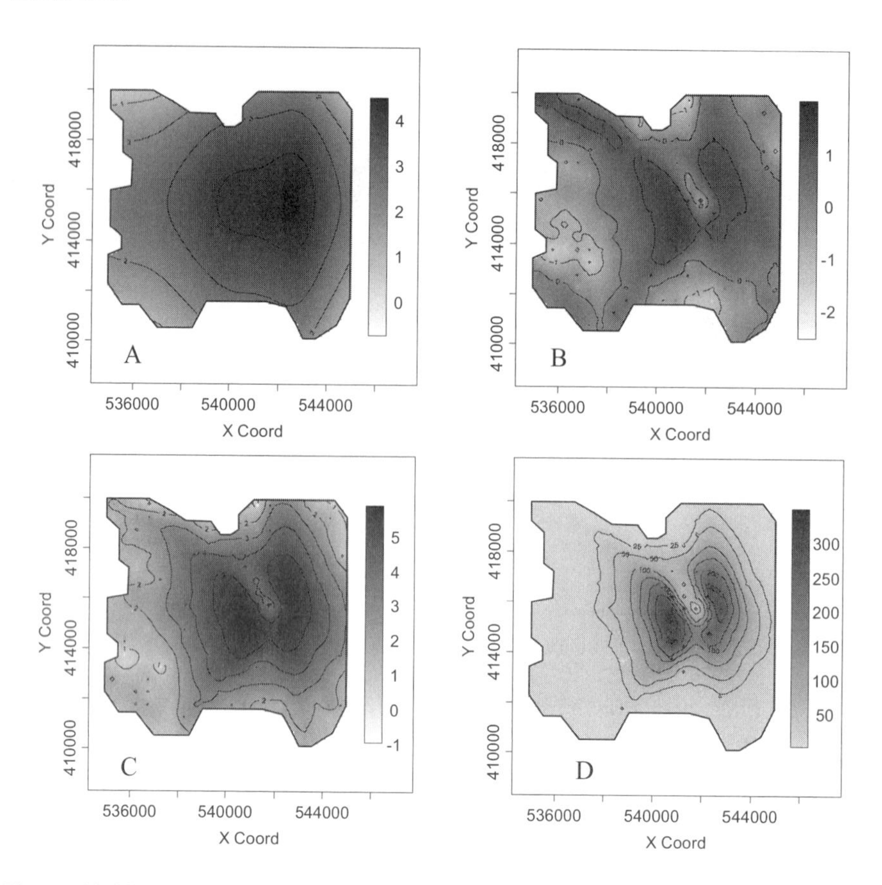

Figure 19.23. Maps of predicted bird counts. A: Prediction based on AM trend model. B: Prediction based on kriged AM residuals. C: Prediction based on log(bird counts) as a sum of trend and residuals maps. D: Prediction obtained by back-transformation of log-counts in map C. A–C: contour level step 1. D: contour level step 50.

20 Univariate methods to analyse abundance of decapod larvae

Pan, M., Gallego, A., Hay, S., Ieno, E.N., Pierce, G.J., Zuur, A.F. and Smith, G.M.

20.1 Introduction

This chapter illustrates how to decide between the application of parametric models (linear regression models) and non-parametric methods (additive models). The techniques applied in this chapter will use as explanatory variables some abiotic (temperature, salinity) and biotic (algal food biomass, as indicated by chlorophyll a) factors that affect the meroplanktonic larvae. We aim to provide preliminary information about some of the pre-settlement processes and the relative influences of environmental factors and variability. Also, the taxonomic identification of some decapod larvae is often difficult, and processing the samples is time consuming. By using information from the samples already analysed and the other available data, such as how many samples per year we could analyse, we may optimise the number of samples examined to achieve the best outcomes and interpretations. In this chapter we therefore also discuss how some models can be used to optimise the number of samples for further sample analysis in other years.

Decapod crustaceans are an important fishery resource in many countries around the world. In Europe alone there are approximately 22 species that are actively exploited. The fisheries for some of these species are very important economically, and their landings value represents nearly 30% of all fish and shellfish landings value in the world (Smith and Addison 2003). The most valuable species landed into Scotland (UK) is the decapod crustacean species *Nephrops norvegicus* (the Norway lobster).

Aside from their economic importance, decapod crustaceans are also of significant ecological importance. They are diverse and abundant in coastal ecosystems, with complex life histories and population dynamics. Many decapods produce pelagic larvae that remain and develop in the water column for days, weeks or months before settling as juveniles into the benthos. After metamorphosis from the pelagic stages, they feed, grow and are fed upon in the benthic community, with survivors recruited to the adult population. For species with planktonic larval stages and then benthic juveniles and adults, there are many factors, biotic and abiotic, which influence the abundance, growth and survival of a cohort. Traditionally, these factors are divided between: (i) pre-settlement processes, operating after the larvae are hatched until they settle into the benthos and (ii) post-settlement process, operating throughout the whole benthic period (Wahle 2003).

Although there are many published studies on decapod crustacea, these are usually related to benthic adults and juveniles and knowledge of their larval phases remains poor. It is necessary to identify and understand the biology of early stages and the environmental factors that affect them, to understand fully the population dynamics of decapods. Improving our understanding of their seasonal population dynamics will contribute to increased understanding of the role of decapods in marine ecosystems. This will also contribute to future developments of stock assessment and management methods for sustainable exploitation of harvestable resources.

As part of a PhD project, this case study chapter is a first approach to ongoing investigations into the ecological role and recruitment of decapod larvae. One aim is to study differences and temporal patterns in the decapod larval communities in the east and the west coast of Scotland. These patterns we may then relate to some or all of several environmental variables known to effect larval production and development. To solve such questions we aim eventually to analyse up to five years of plankton samples from the east and west coast of Scotland. In this case study we only use the abundance of 12 decapod crustacean families studied over two years.

The two sampling points selected in this study of decapod larvae are part of an FRS (Fisheries Research Services) long-term weekly monitoring programme. Sites are located in the northeast (Stonehaven, sampled since 1997) and in the northwest of Scotland (Loch Ewe, since April 2002). Samples analysed for this case study were taken during 2002 and 2003. Both locations are ecologically different, and some of the variables that could influence the larval dynamics are as follows. Due to the shallow depth (50 m), strong tidal flows and the wind effects (Otto et al. 1990; Svendsen et al. 1991), Stonehaven (Scottish northeast coast) is a very dynamic site and the water column remains well mixed for most of the year, with some thermal stratification in calmer summer weather. The fjordic sea loch, Loch Ewe, is a more enclosed sea area and is subject to different environmental conditions and influences. The difference between these reference sites is likely to play an important role in the settlement and recruitment of decapod larvae and could influence, along with other environmental variables, the species diversity and composition. The data show that seabed features differ between locations with hard sand and rocky bottoms off Stonehaven and muddy sand in Loch Ewe.

20.2 The data

The samples were collected between April 2002 and April 2003 from two sampling stations. One is located approximately 3 km offshore of Stonehaven, northeast of Scotland, UK (56° 57.8' N, 02° 06.2' W), whereas the second is located in a sea loch in the northwest of Scotland, Loch Ewe (57°50.14'N, 05°36.61'W) (Figure 20.1).

The plankton samples analysed were taken with a bongo net of 200-μm mesh size towed vertically from near the bottom and immediately preserved. Environmental variables were also sampled: temperature and salinity, measured near the surface and near the bottom (at 1 m and 45 m in Stonehaven and 1 m and 35 m in Loch Ewe), several nutrient chemical concentrations (not used in this case study) and chlorophyll a measured fluorimetrically from surface water sampled over 0–10 m. The decapod larvae present in the samples were counted and taxonomically identified where possible (following dos Santos and Gonzalez-Gordillo (2004), Fincham and Williamson (1978), Pike and Williamson (1958, 1964, 1972), Williamson (1957a, 1957b, 1960, 1962, 1967 and 1983), among others).

The data used in this chapter consist of abundance values for families expressed as numbers of individuals per m^2 (Table 20.1). A total of 24 plankton samples was analysed in the case of Stonehaven, and 21 for Loch Ewe. This ranged from three or four samples analysed per month in the spring/summer to one or two per month in the winter, when decapod larvae are essentially absent from the plankton samples.

The explanatory variables used in these analyses (Table 20.2) were temperature, salinity and chlorophyll a values. Also used as explanatory variables were the locations where the decapod families were found (Stonehaven vs. Loch Ewe) and the collection year (2002 vs. 2003).

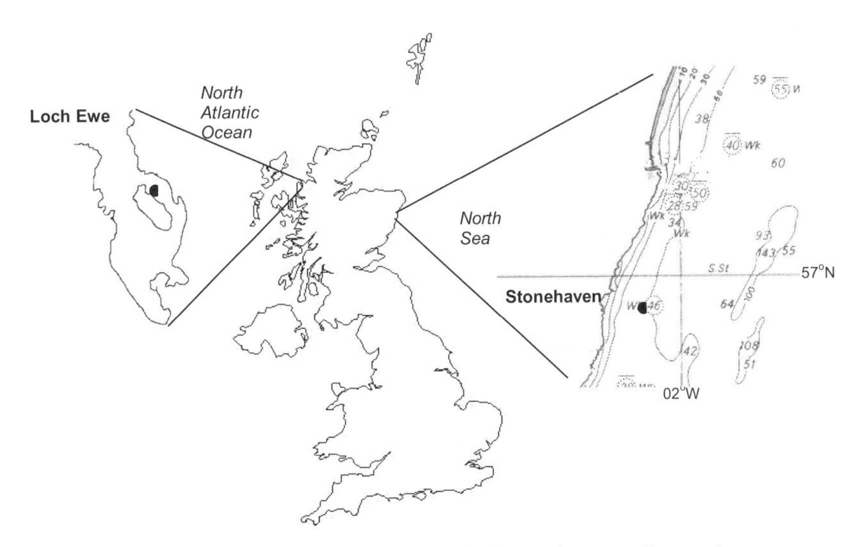

Figure 20.1. Map of the study area. The dots indicate the sampling points.

Table 20.1. List of families used to calculate the species index function.

Families	
Galatheidae	Pandalidae
Paguridae	Alpheidae
Porcellanidae	Callianassidae
Hippolytidae	Upogebiidae
Crangonidae	Laomediidae
Processidae	Nephropidae

Table 20.2. Available explanatory variables.

Explanatory Variable	Notation
Temperature at 1 m	T1m
Temperature at 45–35 m	T45.35m
Salinity at 1 m	S1m
Salinity at 45-35 m	S45.35m
Chlorophyll a	Cho.10m
Location	Location
Year	Year

Options for the analyses of these data include multivariate methods such as principal component analysis, redundancy analysis, canonical correspondence analysis, and even methods such as the Mantel test and ANOSIM. These techniques can be used to detect whether there are relationships between the abundance of the 12 decapod families and the chosen explanatory variables. Examples of such analysis are given in other case study chapters. Here, the aim is different; we want to show how to decide between using parametric models, such as linear regression and generalised linear modelling, versus using non-parametric methods such as additive modelling and generalised additive modelling. All these methods use one response variable, and describe it in terms of the explanatory variables. To obtain one response variable, we can either work on data from one family or convert the data on the 12 families into a diversity index. Obvious candidates are the Shannon–Weaver index, richness index, total abundance, Simpson index, Berger–Parker index and Macintosh index. A detailed discussion on diversity indices can be found in Magurran (2003). In practise, most diversity indices are highly correlated with each other and therefore your choice of index should be based on ecological rather on than statistical arguments. In the remaining part of this chapter, we use richness as the response variable because we are interested in the numbers of families rather than in species abundances. The list of families used to calculate richness is presented in Table 20.1. The total number of sites was 45. The first seven families were measured at 10 or more sites. These families were also more abundant. The available explanatory variables are in Table 20.2. The variables chlorophyll a, salinity at 45–35 m and temperature at 45–35 m have four missing values in the year 2002.

20.3 Data exploration

Figure 20.2 shows a pairplot of the species richness (response variable) and all explanatory variables. The first row and column shows the relationship between richness and each of the explanatory variables. The correlations between richness and explanatory variables (first column) indicate that there might be temperature and chlorophyll a effects. All other panels can be used to asses collinearity: the correlation between explanatory variables. It is clear that temperature (T1m) at the surface and at 45–35 m (T45-35m) cannot be used jointly in any analysis, as the correlation is 0.98. Because (i) T1m has a higher correlation with species richness and (ii) T1m has no missing values (T45-35m has 4 missing values) we decided to use T1m. The correlation between salinity at the surface (S1m) and at 45–35 m (S45.35m) is not as high as for temperature, but we decided to use only S1m. Note that temperature and the nominal variable year have a correlation of 0.74. If a nominal variable has more than two classes, the correlation coefficient is meaningless. However, in this case there are only two classes, which are coded as 2002 and 2003. This means that a correlation of 0.74 indicates that temperature in the second year was higher.

Month and temperature

Note that there is a clear pattern in the month-temperature panel in Figure 20.2. It shows a nice sinusoidal pattern reflecting the seasonal temperature pattern. In the summer months, temperature is higher than in the winter, but we also have higher richness when the temperature is high. This means that we can either use month as an explanatory variable or temperature, but not both, as they both reflect the same ecological effect. Because temperature has a linear effect on richness, we decided to use it instead of month, which clearly has a non-linear relationship with richness. Another reason to avoid using month is that in some months only one observation was taken, whereas in other months we have three of four observations. This would cause problems if month were treated as a nominal variable.

In practice, the situation is a bit more complicated than this as we are actually dealing with a time series. However, for the moment we ignore the time series aspect and assume that temperature represents the seasonal pattern. We return to this aspect in the discussion.

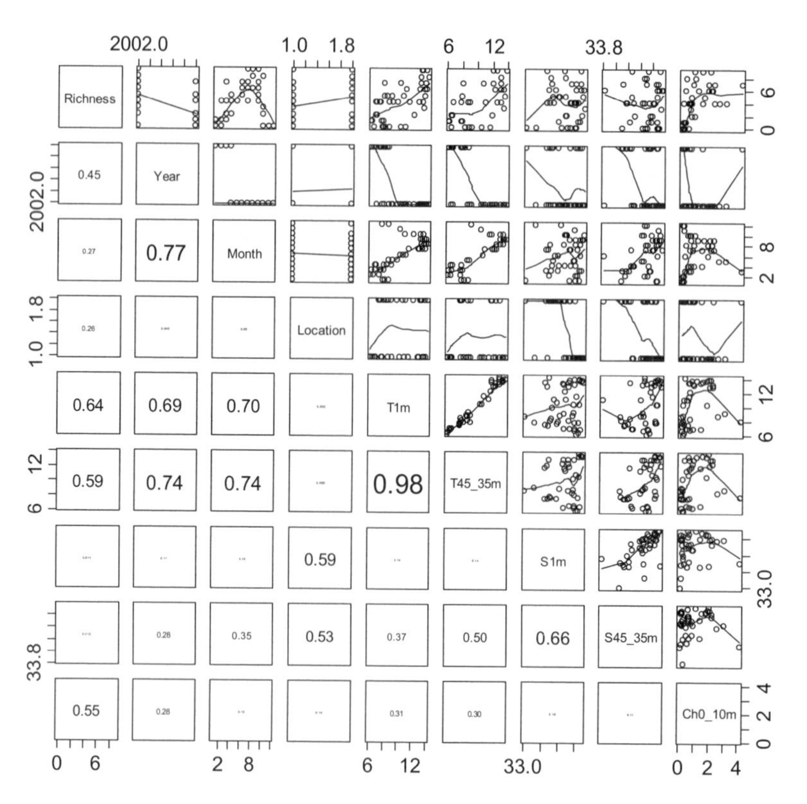

Figure 20.2. Pairplot of species richness and all explanatory variables. The lower diagonal part of the pairplot shows the (absolute) correlation coefficient in which the font size is proportional to the value of the correlation. The upper diagonal shows the scatterplots with smoothing curves.

Transformation

Cleveland dotplots (Figure 20.3) indicated that there are a few samples with slightly higher chlorophyll a values. A first round of application of statistical methods indicated that these observations were influential (e.g., resulting in wide confidence bands for the smoothing techniques), and therefore, we decided to apply a square root transformation to chlorophyll a. The original and transformed chlorophyll a data are presented in Figure 20.3.

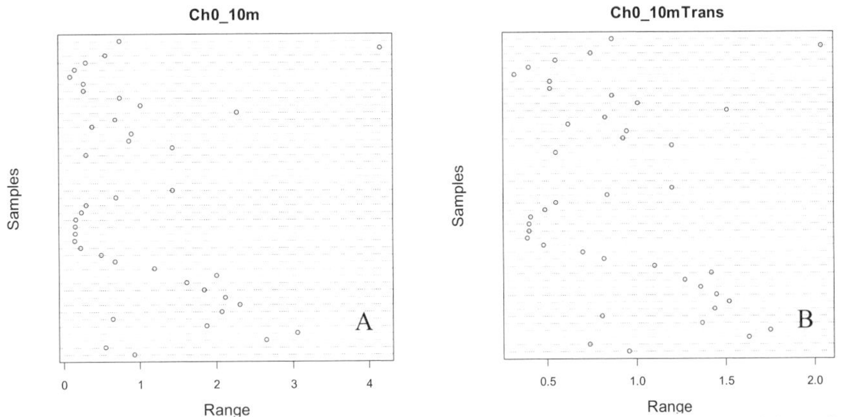

Figure 20.3. Dotplot of chlorophyll a (A) and square root transformed chlorophyll a (B).

20.4 Linear regression results

A linear regression model was applied to explain the variation in the richness index as a function of temperature, salinity, chlorophyll a (transformed), location and year. A forward and backward selection using the AIC was applied and the optimal model, as judged by the AIC, contains temperature, chlorophyll a and location. The estimated parameters, standard errors, *t*-values, etc. are given below.

	Estimate	Std. Error	*t*-value	*p*-value
Intercept)	−4.23	1.06	−3.97	<0.001
T1m	0.43	0.10	3.94	<0.001
chlorofilatrans	3.34	0.71	4.70	<0.001
factor(Location)2	1.55	0.58	2.68	0.011

Residual standard error: 1.76 on 37 degrees of freedom
Multiple R^2: 0.66, Adjusted R^2: 0.63
F-statistic: 24.19 on 3 and 37 df, *p*-value: <0.001

Temperature and chlorophyll a have a positive relationship with richness: the higher the temperature or the higher the chlorophyll a, the more decapod families are present. The results for the nominal variable location show that there are more decapod families found at Loch Ewe (location = 2) than at Stonehaven, which is the baseline value. Panel A in Figure 20.4 shows the standard graphical output of linear regression: residuals versus fitted values to verify the homogeneity assumption, a QQ-plot for normality, the scale-location plot for homogeneity and the Cook distance function for influential observations. None of the panels shows serious problems. The other three panels are residuals versus individual explanatory variables. Any patterns in these graphs are an indication of serious model misspecification. The residuals plotted versus temperature (Figure 20.4-B) show that all samples with lower temperature have a positive residual. Meaning these samples are all under-fitted, as residuals are calculated as observed minus fitted values. Figure 20.4-C shows a graph of residuals versus transformed chlorophyll a. The samples with the six highest chlorophyll a values are all over-fitted (negative residuals). There is no clear residual pattern for location (Figure 20.4-D).

If a graph of residuals versus individual explanatory variables shows any pattern, then model improvement is required. One option is to include quadratic terms in the model, for example:

$$\text{Richness}_i = \alpha + \beta_1 \text{ Temperature} + \beta_2 \text{ Temperature}^2 + \ldots + \varepsilon_i$$

However, the linear and quadratic terms might be highly collinear, and the fit of such quadratic models might be poor. Furthermore, based on prior knowledge of the study system there is no reason to expect a quadratic relationship. Another way to improve the model is by including interaction terms. We added two-way interactions between all possible combinations, but none of the interactions resulted in a significant model improvement as judged by F-statistics of nested models and individual t-values. An alternative approach is a smoothing method like additive modelling or generalised additive modelling.

Additive modelling is based on the Gaussian distribution, and the GAM can use other distributions like the Poisson distribution. The choice of distribution to use can also be inferred from the linear regression results, namely from the plot of residuals versus fitted values; see Chapters 5–7 for further details. If an increase in spread for larger fitted values can be seen, the application of GLM or GAM with a Poisson distribution and log-link function should be considered. The choice between GLM and GAM can be made based on the graphs of residuals versus each explanatory variable. However, for these data, no clear violation of homogeneity can be seen (Figure 20.4), and therefore, the application of additive modelling seems to be appropriate. This decision process is illustrated with a flowchart in Figure 3.1.

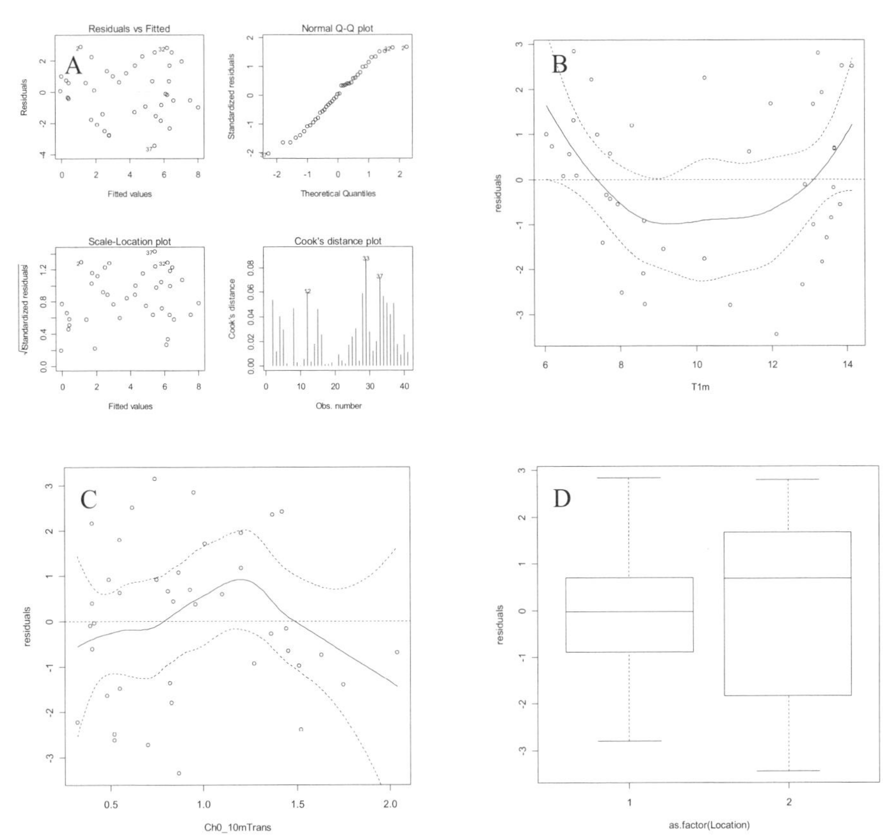

Figure 20.4. Graphical output of the linear regression model. Panel A: standard regression output: residuals versus fitted values, QQ-plot of residuals, a scale location diagram and the Cook distances. The other panels show the residuals versus fitted values for temperature (B), chlorophyll a (C) and location (D). In panels B and C Loess smoothing lines and corresponding point-wise 95% confidence bands were added to aid visual interpretation.

20.5 Additive modelling results

The following additive model (Chapter 7) was applied:

$$\text{Richness}_i = \alpha + f_1(\text{T1m}_i) + f_2(\text{S1m}_i) + \text{f}_3(\text{Ch0.10mTrans}_i) + \text{Year}_i + \text{Location}_i + \varepsilon_i$$

Year and location are fitted as nominal variables. Results of this model indicated that various smoothers can be dropped from the model. Hence a model selection is required. We used the AIC method to choose the optimal model. The AIC of the model containing all explanatory variables is 160.37 (Table 20.3).

Dropping only year gives AIC = 160.04, dropping only Location results in AIC = 165.74, etc. The lower the AIC, the better is the model. In this case, it suggests dropping the nominal variable Year as it is the least important variable. The new model is of the form:

$$\text{Richness}_i = \alpha + f_1(\text{T1m}_i) + f_2(\text{S1m}_i) + f_3(\text{Ch0.10mTrans}_i) + \text{Location}_i + \varepsilon_i$$

The process of dropping variables then repeats, but the AIC indicates that this is the optimal model (Table 20.4). However, the confidence bands and *p*-value ($p = 0.14$) for the salinity smoother were rather large. This is a typical example in which the AIC is too conservative and some extra model selection steps are required. Based on the width of the confidence bands we decided to remove salinity from the model. All remaining variables were significant at the 5% level.

Table 20.3. The results of the first step in the backward selection process in additive modelling. An 'X' means that the explanatory variable was used in the model. In each model cross-validation was applied to estimate the optimal degrees of freedom for each smoother. Bold type indicates the lower AIC and for hence the optimal model.

Variables	Selected Explanatory Variables					
Year	X	-	X	X	X	X
Location	X	X	-	X	X	X
T1m	X	X	X	-	X	X
S1m	X	X	X	X	-	X
Ch0.10mTrans	X	X	X	X	X	-
AIC	160.37	**160.04**	165.74	164.6	164.25	196.13

Table 20.4. The results of the second step in the backward selection process in additive modelling. An 'X' means that the explanatory variable was used in the model. In each model cross-validation was applied to estimate the optimal degrees of freedom for each smoother. Bold type indicates the lower AIC and for hence the optimal model.

Variables	Selected Explanatory Variables				
Location	X	-	X	X	X
T1m	X	X	-	X	X
S1m	X	X	X	-	X
Ch0.10mTrans	X	X	X	X	-
AIC	**160.04**	163.94	169.74	162.4	197.06

The final model is of the form:

$$\text{Richness}_i = \alpha + f_1(\text{T1m}_i) + f_2(\text{Ch0.10mTrans}_i) + \text{Location}_i + \varepsilon_i$$

In each step of the backwards selection, cross-validation was applied to estimate the optimal degrees of freedom for each smoother. The numerical output of the optimal model is given by

	Estimate	std. err.	*t*-ratio	*p*-value
Intercept	3.16	0.33	9.352	<0.001
factor(Location)2	1.78	0.54	3.265	0.002

Approximate significance of smooth terms:

	edf	chi.sq	p-value
s(T1m)	2.15	3.36	0.005
s(Ch0_10mTrans)	1.32	3.82	0.002

R-sq.(adj)=0.697. Deviance explained=73.1%. Variance=2.5872. n = 41

These results indicate that the explanatory variable Location is significantly different from 0 at the 5% level. At location two, 1.78 more families are expected compared with location one (Stonehaven). The degrees of freedom for the smoothers of temperature and chlorophyll a are rather small, indicating nearly linear relationships. The smoothers are given in Figure 20.5, and the shape of the line also indicates a nearly linear relationship. Figure 20.6 shows the residuals of the optimal additive model plotted versus the explanatory variables. As there are no clear patterns now, we can conclude that the additive model is better than the regression model. A more formal approach that comes to the same conclusion is an analysis using the F-test. The output is given by

Model 1: Y ~ s(T1m) + s(Ch0_10mTrans) + factor(Location)
Model 2: Y ~ T1m + Ch0_10mTrans + factor(Location)

	Resid. df	Resid. Dev	df	Deviance	*F*	*p*-value
1	35.53	91.92				
2	37.00	115.35	1.470	−23.430	6.162	0.009

Model one is the additive model in which cross-validation is applied to estimate the optimal degrees of freedom for each smoother. Model two is the (nested) linear regression model. The change in deviance is 23.43. The corresponding F-statistic is equal to 6.16 (p = 0.009) and indicates that the more complicated additive model performs better.

20.6 How many samples to take?

In the previous section, backwards selection was used to find the optimal additive model. It contained temperature, chlorophyll a (transformed) and a location effect. The model was based on 41 samples collected over two years, and only analyses part of a much larger dataset held at at the FRS Marine Laboratory in Aberdeen (UK). At the time of writing, data from several other years are still waiting to be analysed. A wide range of measurements, in addition to those discussed here, are taken for each sample (e.g., DNA information), and a full analysis will be time consuming and costly. As this is a problem common in ecology, it is important to know how many samples per year should be collected.

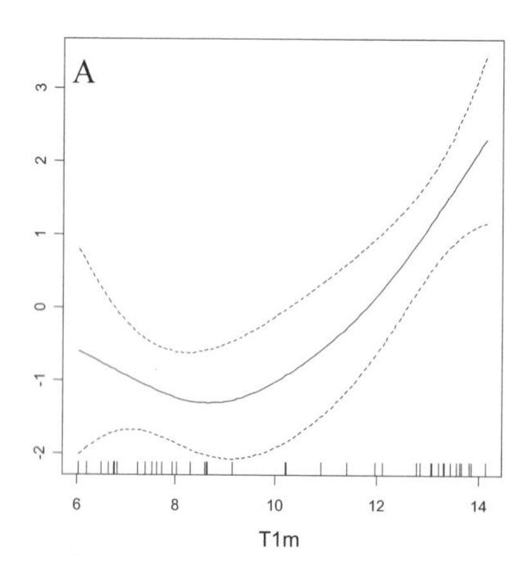

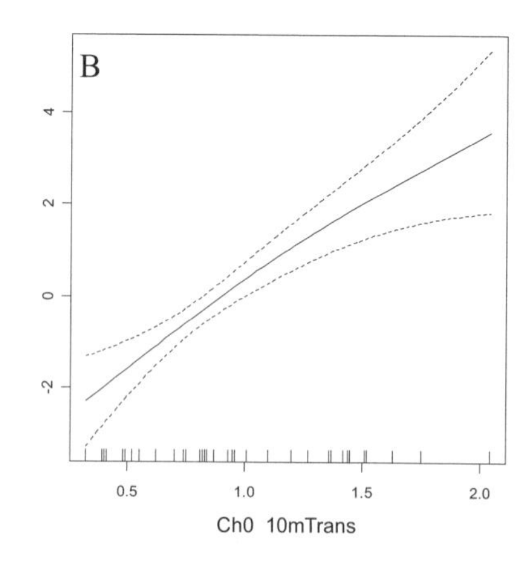

Figure 20.5. Smoothed curves obtained by the additive model. A: The temperature effect on species richness. B: The chlorophyll a effect. Dotted lines are 95% confidence bands. The horizontal axes show the environmental gradient and the vertical axes the contribution of the smoother to the fitted values.

One option is to apply a power analysis using the linear regression results. However, the regression model was not optimal, and therefore, it is less appropriate to use its results for a power analysis (Zar 1999). The use of power analysis for additive modelling is less well developed than it is for regression modelling. However, in this section, we show how we can still get a reasonable idea of the optimal sample size. One approach is to ask the question, what would happen if we had only analysed 35 samples instead of 41: a reduction in sample size of approximately 10%? Will we still find the same relationships as in Figure 20.5? If the answer is yes, then perhaps we can take 10% fewer samples for the remaining years. If the relationship is changed, then clearly we should not omit the samples. One way of assessing what happens if 10% of the data are omitted, is as follows: (i) apply the additive model using all 41 samples, (ii) remove a random 10% of the data and reapply the additive model on the remaining data, and (iii) repeat step (ii) a few times (say 20) so that the choice of the actual omitted samples does not have any influence. This process can then be repeated for 15%, 20%, 25%, 30%, etc. of omitted data. The effect of omitting data on the temperature effect is presented in Figure 20.7. The original smoothing curve and corresponding 95% point-wise confidence bands are plotted (thick lines) plus 20 smoothing curves obtained by omitting x% of the data. Figure 20.7-A shows the smoothing curves if 10% of the data are omitted. The variation among the 20 curves is relatively small compared with the 95% confidence bands of the original smoother. If 25% of the data are omitted, this variation increases, and once 50% of the data are removed, it goes beyond the original range of confidence bands. The same set of graphs is presented in Figure 20.8 for the effects on chlorophyll a results when sample sizes are reduced.

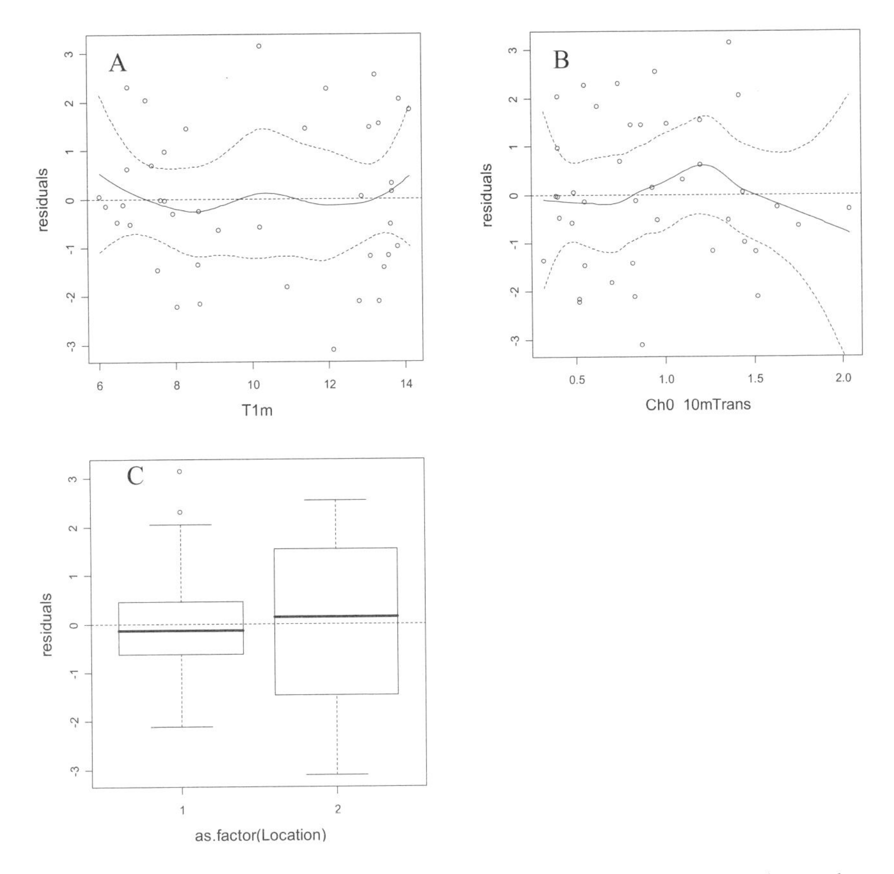

Figure 20.6. Residuals of the optimal additive model plotted versus the explanatory variables temperature (A), chlorophyll a (B) and location (C).

Both graphs seem to suggest that if 10% of the data are omitted, we end up with similar temperature and chlorophyll a relationships compared with using all data. Removing 25% still gives similar curves, but the variation between the fitted curves is higher. These results would suggest that if 90%, or perhaps even 85% of the available samples are processed and included in the analysis, similar temperature and chlorophyll a effects could be expected.

20.7 Discussion

The additive model performed better than a linear regression model and showed that all the major factors analysed (temperature, food and location) are influential in determining the "species" richness (at a family level) of the decapod larval communities examined at the Loch Ewe and Stonehaven sites.

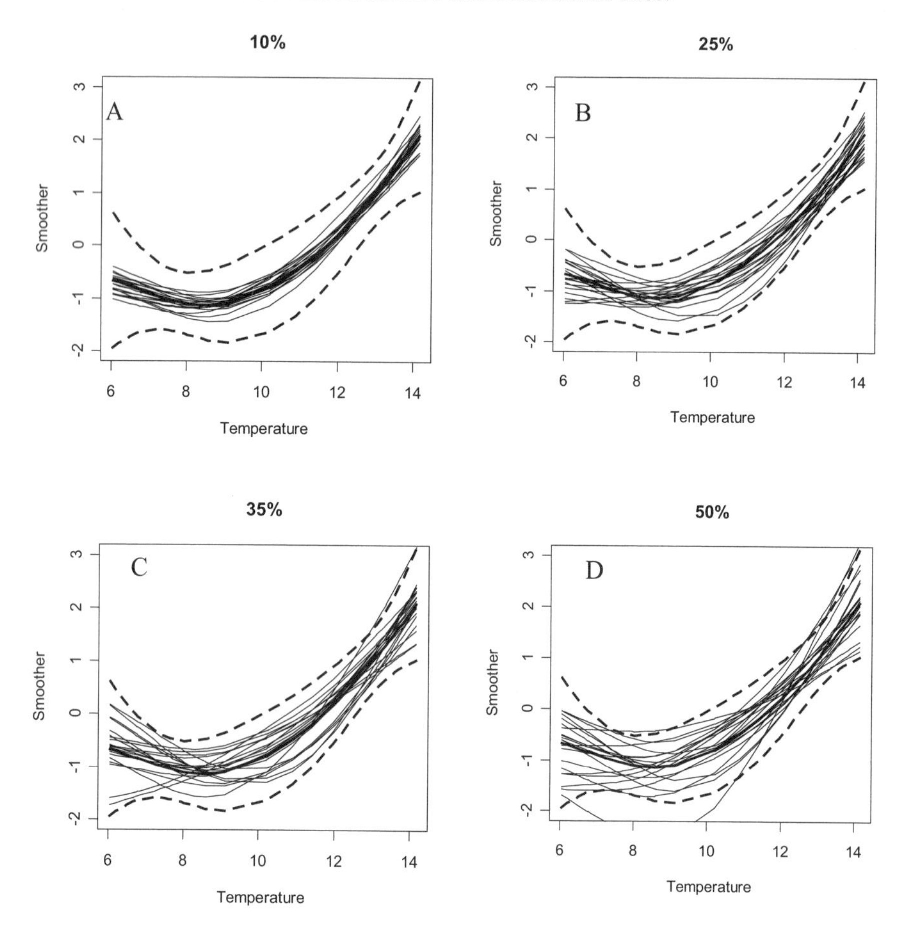

Figure 20.7. Original temperature smoothing curves and 95% point-wise confidence bands and 20 smoothing curves obtained by omitting x% of the data, where x = 15 (upper left), x = 25 (upper right), x = 35 (lower left) and x = 50 (lower right).

It should be noted however that measuring true species richness involves discrimination to species level, which may yield different results. However the results obtained here are at least compatible with intuitive expectations. The major determinant of species richness, over the time periods examined for both sites, appears to be temperature. This is not unexpected as Loch Ewe generally has a higher annual average temperature (10.4°C) than Stonehaven (9.8°C).

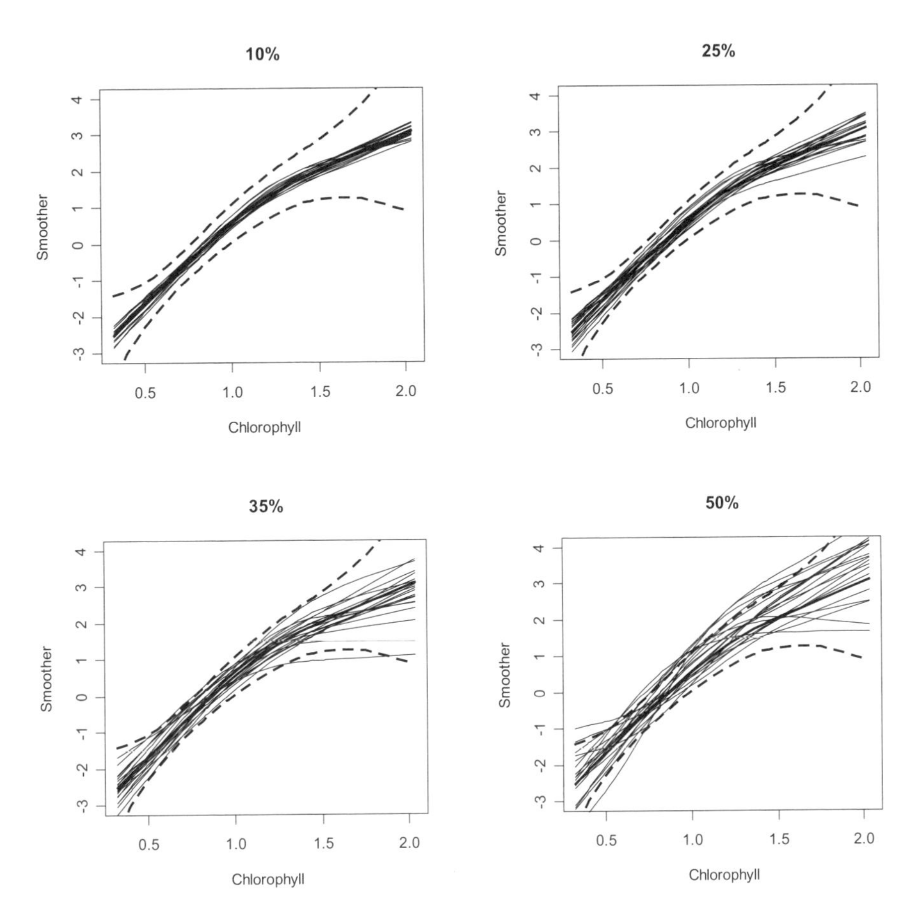

Figure 20.8. Original chlorophyll a smoothing curves and 95% point-wise confidence bands and 20 smoothing curves obtained by omitting x% of the data, where $x = 15$, $x = 25$, x =35 and $x = 50$.

Many marine species groups show increased diversity with increasing temperature, at least when comparing sub-tropical with northern temperate and boreal communities along latitudinal gradients (Raymont 1980). Temperature is the prime environmental variable affecting the physiological rates of growth and reproduction. Temperature also indirectly affects the productivity of food organisms, and it has a major influence on larval development (Costlow and Bookhout 1969). Chlorophyll a is also important, and there were obvious differences between the two sites. The importance of chlorophyll a in the relationships is also expected, as it indicates the availability of decapod food availability. We can therefore conclude that higher field temperatures, increased food levels or extended seasonal food production periods are all likely to be beneficial to the development of decapod larval communities.

The analysis here, therefore, confirms our expectations in a quantitative and objective manner, and allows the derivation of relationships to be applied to wider or more detailed studies.

The time aspect

The decapod data used in this chapter were measured monthly during a period of two years. In some months there are four observations and in other months only one. Once more data become available, time series analysis might be a more appropriate tool to analyse these data. However, this will not be easy as the data are irregularly spaced and most time series methods require equidistant observations.

In Chapter 16 we discussed how generalised least squares (GLS) is used to extend the linear regression model with an auto-correlation structure on the residuals. The same can be done for smoothing methods but confusingly, it is called additive mixed modelling. This combination of linear regression or additive modelling and auto-correlation structures on the noise is demonstrated in several case study chapters. The key to deciding when to apply linear regression and when to apply GLS is: 'do the residuals show any clear auto-correlation?' If they do, then GLS (or the smoothing equivalent) should be applied. For these data, we were on the borderline between the two approaches. Residuals of the optimal models were plotted against the explanatory variable Month (which was not used in the model selection procedure), and there was some small indication that winter months had lower residuals than summer months. We reapplied the smoothing models replacing temperature with month (as a continuous smoother), and the explained variance and model fit were similar to the additive models with temperature. We also defined a new nominal explanatory variable 'season' with values 1 in the summer months and 0 in the winter months and used it as an explanatory variable together with temperature, chlorophyll a and location. This gave a slightly better performance for the smoothing model in terms of residual patterns. But it also complicates the model's interpretation. We expect that once more data become available, GLS and/or its smoothing equivalent will be the more appropriate methods to apply on these data. GLS and its smoothing equivalent is applied in several case study chapters.

Acknowledgements

Part of this work was carried out as part of a Marie Curie Fellowship (QLK5-GH-99-50530-13). Thanks to Alain Zuur, Elena Ieno and Graham Smith for their invitation to contribute with this case study and thanks to all the people of the FRS Marine Laboratory and the University of Aberdeen who were involved in this project. Alain Zuur would like to thank Rob Fryer for a short discussion on the bootstrap approach.

21 Analysing presence and absence data for flatfish distribution in the Tagus estuary, Portugal

Cabral, H., Zuur, A.F., Ieno, E.N. and Smith, G.M.

21.1 Introduction

Understanding the spatial and temporal distribution and abundance patterns of species and their relationships with environmental variables is a key issue in ecology. All too often, despite best efforts, the quality of the collected data forces us to reduce it to presence–absence data, which presents the ecologist with several specific statistical problems. In this chapter, using fish data as an example, we look at several approaches for analysing presence–absence data.

Several statistical techniques have been used to relate fish abundance with abiotic and biotic conditions. These include regression methods, general linear models, ordination methods, discriminant analysis, and several others (Jager et al. 1993; Eastwood et al. 2003; Thiel et al. 2003; Amaral and Cabral 2004; França et al. 2004). However, methods that evaluate the performance of different techniques applied to the same datasets are scarce. And the choice of analytical approach can strongly influence the statistical conclusions and, by implication, the ecological conclusions. This case study compares the adequacy and the performance of several statistical tools when used to analyse fish abundance data and its relationships with environmental factors.

The data used in this example are the abundance of sole, *Solea solea* in the Tagus estuary (Portugal), with the principal ecological question of identifying which environmental factors influence the choice of nursery grounds by this species, in this estuary. The sole is a marine fish that occurs and spawns throughout the continental shelf. Larvae and juveniles tend to migrate to coastal nursery areas using both passive and active transport processes (e.g., Rijnsdorp et al. 1985). The juveniles concentrate in estuaries and bays for a period of about two years (Koutsikopoulos et al. 1989).

The Tagus estuary (Figure 21.1) has long been recognized as an important nursery area for *S. solea*. This estuarine system, located in the centre of the west coast of Portugal, has an area of 325 km^2 and is a partially mixed estuary with a tidal range of about 4 m. The Tagus estuary has long been subjected to industrial development, urbanization and port and fishing activities. Over two million people live around the estuary, mainly in the lower part, where, important industrial complexes such as chemical, petrochemical, food and smelting are found. The upper part (more than 50% of the shoreline) is bordered by land used intensively for agriculture (Fernandes et al. 1995). The *S. solea* abundance data used in this case study were collected during monthly sampling surveys conducted in four areas during 1995 and 1996. Fish were captured using a 4 m beam trawl with 10 mm mesh and one tickler chain. Sampling areas were selected for their importance to juvenile sole, based on results from earlier studies (Costa and Bruxelas 1989). Several environmental variables were also measured: depth, salinity, water transparency, temperature and sediment composition (percentages of mud, medium and fine sand, large sand and gravel).

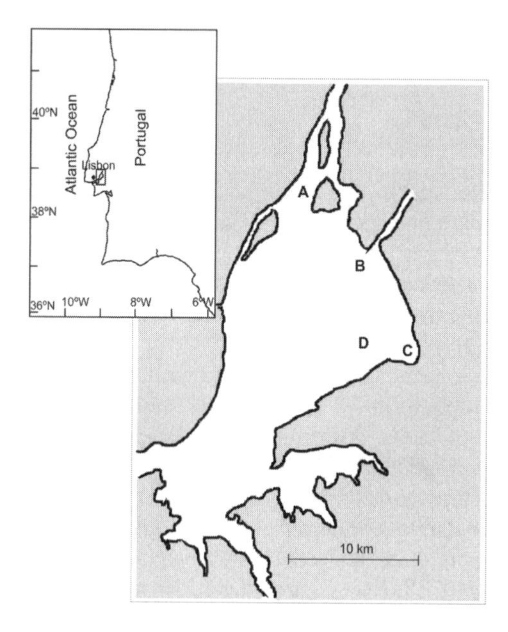

Figure 21.1. The sampling area in the Tagus estuary, Portugal. A, B, C and D indicate the location of the stations.

21.2 Data and materials

The *S. solea* data used in this chapter consist of density values (numbers of individuals per 1000 m^2), recorded from 65 samples in 1995 and 1996. The samples were collected from four areas (Figure 21.1). Sixty percent of the *S. solea* samples were zero, and a few observations had high abundance. This extreme distribution of the data makes it difficult for analysis, as it does not match many models available from standard statistical tools — a common problem with abundance data. It is too extreme for transformation to be successful, and although several methods such as generalised linear models (Chapter 6) have the capacity to deal with overdispersed data, there is a limit to the amount of overdispersion that can be compensated for. In this instance, we decided there was little choice but to reduce it to presence and absence data. Accept the loss of the quantitative information for the sample points where the fish were present. This makes the presence or absence of fish the nominal response variable we are trying to explain using a range of available explanatory variables.

Table 21.1 shows the available explanatory variables. The variables season, month and station are nominal. From the statistical point of view, we have only one response variable, the presence or absence of *S. solea*, and multiple explanatory variables. The question is whether there is a relationship between the two. Because of the presence–absence nature of the response variable, the most appropriate techniques for this analysis are generalised linear models (GLM), generalised additive models (GAM) using a binomial distribution, and classification trees. The number of explanatory variables is more than 10 (8 continuous and 3 nominal). This can be considered as a large number of explanatory variables and might make it difficult to find the optimal model, especially with explanatory variables like mud and the fine sand percentage of the sediment, which have a degree of collinearity. To simplify the problem, our strategy was to:

1. Identify and remove some of the highly correlated (continuous) explanatory variables. Pairplots are a useful tool for this task.
2. Visualise the relationship between the nominal explanatory variables and *S. solea*. Design and interaction plots will be used for this.
3. Visualise the relationship between *S. solea* and the explanatory variables by using coplots.
4. Use classification trees, GAM and GLM to model the relationship between *S. solea* and the selected explanatory variables, and compare the results among all three techniques.

The selection of the explanatory variables in the first step should be based on ecological and statistical considerations.

Table 21.1. Available explanatory variables.

Variable	Nominal	Remark
season	Yes	1 = spring, 2 = summer
month	Yes	
station	Yes	sampling station
depth	No	Depth (m)
temp	No	temperature (°C)
sal	No	salinity (ppt)
transp	No	water transparency (cm)
gravel	No	% gravel in the sediment
large sand	No	% large sand in the sediment
med fine sand	No	% medium and fine in the sediment
mud	No	% mud in the sediment

21.3 Data exploration

Figure 21.2 shows the pairplot for the explanatory variables gravel, large sand, medium fine sand and mud content. The nearly straight, close to 45-degree, straight lines in this plot indicate a strong linear relationship among mud, medium fine sand and large sand content of the sediment. Using all three explanatory variables would lead to serious problems with forward and backward selection procedures in GLM or GAM. Presenting a model that includes all these variables is not a good solution either. The reason for this is that it does not make sense to use a model where some of the explanatory variables mirror the same information. The strong linearity among these three factors allows two to be dropped from the analysis. This reduces the number of parameters in the analysis, and generally the fewer parameters the better, as long as you do not lose too much information. In this instance, we omitted the medium fine sand and the large sand content, using only the mud content and the gravel content for analysis. Note this is an ecological choice rather than a statistical one.

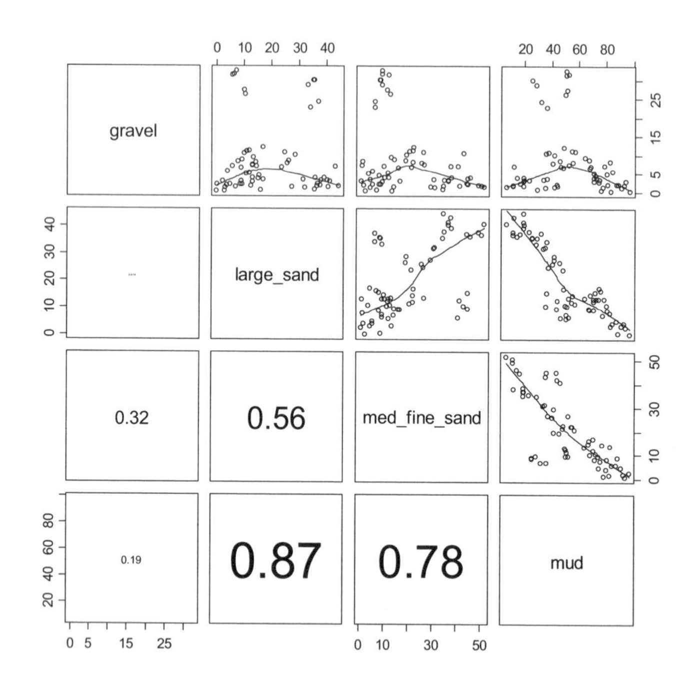

Figure 21.2. Pairplot of four explanatory variables indicating collinearity. The number below the diagonal are (absolute) correlations coefficients. The font size of the cross-correlation is proportional to its strength. The lines in the upper diagonal panels are LOESS smoothers.

A pairplot (Figure 21.3) of the remaining continuous variables gives no immediate indications of further collinearity between the explanatory variables. The first row of graphs shows the relationship between *S. solea* and each of the continuous explanatory variables. Some of the shapes (e.g., for depth and salinity) show the typical fit of a logistic generalised linear model (Chapter 6), and this is a useful guide for choosing an appropriate analysis. Looking at possible relationships between the response variable and nominal variables, a design plot (Figure 21.4) indicates that the mean value in the first season (labelled as 0) is higher than in the second season (labelled as 1), month nine has a considerably lower mean value, and the mean value in area one is nearly twice as high as in the other areas. As season is just a coarser version of month, both these variables should not be used in the same GAM or GLM model (collinearity). It was decided to use month instead of season as it provides more information.

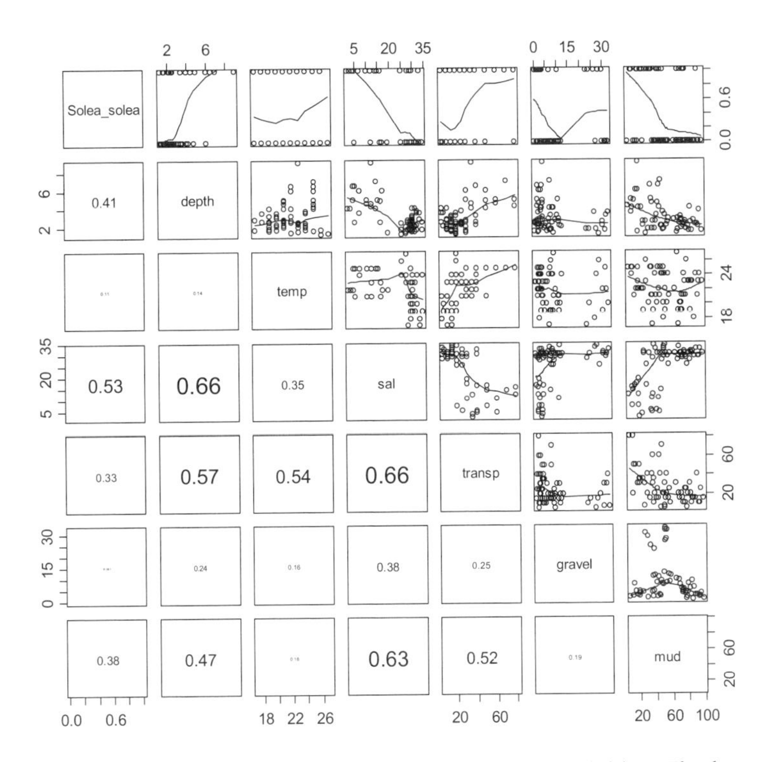

Figure 21.3. Pairplot of *S. solea* and selected explanatory variables. The lower diagonal panels contain the (absolute) correlations. The font size is proportional to the value. The upper diagonal panels show the pair-wise scatterplots. A smoothing line was added.

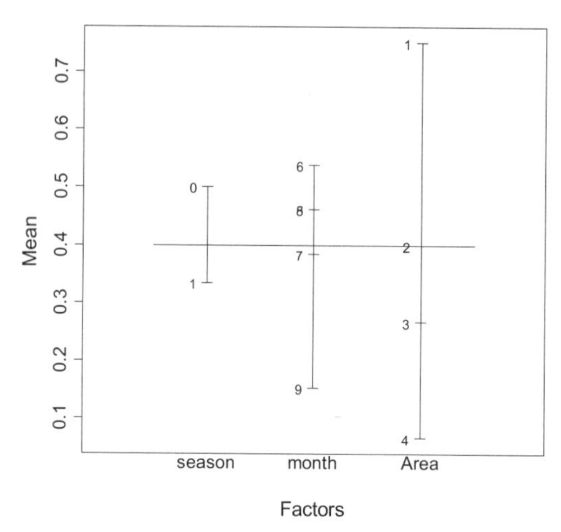

Figure 21.4. Design plot for *S. solea* presence–absence data. The *y*-axis shows mean values per class of the nominal variable. Highest mean values are in month 6 and in area 1.

Figure 21.5 shows a coplot of *S. solea* versus salinity conditional on the nominal variable month. We deliberately selected month as a non-nominal variable in the coplot. As a result, the software groups data from different months and makes a scatterplot between *S. solea* and salinity for those months. For example, the lower left panel in Figure 21.5 shows a scatterplot of *S. solea* and salinity for the samples measured in the months 5 and 6. The lower right visualises the same, expect for samples from months 6 and 7. All panels show a similar *S. solea*-salinity relationship, except for the panel that includes samples from month 9. This is another indication (month 9 also had a particularly low mean in the design plot) that we need to give special attention to month 9 in the GLM and GAM models. A coplot in which month was used as a nominal variable resulted in five panels with less data per panel.

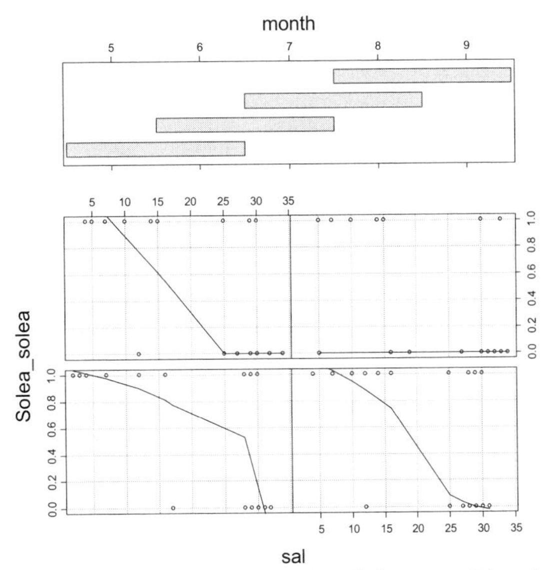

Figure 21.5. Coplot between *S. solea* and salinity, conditional on month. A smoother was fitted in each panel. The lower left panel contain the data from months 5 and 6, the lower right from months 6 and 7, etc. A smoothing curve was added.

21.4 Classification trees

Even though the above exploratory tools allow some variables to be dropped from the analysis, the number of explanatory variables is still relatively large. A classification tree allows a more detailed investigation into the relative importance of these remaining variables. The classification tree (Figure 21.6) indicates that salinity is the most important explanatory variable. A high probability of finding *S. solea* is obtained for samples with salinity smaller than 15.5 and a gravel content larger than 1.34. Using a pruning diagram (Figure 21.7) indicates that the tree presented in Figure 21.6 is sub-optimal (Chapter 8), and that a tree of size two would be optimal. Due to the small sample size, it may be wise to reapply the cross-validation procedure several times using different starting values, and this indicated that one should either use a tree of size two or six. Selecting which one is best is subjective. However, all indicate that salinity is the most important variable, with the importance of gravel, and the variables further down the tree open to argument.

As the same variable appears more than once in the tree, it indicates a weak non-linear relationship between the response variable and the explanatory variables. This suggests that the next step should be a GAM (Chapter 7) as this can deal with non-linearity, and the results from the tree suggest only one important variable to be identified by the GAM (the optimal tree size is 2).

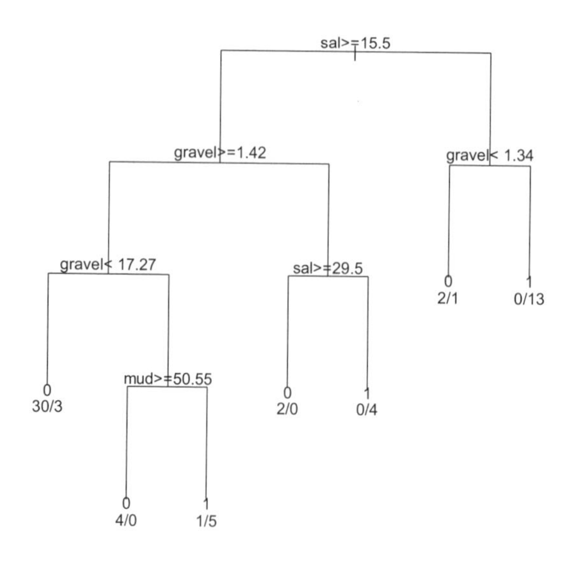

Figure 21.6. Classification tree for *S. solea*. If a statement is true, follow the left branch. Numbers at the end of a branch are the predicted group (1 = presence, 0 = absence) and the classifications per group.

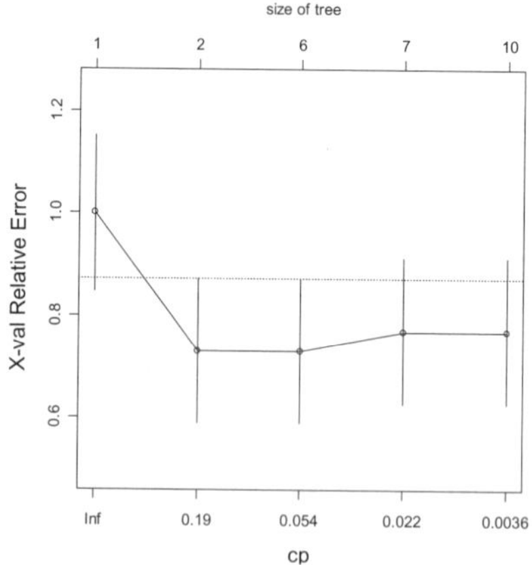

Figure 21.7. Results of cross-validation. The 1-SE rule dictates to select the left-most tree for which the mean relative error is below the dotted line, in this case a tree of size 2.

21.5 Generalised additive modelling

A GAM using the binomial distribution and logistic link function (Chapter 7) was used to relate *S. solea* presence–absence data and the explanatory variables. Strictly speaking, the model is not a GAM with a binomial distribution but rather one with a Bernoulli distribution. With Bernoulli models overdispersion cannot occur. To find the optimal set of explanatory variables, a forward selection was applied. Table 21.2 gives the AIC using each of the continuous explanatory variables as a single explanatory variable. The lower the AIC value, the better it is, and cross-validation was used to estimate the optimal degrees of freedom for each smoother. The GAM identified salinity as the best single explanatory variable. In the next step of the forward selection procedure, two explanatory variables were used, one of them salinity. None of the combinations led to a model with a smaller AIC value (compared with the one with salinity) with significant smoothers. These results indicate that the GAM model using only salinity is the best model. The effect of salinity is presented in Figure 21.8. The dotted lines represent a 95% confidence interval. The cross-validation estimated 1 degree of freedom for the smoother. This means that a GLM should be applied instead of a GAM (GAM with a smoother with 1 degrees of freedom is equal to a GLM). For reasons of completeness, the numerical output for the GAM is given below.

Parametric coefficients:

	Estimate	std. err.	*t*-ratio	*p*-value
Intercept	−0.42	0.29	−1.40	0.16

Approximate significance of smooth terms:

	edf	chi.sq	*p*-value
s(sal)	1	13.823	<0.001

R-sq.(adj)=0.26, deviance explained = 21.6%, n = 65, Deviance = 68.56

Table 21.2. AIC values for first step in the forward selection procedure. Cross-validation was applied in each step. A Binomial GAM was used. The row in bold typeface is the most optimal model.

Single explanatory variable in GAM	AIC	edf
Depth	80.14	1
Salinity	**72.56**	**1**
Temperature	90.63	1
Transportation	83.92	2.7
Gravel	85.35	3.5
Mud	80.4	4.1

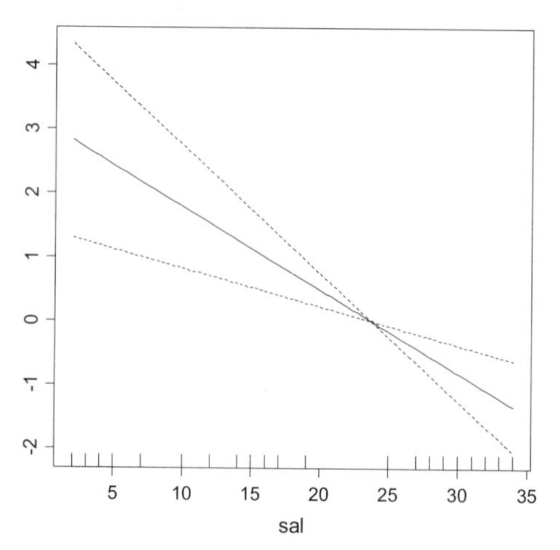

Figure 21.8. Partial fit of salinity. The x-axis shows the salinity gradient and the y-axis is the contribution of the smoothing function f(salinity) in the model logit(Y) = intercept + f(salinity).

21.6 Generalised linear modelling

The advantage of GLM over GAM is that GLM is parametric. Using the same explanatory variables as in the GAM, a backward selection, and a combination of a forward and backward selection was applied. Both approaches gave a model where salinity, month, temperature and gravel (in order of importance) were selected. The estimated parameters, standard errors, z-values and p-values are given below. Note that salinity is highly significant. Other parameters are weakly significant. Instead of assessing the importance of individual explanatory variables using the z-values (or t-values) in Table 21.3, we drop one term in turn and compare deviances of the full and nested model with each other using the Chi-square (Chapter 6). The output of this approach is given in Table 21.4 and shows that all variables are significant at the 5% level, except for gravel.

Table 21.3. Estimated parameters obtained by the GLM model containing temperature, salinity, gravel and month. The AIC is 70.145.

	Estimate	Std. Error	*t*-value	*p*-value
(Intercept)	23.35	9.69	2.41	0.16
Temperature	−0.89	0.42	−2.09	0.03
Salinity	−0.26	0.07	−3.78	0.00
Gravel	0.06	0.03	1.68	0.09
factor(month)6	2.01	1.21	1.66	0.09
factor(month)7	3.37	2.14	1.57	0.11
factor(month)8	3.86	2.01	1.92	0.05
factor(month)9	−2.20	1.32	−1.66	0.09

Table 21.4. Change in deviance and corresponding Chi-square values if one variable is dropped from the model.

Variable to be dropped	df	Deviance	AIC	Chi-square value	*p*-value
<none>		54.14	70.14		
Temperature	1	59.82	74.22	5.68	0.01
Salinity	1	81.33	97.25	27.19	<0.001
Gravel	1	57.08	71.28	2.93	0.08
Factor(Month)	4	65.88	74.71	11.73	0.01

The partial fits (Chapter 5) in Figure 21.9 show the contribution of the individual explanatory variables, while taking into account of the other variables in the model. One of the reasons that gravel is included in the model might be due to the isolated set of samples for high gravel values. Month shows a weak seasonal pattern, with month 9 clearly displaying the lowest values. The variables salinity and temperature have a negative relationship with *S. solea* and a positive relationship with gravel. A detailed model validation included Cook distance values, histograms and QQ-plots of residuals, hat values, changes in fit and parameters after leaving out one variable, and did not indicate any serious problems (Chapter 5).

We now have the output from three different techniques: classification trees, GAM and GLM. All three techniques indicate that salinity is significant, and the most important variable. Gravel was also selected by all techniques, but the *p*-values for this variable indicated a very weak relationship.

Based on the results of the GLM, we re-applied the GAM model using salinity, temperature, gravel and month as explanatory variables. The reason for this is that if the GAM curves obtained by this analysis are straight lines (or can be considered as approximately straight lines within the point-wise confidence bands), then we have a confirmation that the GLM model is indeed the most appropriate method. We used cross-validation in the GAM model to find the optimal degrees of freedom. The output (given below) shows that temperature and salinity are fitted as a linear component, but gravel has a modest non-linear effect.

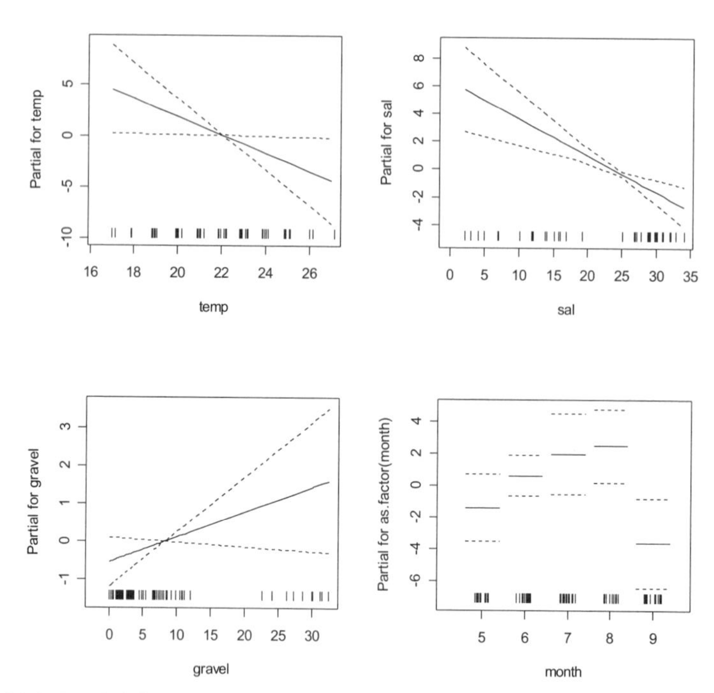

Figure 21.9. Partial fits of GLM model.

Parametric coefficients:

	Estimate	std. err.	*t*-ratio	*p*-value
(Intercept)	−1.71	1.30	−1.38	0.16
factor(month)6	1.72	1.30	1.32	0.18
factor(month)7	2.68	2.26	1.18	0.23
factor(month)8	3.37	2.15	1.56	0.11
factor(month)9	−2.57	1.44	−1.78	0.07

Approximate significance of smooth terms:

	edf	chi.sq	*p*-value
s(temp)	1.11	6.45	0.37
s(sal)	1.07	15.96	0.02
s(gravel)	3.25	10.81	0.28

R-sq.(adj) = 0.42 Deviance explained = 46.8%
n = 65, Deviance = 46.58 , AIC = 67.47

Although the residuals did not show any patterns (which is good), the confidence intervals around gravel (not shown here) were rather large. The p-values for the continuous smoothers indicate a linear salinity effect, but no temperature or gravel effect.

21.7 Discussion

An outline of the analysis workflow is shown in Figure 21.10. The data exploration suggested omitting two continuous and one nominal explanatory variable from the analysis to avoid collinearity problems in the GLM and GAM models. The variables omitted were medium fine sand and large sand content of the sediment and season. Once the optimal models were fitted using the remaining variables, these variables were added back into the analysis as a check, but they did not improve the model.

Classification models were applied next, indicating a strong salinity effect, and possible non-linear relationships, as the same variables occurred at different branches. At this point, you could decide to apply a GLM, but we decided to apply a GAM first as this would help to visualise the type of relationships between *S. solea* and the explanatory variables. The GAM results indicated that the main variable, salinity, was having a linear effect, and therefore we continued with a GLM. A detailed forward and backward selection process in the GLM indicated a month (nominal), temperature, salinity and gravel effect. However, as a GLM is imposing linear relationships (on the predictor scale; Chapter 6), we decided to verify the optimal GLM with a GAM. If the GAM, using month, temperature, and gravel, indicates that the relationships are indeed linear, then we can be confident that the GLM results are correct. We can then present them as our final results. Both temperature and salinity were estimated as linear components by the GAM, but gravel was slightly non-linear. However, in such a model (not presented here) temperature and gravel were not significant.

All the techniques we used showed a significant and strong salinity effect. As to the ecological interpretation, *S. solea* seems to prefer relatively low salinities, and there is a higher probability of catching *S. solea* in months 7 and 8, than in any of the other months.

Obviously, it is difficult to say whether there is a cause-and-effect relationship between salinity and the probability of *S. solea* occurrence. There might be other variables, not measured, but still highly correlated with salinity that are the real driving factors. Despite these difficulties in identifying a cause–effect relationships, the influence of salinity on the abundance of *S. solea* has been reported by other authors. Riley et al. (1981), for the UK waters, and Marchand and Masson (1988) for waters off North France, where it was found that sole less than one year old prefer salinities between 10% and 33%.

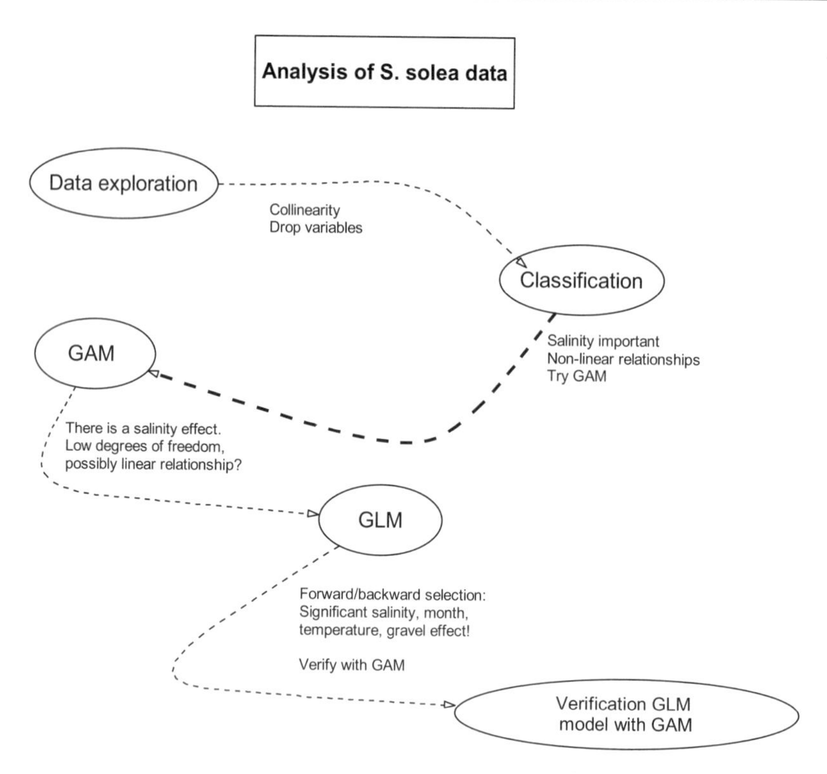

Figure 21.10. Outline of data analysis approach.

The use of several techniques applied to the same dataset allowed us to evaluate the influence on the results due to the selection of a particular statistical technique. Although it is difficult to identify a single best method for comparing different techniques, applied to the same dataset, the approach adopted here should give a useful starting point. The fact that the GLM, GAM and tree models find different optimal models shows the danger of forward selection and relying on only one statistical technique. Even within a particular method, e.g., GAM, one might find different results depending on the model selection strategy, e.g., the use of cross-validation, AIC, or Chi-square deviance tests. However the use of GLM and GAM methods allow a fine-tuning procedure that can lead to marked improvements in the models considered.

Acknowledgement

We would like to thank Ian Tuck and Graham Pierce for useful comments on an earlier draft.

22 Crop pollination by honeybees in Argentina using additive mixed modelling

Basualdo, M., Ieno, E.N., Zuur, A.F. and Smith, G.M.

22.1 Introduction

Throughout the world, honeybees (*Apis mellifera L.*) are used as a pollinator in more than 40 types of commercial crops. This practise not only increases the crop yield, but also the quality of fruited varieties of commercial interest. The use of honeybees may improve crop production as a consequence of cross- pollination or due to the physical contact by foraging behaviour on the flowers.

Darwin formerly described the effect of cross-pollination in 1877. Further work was done by Waite in 1895, with studies carried out on pears showing the value of inter-plantation of cultivars and the roll of honeybees in transferring the pollen among them (Waite 1895).

In seed production systems, the pollination process also has a direct impact on crop production, and consequently on foraging legume or oilseed production such as the sunflower (*Helianthus annuus L.*). The sunflower is the second most important oilseed crop in the world, after soybean, as it is cholesterol-free and also has anti-cholesterol properties. In Argentina, the extension of sunflower cultivated areas for oilseed production has increased in recent years.

Seed from commercial hybrid sunflowers is produced using cytoplasmic Male Sterility. These lines, known as male-sterile (MS) or female, are fertilised with male lines denoted as male-fertile (MF) or restorer lines, allowing the recovery of fertility in the hybrid F1 generation. In the seed fields, the MS–MF lines are planted in separate rows in ratios ranging from 2:1 to 10:1 in order to optimise the number of seed-producing MS plants (Dedio and Putt 1980). To obtain adequate seed production, pollen has to be transferred from the MF lines to the MS lines. The honeybee is considered the most important pollinator of sunflowers (McGregor 1976), and at present, honeybee colonies are placed in sunflower fields to ensure adequate pollination. When honeybees move from MF to MS rows, cross-pollination should occur (Ribbands 1964; Delaude et al. 1978; Radford and Rhodes 1978; Drane et al. 1982). To ensure an adequate seed production, honeybees should visit both parental lines and be located uniformly within the MS. There is a direct linear relationship between the amount of seed produced and the number of visits that the sunflower head (capitulum) receives.

The aim of this case study chapter is to determine how honeybee foraging activity changes in response to days, state of flowering, time of day, temperature and visitation between the MF and the MS lines.

22.2 Experimental setup

The sunflowers used in this study were grown at the Universidad Nacional del Centro de la Provincia de Buenos Aires Campus in Tandil, Argentina. The Tandil district lies in the centre and southeast area of Buenos Aires province, at an altitude of 178 m (see Chapter 28 for a map reference). The area consists of a flat pampas phytogeography region surrounded by rounded slopes. The climatic conditions and soil development make this a particularly warm temperate pampas system and is considered one of the most productive areas in the world for agriculture and cattle rearing (Burkart et al. 1999).

For the experiment, 2 ha were sown with MF and MS sunflower transects at a density of 5 seeds per linear metre. The study plots contained six rows of MS plants followed by four rows of MF, in accordance with the seed company instructions, with a distance of 0.7 m between rows. Flowering extended from February 7 (15% of flowering) to February 16. Honeybee colonies composed of nine standard Langstroth frames were used. The colonies were placed at the edge of each experimental plot with their entrances facing the MS and MF rows, when 15% of the flowers were open (Basualdo et al. 2000).

Before flowering started, the state of flowering on MS and MF plants was estimated by tagging five capitula in five MS rows and five capitula in two MF rows. Each day, the total area of the tagged sunflower head with opened florets was estimated by measuring the maximum diameter of the total capitulum and the minimum diameter of unopened florets. The area of the capitulum with opened florets (OPFL) was estimated (De Grandi-Hoffman and Martin 1993). The number of honeybees collecting nectar on the tagged capitulum was counted daily throughout flowering period, twice in the morning (09.00–11.00 h) and twice in the afternoon (14.00–17.00 h). Daily air temperature was also recorded.

22.3 Abstracting the information

We now discuss how we can summarise the information above in terms of response and explanatory variables.

Quantifying the information

A visual representation of the experiment in shown in Figure 22.1, although it must be emphasised that the real setup contained more transects (see above) that were not sampled. The five male-sterile transects are represented by open triangles and two male-fertile transects by filled triangles. Each transect contained five capitulum. The number of honeybees on each of the 35 capitulum were counted twice in the morning and twice in the afternoon. In this chapter, we will use averages of the two morning and of the two afternoon counts. The reason for this will be explained later. Counts were started on day 7, and further measurements were taken on day 8, 9, 10, 11, 12, 13, 15 and 16. No measurements were taken on day 14. Therefore, the number of honeybees will be our response variable.

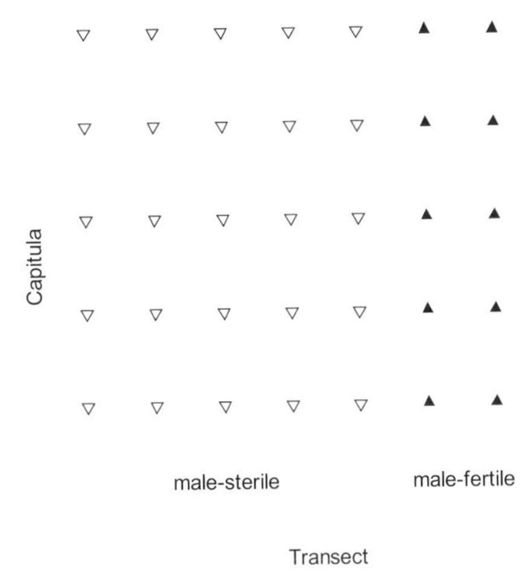

Figure 22.1. Sampling scheme. The graph is slightly misleading in the sense that the capitulum is not in a row. Male-sterile capitulum are labelled as Sex = 1 and male-fertile as Sex = 0.

We have an explanatory variable 'Transect' with values 1 to 7 denoting to which transect an observation belongs. This is a nominal variable and will allow us to check whether there are differences among the seven transects. It is also possible to quantify the difference between MS and MF more directly by using a nominal variable 'Sex' with a value of 0 if a sample was taken from an MF transect and 1 from a MS transect. Note that transects are nested within Sex, and we want to treat both nominal variables as fixed. Similarly, we created a nominal variable 'AMPM' with values 0 (AM) and 1 (PM) indicating at what time of the day the observation was made. All these nominal variables are used as explanatory variables. The day of sampling can also be used as a nominal variable. However, we coded the morning on day 7 as '1', the afternoon on day 7 as '2', the morning of day 8 as '3', etc. and named this explanatory variable 'Time'.

There are also a number of continuous explanatory variables. Air temperature was measured once per day. Another explanatory variable is the area of open florets (cm^2), denoted by 'PercFlower'. A pairplot (Chapter 4) of honeybees, temperature, PercFlower and time is shown in Figure 22.2. From the plot it can be seen that at time 7–10 (days 10 and 11), the temperature was considerably lower. PercFlower had the highest values around day 11, which coincides with the lowest temperature. This might be an indication of collinearity (Chapter 4) between these two variables. However, the correlation was only 0.27, which does not provide enough justification to omit one of them. Figure 22.2 also shows that on the first day of the experiment, not all capitulum had the same area of open florets.

A summary of all explanatory variables is given in Table 22.1. As stated above, we wish to know the relationship between honeybee numbers and the explanatory variables. Therefore, the model we are after is of the form:

honeybees = *F*(Temperature, PercFlower, Transect, Sex, AMPM)

Where *F*() stands for 'function of''. Using Transect and Sex in the same model caused numerical instability, and therefore, we used Transect in first instance, as it is more informative.

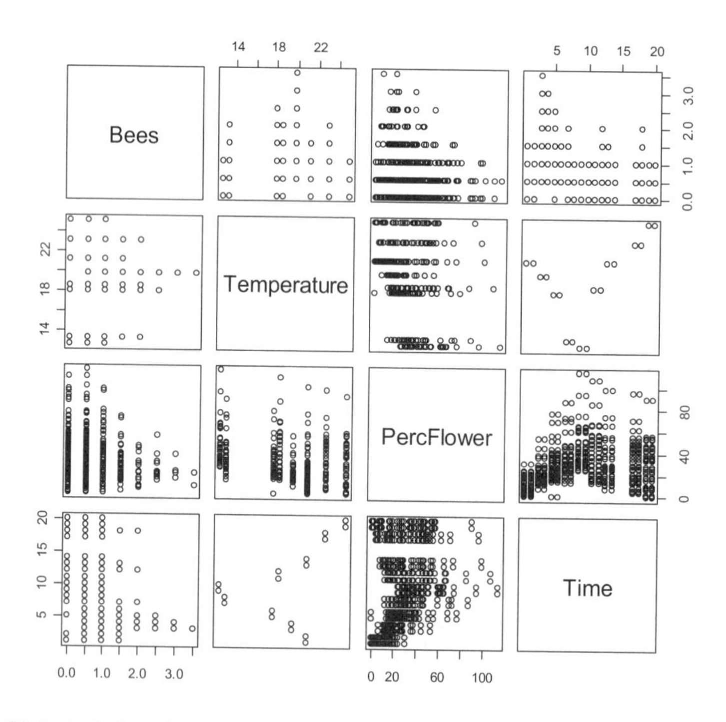

Figure 22.2. Pairplot showing the relationship among number of honeybees, temperature, PercFlower (the area of open florets) and Time.

Table 22.1. A summary of available explanatory variables.

Explanatory Variable	Remarks
Temperature	Continuous variable, one value per day
Days	Nominal variable which identifies day, with values 7–13, 15 and 16.
Transect	Nominal variable with values 1–7 to identify the seven transects.
Sex	Nominal variable with values 1–0 to identify male-sterile (transects 1–5) and male-fertile (transects 6 and 7).
PercFlower	Continuous variable measuring the area of the capitulum with open florets (cm^2).
AMPM	Nominal variable with values 0 (AM) and 1 (PM) identifying the time of the day that sampling took place.

22.4 First steps of the analyses: Data exploration

The first step in any analysis is to explore the data. Boxplots and Cleveland dotplots (Chapter 4) were made and confirmed that none of the continuous variables had outliers or extreme observations. A graph of the number of honeybees, temperature and PercFlower versus time has already been given in Figure 22.2. The number of honeybees observed is between 0 and 5. Previous studies have indicated that temperature is an important explanatory variable. A coplot of the number of honeybees versus temperature conditional on 'Transect' is given in Figure 22.3. To aid visual interpretation, a LOESS smoothing curve with a span of 0.5 was added. The graph suggests that in nearly all transects, the highest numbers of honeybees were observed at a temperature of approximately 20°C. Transect 7 (and also 6) seems to have a slightly different pattern, suggesting a possible interaction effect between temperature and Sex (or Transect). There seems to be only minor differences between AM and PM samples. The patterns in the graph suggest that if we want to include temperature in the models, we have three main options:

1. Create temperature classes and use this as a nominal explanatory variable in linear models.
2. Allow for an interaction between temperature (as a linear term) and any of the other variables.
3. Allow for a non-linear temperature effect.

The first option, converting temperature into classes and using this new variable as a nominal variable, is difficult as it will require arbitrary choices of which temperature values to combine. Including an interaction between transects and temperature as a linear term did not solve the problem. Therefore, we will model the honeybee data as a non-linear temperature effect using smoothing techniques (Chapter 7) together with a smoother-Transect or smoother-Sex interaction.

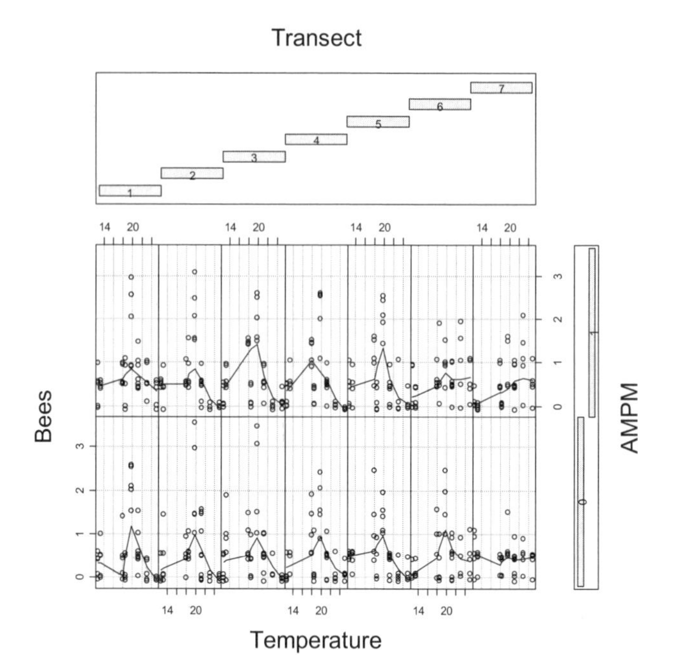

Figure 22.3. Coplot of the number of honeybees (vertical axis) and temperature (horizontal axis) conditional on Transect (blocks at the top) and time of day (blocks at the right hand side). The lower panels correspond to the AM data, and the upper row to the PM data. The left column contains the data from transect 1, and the rightmost column from transect 7. A LOESS smoothing line with a span of 0.5 was added to aid visual interpretation.

22.5 Additive mixed modelling

Because the data exploration indicated that non-linear relationships might be expected, an additive or generalised additive model can be used. The question is whether a Gaussian or Poisson model should be used. In other words, is it additive modelling or generalised additive modelling with a Poisson distribution we want to work with? Theoretically, the Poisson distribution should be applied, as the data are counts. However, additive models are slightly easier to understand and work with. On top of this, we took averages for the morning and for the afternoon data, resulting in non-integer data. Therefore, we will start simple (Gaussian distribution) and do more complicated things (e.g., Poisson) if necessary. The first additive model we could apply is of the form:

$$\text{Honeybees} = \alpha + f_1(\text{Temperature}) + f_2(\text{PercFlower}) + \text{Transect} + \text{AMPM} + \varepsilon \qquad (22.1)$$

The noise component ε in equation (22.1) is assumed to be normally distributed with expectation 0 and variance σ^2. Equation (22.1) models honeybees as a smoothing function of temperature, a smoothing function of PercFlower, and the nominal variables Transect and AMPM. The temperature and PercFlower effects are assumed to be the same for each transect, but Figure 22.3 indicated that this might be unrealistic. There are a couple of other potential problems. Basically we have a time series at each capitulum in each transect. This gives us 35 time series of length 20 with two missing values (day 14), and therefore we have to allow for auto-correlation. Another problem is that the variance might be different per transect, or per sex, or per day, or per morning-afternoon. It is even possible to allow for an increase in spread for increasing temperature (Pinheiro and Bates 2000), but we will not go that far. This brings us into the world of mixed modelling, or since smoothers are involved: additive mixed modelling (Chapter 8). First of all, we have to improve the notation of the model:

$$\text{Honeybees}_{ijs} = \alpha + f_1(\text{Temperature}_s) + f_2(\text{PercFlower}_{ijs}) + \text{Transect}_j + \text{AMPM}_s + \varepsilon_{ijs} \tag{22.2}$$

Honeybees_{ijs} is the value of honeybees at time s in capitulum i in transect j. In additive modelling we assume that the ε_{ijs} are independently normally distributed. We can allow for auto-correlation between honeybees at time s and t (in the same capitulum) by assuming that

$$\varepsilon_{ijs} = \rho\varepsilon_{ij,s-1} + \eta_{ijs} \qquad \text{or equivalently:} \quad cor(\varepsilon_{ijs}\varepsilon_{ijt}) = \rho^{|s-t|} \tag{22.3}$$

The term η_{ijs} is normally distributed. This is one of the most simple auto-correlations structures and is called the auto-regressive correlation of order 1. The further two time points are apart, the lower the correlation. If the difference between two time points is $|s - t| = 1$, then the correlation between the two observations is ρ, if $|s - t| = 2$, then it is ρ^2, etc. In order to implement such a model in a software package, sequential observations need to be identified by a unique number (Pinheiro and Bates 2000, Crawley 2002, mgcv helpfile in R). This was the motivation to take averages per morning and per afternoon. Other auto-correlation structures are discussed in Chapters 16, 26 and 35. We can also allow for different variances per transect by assuming

$$\varepsilon_{ijs} \sim N(0, \sigma_j^2) \tag{22.4}$$

Instead of different variances per transect, we can try different variances per sex; just replace the index j for σ by a k, where $k = 1, 2$.

If you thought that this was complicated, well, there is more to come. We have now specified a model that contains fixed components (the smoothers, intercept and nominal variables) and random components (auto-correlation, different variances). The model selection process for such a model is similar as for mixed modelling (Chapter 8) and consists of three steps: (i) Start with a model that contains as many fixed terms as possible (a just beyond optimal model), (ii) using these fixed terms, find the optimal random structure, and (iii) for the optimal random structure, find the optimal fixed structure. A motivation for this approach was given in Chapter 8. Starting with a 'just beyond optimal model' in terms of fixed components is simple for linear regression, just add as many main terms and interactions as possible, but for smoothing models this is slightly more complicated due to numerical instability. Based on the data exploration, it seems sensible to allow for a different non-linear temperature effect for each transect, and then we can determine in step 3 whether this is really necessary. We can do the same for PercFlower. Hence, our starting model for step 2 is

$$\begin{aligned} \text{Honeybees}_{ijs} = \alpha &+ f_1(\text{Temperature}_s) + f_1(\text{PercFlower}_{ijs}) + \\ &+ f_2(\text{Temperature}_s) + f_2(\text{PercFlower}_{ijs}) + \\ &+ f_3(\text{Temperature}_s) + f_3(\text{PercFlower}_{ijs}) + \\ &+ f_4(\text{Temperature}_s) + f_4(\text{PercFlower}_{ijs}) + \\ &+ f_5(\text{Temperature}_s) + f_5(\text{PercFlower}_{ijs}) + \\ &+ f_6(\text{Temperature}_s) + f_6(\text{PercFlower}_{ijs}) + \\ &+ f_7(\text{Temperature}_s) + f_7(\text{PercFlower}_{ijs}) + \\ &+ \text{Transect}_j \times \text{AMPM}_s + \varepsilon_{ijs} \\ &\varepsilon_{ijs} \sim N(0, \sigma^2) \quad \text{and} \quad cor(\varepsilon_{ijs}\varepsilon_{ijt}) = \rho^{|s-t|} \end{aligned} \tag{22.5}$$

The notation f_1(Temperature) means that a smoother of temperature is used for the data of transect 1. Technically, this is done using the 'by' command in the mgcv library in R; see also Chapter 35. Later, we will allow for different variances. A Transect × AMPM interaction term (plus main terms) was added.

To test whether we indeed need the auto-correlation coefficient, we applied a likelihood ratio test (Chapter 8) on a model with and without auto-correlation. It gave a test statistic of $L = 8.83$ with a p-value of 0.003, indicating that we need the auto-correlation. Note that we are not testing on the boundary (Chapter 8). Allowing for different variances per transect or per morning-afternoon caused numerical problems. As an alternative to check whether we need different variances in the model, we took the residuals from model (22.5) and plotted them against the transect. There were no clear differences in spread. The same holds for sex and AMPM. For this reason, we continue to step 3 with a random component that contains auto-correlation and one variance term σ^2.

In the third step, we search for the most optimal model in terms of fixed components. Most temperature smoothers in equation (22.5) were significantly different from 0 at the 5% level. The amount of smoothing was estimated by cross-validation (Chapter 7), and most temperature smoothers had 4 degrees of freedom (the maximum we allowed for due to the relatively small number of unique temperature values). However, all PercFlower smoothers had only one degree of freedom, and therefore we refitted the model using a linear PercFlower effect, and a PercFlower × Transect interaction term. Using a likelihood ratio test, the interaction term was not significant ($p = 0.19$), and the main term PercFlower was borderline significance ($p = 0.05$). The likelihood ratio test indicated that the AMPM × Transect interaction term was not significant ($p = 0.97$). We then dropped Transect (as a main term) from the model as it was the least significant term, followed by PercFlower and the smoother for transect 7. The remaining temperature smoothers and the AMPM effect were all significantly different from 0 at the 5% level. Hence, the optimal model is given by

$$\begin{aligned}\text{Honeybees}_{ijs} = \alpha &+ f_1(\text{Temperature}_s) + f_2(\text{Temperature}_s) \\ &+ f_3(\text{Temperature}_s) + f_4(\text{Temperature}_s) + \\ &+ f_5(\text{Temperature}_s) + f_6(\text{Temperature}_s) + \text{AMPM}_s + \varepsilon_{ijs} \\ &\varepsilon_{ijs} \sim N(0,\sigma^2) \quad \text{and} \quad cor(\varepsilon_{ijs}\varepsilon_{ijt}) = \rho^{|s-t|}\end{aligned} \tag{22.6}$$

However, the shapes of the smoothers were all similar, and this raises the question of whether we should indeed use a smoother for each transect, or whether we can replace them by one overall temperature smoother, or two temperature smoothers for Sex (Table 22.1), or two smoothers for AMPM, or four smoothers for a Sex–AMPM combination. The model in which two temperature smoothers conditional of AMPM were used, was not better than the one in equation (22.6) as the likelihood ratio test gave a p-value of 0.53. The other comparisons gave:

Model	df	AIC	BIC	logLik	Test	L Ratio	p-value
1	16	1078.92	1150.05	−523.46			
2	6	1039.80	1066.48	−513.90	1 vs 2	19.11	0.04
3	12	1030.01	1083.36	−503.00	2 vs 3	21.79	<0.001
4	8	1010.83	1046.39	−497.41	3 vs 4	11.17	0.02

Model 1 is the model in equation (22.6). In model 2, we only use one temperature smoother for all transects. It is an improvement, but not by much. In model 3, we used four temperature smoothers, one for Sex = 0 and AMPM = 0, one for Sex = 1 and AMPM = 0, one for Sex = 0 and AMPM = 1 and one for Sex = 1 and AMPM = 1. Its AIC is lower than that of models 1 and 2. In model 4, we used two smoothers for temperature, one for Sex = 0 and one for Sex = 1. The likelihood ratio test shows that it is a significant improvement compared with model 3, and its AIC indicates that it is the most optimal model. It is given by

$$\begin{aligned}\text{Honeybees}_{ijs} = \alpha &+ f_{sex=0}(\text{Temperature}_s) + f_{sex=1}(\text{Temperature}_s) \\ &+ \text{AMPM}_s + \varepsilon_{ijs} \\ &\varepsilon_{ijs} \sim N(0,\sigma^2) \quad \text{and} \quad cor(\varepsilon_{ijs}\varepsilon_{ijt}) = \rho^{|s-t|}\end{aligned} \tag{22.7}$$

The numerical output for this model is as follows.

```
                 Estimate      Std. Error      t-value  p-value
Intercept        0.530         0.030           17.391   <0.001
factor(AMPM)1    0.078         0.036           2.152    0.031

Approximate significance of smooth terms:
                          edf      F           p-value
s(Temperature):Sex0       1.703    3.257       0.011
s(Temperature):Sex1       3.949    48.748      0.001

 R-sq.(adj) =  0.292  Scale est. = 0.281   n = 630
```

The scale estimator is the variance of the noise. Because temperature has only a few unique values, we set the amount of smoothing for both smoothers equal to 4 degrees of freedom. Cross-validation (Chapter 7) was used to obtain the amount of smoothing, and it was estimated as 1.7 for the temperature smoother for the male fertile data (Sex = 0) and 3.9 for the male-sterile data (Sex = 1). The smoothing curves are given in Figure 22.4. For the male-sterile data, the largest numbers of honeybees are obtained for temperatures between 18°C and 21°C.

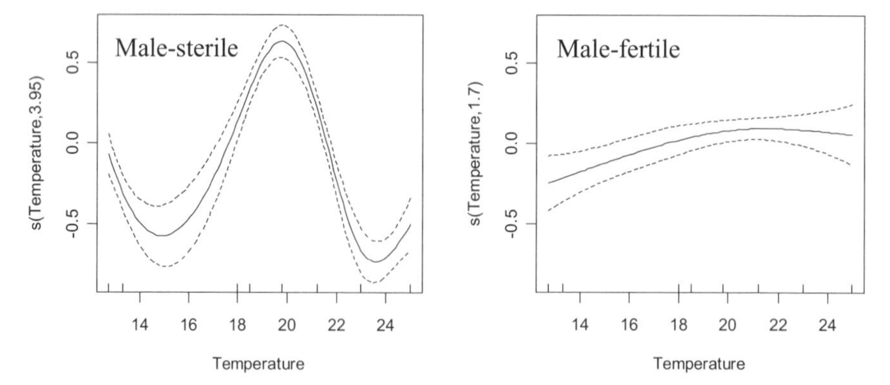

Figure 22.4. Smoothing curves for temperature for the male-sterile (Sex = 1) and male-fertile (Sex = 0) data. Dotted lines are 95% point-wise confidence bands.

The numerical output of our model presented above also indicates that AMPM (time of sampling) is significant but with a p-value close to 0.05. Hence, we should not consider this explanatory variable as important. The relevant model validation graphs to assess the assumptions of normality and homogeneity are given in Figures 22.5 and 22.6. Normality for the residuals of the male-sterile (Sex = 1) seems to be OK, although one can argue about normality of the male-fertile (Sex = 0) residuals. As to homogeneity (Figure 22.6), one can see patterns in these residuals, but this is due to the observed honeybees values only taking on a certain number of unique values, and this results in typical banding of the residuals. However, one can see a small increase in spread for larger fitted values. To understand why this is the case, we plotted residuals versus each nominal explanatory variable. There was no indication that for some levels of transects, sex or AMPM, the spread of residuals was larger. Figure 22.7 gives two examples. Figure 22.8 shows residuals versus temperature conditional on Sex. The larger residuals for the male-sterile data are obtained for temperature around 20°C. This is also the range where higher honeybee values were measured. It may be an option to allow for different residual spread per temperature regime, or use a Poisson distribution.

The model in equation (22.7) gave $\rho = 0.16$. This means that the correlation between two sequential samples is $\rho = 0.16$. If the gap is two units, then the correlation is $\rho^2 = 0.16^2 = 0.03$. These values are rather low.

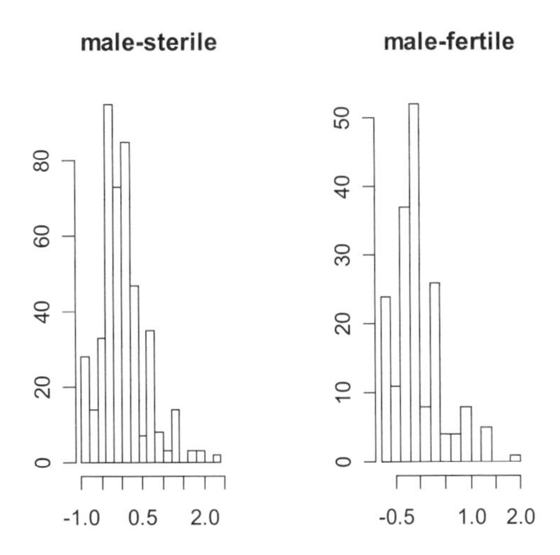

Figure 22.5. Histogram of residuals conditional on male-sterile (Sex = 1) and male-fertile (Sex = 0).

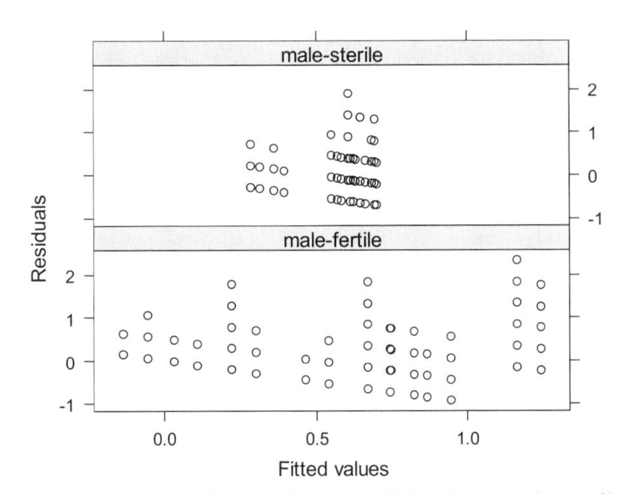

Figure 22.6. Residuals versus fitted values conditional on male-sterile (Sex = 1) and male-fertile (Sex = 0).

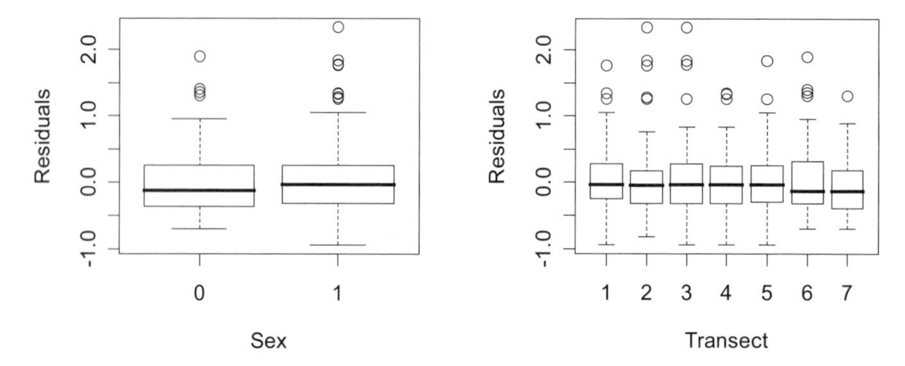

Figure 22.7. Model validation graphs showing that the spread in residuals is homogenous per level of Sex and Transect.

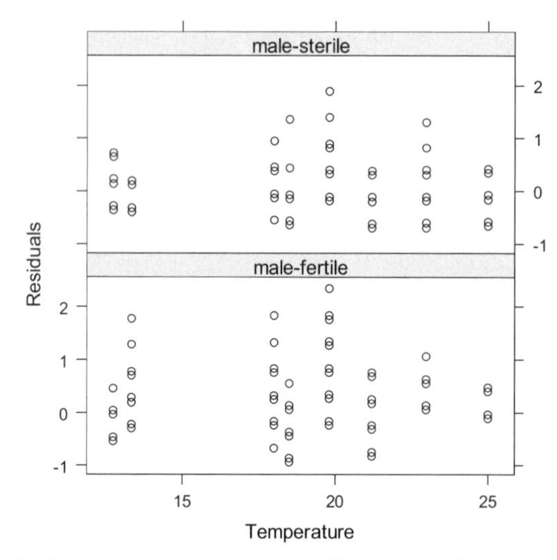

Figure 22.8. Residuals versus temperature conditional on Sex.

22.6 Discussion and conclusions

Initial exploration of the honeybee data indicated that non-linear relationships were evident for some explanatory variables and that the inclusion of linear interaction terms did not solve this problem. Therefore, we applied additive modelling. Indeed, this method showed a non-linear relationship of temperature.

The optimal model contained a temperature smoother for male-sterile (Sex = 1) and male-fertile (Sex = 0) data. However, the model showed a certain amount of heterogeneity, and this means that we have to be careful with interpreting p-values close to borderline significance, as is the case for AMPM. Further model improvement may be obtained by allowing for different variances per temperature regime. Alternatively, a Poisson model can be tried. One might wonder why we should make a lot of fuss about using different variances and correlation structures. The simple answer is, that by including these factors in our model, we can reduce the probability of obtaining a type I error.

The remainder of the discussion will concentrate on whether the additive model was the most optimal model or whether a generalised additive modelling approach should have been used. Interestingly, we did apply the GAM with a log link function and Poisson distribution, and we also converted the honeybee data to presence–absence data and applied a GAM with a binomial distribution and logistic link during the course of our data analysis. Both types of models can be extended with auto-correlation resulting in generalised additive mixed modelling. However, the conclusions resulting from these models were similar to the ones obtained by the additive mixed model.

The selected model indicated that the temperature effect on honeybees was different between MS and MF lines. Within the MS block, a uniform distribution of honeybees was found (different temperature patterns per sex were better compared to difference patterns per transect), which supports the finding of similar studies carried out by Skinner (1987). In this respect, planting schemes used in this study do not affect the seed production.

For the MS block, the maximum number of honeybees was estimated to occur at temperatures between 18°C and 21°C. Foraging activity declined outside this temperature range for the MS block, and temperature seems to be an important factor for honeybee activity. This observation has also been made in previous studies conducted both in this area and also in other areas where a decrease in foraging activity was observed when the temperature dropped by 2°C and 4°C respectively (Núñez 1982; Skinner 1987). At higher temperatures (> 22°C) honeybee activity decreased. However, we cannot say for sure that temperature is the driving factor for honeybee activity as a certain amount of higher dimensional collinearity between temperature, PercFlower, Sex and time exists. Repeating this experiment under different temperature regimes may give more information about the cause–effect relationship.

Acknowledgement

This work was part of the PhD project carried out by the first author who was particularly indebted to Enrique Bedascarrasbure and David de Jong for useful suggestions towards her guidance. Alfonso Lorenzo and Paul Ens kindly provided assistance in data gathering. This work was supported by SECYT-UNCPBA, and Don Atilio Company was acknowledged for supplying the seeds. We would like to thank Alex Douglas for valuable comments on an earlier draft.

23 Investigating the effects of rice farming on aquatic birds with mixed modelling

Elphick, C.S., Zuur, A.F., Ieno, E.N. and Smith, G.M.

23.1 Introduction

Ecologists are frequently interested in describing differences among the ecological communities that occur in habitats with different characteristics. In an ideal world, experimental methods would standardise situations such that each habitat variable could be altered separately in order to investigate their individual effects. This approach works well in simple ecosystems that can be replicated at small spatial scales. Unfortunately, the world is not always simple and many situations cannot be experimentally manipulated. Investigating specific applied questions, in particular, often can be done only at the spatial scales at which the applied phenomena occur and within the logistical constraints imposed by the system under study. In such cases, one is often left with the choice between collecting "messy" data that are difficult to analyse or avoiding the research questions entirely. In this chapter, we investigate just such a case, in which applied ecological questions were of interest, but experimental influence over the system was not possible.

Agricultural ecosystems dominate large areas of the Earth, and they are used by a wide variety of species. Consequently, conservation biologists have become increasingly interested in understanding how farmland can contribute to biodiversity protection (McNeely and Scherr 2002; Donald 2004). Rice agriculture, in particular, has great potential to provide habitat because it is a dominant crop worldwide, and because the flooded conditions under which rice is typically grown, simulate —if only approximately— wetland habitats, which have been widely drained in agricultural areas. In the Central Valley of California, USA, rice is grown in an area of extremely intensive agriculture that historically was a vast complex of seasonally flooded wetland and grassland habitat. In the past, these wetlands supported millions of migratory waterbirds during winter, and the area remains one of North America's most important areas for wintering waterbirds (Heitmeyer et al. 1989).

Traditionally, most California rice fields have been flooded only in the summer months when rice is being grown. Recent legislative changes, designed to improve air quality by phasing out the practice of burning residual straw left over after harvest, have resulted in many farmers flooding their fields during winter in order to speed up the decomposition of rice straw before the subsequent growing season. This activity more closely simulates the historic flooding pattern and has been viewed as a potential boon for wetland birds.

In the original study, the main goals were to determine whether flooding fields result in greater use by aquatic birds, whether different methods of manipulating the straw in conjunction with flooding influences how much fields are used, and whether the depth to which fields are flooded is important. Various straw manipulation methods are used by farmers in the belief that they enhance decomposition rates, and usually involve cutting up the straw to increase the surface area, increasing the contact between the straw and the soil, or both. The effects of water depth are of interest because water is expensive and there are economic benefits to minimizing water use. Limiting water consumption is also important to society because there are many competing demands for a limited water supply. Reduced water levels may have agronomic benefits too, because anoxic soil conditions, which would slow straw decomposition, are less likely with shallow flooding.

The data used in this chapter were collected during 1993–1995, and come from winter surveys of a large number of fields in which waterbirds were counted and identified to species. For the purposes of this study "waterbirds" are defined as Anseriformes (swans, geese and ducks), Podicipediformes (grebes), Ciconiiformes (herons, ibises, and allies), Gruiformes (rails, cranes and allies) and Charadriiformes (shorebirds, gulls, and allies). The data have been analysed previously, both to look for effects on individual species (Elphick and Oring 1998) and to examine composite descriptors of the waterbird community (Elphick and Oring 2003).

Although the study was planned with experimental design principles in mind, the reality of the situation limited the rigour with which these ideals could be followed. Different straw management treatments could not be assigned randomly because their application depended on the separate decisions made by individual farmers. Replication was possible but was similarly constrained by the actions of different growers. Moreover, because all fieldwork was conducted on privately owned land, data collection was subject to the constraints placed by the landowners. Eventually, farms were selected such that each management technique was represented by as many fields as possible and that fields with different treatments were interspersed among each other. Layered on top of these constraints, the nature of the data created various problems for common statistical analyses. In short, from an ecologist's perspective the dataset was an analysts' nightmare! Because of all these issues, the original analyses relied primarily on non-parametric statistical tests and were not well suited to examining different explanatory variables simultaneously or for testing interactions among explanatory variables. In this case study chapter we will apply mixed modelling techniques to deal with the complicated structure of the data.

In the initial analysis, flooding fields clearly had a strong effect on waterbird use, but the effects of the way in which flooding occurred were more ambiguous. Our main questions here, then, concern whether the method of straw manipulation and the water depth in flooded fields influenced waterbird use of the fields. Additionally, we test whether there was geographic and temporal variation in bird numbers. To fully understand all the steps carried out in this case study chapter, it might be helpful to review the theory on mixed models (Chapter 8) afterwards.

23.2 The data

The basic experimental design for this study involved collecting data that were nested at three spatial scales. The main units of study were fields, and it was at this scale that straw management treatments were applied. Some (but not all) fields were subdivided into units referred to as 'checks'. Checks are simply sub-sections of a large field that are separated from each other by narrow earthen levees in order to help farmers control water depth. In this study, data were collected separately for individual checks because water depth typically varied a great deal among the checks within a field, but very little within checks. Collecting data for each check separately, therefore, makes it easier to test for depth effects by ensuring uniform depth within each sampled unit. Fields were also spread over a large portion of the rice-growing region, but were clustered into three geographic 'blocks'. Sampling over a large area was considered important to ensure that the results were broadly applicable. But clustering sampled fields were also necessary in order to facilitate data collection. In our analysis, we wanted to test for differences among these blocks both to determine whether there were geographic differences in the use of fields by birds, and to determine whether the main results applied throughout the rice-growing region. The overall design, then, has checks nested within fields, which are nested within geographic blocks. In all there were 26 fields in block 1, 10 fields in block 2 and 6 fields in block 3 (note that depth data were lacking for a few fields included in the original study, and so the sample sizes here differ slightly from the totals given in the previous papers). The number of checks in a field varied from only 1 in the smallest fields to as many as 16 in the largest. A schematic overview of the data is presented in Figure 23.1.

Birds were counted during surveys that were conducted at approximately 10-day intervals (see Elphick and Oring 1998 for details). A total of 25 surveys were made, 12 between Nov 1993 and Mar 1994 and another 13 between Nov 1994 and Mar 1995. In principle we could make the schematic overview in Figure 23.1 considerably more complex by adding "Survey" as another level in the structural hierarchy. With a few exceptions, each field was visited during every survey. Blocks 1, 2 and 3 had a total of 446, 750 and 836 observations, respectively. We omitted three observations that had extreme water depths (> 60 cm). Possible explanatory variables are water depth (which varies at the check level), straw treatment (which varies at the field level and is labelled as Sptreat in this chapter; see Table 23.1), Block, Field, Survey, and Year.

A complicating factor is that both fields and checks varied in size. The response variables are either numbers or densities of 45 different bird species. If we wanted to work with numbers, then the size of the area would have to be used as an offset or weighting variable; for simplicity in this analysis we used densities.

From an analyst's perspective a major problem with the dataset is that 95% of the observations are equal to zero, which might cause all kinds of statistical problems, including violation of homogeneity (Chapter 5) or overdispersion (Chapter 6), if GLM models were to be applied. The specific aim of this case study chapter is to demonstrate how univariate methods (especially mixed modelling) can be used. Therefore, we pooled the data for all waterbird species to give a single variable (called Aqbirds) and tested models that explain patterns of overall waterbird abundance. Hence, we will not consider the application of multivariate methods like ANOSIM and NMDS (Chapters 10 and 15), which might also help explain the relationships between waterbird use and the various explanatory variables.

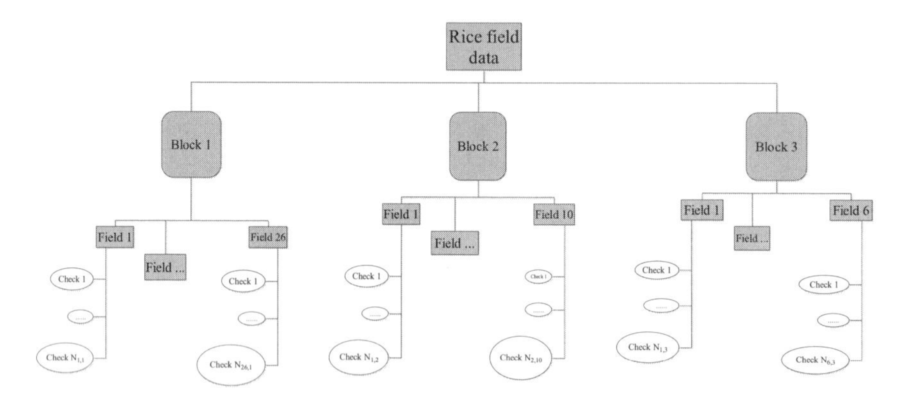

Figure 23.1. Schematic overview of the rice field data. In principle we could make the schematic overview considerably more complex by adding 'Year' and 'Survey as additional levels at the bottom of the structural hierarchy. The surveys were conducted from November to March in each winter.

Table 23.1. Levels of the treatment variable SPTREAT and the corresponding numerical codes used in this chapter.

Treatment	Numerical Code In This Chapter	Description
fld	1	Field flooded with no straw manipulation
rl+fld	2	Field rolled to mix straw and soil, and then flooded
fld+rl	3	Field flooded, and then rolled to mix straw and soil
chp+fld	4	Straw chopped up, and field then flooded
inc+fld	5	Straw incorporated into soil, and field then flooded
rmv+fld	6	Straw removed entirely, and field then flooded

23.3 Getting familiar with the data: Exploration

When deciding how to analyse the rice field bird data, we first had to make a series of decisions about how to structure our models. The first problem was that the data have a hierarchical structure, with checks nested within fields, and fields within blocks. A possible model is

Aqbirds = constant + Depth + factor(Block) + factor(Field) +....

where factor() represents a nominal variable. In the initial dataset, 'Block' was labelled as 1 to 3 and 'Field' was labelled as 1 to 26 for the fields within block 1, 1 to 10 for the fields within block 2 and 1 to 6 for the fields from block 3. This labelling was a problem because field 1 in block 1 is not the same as field 1 in blocks 2 or 3; yet the function factor() would erroneously aggregate all observations with the same field value. Consequently, we re-labelled the fields so that each has a unique label with levels 1 to 42. Using this nominal explanatory variable in the model above means that we must estimate 41 (42 levels minus the baseline) separate intercepts for the field variable alone, which is a lot of parameters to estimate, especially as we are not actually interested in the effects of individual fields. We cannot ignore these effects totally because some of the variation in the data might be due to between-field variation, and accounting for this variation will help us to test for other patterns. To reduce the number of parameters that must be estimated, therefore, we treat Field as a random component rather than as a fixed effect in the model. This approach basically means that we assume that the fields in our study are a random sample from a large number of possible fields, and that we can model all of the variation among fields as a random process that can be described in simple mathematical terms. To be more precise, instead of paying the price of 41 degrees of freedom in order to estimate separate intercepts for every field, we assume that all fields have intercepts that are normally distributed with mean 0 and a certain variance. With this formulation, the only unknown component that must be estimated is this variance, which 'costs' us only one degree of freedom.

We can make the variance structure a bit more complicated and account for the nested design (see Quinn and Keough 2002 for examples) with something like:

Aqbirds = constant + Depth + Sptreat + Block/Field/Year/Survey + noise

The notation Block/Field is a computer notation for nested random effects, and in this study we can extend this approach to incorporate four nested levels. In theory, we could use random components for Field, Block, Survey and Year. We are, however, interested in the specific geographic differences among the blocks, and in any seasonal pattern exhibited by Survey. Consequently, these variables are better left as fixed effects. It would be suitable to model Year as a random component, but there were only two years studied. Generally it is best to have at least four or five levels if one is to treat a variable as a random component instead of fixed because this results in more stable variance estimation and the degree of freedom savings are minimal otherwise. So, we keep the models simple and use something of the form:

Aqbirds = constant + Depth + Sptreat + Year + Survey + Block + Field + noise (23.1)

where only Field and noise are random components, and the rest are considered to be fixed. We call this model a mixed effects model (Chapter 8). Deciding whether to view variables as random or fixed was relatively easy, mainly driven by the underlying questions and the nature of the variables.

The next problem was to determine how to test for temporal patterns. Although each year's data were collected in the same period, the timing of each round of surveys differed somewhat between years due to unexpected restrictions on access to certain sites. For example, although dates for the first survey in each winter coincide, by the fourth survey the timing was quite different (survey 4 occurred from 4–7 January in the first winter and from 19–21 December in the second). This mismatch means that it is misleading to treat a given survey i as temporally equivalent in the two years. Consequently, we decided to analyse each year's data separately. Hence, we removed Year from our previous model and planned to apply the following model separately to data from each year:

$$\text{Aqbirds} = \text{constant} + \text{Depth} + \text{Sptreat} + \text{Survey} + \text{Block} + \text{Field} + \text{noise} \quad (23.12)$$

This model implies a linear relationship between bird densities and water depth, which would not necessarily be expected from a biological perspective. So, we next need to determine whether the effect of depth is indeed linear, and the best way to do this is through data exploration. We should also verify that there are no outliers in these data. Figure 23.2 shows a Cleveland dotplot (Chapter 4) of the response variable Aqbirds for both years combined. Although there are no single extremely large observations, the data would benefit from a transformation as approximately 25 observations (from the second winter) are at least twice as large as the remaining values. QQ-plots (not presented here) show clearly that a $\log_{10}$ transformation is better than a square root or no transformation (see Chapter 4 for a discussion on transformations and how to choose among them). A constant value of 1 was added to each bird density value to avoid problems caused by taking the log of zero. After pooling data for all species the percentage of zero observations in the response variable was much reduced, but still accounted for 35% of the observations.

Figure 23.3 shows a coplot of $\log_{10}(Y+1)$ transformed waterbird densities versus Depth conditional on the variables Sptreat and Year. To aid visual interpretation, a LOESS smoother with a span of 0.7 (Chapter 7) has been added. Note that there appears to be evidence for a non-linear relationship between depth and waterbird densities.

A boxplot of the log transformed waterbird data conditional on survey and year is given in Figure 23.4. The 12 boxplots on the left-hand side are from the winter of 1993/94, the 13 on the right from 1994/95. Note that there is a clear seasonal pattern in the first winter, but that this pattern is less clear in the second winter. This difference may relate to very different weather patterns in the two years, with extensive rainfall and natural flooding in the second year that caused many fields to stay wet longer than normal (Elphick and Oring 1998). If the data from both years were to be analysed together, the patterns in the boxplots indicate that interaction terms between Year and other variables may be required.

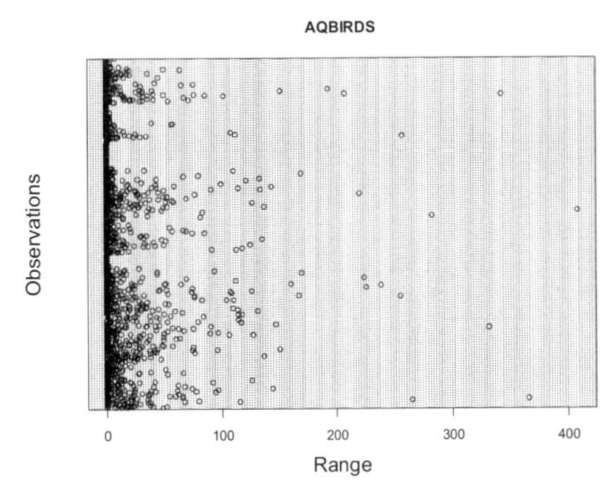

Figure 23.2. Cleveland dotplot of waterbird densities. The horizontal axis represents the values of the observations, and the vertical axis corresponds to the order of the data. The observation at the bottom is the first row in the spreadsheet, and the observation at the top the last.

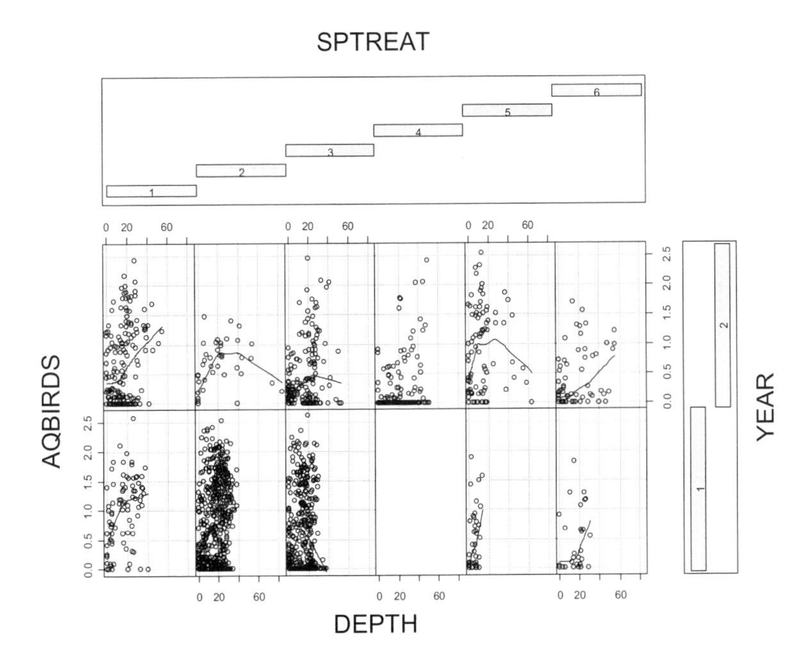

Figure 23.3. Coplot of waterbird densities versus depth conditional on Sptreat and Year. The lower six panels correspond to treatments 1 to 6 (see Table 23.1) in year 1; the upper six panels to the same treatments in year 2. Smoothing curves were added to aid visual interpretation. Treatment 4 was not measured in year 1.

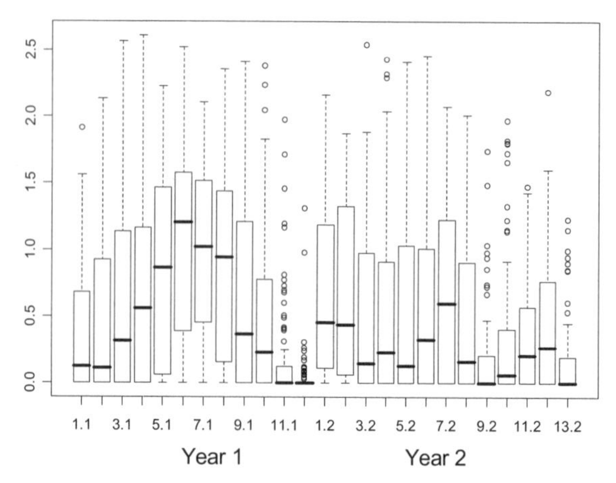

Figure 23.4. Boxplots of log transformed waterbird data conditional on survey and year. The first 12 boxplots from the left are for year 1, the others for year 2. The labelling '*i.j*' refers to year '*j*' and Survey *i*.

23.4 Building a mixed model

Now that we have a sense of what the data look like, we can start building an initial model. First we included Sptreat, Depth and Survey as explanatory variables. Data exploration indicated that there might be a non-linear relationship between waterbird density and water depth, and our initial attempt to use an additive model showed that the relationship is clearly non-linear in the first winter and nearly linear in the second winter (Figure 23.5-A and B). This difference is indicated by the estimated degrees of freedom for the smoothers, which are smaller in the second year, and closer to 1, which would indicate a linear relationship (Chapter 7). With this model, all explanatory variables were significant at the 5% level. This analysis ignores variation among fields, though, accounting for it only as part of the error term.

From this initial analysis, there are at least two ways to proceed: (i) Apply additive mixed modelling to allow for a non-linear depth effect and the between-field variation using a random component or (ii) apply linear mixed modelling using a polynomial function of depth and between-field variation using a random component. So, the problem is how to model the non-linear depth effect. Because (i) additive mixed modelling has already been used in two other case study chapters, and (ii) software for linear mixed modelling is better developed than for additive mixed modelling, we use the former here.

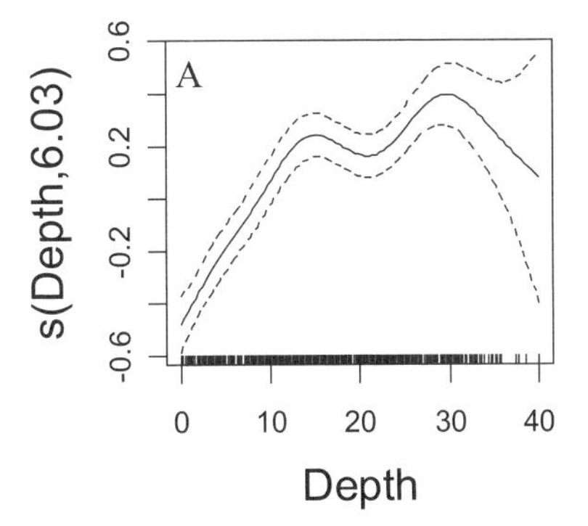

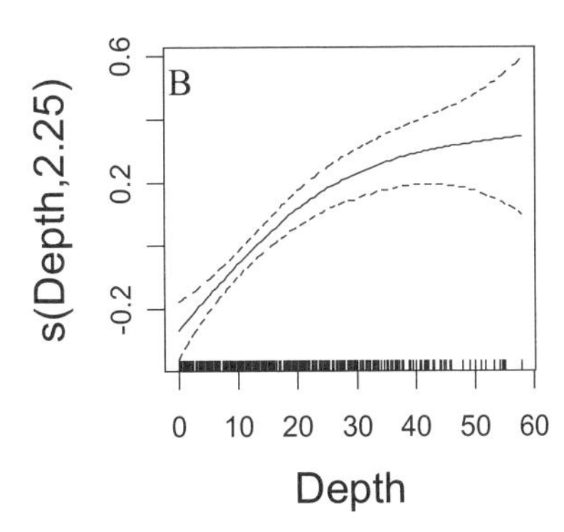

Figure 23.5. Smoothing curve obtained by an additive model using the year 1 data (panel A) and the year 2 data (panel B), showing the effect of water depth on waterbird densities. Dotted lines are 95% confidence bands. Vertical lines at the bottom indicate the values of depth at the corresponding site. The vertical axes represent the contribution of Depth to the fitted bird values (Chapter 7). Cross-validation was used to estimate the degrees of freedom for each smoother.

The limited availability of software for (generalised) additive mixed modelling is because the technique is reasonably new (Ruppert et al. 2003, Wood 2004, 2006). Application of these methods at present requires expert knowledge and programming skills. As shown in other case study chapters in this book, (generalised) additive mixed modelling has huge potential and its use in ecology can be expected to grow.

Linear mixed models

The first linear mixed model we applied to each year's dataset was as follows:

M_0: Aqbirds = constant + Depth + Depth^2 + Sptreat + Block + Field + noise

The Depth^2 term was included to model the non-linear relationship between waterbird density and depth. Later, we will test whether it is really necessary to include this term. All explanatory variables in model M_0 are nominal except for Depth and Depth^2 and only Field was treated as a random component. Unfortunately, this model crashed, producing the message: 'NA in data'. The reason for this error message was that some straw treatments only occurred in one block causing a perfect collinearity between Block and Sptreat. To solve this problem we would either have to eliminate certain straw treatments or drop one of the confounded terms from the analysis. As we are primarily interested in the straw effect, we chose to omit Block. The new model becomes

M_1: Aqbirds = constant + Depth + Depth^2 + Sptreat + Field + noise

The next step was to assess whether a seasonal effect should be included in the model. To do this we plotted standardised residuals from model M_1 versus Survey (Figure 23.6). The resulting boxplot shows a clear seasonal pattern, suggesting that adding a time component is important:

M_2: Aqbirds = constant + Depth + Depth^2 + Sptreat + Survey +
Field + noise

The residuals from this model did not show a seasonal pattern when medians were compared, but the amount of variation still varied among surveys, which violates the homogeneity assumption (Figure 23.7). Because heterogeneity of variance is a serious problem, the *p*-values produced by this model should not be used. Mixed modelling provides a solution, especially if we use separate variances to different portions of the data.

To clarify what we have done so far, the terms Depth, Depth^2, Sptreat and Survey were all used as fixed components in model M_2. Field was used as a random component and is an important part of the model because it reduces the portion of the variation that cannot be explained at all. The remaining unexplained variance was treated as another random component named 'noise', although this notation is a bit sloppy. Elsewhere in the book the residuals that describe this 'noise' have been referred to as ε. The homogeneity assumption applies to these residuals, which are assumed to be normally distributed with expectation 0 and variance σ^2. With mixed modelling, we can account for the heterogeneity by assuming a different value for the residual variance associated with each survey (labelled σ_1^2 to σ_{12}^2 in year 1), rather than just assuming that all residuals are distributed with a mean of 0 and a single variance of σ^2. This gives the following model:

M_3: Aqbirds$_j$ = constant + Depth + Depth^2 + Sptreat + Survey + Field + ε_j

where $\varepsilon_j \sim N(0, \sigma_j^2)$ and Field $\sim N(0, \sigma_{Field}^2)$. Incidentally, instead of using separate variances σ_j^2 for each survey, we could try to account for heterogeneity by using different variances for each straw treatment. In the next section, we will discuss model M_3 for the year 1 data and once we have found the optimal model in terms of random terms we will try to identify unnecessary fixed terms.

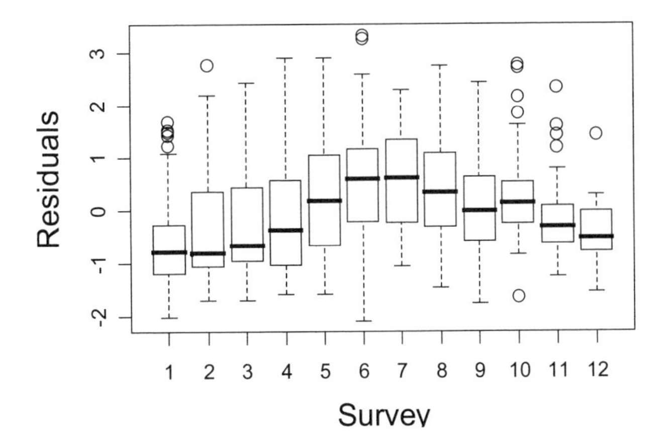

Figure 23.6. Standardised residuals of model M_1 plotted versus Survey for year 1. A strong seasonal trend occurs in both the median values and in the amount of variation within each survey.

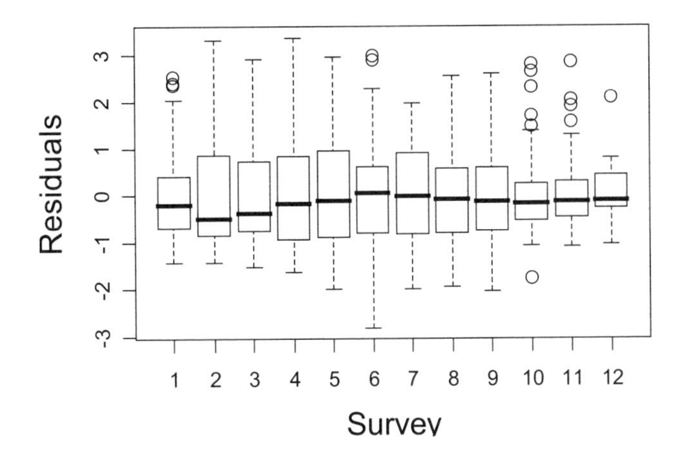

Figure 23.7. Standardised residuals of model M_2 plotted versus Survey for year 1. There is no clear seasonal variation in the median residuals values per survey anymore, although the variance heterogeneity remains, with especially smaller residuals in the later surveys.

23.5 The optimal model in terms of random components

In the previous section we developed a model that contained the random component Field and, in model M_3 also introduced 12 different variance components, one for each survey. Although the modifications introduced in model M_3 solve the variance heterogeneity problem, we are paying a high price in terms of degrees of freedom, because we now have to estimate an extra 11 parameters. Using the AIC or the likelihood ratio test, we can compare models M_2 and M_3 and assess whether this price results in significant model improvement:

Model	df	AIC	logLik	L.Ratio	p value
M_2	20	2146.58	−1053.29		
M_3	31	2006.26	−972.13	162.32	<0.001

In this comparison the smaller AIC for M_3 indicates that the more complex model is the better one and the likelihood ratio test allows us to go further and say that using the 12 variances is a significant ($p < 0.001$) improvement. We can also determine whether we need such a construction for Sptreat as well. In this case, we compared model M_2 with a model M_4 that containing six separate variances, one for each straw treatment. In this case, however, the likelihood ratio test gave $p = 0.094$ indicating that M_2 is the better of these two models. If M_4 had proven better than M_2, then we would have had to choose between this model and M_3. Because these models are not nested, a likelihood ratio test would have been inappropriate and we would have had to choose by comparing AIC values.

So far, our best model is M_3, but this model assumes independent residuals ε. Because the data form a time series, this assumption may be invalid, so we tested the effect of adding a simple auto-correlation structure in which each survey is treated as a point in time. We used an ARMA(1,0) structure for the noise, see also Chapters 16, 27 and 36. In such a model we allow for a relationship over time in the error term: $\varepsilon_t = \theta\varepsilon_{t-1} + \eta_t$ where η_t is independently and normally distributed noise. In the first year, we have up to 12 sequential observations for each check in a particular field. The model will impose the same auto-correlation structure for all checks. Hence, it will give one value for θ describing the overall auto-correlation in the error term for all checks. For longer time series, more complicated auto-correlation structures might be used. Model M_5 is the same as M_3, except that the error term is now allowed to be ARMA(1,0). When we tried to run this model we initially got error messages, caused by some checks with very short time series (only four or five surveys). This problem arose because there were a few checks in which depth measurements could not be made on every survey because depth stakes that had been placed in fields at the start of the field season were removed or chewed through at some sites. Vandalism by muskrats and theft were not accounted for in the study design. Waterbird densities were still obtained for these checks on every survey, but the data could not be used in our analysis because of the missing depth data. Luckily, this problem affected few checks, but the reader who wants to carry out the analysis should remove checks 5 and 6 of field 302, and checks 2–5 of field 305 for the year 1 data. Given these data discrepancies, we refitted M_3 with the reduced dataset in order to compare it to M_5:

Model	df	AIC	logLik	L.Ratio	p value
M_3	31	1925.33	−931.66		
M_5	32	1866.60	−901.30	60.2	<0.001

The likelihood ratio test shows that allowing for auto-correlation between the residuals gives a significant improvement. The estimated value of θ was 0.274. We will discuss the significance levels of the fixed terms later, but it is interesting to mention one of them here and emphasise the implications of auto-correlation. In model M_3, the straw treatment effect produced a p-value of 0.007, but after auto-correlation was accounted for in model M_5, the p-value for this variable was 0.045. In other words, if we had ignored auto-correlation and used model M_3 we would have concluded, wrongly, that the effect of straw management is highly significant when in fact it only just approaches the $p = 0.05$ significance threshold.

Dropping fixed terms from the linear mixed model

Having found the optimal model for the year 1 data in terms of random components, we now investigate whether any of the fixed components can be dropped from the model or if interaction terms should be added. Our starting model is M_5, and we consider each of the following alternatives:

- Drop only Survey (M_6).
- Drop only Sptreat (M_7).
- Add an interaction between Depth and Sptreat (M_8).
- Drop the quadratic Depth term to test whether the depth relationship is linear or non-linear (M_9).
- Drop both the linear and quadratic Depth terms to test whether depth is related to waterbird density at all (M_{10}).

This leads to the following models:

M_6: Aqbirds = constant + Depth + Depth^2 + Sptreat +
Field + Variance for each survey + auto-correlated noise

M_7: Aqbirds = constant + Depth + Depth^2 + Survey +
Field + Variance for each survey + auto-correlated noise

M_8: Aqbirds = constant + Depth + Depth^2 + Sptreat + Survey +
Depth* Sptreat +
Field + Variance for each survey + auto-correlated noise

M_9: Aqbirds = constant + Depth + Sptreat + Survey +
Field + Variance for each survey + auto-correlated noise

M_{10}: Aqbirds = constant + Sptreat + Survey +
Field + Variance for each survey + auto-correlated noise

To compare mixed models with different fixed components, the estimation routine should use maximum likelihood estimation instead of restricted maximum likelihood estimation (REML). See Chapter 8 for details; note that AIC values obtained with these methods should not be compared with each other. The results are given in Table 23.2. First, M_5 and M_6 are compared to test the null hypothesis that the regression parameters for Survey equal zero. The likelihood ratio test shows that this hypothesis can be rejected ($p < 0.001$) and that dropping Survey results in a significantly worse model. The next comparison also shows that reducing the model causes a significant worsening, and that straw treatment affects waterbird densities. This result, however, is only weakly significant ($p = 0.045$), indicating that the Sptreat effect is much smaller than the Survey effect.

The last three models all examine the effects of water depth. Adding an interaction between Depth and Sptreat slightly improves the model (i.e., it produces a smaller AIC value). But this difference gives a *p*-value of 0.055, suggesting that it is of borderline significance. The quadratic Depth term also produces only a marginal difference, although there is significant improvement when this term is included in the model. In contrast, removing depth entirely, by dropping both the Depth and $Depth^2$ terms, results in a highly significant worsening of the model.

So, which model should we choose? Clearly Survey and Depth should be in the chosen model, as dropping either leads to a substantially worse model. Including Sptreat, $Depth^2$, or the Sptreat × Depth interaction all cause marginal improvements in the model, but none of these terms is as important as the other two. Whether they should all be included, then, depends on how rigidly you want to stick to the 0.05 significance criterion. Formally, choosing the final model should involve an iterative process in which the entire process outlined here is repeated after each variable is dropped, in a backward selection style.

Table 23.2. Comparing models with different fixed effects. The AIC for M_5 is 1778.92. Either the AIC or (better) the likelihood ratio test can be used to compare the nested models.

Test models	Term deleted	AIC	*L*-Ratio	*p*-value
M_5 versus M_6	Survey	1956.830	199.904	<0.001
M_5 versus M_7	Sptreat	1780.620	9.694	0.045
M_5 versus M_8	Add interaction Sptreat and depth	1777.645	9.280	0.055
M_5 versus M_9	$Depth^2$	1781.325	4.399	0.035
M_5 versus M_{10}	Depth and $Depth^2$	1793.191	18.266	<0.001

23.6 Validating the optimal linear mixed model

Assuming that we accept model M_8, the next thing we have to do is to inspect the residuals and check whether normality and homogeneity assumptions hold. Figure 23.8-A shows the fitted values versus standardised residuals. Note that the residuals show a pattern of bands that is characteristic to all linear regression, GLM, mixed models and GAM models when there are lots of observations with the same values; see Draper and Smith (1998) for an explanation. Given that our dataset has so many zeros, this pattern is not surprising and we should not worry about straight lines of residuals. The residuals are standardised, hence, values larger than 2 or –2 are suspect (Chapter 5). There is no clear violation of the homogeneity assumption, and only a few observations are larger than 2 given the size of the dataset. Figure 23.8-B shows a histogram of these residuals and indicates that they are slightly skewed.

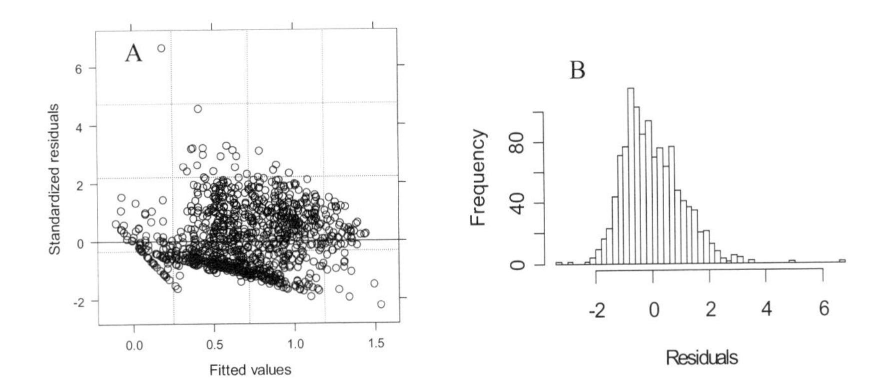

Figure 23.8. A: Fitted values versus standardised residuals obtained by model M_8. B: Histogram of residuals obtained by model M_8.

23.7 More numerical output for the optimal model

The last thing we have to do is to discuss the numerical output for model M_8 (obtained by REML estimation).

```
Random effects:
Formula: ~1 | Field
          (Intercept)    Residual
StdDev:    0.099         0.479

Correlation Structure: ARMA(1,0)
Formula: ~Time | Field/Check
Parameter estimate(s):
Phi1
0.278
```

This information tells us that variation among fields is quite small. The variance between fields and the residual variance are 0.0099^2 and 0.479^2, respectively. Hence, the variation among fields is considerably less than the residual variance. Because the Field effect is small, we also compared various mixed models with corresponding GLS models (a GLS is a mixed model without the random intercept). Comparing AIC values suggested that the mixed models were better, indicating that the random Field component is needed even though the effect is small. The auto-regressive parameter has a value of 0.274. The following lines give the different monthly variances.

```
Variance function:
 Structure: Different standard deviations per stratum
 Formula: ~1 | Survey
 Parameter estimates:
     1      2      3      4      5      6      7      8
 1.000  1.390  1.343  1.407  1.320  1.439  1.209  1.277
     9     10     11     12
 1.289  1.073  0.717  0.350
```

In the analysis, survey 1 is arbitrarily set as the baseline survey, and the variance parameters for the other surveys are scaled relative to this baseline. Examining the pattern in these variance estimates shows that the first survey and the last three surveys have a relatively lower spread than the others. In other words, variation is lower at the start and end of the winter season when bird numbers are low, and variation increases mid-winter when there are large numbers of wintering waterbirds present in the region (see Figure 23.4).

Let us now focus on the estimated parameters and standard errors (obtained by REML estimation) of the fixed components.

	Value	Std. Error	Df	*t*-value	*p*-value
(Intercept)	0.139	0.126	1118	1.097	0.273
Depth	0.037	0.010	1118	3.733	<0.001
I(Depth^2)	-0.001	0.000	1118	-2.483	0.013
Sptreatfldrl	-0.129	0.094	31	-1.363	0.183
Sptreatincfld	-0.160	0.147	31	-1.088	0.285
Sptreatrlfld	-0.074	0.081	31	-0.905	0.372
Sptreatrmvfld	-0.149	0.141	31	-1.053	0.301
Survey2	0.137	0.078	1118	1.768	0.077
Survey3	0.186	0.076	1118	2.450	0.015
Survey4	0.278	0.084	1118	3.317	0.001
Survey5	0.480	0.081	1118	5.930	<0.001
Survey6	0.697	0.087	1118	8.015	<0.001
Survey7	0.656	0.079	1118	8.287	<0.001
Survey8	0.622	0.083	1118	7.517	<0.001
Survey9	0.417	0.087	1118	4.785	<0.001
Survey10	0.427	0.104	1118	4.107	<0.001
Survey11	0.117	0.098	1118	1.199	0.231
Survey12	-0.032	0.093	1118	-0.341	0.733
Depth:Sptreatfldrl	-0.011	0.006	1118	-1.797	0.073
Depth:Sptreatincfld	0.026	0.020	1118	1.312	0.190
Depth:Sptreatrlfld	-0.006	0.005	1118	-1.003	0.316
Depth:Sptreatrmvfld	-0.021	0.011	1118	-1.953	0.051

Just as in linear regression, one should interpret the *p*-values with great care, and the *p*-values in Table 23.2 are better for assessing the importance of each fixed component. It is, however, useful to look at the estimated parameters and their signs. In addition, this output provides some insight into the nature of the differences between levels for the nominal variables. According to this output, the linear depth term is positive and highly significant, indicating that waterbird densities increase with water depth, and the quadratic term is negative and weakly significant, suggesting that this pattern may weaken, or even reverse, in deeper conditions (see also Figure 23.3). The parameter estimates for Sptreat arbitrarily use the first treatment (Sptreatfld) as a baseline, and the negative values for the other treatments indicate that densities were highest for the baseline treatment. The standard errors and p-values for these parameter estimates, however, indicate that differences among treatments are relatively slight, which accords with the earlier conclusion that this effect is only marginally significant. (Note that the need to exclude a few checks due to missing data, resulted in one straw treatment being dropped from the final analysis.) Parameter estimates for Survey are also scaled relative to the first level for the variable (survey 1). In this case, most parameters are positive, reflecting the higher densities of birds in subsequent surveys. More careful inspection shows that Survey parameter estimates steadily increase until survey 6, and then decline. This change matches the buildup of birds during early winter, as migrants arrive from their northern breeding grounds, followed by a gradual decline in late winter as birds begin to move north again (Figure 23.4). Lastly, the parameters that describe the Sptreat × Depth interaction indicate that the effect of depth in fields where straw has been removed (Sptreatrmvfld) is somewhat different from that in the other straw treatments.

Year 2 data

Results for the year 2 data were very similar to those presented for year 1. This analysis needed a separate variance for each survey, a random field component and an auto-correlation structure ($\theta = 0.183$). There was a significant survey effect ($p < 0.001$), a significant straw treatment effect ($p = 0.010$), a weak interaction between straw treatment and depth ($p = 0.056$), and a strong non-linear depth effect ($p < 0.001$ for the linear term and $p = 0.001$ for the quadratic term). We leave it as an exercise for the user to validate the model.

23.8 Discussion

In this chapter, linear mixed modelling techniques were used to analyse the waterbird data. For the year 1 data, detailed model selection indicated that there were highly significant survey and depth effects. The pattern of spread in the residuals varied among surveys, and a random field component was needed in the model. A likelihood ratio test also indicated that we should include an auto-correlation term. Without accounting for auto-correlation, the model indicated a highly significant straw management effect, but once the auto-correlation was added, p-values for straw management and its interaction with depth indicated only weak significance.

This analysis improved on the earlier attempt to analyse these data (see Elphick and Oring 1998, 2003) in several ways. In the previous analyses, the problems created by the hierarchical structure of the data collection, the variance heterogeneity, and the large number of zeros were dealt with by simply testing each explanatory variable separately using non-parametric tests. This approach was not very satisfactory because it meant that the different variables were not tested simultaneously, which made it difficult to parse out their relative importance. In addition, the earlier analysis side-stepped the temporal pattern in the data, by using mean values calculated across each winter for each check. By using data from each survey, separately, the new analysis presented here provides a picture of how waterbird densities change over time, while simultaneously describing and accounting for the depth and straw treatment effects addressed in the earlier analysis. The new analysis could not, however, solve all of the recognized limitations of the earlier attempt. For example, we had to drop the Block variable here because it was confounded with certain straw treatments. Consequently, we were unable to improve on the earlier treatment of geographic effects.

Overall the re-analysis of these data did not radically change our understanding of the dataset. The new analysis, however, does provide stronger support for some of the inferences made in the earlier papers and a deeper appreciation of how the different explanatory variables interact.

In our analysis, we used waterbird densities per check as our response variable. Another approach would have been to use counts as the response variable, but to do that we would need to take into account the variation in check size. One option would be to use the size of the check as a weighting factor, for example checks with small areas could be down-weighted. Alternatively we could use Area as an explanatory variable in the model. Thirdly, we could use a model in which the size of the check (labelled as 'Area') is used as a so-called offset variable. This means that it has no regression parameter. It can be used in a GLM model with a Poisson distribution, as follows:

$$\text{bird number} = e^{\log(Area)+Depth+Depth^2+Sptreat+....}$$

The term log(Area) is fitted as an offset variable, and all other explanatory variables are treated as usual. To do this, however, we would also need to account for the nested structure of the data, which requires generalised linear mixed modelling techniques that are outside the scope of this book.

Acknowledgement

Data collection for this chapter was made possible by the generous help of many California rice growers, and financial support from the Nature Conservancy, Bureau of Reclamation, Central Valley Habitat Joint Venture, Ducks Unlimited, Inc., and the Institute for Wetland and Waterfowl Research.

24 Classification trees and radar detection of birds for North Sea wind farms

Meesters, H.W.G., Krijgsveld, K.L., Zuur, A.F., Ieno, E.N. and Smith, G.M.

24.1 Introduction

In Chapter 9, we introduced univariate regression trees and briefly discussed classification trees. In this chapter we expand on Chapter 9 and provide a detailed explanation of classification trees applied to using radar records to identify bird movements important in choosing sites for offshore wind farms. Tree model software tends to produce large amounts of numerical output, which can be difficult to interpret, and this chapter provides a detailed discussion on interpreting this output.

To increase the supply of renewable energy in the Netherlands, the Dutch government is supporting the construction of a 36-turbine Near Shore Wind farm (NSW) located 10–15 km off the coast of Egmond in the Netherlands (Figure 24.1). This project serves as a pilot study to build knowledge and experience of the construction and exploitation of large-scale offshore wind farms. An extensive monitoring and evaluation programme has been designed to gather information on economic, technical, ecological and social effects of the NSW. This evaluation programme will give information for future offshore wind farm projects as well as an assessment of the current NSW.

Derived from land-based studies, the ecological monitoring programme requires an analysis of three types of possible effects of wind farms on birds: the collisions of birds with turbines, the disturbance of flight paths and possible barrier effects and the disturbance of resting and feeding birds. The project discussed here focused on flight paths, mass movements and flight altitudes of flying birds. It was carried out by Bureau Waardenburg and Alterra Texel and was commissioned by the Dutch National Institute for Coastal and Marine Management. The full results of the study are published in Krijgsveld et al. (2005).

To assess the risks from collision and disturbance of flight paths on birds, it is necessary to identify and quantify the flight patterns of birds in the area, prior to building of the wind farm. Flight patterns were quantified using a combination of automated and field observation techniques. All birds flying through the study area were recorded by means of an automated system using two radars that processed and stored signals in two databases.

One radar rotated horizontally and recorded the direction and speed of all birds flying through the study area. Another radar rotated vertically and recorded the movement and altitude of birds flying through an imaginary line suspended vertically above the radar. This automated system was operated continuously and collected flight data every day of the year, both day and night. In addition, field observations were used to obtain detailed information on bird species composition and behaviour to validate the automated measurements. These field observations were made from an observation platform at sea, Meetpost Noordwijk (Figure 24.1), and comprised observations of mass movements, flight paths and flight altitudes of birds during the day and to a lesser extent during the night. The two radars operated from the same location and were equipped with software designed to distinguish bird echoes from other types of echoes, such as ships and clutter (echoes resulting from radar energy reflected by waves, clouds or other atmospheric conditions), but this was ineffective as a lot of the clutter was still recorded.

It is obviously important to be able to distinguish bird echoes from non-bird echoes if we hope to get an accurate picture of bird movements at sea. We therefore investigated how the characteristics recorded for each echo could be used to identify different classes of birds, or at least help to distinguish echoes from birds from echoes of clutter and ships. In this chapter we show how classification trees were used to classify the echoes from the horizontal radar.

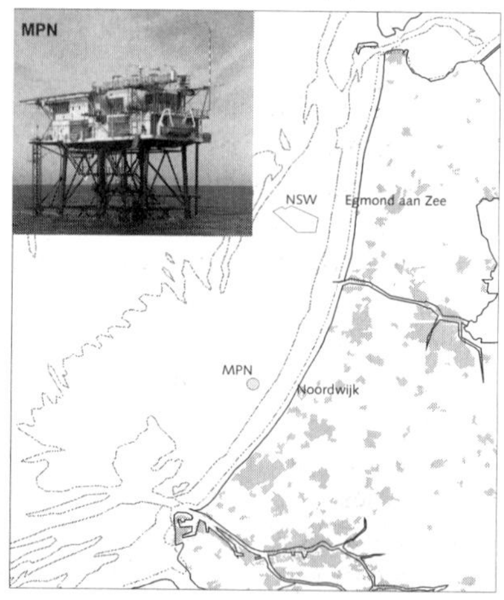

Figure 24.1. Location of planned Near Shore Wind park (NSW) and the observation platform, Meetpost Noordwijk (MPN), off the west coast of the Netherlands. Top left is a picture of the observation platform.

24.2 From radars to data

Observation platform 'Meetpost Noordwijk' is situated at approximately the same distance from the coast as the NSW area (Figure 24.1). It has three decks, the highest of which is 19 m above sea level, giving a good height to observe birds. Two types of radars were used. One, for the observation of flight paths, was a 30-kW Furuno S-band horizontal surveillance radar. This is a standard radar, as used on ships, and scans the area in the horizontal plane around the radar. With this radar, the flight paths of birds flying through the radar beam were tracked and flight speeds (ground speed) and directions were recorded. The second type of radar was a 25-kW Furuno X-band surveillance radar tilted vertically, which was used for the observation of mass movements of birds and flight altitudes. Both radar systems with associated software were installed on the observation platform to continuously measure all bird movements in an area up to 11 km from the platform in every horizontal direction and up to 3 km above the platform. These radars scanned the area above the sea continuously throughout the year, both day and night, and automatically recorded the exact location, direction, speed, and altitude of all flying objects within the scanned area. These data provided the principle dataset on flight patterns, and far more extensive than the field observations because it is continuous, it can record at night and can still record bird movements in poor weather conditions.

In this chapter we only discuss the analyses of data from the horizontal radar. Echoes originate not only from birds, but also from ships, planes, helicopters, or can just be reflections (clutter) from waves, termed wave/sea clutter, or from atmospheric phenomena, called air clutter. The radar equipment records and stores all echoes that it detects unless filtered out by the software as belonging to objects other than birds. Although this should only leave birds in the database, a large percentage of stored echoes still belong to other objects, mainly waves. To separate bird echoes from echoes of other objects (ships, clutter etc), the echoes need to have key identifying characteristics that can be assigned to a certain group, *a posteriori*. To help identify these key characteristics, a fairly large number of characteristics were recorded for each echo (Table 24.1).

Determining the characteristics of radar echoes from different objects requires a training set, a dataset of stored echoes from known objects and these were assembled during fieldwork at Meetpost Noordwijk. During training sessions, field observers and radar operators were in radio contact, so that radar detected objects could be located by the observers and the relevant object visually identified. The resulting set of verified radar echoes were analysed by classification trees to obtain relationships by which *a posteriori* radar images could be identified as belonging to certain groups. The software was set up to follow the flight path of each bird (object) or group of birds and record this as one track. If a bird was lost during several rotations of the radar, for example, due to a sudden movement of the bird, or because it landed on the water, the bird may have been recorded in a second track as well. We assumed that this would generally be the exception in our data and that records (i.e., tracks) were uncorrelated and were from different birds.

The original data consisted of echoes with their associated characteristics as stored by the radar software, plus information from the observer who verified the type of object associated with that the echo. This information was divided into different categories:

- General information (e.g., date, time).

- Information recorded by the observer at the same moment as the echo was made (e.g., species, flock size, flight altitude).
- Echo appearance information (e.g., echo dimensions, reflectivity).
- Echo position (e.g., *x* and *y* coordinates in the radar plane, distance from radar).
- Echo movement (e.g., speed).

Some of these variables represent the same ecological information and had high (>0.8) correlations. Using common sense and statistical tools like correlations and principal component analysis, we condensed the number of original variables into a subset of important characteristics (Table 24.1).

Table 24.1. Variables and abbreviations used in the analyses.

Variable	Description
EPT	Echoes per track.
TKQ	Track quality defined as STT/EPT. STT is the sum of all the TKT within 1 Track..
TKT	Track type, measure for the consistency with which the track was seen by the radar.
AVV	Average velocity of object based on all echoes of one track.
VEL	Velocity of object.
MXA	Maximum echo area (in pixels) of all echoes belonging to one track.
AREA	Area of the target in pixels.
MAXREF	Maximum reflectivity of all echoes in a track.
TRKDIS	Distance covered by the whole track.
MAXSEG	Longest length across the target.
ORIENT	The angle of the longest axis of a target with respect to the horizontal axis. This value is between 0 and 180 degrees.
ELLRATIO	Ratio of Ellipse Major to Ellipse Minor. Ellipse Major/Minor is the length of the major/minor axis of an ellipse that has the same area and perimeter as the target.
ELONG	A measure of the elongation of a target, the higher the value the more elongated the target.
COMPACT	Compactness, defined as the ratio of the target's area to the area of the smallest rectangle.
CHY	The mean length, in pixels, of the vertical segments of a target.
MAXREF	Maximum reflectivity over the entire target area.
MINREF	Minimum reflectivity over the entire target area.
SDREF	Standard deviation in reflectivity over the entire target area.

24.3 Classification trees

Univariate regression and classification trees (Chapter 9, Breiman et al. 1983; Therneau 1983; De'ath and Fabricus 2000) can be used to model the relationship between one response variable and multiple explanatory variables. The tree is constructed by repeatedly splitting the data using a rule based on a single explanatory variable. At each split the data are partitioned into two mutually exclusive groups, each of which is as homogeneous as possible. Splitting is continued until an overly large tree is grown, which is then pruned back to the desired size. This is equivalent to the model selection procedure in linear regression (Chapter 5). Tree models can deal better with non-linearity and interaction between explanatory variables than regression, GLM, GAM or discriminant analysis. Classification trees are used for the analysis of a nominal response variable with two or more classes, and regression trees for a non-nominal/numeric response variable. With classification trees, a transformation of the explanatory variables does not affect the results.

The classification tree tries to assign each observation to one of the predefined groups based on a specific value from one of the variables, thereby maximizing the variation between the groups while minimizing the variation within each group.

The main problem with tree models is determining the optimal tree size: A full tree with lots of splitting rules is likely to overfit the data, but a tree of size of only two or three might give a poor fit. The process of determining the best tree size is called 'pruning the tree'. An AIC type criterium is used to determine how good or bad is the tree. Recall from Chapter 5 that the AIC was defined as

$$\text{AIC} = \text{measure of fit} + 2 \times \text{number of parameters}$$

As measure of fit we used the total sum of residual squares in linear regression and the deviance in GLM and GAM. For a tree model, we have

$$\text{RSS}_{\text{cp}} = \text{RSS} + \text{cp} \times \text{size of tree} \tag{24.1}$$

Where RSS stands for residual sum of squares and cp is a constant. The size of the tree is defined as the number of splits plus one. The RSS component in equation (24.1) is equivalent to the measure of fit in the AIC, size of the tree equivalent to the number of parameters and cp to the × 2 multiplier. In linear regression, we just change the number of parameters and select the model with the lowest AIC. Here, things are slightly more difficult as we do not know the value of cp. If we knew the cp value, then we could do the same as with the AIC and just calculate the RSS_{cp} for different sizes of the tree and choose the tree with the lowest RSS_{cp} value. However, as we do not know the cp value, we need to estimate it, together with the optimal tree size, and use a cross-validation process, which is discussed later. First, we need to discuss how to calculate the residual sum of squares.

Consequently, the next two paragraphs are slightly more technical and the rest of this section may be skipped by readers not interested in this.

Estimating RSS in equation (24.1)

It is easiest to look at 0–1 data first. A leaf represents a group of observations that are deemed similar by the tree and are plotted at the end of a branch. The RSS component, also called the deviance D, at a particular leaf j is defined as

$$D_j = -2[n_{1j} \log \mu_j + n_{0j} \log(1 - \mu_j)]$$

where n_{1j} is the number of observations in leaf j for which $y = 1$ and n_{0j} is the number of observations for which $y = 0$. The fitted value at leaf j, μ_j, is the proportion $n_{1j}/(n_{1j} + n_{0j})$. The overall deviance of a tree is the sum of the deviances over all leaves. If the response variable has more than two classes (e.g., five), the deviance at leaf j is defined as

$$D_j = -2\sum_{i=1}^{5} n_{ij} \log \mu_{ij}$$

$$D_j = -2[n_{1j} \log \mu_{1j} + n_{2j} \log \mu_2 + n_{3j} \log \mu_{3j} + n_{4j} \log \mu_{4j} + n_{5j} \log \mu_{5j}]$$

where n_{1j} is the number of observations at leaf j for which $y = 1$, and $\mu_{1j} = n_{1j}/(n_{1j} + n_{2j} + n_{3j} + n_{4j} + n_{5j})$.

Estimating cp in equation (24.1)

The parameter cp in equation (24.1) is a constant. For a given value, the optimal tree size can be determined in a similar way to choosing the optimal number of regression parameters in a regression model. Setting $cp = 0$ in equation (24.1) results in a very large tree as there is no penalty for its size. The other extreme is $cp = 1$, which will result in a tree with no leaves. To choose the optimal cp value, cross-validation can be applied. With this approach, if the data are split up into, say, 10 parts, a tree is fitted using data of 10 parts, and the tenth part is used for prediction. The underlying principle of this approach is simple; leave out a certain percentage of the data and calculate the tree. Once the tree is available, its structure is used to predict in the group that the omitted data belongs to. As we know which groups the omitted data belong, the actual and predicted values can be compared, and a measure of the error calculated: the prediction error. This process is applied for each of the $k = 10$ cross-validations, giving 10 replicate values for the prediction error. Using those 10 error values, we can calculate an average and standard deviation (an illustration of this process is given later). The entire process is then repeated for different cp values in a 'back-ward selection type' approach. Examples are provided in the next two sections.

24.4 A tree for the birds

The verified dataset consists of 659 cases divided over 16 groups (Table 24.2). In this section we try to classify 9 different groups with at least 10 observations resulting in 629 observations (this excludes cormorants, gannets, land birds, skuas, unidentified birds, ducks, and waders). The nine groups in the analysis were auks, air clutter, water clutter, divers, geese and swans, gulls, sea ducks, ships and terns.

The graphical output of the classification tree for these data is given in Figure 24.2 and Figure 24.3.

We discuss the cross-validation graph (Figure 24.2) first. This graph was obtained with a default value of *cp* = 0.001. The average and the standard deviation of the 10 cross-validations are plotted versus cp and the tree size. The complexity parameter is labelled along the lower *x*-axis, and the tree size is labelled along the upper *x*-axis. The *y*-axis gives the relative error in the predictions, obtained by cross-validation, and the vertical lines represent the variation within the cross-validations (standard deviation). This graph is used to select the most optimal cp value. A good choice of cp is the leftmost value for which the mean (dot) of the cross-validations lies below the horizontal line. This rule is called the one standard deviation rule (1-SE). The dotted line is obtained by the mean value of the errors (x-error) of the cross-validations plus the standard deviation (X-std) of the cross-validations upon convergence. Figure 24.2 indicates that a tree with one split (size of tree = 2) would be the best size of tree. Little would be gained from a tree with more splits.

The final tree, calculated with a *cp*-value of 0.0323 (as suggested by the lowest error in Figure 24.2), is presented in Figure 24.3. The tree is arranged so that the branches with the largest class go to the right. Branch length is proportional to the improvement in the fit. Below each branch the predicted class (or group) and the number of observations in each class are given. The results show that both types of clutter are completely grouped in the left branch predicting that records will be from wave or air clutter (class = 3). Note that this branch includes four ships. The other branch includes all the birds, and as class seven (gulls) had the largest number of cases, all echoes from gulls are included in this leaf. Apart from the wrongly classified non-gulls, 30 ships are also classified as gulls. The variable that can best be used for the first (and only) split is EPT, the number of echoes per track. As EPT are integers, this means that everything with a single EPT is classified as clutter and the rest is classified with the gulls.

Table 24.2. Number of echoes per group. Groups in italics were not used in the tree analysis, which aimed at separating the 9 groups with more than 10 observations per group.

Group	Number of Obs.	Group	Number of Obs.
Auks	11	*Land birds*	8
Clutter air	37	Sea ducks	18
Clutter water	52	Ship	34
Cormorants	2	*Skuas*	1
Divers	22	Terns	16
Gannets	5	*Unidentified Bird*	7
Geese and swans	27	*Unidentified duck*	5
Gulls	412	*Waders*	2

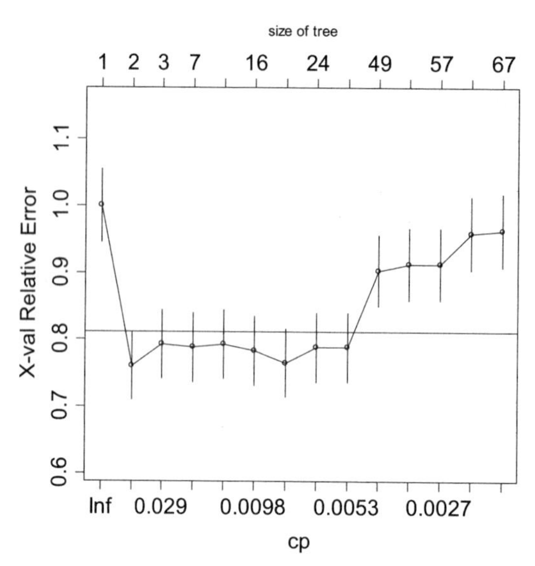

Figure 24.2. Cross-validation results for radar data separated into nine groups. Optimal tree size is 2.

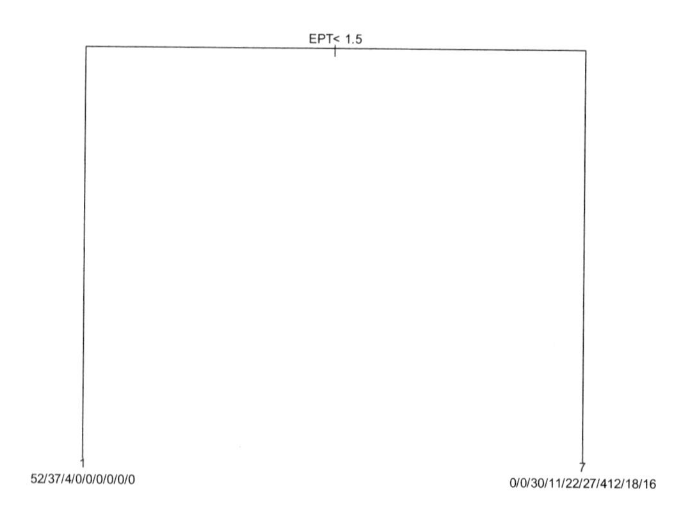

Figure 24.3. Tree for nine groups of radar data using a cp setting of 0.0323. Principal division is based on Echoes per Track (EPT). Groups are from 1 to 9: water clutter, air clutter, ships, auks, divers, geese and swans, gulls, sea ducks, and terns. Branch length is proportional to the improvement in the fit. Below each branch the predicted class is given first, followed by the number of observations in each class.

The numerical output of the tree

Readers not interested in the explanation of the numerical output might want to skip this paragraph in their first reading of this chapter. A lot of numerical output is produced by tree software, and some of the output is presented in Table 24.3. This printout shows that the cross-validation mean value at the largest tree is 1.00 (this is a percentage of the root node error), and the standard deviation is 0.055, which added together gives 1.055. Normally, this value would be taken for the 1-SE rule (representing the horizontal line in Figure 24.2); however, in this case, the tree is not decreasing regularly and the value used for the 1-SE rule is the lowest error value and its standard deviation (0.76 + 0.051 = 0.81) is already reached after the first split. The smallest tree that has a smaller mean cross-validation error (0.76) has one split and is therefore of size two with a *cp*-value of 0.0323 (Figure 24.2). The error columns have been scaled so that the first node has an error of one (multiply columns 3–5 by 217 to get a result in terms of absolute error).

The root node error for a classification tree is the classification error before any splits have been made. Because most samples in the dataset were from group seven (412 gull records), the algorithm classified the entire dataset as group seven. Therefore, samples of all other groups, 217 in total, are wrongly classified, and the root node error is 217 out of 629 (= total number of samples), which is 0.34. Using one split, which corresponds to a tree size of two, results in an error of 76% of the root node error. A tree of size five (Nsplit = 6) has an error of 63% of the root error (Table 24.3).

Table 24.3. Results of cross validation. cp, complexity parameter; nsplit, number of splits; rel error, relative error in the predictions; xerror, the mean value of the errors of the cross-validations; xstd, the standard deviation of the cross-validations. Root node error: 217/629 = 0.34.

	Cp	Nsplit	Rel-error	x-error	x-std
1	0.2396	0	1	1	0.055
2	0.0323	1	0.76	0.76	0.051
3	0.0253	2	0.73	0.78	0.051
4	0.0138	6	0.63	0.78	0.051
5	0.0104	9	0.59	0.78	0.051
6	0.0092	15	0.51	0.81	0.052
7	0.0069	21	0.46	0.82	0.052
8	0.0061	23	0.44	0.83	0.052
9	0.0046	26	0.42	0.83	0.052
10	0.0037	48	0.32	0.89	0.053
11	0.0031	53	0.3	0.94	0.054
12	0.0023	56	0.29	0.94	0.054
13	0.0018	58	0.29	0.97	0.055
14	0.001	66	0.27	1.00	0.055

The numerical output for the final tree constructed with the *cp*-value of 0.0323 (Figure 24.2) was:

```
Node number 1: 629 observations, complexity param = 0.24
  predicted class = 7  expected loss = 0.34
  class counts:   52   37   34   11   22   27  412   18   16
  probabilities: 0.083 0.059 0.054 0.017 0.035 0.043 0.655 0.029 0.025
  left son = 2 (93 obs) right son = 3 (536 obs)
  Primary splits:
    EPT    < 1.5 to the left,  improve=85, (0 missing)
    TKT    < 2.5 to the right, improve=49, (0 missing)
    MAXREF < 940 to the left,  improve=44, (1 missing)
    MXA    < 8.5 to the left,  improve=43, (0 missing)
    AVV    < 57  to the right, improve=29, (0 missing)
  Surrogate splits:
    MXA    < 8.5 to the left,  agree=0.933, adj=0.548, (0 split)
    MAXREF < 940 to the left,  agree=0.930, adj=0.527, (0 split)
    TKT    < 2.5 to the right, agree=0.908, adj=0.376, (0 split)
    TKQ    < 3.9 to the right, agree=0.879, adj=0.183, (0 split)
    AVV    < 66  to the right, agree=0.876, adj=0.161, (0 split)

Node number 2: 93 observations
  predicted class=1  expected loss=0.44
  class counts:   52   37    4    0    0    0    0    0    0
  probabilities: 0.559 0.398 0.043 0.000 0.000 0.000 0.000 0.000 0.000

Node number 3: 536 observations
  predicted class=7  expected loss=0.23
  class counts:    0    0   30   11   22   27  412   18   16

  probabilities: 0.000 0.000 0.056 0.021 0.041 0.050 0.769 0.034 0.030
n= 629
node), split, n, loss, yval, (yprob)
    * denotes terminal node

1) root 629 217 7 (0.083 0.059 0.054 0.017 0.035 0.043 0.66 0.029 0.025)
  2) EPT< 1.5 93  41 1 (0.56 0.4 0.043 0 0 0 0 0 0) *
  3) EPT>=1.5 536 124 7 (0 0 0.056 0.021 0.041 0.05 0.77 0.034 0.03) *
```

The first node is the root of the tree, representing the undivided data, and has 629 observations. The complexity parameter indicates that any *cp*-value between 0.0323 (as suggested in Figure 24.2) and 0.24 would have given the same result. The predicted class for the first split is based entirely on the number of observations in each class. As class seven (gulls) has the highest number of observations the predicted outcome for the first split is class seven. As there are 217 observations in other classes the expected loss (samples incorrectly classified) is 217/629 = 0.34. After this the number of observations in each class together with the relative probabilities for each class are given, followed by the number of observations in the left and right part of the split. Then, the variables available for each split are given. These are ordered by the degree of improvement, with the variable that results in the highest classification score presented first. The actual values of the improvement are not so important, but their relative size gives an indication of the comparative importance of the variables. Clearly, using variables other than EPT results in a serious loss of fit. The primary variables are followed by surrogate variables. These are variables that can be used instead of the first primary variable for cases where a value for the primary variable is missing. The percentage of agreement with the classification of the primary split (for both directions of the split) is given under "agree" (0.933 for MXA). Next, 'adj' gives the adjusted concordance for surrogate splits with the primary split, meaning how much is gained beyond 'go with the majority; rule (calculated as adj = (agree − 536/629)/(1 − 536/629)).

At the end of the detailed numerical output, a summary of the tree analysis is given based on the best variable for each split. The root is always node number one, and the following nodes are defined as twice the 'previous-node-number' for the left split, and twice the 'previous-node-number'+1 for the right split. The splitting rule is given ('split'), the number of cases (n), the number that does not follow the rule and thus are incorrectly classified (loss), the predicted class (yval), and the numbers in each class as a fraction of the total (yprob).

Summary

Classification tree analysis was used to find out how to distinguish among nine different groups of echoes originating from air clutter, wave clutter, ships, auks, divers, geese and swans, gulls, sea ducks, and terns. The tree cannot discriminate among the different groups of birds and wrongly classifies 88% of the ships as birds. However, it classifies 100% of the clutter and 100% of the birds correctly. Because the groups are rather diverse and the analysis indicated that we could not reliably classify all nine groups, we re-applied the tree analysis on the same data but this time grouping them into four group: Birds, ships, air clutter and wave clutter. This is discussed in the next section.

24.5 A tree for birds, clutter and more clutter

For this analysis we grouped all birds into a single group. This gave four groups for the tree to classify: air clutter in air, wave clutter ships and birds. The resulting classification tree is presented in Figure 24.4. The cross-validation plot (Figure 24.5) indicates that a tree with two branches is the best, but this still classifies ships and birds as belonging to the same group. To see whether there was a variable that could separate ships from birds, we used a *cp* value for the tree that was smaller than suggested by the *cp*-plot in Figure 24.5. The resulting tree (Figure 24.4) indicates that by using an extra variable, namely AREA, 41% of the ships can be correctly identified at the cost of misclassifying 2 of the 506 birds. In total, only 3.5% of the data were wrongly classified (taking the two types of clutter together). The tree also indicates that most birds can be discriminated from ships by using AREA (area of the target in pixels). Evidently, ships have quite a different size to the average bird, and this is reflected in the number of pixels on the radar screen. Birds that overlap in area with ships may be (partly) separated by TKQ, track quality, which is generally smaller for birds.

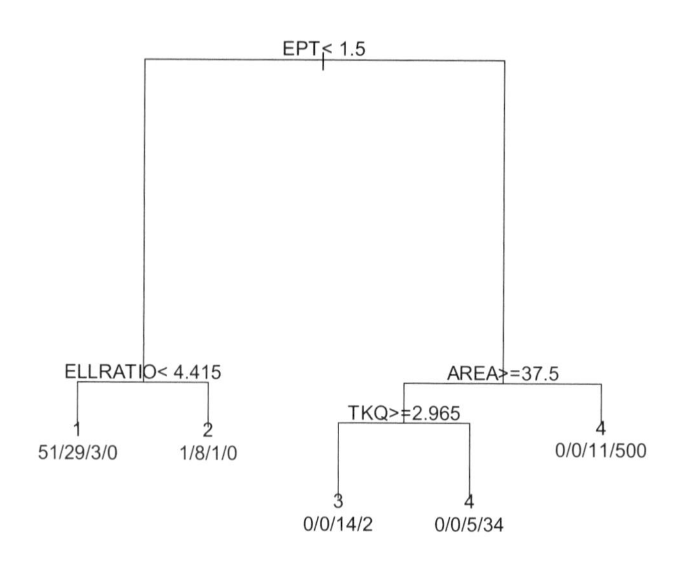

Figure 24.4. Tree for horizontal radar data using four groups and a cp setting of 0.035. Groups are from 1 to 4: air clutter, water clutter, ships, and birds. Branch length is proportional to the improvement in the fit. Below each branch first the predicted class is given, followed by the number of observations in each class.

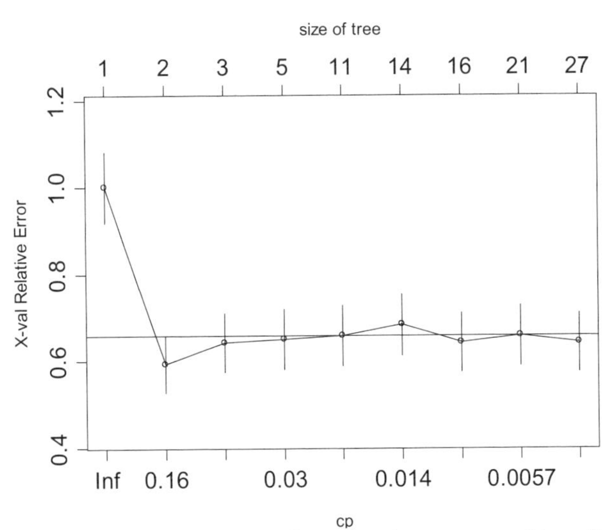

Figure 24.5. Cross-validation results for radar data separated into four groups. Optimal tree size is 2.

A summary of the tree analysis of horizontal radar data using a *cp* value of 0.035 is given below. A '*' denotes a terminal node. The classes in yprob are air clutter, wave clutter, ships and birds.

node)	split	n	loss	yval	(yprob)
1)	root	659	123	4	(0.078, 0.056, 0.051, 0.813)
2)	EPT< 1.5	93	41	1	(0.559, 0.397, 0.043, 0.000)
4)	ELLRATIO< 4.42	83	32	1	(0.614, 0.349, 0.036, 0.000) *
5)	ELLRATIO>=4.42	10	2	2	(0.100, 0.800, 0.100, 0.000)*
3)	EPT>=1.5	566	30	4	(0.000, 0.000, 0.053, 0.947)
6)	AREA>=37.5	55	19	4	(0.000, 0.000, 0.345, 0.654)
12)	TKQ>=2.965	16	2	3	(0.000, 0.000, 0.875, 0.125)*
13)	TKQ< 2.965	39	5	4	(0.000, 0.000, 0.128, 0.871)*
7)	AREA< 37.5	511	11	4	(0.000, 0.000, 0.021, 0.978)*

24.6 Discussion and conclusions

Trees can be used for both description and prediction. Statistical analysis of data is often separated into two phases: exploration and modelling, with the former preceding the latter. Trees can be used for both as shown in this chapter. As an exploratory tool, trees create structure in the data and lead to model building. As a modelling tool, trees may generate models that represent the systematic structure of the data as simply as possible and they can also be used as a prediction tool by accurately predicting unobserved data.

We have shown how classification trees can be used to split echo images from radar at a platform in the North Sea into different classes. Aiming high we first tried to separate the data into nine different classes, which included clutter from waves, clutter from air disturbances, ships and six classes of birds. Starting with 17 different radar variables, the first analysis generated an optimal solution with only one variable: the number of echoes per track. For the sampled data the tree accurately distinguished the two clutter groups from the rest, but was unable to separate the different bird classes from ships nor to distinguish between the two forms of clutter. This led us to try a different approach of lumping all the bird groups into one class called 'birds'. This new analysis again suggested a tree with only one split, but increasing the tree size to one with four splits allowed us to separate a large proportion of the ships from the birds. This new tree correctly classified 96.5% of the data (with wave clutter and air clutter put into one class because our goal was mainly to get rid of the clutter regardless of its origin). Birds (and ships) were mostly followed for several radar rotations and generally had more echoes per track than clutter. This led to the number of echoes per track (EPT) being the main variable in the trees (also indicated by the greater length of the branches following the EPT split, (Figure 24.3 and Figure 24.4). However, there appeared to be a substantial amount of overlap between the different variables, which made it impossible to separate all ships from birds. However, the tree (Figure 24.4) indicates that area can sometimes be used to differentiate ships from birds, but with 36 birds with areas comparable with ships. It is difficult to reconcile how birds can be as big as ships, but looking closely at the data we found that of these 36, 17 consisted of more than one individual (2 to 750 birds), indicating a possible effect of flock size. In addition, AREA is measured as pixels on the radar screen, and depending on the detection distance and detection limit of the radar, an echo of a bird or a ship may be indicated with an identical number of pixels. Using track quality brought about some further improvements, but there remained a small overlap between birds and ships. Also in the branch with smaller areas, a number of ships can be found. This is also clear from dotplots (not shown here).

In conclusion tree analysis proved itself a useful tool to separate clutter from the rest of the data, but it did not succeed fully in separating birds from ships.

Acknowledgement

We would like to thank S. Dirksen, R. Lensink, M.J.M. Poot, and P. Wiersma of Bureau Waardenburg, and H. Schekkerman of Alterra for their help at various stages of the research. The radars and accompanied software were supplied by DeTect Inc. (Florida, USA). We would also like to thank Neil Cambell for comments on an earlier draft.

25 Fish stock identification through neural network analysis of parasite fauna

Campbell, N., MacKenzie, K., Zuur, A.F., Ieno, E.N. and Smith, G.M.

25.1 Introduction

The main aim of fisheries science is to interpret relevant information on the biology of the species in question, records of fishing effort and size of catches, in order to predict the future size of the population under different fishing regimes, allowing fishery managers to make decisions on future fishing efforts. The common approaches to evaluation, modelling and management of fish stocks assume discrete populations for which birth and death are the significant factors in determining population size, and immigration and emigration are not (Haddon 2001). Consequently, for successful management of fisheries, it is vital that populations that conform to these assumptions are identified. In areas where two stocks mix, it is useful to be able to quantify this so that catches can be assigned to spawning populations in the correct proportions.

Several techniques have been used to identify discrete fish stocks and quantify their mixing, such as physical tags, microchemistry of hard parts and a range of genetic markers. See Cadrin et al. (2005) for a comprehensive review. There is no single "correct" approach to stock identification, the trend being towards multidisciplinary studies that apply a range of methods to the same set of fish to allow cross-validation of findings. One of the more popular methods involves the use of parasites as biological tags, and has been used for over 60 years (Herrington et al. 1939). This technique has several advantages over other methods, such as low cost, suitability for delicate species and straightforward sampling procedures. Its main disadvantage is the limited knowledge available on the life cycles and ecology of many marine parasites, but as research in these areas results in more and more information becoming available, the efficiency of the method increases accordingly (MacKenzie 1983; Lester 1990).

The theory behind this technique is that geographical variations in the conditions that a parasite needs to successfully complete its life cycle occur between areas and so between fish stocks (factors such as distribution of obligatory hosts in the life cycle, environmental conditions or host feeding behaviour). This leads to differences in parasite prevalence (the proportion of a host population infected with a particular parasite species), abundance (the average number of a particular parasite species found per host) or intensity (the average number of parasites found in infected individuals) between areas.

A "classical" parasites-as-tags study involves carrying out a preliminary study to identify parasite species that vary in prevalence, abundance or intensity within the study area, followed by the collection of data from a larger number of fish over several years to produce conclusive evidence of a lack of mixing between different parts within the study area. Note that the absence of a difference in parasite prevalence, abundance or intensity between samples is not necessarily indicative of homogeneous mixing. This approach allows migrations, recruitment of juveniles or mixing of different spawning populations to be observed and quantified. The practical application of this method was modelled and verified by Mosquera et al. (2000). It is particularly useful in areas where a small but significant degree of mixing between two populations occurs, obscuring genetic differences between populations.

Several more recent studies have taken a more complex approach to the statistical treatment of parasites as tags of their host populations. These studies have considered each fish as a habitat and treated the entire parasite fauna in that individual as a community. Discriminant analysis (DA) is applied to the parasite abundance data of the community, in order to identify groups of similar fishes (Lester et al. 1985). Moore et al. (2003) had some success with the application of DA to the parasite fauna of narrow-barred Spanish mackerel (*Scomberomorus commersoni*) around the coast of Australia, in order to quantify movement and mixing of stocks.

Discriminant analysis, however, makes several assumptions that make it less suitable for this sort of analysis. First, for testing of hypotheses, DA assumes normality and homogeneity within each group of observations per variable (in this case, parasite abundances). Second, DA works best with roughly equal sample sizes and requires the number of variables to be less than the smallest sample size minus two.

25.2 Horse mackerel in the northeast Atlantic

The Atlantic horse mackerel (*Trachurus trachurus*) is a small pelagic species of fish, with a maximum size of about 40 cm. They are the most northerly distributed species of the jack-mackerels (family Carangidae) (FAO 2000), and they support a sizeable fishery in the northeast Atlantic, both for human consumption and for industrial processing. Catches in the region have been over 500,000 tonnes per year in recent times. They feed at a slightly higher trophic level than many small pelagic fishes, their diet consisting of planktonic copepods, small fishes and benthic invertebrates. This diverse diet is reflected in a diverse parasite community, and 68 taxa have been reported to infect *T. trachurus* (MacKenzie et al. 2004).

There has been uncertainty over the identity of stocks in the northeast Atlantic for over a decade. In this area, the International Council for the Exploration of the Seas (ICES) issues management advice for the horse mackerel, which assumes the existence of three stocks. These are a (i) Western stock, (ii) a North Sea stock and (iii) a Southern stock (Figure 25.1).

These stocks have been defined mainly from observations of the distribution of eggs in regular plankton surveys, and on historical records of the distribution of catches (Eltink 1992; ICES 1992). Until recently, publications dealing with the definition of stock structure in horse mackerel were rare and covered only a small part of the species distribution. In the southern stock there are some works dealing with differences in anisakid infestation levels (Abaunza et al. 1995; Murta et al. 1995); whilst using allozymes, some authors found differences between areas in the northeast Atlantic (Nefedov et al. 1978); whereas others did not (Borges et al. 1993). A recent, EU-funded multidisciplinary study, HOMSIR, has resolved some of the problems with stock identity, but there are still unanswered questions.

One of the problems with stock definition for *T. trachurus* is that it is a highly migratory species, spawning along the edge of the continental shelf, in water around 200 m deep, then dispersing to feed over a wider area. It is thought that the Western and North Sea stocks overlap at certain seasons in the English Channel (Macer 1977), which may cause some degree of mixing between these stocks. Mixing between the Western and Southern stocks remains an unknown quantity, and there has been particular concern about the boundary between these stocks.

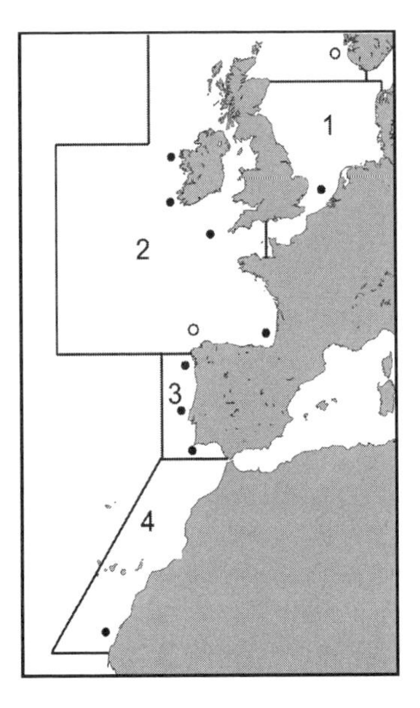

Figure 25.1. A graphical representation of the distribution of horse mackerel stock in the northeast Atlantic. 1. North Sea stock. 2. Western stock. 3. Southern stock. 4. African stocks, after ICES (1992, 2004). Locations of samples collected for verification of stock identity are marked with filled circles and those collected to investigate stock mixing with empty circles.

25.3 Neural networks

The original 'neural network' model was proposed in the 1940s (McCulloch and Pitts 1943), although it is only with the advent of cheap, powerful computers that this technique has begun to be applied widely. There has been a great deal of hype surrounding neural networks. They are, however, simply a non-linear statistical approach to classification. One of the attractions of neural networks to their users is that they promise to avoid the need to learn other more complex methods — in short, they promise to avoid the need for statistics! But this is a misconception: For example, extreme outliers should be removed, collinearity of variables should be investigated before training neural networks, and it would be foolish to ignore obvious features of the data distributions and summaries such as the mean or standard error. The neural network promise of 'easy statistics' is, however, partly true. Neural networks do not have implicit assumptions of linearity, normality or homogeneity, as many statistical methods do, and the sigmoid functions that they contain appear to be much more resistant to the effects of extreme values than regression based methods. Many of the claims made about neural networks are exaggerated, but they are proving to be a useful tool and have solved many problems where other methods have failed.

The name 'neural network' derives from the fact that it was initially conceived of as a model of the functioning of neurons in the brain — the components of the network represent nerve cells and the connections between them, synapses, with the output of the nerve switching from 0 to 1 when the synapses linking to it reach a 'threshold value'.

For the purposes of this chapter, a neural network can be thought of as a classification model of a real world system, which is constructed from the processing units ('neurons') and fitted by training a set of parameters, or weights, which describe a model that forms a mapping from a set of given values known as inputs to an associated set of values known as outputs (Saila 2005). The weights are trained by passing sets of input–output pairs through the model and adjusting the weights to minimize the error between the answer provided by the network and the true answer. A problem can occur if the number of training iterations, or 'epochs' is too large. This reduction in classification success of the data not used in training is known as over-fitting. Once the weights have been set by a suitable training procedure, the model is able to provide output predictions for inputs not included in the training set. The neural network takes all the input variables presented in the data and linearly combines them into a derived value, in a so-called 'hidden layer object' or node (Smith 1993). It then performs a nonlinear transformation of this derived value (Figure 25.2). The use of multiple hidden layer objects in a neural network allows different non-linear transforms of data, with each neuron (node) having its own linear combination, increasing the classifying power of the network.

Originally, the neuron was activated with a step function (represented as the dashed line in Figure 25.2), when the combined input values exceeded a certain value; however, this more flexible sigmoid function allows differentiation and least squares fitting, leading to the back propagation algorithm, making it possible to tune the weights more finely.

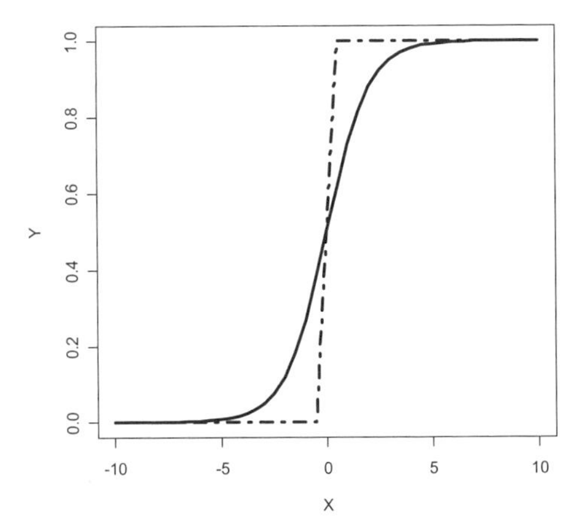

Figure 25.2. An example of the transformation that the hidden layer applies to linear combinations of the input data: $y = 1/(1 + \exp(-x))$. An example of the step function used in early neural network models is shown in the dashed line.

A possible example is shown in Figure 25.3. In this case, we are interested in knowing the stock composition of a mixed sample of fish, and we have count data on six species of parasites from these fish. These counts are treated as the six input variables to our network. The network has four units in the hidden layer. The neural network in such an example would need to have been trained with data from fish that we knew belonged to stocks X and Y beforehand if it was to work successfully. This type of neural network is referred to in the literature by many names, such as feed-forward network, multilayer perceptron, or simply vanilla neural network, named for the generic ice-cream flavour.

The number of units in a hidden layer is variable. Problems can arise if too few or too many units are used. If the network has too few units, it will not be flexible enough to correctly classify the data with which it is presented. On the other hand, if it has too many units, a problem known as over-parameterisation occurs; this reduces the chances of successful classification. Having many hidden layer objects also increases the computing power required to run the function. Often, a trial-and-error approach is used to determine the optimum number of hidden layer units for a particular dataset; however there are other methods that take a more considered approach, such as cross-validation, bootstrapping, and early stop.

Neural networks differ in philosophy from most statistical methods in several ways. A network typically has many more inputs than a typical regression model. Because these are so numerous, and because so many combinations of parameters result in similar predictions, the parameters can quickly become difficult to interpret and the network is most simply considered as a classifying 'black box'. This means that areas where a neural network approach can be applied in ecology are widespread. They are less useful when used to investigate or explain the physical process that generated the data in the first place. In general, the difficulty in interpreting what the functions contained within these networks mean has limited their usefulness in fields like medicine, where the interpretation of the model is vital.

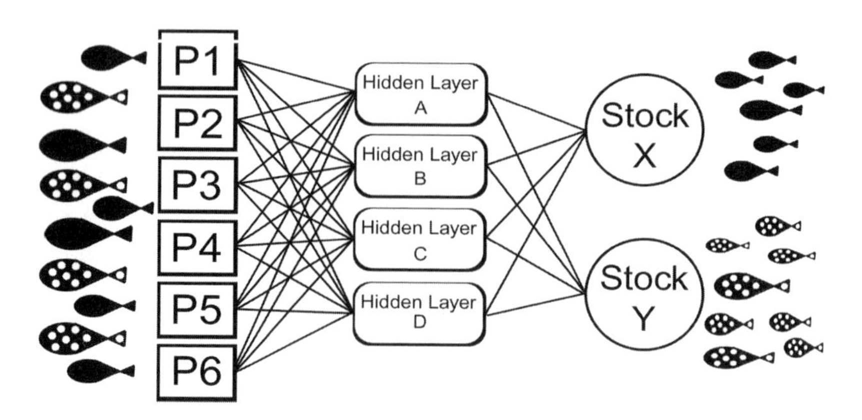

Figure 25.3. Graphical representation of the structure of a neural network. This one has six inputs (P1 to P6), one hidden layer with four neurons or units (A to D) and two output neurons into which it will classify the data (stocks *X* and *Y*).

There have been several uses of neural networks in fisheries science (Huse and Grøjsæter 1999; Maravelias et al. 2003; Engelhaard and Heino 2004), however, these have mainly attempted to predict changes in fish abundance, recruitment or distribution, based on environmental and ecological inputs. The use of neural networks in recognising fish stocks is a relatively new development, and it has been restricted to analyses of morphometry or of otolith microchemistry (Murta 2000; Hanson et al. 2004). These were reviewed and summarised by Saila (2005). A more specific introduction to neural network architectures can be found in Dayhoff (1990) and Smith (1993).

These techniques produce continuously distributed values, such as the distance between two points on the body of a fish, or the quantity of a particular element in an otolith. Parasitological studies, on the other hand, are characterised by relatively large number of fish with a low number of parasites, and a small percentage of observations with high numbers. This tends to give problems with classical statistical techniques that require normality assumptions, such as classical discriminant analysis.

Furthermore, some forms of parasite, such as metazoan species, are too numerous to count, therefore a fish is either classed as infected or uninfected. Neural networks are able to cope with both forms of numeric data, as well as with presence–absence data, in any combination.

Note that the question that is addressed by a neural network approach to parasite data is not 'do these fish belong to different stocks', but, 'based on the parasitological data available, is it possible to successfully assign these fish to a stock'? The difference is subtle, but it should be apparent that low levels of successful classification between two sets of observations would not suggest that they belong to two different stocks, whereas high success would support a hypothesis that the samples were drawn from different populations.

25.4 Collection of data

This work is based on samples of *T. trachurus* collected as part of the HOMSIR stock identification project (see Abaunza et al. In press) for a detailed explanation of the theory behind sample collection). Fish were collected with a pelagic trawl by several research vessels at 11 locations in the northeast Atlantic (see Figure 25.1) in 2001 and were immediately frozen and returned to the laboratory for examination. Between 34 and 100 fish were collected from each location (Table 25.1).

To investigate spawning stock identity, three samples each from the Western and Southern stocks, and one each from the North Sea and African stocks, were examined. For estimation of stock mixing, one sample from a non-spawning seasonal fishery from the Norwegian coast and one spawning sample from the boundary between the Western and Southern samples were examined.

Fish were examined externally for parasites, before opening the visceral cavity. All organs were separated, irrigated with physiological saline and examined for the presence of parasites under a stereo-microscope (6–50×). The opercula (gill covers) were removed along with the individual gill arches, irrigated with physiological saline and examined for the presence of monogenean and copepod parasites under a stereo-microscope. Smears of liver and gall bladder were examined for protozoan and myxozoan infections at a magnification of 325× using phase contrast microscopy.

Table 25.1. Location and size of samples collected for stock identification.

Stock	Lat.	Long.	Sample Size
North Sea	54.45N	06.00E	50
Western	52.53N	12.03W	34
	48.45N	09.29W	50
	51.35N	11.06W	50
	44.00N	01.38W	50
Southern	41.00N	08.50W	100
	38.30N	09.20W	52
	37.00N	08.30W	100
African	19.58N	17.28W	50
Mixed Norwegian	57.41N	05.10E	50
Mixed Spanish	43.35N	08.52W	50

25.5 Data exploration

A total of 636 fish from 11 sites were examined. Parasites that infected less than 2% of fish were deemed to represent either rare species or "accidental" infections and were discounted. Eleven species of parasites were found to be commonly present (Table 25.2).

Exploration of any set of data is an essential first step in carrying out an appropriate analysis. One of the problems with this sort of study is that because fish vary in size, age or sex between samples, the examinations cannot be thought of strictly as replications of observations on a homogeneous population. It is therefore important to look for variation caused by these factors and remove it from further analysis.

Plots were made of the abundance of different parasites against fish length, sex and age (Figure 25.4). No relationships were apparent, with the exception of the nematode, *Anisakis* spp. This species encysts in the body cavity of the horse mackerel and therefore is cumulative with age. A $\ln(y + 1)$ transformation was performed on the *Anisakis* abundance values, and a linear regression of the transformed value was carried out against fish length. The residual values of this regression were then taken forwards for use in the later analysis. This is one way of reducing the bias caused by differing lengths between samples.

Table 25.2. Commonly encountered parasite species used for stock identification analysis.

Class	Species	Location	Data Type
Myxosporea			
	Alataspora serenum (Gaevskaya and Kovaleva 1979)	Gall Bladder	Presence/ Absence
Apicomplexa			
	Goussia cruciata (Theolan 1892)	Liver	Presence/ Absence
Nematoda			
	Anisakis spp.	Body Cavity	Abundance
	Hysterothylacium sp. (larval forms)	Body Cavity	Abundance
	Hysterothylacium sp. (adult forms)	Intestine	Abundance
Digenea			
	Tergestia laticollis (Rudolphi 1819)	Intestine	Abundance
	Derogenes varicus (Müller 1784)	Stomach	Abundance
	Ectenurus lepidus (Loos 1909)	Stomach	Abundance
Monogenea			
	Pseudaxine trachuri (Parona and Perugia, 1889)	Gills	Abundance
	Gastrocotyle trachuri (van Beneden and Hesse 1863)	Gills	Abundance
	Heteraxinoides atlanticus (Gaevskaya and Kovaleva 1979)	Gills	Abundance

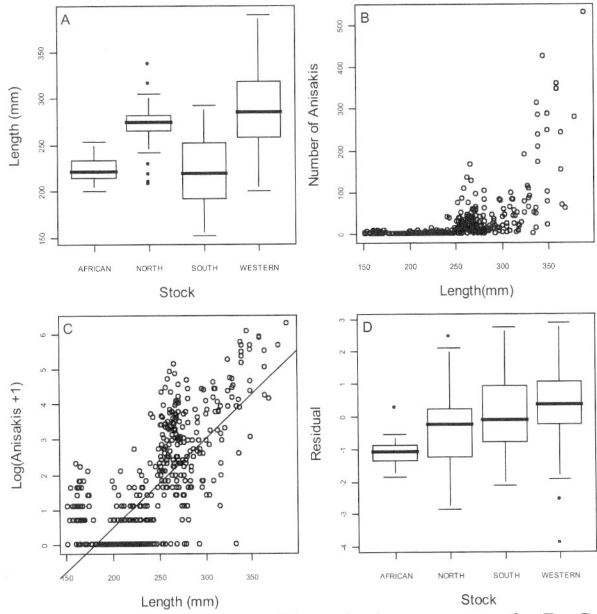

Figure 25.4. A: Fish lengths vary significantly between stock. B: Scatterplot of fish length and *Anisakis* spp. abundance. C: Natural logarithm transformation of *Anisakis* abundance produces a linear relationship. D: The residuals of this relationship are taken forward for use in the analysis after removal of length dependency. The values for the Southern stock are generally lower as the parasite occurred less frequently in this area.

25.6 Neural network results

The first problem in a neural network approach to a classification problem is to select an appropriate structure for the network. This is often done on an *ad hoc* basis. A single hidden-layer, feed-forward neural network was constructed using the *nnet* function (Venables and Ripley 2002) in the R statistical software environment v2.1.1 (R Development Core Team, 2005). This network was provided with half the fish from all samples on a random basis, for use as a training set, then used to reclassify the remaining fish to a stock. In the first instance, the hidden layer contained one object, and to remove chance results caused by selection of fish at random, the process was repeated 100 times with different random training sets. The mean percentage of successful classifications over all simulations is then taken and stored. We then increased the number of units in the hidden layer and repeated the process. Finally, mean successful classification is plotted against number of hidden layer units (Figure 25.5). This allows an educated guess to be made as to the most suitable number of hidden layer objects, which is sufficiently flexible to reclassify data successfully, but not so many that over-fitting occurs and excessive processing time is required. All other settings remained at their default values.

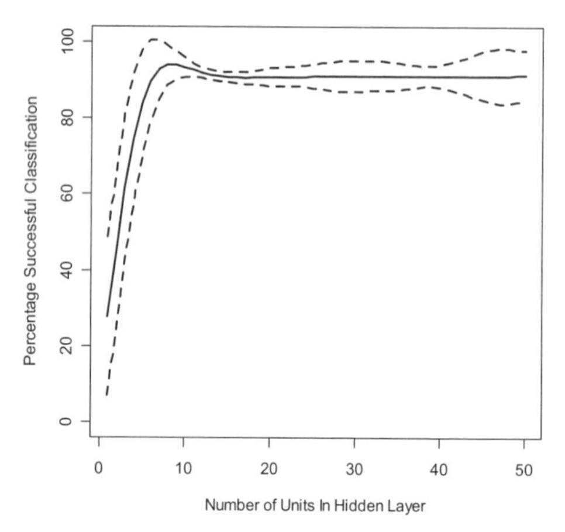

Figure 25.5. Estimation of the optimum number of units in the hidden layer. Mean successful classification (±1 standard deviation) reaches over 90% with eight nodes. Further increases in the number of nodes cause a decrease in success, and an increase in variability, at the cost of increased computing time.

From these data it would appear that the optimum number of hidden layer units is around eight. We decided to use eight hidden layer units for the later neural networks, as there was some decrease in performance of the network at higher values.

To investigate stock identity, the same method of selecting half of the fish in a sample as a training set and reclassifying remaining fish (the "test set") was used. The percentage of fish correctly classified was recorded, along with the numbers from each stock misclassified, and the stock to which they were assigned. This network had 11 inputs, 8 objects in the single hidden layer, and four outputs. To obtain a measure of the error inherent in selecting a training sample at random, and allowing the starting weights of the network to be selected at random, the process was repeated 1000 times. Once outcomes were sorted, median successful classification was represented by the 500th value, and 95% confidence limits by the 50th and 950th values (Figure 25.6).

From these results it is apparent that the neural network is able to correctly classify fish to a stock with a high degree of accuracy — median successful classification for the Southern stock is 95%. These findings support the ICES stock definitions as they are currently applied and suggest that the application of this neural network to mixed stock analysis will give an accurate picture of stock composition. For investigation of the two mixed stock samples, the whole of the spawning dataset was used to train the neural network. This was then used to reclassify the mixed data in question. The neural network was allowed to choose random starting weights; hence, outputs are still variable, even when considering the same set of data. Consequently, this process was also repeated 1000 times in order to produce an estimate of inherent variability.

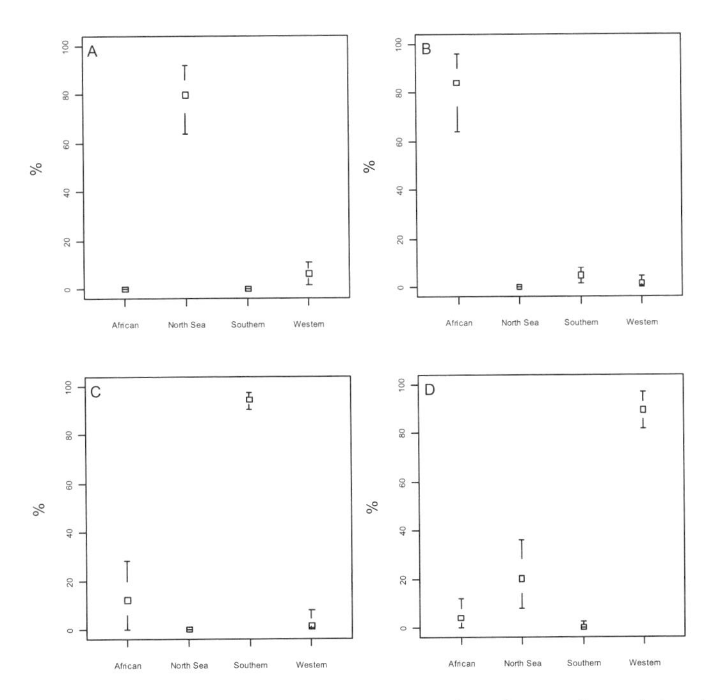

Figure 25.6. Percentage of fish from the test set assigned to each stock by the neural network; A: fish from the North Sea. B: fish from African waters. C: fish from the Southern stock. D: fish collected in the Western stock area.

The Norwegian seasonal fishery shows a more mixed composition than any of the spawning samples, suggesting it may be made up of fish from more than one area. The neural network classifies around 65% of fish as belonging to the Western stock. The remaining 35% are a mixture of Southern and African stocks. Very few fish are assigned to the North Sea stock (Figure 25.7).

The spawning sample from the area of stock uncertainty to the north of Spain is much less conclusive. The neural network assigns around 40% of the sample to the Western stock, 30% to the Southern stock and 20% to the African stock (Figure 25.8).

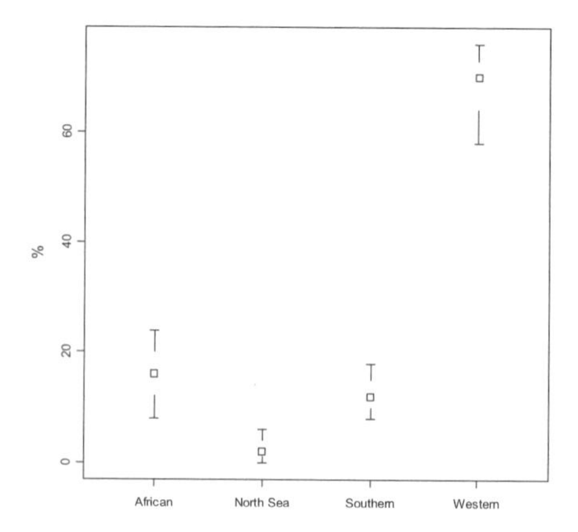

Figure 25.7. Stock membership of horse mackerel from a mixed, non-spawning seasonal fishery that develops in the summer months off the Norwegian coast.

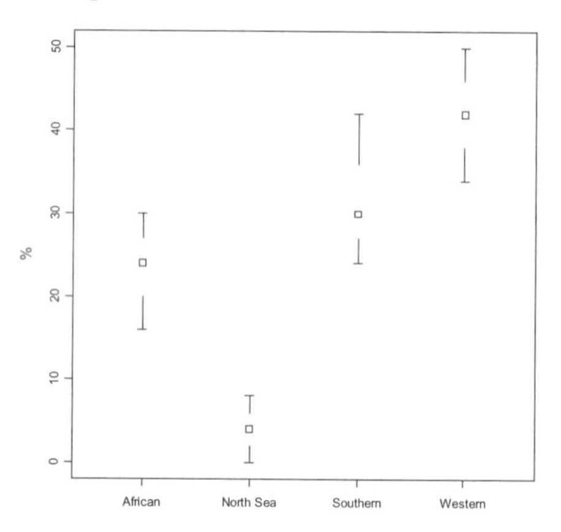

Figure 25.8. Classification of horse mackerel in spawning sample taken from area of stock uncertainty on the northwest coast of Spain.

25.7 Discussion

Stock identification

This study represents the first successful attempt to validate the existence of separate fish stocks by the application of a neural network to parasite abundance data. This technique successfully reclassifies over 90% of fish in the test set, from Western and Southern stocks. Success in reclassifying fish from the North Sea and African stocks is lower, although still over 80%.

The neural network is good at distinguishing between members of the Western and Southern stocks. Median misclassification of fish from the Western stock to the Southern stock is 2%, and from the Southern stock to the Western is 0%. Using parasites as biological tags and multivariate analysis of morphometric distances, it is possible to distinguish between fish from the Western and North Sea stocks, but it has previously been impossible to conclusively distinguish between fish from the Western and Southern areas using any method (ICES 2004).

Although neural networks are robust enough to deal with differences in sample size, it is apparent that the areas with larger sample sizes (western and southern) have higher success rates than the two areas with only 50 fish (North Sea and African stocks). The confidence ranges for these two areas are also much wider than for the larger samples.

It is interesting to note that where misclassification occurs, it tends to be towards stocks that are already believed to mix, rather than to those with which mixing is not regarded as possible. This might suggest that a small number of 'alien' fish from adjacent stocks are present in our 'discrete' spawning samples. These fish would produce such misclassifications, although the difficulties in investigating the processes that go on inside the neural network make this impossible to verify.

It has been suggested that an element of stock mixing can take place between the Western and North Sea stocks while fish over-winter in the English Channel (Macer 1977). This has been supported by recent studies (MacKenzie et al. In press). If stock mixing is occurring in this area, it is likely that this effect is being reflected in the results obtained from the neural network, and that there are a number of fish that belong to the Western stock in the North Sea, and vice-versa. This will slightly confuse the picture that the neural network gives and could explain the tendency of North Sea fish to be misclassified into the Western stock.

A degree of mixing has also been proposed between the Southern and the African stocks (Murta 2000; MacDonald 2005). This is supported by our findings. Although successful classification of fish from the Southern stock is over 95%, the neural network classifies around 20% of fish from the African sample as belonging to the Southern stock. Very few fish from the southern samples are classified as belonging to the African stock. This could suggest that mixing between these areas is a one-way process.

Norwegian non-spawning sample

Having established the utility of a neural network approach to assign fish to spawning stocks based on host-parasite data, it is a straightforward matter to apply this to a non-spawning stock to investigate its composition. These findings suggest that fish in the Norwegian sample do not come from a single stock, but rather are drawn mainly from the Western stock, with a sizeable proportion from much more southerly stocks. This finding is in line with work by Abaunza (In press) who measured growth rates of horse mackerel and found variability from stock to stock, with fishes from warmer waters growing more quickly than their more northerly conspecifics. When examining fish from the Norwegian area, growth rates were noticeably higher than that in neighbouring fisheries, to the west of Ireland and the southern North Sea. Abaunza proposed the existence of a highly migratory 'infra-stock' that spawned and over-wintered in the Southern stock area, then migrated northwards to feed in Norwegian waters. Evidence supporting this hypothesis came from the discovery of characteristically "southern" parasites in fish from the Norwegian area (MacKenzie et al. In press).

The neural network classified few fish from this sample as belonging to the North Sea stock. This is interesting when considering how close it is to this stock. It suggests that very little mixing of North Sea and other fish occurs in this area. This finding is important for fishery managers to consider when allocating catches of fish from the Norwegian fishery to particular spawning populations.

La Coruña spawning sample

The neural network was not able to classify fish from the north west coast of Spain to a particular stock with any great certainty. No one stock appears to dominate this area, and the 95% confidence limits are relatively small. The boundary between the Western and the Southern stocks was recently moved from the north to the west coast of the Iberian peninsula (ICES 2004). These findings suggest that this change was appropriate, in that the sample is classed as more "Western" than 'Southern', but also suggest that a high degree of mixing takes place in this area, and that more intensive sampling in this area would be a worthwhile contribution to stock identification of the horse mackerel.

Conclusions

It is apparent that a neural network approach to classifying individuals into presupposed groups is a powerful tool for problems such as this.

The ease with which it is possible to use neural networks, their lack of restrictive assumptions and their ability to cope with combinations of different types of data make them extremely useful for dealing with problems of classification in ecology.

Acknowledgements

We are grateful to all our partners in the HOMSIR stock identification project for their helpful advice and provision of samples. The work that this chapter was based on was funded by the European Commission, under the 5th Framework, contract no. QLRT-PL1999-01438. Visit www.homsir.com for more details on the project outcomes. We would like to thank Anatoly Saveliev for valuable comments on the statistical aspects of this chapter.

26 Monitoring for change: Using generalised least squares, non-metric multidimensional scaling, and the Mantel test on western Montana grasslands

Sikkink, P.G., Zuur, A.F., Ieno, E.N. and Smith, G.M.

26.1 Introduction

Monitoring programs are vital to assess how plant community succession is affected by environmental change. Each plant community has many biologic, climatic, and abiotic interactions that affect its species differently over time. In temperate grasslands, plant community composition and species dominance can change rapidly in response to changes in the timing and amount of precipitation (Fay et al. 2002; Knapp and Smith 2001). In fact, some of these grasslands are so sensitive to variations in precipitation that they have been dubbed "early warning systems" for global climate change (Kaiser 2001). However, these ecosystems are also sensitive to temperature fluctuations (Alward et al. 1999), the timing and intensity of grazing and fire (Fuhlendorf et al. 2001; Geiger and McPherson 2005; Jacobs and Schloeder 2002), fire exclusion (Leach and Givnish 1996), and invasion of non-native species (Abbott et al. 2000). Monitoring programs that span several decades are critical to determining which of these environmental stresses are important to compositional change within a particular grassland and how rapidly that change occurs.

Our goal in this case study chapter was to analyse monitoring data from two temperate grassland communities in Montana, USA, to determine whether their composition changed over three to five decades and, if so, whether any of these environmental factors can be related to the changes. Our two study areas are protected reserves that have been sites for long-term natural experiments since the late 1950s and 1960s. Both are dominated by native bunchgrasses, including *Pseudoroegneria spicata* (Pursh) A. Love, *Festuca idahoensis* Elmer, and *F. altaica* Trin.; and both have been affected by climatic fluctuations, grazing, and invasion of non-native species throughout their monitoring history. These particular grasslands are unique because they exist in cool, semi-arid northern landscapes that have not been highly fragmented by increases in human population, converted to agricultural fields, or manipulated in experimental studies.

Our primary questions in this case study address the following: Does the biodiversity of these bunchgrass communities change over time? If so, do changes in biodiversity relate with any particular environmental factor? Does species' membership within the bunchgrass community change significantly over 30–50 years? If so, what environmental factors promote changes in the community over time?

The underlying statistical question for these data is simple: 'Is there a relationship between species present in the community at a particular point in time and the environmental variables?' Answering this question is far from simple, however, because the number of environmental variables is relatively large, the plant data have a large percentage of observations with zero species abundance, and the data have an underlying time component. To address how the environmental variables correlate with (i) biodiversity change and (ii) community change, we adopt a multi-step strategy that ultimately provides two different perspectives on the relationship between species and environmental (i.e., explanatory) variables. First, we will reduce the large number of species into an index that represents biodiversity and relate the diversity index to the explanatory variables. Second, we will reduce the number of explanatory variables using variance inflation factor (VIF) analysis. VIF eliminates environmental variables that are collinear and leaves only the variables that contain unique information. Linear regression is then applied to the remaining explanatory variables. Within the linear regression analysis, we must address a crucial question: 'Do the residuals from this linear regression model show any temporal patterns?' If they do, we will need to address residual temporal structure by comparing the results of the linear regression model with a model created using generalised least squares (GLS). For a second perspective on how environmental variables relate to community change, we apply a multivariate analysis, which incorporates the frequency of abundance of all of the species present at each site. Instead of using a diversity index that represents a site's diversity, we compare similarities among and between communities with time based on similarity values computed for each sample period. Because this dataset has many zeros, we will use classical methods like the Jaccard or Sørensen similarity indices combined with NMDS, the Mantel test, and BVSTEP to look at the relationships between species presence and the environmental variables.

26.2 The data

The sites examined for this case study are located in the northern Rocky Mountains of Montana, USA, within Yellowstone National Park and the National Bison Range (Figure 26.1). Both areas are sites for natural experiments that have recorded the response of bunchgrass communities to fluctuating climate, disturbance, and invasion for the past 30–50 years. The communities of interest are the native *Pseudoroegneria spicata* (Pursh) A. Love, *Festuca idahoensis* Elmer, and *F. altaica* Trin. Yellowstone National Park (YNP) is located at Montana's southern border (Figure 26.1). The area is mountainous and an all-season tourist destination. Although the central and southern areas of the park are famous for geysers and volcanic hot-water features, much of the northern boundary area is dominated by native grass and shrub communities and is important winter range for several big-game species including wapiti (*Cervus canadensis*) and bison (*Bison bison*). In 1957, YNP initiated a natural experiment to determine how the grass and shrub communities changed with increasing populations of wildlife in the park (Edwards 1957). Park personnel constructed several five-acre exclosures in the northern winter range that eliminated all big-game grazing within the fenced boundaries. They established multiple permanent transects inside each exclosure. Each transect measured 33.3 m (100 ft). Just outside the exclosures, complementary transects were established that remained open to big-game grazing year-round. The transects selected for analysis in this case study lie between 1650 and 2050 m in elevation and have been resampled six to eight times in the last 50 years.

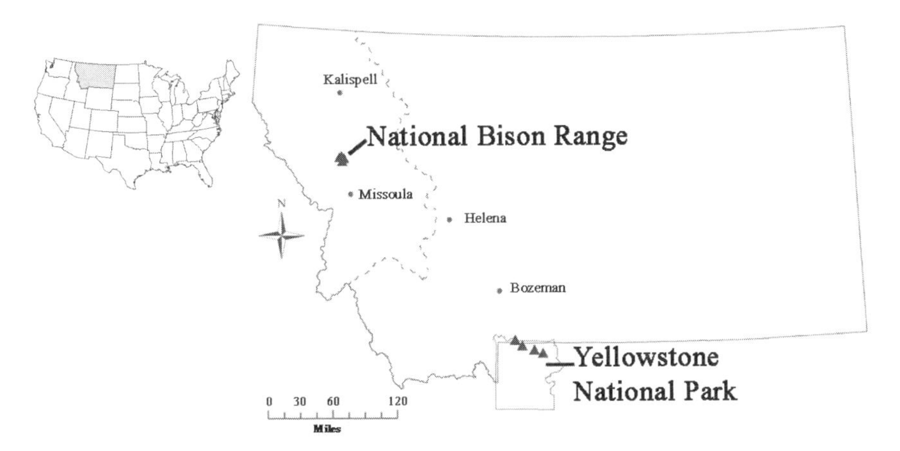

Figure 26.1. Transect locations in Montana are shown by triangles. The dashed line designates the location of the continental divide.

The National Bison Range (NBR) is located near Montana's western boundary (Figure 26.1). It covers approximately 7500 ha. Like YNP, the bunchgrass communities exist at lower elevations within inter-mountain areas. The wildlife refuge is managed to preserve remnants of the once legendary bison population in the western United States and to provide tourists with wildlife viewing and educational opportunities. All areas of the refuge are open to grazing by the same native ungulate species that live in YNP. However, unlike YNP where bison are free to wander outside park boundaries, bison in the NBR are restricted to the refuge and managed within it by several fenced pastures and a rotational grazing system. In the late 1960s, a natural experiment similar to YNP's began in the NBR to monitor range trend and condition. Initially, the experiment began as a way to measure the effects of managed grazing and monitor its trends over very short time frames, but data have continued to be collected for more than 30 years. The transect lines selected for this study have been resampled an average of 10 times since the late 1960s and range from 875 to 950 m in elevation.

Throughout the historic record, sampling intervals have been irregular in both areas (Figure 26.2), but all of the transect lines have been sampled using exactly the same methods. Along each line, vegetation or substrate encountered at each 0.33 m mark was recorded. Vegetation hits were identified to species and recorded as either overstory or understory in the canopy. Substrate hits were recorded as bare ground, rock, pavement, litter, or moss/lichen. Each line had a total of 100 hits per line, so all species and substrate data in this case study represent frequency of occurrence in each sample year. The timing for each re-sampling was matched as closely as possible to the timing of historic samplings so changes in species' frequency over the monitoring period were not confused with seasonal physiologic changes.

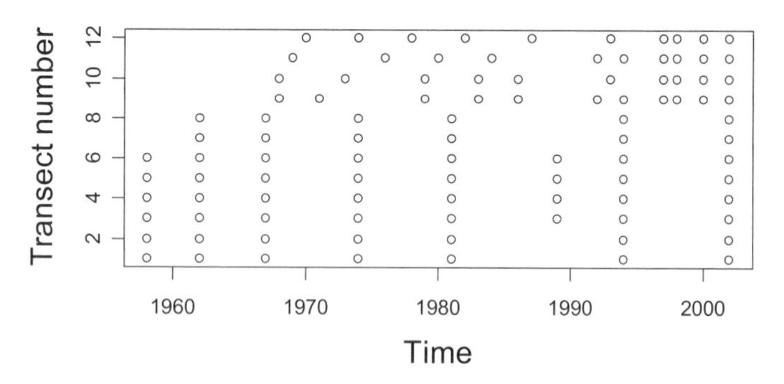

Figure 26.2. Years each transect was sampled (indicated by a circle).

Twenty-one environmental variables are used in this study. They include substrate characteristics, soil texture, and seasonal weather conditions during sampling. Substrate variables include the following surface characteristics at the time of sampling: rock that is greater than 1.9 cm (0.75 in) in diameter (ROCK); exposed soil (BARESOIL); litter (LITTER); and moss or lichen (M/L). Soil samples were analysed for the proportion of sand (PCTSAND), silt (PCTSILT), and clay (PCTCLAY) in their matrix using a combination of sieving and standard hydrometer methods (Bouyoucos 1936). Organic carbon content of soil samples (PCTOrgC) was measured using Loss-On-Ignition procedures (Nelson and Sommers 1982). Soil and organic carbon samples were only collected in 2002, not at each historic sampling.

In general, the climate of both sample areas is semi-arid. YNP has a 38 cm mean annual precipitation and a mean annual temperature of 4.4°C. The NBR is slightly warmer and moister with a 40 cm mean annual precipitation and a mean annual temperature of 7.5°C (Western Regional Climatic Center 2002). To examine how environmental variables correlated with species changes, however, temperature and precipitation values had to be specific to transect site. Specific seasonal values were assigned to each transect for each year from 1958 to 2002 using a surface observation gridding system (SOGS). In the SOGS process, each of the monitoring sites received seasonal climate values that were interpolated from the nearest climate station and adjusted for each site's unique elevation, slope, aspect and location on the landscape (Jolly et al. 2004). The seasonal divisions included fall (September–October prior to sampling year), winter (November–March), spring (April–May), and summer (June–August). The resulting environmental variables include mean minimum temperatures for each season (FallTmin, WinTmin, SprTmin, SumTmin), mean maximum temperatures for each season (FallTmax, WinTmax, SprTmax, SumTmax) and total precipitation for each season (FallPrec, WinPrec, SprPrec, and SumPrec).

For the following data exploration and analysis, transect lines are numerically coded to make interpretations easier. The numeric codes, their corresponding locations, and current management regimes are shown in Table 26.1.

Table 26.1. Transect characteristics in Yellowstone National Park (YNP) and the National Bison Range (NBR), USA.

Site	Transect Number	Location	Management
YNP	1,3,5,7	Inside exclosures	No grazing
YNP	2,4,6,8	Outside exclosures	Open grazing
NBR	9,10,11, 12	North, South, West, East boundaries, respectively	Confined bison grazing, Open wildlife grazing

26.3 Data exploration

The 12 locations from YNP and the NBR contained 93 species, 99 sets of transect data, and 21 environmental variables. As we explore these datasets to find relationships between the species data and the explanatory variables, we are faced with an immediate problem. Many of the explanatory variables are collinear. In previous chapters, we encountered similar collinearity problems, but in each case a simple pairplot gave clear indications of which variables could be dropped. In this study, there are too many explanatory variables to easily use a pairplot. When pairplots become impractical, the PCA biplot is often more appropriate to identify collinearity among variables (Figure 26.3).

Recall from Chapter 12 that lines pointing in the same direction in a PCA biplot indicate that the corresponding variables are correlated with each other, lines pointing in opposite directions are negatively correlated, and lines with an angle of 90^0 are uncorrelated. The directions of lines in Figure 26.3 indicate that many of the climate variables are highly (either positively or negatively) correlated. All seasonal temperatures are highly correlated with each other, and all seasonal precipitation is highly correlated. However, precipitation variables are uncorrelated with temperature variables because they are at 90° angles. Soil texture variables are, for the most part, uncorrelated with climate variables.

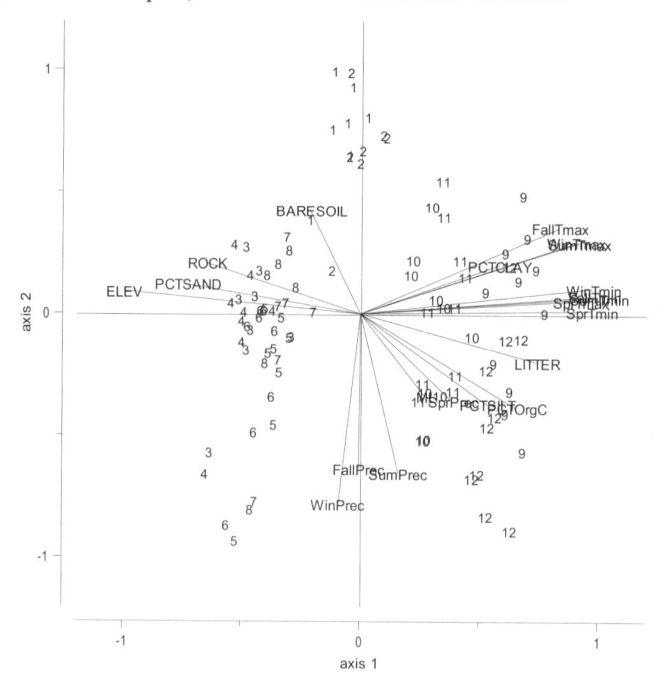

Figure 26.3. PCA correlation biplot for all explanatory variables. The first two eigenvalues are 0.49 and 0.12, which means that these two axes explain 61% of the variation in the data. The numbers in the graph are the PCA scores and represent the transects. Transects 1–8 are from Yellowstone National Park and 9–12 are from the National Bison Range.

Some substrate variables correlate with climate variables (LITTER AND ML), but others do not (ROCK, BARESOIL). In the biplot, notice that most observations from the same transect tend to be close to each other, indicating that the composition of each transect over time is more similar to itself than to other transects. The biplot also shows that some of these transects are more variable over time than others (e.g., transects 7 and 9). Finally, the biplot shows distinct separation between transects 1–8 from YNP and transects 9–12 from the NBR, indicating that environmental conditions differ dramatically between these areas. This interpretation must be used with reservation, however, because some of the explanatory variables (especially soil texture) were assigned the same value for all sampling years. Assigning each site's transects the same value eliminated temporal and spatial variation. Nevertheless, the separations between areas are distinct even considering the few identical soil variables.

Although the biplot suggests that many of the climate variables are correlated with each other, using Figure 26.3 alone makes it difficult to decide which variables are highly correlated and should be eliminated from further analysis. To make the decision more objective, we use another tool called the VIF to assess which variables are highly related. In VIF, one explanatory variable is selected as response variable and all others are set as explanatory variables within a linear regression (Montgomery and Peck 1992). The VIF value for the selected variable is given by $1/(1 - R^2)$ where R^2 is the usual R-squared from a linear regression that gives the amount of variation explained by the regression model (Chapter 5). For each analysis, a different explanatory variable is set as response variable in the regression. If there are 20 explanatory variables, this process is carried out 20 times. The underlying idea is that if three explanatory variables Z_1, Z_2 and Z_3 are highly correlated, and one of them is set as the response variable (say Z_1) in the regression model while all others (Z_2 and Z_3) are set as explanatory variables, then the R^2 of the linear regression model will be high and the VIF value from the model will be large. We do the same calculations with Z_2 set as a response model and Z_1 and Z_3 as explanatory variables to get a VIF value. For Z_3, we repeat the process again. The VIF values obtained when Z_1, Z_2 or Z_3 are used as response variables are compared in a table. A high VIF value is a serious indication of collinearity because it means that the variation in the response variable is explained well by the other variables. For the 21 explanatory variables in this study, 9 have VIF values that make them candidates for elimination (Table 26.2). Because there is no real cut off level for the VIF, this decision is subjective; but some statisticians suggest that values higher than 5 or 10 are too high (Montgomery and Peck 1992). Values greater than 50 definitely require elimination from further analysis.

So, where do we start? If the variables can be combined into related groups, we could subjectively remove a variable (or variables) from each group. For example, the variables here are grouped by substrate, temperature, precipitation, and soil characteristics so we could arbitrarily decide to remove one or more variables from each group. Alternatively, we could use a backward selection method to remove one variable at a time, and recalculate VIF after each iteration until we obtain a set of variables that are not collinear anymore. We should keep in mind that if we remove one variable, and repeat the analysis, all VIF values will change. With this dataset, we decided to use the backward selection process. The variable with the highest VIF value was eliminated first. VIF values were then recalculated for the remaining variables, and the process was repeated until all VIF values were smaller than five. The following variables were omitted from further analysis in order from first to last: FallTmin, ELEV, SumTmin, SprTmin, WinTmin, SumTmax and WinTmax. The VIF values of all remaining variables are reasonably low (Table 26.3) indicating that there is not strong collinearity among them. We will work with these variables in the remaining analyses using the richness.

In addition to collinearity, we need to check whether this dataset needs any transformations (Chapter 4). Cleveland dotplots (Chapter 4) of the selected explanatory variables in (Table 26.3) showed that none of the explanatory variables had extreme observations. Therefore, we do not need to transform any of the explanatory variables.

Table 26.2. VIF values for the explanatory variables. The VIF value for PCTCLAY could not be calculated due to perfect collinearity.

Variable	VIF	Variable	VIF
ROCK	2.929	SumTmax	24.978
LITTER	3.924	WinTmax	46.746
ML	1.832	FallTmin	149.896
BARESOIL	1.953	SprTmin	77.184
FallPrec	7.322	SumTmin	135.605
SprPrec	2.519	WinTmin	56.734
SumPrec	4.570	PCTSAND	4.773
WinPrec	4.052	PCTSILT	8.752
FallTmax	43.045	PCTOrgC	2.438
SprTmax	40.199	ELEV	91.398

Table 26.3. Final VIF values for the selected explanatory variables.

Variable	VIF	Variable	VIF
ROCK	2.522	WinPrec	2.087
LITTER	2.943	FallTmax	3.233
ML	1.599	SprTmax	3.064
BARESOIL	1.526	PCTSAND	4.046
FallPrec	2.171	PCTSILT	3.311
SprPrec	1.502	PCTOrgC	2.210
SumPrec	1.560		

With collinearity and transformation issues addressed, the next step is to develop a diversity index that can be used to model general relationships between species in the bunchgrass community and the environmental variables. To begin this process, we need to address a very common problem with plant data; namely, a species dataset with a high percentage of zeros. The bunchgrasss communities in these two areas have 93 species over time and 99 observations per species. Each transect has relatively few species compared with the large number of species in the overall study, so over 90% of the species observations are equal to zero (Figure 26.4). A large number of zeros creates both statistical and ecological problems. Double zeros, or species absent from two transects, does not necessarily mean that the communities are more similar than two others with and without the species. Zeros during sampling also do not indicate whether an environment is unfavourable for a certain species or whether it was just not found at a particular location. This is the classic "zero truncation" problem in ecology. With only 10% of our total observations having a species present, multivariate methods like principal component analysis, redundancy analysis, (canonical) correspondence analysis, and discriminant analysis are not very useful because each of these analyses is sensitive to double zeros (Chapters 12 and 13). For the data that actually do have values in this study, over 32% have very small values (i.e., frequency <1.0%). With data that have many small values and a few large values, a diversity index like the Shannon-Weaver or total abundance might not be appropriate because each is highly influenced by a few observations. A more appropriate measure of association for these data is a measure that converts all values to presence–absence. Both the Jaccard and the Sørensen indices convert data to presence–absence. We tested both measures on our data and found very little difference in the results, which we expected because these indices have similar definitions (see Chapter 10). Therefore, we decided to use the Jaccard index. For our later multivariate analyses, the Jaccard or Sørensen indices are also more suitable measures of association than either the Euclidean or the Chi-square distance function because they handle data with many zeros better.

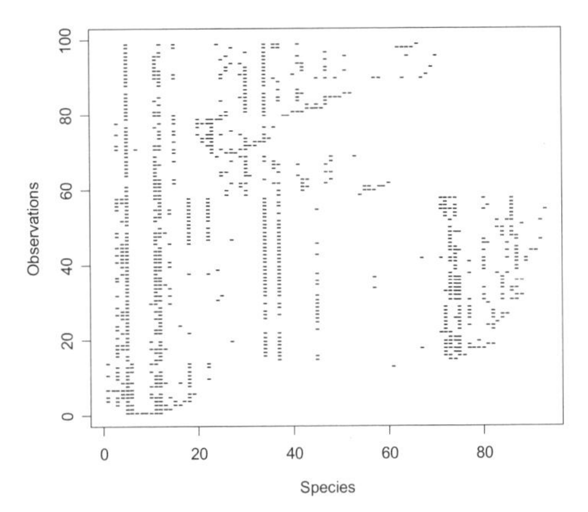

Figure 26.4. Species presence in the dataset. The horizontal axis shows the 93 species in their spreadsheet order and the vertical axis the 99 observations within the transects. A '-' indicates a non-zero value for the species. If all observations had values greater than zero, the graph would be completely filled with dashes.

We are now to a point where we can start exploring whether species richness (i.e., our representation of biodiversity) relates to the explanatory variables and whether time factors into the relationship at all. Pairplots are a good tool to visually look for patterns between our diversity index (richness) and the environmental variables that we selected with the VIF (Table 26.3). The patterns in the top row of the pairplot in Figure 26.5 represent the relationship between richness and the explanatory variables. There are no clear patterns. The other panels can be used to visually assess collinearity, and the lower diagonal panels of the pairplot give the corresponding correlation coefficients. A high (absolute) value (>0.8) indicates that two explanatory variables are linearly related. Whereas these correlations and the previously calculated VIF values measure *linear* relationships among explanatory variables, the upper part of this pairplot can be used to assess whether there are *non-linear* relationships among the explanatory variables. In this case, ROCK and LITTER might have a non-linear relationship. Note that a correlation coefficient is only measuring linear relationships.

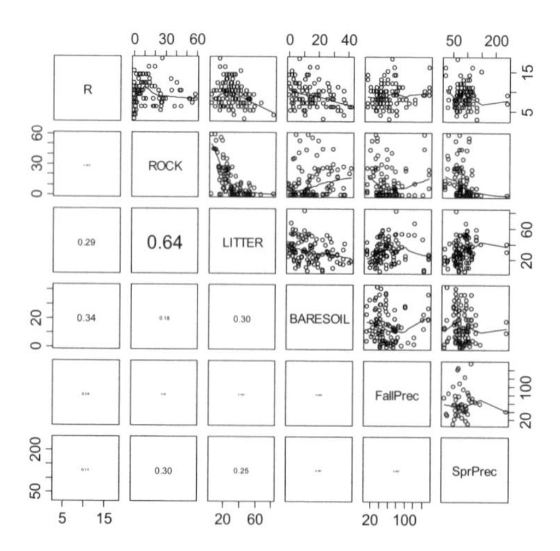

Figure 26.5. Pairplot for species richness and selected explanatory variables. A smoothing curve was added to aid visual interpretation. Other environmental variables are not shown to conserve space. *R* is the species richness. Values on edges represent the units of the variables. The font size of the (absolute) correlation coefficients in the lower diagonal is proportional to their value.

26.4 Linear regression results

In this section, we relate species richness to the explanatory variables using linear regression and try to find the subset of explanatory variables that best explains richness. We start by using the AIC to find the most optimal model. Then we validate the model and explore whether it can explain temporal patterns when 'time' is added as a factor.

Model selection

Recall from Chapter 5 that the AIC measures the goodness of fit but, at the same time, uses the number of explanatory variables as a penalty. The lower the AIC is, the better the model. To find the model with the best AIC, we can either use (i) a backward selection where we start with all explanatory variables and drop one at a time, (ii) a forward selection where we start with models containing only one explanatory variables and add one at a time, or (iii) a combination of forward and backward selections that drops variables at any point in the analysis but can add each back in again at a later stage in the analysis. It could be argued that if the dataset has less than 20 explanatory variables, a fourth option would be to use every possible combination of explanatory variables. This option is not available if analyses are performed using the major statistical software packages, although programming such an algorithm is not difficult if you choose to use this option. We decided to use option three. Our resulting optimal linear regression model has only five environmental variables that relate significantly to richness at the 5% level. The explained variation in the species richness is 54% (R^2), which is reasonably good. Note that sometimes the AIC comes up with a final model in which some of the explanatory variables are not significantly different from 0 at the 5% level. If this were the case, we would have to further remove explanatory variables using the F-test for comparing nested models (which means that we would have used two different selection criteria to find the optimal model). As all explanatory variables are significant in our analysis (see below), we do not have to do any further model selections.

Variable	Estimate	Standard error	t-value	p-value
ROCK	−0.081	0.017	−4.523	<0.001
LITTER	−0.066	0.022	−2.945	0.004
BARESOIL	−0.102	0.020	−5.022	<0.001
FallPrec	0.015	0.006	2.364	0.020
SprTmax	−0.534	0.089	−5.959	<0.001

Model validation

We now move on to the model validation process, which was outlined in Chapter 5. Here we verify the homogeneity, normality and independence assumptions, and check for model misspecification. The residuals of the optimal linear regression model are plotted against the fitted values (Figure 26.6-A). Recall from Chapter 5 that Figure 26.6-A and Figure 26.6-C can be used to assess homogeneity. In this case, there is no violation of homogeneity because the spread of the residuals is nearly the same everywhere. So there is no need to consider more complicated models like generalised linear modelling with a Poisson distribution. We also have normality (Figure 26.6-B). There are no influential observations as judged by the Cook distance function (Figure 26.6-D).

Next we check for misspecification in our optimal model. Misspecification is indicated when graphs of the residuals versus individual explanatory variables show distinct patterns. For most of the explanatory variables used in our model, there is no obvious pattern. However, the residuals versus ROCK and SprTmax both show pattern. The question is, how serious this is? We fitted a smoother with 95% confidence bands to the residuals (Figure 26.7). The width of the confidence bands indicate that '0' is almost entirely within the 95% bands, so the smoother is not significantly different from 0. Only a few samples are different from 0, and one might argue that this could happen by chance. In this case, the patterns seem weak enough for us to ignore, but others might consider them borderline for taking further action. If the residual patterns were stronger, we would address the problem by (i) adding a quadratic term for ROCK and SprTmax, (ii) using a smoothing function for ROCK to get an additive model (Chapter 7), or (iii) adding interactions (Chapter 5) to our model. Some of these options would require generalised linear modelling or generalised additive modelling procedures, which are covered in Chapters 6 and 7. As stated, though, we do not consider the residual patterns strong enough to warrant these techniques, so we will proceed to our most important question. Do the residuals show any patterns over time? If so, then we cannot use the linear regression model because standard errors, *t*-values and *p*-values might be seriously inflated (Chapter 16).

Time effects

The plant communities at each transect are not only affected by environmental factors, but they are also affected by changes in biotic and abiotic conditions over time. Because the data are taken at irregular intervals, they are not as conducive to time series analysis as other datasets in this book. There are ways, however, to evaluate the effects of time on this regression model. A scatterplot of species richness versus time (Figure 26.8) shows some clear patterns that the linear regression model will hopefully explain. If it does not, we violate the assumption of independence and we may commit a type I error (Chapter 16).

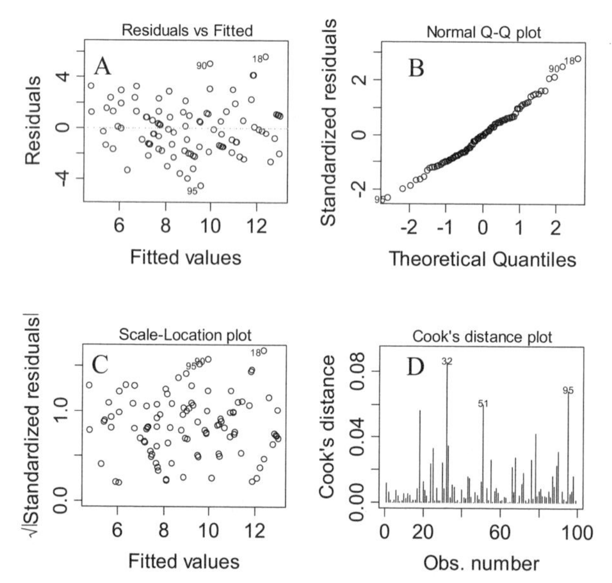

Figure 26.6. Model validation for the optimal linear regression model. Panels A and C indicate that there is no violation of homogeneity. Panel B indicates normal distribution. Panel D shows no outliers using Cook's distance function.

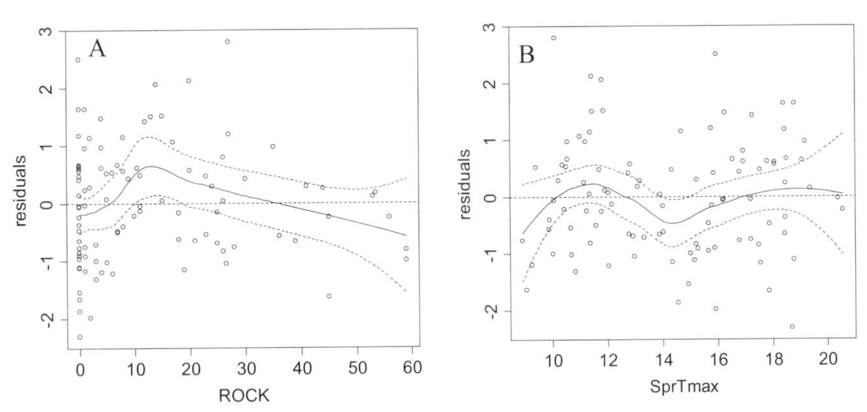

Figure 26.7. Distribution patterns of standardised residuals for the optimal linear regression model versus ROCK (A) and SprTmax (B). The LOESS smoother (solid line) and 95% confidence bands (dotted lines) are shown. All other explanatory variables had smoother lines that were straight horizontal at 0.

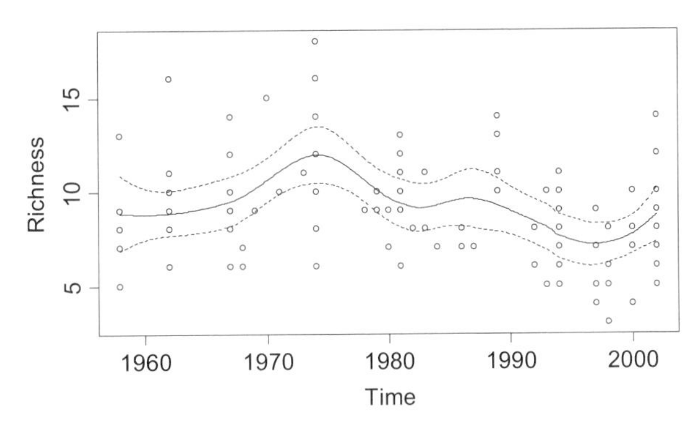

Figure 26.8. Scatterplot for species richness versus time. A LOESS curve with a span width of 0.5 (solid line) was added to aid visual interpretation. The dotted lines represent 95% confidence bands for the smoother.

The easiest way to assess whether the linear regression model explains the temporal pattern in the species richness is to make a plot of the standardised residuals versus time. In our case, this plot does not show a clear temporal pattern (Figure 26.9). A LOESS smoother with a span of 0.75 added to the diagram is almost a straight line at zero indicating that time is not important in the linear regression. Various other tools are available to objectively test whether the residuals contain any temporal pattern. One such tool, GLS, is the subject of the next section.

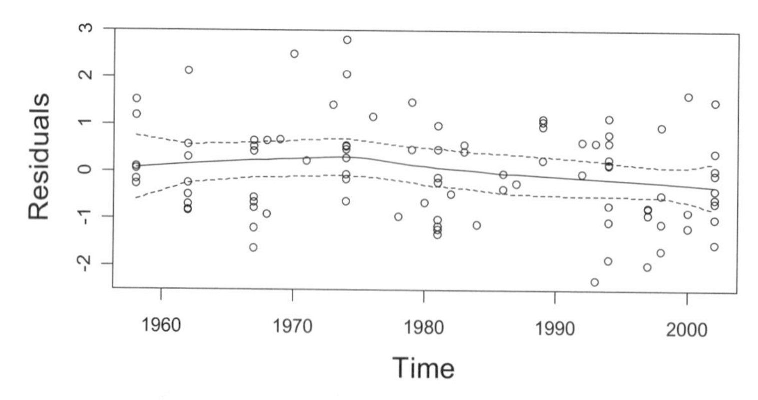

Figure 26.9. Scatterplot of standardised residuals versus year. A LOESS smoother with a span of 0.7 was added. The dotted lines are 95% confidence bands of the smoother.

26.5 Generalised least squares results

In this section, we explore tools that test for a temporal pattern in the residuals. First, we added 'TIME' as a covariate within the linear regression model and tested for its effects. With the YNP and NBR data, 'TIME' was not significant at the 5% level. Alternatively, we could extend the linear regression model so that the residuals are *allowed to* have a temporal pattern, which means that we relax the independence assumption in linear regression. This is done using GLS. When we use GLS, however, there are several ways we can impose a correlation structure on the error terms, and these are discussed below.

ARMA error structure

The auto-regressive moving average (ARMA) adds an auto-correlation structure to the noise component of the linear regression. This technique goes beyond the GLS theory explained in Chapter 16, so we will explain it in more detail here. We start with our optimal linear regression model from the previous section that had five selected explanatory variables:

$$R = \alpha + \beta_1 \text{Rock} + \beta_2 \text{Litter} + \beta_3 \text{Baresoil} + \beta_4 \text{FallPrec} + \beta_5 \text{SprTmax} + N_t \qquad (26.1)$$

In linear regression, we assume that the noise component N_t is independent and normally distributed. In GLS, we use the same equation as in (26.1), but the noise component is no longer restricted to be independent. Recall from Chapter 16 that an ARMA model of order (p,q) for a time series Y_t is defined as:

$$Y_t = a_1 Y_{t-1} + a_1 Y_{t-1} + ... + a_p Y_{t-p} + \varepsilon_t + b_1 \varepsilon_{t-1} + b_2 \varepsilon_{t-2} + ... + b_q \varepsilon_{t-q} \quad (26.2)$$

where ε_t is independently and normally distributed noise. In this model, a time series Y_t is modelled as a function of its past values, its error component, and its past error components. The p defines the number of auto-regressive (past Y values) components and q the number of past error terms. In GLS, this modelling framework is applied on the error component N_t. Hence, Y_t is replaced by N_t in equation (26.2). So, in linear regression we are imposing the condition that N_t is independent normal noise. In GLS, we are decomposing N_t into small building blocks. These building blocks consist of p past terms of N_t, and q building blocks that are independently normally distributed. The question then becomes 'how many of these building blocks do we need?' Or more formally, how many auto-regressive (p) and moving average (q) terms do we need? The easiest way to determine p and q is to apply the GLS with different values of p and q, and choose the one with the lowest AIC (Table 26.4). The AIC values indicate that the model with $p = 1$ and $q = 0$ has the lowest AIC so it is the 'best' model. However, the AIC of the model with $p = q = 0$, which is the linear regression model, is only marginally larger. The question then becomes whether the decrease in AIC is enough to discard the linear regression model in favour of a more complex GLS model. If the difference in AIC values of two models is smaller than 2 (as is the case in Table 26.4), general statistical consensus dictates using the simpler linear model.

Another more formal way to compare the two models uses a likelihood ratio test (see also Chapter 8 for comparing mixed effects models with different random components). The likelihood ratio test indicates that both the ARMA(1,0) and the model without auto-correlation perform equally well (see below), so there is no need to adopt a complicated GLS model over our original linear regression. Before fully committing to the simpler linear regression model, we will explore two additional GLS error structures to see whether they give different effects. The notation GLS(p,q) refers to the ARMA error structure in the GLS.

Model	df	AIC	logLik	*L*-Ratio	*p*-value
GLS(1,0)	8	461.60	−222.80		
GLS(0,0)	7	462.95	−224.47	3.34	0.06

Table 26.4. AIC values obtained by using an ARMA(p,q) structure on the error component in the GLS model ($p = q = 0$ is the linear regression model). Note that the ARMA(1,0) is an auto-regressive model of order 1. The correlation structure in the GLS function within R software can cope with missing values.

	$q = 0$	$q = 1$	$q = 2$	$q = 3$
$p = 0$	462.95	463.69	464.13	466.12
$p = 1$	**461.60**	463.43	465.26	468.16
$p = 2$	463.36	465.24	467.21	469.11
$p = 3$	465.29	466.13	467.66	463.67

Other error structures

In Chapter 35, we examined a salt marsh time series using a special case of GLS error structure, namely the AR(1,0) structure. In the previous section, we examined the effect of time on the residuals using ARMA. In this section, we examine two additional approaches that test the relationships of time lags and covariance between years to the residuals in GLS. The so-called 'compound symmetry approach' is used if it can be assumed that the correlation (of the error) between two different years (within a transect) is equal to ρ whatever the time lag. It is defined as

$$cor(N_t, N_{t+j}) = \rho \text{ for any value of } j \qquad (26.3)$$

The indices t and j stand for time and time lags, respectively. So the correlation between N_t and N_{t+1} is ρ, and the same holds for the correlation between N_t and N_{t+2}, N_t and N_{t+3}, etc. For the YNP and NBR data, the estimated value of ρ using the compound symmetry approach is −0.03 and the AIC is 464.66. This AIC value does not improve on the AIC for the linear regression model. The compound symmetry correlation structure tends to be too rigid and simplistic for most time series datasets (Pinheiro and Bates 2000).

GLS error structure can also be defined using a general correlation matrix. A correlation matrix allows for more flexibility than compound symmetry but at a much higher price in terms of estimated parameters. If there are N different years, the general covariance matrix for the error component is of the form:

$$V = \begin{pmatrix} v_{1,1} & v_{2,1} & \cdots & v_{N-1,1} & v_{N,1} \\ v_{2,1} & v_{2,2} & & \vdots & \vdots \\ \vdots & & \ddots & \vdots & \vdots \\ v_{N-1,1} & v_{N-1,2} & \cdots & v_{N-1,N-1} & v_{N-1,N-1} \\ v_{N,1} & v_{N,2} & \cdots & v_{N,N-1} & v_{N,N} \end{pmatrix} \qquad (26.4)$$

The component $v_{i,j}$ represents the covariance between two years i and j (see Chapter 16). If we use this structure, we allow the residuals to covary in different ways depending on the time lag between them. To use the correlation matrix, however, the sampling protocol must meet fairly strict guidelines. All sample years should be equidistant apart, and each transect should be measured in each sample year. In this case study, the transects were not sampled regularly in either area nor were the sample intervals the same between the two areas, so the software was not able to converge. All other error structures discussed in this section were able to deal with irregular sampling and missing years.

After trying many options within the GLS model, we still do not have models that relate species richness and environmental variables any better than our original linear regression model. We accept the linear regression model then as the best and simplest model to address which environmental variables best correlate with changes in biodiversity in these plant communities.

26.6 Multivariate analysis results

Analysing the relationship between compositional and environmental changes for bunchgrass communities requires a multivariate analysis approach. In a particular type of multivariate analysis called ordination, the arrangement of these values within a multidimensional graph can help us see whether the community data are structured or contain patterns. The patterns may reflect a community's response to multiple environmental changes over time or more subtle biological interactions. Deciding which multivariate analysis to use, however, depends on the data and focus of the study. We could apply principal component analysis or redundancy analysis on these data (Chapter 12), but these methods are sensitive to double zeros (Chapters 10 and 12). Our data matrix contains over 90% zeros, and therefore, we chose an asymmetric measure of association (i.e., joint absence of species at two sites is not contributing towards similarity), namely the Jaccard index (Chapter 10) in combination with non-metric multidimensional scaling (NMDS).

Using NMDS does make it more difficult to produce triplots and visualise the relationships among observations, species and explanatory variables. These relationships must be informally inferred by comparing the transect positions within the NMDS diagram. Determining how time relates to species gradients is also more difficult using NMDS. For some datasets, it might be more appropriate to use an alternative method, such as db-RDA, to relate distance matrices directly with environmental variables (Legendre and Anderson 1999).

NMDS

NMDS calculates a distance matrix from the species data based on a chosen measure of association. It then uses an iterative process to order the distance matrix in *n*-dimensional space to find a configuration that matches the distance matrix as closely as possible (Chapter 15). The distance matrix and the position assigned in the *n*-dimensional space never match perfectly. If they did, a plot of the dissimilarity of each point in the original distance matrix versus the distances between them in the calculated ordination space (also called a Shepard diagram) should be a straight line (or curved for NMDS). To measure the discrepancy of the points from this line, various measures of goodness-of-fit have been proposed. They all use quadratic functions of the original distance and calculated distance in the ordination space (see Legendre and Legendre (1998) or Chapter 15 for details). These measures of goodness of fit are also known as 'STRESS'. High STRESS values indicate a poor fit between the original data structure portrayed by the distance matrix and the ordered positions in *n*-dimensional ordination space.

In this section, we have two options. We can either apply the multivariate analysis on each of our areas separately or on both areas together. Each option gives different perspectives on community and environmental change. In the previous sections, the univariate analysis clearly revealed a difference in composition and environmental conditions between Yellowstone National Park and the

National Bison Range. Combining the areas into one NMDS analysis will likely have the same result. Analysing the two areas separately, however, has the potential of providing more information on how similar local communities are and enables us to track how composition varies within each area over time. It also gives a different perspective on whether climatic and soil variables are important to species presence in the community locally. Although analysing two areas has double the amount of information, graphs, and numerical output to interpret, the benefit for this study is a clearer perspective on the variations within each transect over time.

If NMDS is used, one has to choose the number of axes. In an initial analysis, we used two axes but the STRESS was 0.214 for the YNP data and 0.254 for the NBR data. These values are rather large (Chapter15), and therefore, we present results obtained by using three axes. The problem with three axes is how to present them. We tried three-dimensional scatterplots and even used spin graphs (these are graphs in which the mouse can be used to rotate the three-dimensional scatterplot) but found three separate two-dimensional scatterplots presented the results most clearly.

When NMDS is applied to data from YNP, the communities of most individual transects form tight clusters and the grazed and ungrazed transects at each area are separated (Figure 26.10). Transects from the warm, dry, low elevations (1-2) also separate from all others. Figures 26.10-A and 26.10-C show that transects 1 and 2 are the most dissimilar from the other transects. Along the third axis we can see a difference between transects 5 and 6 versus the other transects. The STRESS using three axes was 0.15, which is still relatively large. We felt the STRESS was too large to make any statement on changes within transects over time.

For the NBR data (Figure 26.11), transects differ much more among themselves (i.e., same numbers are more spread out) than at YNP over time. Panel C seems to indicate differences among transects 9, 10 and 11, which are all in different pastures of the NBR. Again, the STRESS is relatively large using three axes (0.19), so care is needed with the interpretation. Conclusions about change over time are again probably impractical. Our next problem is to find out whether there is a relationship between the environmental variables and the species that could explain these patterns. This requires a Mantel test (Chapter 10).

The Mantel test

The Mantel test evaluates correlations between two distance matrices of the same size. In this study, we have two data matrices. One matrix, called the '**Y**' matrix, contains the species data. It is (number of sites)-by-(number of species) in size. The second matrix is called the '**X**' matrix, and it contains the explanatory variables. Its size is the (number of sites)-by-(number of environmental variables). When a distance matrix is created from each of these matrices using (for example) the Jaccard index for the similarity measure and Euclidean distance measurements, two equal-sized matrices result ($\mathbf{D}_Y$ and $\mathbf{D}_X$).

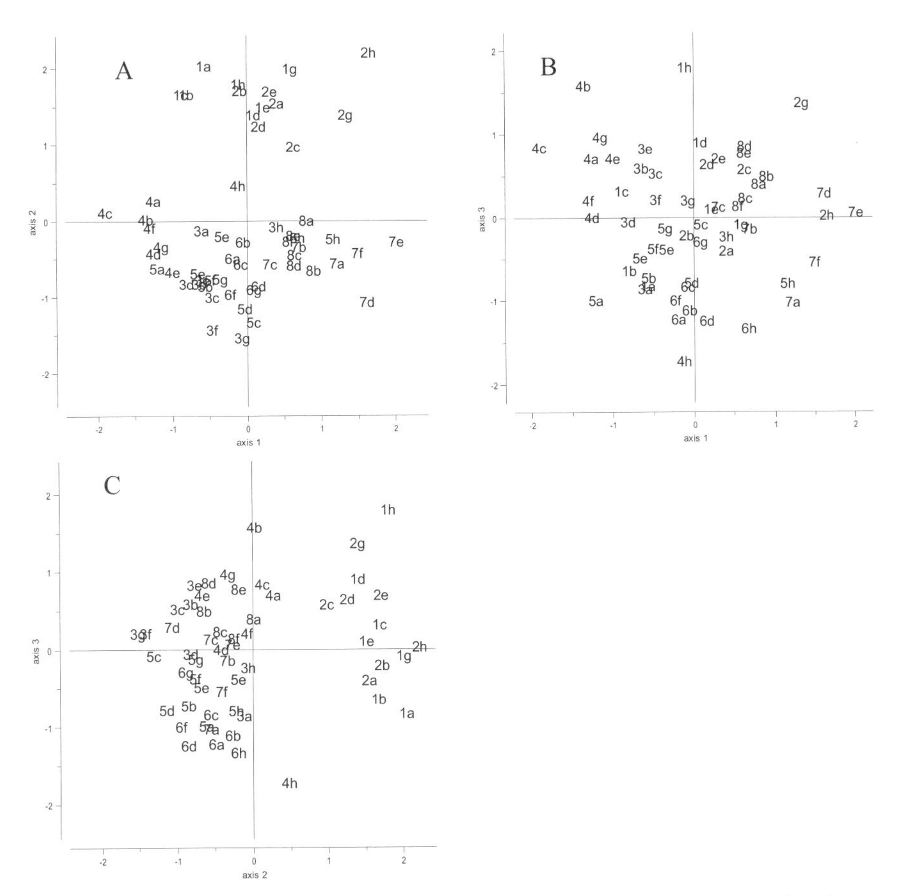

Figure 26.10. NMDS using the Jaccard index for the Yellowstone National Park data. The STRESS value equals 0.15. A: Axis 1 versus 2. B: Axis 1 versus 3. C: Axis 2 versus 3. Alphabetical designations on transect numbers indicate order of sampling (a = first sample).

The ij^{th} element in $\mathbf{D_Y}$ represents the dissimilarity between two observations in terms of species composition, and the ij^{th} element in $\mathbf{D_X}$ the dissimilarity between the same two observations in terms of environmental condition. The Mantel test evaluates how these two distance matrices correlate with each other. To assess the significance of the correlation coefficient, a randomisation process is applied (for details see Chapter 10). It is important to realise what exactly we are testing with the Mantel test. The underlying assumptions are as follows. H_0: distances between the observations in terms of species composition are not linearly related to distances between the observations in terms of environmental conditions. H_1: distances among the observations in terms of species composition are linearly correlated to the distances between the observations in terms of environmental conditions.

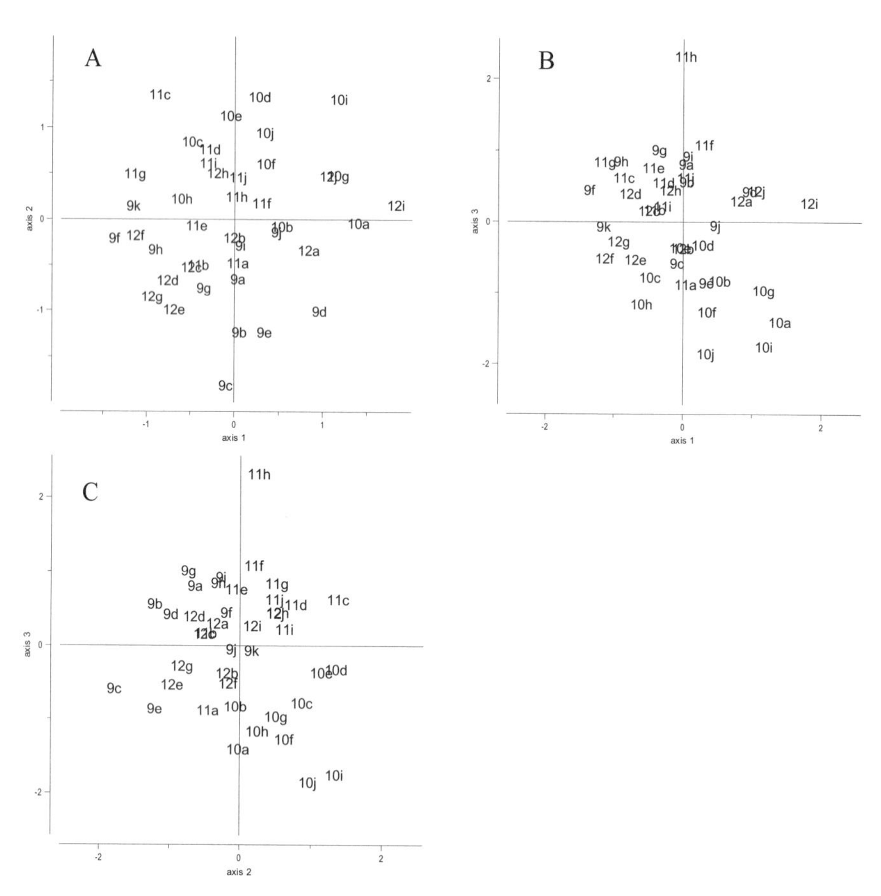

Figure 26.11. NMDS using the Jaccard index for the National Bison Range data. The STRESS value equals 0.19. A: Axis 1 versus 2. B: Axis 1 versus 3. C: Axis 2 versus 3. Alphabetical designations on transect numbers indicate order of sampling (a = first sample).

We applied the Mantel test on the analyses of the separate areas and tested the significance of our correlations between the two distance matrices with a permutation test. The correlation between the elements of $\mathbf{D}_Y$ and $\mathbf{D}_X$ and their significances differed for YNP and the NBR. For YNP, the correlation coefficient was $r_M = 0.378$ ($p < 0.001$). For the NBR, it was $r_M = 0.116$ ($p = 0.06$). For the YNP data, we reject H_0 at the 5% level and conclude that there is a linear relationship between the two distance matrices. For NBR there is no evidence to reject the null hypothesis at the 5% level.

In the previous paragraphs we compared a matrix consisting of distances between observations representing species composition with a matrix reflecting differences in terms of environmental conditions. The question now is 'which explanatory variables should be used when we obtain $\mathbf{D}_X$ so that the correlation between $\mathbf{D_X}$ and $\mathbf{D_Y}$ is as high as possible?' To answer this question, we first calculate (i) a matrix $\mathbf{D_X}$ and (ii) the correlation with $\mathbf{D_Y}$ for all possible combinations of explanatory variables. We then select the combination of explanatory variables that gives the highest correlation coefficient r_M. This process was called BIOENV by Clarke and Warwick (1994). If there are numerous explanatory variables, the computing time for the BIOENV procedure can be long. Sometimes an automatic forward selection procedure called BVSTEP is preferred to speed the process up (Chapter 10). We applied BVSTEP on data from the two different areas. For YNP, the forward selection procedure first selected WinPrec and then PCTSILT. No other variables were needed to explain the correlations. The correlation coefficient between the elements in $\mathbf{D_Y}$ and $\mathbf{D_X}$, in which $\mathbf{D_X}$ was only determined by WinPrec and PCRSILT, was $r_M = 0.408$. For the NBR, the selected variables, in order of importance, were SprPrec, ROCK, and PCTOrgC. The correlation coefficient using these three variables was $r_M = 0.249$. For the National Bison range, we could not use the explanatory variable PCTSILT because it had the same value in all transects.

26.7 Discussion

Many environmental stresses shape the bunchgrass communities of Montana's temperate grasslands during half a century. In this case study, climatic variation, substrate changes, and soil texture are all important to explaining species composition within these communities. When composition is portrayed as a single biodiversity measure (richness) within linear regression, both substrate and climatic factors relate to its variation. Changes in the frequencies of rock, litter, and bare soil at each transect are significant in the linear regression model and relate inversely to richness. Differences in substrate characteristics between the two study areas are not surprising. Yellowstone National Park and the National Bison Range differ in bedrock and soil textures. The YNP transects are located within lower Tertiary igneous and metamorphic rocks. The NBR exists within fine-grained quartzite, argillite, banded slate, and inter-bedded sandstone of the lower pre-Cambrian. Differences in bedrock affect soil textures and the amount of rock at each site. They also affect soil permeability and moisture retention throughout the growing season. The amounts of litter or bare soil affect moisture retention, surface temperature, and nutrients available for plant growth. The more bare soil, the more rapidly moisture is lost from the surface and the higher the surface temperatures during the growing season. As litter increases, moisture is lost more slowly from the soil surface and nutrients are added to the soil.

Biodiversity, as measured by species richness, is also negatively related to maximum spring temperatures and positively related to the fall precipitation. In this area of the northern Rocky Mountains, fall and winter precipitation are very important to re-charging soil moisture for plant growth to begin in spring.

Although time should be significant in studies on community change, neither adding time as a covariate in the linear regression model nor including different error structures improved the model.

The NMDS shows how similar (or dissimilar) each of the communities are to each other; but because of the relatively large STRESS, it cannot be used in this particular study to assess how much the repeat samplings vary at each location.

The three-dimensional YNP ordination graph has several interesting patterns to explore. The grazed plots form distinct groups among themselves, but the groups are also well inter-mixed with clusters of non-grazed plots. This suggests that management type makes some difference in composition in YNP, but that it is not as important as some other factors, such as elevation and moisture, to park grasslands as a whole. The NBR data, which is all grazed on a rotational schedule, varies much more between individual transects.

A more formal approach of determining which environmental variables affect communities in the NMDS species (BVSTEP) shows that winter precipitation and percentage of silt in the soil are significantly related to species presence in YNP while spring precipitation, frequency of rock, and percentage of organic carbon in the soil are most significant in the NBR.

The univariate and multivariate analyses used in this study give slightly different perspectives on how environmental variables affect these bunchgrass communities. Analysing the long-term monitoring records with univariate analyses shows that substrate conditions and precipitation are very important controls on species presence in these grasslands. Analysing the records with the multivariate NMDS shows that temperature, grazing, and non-native species also affect these communities. Whether the sites are analysed together or separately, the factors that affect these bunchgrass communities over several decades are complex and multiple perspectives are essential to explore that complexity.

Acknowledgement

Research was conducted in Yellowstone National Park under research permit Yell-2002-SCI-5252. Roy Renkin and the Yellowstone archive staff helped access historic transect data and historic photographs. Lindy Garner, Lynn Verlanic, and Bill West of the U.S. Fish & Wildlife Service were instrumental in locating historic records and resampling transects in 2002 on the National Bison Range. Dr. Paul B. Alaback, professor of forest ecology at The University of Montana, provided guidance and advice throughout this project. SOGS interpolations were run by Matt Jolly, Numerical Terradynamic Simulation Group, The University of Montana.

27 Univariate and multivariate analysis applied on a Dutch sandy beach community

Janssen, G.M., Mulder, S., Zuur, A.F., Ieno, E.N. and Smith, G.M.

27.1 Introduction

Climate change is, beyond doubt, the most important threat facing the world's coastline and has been accompanied by intensive debate. Marine coastal ecosystems are extremely vulnerable, as they constitute the most productive and diverse communities on Earth. Coastal areas are, however, not only subject to climate change but also to many other forms of human activities. Among them, land-claim, pollution, recreation purposes and dredging activities have been threatening most of the European coasts resulting in many cases in inter-tidal habitat fragmentation and/or degradation (Raffaelli and Hawkins 1996). The consequences of these changes have been well documented in a considerable number of studies that addressed the impact and reported decreased ecosystem performance.

On a more local scale, the Dutch have fought great battles with the North Sea in order to extend their landmass as can be witnessed by the presence of dykes and sophisticated coastal defence systems. The effect of sea level rise on the ecology of the Dutch coastal system constitutes a serious issue that should not be ignored in the short term.

The Dutch governmental institute RIKZ therefore started a research project on the relationship between some abiotic aspects (e.g., sediment composition, slope of the beach) as these might affect benthic fauna. Mulder (2000) described the results of a pilot study that looked at the effects of differences in slope and grain size on fauna in the coastal zone. Using the data from this pilot study and statistical experimental design techniques, a sampling design was developed in which nine beaches were chosen by stratifying levels of exposure: three beaches with high exposure, three beaches with medium exposure and three beaches with low exposure. Sampling was carried out in June 2002, and at each beach, five stations were selected. Effort at each station was low (Van der Meer 1997).

The aim of the project was to find relationships between macrofauna of the intertidal area and abiotic variables. In this case study chapter, univariate and multivariate tools are applied in order to obtain as much information as possible. The results of the combined analyses are then used to answer the underlying question. Instead of presenting the results of the final models, we show how we got to them, and which steps we applied (especially for the multivariate analysis). The outcome of this research will have immediate relevance for assessing and managing disturbance in marine benthic systems, with respect to degradation and biodiversity lost.

27.2 The variables

Table 27.1 gives a list of the available explanatory variables. NAP is the height of the sampling station relative to the mean tidal level. Exposure is an index that is composed of the following elements: wave action, length of the surf zone, slope, grain size and the depth of the anaerobic layer. Humus constitutes the amount of organic material. Sampling took place in June 2002. A nominal variable 'week' was introduced for each sample, which has the values 1, 2, 3 and 4, indicating in which week of June a beach was monitored. The following rules were used. Sampling between 1 and 7 June: $Week_i = 1$. Sampling between 8 and 14 June: $Week_i = 2$. Sampling between 15 and 22 June: $Week_i = 3$ and sampling between 23 and 29 June: $Week_i = 4$. The index i is the station index and runs from 1 to 45. There were nine beaches, and on each beach five stations were sampled (hence, 45 observations). Ten sub-samples were taken per station, but in this chapter we will use totals per station. $Angle_1$ represents the angle of each station, whereas $angle_2$ is the angle of the entire sampling area on the beach. Both variables were used. The variables $angle_2$, exposure, salinity and temperature were available at beach level. This means that it is assumed that each station on a beach has the same value. For $angle_2$ this assumption does not hold, and its inclusion in some of the statistical models should be done with care. A few explanatory variables contained four missing values. Most of the statistical techniques used in the analysis cannot cope with missing values, and therefore, missing values were replaced by averages.

Table 27.2 shows the Pearson correlation among the 12 explanatory variables. Only correlations significant at the 5% level are given. Except for a few variables (chalk and sorting and exposure and temperature), the correlations among the explanatory variables are relatively low, indicating that there is no serious collinearity.

As to the species, in total 75 species were measured. To simplify interpretation of graphical plots, species were grouped in the following five taxa: Chaetognatha, Polychaeta, Crustacea, Mollusca, and Insecta. Within each taxa, between 1 and 28 species were available. Species names were replaced by names taking the following form: P_1, P_2, P_3, etc. for Polychaeta species, CR_1, CR_2, CR_3, etc. for Crustacea species, M_1, M_2, M_3, etc. for Mollusca species, I_1, I_2, I_3 for Insecta, and C_1 for the Chaetognatha species *Sagitta* spec. (only one Chaetognatha was measured). A list of species and notation used in this chapter are available as an online supplement to this book.

Table 27.1. List of available explanatory variables. The columns labelled "Level" indicate whether different values are available for each station on a beach (Beach) or one value for all five stations on a beach (Station).

Number	Variable	Level	Units
1	Week	Beach	-
2	$Angle_1$	Station	-
3	$Angle_2$	Beach	-
4	Exposure	Beach	-
5	Salinity	Beach	‰
6	Temperature	Beach	°C
7	NAP	Station	m
8	Penetrability	Station	N/cm^2
9	Grain size	Station	mm
10	Humus	Station	%
11	Chalk	Station	%
12	Sorting	Station	Mm

Table 27.2. Significant correlations ($\alpha = 0.05$) among explanatory variables.

	Week	$Angle_1$	$Angle_2$	Exp.	Sal.	Temp.	NAP	Pen.	Grains	Hum	Chalk	Sort.
Week	1		−0.61		−0.31				0.41	−0.36		0.31
$Angle_1$		1	0.39						−0.32			
$Angle_2$			1	−0.45					−0.77	0.39		−0.48
Exp.				1		−0.53	−0.72		0.60	−0.32	0.40	0.56
Sal.					1			−0.40	−0.58	0.38	−0.33	−0.49
Temp.						1			−0.40			
NAP							1					
Pen.								1				
Grains									1	−0.35	0.45	0.75
Hum										1		−0.37
Chalk											1	0.79

27.3 Analysing the data using univariate methods

In this section, the species data are analysed by converting them into a diversity index (Magurran 2004). We used the Shannon–Weaver index (with base $\log_{10}$). The shape of the Shannon–Weaver index was similar to that of the species richness Because of this similarity the Shannon–Weaver index can also be seen as an indicator for the number of different species (at least for these data). Figure 27.1 shows a dotplot for the Shannon–Weaver index. Stations are grouped by beach. There are no stations with considerably larger or smaller values.

We now compare the Shannon–Weaver index with the explanatory variables. Quinn and Keough (2002) followed a similar approach and applied regression techniques on the diversity index. Table 27.3 shows the correlation between the Shannon–Weaver index and each of the 12 explanatory variables.

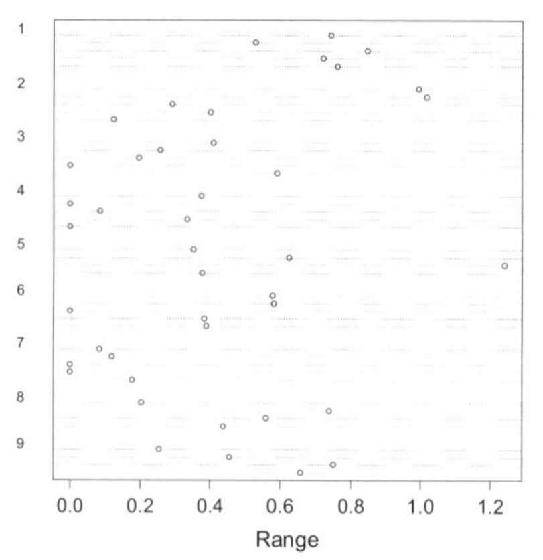

Figure 27.1. Cleveland dotplot of the Shannon–Weaver index. The horizontal axis shows the value at each site, and the vertical axis shows the 45 stations, grouped by beach.

The Pearson correlation coefficients indicate that exposure, salinity and NAP were significantly related to the Shannon–Weaver index. This correlation coefficient does not take into account that the 45 samples come from nine beaches (each beach had five stations). To take into account this two-way data pattern, between-beach and within-beach correlations can be calculated (Snijders and Bosker 1999). Results are presented in the third and fourth column of Table 27.3. The between-beach correlation is the correlation between the beach averages. Humus, chalk and sorting were significantly correlated to the diversity index, and exposure and NAP have a correlation just below the significance level. The within-beach correlation is a correlation coefficient corrected for the beach differences. NAP had by far the largest correlation, followed by grain size, chalk and sorting. For some of the variables, within-beach correlations could not be calculated because these explanatory variables had the same value at all five stations on a beach.

Figure 27.2 shows a pairplot of the explanatory variables. The variables $angle_1$, humus, chalk and sorting all have an observation with a large value. Because this might cause problems in some of the statistical analyses, a square root transformation was applied on these variables. In fact, the correlations in Table 27.3 were obtained after transforming these four variables. Using different transformations for explanatory variables is generally not recommended as it complicates the interpretation. However, in this case it seems justified as the explanatory variables represent different things, also measured in different units.

Pearson correlations between all explanatory variables were calculated and all were smaller than 0.8 in the absolute sense. However, the correlations between (i) $angle_2$ and grain size, and (ii) sorting and chalk were between 0.75 and 0.8, indicating a certain degree of collineairty.

Table 27.3. Pearson correlation, between-beach correlation and within-beach correlation coefficients between Shannon–Weaver index and each of the explanatory variables. Variables in bold typeface are significant at the 5% level. Angle$_1$, humus, chalk and sorting were square root transformed; see the text.

Variable	Cross-Correlation	Between-Beach	Within-Beach
Week	−0.08	−0.14	
Anglc$_1$	−0.03	0.03	0.05
Angle$_2$	0.20	0.33	
Exposure	**−0.39**	−0.64	
Salinity	**0.34**	0.57	
Temperature	0.12	0.20	
NAP	**−0.70**	−0.66	**−0.75**
Penetrability	0.00	0.10	**−0.29**
Grain size	−0.18	−0.50	**0.34**
Humus	0.30	**0.74**	0.06
Chalk	−0.23	**−0.75**	**0.31**
Sorting	−0.21	**0.72**	**0.30**

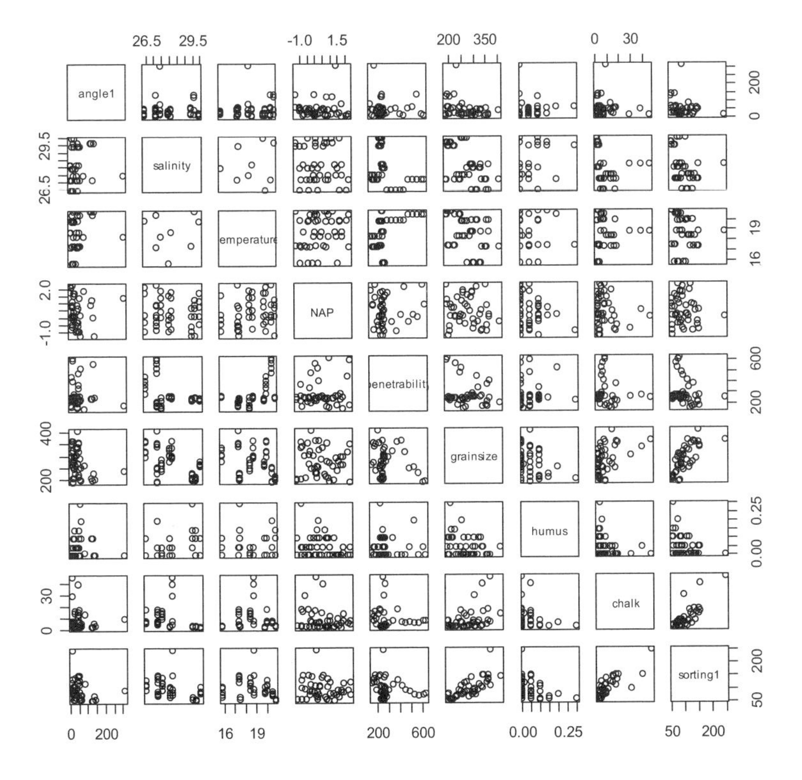

Figure 27.2. Pairplot of some of the explanatory variables.

Additive modelling

We now model the relationship between the Shannon–Weaver index and the 12 explanatory variables. One possible candidate technique is linear regression, but this means that we are imposing linear relationships on the data, whereas in ecology, most relationships are non-linear. We will use a more flexible technique, namely additive modelling. In this technique, the data will tell us the shape of the relationship, and the linear relationship is a special case (Crawley 2002). We will also use regression trees as it is even more flexible in finding relationships.

In a linear regression model, the following mathematical formulation is used to relate the response variable and explanatory variables:

$$H_i = a + b_1 X_{1i} + \ldots + b_p X_{ip} + e_i$$

where b_j are regression coefficients, X_{ij} is the value of the jth explanatory variable (j = 1,..,12) at the ith sample (i = 1,..,45) and H_i is the Shannon–Weaver at site i. In additive modelling the following mathematical formulation is used:

$$H_i = a + f_1(X_{1i}) + \ldots + f_p(X_{pi}) + e_i \qquad (27.1)$$

where $f_j(X_j)$ is a smoothing function (e.g., a LOESS curve) along the jth explanatory variable. Because the response variable is modelled as a sum of smoothing curves, the model in equation (27.1) is called an additive model. An example of a semi-parametric additive model is:

$$H_i = a + b_1 X_{1i} + f_2(X_{2i}) + \ldots + f_p(X_{pi}) + e_i$$

X_1 is fitted parametrically whereas $f_j(.)$ are smoothing functions. We used the additive modelling approach, to find a relationship between the Shannon–Weaver index and the 12 explanatory variables. Because week and exposure are factors, the following initial model was used:

$$Y_i = a + \text{week}_i + f_2(\text{angle}_i) + \text{exposure}_i + f_4(\text{salinity}_i) + \ldots + f_{12}(\text{sorting}_i) + e_i$$

This model resulted in wide point-wise confidence bands for the smoothers, which is probably due to the collinearity we discussed earlier. Perhaps it would have been better to omit one of the explanatory variables angle and grain size. The same holds for sorting and chalk and week and exposure. Besides, in our experience it is better to do a forward selection instead of a backwards selection in additive modelling, as the technique cannot deal well with collinearity. The results of the forward selection are presented in Table 27.4. The AIC was used as model selection tool; the lower the AIC, the better the model. The second column in Table 27.4 contains the AIC values of the additive models with only one explanatory variable. To simplify the analysis, we only used four degrees of freedom for each smoother. The model with NAP was the best model containing a single explanatory variable. The third column (labelled 'NAP +...') contains the AIC value of a model with NAP and each of the remaining variables. Week and exposure were fitted as nominal variables. The model with NAP and week was the best, although differences in the AIC with the model containing NAP and exposure were small. The last column (labelled 'NAP + week +...') contains the AIC of models with three explanatory variables; two of them were NAP and Week. No other combination gave a lower AIC than the model with NAP and Week. Hence, the most optimal model, as judged by the AIC, contained NAP and Week. Competing models are (i) NAP and exposure, and (ii) NAP, week and humus. The last model has large confidence bands for the smoother of humus and is therefore not a serious competing model. However, the model with NAP and exposure has an AIC of −11.02 whereas the model with NAP and Week has an AIC of −13.46. This is a small difference. The explained variance of both models is similar as well (71% for NAP + Week and 68% for NAP + Exposure). A model validation indicated that the model NAP + Week contained some evidence of violation of homogeneity, which was not the case for the NAP + Exposure model. The ecological interpretation of the NAP + Exposure model is easier as well. After all, the interpretation of Week is difficult. Furthermore, there is a certain amount of collinearity between them as the lowest exposure values were measured in the first week. These arguments are all in favour of the NAP + Exposure model, and we therefore present the results of this model. The model is of the form:

$$\mathrm{H}_i = a + f_2(\mathrm{NAP}_i) + \mathrm{Exposure}_i + e_i$$

A useful aspect of additive modelling is that the partial fit of smoothers can be visualised (Figure 27.3). NAP showed a general downwards pattern. In fact, a cross-validation to estimate the degrees of freedom of the smoother (Chapter 7) shows that the smoother can be replaced by a parametric term. This also means that the additive model can be replaced by a linear regression. It should be noted that NAP was one of the few explanatory variables having a linear effect! Hence, starting with additive modelling was not a waste of time. Furthermore, one should also keep in mind that we are applying a model with a normal distribution. So, all the potential problems for linear regression discussed in Chapter 5 apply here as well. A model validation indicated normality and homogeneity of residuals, but if the model is used for prediction, it is theoretically possible to obtain negative Shannon–Weaver values. Switching to a Poisson distribution is not an option here as the diversity index takes on small non-integer values. The numerical output of the additive model is as follows:

	Estimate	std. err.	*t* ratio	*p*-value
Intercept	0.55	0.09	5.86	<0.001
factor(exposure)10	−0.05	0.10	−0.44	0.66
factor(exposure)11	−0.31	0.11	−2.87	0.007

Other numerical information is given by

	edf	Chi-sq	*p*-value
s(NAP)	4	53.94	<0.001

R-sq.(adj) = 0.627. Deviance explained = 67.8%.
Variance=0.038. $n = 45$

Table 27.4. AIC values obtained by forward selection in the additive model. The second column contains the AIC for a model with one variable. The third column contains the AIC value of a model with NAP and each of the remaining variables. The last column contains the AIC of models with three explanatory variables; two of them were NAP and Week.

Variable	One Variable	NAP +...	NAP + Week +...
Week		**−13.46**	
$Angle_1$	30.89	10.29	−6.38
$Angle_2$	25.00	−2.91	−6.06
Exposure		−11.02	−10.26
Salinity	20.99	−8.68	−5.6
Temperature	23.44	−6.51	−6.15
NAP	**3.70**		
Penetrability	32.58	10.42	5.79
Grain size	30.77	6.11	−9.51
Humus	29.70	4.23	−13.31
Chalk	30.94	−0.30	−1.58
Sorting	31.42	6.07	−9.24

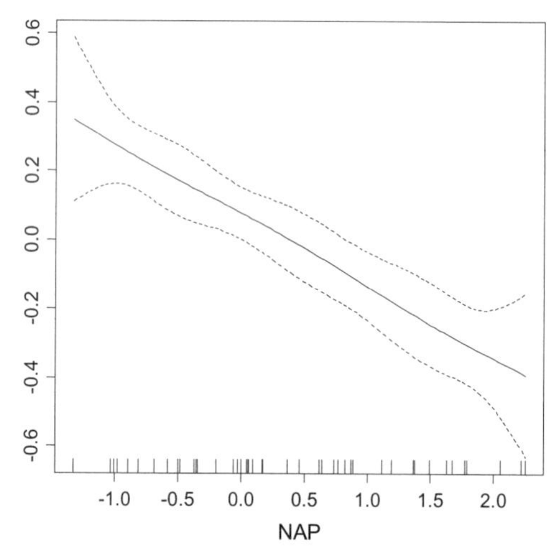

Figure 27.3. Partial fit of NAP in the additive model with exposure. Dotted lines represent 95% confidence bands. The symbols "|" indicate the value of NAP at each station. The smoother has 4 degrees of freedom although further model improvement is possible by decreasing the degrees of freedom. The horizontal axis shows the NAP gradient, and the vertical axis is the contribution of the smoother to the fitted values.

Regression trees

Regression trees (Chapter 9) are useful exploratory tools for uncovering structure in the data. They can be used for screening variables and summarising large datasets. These models result in classification rules of the form: if $x_1 < 0.5$ and $x_2 > 20.7$ and $x_3 < 0$ then the predicted value of y is 10, where x_1, x_2 and x_3 are explanatory variables and y is the response variable. Regression trees deal better with non-linearity and interaction between explanatory variables than regression and GAM models.

The regression tree was applied on the Shannon–Weaver index values. The optimal tree after pruning (Chapter 9) is presented in Figure 27.4. In order to predict the (Shannon–Weaver) index values, one follows the path from the top (root) to the bottom (leaf). The diversity index is first split according to whether the values of NAP are larger (left branch) or smaller than 0.1685 (right branch). Hence, all stations with NAP larger than 0.1865 are in the left branch (these stations are close to the dunes), and the right branch contains the samples with NAP smaller than 0.1685 (these are close to the water line). Observations in the left branch can again be divided into two groups, namely those with very high NAP values (>1.154) and between 1.154 and 0.1685. Values at the end of the tree contain the average value of a group of observations. Observations with a high NAP value (the leftmost leaf) have an average Shannon–Weaver value of 0.051. Week 2 is important for the observations with intermediate NAP values. Observations in the main right branch have higher Shannon–Weaver values, especially if exposure is not 11 (level 3). These results are in line with those of the additive model.

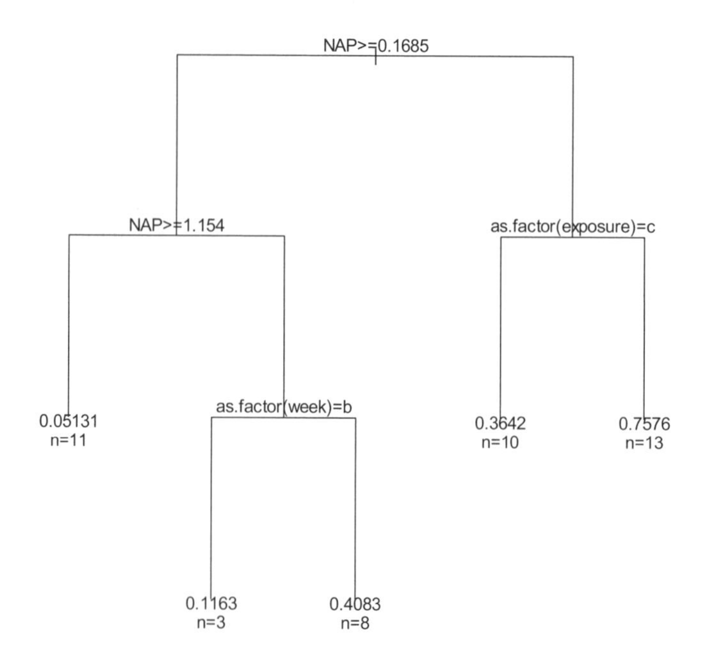

Figure 27.4. Regression tree for the Shannon–Weaver index.

27.4 Analysing the data using multivariate methods

In this section, multivariate analysis is used to analyse whether there are relationships between the species and environmental variables. The species data contain many zeros, and therefore, the application of canonical correspondence analysis and redundancy analysis is inappropriate because of patchy species and double zeros. One option is to take totals per beach and use that in a CCA or RDA. The advantage of this approach is that it reduces the percentage of zeros. But it also reduces the size of the data as there are only nine beaches. An alternative is to apply a special transformation before applying the RDA and to visualise either chord or Hellinger distances (Chapters 12, 28). Legendre and Gallagher (2001) showed that this approach is less sensitive to double zeros and therefore to the arch effect. Here, we present the results of the RDA. Various studies have shown that the Chord distance performs well for ecological data, and therefore, we use the Chord transformation. This means that the RDA software applies a particular transformation on the species data prior to the actual RDA algorithm. As a result, the distances between observations are two-dimensional approximations of Chord distances (Chapters 10 and 12). The triplot is given in Figure 27.5. We wanted to make all species equally important in the analysis, and therefore, the correlation matrix was used (Chapters 12 and 29). A square root transformation was applied to the species data to down-weight the effect of abundant species. Note that this transformation is not related to the Chord transformation; the square root transformation reduces the influences of large values, and the Chord transformation rescales the data of each station.

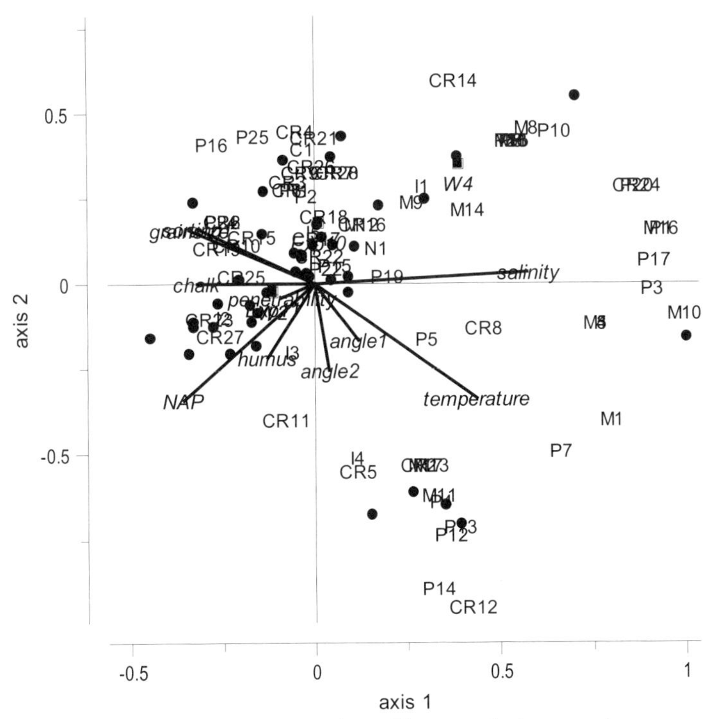

Figure 27.5. RDA triplot for the RIKZ data. The correlation matrix was used. The first two axes explain 45% of the total sum of all canonical eigenvalues (0.40) which corresponds to 18% of the variation in the species data. All explanatory variables were used.

To determine which explanatory variables are important, a forward selection procedure was applied. It uses explained variance as criteria to indicate which explanatory variable should be included/excluded in the model. The forward selection procedure indicated that week, salinity, NAP, humus, temperature and exposure are important (Table 27.5) and that all other variables can be dropped posure are important (Table 27.5) and that all other variables can be dropped from the model. The RDA was refitted using the same Chord transformation, but now only the significant explanatory variables are used. It was felt that if one of the levels of a nominal variable was significant, all levels should be kept. Therefore, all week and exposure levels were included and not just those that were significant in the forward selection. The resulting triplot is presented in Figure 27.6. The amount of explained variance by the first two axes is similar as in Figure 27.5, namely 17%. Such a low number is common in ecological studies. Full details of the numerical output (using only the significant explanatory variables) are given in Table 27.6 and show that the first two eigenvalues are approximately similar. The first two axes explain 33% of the variation in the species data that can be explained with the six explanatory variables. The first two axes explain 52% of this, which works out as 17% of the species variation. The small difference between this percentage and the 18% using all 12 variables means that no important variables were omitted in the analysis. If the first eigenvalue is considerably larger than the second, then interpretation of the triplot should first be done along the horizontal axis. However, this is not the case here as the second eigenvalue is only marginally smaller than the first.

Table 27.5. Order of importance, *F*-statistic and *p*-values for the RDA analysis. W_4, W_2, $Expo_{10}$ and $Expo_{11}$ are nominal variables representing week and exposure levels. We had to drop W_3 as it was collinear with exposure.

Variable	*F*-statistic	*p*-value
W_4	**2.648**	**0.021**
Salinity	**1.955**	**0.004**
NAP	**2.303**	**0.007**
Humus	**1.568**	**0.064**
Temperature	**1.689**	**0.035**
$Expo_{10}$	**1.693**	**0.014**
Chalk	0.864	0.572
penetrability	0.364	0.971
Exp11	0.031	1.000
Angle1	0.682	0.713
Sortin	0.108	1.000
Angle2	0.223	1.000
Grain size	0.037	1.000
W2	0.137	1.000

The triplot indicates that NAP and humus are correlated with each other, but negatively with temperature. A group of *Crustacea* species appear at high values of NAP. Most *Mollusca* species are on the right-hand side of the triplot, which corresponds to low NAP values and high salinity and temperature and in week 4. It is possible that week, salinity and temperature have a certain degree of collinearity as temperature and salinity have the same value for all stations on a beach, and sampling on a particular beach is carried out on the same week. This can easily be verified by replacing the dots by letters A, B, C, D, E, F, G, H and J to visualise which observations are from the same beach (Figure 27.7). We avoided the letter 'I' as it was not clear in the graph. Results indicate that stations of the same beach are close to each other in the triplot. This means that these stations have similar species composition and environmental conditions. In particular beach B seems to have high temperature.

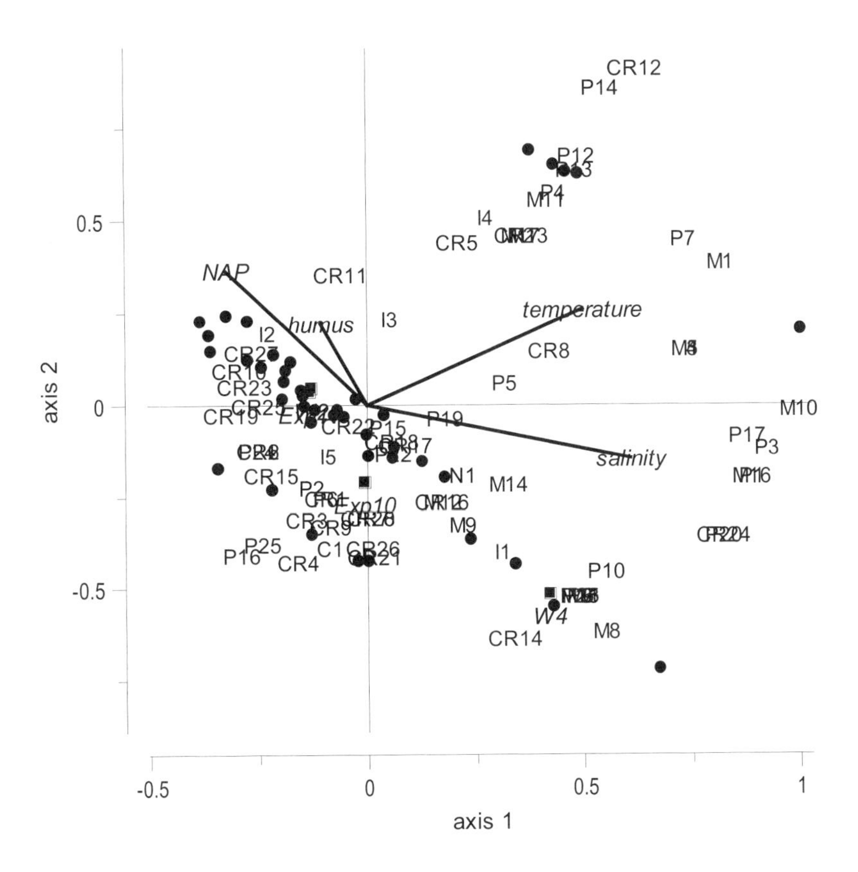

Figure 27.6. RDA triplot for the RIKZ data. The correlation matrix was used. The first two axes explain 45% of the total sum of all canonical eigenvalues (0.40) which corresponds to 18% of the variation in the species data. Only significant explanatory variables were used, namely week, salinity, NAP, humus, temperature and exposure.

Table 27.6. Numerical output for RDA using only the significant explanatory variables. The total variation is 1 and the sum of the canonical eigenvalues is 0.33.

	Axis 1	Axis 2
Eigenvalue	0.10	0.07
Eigenvalue as % of total variation	10%	7%
Eigenvalue as cumulative % of total variation	10%	17%
Eigenvalue as % sum of all canonical eigenvalues	30%	22%
Eigenvalue as cumulative % sum of all canonical eigenvalues	30%	52%

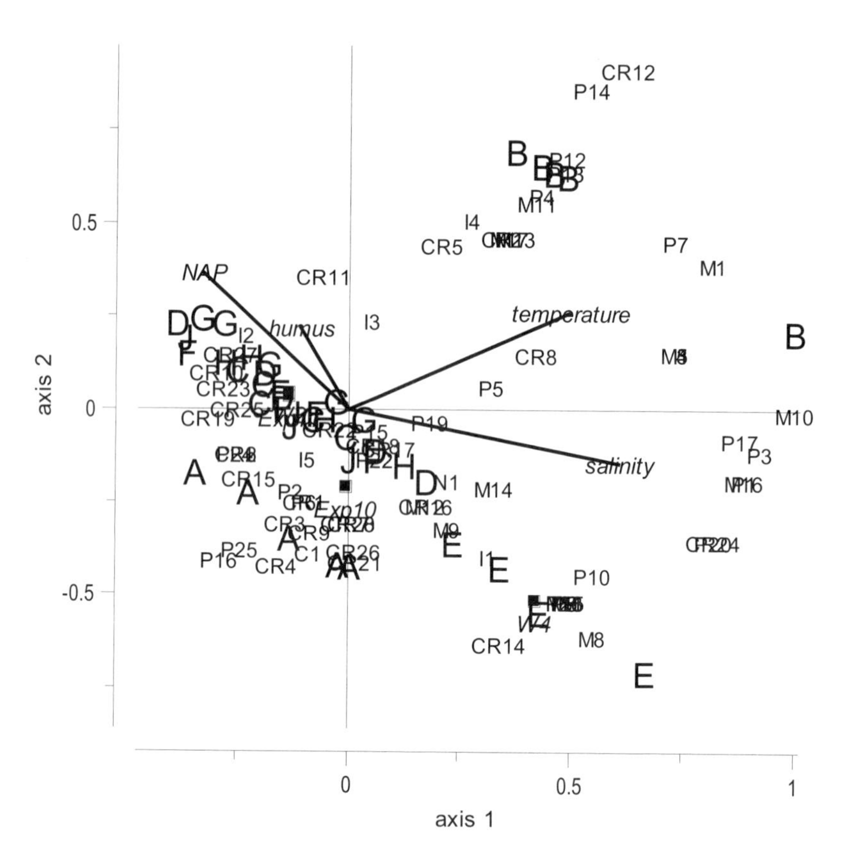

Figure 27.7. Same as Figure 27.6 except that dots are replaced by letters A, B, C, D, E, F, G, H and J for the nine beaches.

Variance partitioning in RDA

We now address the question how much variation in the species data is purely related to exposure. Borcard et al. (1992) showed how variance partitioning is used in CCA and RDA to estimate the contribution of particular groups of explanatory variables. Here, one group is exposure (the two dummy variables) and the other group is defined by all other explanatory variables. Let us call them 'the others'. To estimate the pure exposure effect, the following calculations were carried out (the same transformation as above was used):

1. Apply RDA with exposure as the only explanatory variable.
2. Apply RDA with all other explanatory variables.
3. Apply partial RDA with exposure and 'the others' as the explanatory variables.
4. Apply partial RDA in which exposure is used as explanatory variable and 'the others' as covariables.
5. Apply RDA with all explanatory variables.

Using the eigenvalues of each analysis (Table 27.7), the variance partitioning can be made (Table 27.8). Results indicate that the variation in the species data purely related to exposure is 5%. The other explanatory variables explain 29%. There is also a certain degree of shared information (collinearity) between them: 6%.

Table 27.7. Results of various RDA analysis.

RDA	Explanatory Variables	Sum of All Canonical Eigenvalues
A	Exposure	0.12
B	The others	0.35
C	Exposure with 'the others as covariate	0.05
D	The others with exposure as covariate	0.29
E	All explanatory variables	0.40

Table 27.8. Variance partitioning in RDA. The total variation of the species data in RDA is scaled to 1.

Component	Formula	Variance	%
Pure Exposure	C	0.05	5
Pure 'Others'	D	0.29	29
Shared	A-C or B-D or 1-E-C-D	0.06	6
Residual	1-E	0.60	60
Total		1	100

27.5 Discussion and conclusions

In this chapter, various different statistical techniques were applied to the RIKZ data. The reason for applying a large number of statistical techniques is that each method shows something different. By focusing on only one technique, e.g., RDA, vital information is missed. Besides, one can be more confident about the biological interpretation of the results, if all (or most) methods show something similar.

The statistical methods applied in this chapter can basically be divided into two groups, univariate analysis and multivariate analysis.

Univariate analysis

The Shannon–Weaver index was calculated for all 45 observations. For these data, it can also be seen as a measure of the different number of species per station (as it shows a similar pattern).

According to beach studies, elsewhere in the world (Brown and McLahlan 1990), the most diverse stations can be expected on beaches with fine sand and flat slopes. In this study, the flat beaches are B, G and I, and beaches with fine sand are A, B, G and I. Simple graphs can be used to show that the most diverse stations were on beaches A, B and E. These are the most northern sampled beaches indicating a possible regional effect (Janssen and Mulder 2004, 2005).

Application of all statistical techniques on the diversity indices indicated that there is a significant relationship between the diversity indices and NAP. According to Beukema (2002), the maximum number of macrobenthic species is to be found in the area of the intertidal just between mean tidal level and the low water line. Beukema (2002) studied this phenomenon on the intertidal of a tidal flat area in the Dutch Wadden Sea. Janssen and Mulder (2004, 2005) suggest that this also holds for the inter-tidal of sandy beaches.

An ecological explanation can be found in the combined effects of increasing good food conditions from the high water level down to the low level line (the NAP gradient) and the increasing predation pressure, by shrimp and young fish, up from the low water line to the high water line. Furthermore, there is a negative effect of exposure (the higher the exposure, the lower the diversity) and a week effect. The effect of exposure is in line with literature on this subject (Brown and McLahlan 1990). Exposed beaches in the world show low diversity and abundances compared with sheltered beaches. There is no clear (or strong) relationship between the diversity index and any of the other explanatory variables.

Multivariate analysis

RDA using a special data transformation was applied to the square root transformed abundances of 75 species. The explanatory variables week, salinity, NAP, humus, temperature and exposure were significant at the 5% level. The first gradient represents a NAP versus temperature/salinity gradient. The distribution of the stations in the triplot indicates a major NAP gradient.

Combining the results

The results of all analyses indicated that there is a strong relationship among NAP, exposure and diversity index, and also with the multiple species data. The angle of the beach was not important in any of the analyses. This holds for both $angle_1$ and $angle_2$. Applying a square root transformation on $angle_1$ did not improve the results. Most analyses indicated that there was a strong week effect. Because sampling took place only in June, it is likely that seasonal patterns will exist if sampling were to take place in other months as well. Using the Shannon–Weaver index, the variable NAP turned out to be the most important explanatory variable. In the multivariate analysis, it was also important.

To further optimise the sampling protocol, a statistical experimental design on the 2002 data should be applied. The most obvious way to improve the design is to select beaches that are stratified on exposure (three beaches with low exposure, three beaches with medium exposure and three beaches with high exposure). However, this might be difficult for beaches along the Dutch coast. Perhaps an alternative definition of exposure could be established.

Possible further analysis

Besides the environmental variables, spatial co-ordinates of the beaches were available. Using variance partitioning, the contribution of spatial variation can be determined (and filtered out). Let us make a distinction between spatial explanatory variables and all other explanatory variables (including exposure), denoting these as 'the others'. To determine the variation in the species data that is purely related to spatial locations of the beaches, the following calculations need to be carried out:

1. Apply RDA with the spatial explanatory variables.
2. Apply RDA with the other explanatory variables.
3. Apply partial RDA with the spatial explanatory variables, and the others as covariables.
4. Apply partial RDA in which the other variables are used as explanatory variables and the spatial variables as covariables.
5. Apply RDA in which all explanatory variables are used.

The spatial variables consisted of x and y co-ordinates. These are denoted by x and y, respectively. Following Borcard et al. (1992), the following derived spatial variables can be used: x, x^2, x^3, y, y^2, y^3, xy, x^2y, and xy^2. A linear combination of these spatial variables models most spatial gradients. For example, the function xy models a diagonal (Northeast to Southwest) gradient. However, for the multivariate RIKZ data, we could not apply this analysis as some of the explanatory variables have the same value for all five stations per beach. This means that there is a strong collinearity between the spatial variables that also have the same value per beach.

28 Multivariate analyses of South-American zoobenthic species — spoilt for choice

Ieno E.N., Zuur A.F., Bastida R., Martin, J.P., Trassens M. and Smith G.M.

28.1 Introduction and the underlying questions

Defining spatial and temporal distribution patterns of a soft-bottom benthos community and its relationship with environmental factors has been a common task of many coastal marine ecologists. However, the choice of the most appropriate statistical tools for benthic data has been subject to considerable debate among researchers.

Several research programmes aimed at studying the dynamics of benthic species and their environment, have been carried out at South American estuarine and coastal areas during the last few decades (Bemvenuti et al. 1978; Ieno and Bastida 1998; Lana et al.1989; Giménez et al. 2005) focusing not only on the importance of commercial benthic species but also on the rapid habitat fragmentation and deterioration (Elías 1992b; Elías and Bremec 1994) that have resulted from different levels of human impact and that have reduced the available feeding areas for birds.

Samborombón Bay (Buenos Aires, province, Argentina) is an area of major importance in the life cycle of a large number of organisms that play key roles in the food web of the ecosystem (Ieno and Bastida 1998; Martin 2002). Bivalves, crustaceans and especially polychaetes that inhabit the inter-tidal and tidal flats represent an important link in the food chain from primary producers to predators such as resident and migratory birds and fishes. Samborombón Bay is used by migratory nearctic and austral shorebirds from September to April; the main species preying on macrozoobenthos during the annual stop over migrations are the Red Knot (*Calidris canutus rufa*), the White-rumped Sandpiper (*Calidris fuscicollis*), the Hudsonian Godwit (*Limosa haemastica*), the American Golden Plover (*Pluvialis dominica*) and the Two-banded Plover (*Charadrius falklandicus*) (Myers and Myers 1979; Blanco 1998; Ieno et al. 2004). Direct observation and fecal and gizzard analysis have shown that polychaetes along with decapod crabs are the most important items in the diet of these shorebird species during their stay at Samborombón Bay.

The data analysed here, which have been introduced in Chapter 4, come from a benthic-monitoring programme in the autumn-spring period in 1997 at 30 stations from the inter-tidal mudflats of San Clemente Channel in the south of Samborombón Bay. The area is characterised by a benthos displaying high species densities and low species diversity (Ieno and Bastida 1998). The monitoring plots (transects) on San Clemente Channel were selected to represent the major macrobenthic habitats due to the overwhelming abundance of short-lived and fast-growing polychaete species (Ieno and Bastida 1998; Martin 2002). In the original study, the main goal was to determine the relationship between waders and their inter-tidal food supply. The sampling scheme was determined by the topography of San Clemente Creek as well as the feeding behaviour of the secondary consumers.

The underlying question we aim to answer with this particular data set is whether the environmental variables (sediment composition) had any effect on the macrobenthic species data. We also want to determine whether there are differences between transects and seasons and, in particular, whether the two transects close to the eastern part of the study area are different. The main advantage of these data is that only a few species (infaunal data) were monitored at a very low spatial scale. This makes the following statistical explanation and interpretation easier for the reader to understand.

The aim of this chapter

Because we have multiple species, multiple sites and multiple explanatory variables, we are in the world of multivariate analysis. This means that we have to choose from methods such as principal component analysis, redundancy analysis, correspondence analysis, canonical correspondence analysis, non-metric multidimensional scaling (NMDS), the Mantel test, discriminant analysis, etc. This choice is primarily determined by the underlying questions and the type of data. We require a method that can deal with both species and environmental data; hence RDA, CCA, or the Mantel tests are the most obvious candidates. It is then a matter of carrying out a thorough data exploration to identify the appropriate technique. One not only has to choose a particular technique, but also certain settings have to be selected within that technique.

To illustrate the thinking and decision-making process, we present two different analyses. In the first analysis, which we call the 'careless approach', we highlight the pitfalls of not thinking seriously about the underlying questions and demonstrate some of the common mistakes made by inexperienced users of statistical software packages. We also present the way (we think) the data should be analysed. Hence, the aim of this chapter is to show some of the difficulties and dangers of being spoilt for choice.

28.2 Study site and sample collection

Fieldwork was carried out at the extreme southeastern section of the Samborombón Bay, at the narrow navigation San Clemente Channel (Bértola and Ferrante 1996) (Figure 28.1). The area can be described as a typical temperate South American saltmash, characterised by the conspicuous epifaunal burrowing mud crab, *Chasmagnathus granulata*, together with a dominant vegetation composed by *Spartina densiflora*, *S. alterniflora*, *Salicornia ambigua*, *S. virginica* and *Sirpus maritimus*. Infaunal polychaetes worms mostly dominated the inter-tidal area (Ieno and Bastida 1998; Martin 2002). Sediment texture was dominated by fine and very fine sand with a very soft and flocculent mud fraction of 15–30% (< 63 μm). A full description of the area is given in Ieno and Bastida (1998) and Martin (2002). A photograph of the sampling area is given in Figure 28.2.

Two benthic sampling programmes were carried out on the inter-tidal sediments of the San Clemente channel, one in early May and one in mid-December 1997. Three transects were selected, two in the eastern margin and one in the western margin of the channel, and 60 sediment samples were taken in total to determine infauna abundance and composition. Samples were sieved through a 0.5 mm mesh, and infauna was stored in 70% ethanol (Holme and McIntyre 1984).

The use of a small dredge limited adequate sampling and estimation of the big epibenthic grapsid *Chasmagnathus granulata*; therefore, this species was not included in the data analysis. A list of all available explanatory variables is given in Table 28.1.

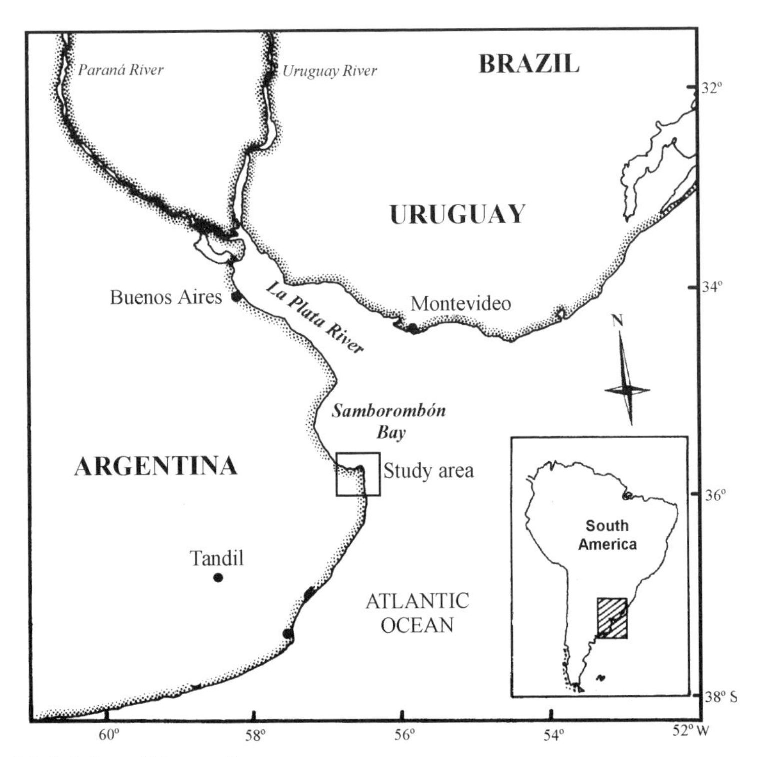

Figure 28.1. Map of the study area.

Figure 28.2. Inter-tidal flats in the study area with the so-called 'Cangrejales' (crab zone) and *Spartina* sea grass bed.

Table 28.1. List of available explanatory variables.

Explanatory Variable	Remarks
MedSand	Continuous variable measured in percentage (%). Medium sand (250–500 μm) (Wentworth 1922)
FineSand	Continuous variable measured in percentage (%). Fine and very finesand (63–250 μm) (Wentworth 1922)
Mud	Continuous variable measured in percentage (%). Particles passing the 64 μm sieve (silt-clay) fraction. (Wentworth 1922).
OrganMat	Continuous variable measured in percentage (%). Organic Matter determined by oxidation method (Walkley and Black 1965)
Transect	Nominal variable: 3 monitored transects (A, B and C), with values 1, 2, and 3.
Season	Nominal variable with values 0 (Southern hemisphere autumn, May) and 1 (Southern hemisphere spring, December) identifying the time of the year that sampling took place.
Channel	Nominal variable with values 0–1 to identify location of transects at both margins of San Clemente channel. (Transect A = Eastern sector) (B and C = Western sector).

28.3 Data exploration

The species data were already used in Chapter 4 to illustrate some of the data exploration techniques. To avoid repeating the same graphs, we will just summarise what we found there. The species data did not contain any large outliers, but two species (*U. uruguayensis* and *N. succinea*) had many observations with zero abundance. The general impression was that a square root transformation on the species would be beneficial as it brings the species within the same range, but this also depends on which statistical method will be applied in the next step.

Except for transect and seasonal effects, we did not look at the explanatory variables in Chapter 4. A pairplot for the continuous explanatory variables shows that mud and fine sand are highly correlated (Figure 28.3). The scatter of points for mud and organic material also indicate a strong linear relationship, although this is not supported by the correlation coefficient, which has a value of only 0.55. This is probably due to one observation, which has high organic material but low mud. Anyway, the scatter of points clearly indicates that mud is collinear with fine sand and with organic material, so we decided to omit mud from any further analyses. One could even argue that fine sand and organic material are negatively related, but it is less obvious as the patterns with mud.

The last question we address in the data exploration is whether the environmental conditions differ by transect. Cleveland dotplots (Figure 28.4) show that transect 2 has considerable higher median sand values and transect 1 has a higher mean organic material. This indicates that the environmental conditions differ considerably per transect, and the implication of this will be discussed in the next section.

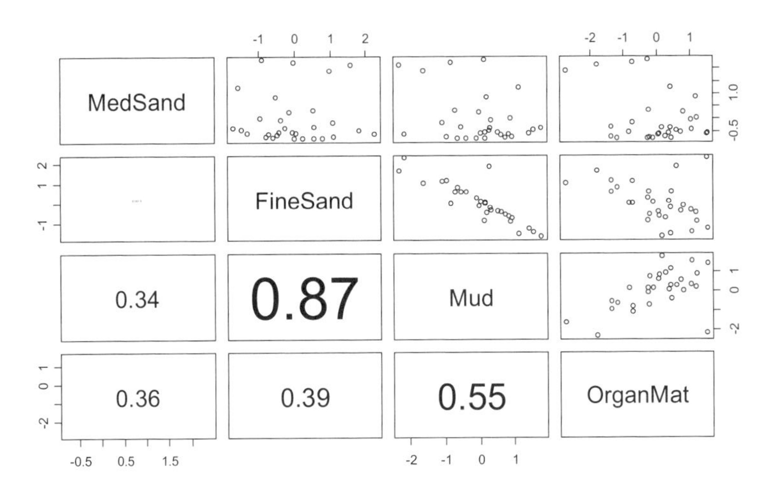

Figure 28.3. Pairplot of all continuous explanatory variables. The lower diagonal panels contain the (absolute) correlation coefficients, and their font size is proportional to the value.

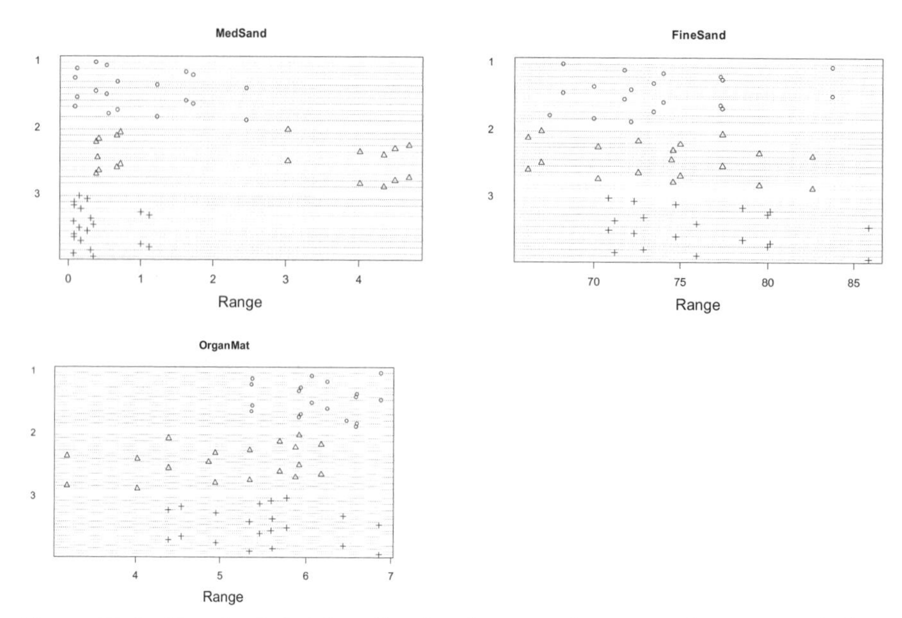

Figure 28.4. Cleveland dotplots for the three continuous explanatory variables medium sand, fine and organic material conditional on transect. The y-axis corresponds to an observation (grouped by transect), and the horizontal axis shows the value for each observation.

The careless approach

The data exploration indicates that two species have lots of observations with zero abundance. To avoid problems with collinearity, mud was removed from the analysis. There is also evidence that the environmental conditions differ per transect. In the careless analysis it is now time to start clicking in a software package. Due to the large number of zeros, PCA, RDA, CA and CCA are less suitable, and probably NMDS combined with the Mantel test would seem to be a good approach.

How-it-should-be-done

Thinking a bit more deeply, we can argue that based on the data exploration, only the Jaccard index or Sørensen index should be used. Owing to the environmental differences among the three transects, the permutation method in the Mantel test needs to be done conditional on transect (i.e., permutations should only be made within transects and not between transects). We also think that a special data transformation combined with an RDA (Chapter 12) might be useful as it visualises Chord distances, a measure of association that is suitable for this type of data.

28.4 The Mantel test approach

Recall from Chapters 10 and 26 that the Mantel test starts with two matrices; one for the species (**Y**) and one for the explanatory variables (**X**). It then calculates a distance matrix for the species data ($\mathbf{D_Y}$) and for the explanatory variables ($\mathbf{D_X}$). These distance matrices are of dimension 58-by-58 as there are 58 observations (two observations were omitted due to absence of all species).

The first problem is to choose a measure of association for the species data and for the explanatory variables. Let us discuss the explanatory variables first. The data matrix **X** is of the form:

	S_1	S_2	S_3		S_{58}
MedSand	2.46	1.23	0.56	.....	0.15
FineSand	72.17	70.60	67.22	.....	70.87
Orgmat	6.59	6.60	6.48	.....	5.78
Season	0	0	0	.....	1
Channel	0	0	0	.....	1

To calculate $\mathbf{D_X}$ we need to define the association between S_i and S_j for every i and j combination between 1 and 58. An obvious choice is to use Euclidean distances, but there is one problem with this; it will be dominated by fine sand as it has the largest variance. Hence, such a distance matrix $\mathbf{D_X}$ would mainly represent differences in sites due to fine sand differences, whereas we want to have a distance matrix that represents all environmental variables. To do this, we have to make sure that each explanatory variable is within the same range. The two easiest ways are either to normalise each explanatory variable (Chapter 4) or apply ranging (divide all the observations of a particular explanatory variable by its maximum observed value). Because the data also contain two nominal variables with values 0 and 1, we decided to apply ranging. As a result all explanatory variables are rescaled between zero and one and the Euclidean distance function now weights each variable in the same way.

The careless approach

In the careless approach we assume that we have access to a software package that allows us to choose from 40 different measures of association.

As to the species data, we picked 10 different measures of association, and each time, we applied the Mantel test. The choices were the Jaccard (S) and Sørensen (S) indices, a variation on these two that give triple weight to joint presence (S), the Ochiai index (S), Euclidean distance (D), Chord distance (D), and Whittaker's index of association (D), the Bray–Curtis index (D), the Chi-square distance (D) and the simple matching coefficient (S). An 'S' indicates that the measure of association is a similarity measure, and a 'D' stands for distance measure. The Mantel test works with distance matrices and the default conversion is distance = 1 − similarity. The results are given in the first two columns of Table 28.2. The R_M statistic is the Pearson correlation between elements of $\mathbf{D_Y}$ and $\mathbf{D_X}$. The permutation tests did not take into that the environmental conditions differed per transect. Imagine the following situation:

	S_1	S_2	S_3	S_4	S_5	S_6	S_7	..	S_{58}
X_1:	1	2	3	96	97	98	1	..	2
X_2:	3	2	1	97	97	97	2	..	3
X_3:	2	2	2	92	93	94	3	..	4
T	1	1	1	2	2	2	3	..	3

Suppose that the *X*s are the environmental variables, S_1 to S_{58} are the observations and *T* identifies the three transects. If we were to permute the observations arbitrarily, the high values of transect 2 would end up in other transects. But the underlying principle of permutation methods and bootstrapping is that we generate data, which are only slightly different compared with the original data. Hence, it is better to permute the data only within transects. The same holds if we work on distance matrices and apply permutation methods to obtain significance values. If we ignore the transect effect, and if there are large differences between transects in terms of environmental conditions, the *p*-values will be too small. There may also be other dangers for the inexperienced user. For example, if the 'similarity to distance conversion method' is changed to

$$d_{ij} = \sqrt{1 - s_{ij}^2}$$

then the R_M statistic involving a similarity coefficient (Jaccard, Sorenson, etc.) becomes considerably larger (Table 28.3). This means stronger relationships! Similarly, care must be taken in quantifying the association between $\mathbf{D_Y}$ and $\mathbf{D_X}$. If instead of using the Pearson correlation coefficient, we had chosen the Spearman correlation coefficient we would have obtained slightly smaller values for R_M. We must also be aware of the importance of carrying out sufficient numbers of permutations. Most software packages permit the user to set the number of permutations. We have used 9999 permutations in Table 28.2 and Table 28.3, but if we had set it to 999, then each time we would have run it, the *p*-values would have been slightly different. If we were unscrupulous, we could have repeated the analysis 5 or 10 times and selected the smallest *p*-value.

We decided to conclude that whatever measure of association we choose, there is a significant relationship between species dissimilarities and environmental differences at the 58 sites. The only exception is the Chi-square distance matrix, but this is probably due to rare species.

Table 28.2. Results of the Mantel test using various different measures of association for the species data. The number of permutations was 9999. The similarity to distance conversion was: $d = 1 - s$, where s is the similarity and d the distance. The p-values marked with * were obtained by permuting the observations only within a transect. The test statistic is the Pearson correlation coefficient.

Index	R_M-statistic	p-value	p-value*
Jaccard	0.242	<0.001	<0.001
Sørensen	0.197	<0.001	0.001
Triple weight to joint presence	0.165	<0.001	0.003
Ochiai	0.179	<0.001	<0.001
Euclidean	0.071	0.040	0.263
Chord	0.078	0.020	0.116
Whittaker's index of association.	0.091	0.007	0.064
Bray–Curtis	0.106	0.004	0.023
Chi-square distance	0.035	0.218	0.018
The simple matching coefficient	0.177	<0.001	0.002

Table 28.3. Results of the Mantel test using various different measures of association for the species data. The number of permutations was 9999. The similarity to distance conversion was: $d = \text{sqrt}(1 - s^2)$, where s is the similarity and d the distance. The p-values marked with * were obtained by permuting the observations only within a transect. The test statistic is the Pearson correlation coefficient.

Index	R_M-statistic	p-value	p-value*
Jaccard	0.287	<0.001	<0.001
Sørensen's	0.281	<0.001	0.001
Triple weight to joint presence	0.272	<0.001	0.003
Ochiai	0.274	<0.001	<0.001
The simple matching coefficient	0.260	<0.001	0.002

How-it-should-be-done

Following the appropriate statistical analysis strategy, we decided to use only the Jaccard index. Permutations conditional on transect were applied, and we used the default conversion of similarity to distance (distance = 1 − similarity). We found $R_M = 0.242$ ($p < 0.001$) using the Pearson correlation coefficient. Table 28.2 also contains p-values obtained by permuting conditional on transect, and these are given in the column labelled p-values*. We also carried out the BVSTEP procedure. Recall from Chapter 26 that this method carries out a forward selection on the explanatory variables and it tries to find the optimal set of explanatory variables so that R_M is maximal. It gave $R_M = 0.265$ using only medium sand, channel and organic material.

As to using the BVSTEP in the careless approach, it would have been very easy to apply the procedure to all 10 measures of association and pick the one we liked best.

28.5 The transformation plus RDA approach

In the 'how-should-it-be-done' approach, we also carried out a special data transformation followed by an RDA. Recall that this approach allows one to visualise Chord distances. The triplot in Figure 28.5 shows a clear zonation of the sites by transect. *L. acuta* was abundant in transect A, *U. uruguayensis* in transect B, which has high medium sand values and the other two species are abundant in transect C. The explained variation by all explanatory variables is 30%, and the first two eigenvalues are 0.21 and 0.7. A forward selection indicated that only medium sand and Channel are important.

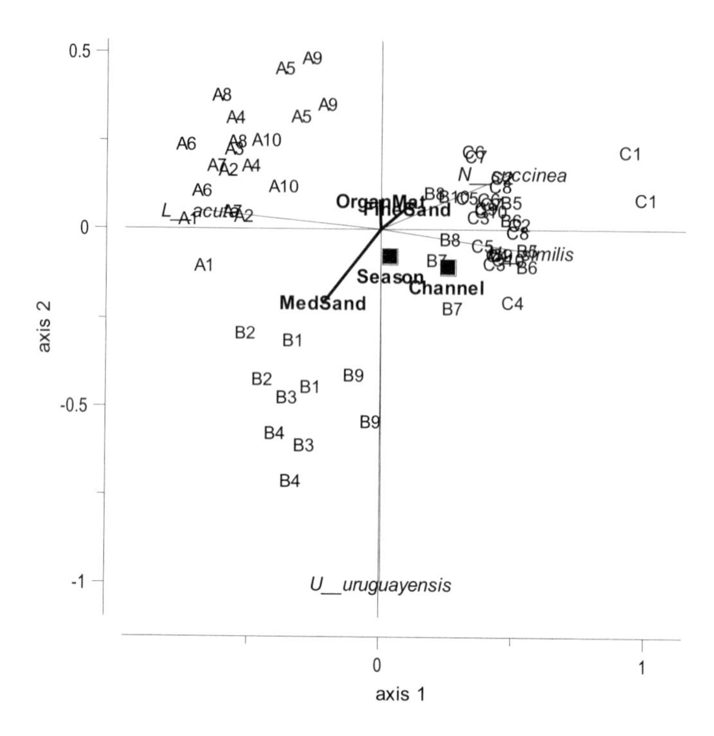

Figure 28.5. RDA triplot obtained using a special data transformation. Distances between observations are two-dimensional approximations of Chord distances.

28.6 Discussion and conclusions

In this chapter we presented two different analysis strategies. In the first we were careless and we showed how we could obtain 'nice' but not necessarily correct results. We found that there is a strong relationship between sites in terms of species composition and environmental conditions, whatever measure of association is used. In the second ('how-it-should-be-done') analysis we carefully considered what to do. We thought carefully about the questions and decided *a priori* (or better: during the data exploration) which methods to use. We found that there was a clear difference among the three transects, and that the most important explanatory variables were medium sand and channel.

As has been highlighted, our research has focused in detail on the relationship between benthic fauna and abiotic factors. For the same transects we determined the average density in which the different birds species foraged and the density in which the different prey species occurred in the soft-bottom sediment layers. However, this complementary information has been left aside as it is not the primary goal of this case study chapter.

Nevertheless, some general ecological implications were raised from benthic food stocks and feeding opportunities. The fact that we found a clear transect effect in the species-environmental relationship may explain why most Golden Plovers feed on the medium sand patches of transect B that are characterized by the presence of the fiddler crab, *Uca uruguayensis*. Plovers defend feeding territories and rely on eyesight for prey detection. Fiddler crabs are particularly active on the surface next to their burrows during low tide and at the same time are vulnerable to predation.

We also detected a channel effect that clearly denoted a higher abundance of the ragworm, *Laeonereis acuta* in transect A, which is likely to be the major prey of the White-rumped sandpiper in the study area. Sandpipers are the most important foragers and have been shown to feed in large social flocks on medium-size *L. acuta* mainly recorded on the eastern sector of San Clemente Channel. Although the explanatory variables in the data seemed to explain part of the variation in the species composition, there are some intriguing conditions that made *H. similis* more abundant in transect C than the others. We suspect that both high turbidity due to the frequent resuspension of softer substratum and sedimentation processes underlie the environmental heterogeneity of the western sector of San Clemente Channel.

The mistakes made in the careless approach may sound silly, but the authors of this book have reviewed various submitted manuscripts in which the number of permutations was only 199 because it was the default number in the software package that was used.

Ultimately, the best way you can make sure that you have applied the most appropriated statistical tool is to think carefully about the underlying questions before applying complicated methods.

Acknowledgements

We would like to thank Fundacion Mundo Marino who kindly allowed us to collect the data and use their experimental facilities. Sergio Moron and Jorge Rebollo provided tremendous assistance in the field. This study was partially supported by AGENCIA and CONICET (Argentina). Special thanks to Barry O'Neill for commenting on an earlier draft.

29 Principal component analysis applied to harbour porpoise fatty acid data

Jolliffe, I.T., Learmonth, J.A., Pierce, G.J., Santos, M.B., Trendafilov, N., Zuur, A.F., Ieno, E.N. and Smith, G.M.

29.1 Introduction

In this chapter we apply principal component analysis (PCA) to data on blubber fatty acid composition in harbour porpoises. Various decisions need to be made when using PCA. As well as showing the usefulness of the dimension reduction achieved by PCA for these data, the implications of these decisions will also be illustrated. The next two sections of the chapter describe the data and the statistical technique respectively. A data exploration section is followed by the main section (Section 29.5), which describes and discusses the results of PCA for the data. Interpretation of principal components can sometimes be difficult and several methods have been suggested for simplifying interpretation. These will be discussed in Section 29.6, and one of them will be illustrated on the fatty acid data. The chapter is completed by a short discussion section.

29.2 The data

The harbour porpoise (Figure 29.1) is probably the most abundant cetacean in British waters (Hammond et al. 2002). However, harbour porpoises, as most cetaceans, are subject to various threats and pressures throughout their range, and increased concern for the status of harbour porpoises has led to the need for more information on the species, including their diet.

Diet is an important aspect of the ecology of marine mammals. Changes in prey type or availability have the potential to affect the distribution, body condition, susceptibility to disease, exposure to contaminants, reproductive success and, ultimately, survival of most marine mammals, including harbour porpoises.

Figure 29.1. Harbour porpoises off Scotland.

Traditional methods used to estimate diets in marine mammals, such as stomach contents analysis, are often limited and estimates can be biased. Fatty acids in predator body tissues have led to the use of fatty acid analysis as a method for understanding the diet of marine mammals. Fatty acids have the potential to act as tracers of diet, with the fatty acid composition of tissues reflecting the average diet ingested over a period of days or months. The influence of prey fatty acid signatures on predator fatty acid profiles has been clearly shown in captive feeding experiments of fish, squid and seals (Kirsch et al. 1998, 2000; Stowasser et al. 2006). Fatty acids in the blubber of different species of free-living marine mammals have also been shown to reflect their diet (for example, Brown et al. 1999; Hooker et al. 2001; Iverson et al. 1995, 1997; Smith et al. 1997; Walton et al. 2000).

However, in addition to diet, other factors also have the potential to influence the fatty acid composition of blubber. Therefore, prior to using fatty acid analysis of blubber samples from stranded porpoises to examine diet, it is important to determine what factors, other than diet, may influence the fatty acid profile. These factors may include decomposition state and blubber thickness.

Blubber samples from 89 harbour porpoises stranded around the Scottish coast between 2001 and 2003 were used for fatty acid analysis, and they form the basis of the present data set. All tissue samples were removed from the left side in front of the dorsal fin during the postmortem examinations by the vets at the Scottish Agricultural College, Inverness. During the postmortem examinations, data were collected on sex, total body length, weight, girth, etc. but these (explanatory) variables were not used in the current analysis because, based on stomach contents analysis, porpoise diet is known to vary in relation to body size (Santos et al. 2004).

Lipids were extracted from the inner blubber layer, and individual fatty acids were identified. The normalised area percentage was calculated for 31 fatty acids: 12:0, 14:0, 14:1n-5, 15:0, 16:0, 16:1n-7, 16:2n-6, 16:3n-6, 16:4n-3, 18:0, 18:1n-9,

18:1n-7, 18:2n-6, 18:3n-6, 18:3n-3, 18:4n-3, 20:0, 20:1n-11, 20:1n-9, 20:2n-6, 20:4n-6, 20:3n-3, 20:4n-3, 20:5n-3, 22:0, 22:1n-11, 22:1n-9, 21:5n-3, 22:5n-3, 22:6n-3 and 24:1n-9. Full details of the lipid extraction method are given in Learmonth (2006).

29.3 Principal component analysis

Principal component analysis was described in Chapter 12. It is a dimension reducing technique, which replaces the original variables Y_1, Y_2, … ,Y_N by a smaller number of linear combinations of those variables, while keeping as much as possible of the variation in the original variables.

Let Y_{ij} be the value of variable j (j = 1,..,N) for observation i (i = 1,..,M). Then the value of the kth principal component (PC) for the ith observation is given by

$$Z_{ik} = c_{k1} Y_{i1} + c_{k2} Y_{i2} + \ldots + c_{kN} Y_{iN}$$

For the current data, j denotes the jth fatty acid, i is the ith porpoise, $N = 31$ and $M = 89$. The numbers c_{1j} (j = 1,..,N) are chosen so that the first PC accounts for as much of the variance in the original variables as possible, the numbers c_{2j} (j = 1,..,N) are chosen so that the second PC accounts for as much of the variation as possible in the original variables, subject to the constraint that the second PC is uncorrelated with the first. Third, fourth, … PCs can be similarly constructed. In theory as many as N PCs can be constructed but, in practice, if the first few account for most of the variation in the original variables, then often only those PCs are of interest.

Various decisions need to be made in conducting a PCA. The first decision is whether to base the PCA on a covariance matrix or correlation matrix. Both analyses will be presented and discussed below. It is fair to say that in most circumstances a correlation-based approach is more appropriate, but there are occasions when a covariance-based analysis is suitable; see Jolliffe (2002, Section 3.3) for more discussion.

Another decision is how many PCs are needed to adequately represent the variation in the original data. Some methods for making this decision are discussed in Chapter 12 and will be illustrated in this chapter. For more details and other methods, see Jolliffe (2002, Chapter 6).

A third decision concerns the values of the c_{kj}. It is their relative values for a particular PC that determines the nature of that component, but for detailed interpretation of a PC it is necessary to decide on a particular 'normalization' of the c_{kj}. The two main normalizations are

$$\sum_{j=1}^{N} c_{kj}^2 = 1 \quad \text{or} \quad \sum_{j=1}^{N} c_{kj}^2 = \text{var}(Z_k)$$

where var(Z_k) is the variance of the kth PC. Each of these normalizations will be illustrated and explained below.

29.4 Data exploration

Boxplot and Cleveland dotplots (Chapter 4) were generated for the 31 fatty acids to identify any extreme values and to determine whether the data required transformation. The boxplots (Figure 29.2) and dotplots (not shown here) indicate a few extreme values, for example, fatty acid 18:1n-7. However, transformation of the data was not required for the fatty acids variables.

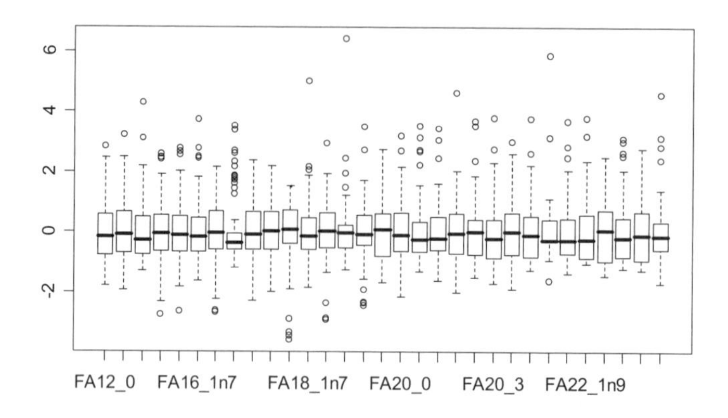

Figure 29.2. Boxplot of 31 fatty acids from the inner blubber layer of 89 harbour porpoises. The data were normalized so that all boxplots are on the same scale. Not all 31 labels are plotted.

29.5 Principal component analysis results

Covariance-based PCA

The first PC accounts for 71% of the total variation in the original variables and the second PC for a further 17%, so that a total of 88% of the variation in the original 31 dimensions can be represented by just the two dimensions defined by the first two PCs. Furthermore, the first two PCs look easy to interpret. Although they are linear combinations of all 31 variables, most of the constants c_{1j} and c_{2j} (the loadings) are small. The first PC is $Z_1 = 0.81Y_6 + 0.41Y_{30} + \dots$, where none of the other 29 terms in the linear combination has a loading greater than 0.21, and only six of them exceed 0.10. Similarly the second PC is $Z_2 = 0.78Y_{11} - 0.49Y_{30} + \dots$, with no other loadings greater than 0.21 and only four greater than 0.10.

However, the massive reduction in dimensionality reduction and easy interpretation of the first two PCs is not as impressive as it might seem. It is caused by large differences in the variances of the original 31 variables. These range from 83.8 for Y_6 (16:1n-7) to 0.00154 for Y_{17} (20:0). When there are big discrepancies between variances, a covariance-based PCA will often tell you little more than which variables have the largest variances. It is no co-incidence that the three variables that dominate the first two PCs are those with the three largest variances. A correlation-based PCA is far more appropriate for these data.

Correlation-based PCA

Table 29.1 provides the variances of, and cumulative proportion of variance accounted for by the first 10 PCs, and Figure 29.3 plots the variances (also known as eigenvalues; see Chapter 12).

Table 29.1. Eigenvalues and eigenvalues expressed as a cumulative proportion.

Axis	Eigenvalue	Cumulative Proportion	Axis	Eigenvalue	Cumulative Proportion
1	13.04	0.42	6	1.28	0.76
2	4.00	0.55	7	0.98	0.79
3	2.23	0.62	8	0.96	0.82
4	1.75	0.68	9	0.82	0.85
5	1.32	0.72	10	0.74	0.87

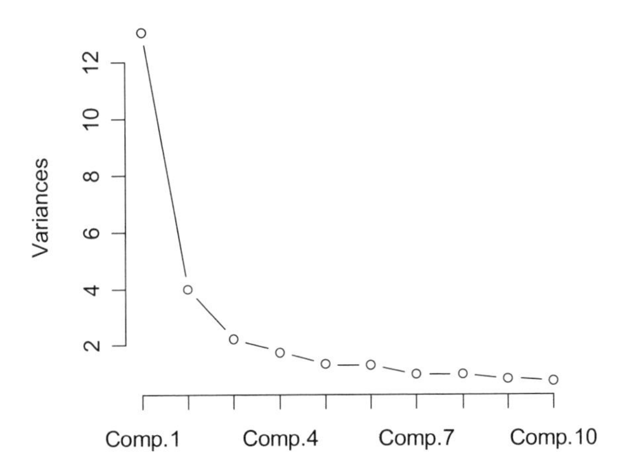

Figure 29.3. First 10 eigenvalues (variances) obtained by the PCA based on the correlation matrix.

There are various proposed rules for deciding how many PCs are needed to adequately represent the variation in the original variables; see Jolliffe (2002, Chapter 6). Some of the simpler ones are described in Chapter 12 of the current text. One method is to plot eigenvalues (variances) against PC number as in Figure 29.3, giving a so-called scree plot, and look for an 'elbow' in the plot. Deciding the location of the 'elbow' is subjective. Here, component 3 probably gives the most plausible 'elbow', implying that only two PCs need to be retained, but other readers looking at the plot might disagree.

A second rule is to retain the minimum number of components needed to account for a certain (high) percentage of the total variation. A common choice for the threshold is 80%, and it can be seen from Table 29.1 that it needs 8 PCs to achieve this. A third popular rule for correlation-based PCA is to retain all PCs whose variances exceed 1, the average variance of PCs for a correlation-based analysis. This rule implies that six PCs are needed.

The three rules are contradictory. The suggestion made after visual examination of the scree plot is almost certainly too small. Two PCs only account for 55% of the total variation, which is distinctly unimpressive. The six PCs suggested by looking at sizes of individual eigenvalues account for 76% of the variation, and this might be acceptably close to 80%.

It is clear that six or more PCs are needed if a substantial part of the original variation is not to be lost. Having said that, what follows concentrates on describing and interpreting the first two components.

The columns of Table 29.2 labelled PC1 and PC2 give the loadings for the first two PCs. The columns labelled SC1, SC2 will be explained in Section 29.6.

From Table 29.2, the first PC is

$$Z_1 = -0.22Y_1 - 0.24Y_2 - 0.23Y_3 - \ldots + 0.25Y_{30} + 0.07Y_{31}$$

where the normalization $\sum_{j=1}^{N} c_{kj}^2 = 1$ is used for the loadings, and var(Z_1) = 13.04. If the normalization $\sum_{j=1}^{N} c_{kj}^2 = \text{var}(Z_k)$ had been used instead, the first PC would be

$$Z_1 = -0.83Y_1 - 0.88Y_2 - 0.85Y_3 - \ldots + 0.93Y_{30} + 0.24Y_{31}$$

This PC is obtained from the first one simply by multiplying it by a constant, the standard deviation of the original Z_1, so that a plot of the values of the observations for a given PC (the PC scores) looks exactly the same for the two normalizations, apart from relabelling the axis to reflect the change of scale. Both normalizations are in common use, and you may see either, depending on which computer software produces your PCs.

The first normalization is fundamental to the derivation of PCs, but the second has the advantage that, for a correlation-based PCA, the loadings are equal to the correlation between the component and each variable. Hence, PC1 for the current data has correlation –0.83 with Y_1, correlation –0.88 with Y_2, and so on.

A third normalization is also sometimes implicitly used. When plotting PC scores, they are sometimes normalized to have unit variance for each PC. This implies that the loadings in the first normalization are *divided* by the standard deviation of the PC (to get the second normalization the loadings in the first were *multiplied* by this standard deviation).

Table 29.2. Loadings for the first two PCs obtained by the PCA based on the correlation matrix (denoted PC1, PC2), and for the first two SCoTLASS components (denoted SC1, SC2) described in Section 29.6.

Variable	PC1	PC2	SC1	SC2	Variable	PC1	PC2	SC1	SC2
Y_1 12:0	−0.23	0.13	−0.15	0.00	Y_{17} 20:0	0.18	0.16	0.00	−0.35
Y_2 14:0	−0.24	0.13	−0.36	0.00	Y_{18} 20:1n-11	0.15	0.22	0.00	−0.46
Y_3 14:1n-5	−0.23	−0.17	−0.17	0.05	Y_{19} 20:1n-9	0.18	−0.18	0.00	0.01
Y_4 15:0	−0.16	0.27	0.00	0.00	Y_{20} 20:2n-6	0.20	−0.14	0.00	−0.12
Y_5 16:0	−0.05	−0.08	0.00	0.00	Y_{21} 20:4n-6	0.23	−0.16	0.29	0.00
Y_6 16:1n-7	−0.24	−0.21	−0.13	0.22	Y_{22} 20:3n-3	0.16	−0.09	0.00	−0.03
Y_7 16:2n-6	−0.08	0.28	0.00	0.00	Y_{23} 20:4n-3	0.25	0.08	0.23	0.00
Y_8 16:3n-6	−0.01	0.05	0.00	0.00	Y_{24} 20:5n-3	0.24	−0.13	0.37	0.00
Y_9 16:4n-3	0.16	0.13	0.00	−0.17	Y_{25} 22:0	0.11	0.06	0.00	−0.05
Y_{10} 18:0	0.24	0.02	0.26	0.00	Y_{26} 22:1n-11	0.12	0.30	0.00	−0.44
Y_{11} 18:1n-9	0.03	0.28	0.00	0.00	Y_{27} 22:1n-9	0.13	0.07	0.00	−0.14
Y_{12} 18:1n-7	0.16	−0.20	0.00	0.00	Y_{28} 21:5n-3	0.25	−0.01	0.29	0.00
Y_{13} 18:2n-6	0.07	0.34	0.00	−0.22	Y_{29} 22:5n-3	0.24	−0.09	0.39	0.00
Y_{14} 18:3n-6	0.16	−0.07	0.00	0.00	Y_{30} 22:6n-3	0.25	−0.08	0.47	0.08
Y_{15} 18:3n-3	0.09	0.33	0.00	−0.30	Y_{31} 24:1n-9	0.07	0.23	0.00	−0.03
Y_{16} 18:4n-3	0.21	0.16	0.00	−0.45					

Turning to the interpretation of the PCs in this example, the loadings of most of the variables are positive, but those for the first 8, in particular Y_1 – Y_4, Y_6, are negative. This implies that the main source of variation in the data is between those porpoises that have greater than average values for the first group of variables coupled with smaller than average values for the remaining variables and those porpoises with the opposite features. Looking at the second PC, the largest positive loadings are for Y_4, Y_7, Y_{11}, Y_{13}, Y_{15}, Y_{18}, Y_{26}, Y_{31}, whereas the largest negative loadings are for Y_6, Y_{12}. This means that the main source of variation that is uncorrelated with the first PC contrasts those porpoises with greater than average values for the first group of variables, and smaller than average values for the second group, with porpoises having the opposite features.

Using knowledge of the nature of the variables, PC1 appears to order fatty acids in relation to carbon chain length, with the shortest chain fatty acids having the most negative correlations with PC1, and vice versa. It may also be noted that short-chain fatty acids are biosynthesised in marine mammals, whereas most long-chain fatty acids are derived from the diet. There appears to be no simple interpretation of PC2, however. Interpretation could be attempted for all the PCs that are retained, but this is by no means straightforward for these data. Further discussion of interpretation will be given in Section 29.6.

Biplots

A complementary way of displaying results from a PCA is by means of a biplot. Such plots are discussed in Chapter 12. Figure 29.4 gives the so-called correlation biplot for the data. From this plot we can glean information about relationships between the variables, about relationships between porpoises, and the values of different variables for different porpoises.

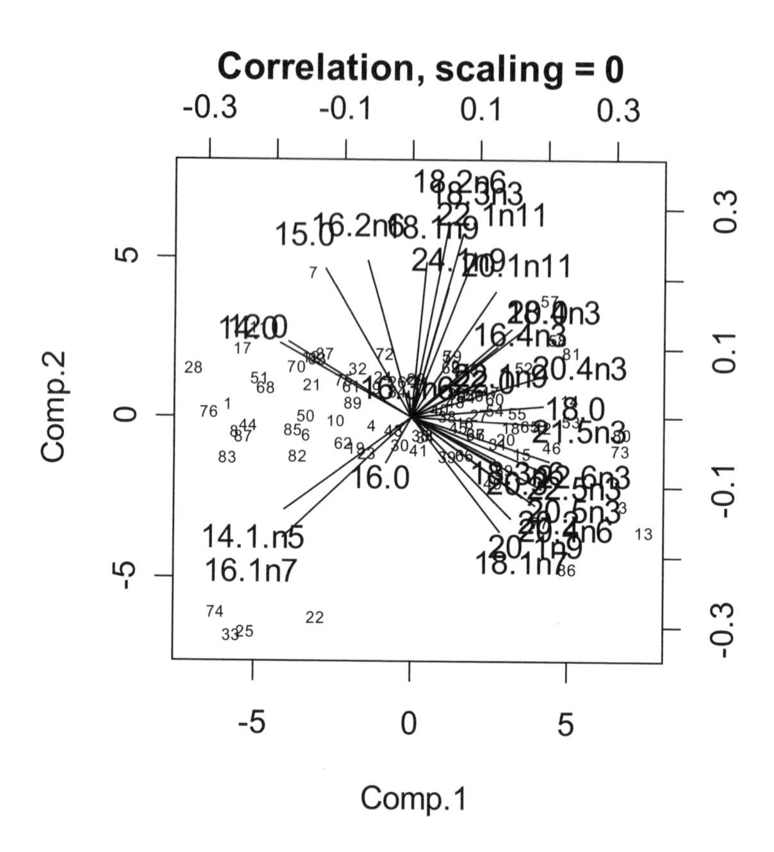

Figure 29.4. PCA biplot. Comp.1 and Comp.2 are the first two principal components.

The angle between lines corresponding to different variables gives an indication of the correlation between the variables (see Chapter 12). Lines pointing in similar directions correspond to variables that are positively correlated, lines pointing in opposite directions to negatively correlated variables, and lines roughly at right angles correspond to almost uncorrelated variables. For example, the lines corresponding to variables Y_3 (14:1n-5) and Y_6 (16:1n-7) point in almost the same direction, so the variables are likely to be highly correlated. Similarly, the pair of variables Y_1 (12:0), Y_2 (14:0) seem likely to be highly correlated with each other, but almost uncorrelated with Y_3, Y_6. The reason for saying 'likely to be' rather than 'are' highly correlated is that the two-dimensional plot only accounts for 55% of total variation in the data, the amount accounted for by the first two PCs. If the first two PCs, and hence the two-dimensional plot, had accounted for more, 90% say, of the variation, the relationship between the angles and the correlations would be more definite. The grouping of fatty acids in the PCA biplot appears to reflect chain length, with all eight of the fatty acids with negative values for PC1 having a carbon chain length of 16 or less and all fatty acids with positive values for PC2, except one Y_9 (16:4n-3), having a carbon chain length of 18 of more.

Turning to the observations, those porpoises that appear close together on the plot are likely to have similar values for the 31 variables. There are two caveats attached to this statement. The first is to note again that the plot displays only 55% of the variation in the data; the second is that the distances approximated on the plot are Mahalanobis distances, not Euclidean distances (see Chapter 12). Despite these caveats, there is still useful information to be gleaned from the plot. For example, porpoises 22, 25, 33, 74 (which had all died due to neonatal death) seem to form a group similar to each other, but separate from the rest, whereas porpoise 7, which had the largest body length out of all the porpoises sampled, seems to be rather different from any of the others.

Finally, the reason for the 'bi' in biplot is that the loadings and scores can be looked at simultaneously. If a line (L_1) is drawn from an observation to the line corresponding to a variable (L_2) so that it intersects L_2 at right angles, the position of the intersection provides information about the value of the variable for that observation. If the intersection with L_1 is a long way from the origin, it suggests that the value of the variable is larger than average for the observation, whereas if the intersection is close to the origin, the variable probably has a fairly average value for that observation. If the line L_2 needs to be extended through the origin in order for L_1 to intersect it at right angles, then the value of the variable is likely to be smaller than average for the observation. For example, porpoise 7 is likely to have larger than average values for variables Y_1, Y_2, Y_4, Y_7, near average values for Y_3, Y_5, Y_6, and below average values for Y_{12}, Y_{19}, Y_{20}, Y_{21} among others. Again the word 'likely' is inserted because the plot represents only 55% of total variation. In fact, examining the original data, porpoise 7 is in the upper quartile for variables Y_4, Y_1, Y_2, in the lower quartile for Y_{19}, Y_{20}, Y_{21}, and in the middle half of values for Y_5, in line with predictions from the biplot. For the other variables mentioned, the predictions are less good, but would be much better if the biplot had represented 90%, say, of the total variation.

29.6 Simpler alternatives to PCA

The linear combinations of variables defined by the PCs are optimal in the sense that they successively account for as much as possible of the total variation in the data. However, they can be difficult to interpret. For example, the loadings of the first two PCs given in Table 29.1 have a wide range of values, which makes them difficult to interpret. In component 2, variables Y_{13} and Y_{15} with loadings 0.34 and 0.33 make a large contribution to the component whereas Y_8 and Y_{10} with loadings 0.05, 0.02 have trivial contributions. But what about Y_1, Y_2? Are their loadings (both 0.13) large enough to consider when interpreting what the component represents?

Although some caution is needed in taking the size of loadings as a definite indication of the importance of variables in a component (Cadima and Jolliffe 1995), it would nevertheless be much easier to interpret a component that had most of its loadings unambiguously large or small with few intermediate values. Several techniques have been developed that make components simpler in this sense. Of course, to achieve this it is necessary to sacrifice something. Simpler alternatives to principal components will typically account for less of the total variation than the PCs and/or they may lose the property of being uncorrelated.

A review of some simpler alternatives to PCA is given in Chapter 11 of Jolliffe (2002), but this is an active area of research and new methods are still in development. The existing methods can be divided into four broad categories:

- Rotation. This is probably the best-known and most popular strategy, and it is borrowed from the related technique of factor analysis (Jolliffe 2002, Chapter 7). The idea is that a decision is made on how many components to retain, perhaps 6 for the current data. This defines a six-dimensional subspace of the 31-dimensional space spanned by the complete data set. Rotation of the axes is then carried out in the six-dimensional space in a way that optimises some 'simplicity' criterion.
- Do a PCA, then simplify the PCs in some way, often by severe rounding of the loadings.
- Find linear combinations that optimise some criterion that simultaneously searches for large variance retention and simplicity.
- Find linear combinations of the variables that successively maximize variance but are subject to additional constraints designed to achieve greater simplicity.

Here, just one technique from the final category will be briefly described and illustrated. As with some other techniques from this category, it can also be formulated as a method in the third category.

The idea for the technique is borrowed from multiple regression, where collinearity between variables can cause difficulties in interpreting a regression equation (Chapter 5). Tibshirani (1996) suggested a technique called the LASSO (Least Absolute Shrinkage and Selection Operator) that addresses this problem. It adds an extra constraint to the usual least squares method of fitting a regression equation.

Jolliffe et al. (2003) adapted the LASSO to the PC context and called the technique SCoTLASS (Simple Components LASSO). As noted in Section 3, PCA finds linear combinations Z_1, Z_2, … of the variables that successively maximize $\text{Var}(Z_k)$ subject to

$$\sum_{j=1}^{N} c_{kj}^2 = 1$$

and subject also to Z_k being uncorrelated with previous Zs. Here c_{kj} is the loading of the jth variable for the kth component. In SCoTLASS an extra constraint is added, namely

$$\sum_{j=1}^{N} | c_{kj} | \leq t ,$$

where t is some threshold, which can lie in the range $1 \leq t \leq \sqrt{N}$, and N is the number of variables. For $t = \sqrt{N}$ the method simply gives the standard principal components, whereas for $t = 1$ it chooses the original variables according to the magnitude of their variances. As t decreases within the range, the components found by SCoTLASS become increasingly simple, with more and more loadings driven towards zero by the extra constraint. At the same time the variance accounted for by the first few components decreases compared with that accounted for in PCA. This behaviour is illustrated in Figure 29.5, which plots the amount of variance retained by the first six SCoTLASS components and the number of zero loadings (to two decimal places) in those components, as t varies.

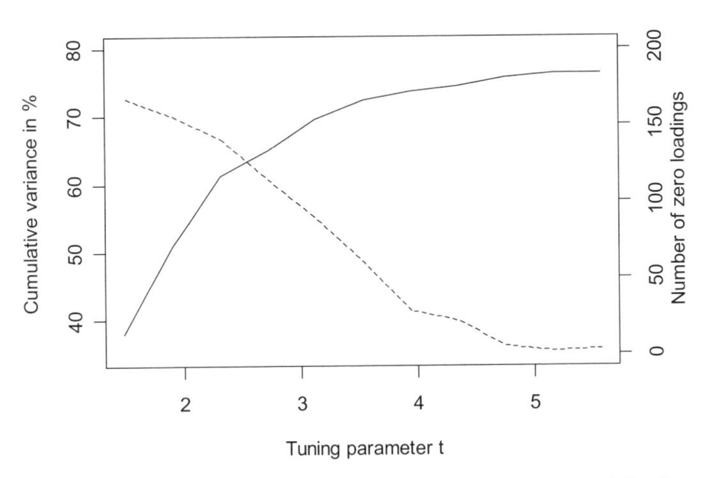

Figure 29.5. For SCoTLASS, the cumulative variance accounted for by the first six components (solid line) and the number of zero loadings in those six components (dotted line) as the tuning parameter varies from 1 to 5.57 (the latter gives PCA).

For example, for PCA (corresponding to t = $\sqrt{N}$ = 5.57) the first six components account for 76% of total variation but there are only three zero loadings. As t decreases to 4.35, 2.72, 1.91, the variance accounted for by six components drops to 74%, 65%, and 51%, respectively. At the same time, the number of zeros increases to 22, 114 and 156. The choice of t is subjective; from a plot such as Figure 29.5, the user can decide at what point the gain in simplicity is more than offset by the loss of variance.

A possible choice here is t = 2.72, and for this value, the loadings produced by SCoTLASS were included in Table 29.2. It can be seen that the numerous zeros in the loadings make the components potentially much simpler to interpret than the PCs. Component 1 is a contrast among variables Y_1, Y_2, Y_3, Y_6 (fatty acids 12:0‡, 14:0^{+}, 14:1n-5‡ and 16:1n-7^{+}, which are ‡primarily from biosynthesis or $^{+}$can be from both biosynthesis and diet, based on Iverson et al. 2004) on the one hand and Y_{10}, Y_{21}, Y_{23}, Y_{24}, Y_{28}, Y_{29}, Y_{30} (fatty acids 18:0^{+}, 20:4n-6*, 20:4n-3*, 20:5n-3*, 21:5n-3*, 22:5n-3^{+} and 22:6n-3*, which are *primarily from direct dietary intake or $^{+}$can be from both biosynthesis and diet, based on Iverson et al. 2004) on the other, with the remaining 20 variables apparently unimportant. The 'cost' of this simplification is a substantial reduction in variance accounted for by the first component, down from 42% to 20%. This is not as bad as it might seem. The reduction in variance for the first six components, from 76% to 65%, is much less, as some of the later SCoTLASS components account for more variance than the corresponding PCs. For example, for the second component, the amount of variance accounted for is 15%, more than the 13% associated with the second PC.

Turning to interpretation of the second component, there is a more straightforward interpretation than for the second PC. Here the second component is mainly measuring the group of variables Y_9, Y_{13}, Y_{15}, Y_{16}, Y_{17}, Y_{18}, Y_{20}, Y_{26}, Y_{27} (fatty acids 16:4n-3*, 18:2n-6*, 18:3n-3*, 18:4n-3*, 20:0^{+}, 20:1n-11*, 20:2n-6*, 22:1n-11* and 22:1n-9*, which are *primarily from direct dietary intake or $^{+}$can be from both biosynthesis and diet, based on Iverson et al. 2004), but contrasted with Y_6 (fatty acid 16:1n-7, which can be from both biosynthesis and diet). There are also a few variables with smaller, but non-zero, loadings whose contributions are equivocal. Fifteen of the 31 variables have zero loadings.

29.7 Discussion

In this chapter, PCA has been illustrated on an interesting data set, and various decisions that are needed in order to implement PCA have been discussed. An alternative to PCA that makes interpretation of the derived variables simpler has also been described and illustrated. More specifically, we looked at the relationship between fatty acids in the inner blubber layer of harbour porpoises stranded around Scotland. This study was used as a preliminary analysis of the fatty acid data prior to using fatty acid analysis as a method to examine diet.

A considerable reduction of dimensionality was achieved for the data set whilst still accounting for a large proportion of the original variation. The reduction was less for correlation-based PCA than for covariance-based PCA, but the latter was clearly inappropriate for these data.

Biological interpretations were found based on some leading PCs and their simpler alternatives. The biplot associated with the PCA also provided useful information, even though the two-dimensional approximation to the data that it provides was not especially good for these data. There was some clear grouping of the fatty acids, which was consistent with known differences in the probable sources of these fatty acids. Long-chain mono- and polyunsaturated fatty acids that were primarily from direct dietary intake were generally separated from fatty acids that are primarily from biosynthesis. Of the fatty acids that can be from both the diet and the biosynthesis, fatty acids with a carbon chain length of 16 or less were grouped with fatty acids that were primarily from biosynthesis.

The four porpoises that had died due to neonatal death were clearly separated in the principal component analysis biplot. This probably reflects the fact that, in neonates, the blubber fatty acids have been obtained through maternal transfer and foetal synthesis prior to birth, as well as the transfer of fatty acids in milk. A study on the fatty acid profile of harbour porpoises from the mid-Atlantic coast of America revealed differences between maternal and foetal blubber, suggesting selective transfer of fatty acids to the foetus (Koopman 2001). Similar observations have also been made in seals, for example, by Iverson et al. (1997).

Although PCA is fundamentally a very simple idea, it is an extremely powerful tool, and new applications and modifications are still being developed. The idea of constructing alternatives to PCA that retain its main objectives while trying to simplify the results is just one manifestation of this. New developments are scattered through the literature of many subject areas. As well as mainstream statistics and ecology, computer science, data mining, genetics and psychology are among the areas of active research in such topics.

Finally, one particular modification of PCA should be mentioned. The data analysed in this chapter are ‘compositional’; that is the sum of the variables is the same (100%) for every porpoise. Special techniques are available for such data (Jolliffe 2002, Section 13.3), although with as many as 31 variables, it is unlikely that conclusions would be changed much by using these methods.

Summarising, the data set studied proved to be unsuitable for covariance-based PCA, but correlation-based PCA revealed informative patterns. The results of this analysis suggest that the relationships between different fatty acids in the inner blubber layer of harbour porpoises, namely the tentative identification of several groups of fatty acids, could reflect differences in their origin, i.e., diet or biosynthesis.

Acknowledgements

This project would not have been possible without the samples, data, help and advice of Bob Reid and Tony Patterson at the SAC in Inverness and the help of various staff members at the FRS Marine Laboratory, Aberdeen. Many thanks to Sarah Canning for the photo of the harbour porpoises.

30 Multivariate analyses of morphometric turtle data — size and shape

Claude, J., Jolliffe, I.T., Zuur, A.F., Ieno, E.N. and Smith, G.M.

30.1 Introduction

Morphometry is the measurement of shape. For most morphometric studies, a large number measurements is required. For example, in Chapter 14, we used six body measurements collected on 1100 sparrows. One option is to analyse each variable separately using univariate methods. But this is a time-consuming process and, moreover, the multivariate nature of the data is not taken into account. A sensible approach is then to apply multivariate techniques, and obvious candidates are often principal component analysis (PCA) and discriminant analysis (DA). These techniques were discussed in Chapters 12 and 14. DA can be used if there is a grouping in the observations, as we had for the sparrow data (different observers, species or sex). Although PCA was successfully applied on the dolphin data in Chapter 29, there is a problem with PCA if it is applied on morphometric data.

Recall that the first axis in a PCA represents the major source of variation in the data. If all morphometric variables are related to the overall size of an organism, then typically the first axis is determined by size. Typically, all morphometric variables have a similar influence on this axis and for that reason it is often called a size axis. Because further axes are required to be orthogonal to this axis, we end up with axes that mostly describe some aspect of the shape of the subject. For this reason, we talk about size and shape axes. In practice, this often means that the first axis explains most of the variation in the data.

The aim of this chapter is to discuss several methods that can be used to deal with the size and shape problem. We will discuss classic approaches based on pre-standardisation and PCA, but we will also present a more recently developed approach that analyses the relative positions of landmarks that can be digitised on a picture or directly on the biological structure of interest (Bookstein 1991). It is referred to as *geometric morphometrics*, and an increasing amount of effort has been devoted to this approach (Bookstein 1991; Dryden and Mardia 1998; Zelditch et al. 2004). To illustrate the methods, a turtle morphometric dataset is used.

The outline of the chapter is as follows. We will first introduce the data and apply a short data exploration. We will then discuss a series of classic approaches, and we present results of some of these approaches for the turtle data. Finally, an introduction into geometric morphometrics is given and a short illustration is provided.

Turtles in the dataset are living aquatic and terrestrial species from Europe, Asia, Africa, South and North America. Most of them are housed in the collections of the Chelonian Research Institute (Oviedo, Florida), some are from the National Museum of Natural History of Paris, and some are from Marc Cheylan, Haiyan Tong and Julien Claude. Institutional numbers of specimens are available in Claude et al. (2003, 2004).

30.2 The turtle data

Since their first occurrence in the late Triassic fossil record (225 million years ago), many families of turtles have appeared and have evolved to marine, freshwater and terrestrial environments. Turtles are unique among living amniotes because they possess an anapsid skull. Despite their primitive appearance their skull exhibits a wide array of morphologies; this has been important to morphologists to establish phylogenetic scenarios (e.g., Gaffney 1975).

The skull is an integrated morphology, and any description should take into account relations between the morphological traits that compose its structure. In addition there is a huge size variation between turtles, something that is not necessarily of interest if we want to focus on shape variation.

Few studies have attempted to describe the morphological variation of the skull in quantitative terms (Claude et al. 2004), although these approaches can demonstrate explicitly the role of evolutionary mechanisms such as selection or constraints.

Twenty-four measurements were obtained from inter-landmark distances on the skulls of 123 species of turtles (Figure 30.1). We will denote these variables as D_1 to D_{24} is this chapter. A description of these variables is presented online. The sample covers all the modern families of turtles (14) and the three main ecological groups (freshwater, marine and terrestrial). In addition, the skull of the more primitive and oldest turtle, *Proganochelys quendstedti* (225 million year old), was added to the data. For some species, several individuals were available to correct for intra-species variation. Hence, for some species the observation corresponds to the average of individual measurements. A list of taxa and effective size for each species is available online.

As we will show in Section 30.4, the variation in size among the 123 species is considerable for this dataset. The overall size of a turtle skull is often described by the midline length of the specimen from the premaxilla to the occipital condyle. This is D_2 in Figure 30.1. Its size ranges from 23 mm for the smallest specimen (*Testudo kleinmani*) to more than 220 mm for the biggest species analysed here.

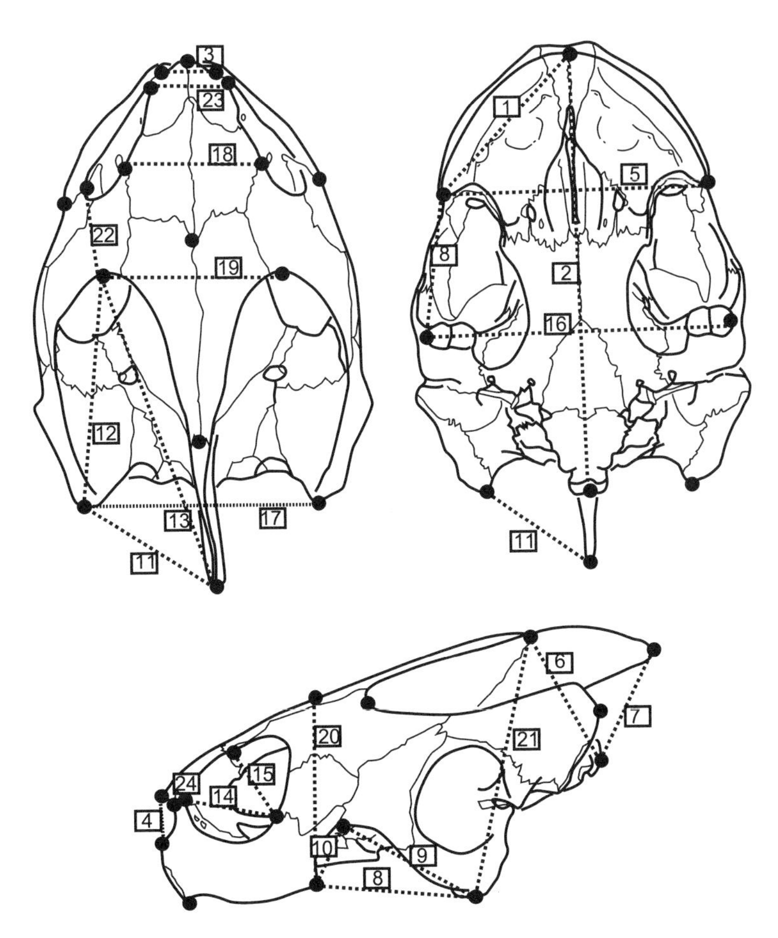

Figure 30.1. Definition of distance measures. The 24 distance variables will be called D_1 to D_{24} in the text. These measures are inter-landmark distance. The 25 landmarks (black dots) will be used in the Section 30.7.

30.3 Data exploration

The first question we address is whether there are any outliers in the data. An outlier is in this case a turtle with an extreme large or small value for one or more distance variables. Graphical tools to identify such observations are the boxplot and Cleveland dotplot; both were discussed in Chapter 4. The boxplot for all distance variables is given in Figure 30.2, and it shows that (i) not all distance variables are within the same range (some vary around 50 mm and others around 10 mm) and (ii) there is considerable variation within each variable. The first point may sound trivial, especially if you look at how these variables were defined in Figure 30.1. For example, D_3 and D_4 are considerably smaller than D_2 and D_{21} and have therefore smaller spread. However, this immediately raises an important question for the PCA; do we consider all distance variables equally important or do we mainly focus on the variables with larger spread? The answer to this question is based on biology and not on statistics. In this case, it was felt that all distance variables were equally important, and hence, we will use a correlation-based PCA.

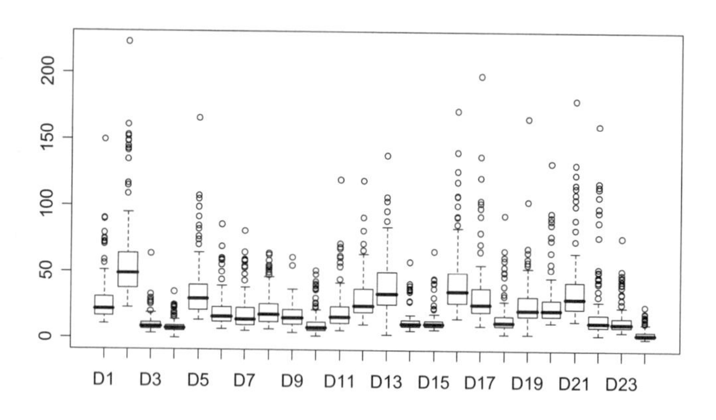

Figure 30.2. Boxplot of the 24 distance variables. Due to limited space along the horizontal axis, not all labels are printed.

To answer the question, whether there are outliers, Cleveland dotplots (not presented here) confirmed the impression already given by the boxplots, namely that some variables would benefit from a transformation. The reason for this is that for some variables there were individual observations, and also groups of observations, that were relatively far away from the centre of the data. Before applying any data transformation, we constructed a pairplot (Figure 30.3), which shows that these observations are not real outliers as they comply with the linear relationship between the variables (see also Section 4.3). For this reason we did not apply a transformation on the distance variables. Note that most of the correlation coefficients (Figure 30.3) between the distance variables are relatively large. This is typical for morphometric data; a turtle with a large skull tends to have large values for all distance variables. Another possible reason for the high correlation coefficients can be a grouping structure in the observations. This is often the case when biological groups segregate the observations. In our example, some turtle species share a similar way of life that may generate resemblance because of convergent evolution. In addition, species belong to different groups (genera, families, superfamilies) that have distinct evolutionary history, and this will directly result in segregation among the observed morphologies. Imagine, for example, that we have two morphologically distinct families in our data. This will produce a correlation structure in the data, even if within a family, variables are unrelated. To illustrate such a grouping structure (and associated problems), suppose we have *artificial* data with values:

Y_1: 1 2 1 3 4 1 2 100 99 80
Y_2: 2 3 4 1 6 3 1 96 90 75
Y_3: 3 2 6 6 3 4 9 99 60 84

Biologically speaking the higher values for the three last observations for the three variables may have evolutionary grounds. As an example, imagine that the three last observations belong to a different taxonomic group (family A versus family B) or to a distinct ecological group (aquatic versus terrestrial forms, for example). For the turtle data we do not have Y_1 to Y_3 with 10 observations each, but D_1 to D_{24} with 123 observations each. However, the following discussion is relevant for the turtle data. The correlation coefficient among Y_1, Y_2 and Y_3 will be close to 1 only because a group of observations has much higher values. We have similar problems in the turtle data; turtles from particular environments have all higher distance values (this can be illustrated with, for example, a Cleveland dotplot).

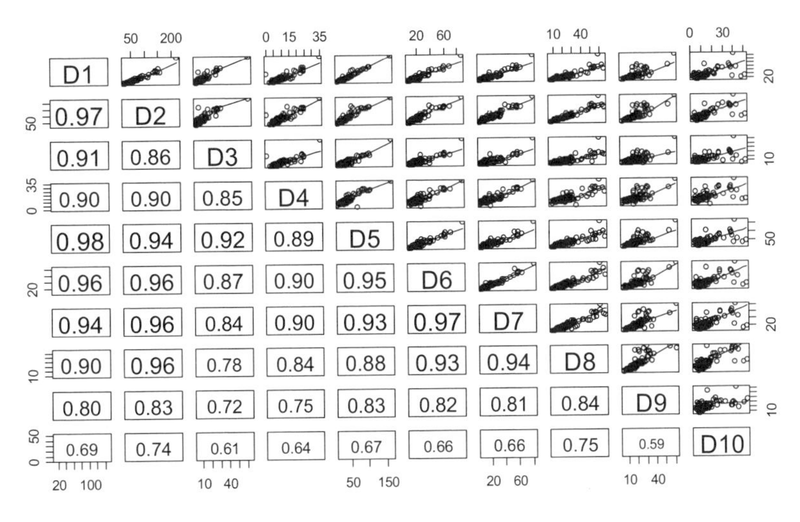

Figure 30.3. Pairplot for selected distance variables (D_1 to D_{10}). The lower panel contains correlation coefficients between distance variables. The font size is proportional to the correlation.

Dependence among observation is more difficult if we want to take phylogeny into account. In this case we have to take into account the shared evolutionary history of pairs of species that will be expressed as the amount of independent morphological divergence. As species sharing a more recent common ancestor are more likely to look similar, the dependence structure among observations can be directly related to the phylogenetic tree structure. Generalised estimating equations (GEE, an extension of GLM) have been used in this context (Paradis and Claude 2002). In addition to GLM, GEE introduces a correlation structure among observation. Paradis and Claude (2002) have explained how a phylogenetic tree can be transformed into this correlation matrix. However, this issue is beyond the scope of this chapter. Other methods have been attempted to correct the phylogenetic effect (Felsenstein 1985; Cheverud et al. 1985; Grafen 1989; Martins and Hansen 1997), following diverse approaches (auto-regression, "contrast" analysis, generalised least squares).

30.4 Overview of classic approaches related to PCA

There is a large literature on the study of size and shape using morphometric data, and various approaches have been suggested for quantifying size and shape. Several of them use PCA-related techniques and we concentrate on these in this section, which is based on Section 13.2 of Jolliffe (2002). By 'classic' we mean that inter-landmark distances are analysed, rather than the co-ordinates of the landmarks themselves. Analyses based on the latter are discussed in Section 30.7.

The idea of using the first principal component (PC) as a measure of size, with subsequent PCs defining various aspects of shape, dates back at least to Jolicoeur (1963). It is fairly conventional, for reasons explained in Jolicoeur (1963), to take logarithms of the data, with PCA then conducted on the covariance matrix of the log-transformed data.

Whether PCA is done on untransformed or log-transformed data, the coefficients or loadings for all variables in the first PC will often be similar, but they will not be identical. Some authors argue that all variables should have *equal* importance in measuring size, so that the first PC should be replaced by an equally weighted combination of all variables. This is known as *isometric size*. Somers (1989), for example, argues that the first PC contains a mixture of size and shape information, and that in order to examine 'shape', an isometric component rather than the first PC should be removed. Several ways of removing isometric size and then quantifying different aspects of shape have been suggested, some of which are discussed below. The problem is that by replacing the first PC by isometric size, at least one desirable property of PCA is lost.

It does not help that none of the terms size, shape, isometric or allometry is uniquely defined, which leaves plenty of scope for vigorous debate on the merits or otherwise of various procedures; see for example Bookstein (1989) or Jungers et al. (1995).

PCA has the two properties that vectors of loadings defining the PCs are orthogonal and that the PCs themselves are uncorrelated. Both of these properties are desirable as they mean that, in two senses, different PCs are measuring clearly different things. Most alternatives that attempt to use the isometric size instead of the PCs lose at least one of these two properties. The covariance matrix S of a dataset consisting of measurements on p variables x, can be written as

$$S = l_1 a_1 a_1' + l_2 a_2 a_2' + \cdots + l_p a_p a_p'$$

where $l_1 \geq l_2 \geq \cdots \geq l_p$ are the eigenvalues of S, and a_1, a_2, ..., a_p are the corresponding eigenvectors. If the first term on the right-hand-side of this equation is removed and a PCA is done on the resulting matrix, the first PC is the second PC of S, the second PC of the new matrix is the third PC of S, and so on. Somers (1986) suggested removing $l_0 a_0 a_0'$ from S and doing a 'PCA' on the resulting matrix to get 'shape' components, where l_0 is the variance of $a_0'x$ and

$$a_0 = \frac{1}{\sqrt{p}}(1,1,\cdots,1)$$

This procedure has several drawbacks (Somers 1989; Sundberg 1989). In particular, the shape components lose the desirable properties mentioned above.

One alternative suggested by Somers (1989) is to find shape components by doing a PCA on a double-centred version of the log-transformed data. Double-centring subtracts the mean of each row in a dataset as well the mean of each column, whereas ordinary PCA only centres the columns (variables). It can be considered to remove size because the isometric vector is one of the eigenvectors of its covariance matrix, but with zero eigenvalue. Hence the vectors of coefficients of the shape components are orthogonal to the isometric vector, but the shape components themselves are correlated with the isometric component. Cadima and Jolliffe (1996) quote an example in which these correlations are as large as 0.92. Double-centring is illustrated for the turtle data in Section 30.6.

By using projections, Ranatunga (1989) formulated a method for which the shape components are uncorrelated with the isometric component, but her technique sacrifices orthogonality of the vectors of coefficients.

Once we move away from eigenvectors of the original covariance or correlation matrix, preserving both properties becomes difficult. Cadima and Jolliffe (1996) succeeded in deriving a procedure, combining aspects of double-centring and Ranatunga's approach, that gives shape components that are both uncorrelated with the isometric component and have vectors of coefficients orthogonal to a_0. However, as pointed out by Mardia et al. (1996), another fundamental property of shape components is lost in this method. If $x_h = cx_i$, where x_h, x_i are two observations of the p variables and c is a constant, then most definitions would say that x_h, x_i have the same shape. However, the scores of the two observations are generally different on the 'shape components' found by Cadima and Jolliffe's (1996) method.

Another approach to the analysis of size and shape is to define a scalar measure of size, and then calculate a shape vector as the original vector x of p measurements divided by the size. This is intuitively reasonable but needs a definition of size. Darroch and Mosimann (1985) list several possibilities but hone in on:

$$g_a(x) = \prod_{k=1}^{p} x_k^a, \qquad \text{where} \quad a = (a_1, a_2, \cdots, a_p) \quad \text{and} \quad \sum_{k=1}^{p} a_k = 1$$

So size is thus a generalization of the geometric mean. We now have a shape *vector* $x/g_a(x)$ but may want to decompose it into scalar shape components. This brings us back to PCA. Darroch and Mosimann (1985) suggest using PCA on the log-transformed shape vector, leading to shape components. They show that the PCs are invariant with respect to the choice of *a*, provided that its elements sum to 1. The covariance matrix of the log shape data has the isometric vector as an eigenvector, with zero eigenvalue, so that all shape components are contrasts between log-transformed variables.

30.5 Applying PCA to the original turtle data

To illustrate what happens if PCA is applied on morphometric data, we apply it to the turtle data. As discussed in Chapter 12, two main questions have to be addressed before applying PCA, namely (i) should all the distance variables have the same influence or should the variables with more variation dominate the analysis, and (ii) are we interested in relationships between observations (turtles) or between distance variables? The first question is directly related to the problem of whether we should apply the PCA on the correlation or covariance matrix, and the second question dictates whether we should use the distance biplot or (confusingly) the correlation biplot. The correlation biplot is also called the species conditional biplot, but that would cause even more confusion, as the turtle species are actually the observations.

As noted in Section 30.3, we decided to use the correlation matrix and correlation biplot. As a result we obtain a biplot in which angles between lines (the distance variables) represent correlations between the distance variables, but distances between observations (turtles) cannot be compared directly with each other; they are approximate Mahalanobis distances that can sometimes be a bit more difficult to interpret. This means that if we obtain the biplot, we should not enthusiastically start to compare different groups of observations (e.g., freshwater and marine turtles) as if distances were Euclidean, but merely concentrate on the distance variables.

The biplot is given in Figure 30.4. Some software packages scale the eigenvalues so that the sum of all eigenvalues is equal to 1, and we use this convention here. The first two eigenvalues are 0.84 and 0.07, respectively, which means that the first two axes explain 91% of the variation. All variables are strongly related to the first axis. This is typical for morphometric data, and the first axis can be thought of as a size axis (Jolliffe 2002).

For most datasets we would be very content with 91% explained variation, but not in this case. Recall from Chapter 12 that this 91% means that the biplot explains the correlation matrix rather well, but the correlation matrix itself is mainly determined by length of the skulls. So the first axis represents a rather trivial outcome, namely that most distance variables are highly correlated.

In the next section we look at several alternatives to straightforward PCA, using the inter-landmark data.

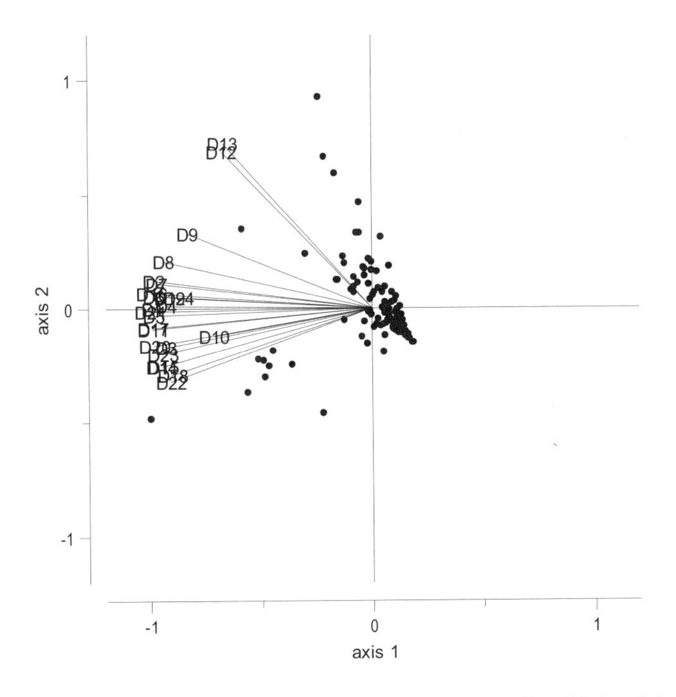

Figure 30.4. PCA biplot; axis 1 versus axis 2. The correlation biplot (also called: species conditional) biplot was made based on the correlation matrix. The first three eigenvalues are 0.84, 0.07 and 0.03, respectively. Eigenvalues were scaled to have a sum of one, which means that the first two axes explain 91% of the variation in the data.

30.6 Classic morphometric data analysis approaches

As noted earlier, by classic we mean analyses based on inter-landmark distances, as opposed to those based on the co-ordinates of the landmarks.

Option 1: Using variable D_2 as a scalar measure of size

The first option we discuss is filtering out the size effect by using a scalar measure of size. For some morphometric datasets, an overall measure of length is available. For the turtle data, the variable D_2 may be considered as the overall measure of size, but this is an arbitrary choice specific to this dataset only. To filter out the overall-size effect, it may be an option to apply a linear regression on each distance variable (except for D_2) and use D_2 as the explanatory variable. The residuals of these regression models are uncorrelated with the general size variable D_2. This process was applied on the turtle data. We used variable D_2 as a proxy for overall length of the skull, although this is a choice can be argued about. The PCA correlation biplot (based on the correlation matrix of residuals from the procedure) is given in Figure 30.5. Note that we no longer have the situation in which all variables point in the same direction.

The first axis represents 46% of the variation and shows distinctions between robust skulls with shallow temporal emargination and skulls with deep emargination (see the opposition between D_{12}, D_{13} and most other distances on the first shape axis). The second axis accounts for 15% of the variation and is also influenced by the degree of temporal emargination (D_{12}, D_{13}, D_{22}), the development of the length of the posterior part of the skull (D_8, D_9), the location of the jugal emargination (D_{10}, D_9), and the relative development of the orbit (D_{14}, D_{15}). However most observations are grouped close to the origin and this first attempt to remove size from shape seems to yield rather little biological information in terms of the structure of the shape space. Allometry (the relationship between size and shape) may drive such a pattern, at least on PC1, because a lot of variables are contributing in the same way to this axis. As for the observations (the turtles), the axes seem to depend greatly on peramorphic (hyper-adult) morphologies of giant species that represent often extreme plots (the narrow-headed giant softshell — maximal score on axis 1— and the giant leatherback turtle —minimum score on axis 1—, the alligator turtle —minimum score on axis 2—). As these species are among the biggest in our sample, this may reflect different allometric relationships. Only groups like marine turtles and trionychid turtles were distinct from each other on these two axes. However, for a statement like this, it is better to use a distance biplot and not a species conditional biplot.

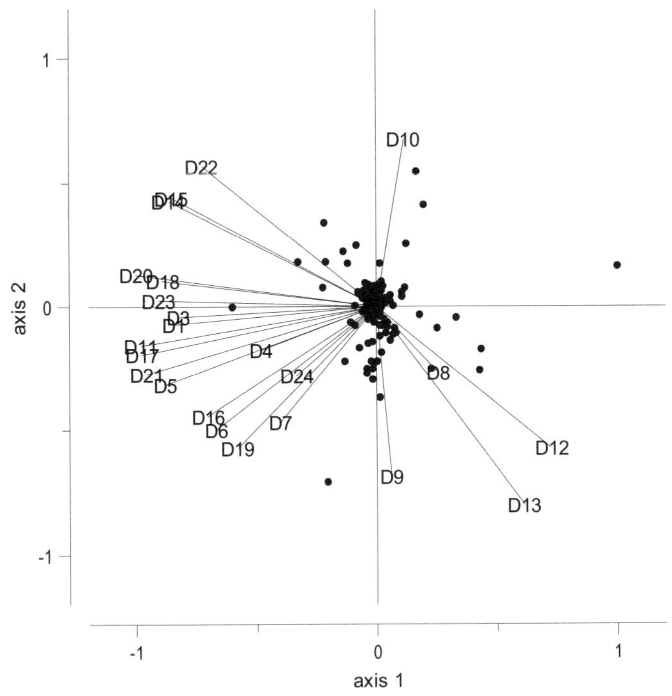

Figure 30.5. Correlation biplot for the data in which D_2 was partialed out. The first three eigenvalues are 0.46, 0.15 and 0.09.

Option 2: Removing isometric size with a pre-standardisation

The next option we discuss is a pre-standardisation of each observation with a size estimator. We choose to use the total squared distance of each observation from the origin as a general measure of size for each observation. This means that for each species, we calculate

$$D_{tot} = \sqrt{\sum_{j=1}^{24} D_j^2} \tag{30.1}$$

The variable D_j is the original distance variable (no transformations were applied on it). Then each of the 24 observations for this species is divided by D_{tot}. This process is repeated for all 123 species. After this pre-standardisation, each distance variable is normalised so that all variables have the same weight in the PCA. In this case, we use the distance biplot as we are interested in how the turtles are projected on the two-dimensional space. Compared with the previous analysis the first two axes contain less variation (39%), and the eigenvalues decrease less rapidly; the first three eigenvalues are 0.24, 0.15 and 0.13. The biplot is given in Figure 30.6-A and Figure 30.6-B. We used two panels, one for the variables and one for the scores as we did not want to clutter the graphs too much. So, one has to imagine these two panels on top of each other. As this is the distance biplot, we cannot compare the distance variables by their angles. All we can say is that variables pointing in the same direction have similar values for certain groups of turtles. Panel A shows that we can indeed see a certain grouping in the variables, namely (i) D_8, D_9, (ii) D_{12}, D_{13}, (iii) D_3, D_4, D_{14}, D_{15}, D_{18}, D_{20} and D_{23}, and (iv) D_{11}, D_{17} and D_{22}. Some of the variables were poorly represented on the first two axes (e.g., D_2, D_{16}, D_{19}, D_{24}). In panels B and C we used super family and environment information as labels. Groups of observations corresponding to either taxonomic families or ecologies appear clearly. The first axis seems to represent a distinction between marine and terrestrial turtles versus all the other freshwater turtles. This axis represents, among other things, a change in the relative middle height of the skull (D_{20}) and relative size of the posterior part of the skull (D_8, D_9), which varies from flat and triangular with more developed posterior part versus higher, more rectangular and shorter posterior part in lateral views along the axis. The second axis opposes marine turtles (bottom) to terrestrial ones (top). Species close to the top of the graph get a shortest relative distance between the occipital condyle and the supraoccipital crest (D_6, D_7), longer distance between supraoccipital crest and ends of squamosals (D_{11}), deeper temporal emargination (D_{12}, D_{13}), longer basioccipital to maxillary length (D_2), and wider orbits and nares (D_3, D_4, D_{14}, D_{15}). Clades were grouped rather well on the first axes; for example, trionychoids (labelled 3) were gathered at the top-left part of the plot, representing flat and deeply emarginated skulls.

The pre-standardisation carried out above was also discussed in Chapter 12. Distances between observations (turtles) in the PCA biplot (applied on the transformed data) are now two-dimensional approximations of the Chord distance (Chapter 10) between two turtles (Legendre and Gallagher 2001).

Option 3: Removing isometric size with double centring

There is another pre-standardisation option, namely double centring, which was described in Section 30.4. If we imagine the data as being from 10 turtles (T_1 by T_{10}), we have:

	T_1	T_2	T_3	T_4	T_5	T_6	T_7	T_8	T_9	T_{10}
Y_1:	1	2	1	3	4	1	2	100	99	80
Y_2:	2	3	4	1	6	3	1	96	90	75
Y_3:	3	2	6	6	3	4	9	99	60	84

Turtles T_8, T_9 and T_{10} are just large species. Because of these three, the correlation coefficients among Y_1, Y_2 and Y_3 are very high. To down-weight their effect on the correlation coefficients (and therefore in the PCA), we could calculate the mean value for each turtle over the three observations (the mean for T_1 is 2, etc.), and subtract this mean value from the corresponding column. For the artificial data, this gives

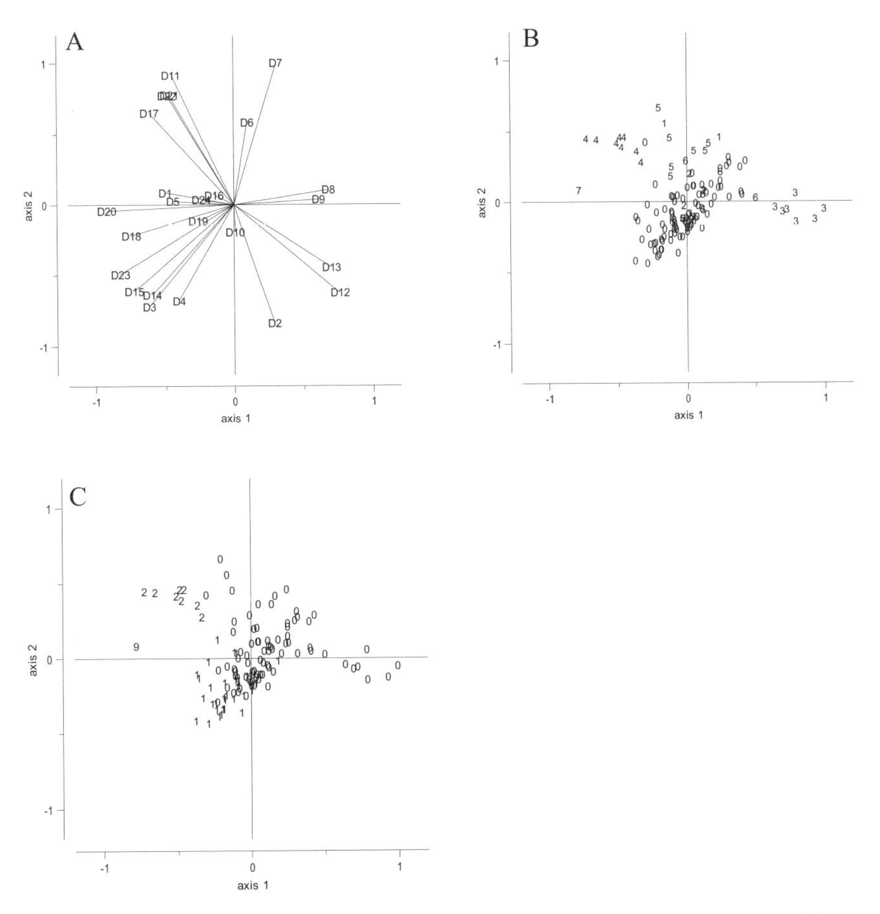

Figure 30.6. A: Axes 1 and 2 for the PCA distance biplot. Only the variables are shown. B: Scores of the PCA distance biplot in panel A. The numbers refer to super-families, and the ‘7’ is the fossil turtle. C: Scores of the PCA distance biplot in panel A. The numbers refer to the environment (0 = freshwater, 1 = terrestrial, 2 = marine and 9 is the fossil turtle).

	T_1	T_2	T_3	T_4	T_5	T_6	T_7	T_8	T_9	T_{10}
Y_1:	−1.00	−0.33	−2.67	−0.33	−0.33	−1.67	−2.00	1.67	16.00	0.33
Y_2:	0.00	0.67	0.33	−2.33	1.67	0.33	−3.00	−2.33	7.00	−4.67
Y_3:	1.00	−0.33	2.33	2.67	−1.33	1.33	5.00	0.67	−23.00	4.33

The data have been centred by columns. Calculating correlations or covariances, which is necessary for PCA, for these column-centred data will automatically centre by rows — hence the name double-centring. We are slightly inconsistent here as double centring in Section 30.4 was explained for data in which the columns represent the variables, whereas in the artificial example above, the variables are in rows (to save space).

Results of the double centring are given in Figure 30.7. Observations on each turtle were first centred, and then a PCA using the covariance matrix was applied. The turtles' scores are plotted as labels using the environment in which they lived (0 = freshwater, 1 = terrestrial, 2 = marine and 9 is the fossil turtle). Note that the marine species are grouped together and that these have higher values for D_2, D_{17} and D_{22} and lower for D_{12} and D_{13} (implying a shallow to absent temporal emargination).

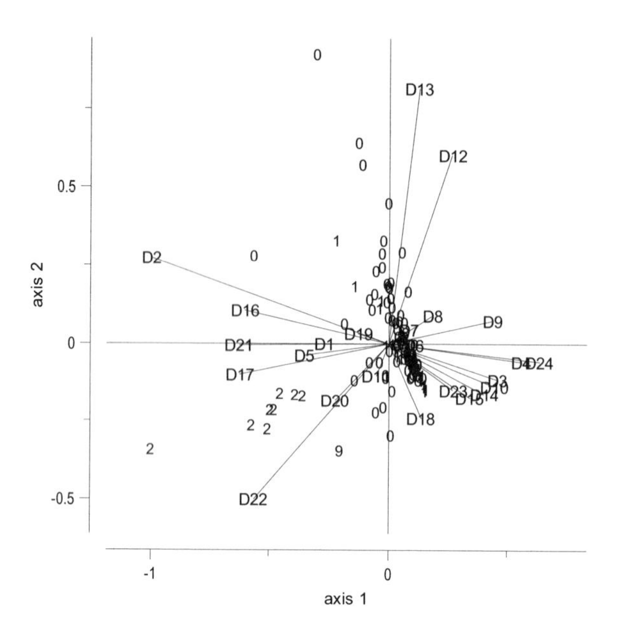

Figure 30.7. PCA biplot based on the covariance matrix after double centring the data of each turtle. The labels for the turtles refer their environment (0 = freshwater, 1 = terrestrial, 2 = marine and 9 is the fossil turtle). The first two eigenvalues are 0.62 and 0.22 corresponding to 62% and 22% of the total variation, respectively.

30.7 A geometric morphometric approach

In the previous analyses, shape components were extracted directly from measurements. These measurements were defined as inter-landmark distances. We have seen that once the size effect is removed, the relationships between these corrected measures and the position of individuals on principal component biplots can be ideally interpreted as shape differences. It is legitimate to think that these shape differences represent, more or less, the relative position of landmarks in each configuration (a configuration here is defined as a set of landmarks). Imagine that rather than taking many measurements, we had scaled, rotated and translated all the configurations on their centroid (the position of the averaged coordinates of a configuration). Differences between configurations can be quantified as differences of position between corresponding landmarks on an optimally superimposed configuration. Indeed shape information is what we have to retain when translation, rotation and scale effect are removed from the original data.

Geometric morphometrics (introduced in the 1970s) are a set of statistical tools devoted to the study of shape variation. They are particularly relevant for isolating the size and shape components from a configuration of landmarks. The more common methods use superimposition procedures to fit the set of configurations, followed by statistical methods applied to the superimposed coordinates of the landmarks of the different configurations. The differences in landmark positions among superimposed configurations reflect shape differences.

The coordinates of 25 landmarks (Figure 30.1) were digitised in three dimensions on the same sample. Each species corresponds to a configuration of 25×3 coordinates. To obtain a shape space, the information about position, size and orientation have to be removed from the original data. For this aim all configurations are superimposed following three steps: The first scales all configurations to the same centroid size (the centroid size is the square root of the sum of squares from the centroid to each landmark of a configuration). The second step translates all configurations to have a common centroid, and the last step rotates all configurations such that distances between corresponding landmarks are minimised (see Figure 30.8). These distances between corresponding landmarks are minimised during the superimposition.

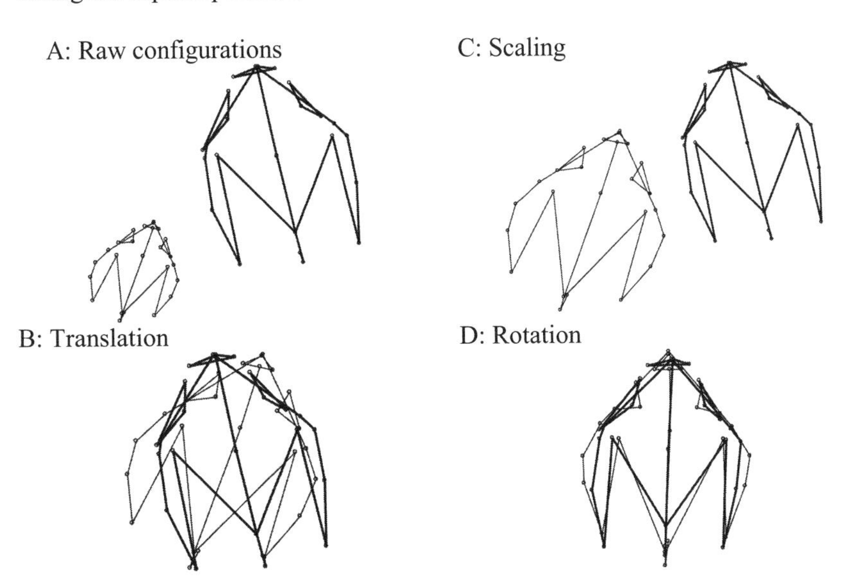

Figure 30.8. The three steps of the superimposition procedure.

For two configurations the procedure is simple. However when there are more than two configurations, finding the best fit uses an iterative procedure: All scaled and centred configurations are first rotated on a reference configuration (for example the first), an average configuration is computed and the rotations are recomputed iteratively until the reference (the average) is stabilized. The details of the transformation are fairly complicated and involve a projection of the shape space (which is non-Euclidian) to a tangent Euclidian one. The interested reader should consult Dryden and Mardia (1998).

Once the superimposition and projection are computed, the resulting transformed coordinates can be used as variables in statistical analysis. As the shape space is multidimensional, principal component analysis on the variance–covariance of coordinates is a way visualizing the position of each shape in this space. Readers interested in other aspects of superimposition methods and shape statistics should again consult Dryden and Mardia (1998).

A shape change along each principal component can be visualised in using the original contribution of each coordinates variables that can be added or subtracted to the average configuration (the origin of the PCA). The score of individuals (here species) can be used for scaling the contribution in change; it is even possible to amplify the eigenvector contribution if shape differences along the axis are not obvious. By adding a few graphical links between landmarks of these extreme shapes, the variation in shape accompanying each axis can be easily described biologically. An illustration is given in Figure 30.9 and Figure 30.10; one illustrates the relation between environments and shape variation and the other the relation between shape variation and super-families.

The first two components accounted for 32.9% and 19.2% of the total shape variance. The different super-families and ecologies are well grouped on the two first PCs, indicating that the geometric shape space directly reflects biological differences. The first component opposes skulls with a deep temporal emargination to skulls with a shallow one. Marine turtles and terrestrial turtles occupy rather opposite locations on this component. Freshwater turtles are more variable, but most of them are close to the origin of the plot. The second principal component opposes high skulls to flatter ones. The flatter skulls have orbits that with a long posterior part of the skull occupy the bottom of the plot, whereas the maxillary part of the skull is more developed and orbits are positioned more laterally for individuals occupying the top of the plot. This PC opposes snapping turtles (chelids, trionychids) to terrestrial ones, with other aquatic and marine ones having an intermediate position. Based on the pattern of variation along two axes, a functional interpretation can be given: The pattern observed on the first axis can be related to skull retraction under the carapace, which is likely related to patterns of temporal emargination — turtles located in the right part of the plot (marine turtles and chelids) have, respectively, no skull retraction, or horizontal skull retraction, whereas those in the left part have a vertical skull retraction. The pattern observed on the second axis can be related to feeding mode — a more elongated anterior part, and shortest posterior part of the skull, is observed in terrestrial turtles that do not need to swallow food as aquatic ones do. See also Claude et al. (2004) for a more detailed analysis. Amazingly, the fossil *Proganochelys* was rather close to the top of the plot (close to terrestrial and marine turtles), which may be interpreted as resulting from a terrestrial feeding mode, that is, a non-retraction possibility of the neck. We may reach similar interpretations based on the approach used in Section 30.4, option 2. But patterns of skull variation are more easy to interpret with the geometric approach.

Geometric morphometrics seemed to be at least as efficient as the approaches to identify patterns of biological variation described in Section 30.4. In using these two methodologies, outliers influenced PCs less than for the previous options. For geometric morphometrics, this is probably due to the superimposition procedure that spreads variation among the different landmarks of the configurations.

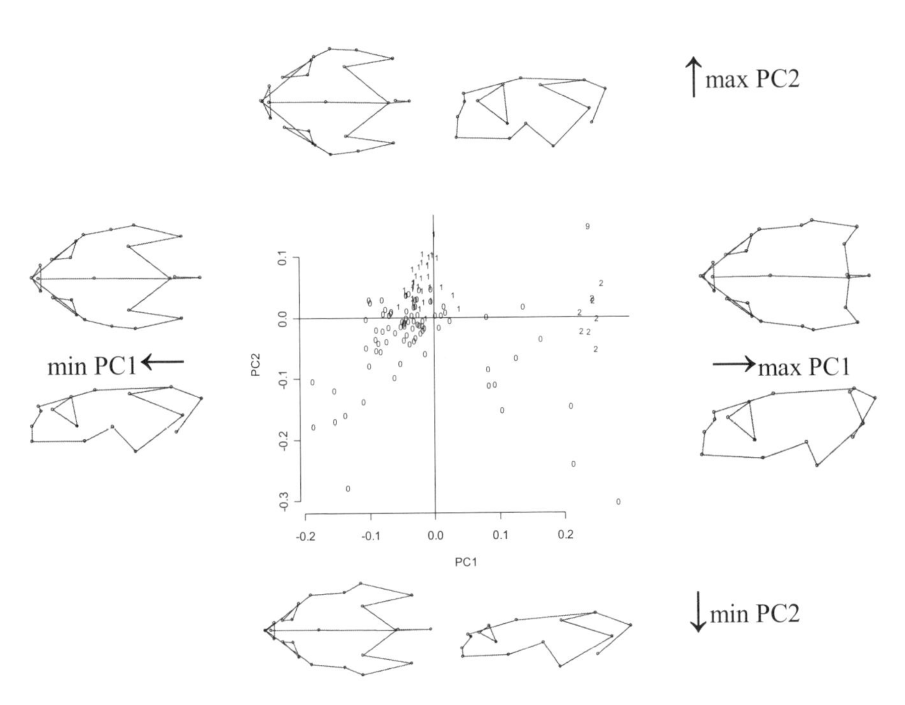

Figure 30.9. Axes 1 and 2 for the PCA of superimposed coordinates. Drawings around the plot represent extreme shapes on each PC. (0 = freshwater, 1 = terrestrial, 2 = marine and 9 is the fossil turtle).

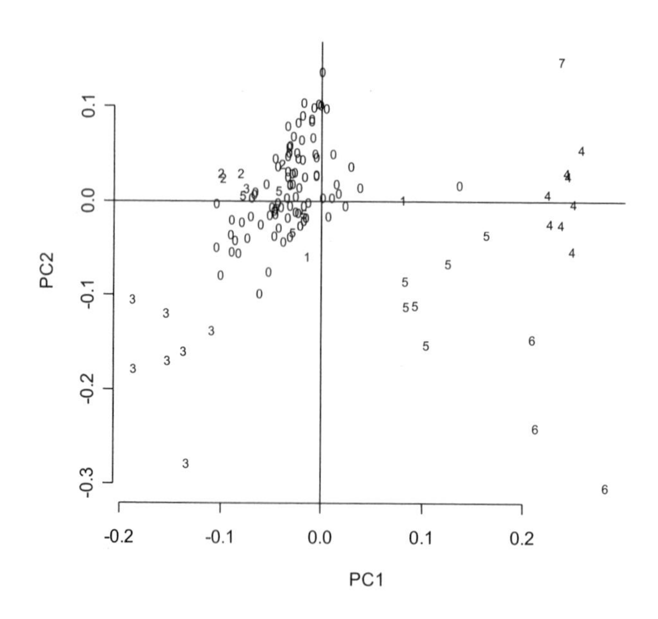

Figure 30.10. Axes 1 and 2 for the PCA of superimposed coordinates. The numbers refer to super-families and the '7' is the fossil turtle.

Many other geometric approaches of shape variation have been developed during the last 30 years: Some methods rely on thin plate spline and deformation grids, others on Fourier analysis, other on angles in configuration, etc. The scope of this chapter is just an introduction; therefore, the reader is invited to read the literature (indicated in the text) if (s)he wants to set a protocol of measurement for analysing shape variation. Geometric morphometrics are now an important alternative approach to a statistical analysis of *ad hoc* inter-landmark distances, and they deal with the size effect efficiently. However, as for any other methods, they can introduce biases and additional problems, the more important one being probably the size estimation that is still subjective.

Acknowledgements

The authors are grateful to P.C.H. Pritchard, H. Tong, M. Cheylan, and F. Renous for providing the material used in this study. This is publication item 2006-046 for J. Claude.

31 Redundancy analysis and additive modelling applied on savanna tree data

Lykke, A.M., Sambou, B., Mbow, C., Zuur, A.F., Ieno, E.N. and Smith, G.M.

31.1 Introduction

Between 1930 and 1970, the colonial administration and the Senegalese state established 213 protected areas aimed at preserving the natural heritage and ensuring a future supply of natural resources for the local population. Today, the woody resources from the protected savannas still provide an important source of firewood, construction materials, food, animal fodder and medicine for the local people. The management of these protected areas has until recently been centralised and directed by the authorities without reference to the views of the local societies. This has often led to a lack of concern and understanding by the local people, and the protected savannas have continued to decline through uncontrolled fires, grazing animals, agriculture and logging. The decline in tree density within the savannas has been drastic during the last decades, and the remaining areas of savanna are under increasing pressure as the demands on their resources continue to grow.

Today a general agreement is emerging that locally based management of protected areas is the way forward, particularly in the light of an increased concern about the state of the natural resources among local people. For integrated sustainable use and conservation of habitats and biodiversity, it is necessary to combine local needs with a scientifically based comprehensive understanding of the biodiversity and ecology of savannas.

The comprehension of ecological processes is complicated because the changes are gradual, dynamic and related to the land use patterns, including fire and grazing by domestic animals. Furthermore, major vegetation changes took place 20–70 years ago, before most scientific investigations started. The ecology and dynamics therefore need to be investigated on the basis of current vegetation data. During the last two decades, satellite images have become available, and several studies have used remote sensing to assess vegetation status; however, the relations between vegetation characteristics and satellite-based indices are poorly understood and no consensus on methods exists. The current study aims to identify whether species composition and vegetation characteristics can be recognized from satellite images or by other simple vegetation parameters.

31.2 Study area

Field data were collected from a typical savanna in western Senegal (Figure 31.1). The area received protection status as classified forest in 1933 under the name Patako Classified Forest; since then, the vegetation has changed from dry forest dominated to savanna dominated: a change that is perceived as negative by local people who rely on its woody resources for subsistence and revenue. Today, Patako Forest is surrounded by agricultural land and functions as an important reserve of natural resources for the local population as well as being commercially logged for firewood. The climate is seasonal tropical with 900 mm annual precipitation falling within a four-month period from July to October. Savanna fires burn the vegetation every year during the dry season.

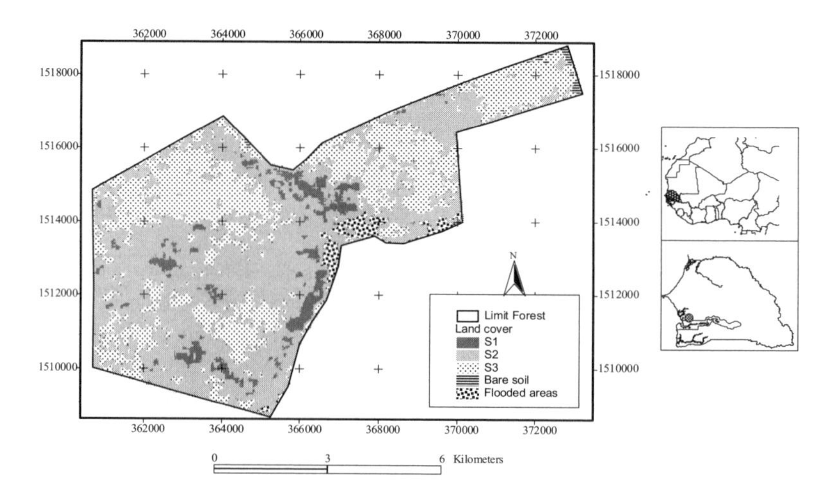

Figure 31.1. Left: Map of study area. Right: West Africa and Senegal.

31.3 Methods

The sampling was designed (i) to cover all habitat types in a vast and heterogeneous area, (ii) to be efficient in the field, and (iii) to allow for statistical analysis. The study area was divided into 250 m × 250 m quadrats and stratified into homogenous zones based on satellite images (Figure 31.2). Across the strata, 22 quadrats were selected randomly for sampling. One sample was taken in each of the selected quadrats. A sample consisted of eight, 20 m × 20 m sub-plots placed at random as follows: A pole was placed at a random point within the selected quadrat and from that point, a 115 m long line was located in the following directions N, NE, E, SE, S, SW, W and NW (Figure 31.3). The sub-plots were placed on each line at a random distance from the pole. In total, 7.04 ha were investigated. All woody plants over 5 cm dbh (diameter at 1.3 m above the ground) within sub-plots were identified to species. Smaller individuals were counted in two groups (less than 1 m and over 1 m tall).

For the statistical analysis, the eight sub-plots in each strata were pooled in order to eliminate fine-scale heterogeneity. To eliminate rare species that were only measured at a few sites, a cut off level of five species was chosen. This reduced the number of species from 50 to 16 (Table 31.1). Thus, the final dataset contained the abundance of 16 woody plant species measured at 22 sites. Several diversity indices were calculated on the basis of the 16 selected species.

There are two types of explanatory variables, namely those derived from the satellite images (Table 31.2) and those derived on the basis of other vegetation parameters (Table 31.3). Because of extreme large cross-correlations (>0.98), some of the satellite variables were not used in the analyses (band 3, 4, 5, 7 and ndvi).

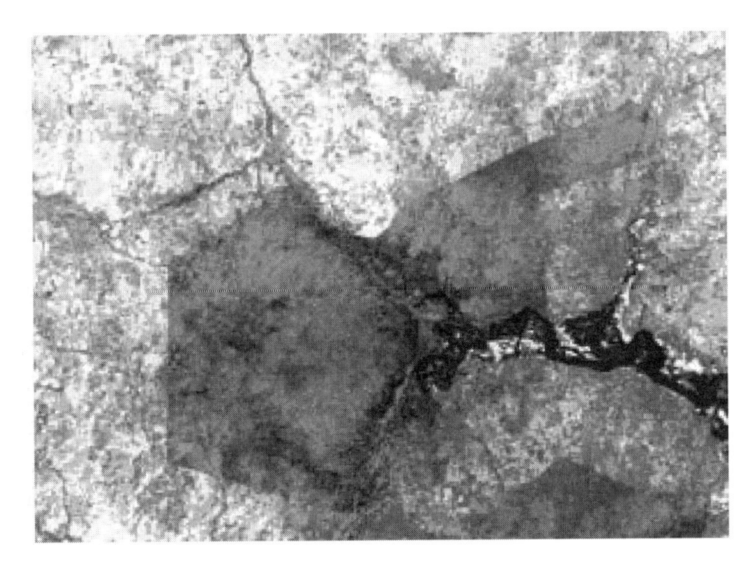

Figure 31.2. Satellite image of study area (LANDSAT — ETM data from 9 December 1999).

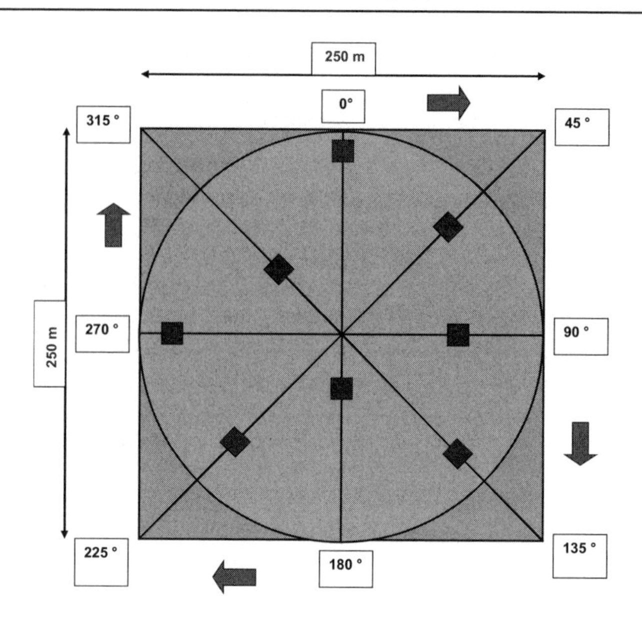

Figure 31.3. Sampling method. One sample consists of eight sub-plots.

Table 31.1. List of 16 species used as response variables in the statistical analysis.

Species	Code	Family	Total Abundance
Acacia macrostachya	Acamac	Mimosaceae	93
Bombax costatum	Bomcos	Bombacaceae	17
Combretum glutinosum	Comglu	Combretaceae	721
Combretum nigricans	Comnig	Combretaceae	208
Cordyla pinnata	Corpin	Caesalpiniaceae	64
Daniellia oliveri	Danoli	Caesalpiniaceae	45
Detarium micrantum	Detmic	Caesalpiniaceae	33
Lannea acida	Lanaci	Anacardiaceae	51
Prosopis africana	Proafr	Mimosaceae	16
Pterocarpus erinaceus	Pteeri	Fabaceae	36
Sclerocarya birrea	Sclbir	Anacardiaceae	8
Securidaca longepedunculata	Seclon	Polygalaceae	39
Sterculia setigera	Steset	Sterculiaceae	11
Struchnos spinosa	Strspi	Loganiaceae	9
Terminalia macroptera	Termac	Combretaceae	142
Xeroderris stuhlmanii	Xerstu	Fabaceae	18

Table 31.2. Explanatory variables based on satellite images (Landsat ETM).

Explanatory Variable	Definition
Strata	
Band 1	band1 (0.45 – 0.52 μm / blue)
Band 2	band2 (0.53 – 0.61 μm / green)
Band 3	band3 (0.63 – 0.69 μm / red)
Band 4	band4 (0.78 – 0.90 μm / near infrared)
Band 5	band5 (1.55 – 1.75 μm / short wave infrared)
Band 7	band7 (2.09 – 2.35 μm / medium infrared)
Brightness	0.2909 × band1 + 0.2493 × band2 + 0.4806 × band3 + 0.5568 × band4 + 0.4438 × band5 + 0.1706 × band7
Greenness	0.2728 × band1 + 0.2174 × band2 + 0.5508 × band3 + 0.7221 × band4 + 0.0733 × band5 + 0.1648 × band7
Wetness	0.1446 × band1 + 0.1761 × band2 + 0.3322 × band3 + 0.3396 × band4 + 0.6210 × band5 + 0.4189 × band7
Ratio72	band7 / band2
Savi	((band4 / band3) / (band4 + band3 + 0.5)) × (1 + 0.5)
Ndvi	(band4 - band3) / (band4 + band3)

Table 31.3. Explanatory variables based on vegetation parameters.

Explanatory variable	Definition
cl_1	regeneration, woody plants ≤ 5 cm dbh and > 1 m tall
cl_2	regeneration, woody plants ≤ 5 cm dbh and ≤ 1 m tall
Dead trunks	No. of standing dead trunks

31.4 Results

Data exploration

A data exploration was carried out to identify extreme observations and the type of relationship between species, between explanatory variables and between species and explanatory variables. Cleveland dotplots (Chapter 4) of various species showed that a data transformation was needed, as there are various extreme observations. Figure 31.4 shows two examples of this where the largest value (on the right-hand side) is considerably larger than the majority of the other observations. Cleveland dotplots of the satellite variables indicated that there are no variables with extreme observations, but there are a few variables that show a strata effect, the two most obvious are greenness and savi (Figure 31.5). It might be an option to subtract the mean in each strata for each variable. This would remove a high correlation between variables due to strata differences. On the other hand, differences between the strata might be important; so there are arguments for and against removing it. We decided not to remove it.

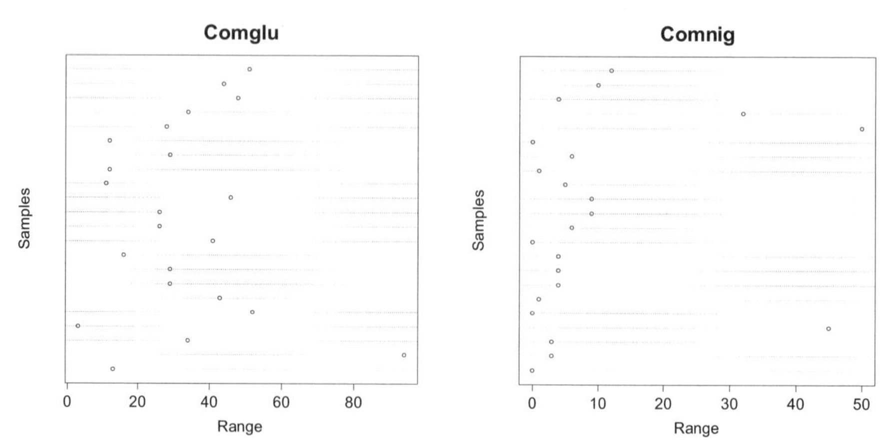

Figure 31.4. Cleveland dotplots of two species showing the need for a data transformation. The horizontal axis shows the value at a site and the vertical axis the sample number, as imported from the spreadsheet. The value at the top is the last value in the spreadsheet and the sample at the bottom the first.

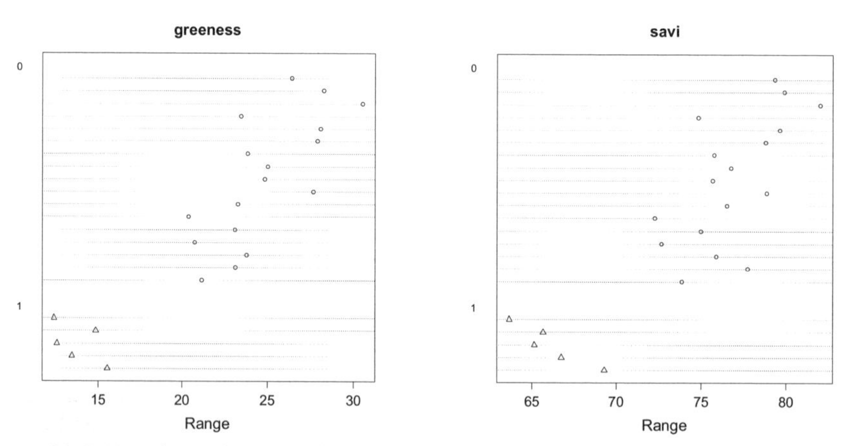

Figure 31.5. Dotplots of two explanatory variables illustrating a strata effect. The circles are observations from strata 0 and the triangles from strata 1. The horizontal axis shows the value at a site and the vertical axis the sample number, grouped by strata.

A data exploration applied on the regeneration variables (cl_1 and cl_2) and the dead trunks indicated that a transformation on these variables is required because they have a few samples with considerably larger values than the rest. We decided to apply a square root transformation on the species and a log(X + 1) transformation on cl_1, cl_2 and dead trunks. The square root transformation also improved the linear relationship between species and (untransformed) satellite variables, as indicted by scatterplots and correlation coefficients. Some cross-correlations between species and satellite variables were around 0.5. The reason we used these two transformations is based on the range of the original data; the log transformation is considerably stronger than a square root transformation.

A scatterplot between the satellite variables (Figure 31.6) indicated serious collinearity, and it was decided to omit the variables band 2, ratio72 and savi. The choice for these variables is based on the correlations in Figure 31.6, histograms of the explanatory variables (showing the coverage of the gradient) and a principal component analysis on the satellite variables (Figure 31.7). Ratio72 is negatively correlated to wetness. Greenness and savi are related. Band 2 and brightness are related. An alternative is to use VIF values (Chapter 26).

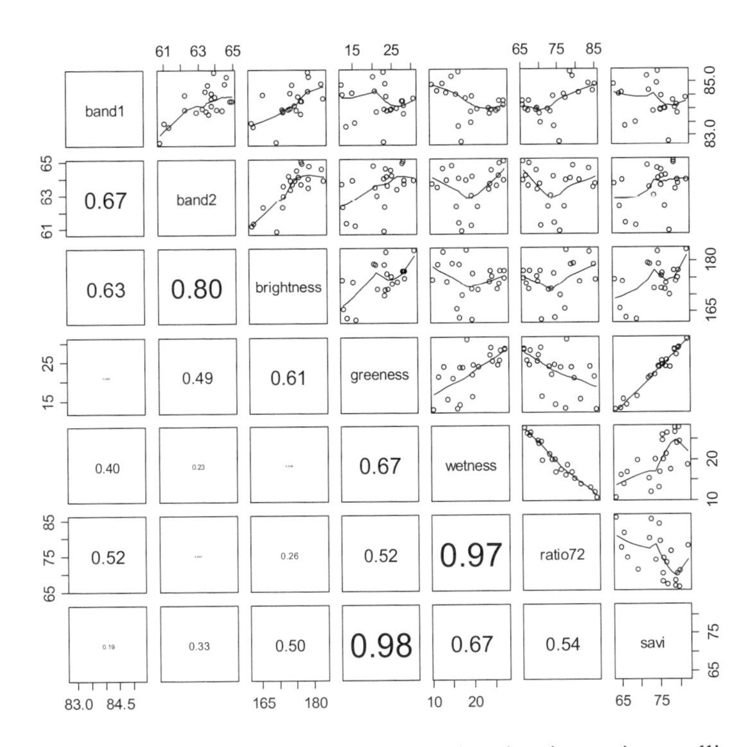

Figure 31.6. Pairplot between the satellite variables showing serious collinearity. The graphs above the diagonal are scatterplots, and numbers below the diagonal represent (absolute) correlations between the variables. Font size is proportional to the value of the correlation.

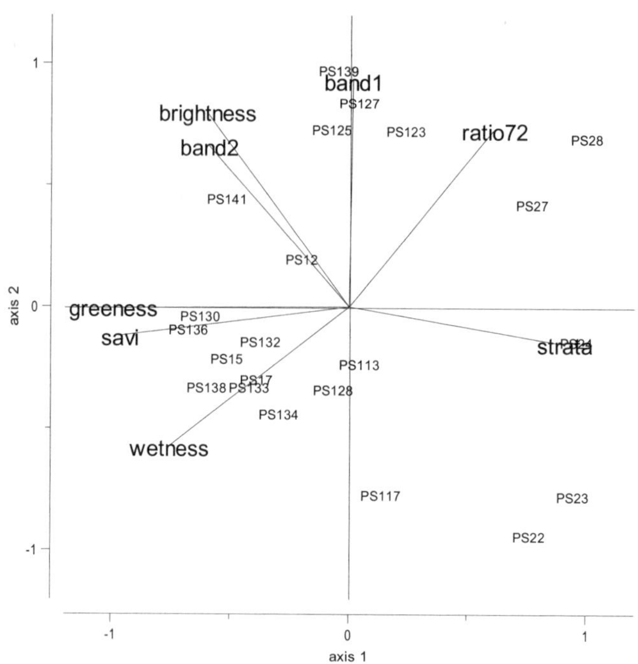

Figure 31.7. PCA applied on the satellite variables, some showing collinearity. The first two axes explain 87% of the variation in the data (53% on axis 1 and 34% on axis 2).

Univariate analysis: Species diversity versus satellite-based explanatory variables

Regression trees were used to analyse the relationship between various diversity indices (richness, Shannon–Weaver, total abundance, Simpson, Berger–Parker, Macintosh) and the selected satellite variables. However, no satisfactory model could be found. Other univariate methods like linear regression and additive modelling were also applied, but these methods did not give any convincing result neither.

Multivariate analysis: Species versus satellite-based explanatory variables

To relate the species data to the satellite variables, either redundancy analysis (RDA) or canonical correspondence analysis (CCA) could be applied. Data exploration, including coenoclines (Chapter 13), showed that most relationships between species and explanatory variables are approximately linear, which justifies the application of PCA and RDA instead of correspondence analysis (CA) and CCA, as the first methods are based on linear relationships and the second on unimodal relationships. The additive modelling, carried out later on, also confirmed that most relationships between species and explanatory variables are approximately linear.

There are two main points to consider when applying RDA, namely (i) the covariance versus correlation and (ii) species conditional or site conditional scaling (Chapters 12 and 29). The choice between covariance and correlation requires some thought. The RDA based on the correlation matrix considers all species to be equally important. Making a small variation in a less abundant species is as important as a larger variation in a more abundant species (if the relative change is the same). The RDA, based on the covariance, focuses more on the abundant species. Here, it was decided to base the analysis on the covariance matrix in order to give the common species more weight in the analysis as this reflects the often more important ecological role of common species and better corresponds to way the vegetation is perceived by the local people. However, the effect of the more extreme values in the common species has been dampened by the square root transformation.

The scaling determines the interpretation of the triplot. The species conditional scaling gives a triplot in which angles between species and satellite variables can be interpreted in terms of correlation or covariance, but distances between samples are more difficult to interpret. In the site conditional scaling, sites can be compared with each other but angles between species do not have any formal interpretation. As we are primarily interested in the relationship between the species and the species and satellite variables, the species conditional scaling was used.

The resulting triplot is presented in Figure 31.8. Before discussing the graphical output, we discuss the numerical output. All five explanatory variables (the four satellite variables and strata) explain 35% of the variation in the species data. The two-dimensional approximation in Figure 31.8 explains 81.49% of this (53.58% on axis 1 and 17.91% on axis 2). Therefore, the first two axes explain 28.35% of the total variation in the species data.

The results of a forward selection and permutation tests, presented in Table 31.4, indicate that brightness is significantly related to the species data ($p <$ 0.001). There is also a weak strata effect ($p = 0.026$). The triplot in Figure 31.8 indicates that strata is related to Comnig, and brightness is negatively related to the species Pteeri, Danoli and Comglu.

Because RDA explained only 35% of the variation, we decided to verify the results with another statistical method. A possible way of doing this is applying additive modelling (Chapter 7) in which each species is used in turn as response variable and brightness and strata as explanatory variables. Although RDA is based on linear relationships, we decided to use additive modelling as it allows for more flexibility than a parametric model. An alternative method is GAM using the Poisson distribution and log link function as the data are count data. However, this would complicate comparing the RDA (which is based on covariance) and GAM results. And as the species were square root transformed, which should stabilise the mean-variance relationship, a GAM was considered unnecessary, and the following additive model was applied on each of the 16 species:

$$\text{Species} = \text{constant} + f(\text{brightness}) + \text{strata} + \text{noise}$$

where $f(.)$ stands for a smoothing function and strata is modelled as a nominal variable. We used cross-validation (Chapter 7) to estimate the optimal amount of smoothing for brightness. Results indicated that brightness was significantly related to seven species, namely: Bomcos, Corpin, Danoli, Pteeri, Sclbir, Termac and Xerstu. Strata was significantly related to Proafr and Termac. The smoothing curves for these species are presented in Figure 31.9.

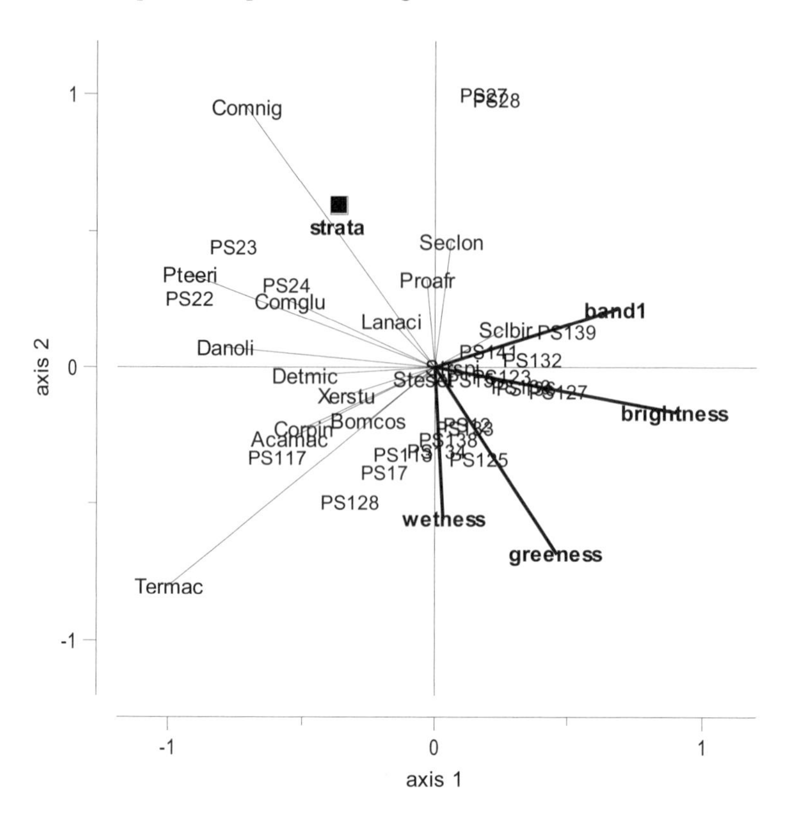

Figure 31.8. Triplot obtained by RDA.

Table 31.4. F-statistic and p-values of conditional effects obtained by a forward selection and permutation test in RDA. The number of permutations was 9999.

Variable	F-statistic	p-value
Brightness	3.685	0.000
Strata	2.248	0.026
Band1	0.828	0.570
Wetness	0.634	0.757
Greenness	1.073	0.365

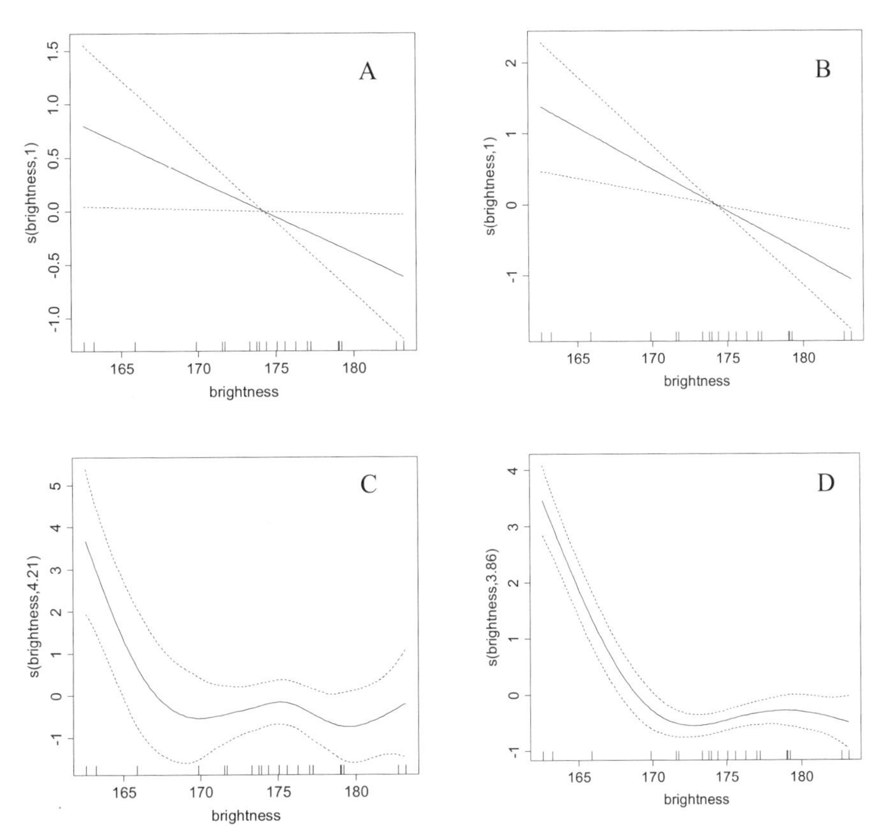

Figure 31.9. Smoothing curves for all additive models in which the smoothing curve for brightness was significant at the 5% level for the species Bomcor (A), Corpin (B), Danol (C) , Pteeri (D), Sclbir (E) and Termac (G).

Multivariate analysis: Species versus onsite explanatory variables

A similar analysis was applied on the species data using the vegetation parameters cl_1, cl_2 and dead trunks as explanatory variables. The triplot is presented in Figure 31.10. All three explanatory variables explain 24% of the variation, and the first two axes represent 89% of this (73.51% on axis 1 and 15.15% on axis 2). A forward selection and permutation test showed that cl_2 is significantly related to the species data. Other explanatory variables were not significant at the 5% level. Just as before, a more detailed analysis was applied using additive modelling in which cl_2 was used as the only explanatory variable. Results indicated that the cl_2 was significantly related to the following three species: Comnig (linear and positive). Danoli (approximately linear and positive) and Pteeri (step function going from low to high, indicating a positive relationship).

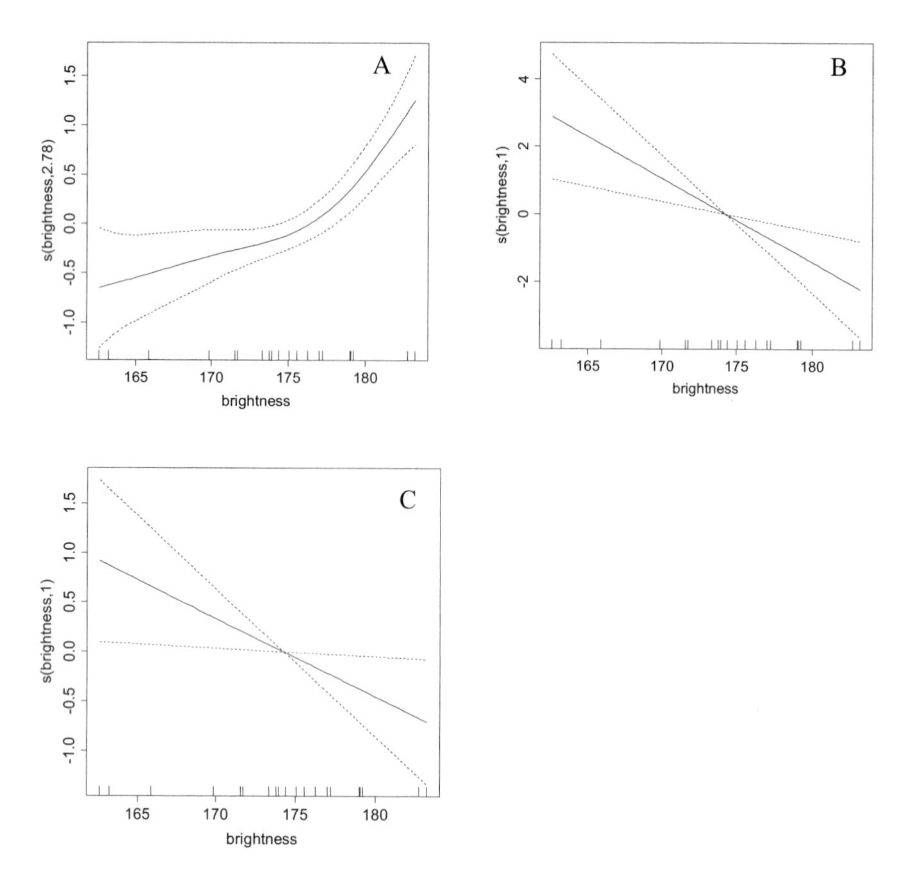

Figure 31.9 (continued). Smoothing curves for all additive models in which the smoothing curve for brightness was significant at the 5% level: Sclbir (A), Termac (B) and Xerstu (C).

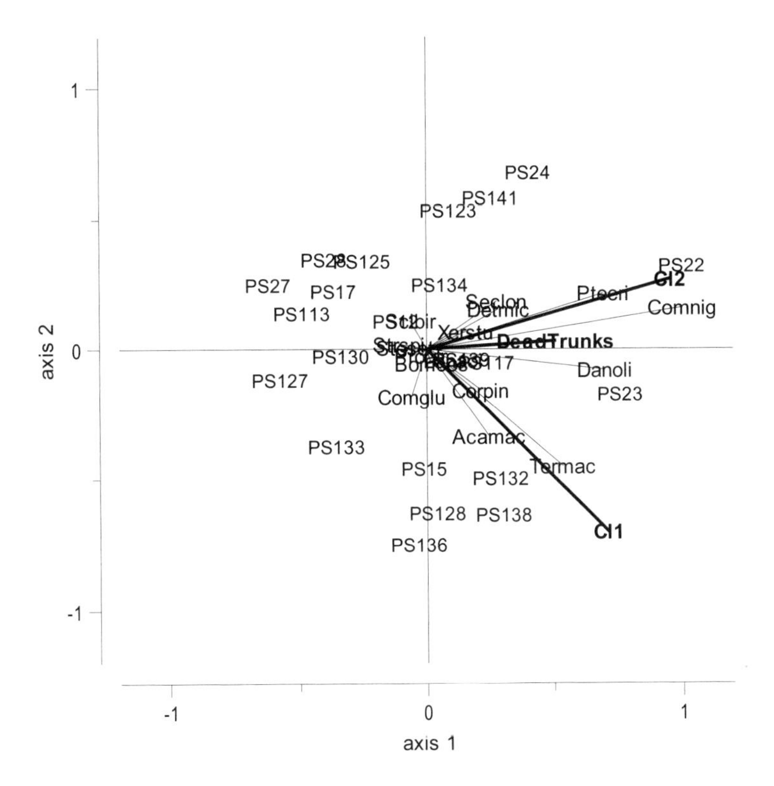

Figure 31.10. RDA triplot using *cl1*, *cl2* and dead trunks as explanatory variables.

31.5 Discussion

In this chapter, we looked at the relationship between woody plant species and satellite-derived variables. The (univariate) diversity indices are useful from an ecological point of view, as they are measures of biodiversity and abundance, but no relationship could be detected between these variables and satellite-based explanatory variables; i.e., satellite images were unable to detect patterns of diversity and density.

The RDA focussed on multiple species and satellite-based variables. We showed by aid of forward selection that two satellite-based variables, brightness and strata, were significantly related to the species data. Using these two variables in an additive model showed that seven species were significantly related to brightness and two species were significantly related to strata. Six species (Bomcos, Corpin, Danoli, Pteeri, Termac and Xerstu), all relatively large tree species, and quality species from local people's point of view, were negatively related to brightness. This indicates that brightness might be a measure of poor woody cover and poor vegetation quality. One species (Sclbir), characteristic for drier and more open environments, was positively related to brightness. Strata was significantly related to Proafr and Termac after the effect of brightness was partialled out.

In a PCA applied on the species data using the covariance matrix, the first two axes explain 53% of the variation. Because RDA is a restricted form of PCA, it is interesting to compare the amount of explained variation in the species data obtained by both methods along the first two axes. If the explained variation by PCA and RDA along the first two axes is similar, then the selected explanatory variables explain the data rather well. On the other hand, a large difference indicates that a poor selection of explanatory variables was used. The first two axes in RDA explain 28% of the total variation in the species data. Hence, by putting restrictions on the PCA axes, we explain 25% less variation. This means that there are other explanatory variables, not used in the analysis, that are responsible for this 25%.

In a second analysis, RDA was applied on the species and regeneration information (cl_1, cl_2, dead trunks) to identify which species were abundant at sites with high (or low) regeneration. Just as before, a more detailed analysis using additive modelling was applied, indicating that cl_2 (regeneration > 1 m tall) was significantly and positively related to the species Comnig, Danoli and Pteeri. This type of regeneration indicates vegetation that has had time to restore without heavy fire impact. In areas that are heavily burned annually, the regeneration is arrested at an early stage (usually less that 1 m tall) or eliminated.

Conclusions

Scientific results play a critical role in the comprehension of vegetation changes and in identifying and evaluating the value of management solutions. The statistical investigations revealed no clear relation between satellite-based variables and typical measures of vegetation quality, such as diversity and density, which makes it risky to build general conclusions about vegetation quality in this environment on the basis of remote sensing.

More detailed statistical investigations based on RDA and additive modelling, however, indicate a relation between some of the larger tree species and some satellite-based variables such as strata and brightness. A more detailed understanding of such relations can improve the use of remote sensing in vegetation management.

The abundance of regeneration >1 m tall was also found to be an indicator of vegetation quality as it was significantly related to a number of large tree species of high quality. This could be explained by the high abundance of regeneration >1 m tall in less fire affected areas.

Acknowledgement

We thank ENRECA/Danida for financing the fieldwork.

32 Canonical correspondence analysis of lowland pasture vegetation in the humid tropics of Mexico

Lira-Noriega, A., Laborde, J., Guevara, S., Sánchez-Ríos, G., Zuur, A.F., Ieno, E.N. and Smith, G.M.

32.1 Introduction

The aim of this chapter is to provide an application of canonical correspondence analysis (CCA), and we will use a lowland tropical vegetation data set. It should be noted that the aim is not to provide a detailed statistical analysis of these data, as other methods may be more appropriate to answer the underlying questions. The reason for this is that the original data set contained a large number of zero abundance for most species, and therefore statistical techniques discussed in Chapters 10, 15, 26 and 28 (non-metric multidimensional scaling and the Mantel test) are more appropriate tools to analyse the original data. However, aggregating the data (using families instead of individual species and averages per pasture instead of individual sampling plots) to reduce the number of zeros gave a data set to which CCA can be applied.

It is well known that from the middle of the twentieth century to the present, vast areas of the tropical rain forest on the American continent have been destroyed, and at an alarming rate. Entire landscapes, previously covered by luxuriant tropical rain forest, are now occupied by extensive man-made pastures where cattle graze, mainly to satisfy the demand of the ever growing urban population of developing countries. Most people, particularly ecologists, think of these man-made pastures as the antithesis of life (or biodiversity). Not only do they destroy the most complex and diverse of vegetation communities, they are extremely poor in species and have become simplified systems. Furthermore, in the Americas, this is exacerbated by the fact that they are grazed by Asian-derived cattle and are commonly dominated by African grasses. However, even though pastures are currently the most common type of vegetation in the lowland humid tropics of the Americas, ecologists seem to have persistently avoided studying them.

The lack of the most basic information about these pastures in the Americas is remarkable when we compare this with the detailed knowledge that exists for the vegetation characteristics and ecology of man-made grasslands in temperate countries, mainly Europe and the U.S.A., and of South American savannas. Another more immediate and local contrast emerges when one considers that the vast body of knowledge and published research on the vegetation structure, composition and dynamics of tropical rain forest has been carried out within biological research stations. The majority of these are in close contact with or surrounded by man-made pastures; however specialised literature, more often than not, regards these pastures as a 'non-habitat'. Even though the replacement of tropical rain forest by pastures is a serious modification of the natural world, we can say little about this anthropogenic system without engaging in a detailed study of its vegetation.

The current study was carried out at two localities adjacent to a tropical rain forest reserve in Mexico, where cattle are currently being raised, although the history of the establishment and management of the pastures differ between localities. Twenty active pastures were sampled and analysed in order to describe the spatial variation of vegetation characteristics and to evaluate whether differences in their management history and practices (i.e., grazing regime, herbicide use, etc.) were able to explain this spatial variation.

32.2 The study area

Vegetation sampling was done in the volcanic mountain range known as Los Tuxtlas, in the state of Veracruz, Mexico (Figure 32.1). The mountain range is 90 km long, oriented in a NW-SE direction, and 40–50 km across. It ranges from sea level up to 1680 m above sea level, emerging from the coastal plain in the southernmost part of the Gulf of Mexico. Los Tuxtlas is the most humid region along the coast of the Gulf of Mexico, with an annual precipitation of 4000 mm. Although it rains all year round, there is a three-month 'dry' season from March to May and a 'wet' season from June to February.

At the beginning of the twentieth century, Los Tuxtlas was covered by more than 300,000 hectares of tropical forest; in 1991 only 15–20% of the original area was still forested, the rest had been transformed into pastures, crops, roads and urban areas. In 1991, pastures covered 160,000 ha of previously forested areas and were mostly located in the lowlands, below 500 m. Currently, the largest tract of tropical rain forest in the lowlands of the region is found in the Los Tuxtlas tropical biology research station (640 ha) and has been managed by the National Autonomous University of Mexico (UNAM) since 1967.

The two localities studied are La Palma and Balzapote. These are adjacent to each other and to the UNAM station (Figure 32.1). La Palma was founded in the 1930s, and since that time, cattle raising has been the main economic activity of its inhabitants. Balzapote was founded 10 years later by farmers who had little or no cattle, but instead practiced 'slash and burn' agriculture, growing mainly maize, beans and squash. During the 1970s and the 1980s, socioeconomic factors discouraged subsistence agriculture, while cattle ranching was favoured. At the beginning of the 1990s, most of Balzapote was covered by pastures that were tended by reluctant ranchers who were previously successful farmers.

In the region there are two different ways of establishing a pasture. In one, cattle that have been grazing in another pasture (where grasses are seeding) are introduced into a crop field with maize stalks. The grass seeds are deposited via the cow dung, and with the help of frequent weeding (by hand and using a machete, or by spraying selective herbicides), the growth of native grass species is favoured. In this case, grasses are not actively sown, but their establishment is induced. This type of pasture is locally known as 'grama pasture'. Alternatively, the grass is directly sown or planted in the ground, and currently the most commonly used species is African Star grass (*Cynodon plectostachyus*). This introduced and improved species does not produce viable seeds naturally in Mexico, and therefore, it is propagated vegetatively in the region; i.e., by planting 10–15 cm long segments of its stolon between crops. This type of pasture is known locally as 'star pasture'.

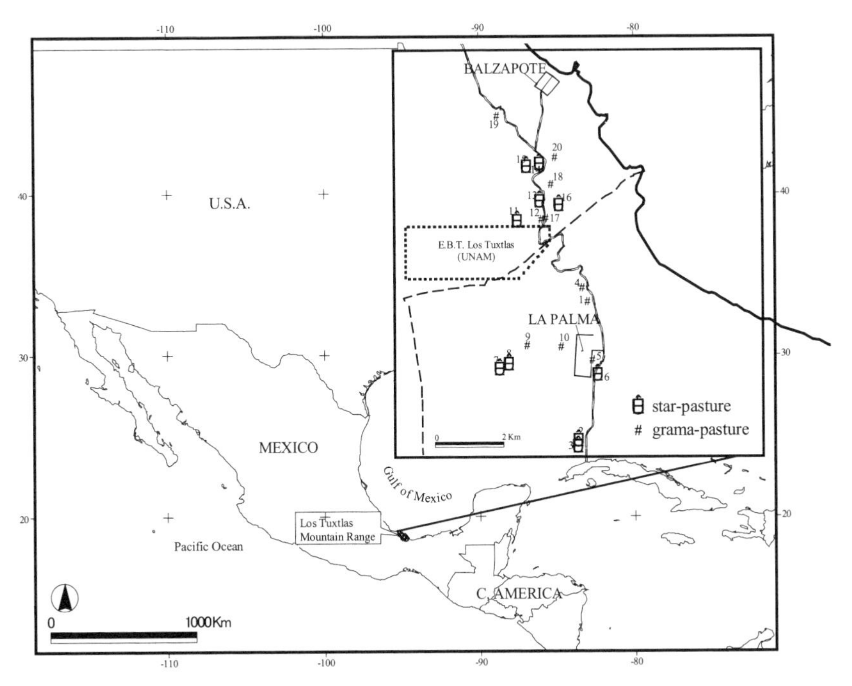

Figure 32.1. Map of study area, showing the 20 pastures sampled.

32.3 The data

Vegetation sampling in the 20 pastures was carried out during the dry season (April 10–May 12) and the rainy season (September 25–October 11) of 1992. However, as temporal changes in vegetation were only related to biomass and not to composition (Lira-Noriega 2003), we only used the dry season data set for the CCA. At each of the two localities, 10 actively grazed and well-tended pastures (hearty herbaceous cover without invading shrubs) were selected: Five were 'star pastures', and five were 'grama pastures' (Figure 32.1). Selected pastures were those in which the owner allowed us access and agreed to be interviewed about pasture history and management details. At each pasture a visually homogeneous and undivided area (no fence subdivisions) larger than one hectare was selected, and within this, a 100 x 100m area was marked with pegs avoiding the shade of trees when possible. Within the marked hectare, ten 2 × 2 m plots were randomly selected for sampling (Figure 32.2). All plant species rooted within the plot were identified. The percent cover of each species within the plot was assigned to one of six possible categories: < 1%; 1–5%; 5–25%; 25–50%; 50–75% and >75%, represented by the numbers 1 to 6, respectively. For each plot the minimum and maximum height of the turf (foliage height) was measured to the nearest centimetre. For each pasture, the altitude above sea level and slope were recorded. In addition, several variables were obtained by interviewing the owners of each pasture. These included the dates the forest was felled, and the pasture was established, in addition to other features related to management practices (number of cows, frequency and method of weeding, etc.; see Table 32.1).

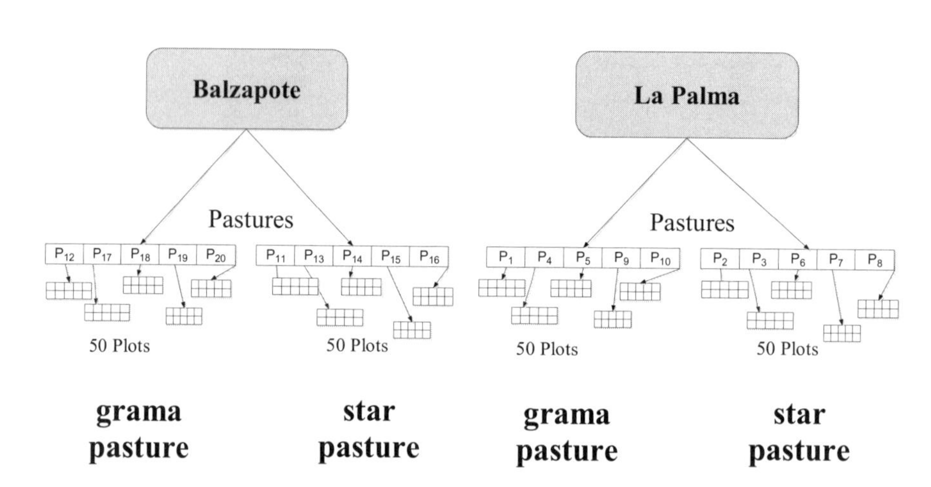

Figure 32.2. The experimental design. The two sampling localities (Balzapote and La Palma) each contain 10 pastures: 5 'grama pastures' with native grass and 5 'star pastures' with introduced grass (see Section 32.2). Each pasture has 10 plots.

The main underlying question in this study is whether there is a difference in species communities among the four groups of pastures in Figure 32.2. Each group is characterized by a different management intensity and history. To quantify these four groups, a nominal variable 'block' is introduced. It has four values identifying the classes (i) grama pastures in Balzapote, (ii) star pastures in Balzapote, (iii) grama pastures in La Palma and (iv) star pastures in La Palma.

The original data contained 171 species, most of them are rare. To reduce the large number of zeros in the data matrix, field data per species were converted into plant cover per family for each plot. As the introduced African star grass (*C. plectostachyus*) is the only species that is directly sown by the ranchers, it is considered a separate family for this analysis. The remaining species of the grass family (Poaceae, formerly Gramineae) are pooled together as they all produce viable seed in Los Tuxtlas and are native to Mexico. We treat cover as a continuous variable because the difference between coverage of 0 and 1 is considered equally important as the differences between coverage of 5 and 6. It is in fact an ordinal variable with six classes (<1%, 1–5%, 5–25%, 25–50%, 50–75% and >75%). A list of all the families used in this analysis is available online.

Table 32.1. A summary of available explanatory variables. *Measured in the field; ** From interviews with the owner of the pasture.

Explanatory Variable	Remarks
Altitude above sea level*	Continuous variable (metres).
Field slope*	Continuous variable (degrees).
Bared soil*	Continuous variable measured in a similar way as vegetation cover (1–6).
Time since forest clearing**	Categorical variable with index values from 1–8, representing ages from approximately 6 to 40 years.
Cattle grazing intensity**	Continuous variable (head of cattle per hectare).
Weeding frequency**	Categorical variable: 1 = no weeding, 2 = weeding with a machete once per year, 3 = weeding with a machete more than once per year.
Herbicide spraying**	Categorical variable: 1 = no spraying, 2 = one spray per year, 3 = more than one spray per year.
Plague**	Nominal variable: 0 = no plague, 1 = plague in the pasture the year before sampling. This was an atypical insect herbivore attack on grass leaves in some pastures the year before sampling.
Minimum vegetation height*	Continuous variable (cm).
Maximum vegetation height*	Continuous variable (cm).

32.4 Data exploration

Observations equal to zero

The first point we look at is how many observations in the family data are equal to zero as this determines which multivariate method should be applied in the next step of the analysis. If there are lots of zeros in the data, then the correlation and covariance coefficients (used by principal component analysis and redundancy analysis), and the Chi-square distance function (used by correspondence analysis and canonical correspondence analysis) are less suitable to define association (Chapters 12, 13, 26, 28). For such data the Jaccard, Sørensen or Bray–Curtis indices might be more appropriate, followed by non-metric multidimensional scaling.

There are different ways to get an idea of the number of zeros in a data set. We can either look at the spreadsheet, express the number of zeros as a percentage, or visualise the zeros in a figure. The last option is carried out here (Figure 32.3). Each symbol '-' means that a particular observation was equal to 0. Note that there is indeed a large number of observations equal to zero.

This indicates that it is not appropriate to apply principal component analysis (PCA) or redundancy analysis (RDA) on these data as two families who are jointly absent at sites are calculated as more similar than families who are jointly present. And correspondence analysis (CA) and canonical correspondence analysis (CCA) will be dominated by patchy families. In an initial analysis, we did apply a CCA on these data and it gave a strong arch effect (Chapters 12 and 13). Detrended CCA removes the arch effect, but the results may still be dominated by patchy species (or families in this case). It may be an option to apply a special data transformation so that Chord distances can be visualised in RDA.

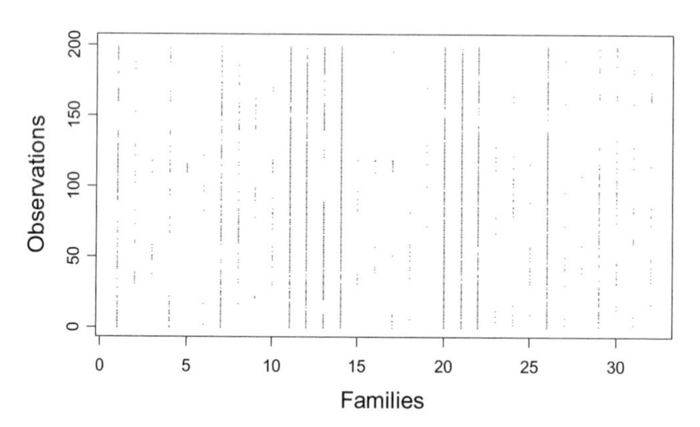

Figure 32.3. Visualisation of the number of observations equal to zero. The horizontal axis represents the different families and the vertical axis the 200 observations (sampling plots). The order of the observations corresponds to the order in the spreadsheet. A '-' indicates an observation equal to 0.

Instead of applying NMDS or RDA (combined with the Chord transformation), there is an alternative. The data contain 10 replicate observations per pasture (Figure 32.2), and we decided to use the average of these 10 replicates. The reasons for this are that (i) the main underlying question for this study is whether there is a block effect, and (ii) most explanatory variables have the same value for all replicates within a pasture.

The average index of family cover per pasture was calculated, and as a result, we have a data set with 32 families and 20 observations (average index per pasture). Most of the explanatory variables had the same value for all replicates in a pasture. For those that differed within pastures, we took the average. Hence, we have a new data set of dimension 20-by-32 with average indices, and nine explanatory variables measured at the same pastures. This new data set has considerably fewer zeros for the family data.

Outliers

Cleveland dotplots and boxplots (not shown here) showed that none of the families had extreme observations. Considering the explanatory variables, bared soil had one observation that was twice as large as the second largest observation and therefore bared soil was log transformed. Weeding frequency had 16 pastures with the same value, and only three unique values overall, so with respect to weeding frequency the data are highly unbalanced and we decided to omit this variable from the analysis. Five families (Boraginaceae, Adiantaceae, Schizaeaceae, Selaginellaceae, Sapindaceae) were measured at less than five pastures and therefore were omitted from the analysis.

Collinearity

In the next step of the data exploration, we investigate the relationship between the explanatory variables. Figure 32.4 shows a pairplot of the continuous explanatory variables, and there are some problems. Minimum and maximum vegetation height have a cross-correlation of 0.82 indicating a strong linear relationship. The scatter of points for these two variables (panel opposite "0.82") confirms the strong linear relationship. This means that we have to omit one of the vegetation height variables (it does not matter which one because both variables are basically representing the same ecological signal). We decide to drop minimum vegetation height.

The fact that the pairplot indicates that there are no other strong two-way interactions does not mean that there is no further collinearity. We have not included the two nominal variables 'block' and 'plague', and there might be three- or multi-way interactions between the explanatory variables. A dotplot or boxplot conditional on 'block' (not shown here) gives the impression that there is a block effect in some of the explanatory variables. We will discuss this further when applying the multivariate analysis.

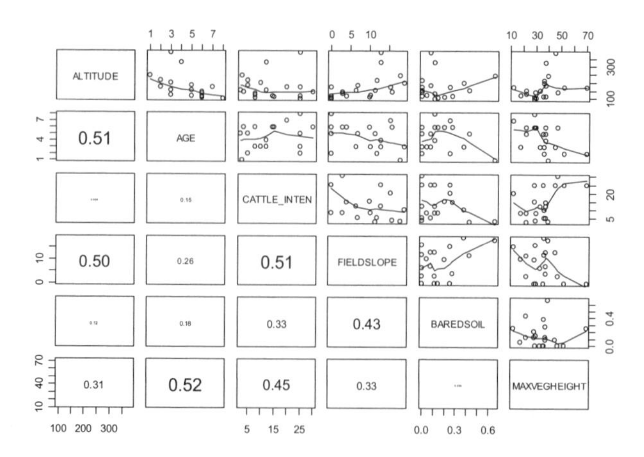

Figure 32.4. Pairplot of the continuous explanatory variables. The lower diagonal panels contain the (absolute) correlation coefficients, and the upper diagonal panels contain scatterplots of each combination. The font size of the correlation coefficient is proportional to its value.

32.5 Canonical correspondence analysis results

We know *a priori* that the gradients are relatively long. For example, altitude ranges from 100 to 400 m and it is unlikely that species-environmental relationships are linear along such a long gradient. For this reason (Chapter 13), we apply canonical correspondence analysis on the pooled family data (20 pastures).

The first question we have to address is which triplot type we want to use: a species-conditional or a site-conditional triplot (Chapters 12 and 13). The underlying question in this study is whether there is a block effect. This means that we want to compare observations with each other and that we are less interested in comparing families with each other. Hence, we should use the site-conditional biplot scaling (also called: distance scaling). In this scaling, distances between observations represent two-dimensional approximations of Chi-square distances, but angles between families cannot directly be interpreted as correlations. All we can say is that lines (families) pointing in the same direction mean that those families appear at the same pastures.

The nominal explanatory variable 'plague' is coded as 0–1 and can be used in the CCA. For 'block' this is slightly more complicated. It has four levels, and therefore we create four new dummy variables B_1, B_2, B_3 and B_4, with B_j equal to 1 if the observation was taken in block j, and 0 otherwise. To avoid 100% collinearity, one of these levels has to be omitted. We select B_4. The CCA triplot is presented in Figure 32.5. The block effect seems to dominate the triplot.

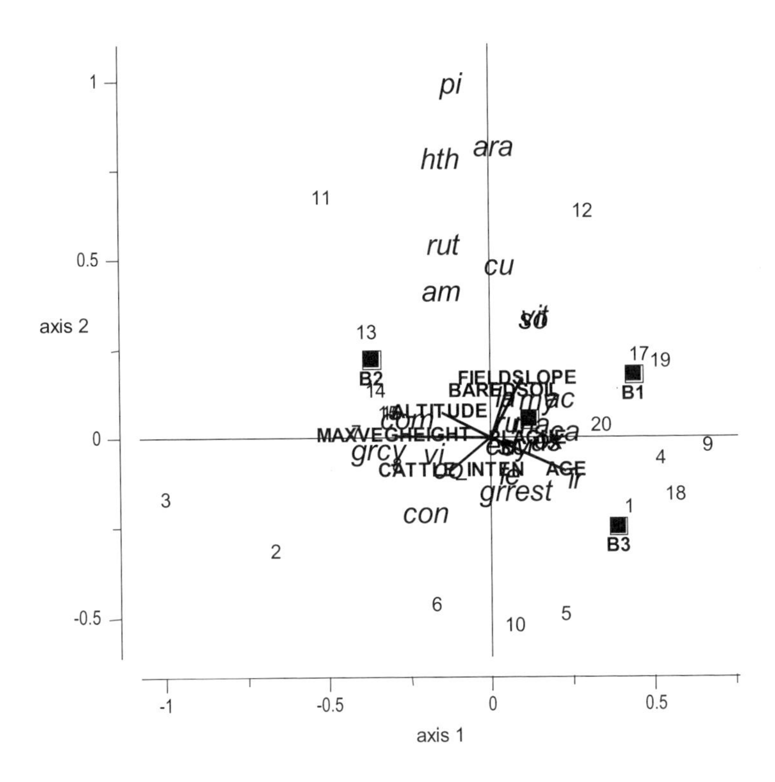

Figure 32.5. CCA triplot for pooled family data. The continuous explanatory variables are represented as lines and the nominal explanatory variables as square blocks. The numbers 1–20 refer to the pastures, and plant families are plotted as labels.

The total inertia (variation) in the family data is 0.45, and the inertia explained by all explanatory variables is 0.29. Hence, all explanatory variables explain $100 \times 0.29/0.45 = 64\%$ of the variation in the data. The first two axes explain 59% of this 64%, which means that they explain 38% of the variation in the family data. This is quite good compared with other ecological studies.

The next question is which explanatory variables are important, and this is typically investigated with a backward selection combined with a permutation test (Chapters 12 and 13). The results in Table 32.2 indicate that maximum vegetation height is highly significant, and that field slope, B_2 and B_1 are significant as well at the 5% level. The p-values for B_1 and B_2 are close to 0.05, which suggests that their role is less important as was inferred from the triplot. However, the situation is slightly more complicated. Recall that in the data exploration, we mentioned that a conditional dotplot showed a block effect in some of the explanatory variables. A good way to assess collinearity between all variables are Variance

Inflation Factors (VIFs). These were discussed in detail in Chapter 26. The VIF values for the explanatory variables are given in Table 32.3 and indicate that a considerable part of the variation in B_1 is explained by the other explanatory variables, and the same holds for B_3 and field slope.

Table 32.2 Results of the forward selection process. The number of permutations was 9999.

Explanatory Variable	F-statistic	p-value
MAXVEGHEIGHT	4.497	<0.001
FIELDSLOPE	2.214	0.013
B_2	1.841	0.047
B_1	1.951	0.037
ALTITUDE	1.637	0.091
BAREDSOIL	1.472	0.149
AGE	1.115	0.326
PLAGUE	0.862	0.531
CATTLE_INTEN	0.563	0.799
B_3	0.362	0.934

Table 32.3.VIF values for all explanatory variables. The higher a VIF value, the more variation in the explanatory variable is explained by the other explanatory variables (collinearity). VIF values larger than 5 can be considered as a problem.

Explanatory variable	VIF
B_1	6.96
B_2	3.22
B_3	7.12
ALTITUDE	4
AGE	3.59
CATTLE_INTEN	2.77
FIELDSLOPE	4.17
PLAGUE	1.78
BAREDSOIL	1.71
MAXVEGHEIGHT	3.39

There are now two ways to proceed. The first option is to remove some of the explanatory variables and identify which are collinear with the block dummy variables. The problem is that this will be a difficult and arbitrary process, and therefore, a more objective tool is needed. The second option is to apply a partial CCA and calculate the pure bock effect.

Partial CCA and variance partitioning

In a partial CCA, five different CCAs are applied on the family data. In each analysis, a different set of explanatory variables is used and the total sum of all canonical eigenvalues of each CCA is used to calculate the pure block effect, the shared information, the (pure) effect of the other explanatory variables and the residual information. The results of these five CCA steps are given in Table 32.4, and Table 32.5 summarises the results. All explanatory variables explain 64% of the inertia (variation) in the family data, and 36% of the variation cannot be explained by these explanatory variables. Decomposing the 64% shows that the pure block effect is 13%, and the variation explained purely by the other variables is 44%. The shared explained variation is 7%, and this is due to collinearity. The percentages 13, 44 and 7 add up to 64. Summarising, the block effect explains between 13% and 20% of the variation in the family data.

Table 32.4. Results of the partial CCA. Total variation is 0.45. Percentages are obtained by dividing the explained variance by total variance. The block variables are B_1, B_2 and B_3 and 'Others' represent the remaining seven explanatory variables.

Step	Explanatory Variables	Explained Inertia	%
1	Block and others	0.29	64%
2	Block	0.15	33%
3	Others	0.23	51%
4	Block with others as covariable	0.06	13%
5	Others with Block as covariable	0.20	44%

Table 32.5. Variance decomposition table showing the effects of block and the other variables. Components A and B are equal to the explained variances in steps 5 and 4, respectively. C is equal to the variance in step 3 minus the variance in step 5, and D is calculated as Total inertia minus the explained inertia in step 1.

Component	Source	Calculation	Inertia	%
A	Pure others		0.20	44%
B	Pure Block		0.06	13%
C	Shared (3–5)	0.23–0.20	0.03	7%
D	Residual	0.45–0.29	0.16	36%
Total				100

32.6 African star grass

In the previous section we analysed whether there was any difference in the species community, and we were particularly interested in the variation in different blocks as this might represent the effect of grcyn (i.e., the cover of African star grass; *C. plectostachyus*). However, the complete family data contained this species (grcyn) as a separate family. A Cleveland dotplot (Figure 32.6) shows that this species is mainly observed in blocks 2 and 4 (see also Figure 32.2). In fact, the CCA applied in the previous section also produced diagnostics for each family (not shown here), and these show that the African star grass is fitted rather well.

Although it is difficult to verify, it might be the case that the CCA was mostly depicting only the grcyn-block relationship. One way to avoid this would be to use grcyn as an explanatory variable, but this would cause trouble with the other explanatory variables as grcyn might be influenced by altitude, field slope, vegetation height, etc. As an alternative, we apply the same analysis as in the previous section but without grcyn and it then becomes interesting to know what the pure block effect is as it may be hypothesised that it represents the grcyn effect. The triplot (not shown here) looks similar as in Figure 32.5, and the results of the variance partitioning (Table 32.6 and Table 32.7) show that the pure block effect is now 15%.

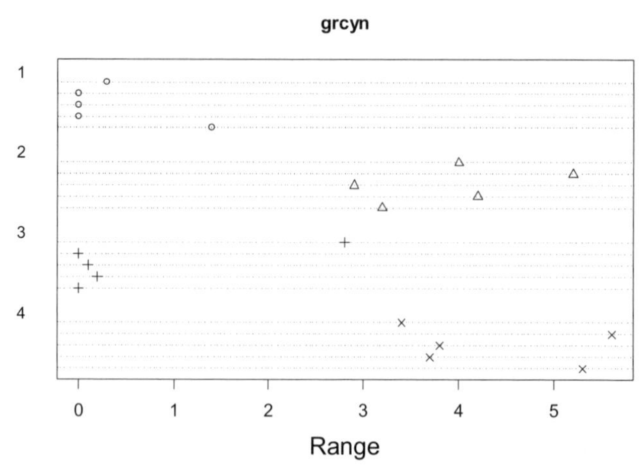

Figure 32.6. Cleveland dotplot of the African star grass *Cynodon plectostachyus* (grcyn) conditional on Block ('o' = grama pasture in Balzapote, 'Δ' = star pasture in Balzapote, '+' = Grama pasture in La Palma, "x" = star pasture in La Palma). The horizontal axis shows the value of the observations and the vertical axis the observations (within groups).

Table 32.6. Results of the partial CCA. The African grass species grcyn was not used in the analysis. Total variation is 0.39. Percentages are obtained by dividing the explained variance by total variance. The Block variables are B_1, B_2 and B_3 and the 'other' variables are the remaining seven explanatory variables.

Step	Explanatory Variables	Explained Inertia	%
1	Block and others	0.23	59%
2	Block	0.08	21%
3	Others	0.17	44%
4	Block with others as covariable	0.06	15%
5	Others with Block as covariable	0.15	38%

Table 32.7. Variance decomposition table showing the effects of Block and the other variables. The African grass species grcyn was not used in the analysis. Components A and B are equal to the explained variances in steps 5 and 4, respectively. C is equal to the variance in step 3 minus the variance in step 5, and D is calculated as Total inertia minus the explained inertia in step 1.

Component	Source	Calculation	Inertia	%
A	Pure others		0.15	38%
B	Pure Block		0.06	15%
C	Shared (3–5)	0.17–0.15	0.02	5%
D	Residual	0.39–0.23	0.16	41%
Total				100

32.7 Discussion and conclusion

In this chapter, CCA was applied to pooled data. The motivation for pooling the data was the large number of zeros in the original data. Alternative analyses would have been non-metric multidimensional scaling using the Jaccard index, or RDA (combined with the Chord transformation).

The results of variance partitioning show that the block effect explains 13–20% of the variation in the family data. The problem with this approach is that the block effect may be caused by the presence of the African star grass *C. plectostachyus* (grcyn) in the data matrix. Hence, the CCA may pick up only the grcyn-block relationships. If the African star grass grcyn is omitted from the analysis, the block effect represents 15–20% of the variation. In this case, the block effect may represent the effect of grcyn. If there are no other factors that differ between the blocks, then grcyn may be the reason for the 15–20%. However, simple dotplots and boxplots give some indication that not only grcyn differs per block as do a few of the other explanatory variables.

The main underlying question was whether there is a block effect, and the answer to this question is positive. However, understanding why there is a block effect is more difficult as the block effect not only represents differences in the presence of the introduced African star grass, but also differences between explanatory variables, as shown by the VIF values in Table 32.3. A large proportion of the variation in the block variables is explained by other variables so it is difficult to hypothesise what exactly the block effect means in terms of ecology.

As suggested by the CCA triplot (Figure 32.5), the method of pasture establishment (induced cover of native grasses or sowing the introduced African grass) has a strong effect on pasture family composition (the block effect), and this not only represents differences in the presence of the introduced African star grass, but also differences between explanatory variables, as shown by the VIF values (Table 32.3). Of all the explanatory variables considered, vegetation height and the slope of the terrain had the strongest effect on family composition. Vegetation height is highly variable and depends directly on the duration and timing of resting vs. grazing in each pasture. Slope is directly linked to grazing intensity; in flat pastures, more cows are left to graze for longer periods than on steep terrain. Unexpectedly, other variables such as the age of the pasture and the number of cows that are kept throughout the year in each pasture, did not have an important effect on family composition. Overall, the analysis shows that a pasture's management intensity and history have a notable effect on its vegetation composition in Los Tuxtlas. Of the dozens of variables originally selected for study, the results of the CCA indicate that in order to understand the spatial variation in family composition and cover in a pasture, cattle density and grazing regime in particular require further study.

Acknowledgement

We thank Bianca Delfosse for translating parts of the manuscript into English and are grateful to all those who participated in the vegetation sampling, and to the specialists who identified the plants collected in this study. This research was funded by the Departamento de Ecología Funcional (902-17) of the Instituto de Ecología, A.C. and by the Consejo Nacional de Ciencia y Tecnología (project CONACYT 0239-N9107). We would like to thank Toby Matthews for valuable comments on an earlier draft.

33 Estimating common trends in Portuguese fisheries landings

Erzini, K., Zuur, A.F., Ieno, E.N., Pierce, G.J., Tuck, I. and Smith, G.M.

33.1 Introduction

Commercial multi-gear fisheries in Portugal, as in most European countries, are multi-species fisheries. For some individual fish species, the effects of environmental conditions on abundance trends have been analysed, e.g.,, the effects of upwelling on recruitment trends in sardine (*Sardina pilchardus*) and horse mackerel (*Trachurus trachurus*) (Santos et al. 2001), the influence of wind and North Atlantic Oscillation (NAO) on sardine abundance (Sousa Reis et al. 2002; Borges et al. 2003) and long-term changes in catches of bluefin tuna (*Thunnus thynnus*) and octopus (*Octopus vulgaris*) in relation to upwelling, NAO and turbulence indices (Sousa Reis et al. 2002). However, to date no multivariate time series analysis has been carried out to compare changes over time in different exploited species and to determine the effects of fishing and environmental conditions on abundance trends.

Reliable data on effort and population dynamics parameters for individual species are scarce. Consequently, age-based methods of assessment or those based on catch per unit effort (CPUE) are not possible for most species. Traditional stock assessment methods also do not take into account environmental variability.

The analysis of long-term trends in fisheries and environmental variables may shed light on factors influencing particular species or species groups and provide a basis for improved assessment and management. Trends that are common to several species may reveal changes in the structure of marine communities and potentially provide useful indicators for ecosystem-based fishery management (c.f. Rochet et al. 2005). The aim of this case study chapter is to identify and estimate common trends in the landings time series for the Algarve area of Portugal, and to estimate possible relationships with environmental variables. Although landings may not provide a reliable indicator of abundance, trends in landings will relate in part to abundance, as well as to fishing effort and the management and regulatory regime under which the fishery operates. Furthermore, although fishing effort data were not available, it was possible to use the numbers of registered boats and fishermen as proxies for effort.

The environmental variables selected for this analysis included the NAO index (Hurrell 1995) and sea surface temperature, both of which have been shown to affect the abundance and/or distribution of many marine resource species (e.g., Drinkwater et al. 2005); also parameters relating to local sea conditions, namely upwelling, river flow and rainfall.

Given the multi-species, multi-gear nature of the Algarve fisheries, the shortness of the available time series and the lack of information on fishing effort for particular species, it was decided to use two complementary methods, namely (i) min/max autocorrelation factor analysis (MAFA) and (ii) dynamic factor analysis (DFA). Of all the techniques discussed in this book, these two are the most suitable for estimating common trends. Although additive mixed modelling or generalised least squares can be used to estimate trends (see for example the salt marsh time series chapter), these methods are less suitable for the estimation of *common* trends.

A more detailed analysis of these data is presented in Erzini (2005). Here, we use a subset of the explanatory variables and focus on the complimentary aspect of MAFA and DFA.

33.2 The time series data

Official landings statistics, for the Algarve region, for 12 species/groups from 1982 to 1999 were selected for the multivariate time series analysis. These include small pelagics (sardine, *Sardina pilchardus*; anchovy, *Engraulis encrascicholus*; horse mackerel, *Trachurus trachurus*; chub or Spanish mackerel, *Scomber japonicus*), deep water or mesopelagic fish (European hake, *Merluccius merluccius*; black spot sea bream, *Pagellus bogaraveo*; Ray's bream, *Brama brama*; silver scabbard fish, *Lepidopterus caudatus*), cephalopods (octopus, *Octopus vulgaris*; cuttlefish, *Sepia officinalis*) and crustaceans (deep water shrimp; mainly *Aristeus antennatus, Aristaeopsis edwardsiana* and *Aristeomorpha foliacea*; spider crab, *Maia squinado*). These species accounted for up to 83% of the total annual landings in the study area. A lattice plot of the standardised fisheries time series is given in Figure 33.1. The motivation for the standardisation is given in the next section. Note that the 12 time series do not all appear to show the same pattern.

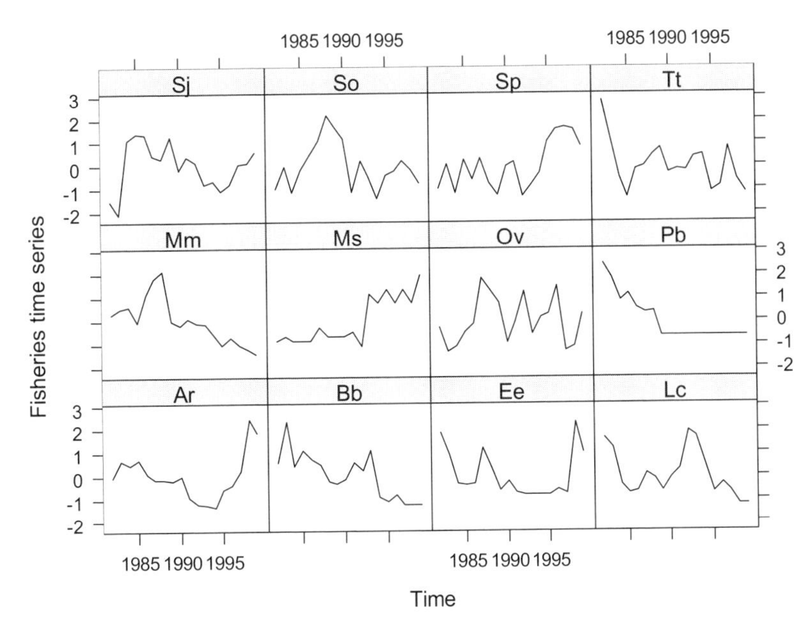

Figure 33.1. Standardised (each series is mean deleted and divided by its standard deviation) and fisheries time series. The abbreviations refer to the 12 fish and invertebrates and are Ar (shrimp), Bb (Ray's bream), Ee (anchovy), Lc (silver scabbard fish), Mm (hake), M (spider crab), Ov (octopus), Pb (black spot sea bream), Sj (Spanish mackerel), So (cuttlefish), Sp (sardine), and Tt (horse mackerel).

The environmental variables were Guadiana River flow or discharge recorded at the Pulo do Lobo station, the NAO index (Hurrell 1995), sea surface temperature, an upwelling index and rainfall. The upwelling index was derived from wind data recorded in Faro (Sousa Reis 2002). Likewise, mean annual and seasonal sea surface temperature at 36°N 8°W and rainfall recorded in Faro for the period from 1982 to 1999 were used as potential explanatory variables. In the absence of reliable fishing effort data, two proxy variables were used. These were the number of licensed fishing boats and the number of licensed fishermen in the Algarve. Erzini (2005) also used seasonal averages of the SST and rainfall, but these are omitted here in order to simplify the analyses. Besides, they were not found to be important in Erzini (2005). A lattice plot of the explanatory variables is given in Figure 33.2. Peak annual Guadiana River flow was recorded in 1997, following a period of severe drought in the first half of the 1990s. The NAO had a contrasting pattern, with maximum values in the first half of the 1990s, decreasing to a low in 1996 and increasing thereafter. Both sea surface temperature and the upwelling index displayed a similar pattern, with low values for most of the time period, reaching a maximum in the mid-1990s. To assess collinearity between the explanatory variables, a pairplot was made (Figure 33.3). The correlations (lower diagonal) indicate that there is no serious collinearity (all values ≤ 0.6; values above 0.9 may be considered problematic) although some of the scatterplots show that obvious relationships do exist, e.g., between the number of boats and fishermen. Nevertheless, the pairplots indicate no immediate problems with the data; hence, we will continue with what we have and without any transformations.

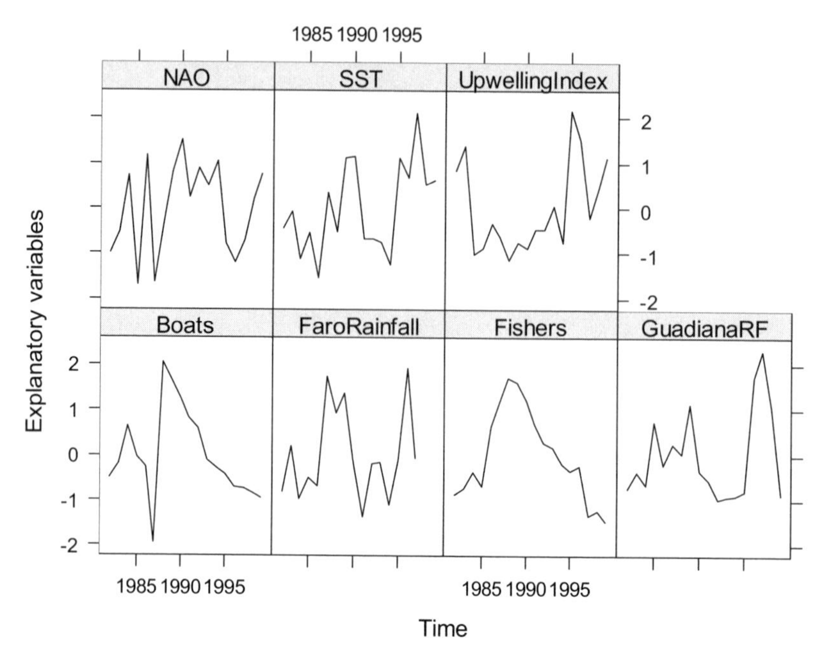

Figure 33.2. Standardised (each series is mean deleted and divided by its standard deviation) environmental variables for 1982–1999. SST stands for sea surface temperature.

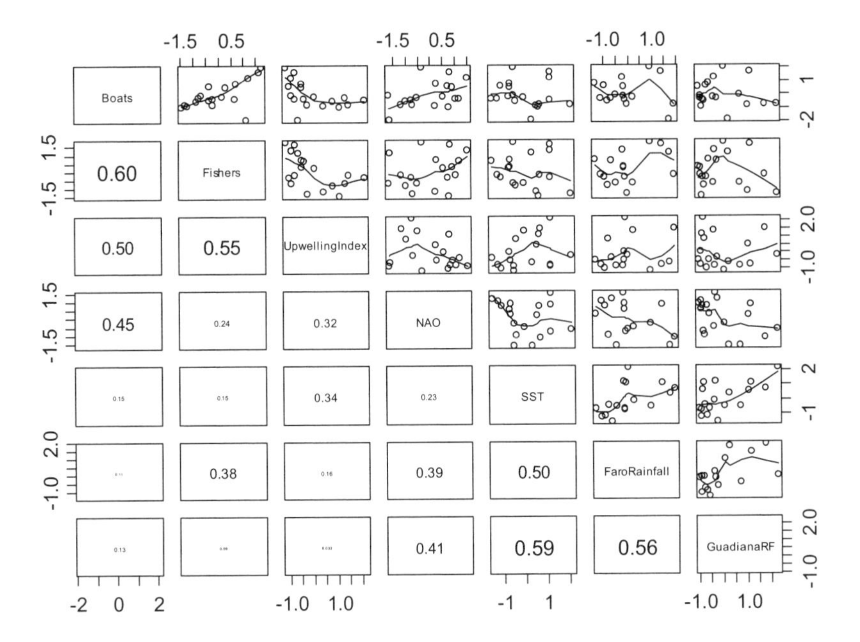

Figure 33.3. Pairplot for the explanatory variables. The upper diagonal shows scatterplots (and smoothing line was added) and the lower diagonal the (absolute) correlations between two explanatory variables. The font size of the correlation is proportional to its value.

33.3 MAFA and DFA

MAFA and DFA were discussed in detail in Chapter 17. Both methods can be used to estimate common trends for the 12 fisheries time series, but they do this in a different way. MAFA works like PCA except that the first axis does not explain maximum variance; rather it accounts for maximum auto-correlation with time lag 1. A high auto-correlation with time lag one is associated with a smooth curve; hence, the first MAFA axis represents the main underlying common trend. The second axis is the second most important trend, etc.

To assess the statistical significance of MAFA axes, a randomisation process was used (Solow 1994, Chapter 17). Correlations between MAFA axes and the landings time series were used to determine the relationship of individual fish species to particular MAFA axes. Correlations between MAFA axes and explanatory variables were also calculated, and these give information on the relationships between the main underlying trends and the environmental variables.

DFA is a completely different modelling approach. The 12 fisheries time series are modelled as a function of a linear combination of common trends, an intercept, one or more explanatory variables and noise (Zuur et al. 2003a,b; Zuur and Pierce 2004; Chapter 17; Chen et al. 2006). DFA was used to analyse the 12 time series of landings, together with various explanatory variables. A series of models were fitted, ranging from the simplest, with one common trend plus noise, to the most complex with up to four common trends, two explanatory variables plus noise. Models were fitted with both a diagonal covariance matrix and a symmetric positive-definite covariance matrix (see Chapter 17 for details) and compared using the Akaike's information criterion (AIC) as a measure of goodness-of-fit. Both fisheries and explanatory variables were standardised to facilitate interpretation of the loadings in DFA.

The main difference between DFA and MAFA is that MAFA cannot incorporate explanatory variables, which makes it a so-called indirect gradient analysis. If the main trends in the data are related to the explanatory variables, then we can pick this up by calculating correlation coefficients between the trends and the explanatory variables once the MAFA analysis is finished. But one will not find strong correlations if the main trends in the species data are not related to the explanatory variables. In DFA we can add the explanatory variables directly in the model and the trends represent the remaining common information.

33.4 MAFA results

Three MAFA axes were significant, and these are presented in Figure 33.4. The first MAFA axis represents a steady declining trend over time. The second MAFA axis shows an increasing trend from 1982 to the early 1990s, followed by a decline. The third MAFA axis decreases to a minimum in 1988, increases to a maximum in 1994 and then decreases steadily thereafter.

The canonical correlations, illustrating the relationship between the species and the three MAFA axes, are shown in Figure 33.5. There were 11 significant cross-correlations, and by chance alone there should only have been two. It can be seen that the first MAFA axis is important for hake, Ray's bream, black spot sea bream, sardine and spider crab. For the first four of these species the positive correlation with the main trend indicates a general decrease in landings, whereas for sardine and spider crab, a tendency for increasing landings is shown by the negative correlation. The plot of canonical correlation for the second MAFA axis showed relatively weak positive correlations only for octopus and cuttlefish. Significant negative correlations for anchovy, black spot sea bream and shrimp indicate trends in landings that are in contrast to the second MAFA axis. The third MAFA axis was significant for silver scabbard (positive correlation) and for chub mackerel and cuttlefish (negative correlation).

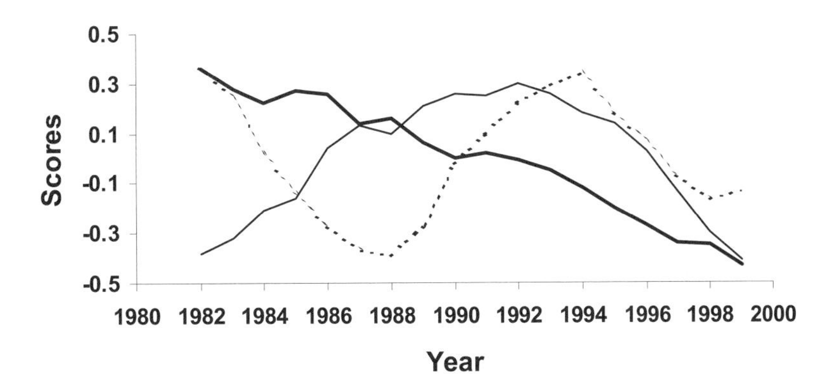

Figure 33.4. Three significant MAFA axes: MAFA 1 (bold), MAFA 2 (continuous line), MAFA 3 (dashed). The vertical axis is unitless.

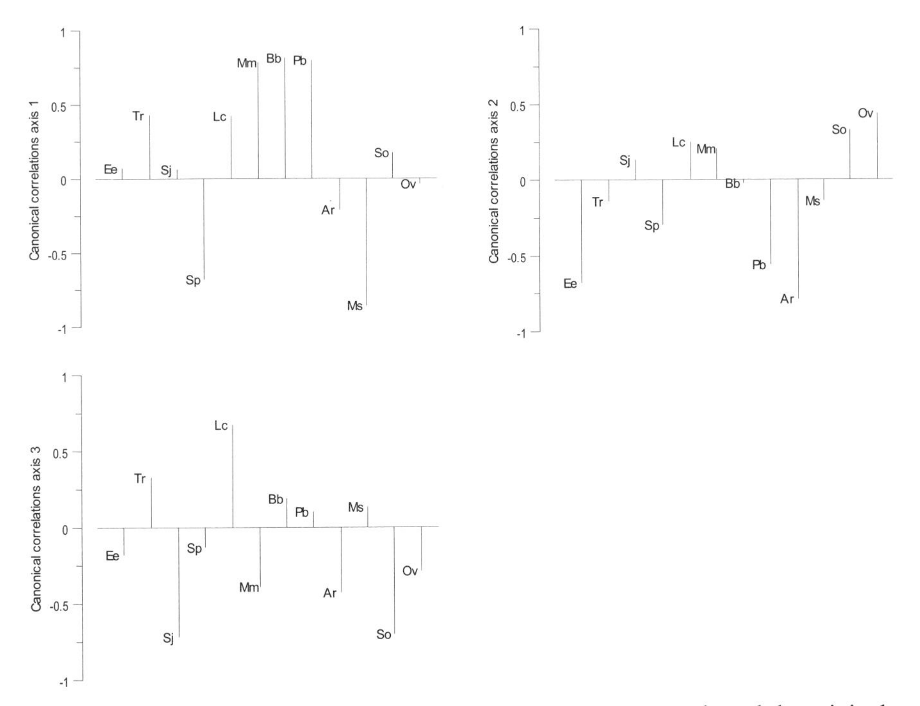

Figure 33.5. Canonical correlations (= correlations between trends and the original time series) for the three MAFA axes. Correlations < -0.47 or > 0.47 are significant at the 5% level. Codes refer to species and are explained in the text.

The correlations between the three significant MAFA axes and the explanatory variables are given in Table 33.1. The majority of the 21 cross-correlations were negative, and 4 were significantly different from 0 at the 5% level. Only one would have been expected to be significant by chance alone, indicating strong relationships between explanatory variables and the significant MAFA axes. Mean SST was significantly negatively correlated with MAF1. Annual rainfall and Guadiana River discharges or flows were significantly negatively correlated with MAF3. The number of licensed fishermen was significantly correlated with MAF2.

Table 33.1. Correlations between explanatory variables and the three significant MAFA axes. Significant correlations are in bold (significance level for correlations: 0.47).

Explanatory Variable	MAF1	MAF2	MAF3
Boats	0.311	0.410	−0.101
Fishers	0.387	**0.757**	−0.391
UpwellingIndex	−0.362	−0.390	0.395
NAO	−0.135	0.299	0.087
SST	**−0.566**	−0.074	−0.215
FaroRainfall	−0.171	0.126	**−0.491**
GuadianaRF	−0.289	−0.130	**−0.474**

33.5 DFA results

Erzini (2005) also tried models with time lags in the explanatory variables, but these were not as good as those with no time lags; hence, only results based on lag 0 data will be presented and discussed here. It should also be pointed out that because of the lag analysis, the DFA results in Erzini (2005) are for the 1984–1999 data, whereas here we use the full time series (1982–1999). Models based on a diagonal covariance matrix generally gave the 'best' fits with three or four common trends, whereas the smallest AIC values corresponded to only one or two common trends for models containing a symmetric non-diagonal covariance matrix. The AIC values of all possible models are given in Table 33.2, and results indicate that the smallest AIC value was for the model with annual rainfall and the number of fishermen as explanatory variables and a symmetric non-diagonal covariance matrix. Results of this model are presented here.

The single common trend (Figure 33.6) of this model shows a decrease over time until the early 1990s, levelling off during the last few years of the time series. The canonical correlations are plotted in Figure 33.7 (factor loadings were similar and are not presented here). The canonical correlations are very similar to those of the first MAFA axis (Figure 33.5), with hake (Mm), Ray's bream (Bb) and black spot sea bream (Pb) strongly positively correlated with the common trend, whereas the spider crab (Ms) was negatively correlated. It is interesting to note that anchovy (Ee), horse mackerel (Tt), shrimp (Ar), cuttlefish (So), octopus (Ov), Spanish mackerel (Sj), and sardine (Sp) landings are poorly or moderately correlated to the common trend. These are species that can be considered to be short-lived, whereas the species that are strongly positively correlated are relatively long-lived.

Table 33.2. AIC values for DFA models with 1 to 4 common trends and 0 to 2 explanatory variables. The lowest AIC value per series of models is in bold. UI stands for upwelling index and RF for rainfall.

	M Number of Trends						
	Diagonal Error Matrix				Non-diagonal Error Matrix		
Explanatory Variables	1	2	3	4	1	2	3
M common trends + noise							
	595	569	539	**539**	570	563	**563**
M common trends + expl. var(s) + noise							
Boats	593	565	548	**539**	565	564	**562**
FaroAnnualRF	**564**	700	887	585	**530**	639	nc
Fishers	552	523	520	**515**	**528**	528	544
GuadianaRF	600	572	550	**546**	565	**556**	560
NAO	596	565	533	**528**	561	554	**554**
SST	603	571	540	**538**	567	**560**	560
UI	586	550	543	**538**	562	**560**	561
Boats-FaroAnnualRF	564	**542**	562	749	**524**	546	563
Boats – Fishers	560	529	518	**515**	523	**521**	535
Boats – GuadianaRF	596	573	557	**542**	560	**555**	559
Boats – NAO	603	568	546	**537**	559	553	**551**
Boats – SST	596	571	542	**534**	562	**559**	560
Boats – UI	594	560	552	**543**	560	**559**	562
FaroAnnualRF - Fishers	525	**501**	521	539	**490**	492	524
FaroAnnualRF-GuadianaRF	595	566	537	**536**	533	**525**	527
FaroAnnualRF - NAO	556	**535**	555	573	517	**515**	566
FaroAnnualRF – SST	559	**528**	548	566	516	**508**	550
FaroAnnualRF – UI	551	**525**	552	563	**521**	545	557
Fishers – GuadianaRF	552	533	525	**523**	**522**	524	538
Fishers – NAO	550	516	**509**	514	520	**518**	532
Fishers – SST	547	519	520	**517**	**520**	522	534
Fishers – UI	556	522	521	**519**	523	**523**	538
GuadianaRF – NAO	599	567	541	**525**	555	**548**	554
GuadianaRF – SST	614	582	557	**554**	564	**559**	560
GuadianaRF – UI	589	561	553	**544**	557	**555**	557
NAO – SST	605	575	530	**523**	557	550	**549**
NAO – UI	592	551	544	**538**	556	558	**553**
SST – UI	592	560	542	**538**	556	**552**	555

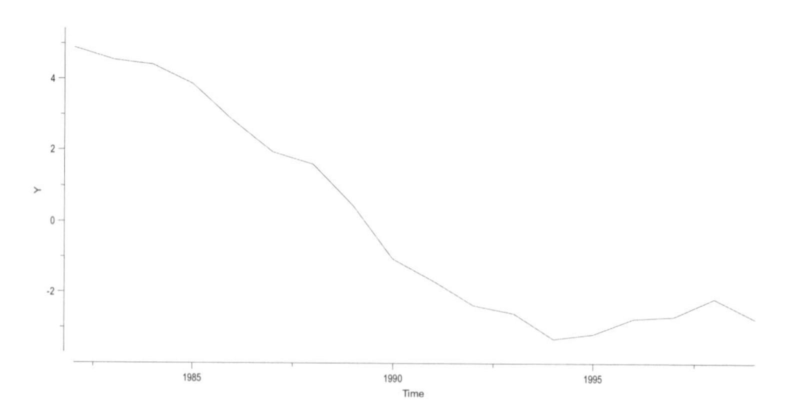

Figure 33.6. The main common trend for the DFA model with two explanatory variables: annual rainfall and the number of licensed commercial fishermen in the Algarve. The vertical axis is unitless.

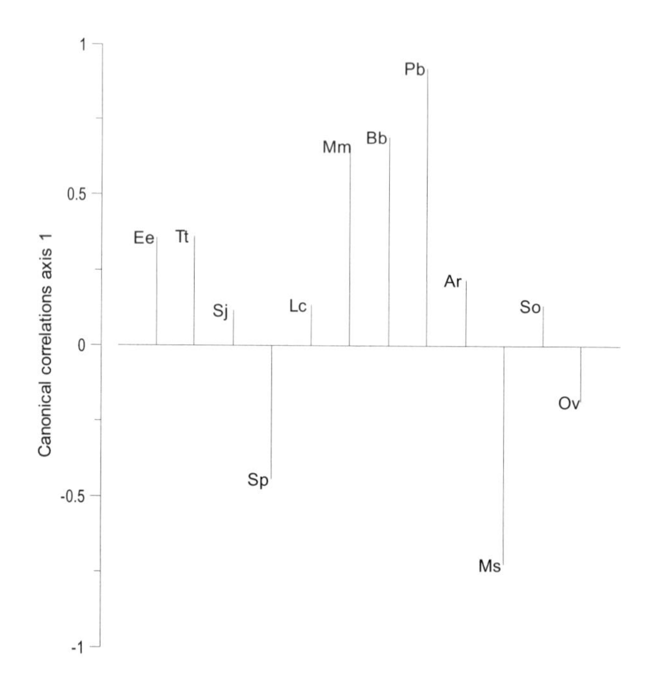

Figure 33.7. Canonical correlations for the DFA model with one common trend and two explanatory variables (annual rainfall and the number of licensed commercial fishermen in the Algarve). Species codes are given in the text and in Table 33.3.

The estimated regression parameters for the two explanatory variables (number of fishermen and rainfall) are given in Table 33.3. Relatively large absolute *t*-values were obtained for hake, shrimp, cuttlefish and octopus for the first explanatory variable, indicating a strong relationship with this proxy measure of effort. On the other hand, only anchovy and shrimp landings seem to be strongly related to rainfall.

The cross-correlation between the two explanatory variables was –0.09. The error covariance matrix *R* (not presented here) of the model showed relatively large diagonal elements for horse mackerel, Spanish mackerel, silver scabbard fish, and octopus, indicating that these landings time series were not fitted well. The correlation coefficients calculated from the error covariance matrix *R* indicated relatively strong positive correlations between the error components of species with similar patterns such as sardine and shrimp (r = 0.721), silver scabbard and Ray's bream (r = 0.614), anchovy and black spot sea bream (r = 0.688) and cuttlefish and shrimp (r = 0.633). On the other hand, relatively strong negative correlations were found between shrimp and silver scabbard fish (r = –0.702), Spanish mackerel and anchovy (r = –0.719) and Spanish mackerel and black spot sea bream (–0.740).

The dissimilarity coefficients calculated from the correlation coefficients following Zuur et al. (2003b) were visualized using multi-dimensional scaling. Results (not shown here) indicated that anchovy (Ee), horse mackerel (Tt), blackspot sea bream (Pb) and Spanish mackerel (Sj) were close to each other and separate from all others, indicating that the response variables in these two groups share a certain amount of information that is not explained by the common trend and the explanatory variables.

Table 33.3. Estimated regression parameters, standard errors (s.e.) and *t*-values for the explanatory variables 'number of fishermen' and annual rainfall. Significant parameters are in bold font.

		Explanatory Variables					
		Fishers			Faro Rainfall		
		Estimate	s.e.	*t*-value	Estimate	s.e.	*t*-value
Anchovy	Ee	0.02	0.18	0.11	0.002	<0.001	**4.84**
Horse mackerel	Tt	−0.11	0.24	−0.46	−0.001	<0.001	−1.13
Mackerel	Sj	0.48	0.24	**2.02**	0.001	<0.001	1.80
Sardine	Sp	−0.43	0.21	−2.04	0.000	<0.001	0.42
Silver scabbard fish	Lc	−0.14	0.23	−0.58	−0.002	<0.001	**−2.27**
Hake	Mm	0.59	0.15	**3.92**	0.000	<0.001	0.09
Ray's bream	Bb	−0.02	0.20	−0.10	−0.001	<0.001	−1.28
Black spot sea bream	Pb	−0.31	0.16	**−2.00**	0.000	<0.001	−0.64
Shrimps	Ar	−0.13	0.13	−0.99	0.003	<0.001	**7.31**
Spider crab	Ms	−0.39	0.17	**−2.27**	0.000	<0.001	0.44
Cuttlefish	So	0.80	0.19	**4.22**	0.001	<0.001	1.48
Octopus	Ov	0.57	0.22	**2.52**	0.000	<0.001	0.25

The observed and fitted landings are shown in Figure 33.8. In most cases the fits were reasonable, with the model adequately describing the trend. Several patterns can be seen for the 12 species. Hake (Mm), black spot sea bream (Pb) and Ray's bream (Bb) show a general decline over time. Anchovy (Ee), chub mackerel (Sj) and shrimp (Ar) are characterized by an initial decrease in landings followed by steady increases in the 1990s. Sardine (Sp) and crab (Ms) landings fluctuate initially and then increase in the second half of the time series. Cuttlefish (So) landings show oscillations over time, with two peaks during the 18-year period, whereas horse mackerel (Tt) and silver scabbard fish (Lc) landings increased initially before decreasing in the late 1990s. The worst fits were for octopus (Ov) landings, which fluctuated considerably, with four peaks in the landings in less than two decades, and for horse mackerel (Tt).

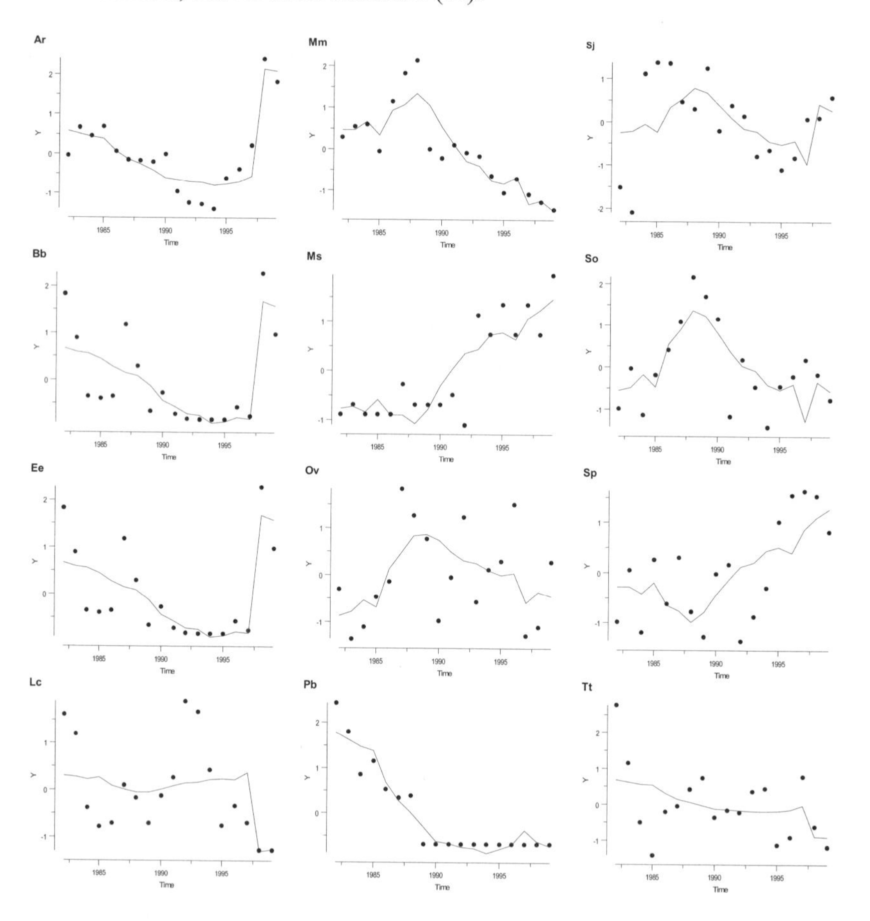

Figure 33.8. Observed (filled circles) and fitted landings (lines) for the DFA model with one common trend and two explanatory variables (annual rainfall and the number of licensed commercial fishermen in the Algarve). Species codes are given in the text and in Table 33.3.

33.6 Discussion

The two techniques produced results that were largely in agreement. The most important MAFA axis showed a steady decline over time, which was similar to the DFA common trend. This strongly suggests that an important underlying pattern in Algarve landings is that of a decrease over time. Indeed, most of the species showed a decline in landings for at least part of the time period. From Figure 33.7 it is evident that, except for the small-size pelagics (anchovy and sardine) and shrimp and crabs, the landings of the remaining large-sized fish declined almost constantly during the study period. This is in agreement with the general trend of sharply declining biomass of large-sized fish observed in the North Atlantic (Christensen et al. 2003; Myers and Worm 2003; Baum et al. 2003).

The AIC indicated that the DFA model with $M = 1$ and two explanatory variables, annual rainfall and the number of fishermen, was the most optimal. Given that the second and third MAFA axes were highly correlated with number of fishermen and annual rainfall, respectively, it is not surprising that the DFA model with $M = 1$ and annual rainfall and number of fishermen as explanatory variables was the best. Thus, both techniques (MAFA and DFA) lead to the same conclusion and their combined use appears to be a useful approach. The results of both techniques could also be coherently interpreted in the light of the life history characteristics of the species. Species such as hake, black spot sea bream and Ray's bream that can be considered relatively long-lived and more susceptible to increasing fishing mortality were more related to the first MAFA axis and the common trend of the DFA, in that both indicate a decline over time. On the other hand, landings of the more short-lived species such as the sardine, anchovy, shrimps and cuttlefish were more related to the other more variable MAFA axes, probably reflecting the greater influence of environmental factors on the recruitment and abundance of these species.

Spawning of most of the commercial species landed in the Algarve takes place in winter and spring, at the end of (or following) the rainy season and the period of maximum river flow (November to February). Since the recruitment of short-lived species such as anchovy and sardine to the fishery takes place in their first year of life, a likely explanation for the influence of rainfall and associated river flow on the landings is that recruitment success of these species is enhanced by increased primary production, larval and juvenile survival and extension of nursery habitat (Kimmerer 2000a,b). Other studies have also found that local environmental variability is important for fisheries landings. In the northwestern Mediterranean Lloret et al. (2001) found that, for most species, catches were significantly positively correlated with river flows and a wind mixing index, with periods of both low discharge and low wind mixing being particularly unfavourable for many of the Mediterranean species studied. Octopus and cuttlefish commercial trawl landings (1982–1999) and artisanal octopus landings (1992–1999) in the Gulf of Cádiz were correlated with sea surface temperature, rainfall and Guadalquivir River discharges (Sobrino et al. 2002).

The NAO and the UI were not found to be as important as rainfall, even though many authors have reported links between long-term trends in climate and different components of the marine ecosystem (Sharp and Csirke 1983; Glantz 1992; Koranteng and McGlade 2001; Parsons and Lear 2001; Ravier and Fromentin 2001; Yanez et al. 2001; Drinkwater et al. 2005).

In conclusion, MAFA and DFA are useful techniques for analysing short, non-stationary time series, with the ability to handle trends, explanatory variables, missing values and interactions between trends, response and explanatory variables. Although both techniques estimate common patterns in multiple time series, they are different statistical techniques, with for example DFA using explanatory variables in a direct way, and allowing for interactions between the response variables using a non-diagonal error covariance matrix. They therefore show different aspects of the data, and their combined use proved to be advantageous, strengthening the conclusions of the analysis.

In this study both techniques were efficient to identify patterns in landings and environmental variables as well as interactions between time series of response and explanatory variables. The importance of local environmental variability for fisheries landings, in particular annual rainfall, rather than large-scale climatic factors, was shown. Once more data are available, other techniques such as chronological clustering (Legendre and Legendre 1998) can be used to see whether breakpoints in the multivariate time series coincide with changes in annual rainfall and river flow.

Acknowledgement

We would like to thank Madalena Salgado and Luís Bentes for compiling the data used in this analysis. We are also very grateful to Dr. Margarida Castro and Dr. Kostas Stergiou for their critical reviews that greatly contributed to improving the manuscript. This work was funded in part by the European Union (AQCESS project, Ref.: QLRT-2000-31151; INCOFISH project, Ref.: PL 003739).

34 Common trends in demersal communities on the Newfoundland-Labrador Shelf

Devine, J.A., Zuur, A.F., Ieno, E.N. and Smith, G.M.

34.1 Introduction

In this chapter another example of dynamic factor analysis (DFA) and min/max auto-correlation factor analysis (MAFA) is presented. The statistical methodology was explained in Chapters 33 and 16 and is not repeated here.

The Newfoundland and Labrador Shelf system supported one of the world's greatest fisheries. Today, many of the stocks are decimated. Annual landings of all groundfish species declined rapidly in 1978, stabilised in the 1980s, and then declined sharply in the early 1990s (Boreman et al. 1997). Many groundfish fisheries, including Atlantic cod *Gadus morhua,* were closed in 1992. Changes in abundance, mean size and biomass are not restricted to commercial species; non-commercial species have also shown declines (Gomes et al. 1995; Haedrich and Barnes 1997; Bianchi et al. 2000; Zwanenburg 2000). Although groundfish populations were declining, seal populations in the Northwest Atlantic were steadily increasing; predation by seals has been suggested as hindering the rebuilding of some important commercial stocks (Morissette et al. 2006).

The Newfoundland-Labrador Shelf is a unique ecosystem due to its topography and circulation patterns. The shelf is broad, ranging from 150 to 400 km wide, overlain by polar waters and contains the deepest shelf region off eastern North America (Helbig et al. 1992; Drinkwater and Mountain 1997). The Labrador Shelf topography is very complex; the shelf contains numerous shallow banks separated by deep channels (Drinkwater and Harding 2001). Inner basins typically reach maximum depths of over 800 m (Drinkwater and Mountain 1997). To the south, the shelf forms the Grand Banks of Newfoundland, a relatively flat area with an average depth of 80 m (Helbig et al. 1992). The Labrador Current, the dominant hydrographic feature of the region, forms a distinctive cold intermediate layer, capped above and below by warmer waters, which effects the distribution and migratory patterns of many fish species (Drinkwater and Harding 2001).

The Newfoundland-Labrador Shelf experienced different environmental conditions beginning in the 1960s. The system experienced below average temperatures in the mid-1980s to mid-1990s, with the early 1990s experiencing the lowest recorded temperature anomalies in sea surface waters (0–176 m) since 1950 (Drinkwater 2002). Great salinity anomalies (GSAs) also occurred in the early 1970s, 1980s and 1990s and may be linked to changes in the North Atlantic Oscillation (NAO) index (Belkin 2004). The NAO, the atmospheric pressure differential between the Azores and Iceland, exerts a strong influence over the ocean and atmosphere of the North Atlantic Ocean. The NAO influences sea ice extent and melt, water temperature, the distribution and fluxes of major water masses and currents, deep water formation in the Greenland Sea and intermediate water formation in the Labrador Sea (Hurrell et al. 2003).

Over-fishing, predation, changes in prey availability and environmental factors have all been pinpointed as possible causes for the observed declines in size and abundance of demersal fish species, and a long and ongoing debate concerns which of the many possibilities has played the greatest role (NRC 1999; Hamilton et al. 2004). Examinations have ranged from the descriptive (e.g., Villagarcía et al. 1999) to the broadly analytical (e.g., Bianchi et al. 2000) and from the application of local ecological knowledge (e.g., Neis et al. 1999) to quantitative ecosystem models based on theory (e.g., Murillo 2001). Most studies, however, have been more traditional and have employed very standard approaches. Furthermore, the great majority of these studies have focused on only one species, Atlantic cod, with little or no consideration of other species in the system. As would be expected, conclusions range across the spectrum as to principal causes, but there is general agreement that the situation is complex with underlying dynamics operating at several scales.

Our objective is to determine whether the complex dynamics involving biomass of the Newfoundland-Labrador Shelf demersal community could be described using multivariate time series analysis. We used MAFA and DFA to analyse trends in relative biomass of commercial and non-commercial species and examine relationships with external factors.

34.2 Data

The ECNASAP (East Coast North American Strategic Assessment Project) dataset was used as the source of records for the Newfoundland-Labrador Shelf (NAFO Divisions 2J3KL) (see Brown et al. 1996 for details). This database consists of records collected from random stratified scientific survey tows for the years 1978 through 1994. A mixture of important commercial, rare and non-commercial demersal teleost and elasmobranch species were chosen for the analysis: Atlantic cod *Gadus morhua* (AC), American plaice *Hippoglossoides platessoides* (AP), onion-eye grenadier *Macrourus berglax* (RHG), rock grenadier *Coryphaenoides rupestris* (RKG), Greenland halibut *Reinhardtius hippoglossoides* (GH), thorny skate *Raja radiata* (TS), deepwater redfish *Sebastes mentella* (DR), golden redfish *Sebastes marinus* (GR), spinytail skate *Bathyraja spinicauda* (SS), Atlantic wolffish *Anarhichas lupus* (AW), northern wolffish *Anarhichas denticulatus* (NW), spotted wolffish *Anarhichas minor* (SW), blue hake *Antimora rostrata* (BH), and black dogfish *Centroscyllium fabricii* (BD). Weight per tow was used as an index of relative biomass over time (Figure 34.1-A). Because trawl survey data often have a skewed distribution, data were $\log_{10}(x + 1)$ transformed. All data were standardised by normalisation to assist with interpretation.

Environmental variables were sea surface temperature recorded to 100 m depth (SST), bottom temperature at 250–1485 m (the maximum depth of the survey, BT), and salinity at 0–250 m (SAL) in NAFO Divisions 2J3KL, 1960–1994 (Figure 34.1-B). Annual anomalies were estimated by subtracting the 1960–1994 mean from the annual mean and dividing by the 1960–1994 standard deviation. The NAO annual index (NAOA) and the NAO winter index (NAOW) were also included (www.cgd.ucar.edu/~jhurrell/nao.stat.ann.html). Fishing effort data (number of days) (EFF) were obtained from the NAFO annual fisheries statistics database (www.nafo.ca), and harp seal abundance (HARP) was obtained from DFO. All data were standardised by normalisation to assist with interpretation.

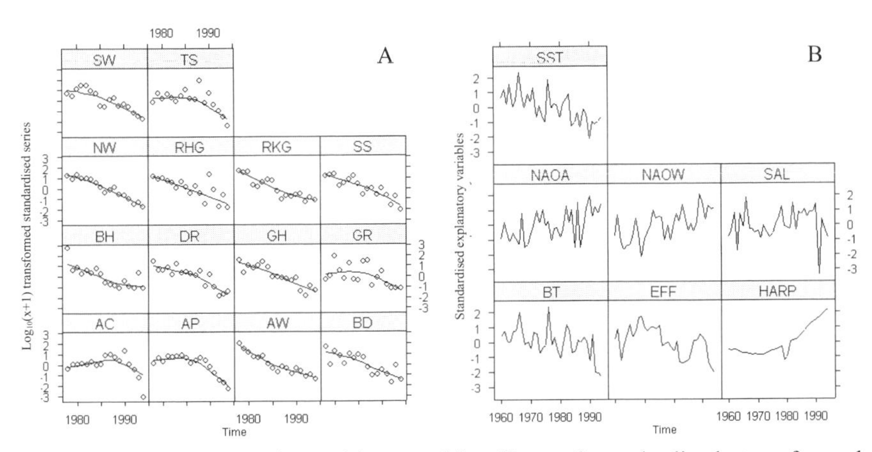

Figure 34.1. A: Lattice plots with smoothing lines of standardised, transformed scientific survey CPUE (weight per tow) time series 1978–1994. B: Standardised explanatory variable time series 1960–1994 (B) in NAFO Divisions 2J3KL. See text for abbreviations.

34.3 Time series analysis

Cross-correlations were used to decide which lags, if any, were important between response and explanatory variables. MAFA and DFA were discussed in detail in Chapter 17, and an example of their use with fisheries landings data was given in Chapter 33. The methodology is the same; therefore, we omit description of the technique here.

Time series and correlations

Cross-correlations between response variables and explanatory variables at lags up to 5 years, 10 years for fishing effort, showed most lagged explanatory variables had higher correlations than variables with no lags (not shown here). Only harp seal abundance (all lags), fishing effort (lags > 6) and sea surface temperature (lags 1 and 2) were significantly correlated with most of the species. Bottom temperature (lags > 0) and the NAO annual index (all lags) were not significantly correlated with many, if any, of the species. If cross-correlations are estimated for many variables ($n > 100$), spurious significant cross-correlations may be estimated (Chatfield 2003). Out of 658 estimated cross-correlations, 209 were significantly different from 0 at the 5% level; by chance alone, 33 could have been significant.

MAFA

MAFA showed two main trends in relative biomass were significant (Figure 34.2). The first MAFA axis (auto-correlation of 0.996, $p = 0.054$) represents a steady decline over time; this is the main pattern underlying the time series. The second MAFA axis (auto-correlation of 0.922, $p = 0.016$) shows an increase from 1978 to 1987, followed by a decline.

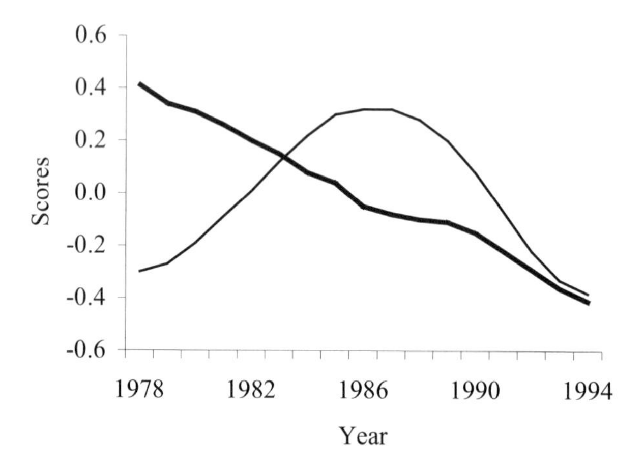

Figure 34.2. The two main trends identified by MAFA in the relative biomass of 11 teleost and 3 elasmobranch species from scientific random stratified surveys in NAFO Divisions 2J3KL, 1978–1994. The thick line indicates main MAFA axis. The *y*-axis is unitless.

Canonical correlations between the species and MAFA axes indicated that the first axis was important for all species except Atlantic cod, whereas the second axis was important only for American plaice, Atlantic cod and thorny skate (Figure 34.3). All significant correlations were positive.

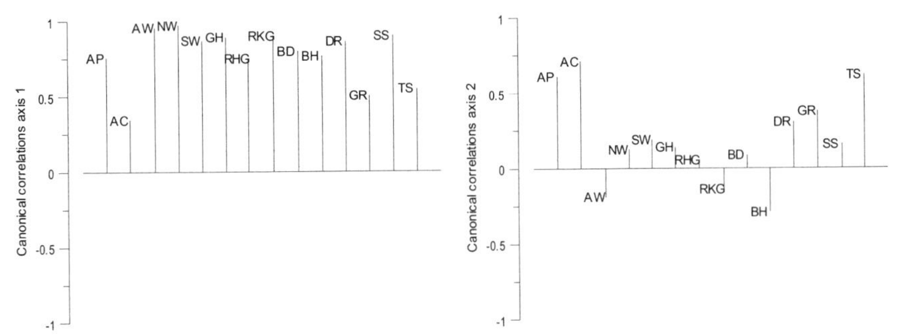

Figure 34.3. Canonical correlations between species and MAFA axes for the two main trends in biomass in NAFO Divisions 2J3KL, 1978–1994. Significance level for correlations is 0.49.

Forty-two explanatory variables and lags were used, but here only the highest correlations between the MAFA axes and the explanatory variables are presented to save space (Table 34.1). Harp seal abundance (no lag and lag 1), NAO winter index (lagged 1 year), and salinity (lagged 4 years) were significantly negatively correlated with MAF1, whereas bottom temperature (no lag), sea surface temperature (lagged 3 years), and fishing effort (lagged 10 years) were significantly positively correlated to MAF1. The NAO annual index, although not significant at the $p = 0.05$ level, was highly negatively correlated with MAF1. Bottom temperature (no lag) was significantly positively related to the second MAFA axis, whereas fishing effort (lagged 4 years) was significantly negatively related to the axis. As fishing effort decreased, biomass for these species increased and vice versa as shown by the high negative correlation between fishing effort and this trend. Out of 84 estimated cross-correlations, 4 could have been significant by chance; 26 were estimated to be significantly different from 0 at the 5% level.

DFA

DFA models with lagged explanatory variables provided a better fit than those with no lags. Additionally, models based on a symmetric non-diagonal covariance matrix provided better fits than those based on a diagonal covariance matrix. Akaike's Information Criterion (AIC) was used initially to determine the most optimal model in terms of goodness-of-fit and the number of parameters; the model with the smallest AIC value was selected as being the best. Additionally, fitted values and residuals were also used to determine goodness-of-fit. Instead of presenting the results of only the most optimal DFA model, we present various models to obtain greater insight into possible causes of changes in groundfish relative biomass. We start with DFA models containing one explanatory variable. The best models with only one explanatory variable, using a symmetric non-diagonal covariance matrix, were harp seal abundance lagged 1 year and fishing effort lagged

10 years (Table 34.1). Although the one trend model with salinity lagged 4 years had a lower AIC value than the models with harp seal abundance or fishing effort, the fit of the model to the data was poor, indicating this was probably not the best model. Plots of residuals and fits indicated the models with two trends were better than those with only one trend. Again, harp seal abundance lagged 1 year and fishing effort lagged 10 years were the best models.

Because of collinearity between harp seal abundance and fishing effort at high lags (in effort), these two explanatory variables could not be combined in the analysis. The cross-correlation between harp seal abundance lagged 1 year and fishing effort lagged 10 years is −0.85. The best model with two explanatory variables, determined using the AIC value and plots of residuals and fits, was harp seal abundance lagged 1 year and salinity lagged 2 years (Table 34.2). Plots of residuals and fits indicated the model with three trends was better than the models with one or two trends, therefore, the model with three common trends was chosen (Figure 34.4). The first common trend is similar to the first MAFA axis; it showed a declining trend. The second trend, similar to the second MAFA axis, showed an increase and then a decrease over time. The third trend was a decrease until 1984 and then an increase until 1991.

Table 34.1. Correlations between the MAFA axes and the explanatory variables and AIC values for DFA models with one and two common trends. Significance level for correlations = 0.49.

Explanatory Variable	MAFA 1	MAFA 2	DFA 1 Trend	DFA 2 Trends
NAOW1	−0.58	0.00	345.3	341.4
NAOA1	−0.47	−0.08	370.6	365.6
BT	0.53	0.49	338.1	359.6
HARP	−0.96	0.04	326.0	340.1
HARP1	−0.95	−0.02	275.0	282.1
EFF4	0.23	−0.84	342.7	351.3
EFF10	0.93	0.12	282.0	291.6
SST3	0.69	−0.02	358.4	348.3
SAL4	−0.64	0.01	245.3	339.7

Table 34.2. AIC values for DFA models with one to three common trends and two explanatory variables; cross-correlations between the two explanatory variables used in the model are also shown. Only models with the lowest AIC values are shown.

Explanatory Variables	1 Trend	2 Trends	3 Trends	Correlation
HARP1, SAL2	172.7	181.7	196.3	−0.15
HARP1, SST2	259.9	266.7	279.5	−0.63
HARP1, NAOW1	178.1	187.0	198.5	0.60
HARP1, NAOA1	269.9	278.2	289.2	0.46
HARP1, BT	217.7	225.0	237.0	−0.43
EFF10, SAL2	212.3	218.8	227.1	0.78
EFF10, SST2	274.2	278.0	290.3	0.61
EFF10, NAOW1	238.2	241.3	255.9	0.14
EFF10, NAOA1	218.5	223.3	236.8	−0.43
EFF10, BT	223.2	228.7	236.5	−0.32

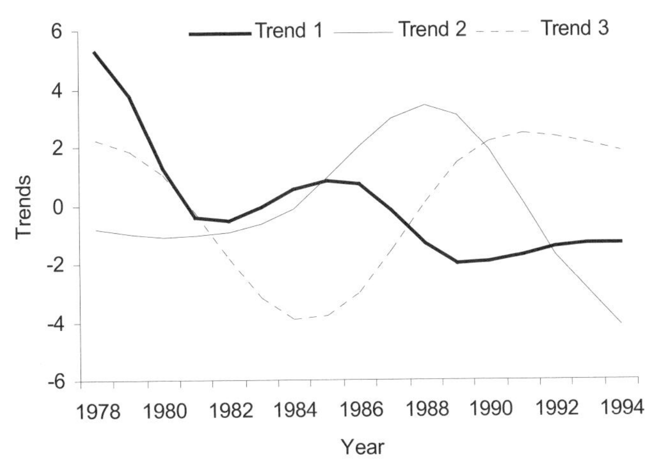

Figure 34.4. The three main common trends in relative biomass for 11 teleost and 3 elasmobranch species for the DFA model with two explanatory variables, salinity lagged 2 years and harp seal abundance lagged 1 year, in NAFO Divisions 2J3KL, 1978–1994.

Canonical correlations were used to determine which species were related to a particular trend (Figure 34.5). All species except Atlantic cod, American plaice, golden redfish and thorny skate were strongly related to the first common trend, a decline over time. The second common trend was strongly and positively related to American plaice, Atlantic cod and thorny skate. Deepwater redfish, golden redfish and spinytail skate were also positively related to the second trend. All species except Atlantic cod were negatively related to the third common trend; thorny skate was only weakly negatively correlated. Species negatively correlated to a trend display a trend exactly the opposite of what is indicated; all species except Atlantic cod increased in biomass until 1984 and then declined until 1991.

The estimated regression parameters for the two explanatory variables show that only black dogfish and blue hake had a strong relationship with salinity lagged two years, which is indicated by the relatively high t-value (Table 34.3). All species except Atlantic cod and blue hake had a strong relationship with harp seal abundance lagged 1 year. The cross-correlation between the two explanatory variables was -0.15. The model with three trends and two explanatory variables (salinity lagged 2 years and harp seal abundance lagged 1 year) improved the fit of the model compared with the model with three trends and one explanatory variable (Figure 34.6). Adding salinity to the model improved the fit for black dogfish, Greenland halibut, blue hake, spinytail skate and onion-eye grenadier for the last years of the time series.

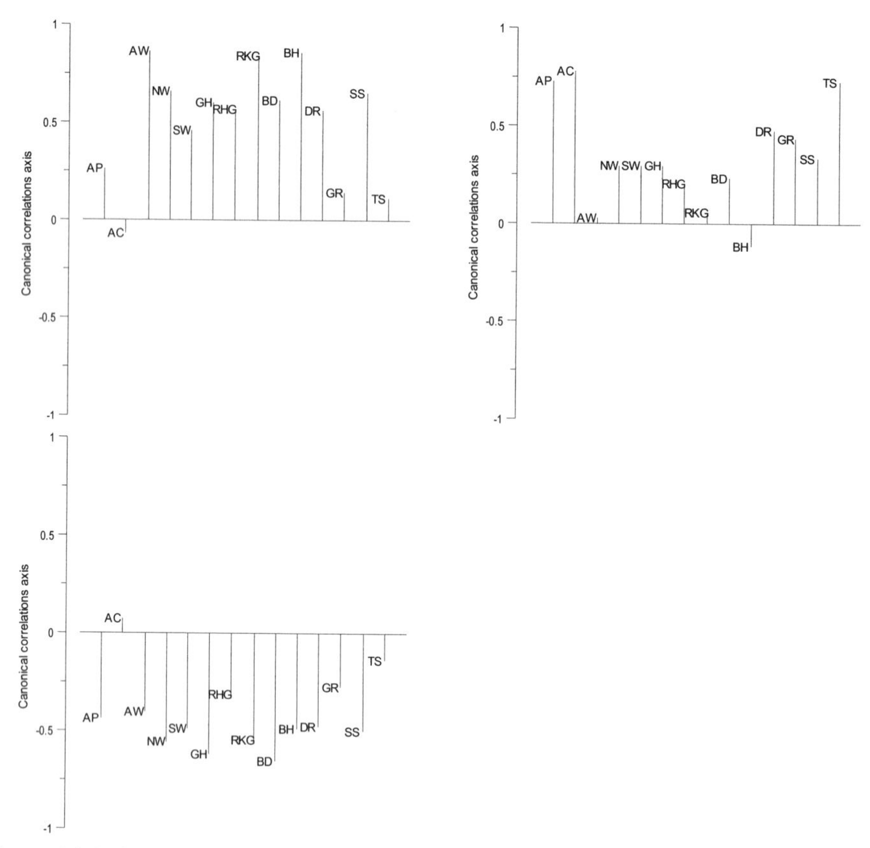

Figure 34.5. Canonical correlations for the DFA model with three common trends and two explanatory variables, harp seal abundance lagged 1 year and salinity lagged 2 years. Top left: trend 1, top right: trend 2, bottom: trend 3.

Table 34.3. Estimated regression parameters, standard errors and *t*-values for the DFA model with three trends and two explanatory variables, salinity lagged 2 years and harp seal abundance lagged 1 year.

		Salinity Lagged 2 years			Seal Abundance Lag 1 year		
Species	Code	Estimated values	S.E.	*t*-values	Estimated values	S.E.	t-values
Hippoglossoides platessoides	AP	−0.05	0.09	−0.61	−0.72	0.18	−4.03
Gadus morhua	AC	−0.04	0.13	−0.29	−0.37	0.27	−1.35
Anarhichas lupus	AW	−0.07	0.09	−0.76	−0.64	0.15	−4.24
Anarhichas denticulatus	NW	−0.07	0.08	−0.87	−0.86	0.09	−9.39
Anarhichas minor	SW	0.02	0.11	0.22	−1.10	0.14	−7.95
Reinhardtius hippoglossoides	GH	−0.15	0.11	−1.35	−0.71	0.13	−5.57
Macrourus berglax	RHG	−0.12	0.17	−0.74	−0.58	0.19	−3.07
Coryphaenoides rupestris	RKG	0.00	0.10	−0.05	−0.44	0.17	−2.65
Centroscyllium fabricii	BD	−0.30	0.10	−3.06	−0.47	0.16	−3.03
Antimora rostrata	BH	−0.27	0.12	−2.27	−0.21	0.21	−0.98
Sebastes mentella	DR	−0.04	0.11	−0.38	−0.55	0.17	−3.27
Sebastes marinus	GR	0.04	0.18	0.24	−0.63	0.20	−3.20
Bathyraja spinicauda	SS	−0.11	0.12	−0.93	−0.60	0.16	−3.79
Raja radiata	TS	0.01	0.13	0.05	−0.54	0.23	−2.33

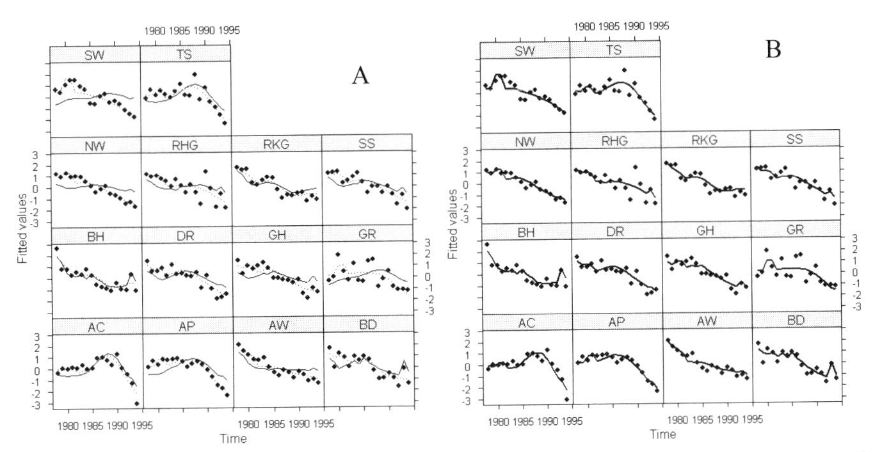

Figure 34.6. A: Fit of three trends and one explanatory variable model: Harp seal abundance lagged 1 year (dotted line) and salinity lagged 2 years (solid thin line) to the relative biomass index (points) in NAFO Areas 2J3KL, 1978–1994. B: Fit of three trends and two explanatory variables model (harp seal abundance lagged one year and salinity lagged 2 years) to the relative biomass index.

34.4 Discussion

Two similar trends were shown by both MAFA and DFA. DFA results indicated that a third trend, not appearing as a result in MAFA, was important and improved the model fit to the data. However, for this chapter, the discussion will focus on the two similar trends identified.

The main trend in biomass of demersal species on the Newfoundland-Labrador Shelf 1978–1994 was a decline over time. MAFA results showed declining biomass was highly correlated to most variables. Declines in biomass may indicate that there are either fewer fish present or that fish are smaller. Paucity of older fish is primarily an effect of fishing effort (Hamilton et al. 2004), whereas smaller fish may be a sign that size-at-age is changing (Bianchi et al. 2000). Reductions in size-at-age can result from both fishing and environmental changes. Declines in the population biomass indices occurred during a period when harp seal abundance was increasing, NAO anomalies were generally positive (indicating more and stronger winter storms in the northwest Atlantic), salinity was higher than average (except for anomalies in the early 1980s and 1990s), bottom temperatures and sea surface temperatures were lower than average, and fishing effort was variable. Additionally, population age-class structure and mean size-at-age for many commercial and non-commercial fish species captured in research surveys and commercial fisheries has declined since the early 1980s (Haedrich 1995; Bowering et al. 1997; Haedrich and Barnes 1997).

The second trend in biomass, an increase until the mid-1980s followed by a steep decline, was strongly related to Atlantic cod, American plaice, thorny skate, deepwater redfish and golden redfish. These five are important commercial species. According to the MAFA results, fishing effort lagged 4 years was negatively correlated to this trend, whereas bottom temperature was positively related. Fishing mortality on most groundfish stocks was low 1977–1980, following the exclusion of foreign fleets within Canada's 200-mile economic management zone and may have allowed a slight rebuilding of stocks. Fishing mortality subsequently increased rapidly due to an increase in the size of domestic fleets until the closure of many fisheries in the 1990s (Boreman et al. 1997). Mean weight, body condition, length and population size structure are often negatively correlated with increasing fishing effort, a sign of size selective exploitation (Pauly and Maclean 2003). Bottom temperature, except for a low around 1984–1985, tended to follow the trajectory of the MAFA trend (increasing in early 1980s, decreasing to lows after 1992). Changes in species abundance or size and temperature are often highly correlated (McGinn 2002).

When single explanatory factors were considered, DFA showed that predation (fishing effort or harp seal abundance) at lags was strongly related to most of the demersal species, but environmental variables were not. Fishing effort has an immediate effect on biomass, reducing the number and size of fish present, but it also has a cumulative effect, reducing the number of fish that will recruit to the population in the future. Fishing effort declined from 1970 until the mid-1980s and increased until the closure of many fisheries in the early 1990s. These closures did not result in decreased predation as these species are captured as bycatch in other fisheries. As the groundfish fisheries were collapsing, fishermen turned to harvesting shrimp and crab; these two fisheries have some of the highest discard rates. Ninety-eight percent of the bycatch from shrimp fisheries in the northwest Atlantic, which is mainly juveniles and sub-adult fish, is discarded (Alverson et al. 1994). Although fishing has an effect on all sizes of fish, harp seals have been shown to select a variety of size ranges including fish that have not yet recruited to the fishery (Stenson et al. 1997; Hammill and Stenson 2000; Morissette et al. 2006).

When two explanatory variables were considered, the best model included lagged salinity and predation. Salinity, except for an anomaly in the early 1980s, was generally higher than average for most of the 1980s. Changes in salinity are associated with changes in primary production in surface waters and thus food availability for pelagic larvae and juveniles (Boreman et al. 1997). Increased primary production could lead to higher survival of larvae and greater recruitment to the adult population. Salinity anomalies tend to lead to sea ice and temperature anomalies (Marsden et al. 1991), which directly influences the timing and extent of harp seal migrations southward (Stenson et al. 1997). Changes in fish biomass may have been related to greater harp seal predation following greater southward extent of ice in the mid-1980s .

Our analysis has illustrated an objective technique to gain insight into the elements of a changing system, and it is unique in being based on fisheries, independent data while including the potential effects of a natural predator. These techniques highlight the complexity of the Newfoundland-Labrador Shelf ecosystem and give insight into a rich system; many dynamics are occurring at the same time, often within populations of the same species. MAFA and DFA methods allow testing for relationships between time series of the demersal community with a variety of explanatory variables. Our analyses show predation, in general, has important effects on biomass of demersal species on the Newfoundland-Labrador Shelf. Many studies have described community changes due to environmental changes and fisheries, a type of selective predation; however, few have looked at predation by multiple causes. Predation is always present in marine systems, and removals due to natural predation can exceed removals by fisheries (Bax 1998). Seal predation has been shown to consume large quantities of fish in Atlantic Canada (Shelton et al. 2006); however, this is only one type of predation. Consumption of fish by other fish can exceed that of marine mammals (Morissette et al. 2006). We have shown that one type of external factor is not responsible for the changes occurring within this fishery ecosystem. Biological and environmental factors, acting in combination, have resulted in the dramatic changes we see today.

Acknowledgements

We thank the Department of Fisheries and Oceans for supplying data and R.L. Haedrich for critical review of the manuscript. This work was funded in part by the US National Science Foundation (NSF) and the Natural Sciences and Engineering Research Council of Canada (NSERC).

35 Sea level change and salt marshes in the Wadden Sea: A time series analysis

Dijkema, K.S., Van Duin, W.E., Meesters, H.W.G., Zuur, A.F., Ieno. E.N. and Smith, G.M.

35.1 Interaction between hydrodynamical and biological factors

Salt marshes are a transitional zone between the sea and land formed by flooding, sedimentation and erosion. This highly specialized zone is characterised by a close interaction of physical and biological processes. The saline plant and animal communities play an important role in the geomorphological development. Because the salt marshes are a sedimentary belt, their potential for recovery is essential to coastal protection (Erchinger 1995).

The Danish-German-Netherlands Wadden Sea harbours substantial areas of salt marshes. With 400 km^2, first place in Europe is shared with the United Kingdom (Dijkema 1990; Bakker et al. 2005). In the Netherlands part of the Wadden Sea, 3.6% of the total tidal area is salt marsh, divided into a barrier island-type and a mainland-type of salt marsh. In the past, embankments have far exceeded the natural accretion rate of the mainland-type of salt marsh. Therefore, the total salt marsh area decreased. By stimulating the sedimentation process through the creation of sedimentation fields sheltered by brushwood groynes and improving vegetation development through artificial drainage (digging creeks) the man-made mainland salt marshes in the provinces of Friesland and Groningen (Figure 35.1) helped to catch up on this backlog over the past decades.

Salt marsh plants play an essential role in the interaction between hydrodynamical and biological factors. In general, salt marsh development is possible on locations with a gently sloping coastline, low wave energy and low water velocity. Therefore, salt marshes are absent on exposed rocky coastlines, but they are usually present along flat coasts and in sheltered bays. If there is sufficient sediment in the water, the tidal flat increases in surface level and the pioneer plants *Spartina anglica* and *Salicornia dolichostachya* appear (Figure 35.2). At around mean high tide (MHT) level, the grass *Puccinellia maritima* is the next step in the marsh building process. This perennial grass provides sufficient cover (i) to produce the highest accretion rate in the entire development of the salt marsh, (ii) to prompt the development of a natural creek system, and (iii) to fix the newly deposited sediments. The development of a creek system (or the digging of ditches in man-made salt marshes) provides a major stimulus for the growth of most salt marsh plants by improving drainage. Several studies have shown that the mud supply in the Wadden Sea is more than sufficient and therefore in no way restricts the rate of accretion of the mainland salt marshes. Due to artificial draining in mainland salt marshes, no major spatial patterns in sedimentation can be found. This contrasts with the barrier island-type of salt marshes where creeks, levees and basins are natural phenomena.

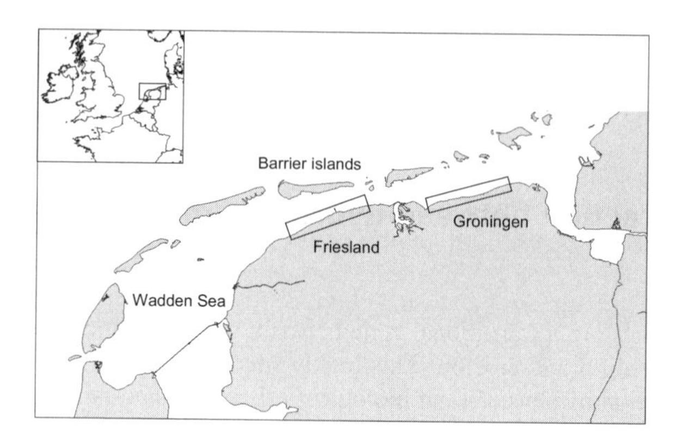

Figure 35.1. Location of mainland salt marsh areas (in rectangles) in Friesland and Groningen.

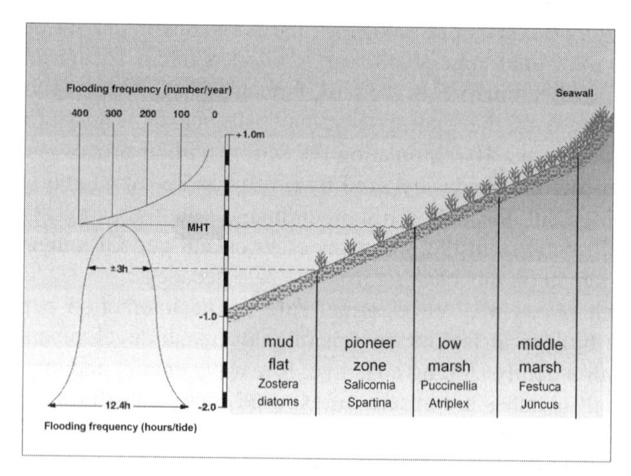

Figure 35.2. Schematic overview of the zonation of the vegetation in a typical salt marsh as a consequence of flooding frequency and MHT. The pre-pioneer zone is the transition between pioneer zone and mudflat and has a scarce vegetation cover (Erchinger 1985).

Salt marsh plants have a characteristic vertical range in relation to sea level. For long-term vegetation development, the accretional balance of sedimentation, erosion, sea level rise and soil subsidence is thought to be the determining factor (Dijkema 1997). Salt marsh plants have a characteristic vertical range in relation to sea level. Year-to-year changes in MHT (Figure 35.3), however, have an impact on the distribution of plants and on the vegetated area of the different salt marsh zones, under the precondition that the drainage pattern does not change (Olff et al. 1988; Dijkema et al. 1990; De Jong and Van der Pluijm 1994). The number of yearly floodings changes with MHT. The lower limits of the vegetation zones may follow trends of changing water level. An increase in flooding frequency may worsen growing conditions and shift the lower limits of vegetation zones to higher grounds that are less frequently inundated. A decrease in flooding frequency may improve growing conditions at lower levels and will stimulate the plants to move to lower levels (= towards the mudflat). An increase in MHT of just 5 to 10 cm in a single year may already result in a shift of some plant species (Beeftink 1987).

The main questions in this case study chapter are (i) do the year-to-year changes in MHT levels affect the development of salt marsh vegetation, (ii) are MHT levels responsible for major shifts in the lower limits of the vegetation zones, and (iii) are the impacts on pioneer zone and salt marsh zone different?

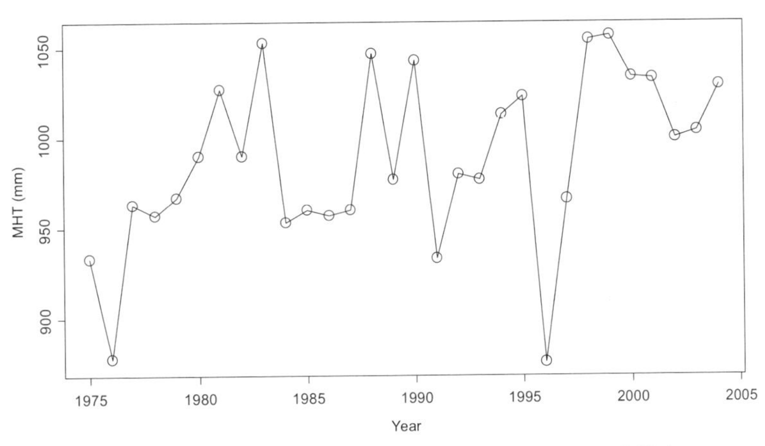

Figure 35.3. Yearly MHT in the Netherlands Wadden Sea 1975–2004.

35.2 The data

The time series used in this chapter are taken from a large dataset on man-made salt marshes along 50 km of the Dutch mainland coast collected by the Department of Public Works (Wadden Sea Unit) between 1937 and 2004. The database includes records on elevation, accretion, soil composition, vegetation composition, vegetation coverage, vegetation type, yearly average MHT, management, etc.

(Dijkema 1997). It is probably the oldest and largest monitoring dataset for a salt marsh site in Europe. We selected 18 stations (Table 35.1) located in the provinces of Groningen and Friesland. For each station, information of plant community relevees were used to calculate the yearly lateral seaward shifts of the lower limits of the pioneer and marsh zones in metres. Hence, PP_GR_East is an annual time series that consists of the distance from the seawall the boundary of the pre-pioneer species at location east in Groningen, relative to 1980. The same holds for PP_GR_Mid, but now for a different location. We use the yearly MHT for all zones (Figure 35.3). Figure 35.4 shows the man-made salt marshes along the mainland coast of Groningen.

Table 35.1. Names of the 18 time series used in this case study chapter. There are three vegetation zones, namely pre-pioneer zone (0–5 % coverage), pioneer zone (> 5 % coverage) and salt marsh. For each vegetation zone, there is an east, mid and west time series referring to the position of the stations.

	Area	
Type	**Groningen**	**Friesland**
Pre-pioneer	PP_GR_East	PP_FR_East
Pre-pioneer	PP_GR_Mid	PP_FR_Mid
Pre-pioneer	PP_GR_West	PP_FR_West
Pioneer	P_GR_East	P_FR_East
Pioneer	P_GR_Mid	P_FR_Mid
Pioneer	P_GR_West	P_FR_West
Salt marsh	K_GR_East	K_FR_East
Salt marsh	K_GR_Mid	K_FR_Mid
Salt marsh	K_GR_West	K_FR_West

Figure 35.4. Left: Salt marsh works in Groningen. The sea wall is the white wall in the middle of the photo. The different vegetation zones (salt marsh, pioneer and pre-pioneer) can be seen by moving from the sea wall towards the left. The land on the right side of the seawall is never flooded and is used for agriculture. Right: Salt marsh works viewed from the sea wall to the mudflat.

35.3 Data exploration

Figure 35.5 shows a principal component analysis (PCA) correlation biplot for the nine time series from Groningen and the nine series from Friesland. Most of the Groningen time series point in the upper right direction, whereas the majority of the Friesland series point towards the lower right. Recall from Chapter 12 that lines pointing in the same direction implies that the corresponding variables are correlated, whereas if the angle is 90 degrees, they are uncorrelated (although keep in mind that the biplot is a two-dimensional approximation of a high-dimensional space). Hence, the biplot indicates that the nine time series of Groningen behaved differently from those in Friesland, and therefore, we will analyse the data in two steps, first for the Groningen series and then for the Friesland series.

We have nine time series for Groningen; three pre-pioneer, three pioneer and three salt marsh time series and a lattice plot are presented in Figure 35.6 (left). Data from 1975 onwards were used as this provides a regular-spaced dataset without missing values. Most of the series exhibit a general decline over time. This means that the lower limits of the three vegetation zones are moving towards the sea wall. Two time series (PP_GR_East and PP_GR_Mid) have a large peak in the mid-1990s but as it is present in more than one time series is unlikely to be a typo. Note that this peak also coincides with the low MHT value in 1996. There is no immediate reason to apply a data transformation. The lattice graph for the Friesland series is presented in Figure 35.6 (right). Again, there is no need for a data transformation. Both lattice graphs of the time series versus Year indicate that there are indeed differences in the temporal patterns in the Groningen and

Friesland series. Lattice graphs of the time series versus MHT are presented in Figure 35.7 and show a weak linear relationship at most stations.

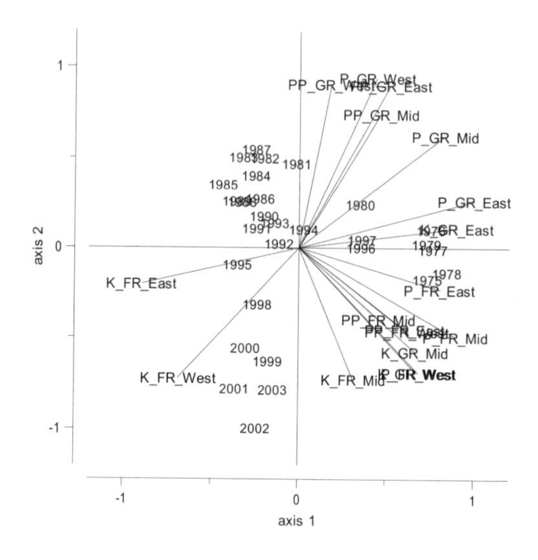

Figure 35.5. PCA correlation biplot of all 18 time series from Groningen and Friesland. Lines pointing in the same direction indicate high correlation. The first two eigenvalues are 0.36 and 0.30, respectively, corresponding to 66% of the variation in the data. The software used scales the eigenvalues such that the sum of all eigenvalues is equal to 1.

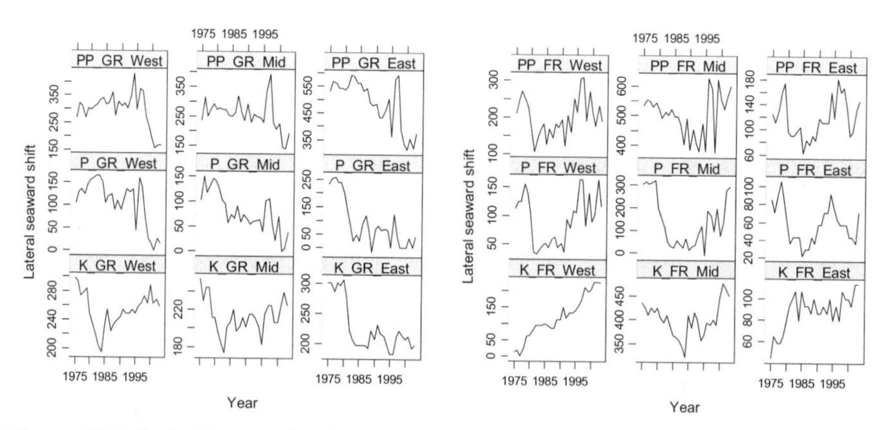

Figure 35.6. Left: Time series for Groningen (GR) based on yearly lateral seaward shifts of pre-pioneer (PP), pioneer (P) and salt marsh zone (K) plotted versus Year. Right: Time series for Friesland (FR) based on yearly lateral seaward shifts of pre-pioneer (PP), pioneer (P) and salt marsh zones (K) plotted versus Year.

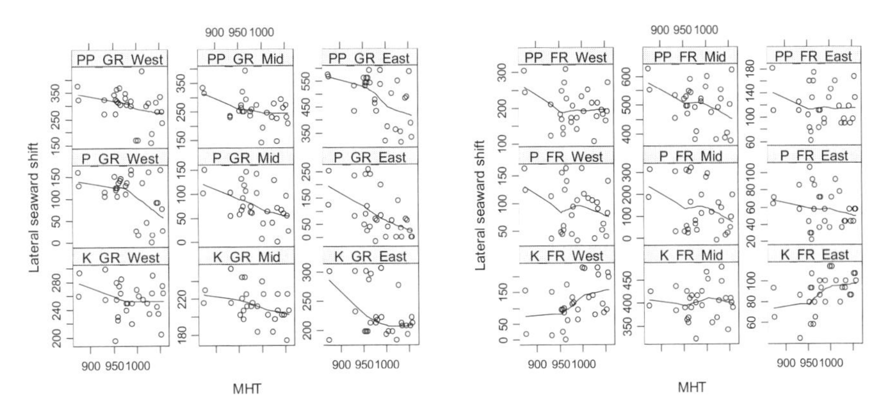

Figure 35.7. Left: Time series for Groningen (GR) based on yearly lateral seaward shifts of pre-pioneer (PP), pioneer (P) and salt marsh zones (K) plotted versus MHT (mean high tide). A smoothing LOESS curve was added. Right: Time series for Friesland (FR) based on yearly lateral seaward shifts of pre-pioneer (PP), pioneer (P) and salt marsh zones (K) plotted versus MHT (mean high tide). A smoothing LOESS curve was added.

35.4 Additive mixed modelling

Based on prior knowledge, we expect to find a linear relationship between MHT and any of the time series. The question is whether this relationship is *indeed* linear for these data, and whether there is also a temporal trend in the time series. A possible modelling approach is a linear regression model of the form:

$$Y_t = \alpha + \beta_1 \text{MHT}_t + \beta_2 t + N_t \qquad (35.1)$$

In equation (35.1), Y_t is the value of the lateral shift of the lower limit of the vegetation zone at a particular station in year t, α is the intercept, MHT_t is the mean high tide in year t, the component $\beta_2 t$ is a trend over time and N_t is independently normally distributed noise. The parameter β_1 tells us whether there is a linear effect of MHT on the time series and the slope β_2 represents the effect of the remaining trend. There are three potential problems with this model: (i) the effect of MHT might not be linear (and the shape of the LOES smoothers in the lattice plots gave some indication that this indeed may not be the case), (ii) the trend might not be linear, and (iii) the noise component might not be independent. The first two issues can be dealt within the additive modelling framework:

$$Y_t = \alpha + f_1(\text{MHT}_t, \lambda_1) + f_2(\text{Year}, \lambda_2) + N_t \qquad (35.2)$$

The function $f_1(\text{MHT}_t, \lambda_1)$ is a smoothing curve for MHT and can have any shape. The amount of smoothing is determined by λ_1. If $\lambda_1 = 1$, then the smoothing curve is a straight line making it identical to the β_lMHT$_t$ component in the linear regression model. The function $f_2(\text{Year}_t, \lambda_2)$ is the trend over time, and again, it can have any shape depending on the amount of smoothing λ_2 (the degrees of freedom of the smoother). So, using smoothing curves in equation (35.2) instead of parametric components in equation (35.1) allows for non-linear relationships. The amount of smoothing for each smoother can be determined with the AIC or with an automatic selection tool like the cross-validation; see Chapter 7 for further details.

So, what do we do with the noise component? As in this study the positions of the lower limits of the various vegetation zones were measured repeatedly at the same locations over a long period, the errors will not be independent. In various chapters (8, 16, 22, 26), we discussed how linear regression can be extended to generalised least squares by imposing for example an ARMA(p,q) structure on the noise component N_t in equations (35.1) or (35.2). Recall that such an error structure is of the form:

$$N_t = \gamma_1 N_t + \ldots + \gamma_p N_p + \varepsilon_t + \phi_1 \varepsilon_{t-1} + \ldots + \phi_{t-q} \varepsilon_{t-q} \qquad (35.3)$$

The same can be done in the additive model in equation (35.2), which is then confusingly called additive mixed modelling (Wood 2006). The parameter p in equation (35.3) defines the number of auto-regressive error terms N_t and q the number of moving average terms. The noise component ε_t is independently normally distributed. The only purpose of equation (35.3) is to break down the *dependently* distributed noise term N_t into smaller building blocks that consist of past noise terms of N_t and something that is indeed independently and normally distributed. The question is then how many of these building blocks we need. The strategy we adopt here is to try different values of p and q and use the one with the lowest AIC. If it turns out that $p = q = 0$ is the most optimal combination, then we are close to assuming that the original error term N_t is independent. We deliberately used the phrase 'close to' as we could also try other types of error structures (Pinheiro and Bates 2000).

More problems

There is an extra problem. We have one smoothing model for each of the 18 time series. Even if we analyse the data from the two provinces separately, we still have nine models per province. To reduce this number, one model with interaction terms can be applied:

$$Y_{ti} = \alpha_i + f_{1i}(\text{MHT}_t, \lambda_{1i}) + f_{2i}(\text{Year}, \lambda_{2i}) + N_{ti} \qquad (35.4)$$

Y_{ti} is the value of the lateral shift of the lower limit of the vegetation zone in year t at station i, where i = 1,..,9. Each station can have its own smoother for MHT and Year. It is also possible to use one MHT smoother or one Year smoother for a couple of stations, e.g., for those close to each other (and showing the same pattern). Technically, the interaction is modelled using the 'by' command in the gamm function in the R library mgcv (Wood 2006). The alternative (although not applied here) is to use one overall smoother for MHT and one for Year, and include a smoother for each station representing the deviation from the overall pattern. Significance levels of the smoothers indicate which stations deviate form the general pattern.

So what about the noise term N_{ti}? The lattice graphs in the data exploration show that the variation between the stations differ considerably. Hence, assuming that N_{ti} is normally distributed with the same variance for each station is likely to be wrong; the model validation would show heterogeneity of residuals. Therefore, we will allow for different variances per station:

$$N_{ti} \sim N(0, \sigma_i^2) \tag{35.5}$$

Note the index i for the variance. To reduce computing time, one auto-correlation structure for all nine stations is used. As a result, the ARMA parameters do not have indices i:

$$N_{ti} = \gamma_1 N_{ti} + \ldots + \gamma_p N_{pi} + \varepsilon_{ti} + \phi_1 \varepsilon_{t-1,i} + \ldots + \phi_{t-q} \varepsilon_{t-q,i} \tag{35.6}$$

This is a practical choice driven by lack of sufficient computing power, but for smaller datasets, one can make auto-correlation graphs of the residuals for each time series and investigate whether the patterns are different.

The question is now how to find the optimal model. What is the model selection strategy? Because this is an extension of linear mixed modelling, we should also adopt the same selection strategy as in Chapter 8:

1. Start with a model that is reasonable optimal in terms of fixed components. This refers to the intercepts and smoothers in equation (35.4). Based on initial model runs with independently normally homogeneously noise, we expect that the final model is not more complex. In fact, most smoothers for MHT were linear lines, so some simplification (in later steps of this strategy protocol) may be needed.
2. Find the optimal model in terms of random components. This refers to equations (35.5) and (35.6). Do we really need the nine variances, and what is the optimal ARMA structure? We can use the AIC and likelihood ratio tests (using REML) to answer both questions
3. For the optimal random structure obtained in the previous step, find the optimal model in terms of fixed terms, i.e., drop non-significant smoothers. Just as in mixed modelling, maximum likelihood estimation instead of REML is needed.

35.5 Additive mixed modelling results

Groningen

We will apply the additive mixed model in equations (35.4)–(35.6) on the nine time series from Groningen. The three-step protocol was followed.

Step 1

Using a normally independently homogeneously distributed error term, the model in equation (35.4) was fitted. It gave a Year and MHT smoother for each station. For some stations, the smoothers were non-linear, but for the majority of stations, they were straight lines.

Step 2

Two questions are of prime interest here: (i) Do we need the nine variances or is one sufficient and (ii) which ARMA structure do we need for the auto-correlation? The first question is quite simple; fit the model in equations (35.4)–(35.6) containing nine variances, and fit a model with only one variance and apply a likelihood ratio test.

Model	df	AIC	BIC	logLik	*L*-Ratio	*p*-value
1	55	2393.12	2583.16	−1141.56		
2	47	2469.36	2631.76	−1187.68	92.2379	<0.001

Model 1 contains the nine variances and model 2 only one; hence, the differences in number of parameters is eight. Both the AIC and the likelihood ratio test indicate that the model with nine variances is preferred. A p-value smaller than 0.001 can be seen as an indication that adding nine variances does improve the model. Now the more difficult question, what values of p and q should we choose for the ARMA(p,q) structure? This is a matter of trying a series of combinations and selecting the combination with the lowest AIC. The reader is warned that this is a time-consuming exercise with about half an hour computing time per p-q combination. And if the number of parameters is increasing, choosing 'good' starting values to avoid non-convergence becomes some sort of art as well. The model with no auto-correlation structure ($p = q = 0$) had AIC = 2413. For $p = 1$ and $q = 0$ we had AIC = 2393, $p = 1$ and $q = 1$ gave AIC = 2395, $p = 2$ and $q = 2$ gave AIC = 2398, $p = 3$ and $q = 0$ gave AIC = 2397, $p = 3$ and $q = 1$ gave AIC = 2399. This clearly shows that an auto-regressive error structure of order 1 ($p = 1$, $q = 0$) should be used. A model with no auto-correlation ($p = q = 0$) is nested within a model with ARMA(1,0), also called AR(1). Hence, we can also apply a likelihood ratio test giving a p-value smaller than 0.001.

Step 3

Using nine different variances and an ARMA(1,0) error structure, we now investigate the optimal model in terms of fixed components. Cross-validation (Chapter 7) was applied to estimate the amount of smoothing for each smoother. We dropped (one by one) the smoothers that were not significant at the 5% level.

Results for the Year smoothers showed that seven of them were significant at the 5% level (with p-values all smaller than 0.004), and these are presented in Figure 35.8. Four stations show a linear decline over time (getting closer to the seawall), whereas three stations show a non-linear pattern. As to the MHT smoothers, they all had 1 degree of freedom, indicating a linear relationship between MHT and the lateral shift of the lower limit of the vegetation zone (Figure 35.8). The MHT smoother was significant for five of the nine time series in Groningen: one salt marsh series (mid), two pioneer series (west and east) and two pre-pioneer series (west and east). The relationship was negative; higher MHT corresponds to lower lateral shift values, as was expected; the lower limit of the vegetation zones moves more towards the sea wall. However, it should be noted that p-values for the significant MHT smoothers were all between 0.01 and 0.05 indicating only a weak relationship. It should be noted that the smoothing curves in Figure 35.8 are partial fits. Each smoother shows the effect, while taking into account the effect of the other smoothers. So, they are not directly comparable with the smoothers in Figure 35.7.

The numerical output for the random components shows that the auto-correlation coefficient is equal to γ_l = 0.5, and the nine variances for the salt marsh, pioneer and pre-pioneer series are, respectively, for west, middle and east for each zone 1.00, 1.09, 1.22, 2.19, 1.92, 3.14, 2.75, 3.54 and 4.06. Note that the salt marsh series (the first three) have the lowest variances and the pre-pioneer series (the last three) the highest. Further model improvements may be obtained by using only three variances, one per vegetation type.

Because the MHT effect was linear, the following model was also applied:

$$Y_{ti} = \alpha_i + \beta_i \text{MHT}_t + f_{2i}(\text{Year}, \lambda_{2i}) + N_{ti} \tag{35.7A}$$

The MHT effect on the lateral shift of the vegetation zone is modelled as a linear term and is allowed to differ per station (this is just the interaction between station and MHT). A competing model is without the interaction:

$$Y_{ti} = \alpha_i + \beta \text{MHT}_t + f_{2i}(\text{Year}, \lambda_{2i}) + N_{ti} \tag{35.7B}$$

A likelihood ratio test gives a p-value of 0.03, indicating that the interaction term is only weakly significant. However, from model (35.7A) it is more difficult to infer for which stations MHT is important.

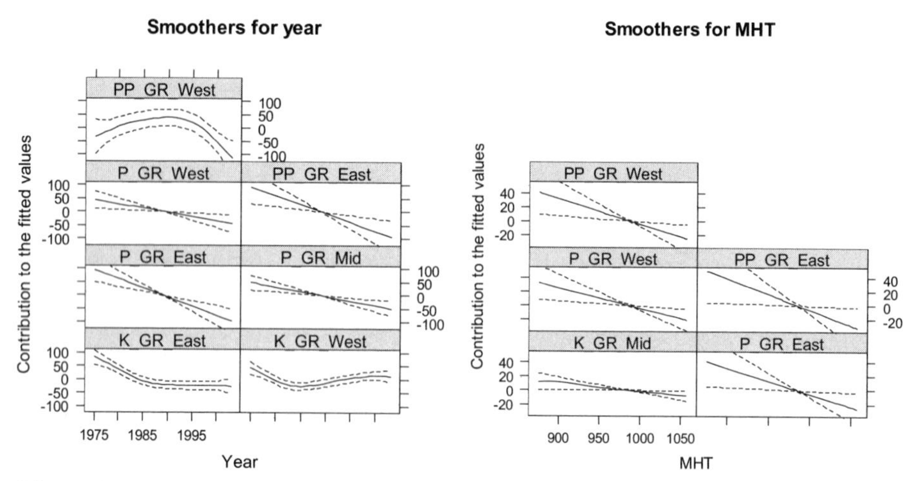

Figure 35.8. Smoothing curves for Year (left) and MHT (right) for the province Groningen. Dotted lines are 95% point-wise confidence bands.

Friesland

The same approach was applied to the nine time series from Friesland. Just as for the Groningen series, the model for the Friesland time series improved significantly by using nine variances; a likelihood ratio test (comparing models with nine and one variances) gave $p < 0.001$. The nine variances for the salt marsh, pioneer and pre-pioneer series were, respectively, for west, middle and east for each zone: 1, 1.62, 0.80, 2.08, 4.49, 1.01, 3.07, 4.10 and 1.80. As to the ARMA(p,q) autocorrelation structure, $p = q = 0$ gave AIC = 2399, $p = 1$ and $q = 0$ gave AIC = 2397, $p = q = 1$ gave AIC = 2389, $p = 2$ and $q = 1$ gave AIC = 2387, $p = 3$ and $q = 0$ gave AIC = 2388, $p = 3$ and $q = 1$ gave AIC = 2389, and $p = 3$ and $q = 2$ gave AIC = 2389. Based on these AIC values it seems that either $p = q = 1$ or $p = 2$ and $q = 1$ is the 'best' choice. We selected ARMA(1,1) as the difference between its AIC and that of the optimal ($p = 2$, $q = 1$), but a more complicated model is smaller than 2. An AIC difference of less than 2 is seen as not important enough to go for a more complicate model.

In the third step of the model selection process, we looked at the smoothers for MHT and Year. All smoothers had 1 degree of freedom, except for one station (however the smoother for this station was not significant). Only 4 from the 18 smoothers were significantly different from 0 at the 5% level, namely the MHT smoothers for stations P_FR_Mid and PP_FR_Mid (showing a linear decreasing trend over time), and the Year smoother for stations K_FR_West and K_FR_East (showing a linear increasing trend over time).

Generalised least squares

A possible criticism on the additive mixed model is that the trend at each station is modelled with a smoother, whereas parametric models are easier to present in terms of numerical and graphical output. The results of the additive mixed model suggest that MHT has a linear effect and that time is slightly non-linear for some of the time series. For those time series where time had a non-linear effect, the shape of the smoothing curve is similar to that of a second order polynomial function. So, if we want to use a parametric model that is capable of similar fitted curves, we could use a model of the form:

$$Y_{it} = \alpha_i + \beta_{1i}\text{MHT}_{ti} + \beta_{2i}\text{t} + \beta_{3i}\text{t}^2 + N_{ti} \tag{35.7}$$

Instead of the smoother for Year, we now use a second-order polynomial function. We have added the index i that takes values from 1 to 9, and it identifies the nine stations series. The interaction allows for different relationships per station. The noise component can again be modelled using an ARMA(p,q) structure as in equation (35.3), and different variances per station can be used. It is advisable to centre the explanatory variables to reduce collinearity. The model selection process follows the same strategy as above; start with a reasonable model in terms of fixed components, find the optimal random structure, and drop all non-significant fixed terms one at a time. The parameters in the model in equation (35.7) can be estimated using generalised least squares. The problem of the model in equation (35.7) is that it results in a large number of estimated parameters for the stations and its interaction terms with time and MHT. In some situations, we might not be interested in station effects. So why should we sacrifice so many precious degrees of freedom? We can avoid this by applying linear mixed modelling and use station as a random component (Chapter 8). We can even apply such a model on all 18 time series at the same time. It would model the 18 plant zones time series as a function of an overall MHT and time effect, and at each of the 18 stations, a random variation of the intercept and slopes is allowed. Obviously, we still have to allow for an auto-correlation structure on the error components N_t, but this can be done with the ARMA(p,q) structure. If the homogeneity assumption is violated in the mixed model, then different variance components can be used. Further nominal explanatory variables identifying the two different provinces (Groningen and Friesland) could be added. We leave this as an exercise for the reader.

35.6 Discussion

Based on the results the answer to the questions (i) do the year-to-year changes in MHT levels affect the development of salt marsh vegetation, (ii) are MHT levels responsible for major shifts in the lower limits of the vegetation zones, and (iii) are the impacts on pioneer and marsh zones different is threefold yes.

On the 9 time series from Groningen, we applied a model that contained 18 smoothers: a Year smoother and an MHT smoother for each station. We then dropped the non-significant smoothers (one at a time) and ended up with a model containing seven significant Year smoothers and five significant MHT smoothers. Using cross-validation, all MHT smoothers had 1 degree of freedom indicating a linear effect. This effect was negative, which means that high MHT values are associated with a vegetation boundary close to the seawall.

The same process was applied on the nine Friesland time series, but results were less 'good' from a statistical point of view. With 'less good' we mean that we could find less significant smoothers. Only two stations had a significant (linear and negative) MHT effect and a significant Year effect.

These results indicate that MHT has a greater influence in Groningen, and this area also exhibits more significant long-term trends (the Year smoother). The terms 'greater' and 'more' are in terms of number of stations. A possible explanation for this is that in Groningen there is more sandy soil, lower sedimentation rate and maintenance problems with the groyne system. These factors offer the vegetation less protection against waves and currents allowing the physical forces to do their job.

For the Groningen series, MHT had a significant negative effect on two pre-pioneer, two pioneer and only one salt marsh series. It is tempting to conclude from this that the effect of MHT levels on the development of the salt marsh zone is minimal (at least in Groningen), due to the stronger resistance of the perennial vegetation. If this is indeed the case, then this means that we have to focus on (protection of) the pioneer zones for coastal defence. This is an important piece of information for our knowledge on the effects of enhanced sea level rise on coastal salt mashes and for coastal defence!

In the models we used a smoother for MHT and Year. The second term represents the long-term trend. A collinearity problem arises if MHT itself contains a strong long-term trend, which is not the case here as can be seen in Figure 35.3.

In this chapter, we applied additive mixed modelling. We could also have applied DFA or MAFA (Chapters 16 and 17) on these data. In fact, we did and the results were similar. We could have added another section to this chapter with the DFA and MAFA results, but then it would have been a copy of the Hawaiian bird time series case study chapter. However, it does show that if one has a multivariate time series dataset, and if the main focus is on trends and the effects of explanatory variables, then the reader of this book has now a series of useful techniques in his or her toolbox. Which tool to use (AMM, GLS, DFA or MAFA) is matter of personal preferences, size of the data and underlying questions. A few points to consider: MAFA cannot cope with missing values, DFA cannot cope well with datasets containing lots of time series (>25), AMM cannot cope well with a large number of interactions, GLS creates lots of estimated parameters, and the model selection process in mixed modelling is complicated. On the other hand, all these methods were applied in the three case study chapters and each time they did the job.

36 Time series analysis of Hawaiian waterbirds

Reed, J.M., Elphick, C.S., Zuur, A.F., Ieno, E.N. and Smith, G.M.

36.1 Introduction

Surveys to monitor changes in population size over time are of interest for a variety of research questions and management goals. For example, population biologists require survey data collected over time to test hypotheses concerning the patterns and mechanisms of population regulation or to evaluate the effects on population size of interactions caused by competition and predation. Resource managers use changes in population size to (i) evaluate the effectiveness of management actions that are designed to increase or decrease numbers, (ii) monitor changes in indicator species, and (iii) quantify the effects of environmental change. Monitoring population size over time is particularly important to species conservation, where population decline is one key to identifying species that are at risk of extinction.

Our goal in this case study chapter was to analyse long-term survey data for three endangered waterbirds that are found only in the Hawaiian Islands: Hawaiian stilt (*Himantopus mexicanus knudeseni*), Hawaiian coot (*Fulica alai*), and Hawaiian moorhen (*Gallinula chloropus sandvicensis*). The survey data come from biannual waterbird counts that are conducted in both winter and late summer. Surveys were initiated for waterfowl in the 1940s on most of the major Hawaiian Islands, and they were modified in the 1960s to better monitor the endangered waterbirds as well (Engilis and Pratt 1993). In the 1970s, surveys were expanded to encompass all of the main Hawaiian Islands. Surveys are coordinated by the Hawaii Division of Forestry and Wildlife, and the goal is to complete each statewide survey in a single day to reduce the risk of double counting individuals.

We analysed winter survey data to reduce the contribution of recently hatched birds to survey numbers, which can increase count variability. There are two previously published analyses of population trends of these species. Reed and Oring (1993) analysed winter and late-summer data for Hawaiian stilts and concluded that there was a statewide increase in population size during the 1970s and 1980s. Engilis and Pratt (1993) analyzed data from 1977–1987 for all three of the species addressed in this chapter, and found that annual rainfall patterns had a strong effect on counts. They concluded that, for this time period, moorhen and coot numbers varied with no clear trend, whereas stilt numbers rose slightly. Trend analyses in these papers were based on visual inspection (Engilis and Pratt 1993) or used linear and non-linear regressions (Reed and Oring 1993). Since these techniques ignore time structure, specific time series analysis tools are applied in this case study chapter. We limited our analyses to the islands of Oahu, Maui and Kauai (Figure 36.1) because these islands have the longest survey history and collectively contain the vast majority of the population for each species.

Our primary questions were to determine whether there are detectable population size changes over time, whether trends are similar among time series, and to evaluate the effect of rainfall described for a shorter time span by Engilis and Pratt (1993).

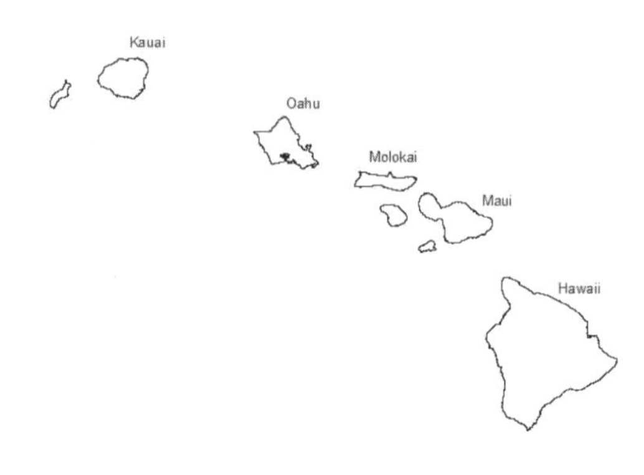

Figure 36.1. Sampling area, showing the main Hawaiian Islands.

36.2 Endangered Hawaiian waterbirds

Hawaiian stilt, coot, and moorhen breed and live year-round in low-elevation wetlands, and they are endangered because of habitat loss and invasion by exotic species that either prey upon them or change the nature of their habitat (Griffin et al. 1989). These waterbirds differ significantly in detectability, which contributes to concerns about how accurately the biannual waterbird surveys enumerate individuals, and whether the number of birds counted is an accurate index of population size (Chang 1990; Engilis and Pratt 1993). Data are considered most accurate for Hawaiian stilts, which spend most of their time in the open where they can be easily seen, and least accurate for moorhen, which live in dense emergent vegetation and have cryptic behaviour (Chang 1990).

Hawaiian stilts and coots are found on all of the major islands (Hawaii, Kauai, Maui, Molokai, Oahu), as well as on Niihau, which shares birds seasonally with Kauai (Engilis and Pratt 1993; Reed et al. 1998). The Hawaiian moorhen's distribution once covered five islands, but now is restricted to Oahu and Kauai. For all three species, population size was estimated only sporadically before the 1950s. Reports, however, tend to describe populations that were fairly large in the 1800s, with declines noticeable by the end of the nineteenth century and continuing until the 1940s (Banko 1987a,b). Since the 1960s, population sizes have increased (Engilis and Pratt 1993; Reed and Oring 1993), with current statewide estimates of approximately 1500 stilts, 3000–5000 coots, and 300–700 moorhen. Because Hawaiian moorhen are secretive, their survey results are considered to be an underestimate of the true number (Chang 1990). The time series used in our analyses are given in Table 36.1. Rainfall data came from the National Climate Data Center (http://cdo.ncdc.noaa.gov/ancsum/ACS) for the Kahului airport on Maui.

Table 36.1. Species, islands, and names used in this case study. Stilts and coots move seasonally between Kauai and Niihau, and so for this time series, we only used data from years in which both islands were surveyed for these two species. Moorhen do not occur on Maui, so there is no time series for that combination.

	Species	Island	Name in this chapter
1	Hawaiian Stilt	Oahu	Stilt_Oahu
2	Hawaiian Stilt	Maui	Stilt_Maui
3	Hawaiian Stilt	Kauai	Stilt_Kauai_Niihau
4	Hawaiian Coot	Oahu	Coot_Oahu
5	Hawaiian Coot	Maui	Coot_Maui
6	Hawaiian Coot	Kauai	Coot_Kauai_Niihau
7	Hawaiian Moorhen	Oahu	Moorhen_Oahu
8	Hawaiian Moorhen	Kauai	Moorhen_Kauai

36.3 Data exploration

The first question we asked was whether there are any outliers or extreme observations because of their potential undue influence on the estimated trends. Cleveland dotplots and boxplots (Chapter 4) indicated that various time series had observations that were distinctly larger than in other years within that series. We compared Cleveland dotplots and boxplots of square root and $\log_{10}$ transformed data. Both transformations resulted in considerably fewer influential observations, and the choice of which one to use was merely subjective. Consequently, we used the square root transformed data because it changes the real data less than does the stronger $\log_{10}$ transformation. The second important step in a time series analysis is to make a plot of each variable versus time; lattice graphs are a useful tool for this. In making such a graph, the question one should ask is whether the absolute differences between the time series are important or whether an increase in one abundant species is equally important as a proportionally similar increase in a less abundant species. In the second case, normalised data (Chapter 4) should be used. For this analysis, we made all time series equally important and Figure 36.2 shows a lattice plot of all eight normalised time series. The graph highlights the fact that not all series start at the same time, some series fluctuate more than others, and there are various missing values (these are the gaps in the lines). Note that some time series techniques cannot cope well with missing values (e.g., MAFA). The time series for stilts and coots on Kauai_Niihau have many missing values because Niihau surveys were initiated late in the survey period.

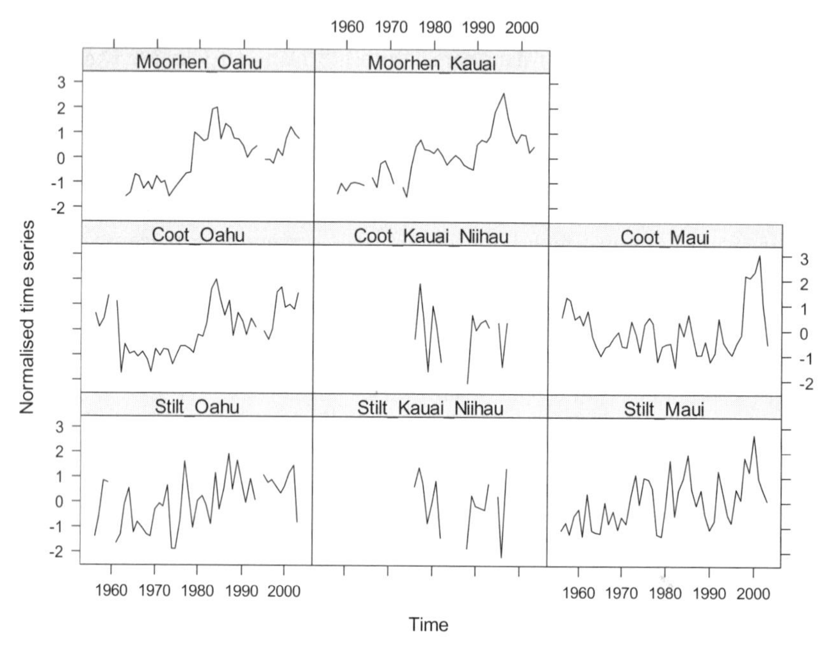

Figure 36.2. Lattice plot of all time series. The horizontal axis shows the time (in years) and the vertical axis the normalised bird abundance.

To obtain insight into the strength of the relationships over time for individual time series, auto-correlation functions were calculated. All series except those for coot and stilt on Kauai_Niihau show an auto-correlation pattern that might indicate the presence of a trend. Recall from Chapter 16 that an auto-correlation function measures the relationship between a time series Y_t and its own past Y_{t-k}. The auto-correlation of a time series with a strong trend will have a high auto-correlation for the first few lags k, compare also Figure 36.3 and Figure 36.2.

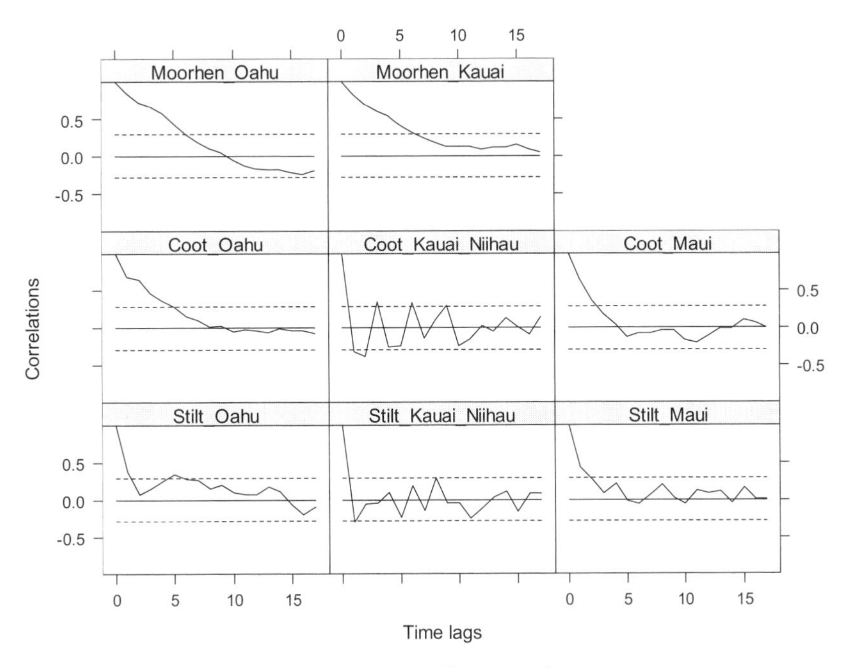

Figure 36.3. Auto-correlation functions for all time series.

36.4 Three ways to estimate trends

LOESS smoothing

The first approach we applied to visualise the possible presence of common trends is LOESS smoothing (Chapter 17). In this method, each (normalised) time series is smoothed (using time as the explanatory variable) and all smoothing curves are plotted in one graph (Figure 36.4). The thick line is the average, which can be calculated because the original series were normalised (Chapters 4 and 17). Note that missing values were omitted from the analysis. If there are only a few missing values, they can be replaced by an appropriate estimate, for example using interpolation or an average value. However, in this case two series have a considerable number of missing values and replacing them by an estimate would be inappropriate. The shape of the smoothing curves does indicate a general increase after about 20 years (i.e., starting in the mid-1970s). However, there is considerable variation among the curves.

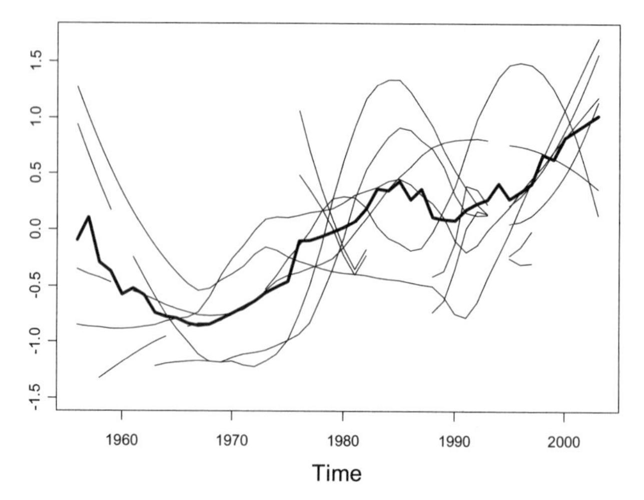

Figure 36.4. Curves obtained by LOESS smoothing. Breaks in lines indicate years in which a time series had missing values. The horizontal axis represents time in years, and the vertical axis shows the values of the smoothers. The time series were normalised (Chapter 4); hence, the vertical axis does not have units. The thick line is the average of the smoothers in each year.

Min/max auto-correlation factor analysis (MAFA)

A more formal way to estimate common trends is min/max auto-correlation factor analysis (Solow 1994). The method was explained in Chapter 17 and in various case study chapters (Chapters 33 and 34). Just like principal component analysis, this method gives axes. However, these axes are derived in such a way that the first axis has the highest auto-correlation with a time lag of one year, which is typically associated with a smooth trend. MAFA extracts the main trends in the time series, and the first MAFA axis can be seen as the most important underlying pattern, or index function in the multivariate time series dataset. This technique cannot cope well with missing values, and because both time series from Kauai_Niihau contained lots of missing values, we decided to omit these time series from the MAFA analyses. MAFA axes are presented in Figure 36.5.

The first MAFA axis looks like a step function: reasonably stable up to the mid-1970s, followed by a rapid increase to the early 1980s, and then stable again. The second axis has high values around 1960, a peak in the mid-1980s and a drop in the mid-1990s. Just as in principal component analysis (PCA), in MAFA we can infer which of the original variables is related to the first axis, to the second axis, and to both, by using each variable's loadings on each axis (Chapter 17). Here, we can address the same question: Which of the original six time series are related to these MAFA axes? One way to answer this question is to calculate Pearson correlations between the original time series and the axes (Table 36.2). A high positive value indicates that the bird time series follows the same pattern as the MAFA axis. A high negative value means that it has the opposite shape.

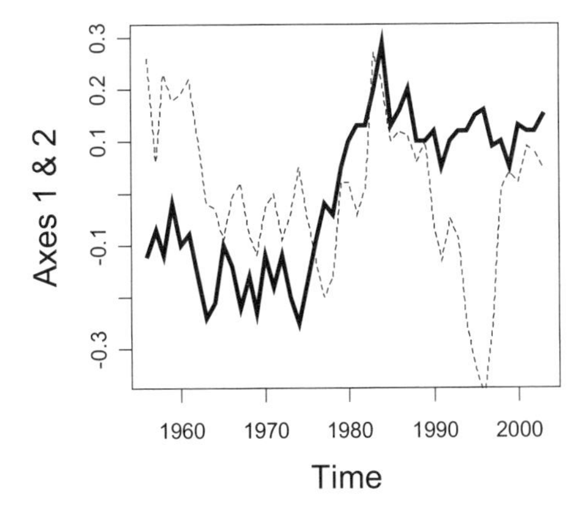

Figure 36.5. Results from the MAFA. The thick line is the first MAFA axis, and the dotted line is the second axis.

Table 36.2. Correlations between the first two MAFA axes and the original time series used for the MAFA.

Time Series	MAFA Axis 1	MAFA Axis 2
Stilt_Oahu	0.56	−0.12
Stilt_Maui	0.41	0.02
Coot_Oahu	0.71	0.47
Coot_Maui	0.05	0.34
Moorhen_Oahu	0.89	0.41
Moorhen_Kauai	0.65	−0.69

Results indicate that the bird time series for stilts, coots and moorhens on Oahu, for moorhens on Kauai, and to a lesser extent for stilts on Maui follow the pattern of the first trend. Indeed, one can recognise this shape in the corresponding time series in Figure 36.2. The second axis is mainly positively related to coot on Oahu and negatively to moorhen on Kauai, with weaker positive relationships with moorhen on Oahu and coot on Maui. Again, one can spot patterns of the second MAFA axis in these time series, especially the big peak (or drop depending on the sign of the correlation) during the mid-1990s. The correlation between the first two MAFA axes and rainfall is not significantly different from 0 at the 5% level. This means that there is no relationship between the two main underlying patterns in the six selected time series (or: MAFA axes) and rainfall. A permutation test indicated that the auto-correlation of the first, and of the second, MAFA axis is significantly different from 0 at the 5% level.

Dynamic factor analysis (DFA)

A more formal tool to estimate common trends and the effects of explanatory variables is dynamic factor analysis (Chapters 17, 33, 34). The DFA model decomposes the eight bird time series into:

- A linear combination of common trends.
- Effects of explanatory variables.
- An intercept for each time series.
- Noise.

The difference between MAFA and DFA is that in MAFA all common information is extracted in the form of trends. In DFA, the common trends represent the information that is common in a set of time series, and that cannot be explained by the explanatory variables. Information that is 'not common' enough is routed to the noise component. As to this noise component, it can be modelled as something that is independent for each bird time series, or alternatively, the noise term of one bird time series can be allowed to interact with that of another bird time series. This brings us close to a generalised least squares approach in which an error covariance matrix can be used to model interactions between noise components (Chapter 17). The interactions of the noise associated with two time series can be seen as unexplained information that exists between the two series, but that forms a pattern not strong enough to justify the need for an extra common trend in the model. Technically, a non-diagonal error covariance matrix is used (Zuur 2003a,b, 2004). For these data, DFA models containing such an error matrix performed better (judged by the AIC) compared with a diagonal error matrix.

DFA can deal better with missing values than can MAFA because it makes use of the so-called Kalman filter and smoother (Zuur et al. 2003a), and therefore all eight time series were used. But it should be noted that DFA cannot do magic with this respect. We noticed that the algorithm became unstable if more than one trend was used. This is probably due to the non-random distribution of the missing values. Therefore, DFA models with only one common trend were used.

Explanatory variables in the DFA

Conservation protection laws for coots and stilts came into force in 1970 and for moorhen in 1967. To test whether these laws had any effect on population sizes, nominal variables can be introduced. For example, we created a new variable called C_{1970} that had values of zero up until 1969 and values of one thereafter. The same was done for C_{1967}; zeros up to 1966 and ones from 1967 onwards. The DFA model allows one to include explanatory variables, but it will produce a different regression coefficient for each time series, and therefore describes a separate effect for each of the eight time series. So, using either C_{1967} or C_{1970} will allow the variable to influence all three species, which might not make sense given the different timing of protection. The shape of the first MAFA axis, moreover, suggests that any population increase is delayed by a few years and starts after 1974. This pattern makes biological sense because any response to management is unlikely to be instantaneous, and because the implementation of new management activities is likely to lag a few years behind new legislation. Also, despite their ecological differences, these three species all occur in the same wetlands and face similar threats. Consequently, protection of management for one species is likely to also have benefits for the others. An alternative, therefore, is to test another nominal variable; let us call it C_j. It is defined as

$$C_j = \begin{cases} 0 & \text{for the years 1956 to j - 1} \\ 1 & \text{for the years j to 2003} \end{cases}$$

We then applied the DFA model (in words):

8 bird time series = intercept + 1 common trend + rainfall + C_j + noise

The questions that then arise are whether rainfall is significant, which value for j should be used, whether C_j is significant and how the trend looks for different values of j. Note that the trend is basically a smoothing curve, and that, in this application, an increasing trend would indicate an overall pattern of population recovery after past declines.

Just as in linear regression we obtain t-values for the regression coefficients (rainfall and C_j) and the AIC can be used to find the optimal model and also the optimal value of j. The AIC is a model selection tool, for which the lower the AIC value the better (Chapter 17). Table 36.3 shows the AIC values for various DFA models. The model with no explanatory variables has the highest AIC, indicating that it is the worst model. We also considered models with only rainfall (AIC = 754.08) and only the C_j variables, but these were also sub-optimal models. Models that contained both rainfall and C_j were better.

Figure 36.6 shows the estimated trends for some of these models. The upper right panel (labelled 'none') is the trend obtained by the model with rainfall but no C_j value used as explanatory variables. The trend has the shape of a step function. The models with rainfall and C_j where j is between 1966 and 1974 all have similar AIC values and their trends are similar in shape. At j = 1975, the AIC drops and the trend shape changes.

So, what is the DFA doing? To understand this, it might be useful to compare results with that of the MAFA. The MAFA produced two common trends, the first one showing stability, followed by a period of increase, and followed by stability again. The function C_j in the DFA model with one common trend and rainfall is just picking up either the transition at the lower part of the step function or that at the upper part depending on the value of j. The next question is, then, which model is the best? From a statistical point of view, it is the model with j = 1979 because it has the lowest AIC. The fact that the model with C_{1979} is better than with C_{1975} indicates that the difference between the second period of stability and the other years is more important for explaining the overall pattern in the data than the difference between the first period of stability and the rest. A further model improvement might be obtained if we use two dummy variables simultaneously in the model, C_{1975} and C_{1979} as it would effectively replace the first MAFA axis by two dummy variables. For now, however, we discuss the results obtained by the DFA model with C_{1975} because the timing of the start of the recovery is of more interest to conservation biologists wanting to understand the mechanisms that underlie the reversal of population declines.

Estimated regression coefficients for this model are presented in Table 36.4. In the DFA model, rainfall has a negative and significant effect on several time series. Why is this the case in the DFA when rainfall was not important in the MAFA? Well, let us look in detail at the time series for which it is actually important. Based on the sign, magnitude and significance levels, rainfall is important for both stilt and coot on both Maui and Kauai_Niihau. But two of these time series were not used in the MAFA and the other two time series were not strongly related to the first two MAFA axes!

Table 36.3. DFA results using all eight time series. Only DFA models with one common trend were considered.

Model	Explanatory Variable	AIC	Model	Explanatory Variable	AIC
1	None	770.74	11	Rainfall & C_{1972}	751.70
2	Rainfall	754.08	12	Rainfall & C_{1973}	749.91
3	C_{1966}	768.80	13	Rainfall & C_{1974}	748.68
4	C_{1969}	768.74	14	Rainfall & C_{1975}	730.68
5	Rainfall & C_{1966}	744.52	15	Rainfall & C_{1976}	722.83
6	Rainfall & C_{1967}	751.50	16	Rainfall & C_{1977}	720.61
7	Rainfall & C_{1968}	753.47	17	Rainfall & C_{1978}	725.17
8	Rainfall & C_{1969}	748.58	18	Rainfall & C_{1979}	717.49
9	Rainfall & C_{1970}	753.47	19	Rainfall & C_{1980}	724.29
10	Rainfall & C_{1971}	753.02			

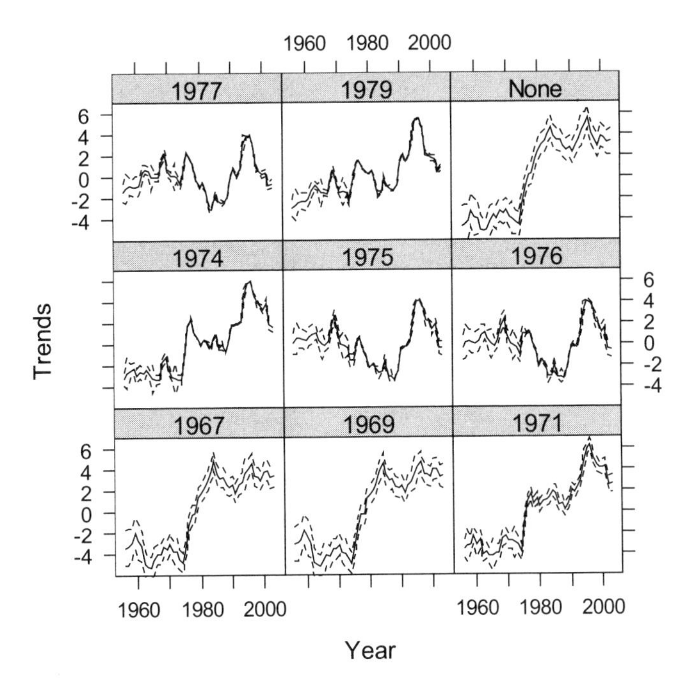

Figure 36.6. Common trend obtained by a DFA model using one common trend, rainfall, and C_j as explanatory variables. The *y*-axis represents the partial fit of the trend (Chapter 17). It shows the standardised population size. Dotted lines are 95% confidence bands for the trend. The numbers above the panels refer to different values for *j*.

The next question is: Why is C_{1975} in the model? The time series for which C_{1975} is important were also strongly related to the first MAFA axis. Hence, C_{1975} takes over part of the role of the first MAFA axis.

Besides the effects of the explanatory variables rainfall and C_{1975}, there is a certain amount of common information in the time series. This is captured by the common trend (see the panel for j = 1975 in Figure 36.6). Just as in MAFA one can ask the question: Which of the original time series show this pattern? Again, correlations between the trend and the original time series can be used. These can either be presented in a graph or tabulated (Table 36.4). In this case, only three time series had correlations that were not close to zero and, hence, contained a pattern similar to the DFA trend. These were Coot_Maui and both moorhen series. All other time series are mainly related to the explanatory variables. Notice that Moorhen_Oahu shows a negative correlation with the trend, whereas the other two time series show positive correlations. Note also that the correlations are calculated for available data only (missing values were not substituted by averages).

One can indeed recognise a bump or a dip (moorhen on Oahu) during the mid-1990s in these three time series; see also Figure 36.2.

Table 36.4. Estimated regression parameters, standard errors and t-values obtained by a DFA model with one common trend, rainfall, and C_{1975} as explanatory variables and a non-diagonal error covariance matrix. The last column contains the cross-correlations between the trend and the original time series. Bold face indicates regression parameters significantly different from 0 at the 5% level.

	Rainfall			C_{1975}			Trend
	Estimate	SE	t-value	Estimate	SE	t-value	
Stilt_Oahu	0.00	0.02	0.21	**1.00**	**0.27**	**3.76**	−0.004
Stilt_Maui	**−0.06**	**0.01**	**−4.56**	**0.95**	**0.22**	**4.29**	−0.079
Stilt_Kauai_Niihau	**−0.06**	**0.02**	**−3.21**	0.00	1.60	0.00	0.139
Coot_Oahu	−0.02	0.02	−1.50	**0.87**	**0.26**	**3.30**	−0.162
Coot_Maui	**−0.06**	**0.02**	**−4.12**	0.14	0.27	0.52	**0.300**
Coot_Kauai_Niihau	**−0.05**	**0.02**	**−2.23**	1.20	1.60	0.75	−0.002
Moorhen_Oahu	0.00	0.01	−0.01	**1.26**	**0.25**	**4.80**	**−0.422**
Moorhen_Kauai	0.01	0.01	1.33	**1.82**	**0.35**	**5.13**	**0.431**

36.5 Additive mixed modelling

The results obtained in the previous section indicate that there is at least one major pattern over time underlying all eight time series. In this section, we show how smoothing techniques can be used to gain similar information. To apply the methods in this section, the data were re-organised such that all eight time series were placed under each other in the same spreadsheet. Hence, there is now one long column of data. A new nominal variable 'ID' was created with values 1 to 8, identifying the time series. The columns with rainfall and year were copied eight times as well. As a result, the re-organised dataset contained only four columns. In order to make the additive modelling (AM) results directly comparable with the other methods that we have used, we first standardised the time series just as we did for the DFA and MAFA analyses. This step is not strictly necessarily as we will discuss later. We can start the AM analysis in different ways. If we had no prior knowledge from the DFA or MAFA, we could start the analysis by applying the AM to each individual time series:

$$\text{Birds}_t = \text{constant} + f_1(\text{Rainfall}_t) + f_2(\text{Time}_t) + \varepsilon_t$$

where $f_1()$ and $f_2()$ are smoothing functions and ε_t is independently normally distributed noise. The index t refers to year. The problem with this model is that the data form time series, and therefore, the independence assumption on ε_t may be incorrect (Chapters 16, 26). Additive *mixed* modelling (AMM) allows one to include an auto-correlation structure, and the simplest choice is an AR-1 structure:

$$\varepsilon_t = \rho\varepsilon_{t-1} + \gamma_t$$

where ε_t is now allowed to be correlated with noise from previous years and γ_t is independently normally distributed noise. Other auto-correlation structures were discussed in Chapters 16 and 26. It is also possible to apply the AMM on the combined data matrix:

$$\text{Birds}_{ti} = \text{constant} + f_{1i}(\text{Rainfall}_{it}) + f_{2i}\,(\text{Time}_{ti}) + \varepsilon_{ti} \qquad (36.1)$$

$$\text{where} \quad \varepsilon_{ti} = \rho\varepsilon_{t-1,\,i} + \gamma_{ti}$$

The index i refers to time series and runs from 1 to 8. The constant does not contain an index i because the time series were standardised and therefore have a mean of zero. Technically, this model is fitted using the 'by' command in the gamm function in the R library mgcv (Wood 2006). The 'by' command allows one to include an interaction between smoothers and nominal variables. The model in equation (36.1) assumes that the rainfall-bird relationship and the long-term trend are different for each time series. This means that a lot of precious degrees of freedom have to be estimated. Yet, another possible AMM, assuming that the rainfall effect and trend is the same for all time series, takes the form:

$$\text{Birds}_{ti} = \text{constant} + f_1(\text{Rainfall}_{ti}) + f_2(\text{Time}_{ti}) + \varepsilon_{ti} \qquad (36.2)$$

$$\text{where} \quad \varepsilon_{ti} = \rho\varepsilon_{t-1,\,i} + \gamma_{ti}$$

Note that we have omitted the index i from the smoothers f_1 and f_2 in equation (36.1). The model in (36.2) is nested within the model in (36.1), and therefore we can use a likelihood ratio test to see which one is better. Results indicated that there was no evidence ($p = 0.84$) that the model in (36.1) is better than the AMM in equation (36.2). This does not necessarily mean that there is only one rainfall pattern, and one long-term trend, but it does mean that a model with two smoothers is better than the model with 16 smoothers.

One can also raise the question of whether the three species have the same trend, and such a model is given by

$$\text{Birds}_{ti} = \text{constant} + f_{1,species}(\text{Rainfall}_{ti}) + f_{2,species}(\text{Time}_{ti}) + \varepsilon_{ti} \qquad (36.3)$$

$$\text{where} \quad \varepsilon_{ti} = \rho\varepsilon_{t-1,i} + \gamma_{ti}$$

In this model, each species is allowed to have a different rainfall effect and a different trend. Because the model in equation (36.1) is nested within (36.3) the likelihood ratio test was applied and there was no evidence to prefer the more complicated model with three trends ($p = 0.17$).

Obviously, one can also ask whether the three islands have the same trend, and such a model is given by

$$\text{Birds}_{ti} = \text{constant} + f_{1,island}(\text{Rainfall}_{ti}) + f_{2,island}\,(\text{Time}_{ti}) + \varepsilon_{ti} \qquad (36.4)$$

$$\text{where} \quad \varepsilon_{ti} = \rho\varepsilon_{t-1,i} + \gamma_{ti}$$

Each island is now allowed to have a different rainfall effect and a different trend. Again, model (36.1) is nested within model (36.4) and the likelihood ratio test gave a p-value of 0.04. Note, however, that these p-values should be interpreted with care due to its approximate nature. We have decided to present the results of the model in equation (36.4). The numerical output is given by

	edf	*F*-statistic	*p*-value
s(Rainfall):S1	2.46	2.47	0.010
s(Rainfall):S2	1.00	13.31	<0.001
s(Rainfall):S3	3.65	4.87	<0.001
s(Year):S1	4.64	4.95	<0.001
s(Year):S2	1.00	20.59	<0.001
s(Year):S3	1.00	0.29	0.59

These are approximate significance values of smooth terms. The notation s(Rainfall):S1 refers to the smoother for rainfall at island 1, whereas (sYear):S1 gives the smoother for the temporal trend on the same island (1 = Oahu, 2 = Maui, 3 = Kauai_Niihau). Except for the trend for Kauai_Niihau, all smoothers are significant. An estimated degree of freedom (edf) of 1 means that the relationship is modelled by a straight line. Hence, the rainfall effect for Maui and the long-term trend at Maui are linear, and all other smoothers are non-linear. The auto-correlation parameter ρ was 0.39. The smoothing curves are plotted in Figure 36.7. The shape of these curves indicates that on Oahu, there was a strong trend over time with an increase from the mid-1970s until the mid-1980s. Layered on top of this pattern, there was a general decrease in bird numbers with rainfall; the more rainfall, the lower the bird numbers. On Maui, there is a decrease in bird numbers with rainfall, but an increasing trend over time. Finally, on Kauii_Niihau, there is a non-linear rainfall effect, with numbers generally decreasing with more rain, but increasing again when there is very high rainfall.

The model can be extended in various ways: (i) We can add dummy variables that measure the effect of conservation protection in the same way as we did for the DFA, and (ii) instead of standardising the time series, we can add a nominal variable ID with values 1 to 8 (allowing for different mean values per time series) and use eight different variances (allowing for different spread per time series). It is also possible to use different spreads (variances) per species or per island. Avoiding standardisation is useful if the model will be used for prediction.

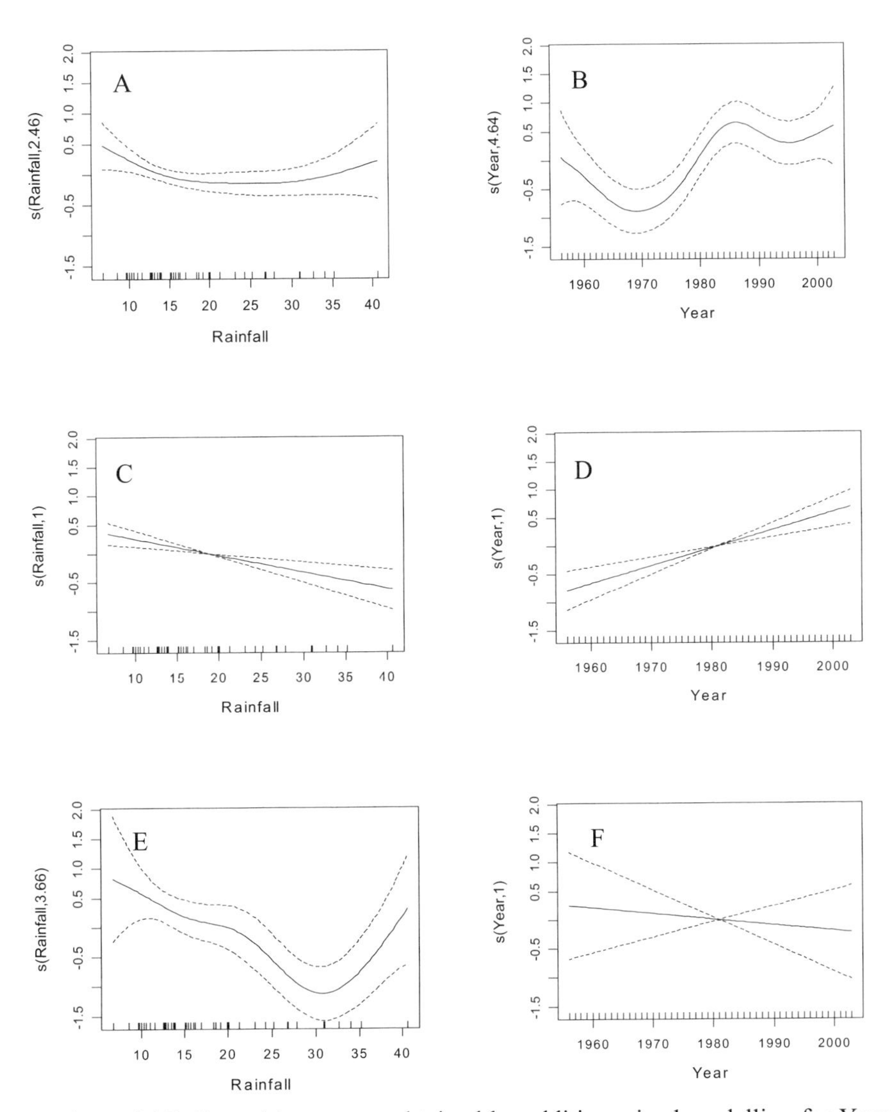

Figure 36.7. Smoothing curves obtained by additive mixed modelling for Year and Rainfall using an additive mixed model. The dotted lines represent the 95% confidence bands. The vertical axis represents the contribution of the smoother to the fitted values. Cubic splines were used. Panels A and B refer to Oahu, C and D to Maui and E and F to Kauai_Niihau.

36.6 Sudden breakpoints

Both the MAFA and DFA indicated that there was a sudden increase in bird abundance in a majority of the time series. To confirm this interpretation, we applied chronological clustering (Chapter 19). This technique calculates a distance matrix between the years. Each element in the matrix represents the association between two years (as measured over all eight variables). We decided to use the Whittaker index (Bell and Legendre 1987; Legendre and Legendre 1998). The problem with simple clustering is that it could produce groups of non-sequential years, ignoring the temporal dependence between years, which is crucial in an analysis involving time series. Instead of applying ordinary clustering on the distance matrix, chronological clustering creates groups that are constrained to include only sequential years. In this method, two clustering parameters have to be set, alpha and the connectedness. Following the recommendations in Legendre et al. (1985), we kept the connectedness fixed at 0.5 and presented results for different values of alpha. Small values of alpha mean that we end up with rather conservative groupings. This can also be seen as a bird's eye overview: Only the major breakpoints are obtained. Increasing the value of alpha gives greater resolution and more groups. Results in Figure 36.8 indicate that there is a clear breakpoint in 1975, which corresponds with our earlier analyses. Other measures of similarity gave identical breakpoints. Two time series (those for Kauai_Niihau) started in 1976 and could have influenced the location of the breakpoint. We repeated the analysis without these time series, however, and a nearly identical picture was obtained except that the main breakpoint shifted slightly to 1978.

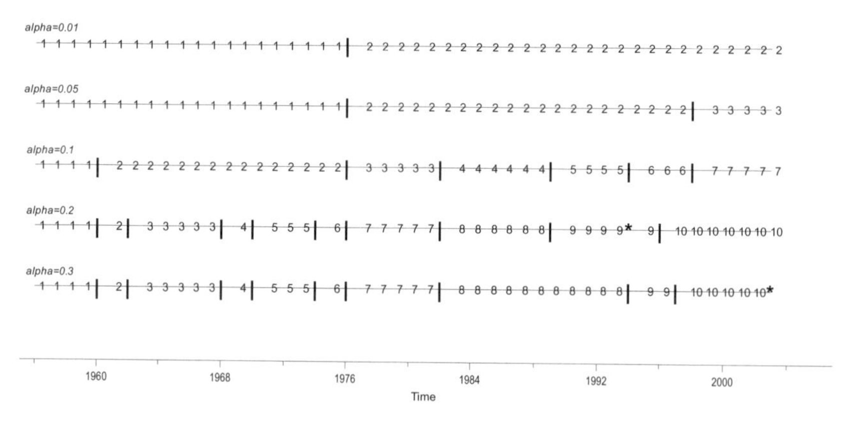

Figure 36.8. Results of chronological clustering applied on all eight bird time series. Vertical bars identify breakpoints for different alpha levels.

36.7 Discussion

The main underlying questions for this study were (i) whether the eight time series have similar trends or patterns over time, and (ii) whether there are any sudden changes. To answer these questions, various time series techniques, discussed earlier in this book, were applied. The time series dataset used here is not large, although it is relatively long for an annual monitoring program in the field of ecology. Nonetheless, it turned out that it was not an 'easy' dataset from a statistical point of view. No single technique was conclusive; however, the case study demonstrates how results from all these techniques combined can give insight into what is going on. Moreover, results from both optimal and sub-optimal models were informative.

The analyses showed that there was a major increase starting in the mid-1970s that continued up to the end of the 1970s. Because the start of two of the time series coincided, more or less, with the beginning of this period of increase, it is reasonable to wonder if the change is an artefact caused by the pattern of data collection. The basic result, however, was the same for both analyses that include these two time series and for those that do not.

We used DFA to identify when the change in population sizes took place. Using dummy variables, the year 1975 was identified as the year in which the increase started. It is unlikely that the two time series starting in 1976 are responsible for the shift as all time series were normalised in the DFA. Besides, the MAFA showed the same breakpoint and it did not use these two time series. The additive mixed modelling indicated an increase from the early 1970 up to the mid-1980s on Oahu.

Chronological clustering also shows the same breakpoint in the mid-1970s, but it does not indicate why the breakpoint occurred. MAFA and DFA gave more details on reasons for the pattern.

To conclude, the analysis suggests that there is a rainfall effect primarily associated with the stilt and coot time series from Maui and Kauai, providing additional support for the idea that fewer birds are recorded in years when there is a lot of rain. On top of this, several time series were affected by a major population increase in the 1970s, which occurred not long after endangered species legislation was introduced to protect the birds under study. Finally, three series contained an additional common pattern: coot on Maui and moorhen on Kauai showed an overall increase, whereas moorhen on Oahu decreased.

Acknowledgements

We thank the Hawaii Division of Forestry and Wildlife, and the Hawaii Natural Heritage Program for access to the biannual waterbird survey data, and the Hawaii Division of Forestry and Wildlife for their long-term coordination of these surveys.

37 Spatial modelling of forest community features in the Volzhsko-Kamsky reserve

Rogova, T.V., Chizhikova, N.A., Lyubina, O.E., Saveliev, A.A., Mukharamova, S.S., Zuur, A.F., Ieno, E.N. and Smith, G.M.

37.1 Introduction

This case study illustrates the application of spatial analysis methods on a boreal forest in Tatarstan, Russia. Using remotely sensed data and spatial statistical methods, we explore the influence of relief, soil and climatic factors on the forests of the Raifa section of Volzhsko-Kamsky State Nature Biosphere.

The Raifa area is a challenging area to study relationships between environmental factors and the spatial distribution of vegetation because, even though its climatic variation is minimal, the heterogeneity of its plant cover is high. Its climatic conditions are more similar to a milder taiga subzone further to the south.

Raifa's forests belong to three distinct biocenoses, including southern taiga, broadleaved forests, and mixed forests. Taiga-type forests consist of spruce, fir, pine and larch. At a large spatial scale, the gradient of climatic factors causes zonal replacement of these forest communities; and at a small scale it affects the community structure. Because the compositions of these communities vary so much on the landscape, deciphering which environmental factors are important to these boreal communities requires special spatial analyses that are sensitive to the characteristics of each type of forest.

Separating the large-scale spatial patterns caused by climatic variation from the small-scale patterns caused by stochastic factors or community interactions requires a spatial approach. As pointed out by Legendre and Legendre (1998), field observations often result from a combination of two different processes that account for large-scale and small-scale spatial structuring. The response variable z is spatially structured at a large scale because the explanatory variables are themselves spatially structured by their own generative processes. On the other hand, the deviation of z from the large-scale trend, or the residuals, can result from some stochastic spatial process involving the variable z itself, which causes a lack of independence among the residual components. This is called spatial autocorrelation, and it is generally a function of the geographic distance between sample locations.

To analyse the relationship between environmental factors and the Raifa boreal forest, we created a boreal forest index that should be responsive to the specific environmental factors acting upon this type of forest. The index is defined to be the number of species that belong to a set of boreal species divided by the total number of species at a site. This variable is hereafter referred to as the *boreality*. The model that will be applied assumes that the value of the dependent variable z (boreality) at site j can be presented as the sum of the effect of explanatory variables at site j, a weighted sum of the residuals of the same variable at sites i that surround site j and a random error ε_j. The response variable (boreality) can thus be modelled (in words) as

Boreality = *f*(Explanatory variables) + Auto-correlated noise + noise

The function of the explanatory variables *f*(Explantory variables) is also called the spatial trend. The last noise component is independently distributed noise. In mathematical terms, the boreality model formula is denoted as

$$z_j = f(\mathbf{X}_j) + \sum_i w_i (z_i - f(\mathbf{X}_i)) + \varepsilon_j \qquad (37.1)$$

The first term $f(\boldsymbol{X}_j)$ represents the effects of the explanatory variables $\boldsymbol{X}$ at a site j. It may be the effect of all factors that we assume influence the modelled variable z. These factors may be climatic, such as temperature or wetness, or topographic, such as slope or exposure. In other words, all available data derived from direct measurements or satellite images can be included in $f(X_j)$. The second term in equation (37.1) is the spatial auto-correlation, and it is a weighted sum of the residuals of neighbouring sites. The residuals are obtained by subtracting the effects of the explanatory variables $\boldsymbol{X}$ measured at sites i that surround site j from z. The weight w_i is a function of the geographic distance between location j where we predict variable z and the surrounding locations i. The impact of the surrounding sites is determined by the *range* of the auto-correlation process. Recall from Chapter 19 that the *range* defines the distance between two sites at which there is no more spatial auto-correlation. Hence, a large range means that sites far away from each other are still interacting. The values of the range and weight function w of an auto-correlation are computed through variography analysis, which was described in Chapter 19. The third term in equation (37.1) is the error term and is assumed to be independently normally distributed.

In the context of this study, we assume that the spatial gradients of environmental factors provide large-scale structure of plant cover characteristics, and that the fine-scale spatial heterogeneity of vegetation cover, which is caused by auto-correlated processes, is an intrinsic feature of vegetation itself. Therefore, the main task in this chapter is to apply spatial analysis methods that detect the large-scale spatial gradient of factors reflected in plant cover (the function *f*()), while taking into account the small-scale spatial interaction (the second term in equation (37.1).

37.2 Study area

The study area is in the Volga Valley on a terrace above the flood-lands of the Volga River, ca. 30 km west of Kazan, Tatarstan, Russia (Tajsin 1969, 1972). The western zone of the Raifa section of Volzhsko-Kamsky State Nature Biosphere Reserve is about 7 km in length from north to south (Figure 37.1), and is located within a sub-taiga coniferous-broadleaved forest biogeographical zone. We restricted this study to the western zone of the Raifa because it is the best sampled (see sample locations in Figure 37.1-C); and it exhibits regular replacements of natural territorial complexes from south to north because of changes in latitude, slope, aspect, lithology and moisture (Tajsin 1972). Although the geographic relief of the Raifa area is approximately 60–100 m above sea level and its topography is basically flat, the region contains pronounced erosion features, including ravines and gullies. The mean temperature in the area is 3.4°C, and the mean annual precipitation is 568 mm.

The 3846 hectare Nature Biospere Reserve, which includes the Raifa forests, was created in 1960 to protect unique boreal forests that contain high biodiversity, rare tree species, and a variety of phytocoenosis. The changes in vegetation types over just 2 km within the Raifa section of this preserve are comparable with what can be observed over a 100-km section of European Russia.

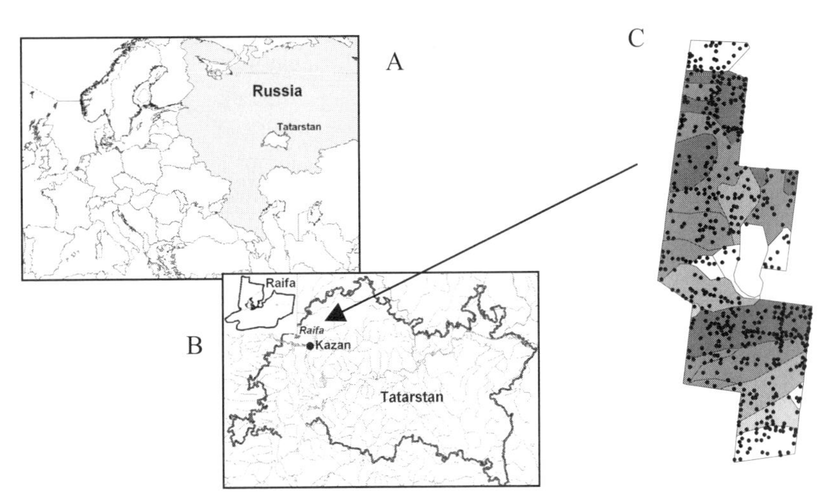

Figure 37.1. A: The location of Tatarstan. B: The location of the Raifa reserve in Tatarstan. C: Study area of forest types, on the basis of species domination (Porfiriev 1968). The extent of the study area is 7.2 km from north to south, and 2.7 km from east to west. Lake Raifskoe is in the centre of the region, with the Sumka River flowing into the lake from the northeast (not plotted on the study area map). Each dot in panel C represents a sample site. The shading corresponds to different forest types. Grey colouring in the study area indicates the degree of boreal species (dark corresponds to higher values).

The predominant species in Raifas' forest community is *Pinus sylvestris*. The communities also includes representatives of taiga and broadleaved zones, such as *Quercus robour, Tilia cordata, Betula pendula, Alnus glutinosa, Pinus silvestris, Piceax fenica* and more rarely, *Abies sibirica*.

37.3 Data exploration

A total of 534 geobotanical sites were sampled for this study (Figure 37.1C). Vascular plant community compositions were sampled using 100-m^2 plots in June of 2000–2004. In all, 485 vascular plant species were recorded, and 327 of them were present in more than three sites. The total number of boreal species found on the plots amounts to 22 species (Bakin et al. 2000). The proportion of these boreal species to species of all coenosis groups at each site, i.e., the site's *boreality*, ranged from 0% to 60%. The Cleveland dotplot (see Chapter 4) shows that no sites have extremely high values of boreality (Figure 37.2), but there are many observations with boreality equal to zero.

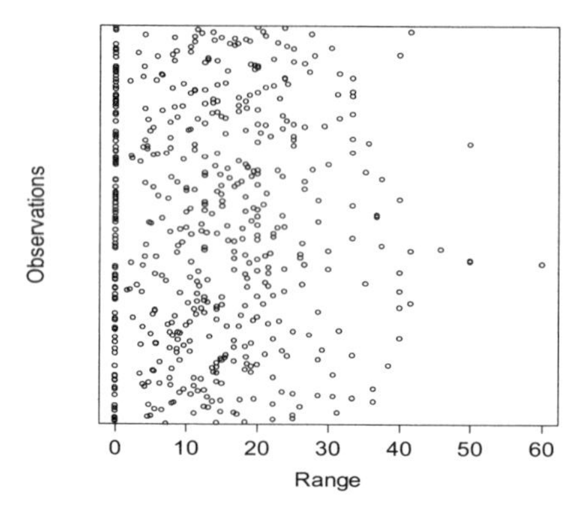

Figure 37.2. Cleveland dotplot for boreality. The vertical axis shows the observations in the order as they were in the spreadsheet, and the horizontal axis gives the value.

Four remotely sensed variables, derived from the LANDSAT 5 satellite images that were taken on June 22, 1987, were used as explanatory variables, namely:

1. The normalised difference vegetation index (NDVI). Positive values of NDVI indicate green vegetation; negative values indicate non-vegetated surface features, such as water, barren rock, ice, or snow.
2. Temperature in degrees Kelvin (Temp).
3. Index of Wetness (Wet). The index of wetness ranges from −1 to 1. Positive values indicate water surfaces; negative values correspond to wet habitats.

4. Index of Greenness (Grn). The index of greenness ranges from −1 to 1. Larger values correspond to more dense vegetation cover.

In addition to the data derived from satellite images, we also used latitude (*X*) and longitude (*Y*) of the sites as explanatory variables, assuming that they represented an indirect relationship with climatic gradients.

A pairplot for the four satellite variables and the two spatial coordinates is presented in Figure 37.3. All variables have some outliers and show collinearity. One site has outliers in nearly all variables. Temp has two additional sites with outlier values. Correlations between the latitude *X*, and longitude *Y* variables and the satellite variables are relatively low (<0.4). The correlations between the remotely sensed variables (see lower diagonal panels in Figure 37.3) all have serious collinearity issues, even after removing the outlier.

To detect collinearity (Chapter 5) between *all* explanatory variables, we calculated pair-wise correlations between the six variables and variance inflation factors (VIFs). Recall that VIF values were used in Chapter 26 to assess collinearity, and the reader is referred to that chapter for a detailed explanation of their calculation and interpretation. Basically, high VIF values, say >5 or >10, are an indication of collinearity. The VIF values for all six variables were *X* (1.365), *Y* (1.450), Temp (1.882), Wet (4.929), NDVI (13.564) and Grn (19.461). The VIF values for index of greenness and NDVI in this study indicate serious collinearity. Both of these variables reflect vegetation cover density. Based on these VIF values (and those obtained by dropping Grn), we decided to omit NDVI and Grn from further analysis. As a result, the remaining variables all have correlations smaller than 0.8.

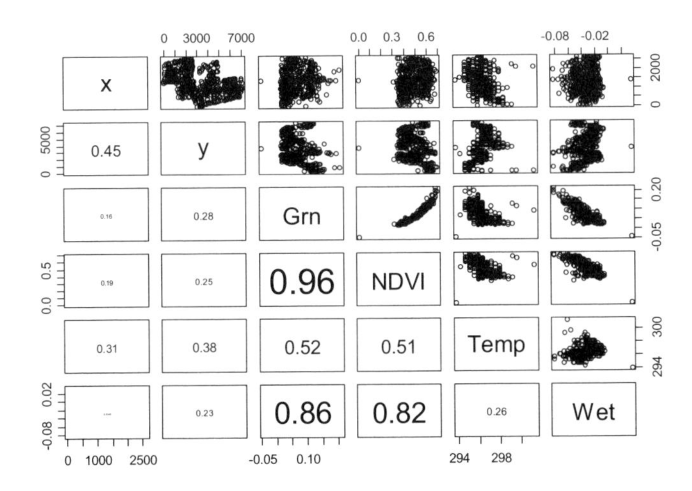

Figure 37.3. Pairplot of the two spatial variables and the four satellite variables. The lower diagonal elements contain the (absolute) correlations, and the font size is proportional to the value. All variables have outliers and show collinearity.

37.4 Models of boreality without spatial auto-correlation

To capture how much the boreality index varies with environmental factors, two different models were applied and their results compared. One was a linear regression model (LM). The other was an additive model (AM) with Gaussian error distribution. We examined the LM and AM models with different sets of predictors (i.e., the set of spatial coordinates and the set of remotely sensed data) to obtain insight into the importance of each explanatory variable. Remotely sensed variables and spatial coordinates were added to each model one at a time, and the relevance of each predictor in the individual models was investigated using an F-test for nested models (Chapters 5 and 7). Figure 37.4 shows one example of an AM model in which X, Y, Temp and Wet were used as explanatory variables. It has a pattern of increasing residual spread for larger fitted values, which indicates a violation of the homogeneity assumption. The same pattern was encountered in all LM and AM models; therefore, all violated assumptions of homogeneity. To proceed further, we have several options, namely:

- Apply a data transformation on boreality. As long as a monotonic transformation is applied, it makes biological sense to do this (or formulated differently, it is not biological nonsense to apply a transformation on boreality percentages).
- Boreality is defined by the numbers of species that belong to a set of boreal species divided by the total number of species at a site. Hence, we have S_i successes out of n_i trials, which can be modelled as a Binomial GLM or GAM (Chapter 6).
- The number of species belonging to the boreal coenosis species can be modelled as a Poisson GLM or GAM with the number of species per site as an offset variable (Chapter 6).
- Model the heterogeneity in terms of spatial covariance of the error component.

There is no perfect solution, so any of these approaches can be applied to a given dataset. Some may work; some may not work. We decided to transform boreality (because it worked) according to the following transformation:

$$z_i = (1000(S_i + 1) / n_i)^{1/2}$$

where z_i is transformed boreality, S_i is the number of species that belong to boreal coenosis species, and n_i is the number of all species at the site i. See Cressie (p. 395, 1993) for a discussion of this transformation, and other types of transformations (i.e., the Freeman–Tukey transformation). The effect of the transformation for the LM and AM can be seen in Figure 37.5. All explanatory variables selected in the data exploration step were used in these models, and except for the coordinate X, all variables were significant at the 99% confidence level (Table 37.1).

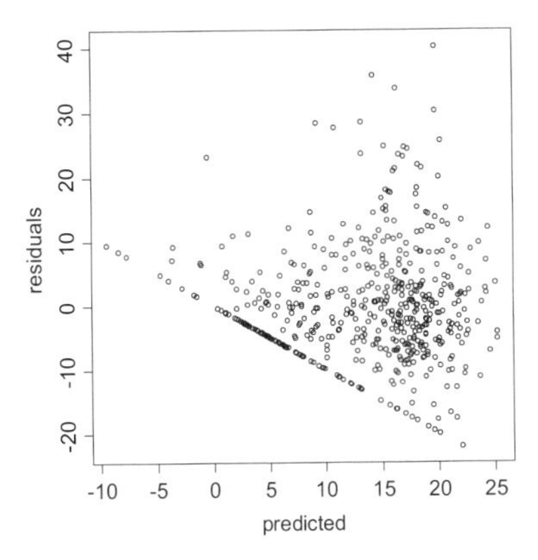

Figure 37.4. Residuals versus fitted values. The residuals were obtained from an AM model containing *X*, *Y*, Temp and Wet as explanatory variables. Cross-validation was used to estimate the optimal amount of smoothing for each term.

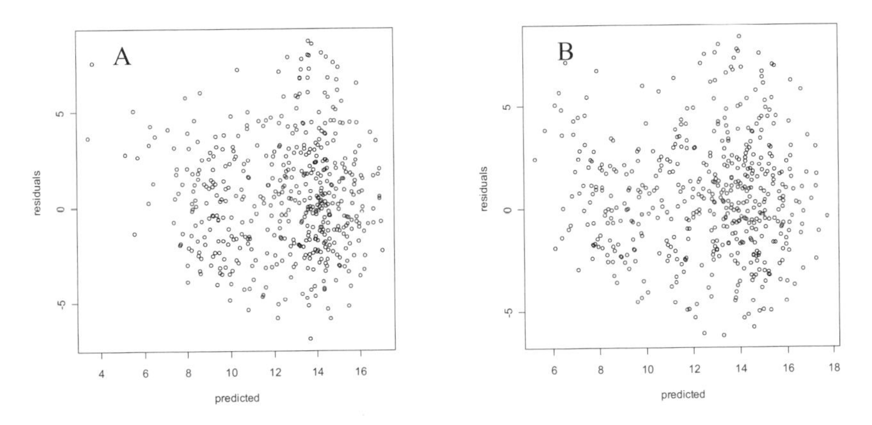

Figure 37.5 Residuals versus fitted values from a model containing Y, Temp and Wet as explanatory variables. A: The residuals were obtained from the LM model. B: The residuals were obtained from the AM model. AM residuals show better accordance with the homogeneity assumption. In both panels boreality was transformed.

Table 37.1. p-values for the explanatory variables. RS stands for remote sensing.

	X	Y	Temp	Wet
LM(X,Y)	0.083	<0.001		
LM(RS data)			<0.001	<0.001
LM(X,Y, RS data)	0.045	<0.001	<0.001	<0.001
AM(X,Y)	0.842	<0.001		
AM(RS data)			<0.001	<0.001
AM(X,Y, RS data)	0.157	<0.001	<0.001	<0.001

We regarded site position on the north-south axis (coordinate Y) as a factor that reflects the spatial (zonal) gradient of environmental factors. Table 37.1 shows that the Y coordinate is highly significant in both the LM and the AM. This result can be considered as a confirmation of our assumption regarding the large-scale spatial trend of boreality. The variable X is not important in any of the models, probably because the extent of explored territory in this direction is too small to reveal any serious environmental trends. We decided to remove the X coordinate from further analyses.

Because all remaining explanatory variables are significant in the models, there is no point in doing any further model selection. In the next section, we will improve upon these models by taking into account spatial auto-correlation. We will use two models, namely (i) the LM based on Y coordinates of site locations, temperature and index of wetness; and (ii) the AM based on Y coordinates of site locations, temperature and index of wetness, and we will add spatial auto-correlation (i.e., values for the second term in our boreality equation 37.1)

37.5 Models of boreality with spatial auto-correlation

The models discussed in the previous section can reveal spatial trends of environmental factors in boreality. Recall from equation 37.1 that the term 'spatial trend' refers to the effect of the explanatory variables. However, not all variability in boreality is explained by the spatial trend. One way to show how much is not explained is to calculate the correlation between observed and fitted values for each model. For the AM, this correlation was 0.697. We can try to improve these models by taking into account spatial auto-correlation. We can add spatial correlation structure in several ways, and we discuss two options here. The first option is to use an iterative scheme, as described in Bailey and Gatrell (1995). It works in the following way:

1. Start with an LM or AM without correlation structure and obtain residuals.
2. Estimate the residual covariance structure from the variography analysis (Chapter 19) and incorporate this in the model using a generalised least squares (GLS) approach (Chapters 5, 16, 26). Recall that GLS is an estimation procedure that allows for a non-diagonal error covariance matrix in linear regression or additive modelling.
3. Get new residuals and re-estimate the covariance structure.

4. Repeat steps 2 and 3 until convergence.

The second option is to estimate the spatial trend and correlation structure simultaneously using numerical optimisation of the likelihood function (Pinheiro and Bates 2000). In this chapter, we will discuss both approaches.

The Bailey and Gatrell scheme

We subtract the LM spatial trends from the observed boreality to obtain the residuals and use these in *variography* or *variogram analysis*. Because the variography analysis is based on the assumption of multivariate normality and stationarity of the residuals (in this case), we need to verify that the residuals meet these assumptions before continuing.

With stationarity, we mean the so-called second-order stationarity (Chapter 19), which can be checked by testing for spatially varying mean and variance, and heterogeneity of spatial covariance. Because we have already subtracted the LM and spatial trend from the observed boreality, the residuals have zero mean.

To test the LM residuals for departure from stationarity and multivariate normality, we use the h-scatterplot. This tool is used to plot the residuals in points s_1 against residuals in points s_2 separated by distance lag $\boldsymbol{h}$. To be more precisely, for each observed site in the study area, we can obtain a list of sites that are *exactly* $\boldsymbol{h}$ units away from it. Call this collection of points E. It may be an option to use $\boldsymbol{h}$ plus/minus a small number or else E only contains a few or no points at all. In an h-scatterplot, we plot the value of the residual at $\mathbf{s}_1$ along the x-axis and the values of the residuals in E along the y-axis. This whole process is then repeated for each site. The point $\mathbf{s}_1$ is also called the head and $(s_1+\boldsymbol{h})$ the tail. If the distribution of the residuals is highly skewed, then small values of residuals in heads correspond to big values of residuals in tails and vice versa. If this is the case, then the h-scatterplot will display a so-called 'butterfly wing', i.e., groups of points that are far away from the diagonal line. h-scatterplots for LM residuals are presented in Figure 37.6. We used two different lags: 200 and 800 m. Both graphs indicate that there is no clear butterfly wing effect. If the points follow the diagonal line, then there is positive correlation (high values of the residuals at both $\mathbf{s}_1$ and $(s_1+\boldsymbol{h})$). In this case, there is no clear trend.

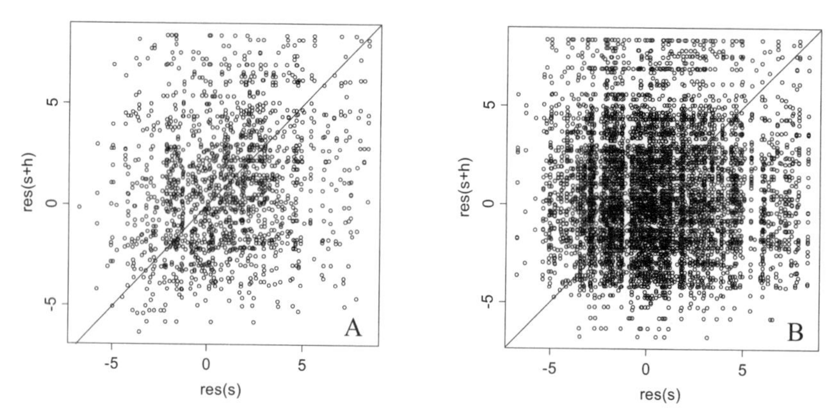

Figure 37.6. *h*-scatterplots. A: Plot for LM(ce) residuals for lag $h = 200$ m. B: Plot for LM(ce) residuals for lag $h = 800$ m. The value of the residual at $\mathbf{s}_1$ is along the horizontal axis, and residuals of all points that are **h** units away from $\mathbf{s}_1$ are plotted along the vertical axis.

Another way to assess the multivariate normality assumption of the residuals is to verify one-dimensional normality by using QQ-plots (Chapter 4). For these data, the *h*-scatterplots and QQ-plot indicate that the stationarity and normality assumptions are valid. Therefore, we can proceed with variography analysis and use geostatistical means to model the boreality residuals.

Recall from Chapter 16, that when we applied GLS on time series data, we had to specify the type of auto-correlation structure. Options were the AR(1), ARMA(*p*,*q*), and compound symmetry structure, among others. A useful tool to decide between the various temporal auto-correlation structures was the auto-correlation function. In spatial statistics, we have the same problem. The spatial equivalent of the auto-correlation function is the variogram, or, more precisely, the *semi-variogram*. It is a measure of spatial dependence between boreality in point s_1 and s_2. These points are *separated* by the distance $\boldsymbol{h}$, with the direction from s_1 to s_2. The estimated variogram (using sample data) is referred to as the *empirical variogram* or sometimes as the *sample variogram*. The mathematical formulation of the variogram and its estimator are given in Chapter 19.

Based on the shape of the variogram we can choose an appropriate error structure for the GLS. Common options are the exponential, Gaussian, linear and spherical models. Figure 19.11 in Chapter 19, Figure 5.9 in Pinheiro and Bates (2000) or Figure 13.8 in Legendre and Legendre (1998) show the corresponding shapes of the variogram for these models. So, we need to make a variogram for the residuals (sample variogram), and decide which model for the error structure is the most appropriate (i.e., which model fits the sample variogram best). In Figure 37.7, the sample variogram graph (dots) for LM residuals shows that two variogram models may be appropriate: spherical and exponential. To decide which of these models fits the sample variogram best, we can fit both models and compare their weighted sum of squares, i.e., the sum of distances between fitted variogram values and sample variogram values. The smaller sum of squares indicates the better fit.

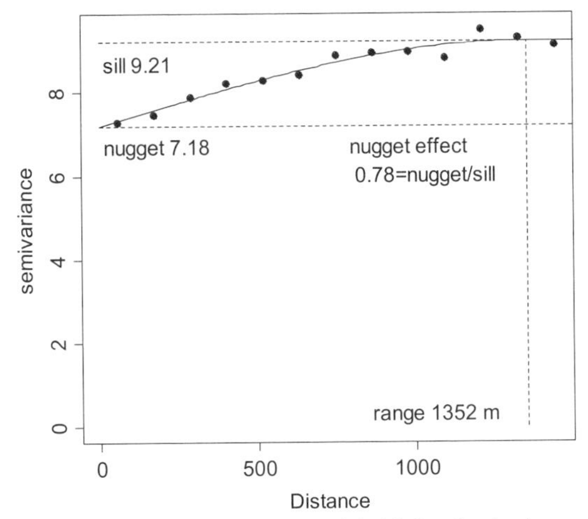

Figure 37.7. Graph of sample variogram and initial spherical correlation structure for LM residuals.

All variograms have a couple of unknown parameters that we need to estimate. The line can be characterised by the following parameters (Figure 37.7): the *range* of the spatial auto-correlation and the *nugget effect*. The *range* of the variogram model gives the maximum distance of site-to-site influence. The *nugget effect* is the variance of the difference between values at the sites that are co-located, or so close in space that we can consider them as one location. It is expressed as a proportion of the sill (but this may depend on software). *Sill* is half of the maximum variance of the difference between values at different locations.

For the residuals obtained from the LM and AM, we fitted both spherical and exponential variogram models. The spherical variogram model had the better fit (its sum of squares was smaller). Therefore, we will only present results obtained with the spherical variogram model.

Based on Figure 37.7, we assume (initially) that the range is 1352 m and the nugget is 0.78 (expressed as a fraction of the sill). The nugget effect plays a role similar to the R^2 in linear models; not more than 28% (= 1 − 0.78) of the variance in the data at any given location can be explained using data in other locations. In other words, only 28% of the variance of boreality residuals can be explained by auto-correlation process. These parameters of the correlation structure are just initial estimates, because we used the residuals of the LM trend model to estimate them. So, following the iterative scheme described by Bailey and Gatrell, we need to (i) incorporate these parameters into the model by specifying the range and nugget effect, and (ii) get new residuals and re-estimate the covariance structure by fitting the variogram model to the newly calculated sample variogram of these residuals. Steps (i) and (ii) are repeated until the spatial correlation parameters converge. The spherical data correlation structure in the final LM(ce) trend model obtained by this iterative scheme has a range of 1549 m, and the nugget effect is 0.66 (nugget is 6.66 and sill is 10.13). Note that these values differ from the initial values.

In time series analysis, time's only direction is forward. In spatial statistics, the lag can be in any direction. So far we have assumed that the relationship between two sites that are separated by a distance $\boldsymbol{h}$ is the same whatever the direction of $\boldsymbol{h}$ is. The term *anisotropy* is used when the strength of the spatial auto-correlation process is not the same in all directions; in some directions the strength is stronger and in some it is weaker. As the GLS model that was used to incorporate the data correlation structure *assumes* isotropic auto-correlation (where the strength *is* the same in each direction), we need to verify this assumption.

To detect anisotropy, we can look at the behaviour of empirical variograms for the separation vector $\boldsymbol{h}$ taken only along selected directions. These are so-called directional variograms. In Figure 37.8 empirical directional variograms of LM(ce) residuals are presented. Four directions were used: 0°, 45°, 90° and 135°. The direction angle is evaluated as the angle between the y-axis and the selected direction.

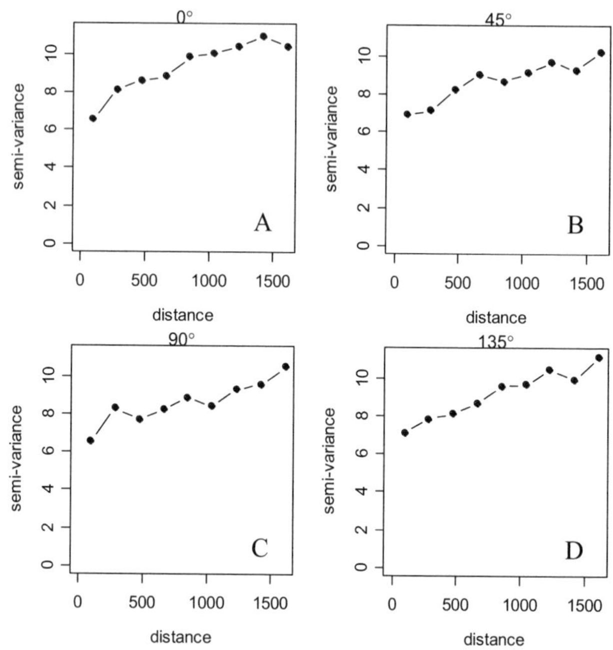

Figure 37.8. Empirical directional variogram for LM(ce) residuals for $\boldsymbol{h}$ distances. A: for the 0° direction, B: for the 45° direction, C: for the 90° direction, D: for the 135° direction.

We can identify anisotropy if there is a difference between slopes of the curves that join values of sample variograms. A visual inspection may reveal anisotropy if the variogram values increase fast in some directions and more slowly in others. In this case, there is some evidence of anisotropy at the 45º and 90º directions, but it is not strong so it is difficult to see a clear difference between these empirical variograms. We can detect anisotropy numerically by fitting variogram models to two directional empirical variograms that have perpendicular directions. The first variogram must have the flattest curve (as judged by eye), and the second variogram must have a perpendicular direction.

Here, we use two directional empirical variograms for the 0º and 90º directions, and Figure 37.9 shows the corresponding empirical variograms (dots) with fitted spherical variogram models. Anisotropy is characterized by two parameters: the anisotropy angle and the anisotropy ratio. The anisotropy angle is the direction of the variogram with the flattest curve and greater range (here it is the 90º one). The anisotropy ratio is the ratio of greater range to the smaller range.

The fitted spherical variogram models for LM(ce) residuals in Figure 37.9 have a nugget effect 0.60 (sill 10.60 and nugget 6.38). It is slightly smaller than the nugget effect determined without taking into account anisotropy (0.66). The direction with the greater range was 90º, and the range was 2185 m (Figure 37.9-A). The smaller range equals 1411 m (direction 0º, Figure 37.9-B). The anisotropy ratio is 2185/1419 ≈ 1.55, so the auto-correlation of sites along the 90º direction diminishes with distance in 1.55 times slower than along the 0º direction.

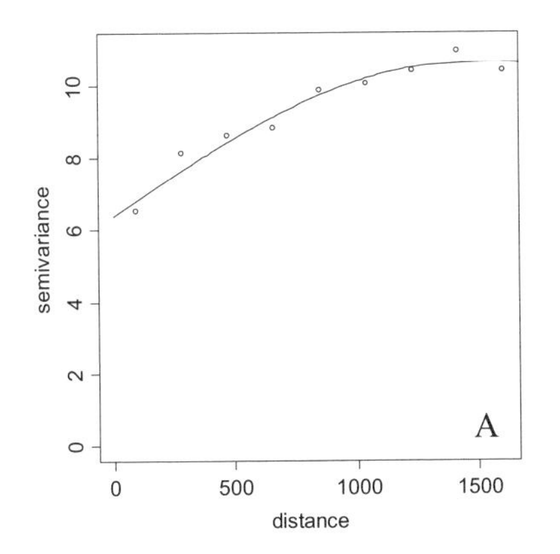

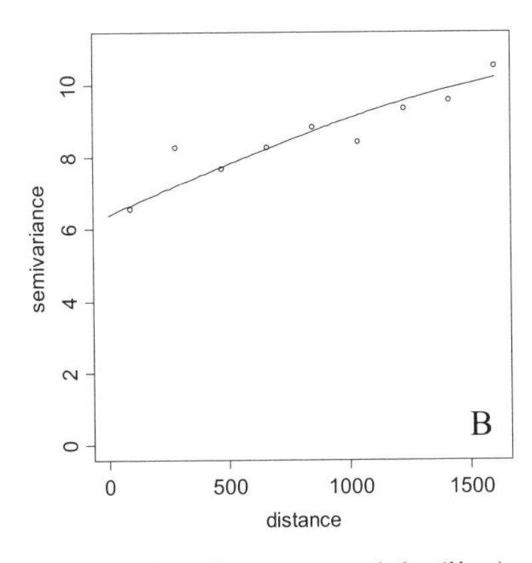

Figure 37.9. Empirical variograms (dots) with fitted variogram models (line). A: Empirical directional variogram for LM(ce) residuals along the 0º direction and fitted exponential variogram model with range 1411 m, sill 10.60 and nugget 6.38. B: Empirical directional variogram for AM(ce) residuals along the 90º direction and fitted exponential variogram model with range 2185 m, sill 10.60 and nugget 6.38. This means that the variogram value of co-located sites is 6.38 and the variogram values of sites separated by more than approximately 1411 m along the 0º direction and more than 2185 m along the 90º direction is assumed equal to 10.60.

Approach 2: Numerical optimisation of the likelihood function

In the approach described above, the Bailey and Gatrell iterative scheme was used to estimate the trend component and the correlation structure. We now briefly describe how to estimate the spatial trend and correlation structure simultaneously using numerical optimisation of the likelihood function (Pinheiro and Bates 2000). We will apply it to the AM trend model, obtain the AM(ce) model and inspect its residuals by variography analysis to detect anisotropy of the auto-correlation process.

Before applying the numerical optimisation routines, we inspected (using the same tools as above, e.g., QQ-plots and *h*-scatterplots) the boreality residuals obtained by the AM trend model to check multivariate normality and stationarity. Both assumptions were valid. We then applied the numerical optimisation routines to estimate the spatial trend component and correlation structure simultaneously. This model uses the same variogram models. We can specify different spatial auto-correlation structures and then compare them using the AIC. The smallest AIC shows the most adequate data correlation model. Just as before, we compared spherical and exponential models. The AIC were nearly equal, but we decided to use the spherical model.

A spherical data correlation structure was specified for the AM(ce) with the following starting values: *range of auto-correlation* 821 m and *nugget effect* 0.70.

The software that we used (nlme library in R) scales the maximum variance (sill) to one, and as a result we can only obtain relative values of the nugget effect. If the aim of the analysis is to model the spatial trend and take into account the spatial auto-correlation, then we need the exact values of nugget and sill.

The models used above, the AM(ce), LM(ce), and the GLS in the Bailey and Gatrell scheme, can only cope with isotropic spatial auto-correlation (due to software implementation). So, after convergence of the estimation methods, we need to check the residuals for anisotropy, and this can be done in the same way as above. For the AM(ce) residuals, anisotropy was detected along the 90° direction (the greater range along the 90° direction is 1227 m, and the smaller range along the 0° direction is 752 m. The anisotropy ratio is $1227/752 \approx 1.63$. The nugget is 6.3 and sill 9.1. So the nugget effect determined by variography analysis ($0.69 = 6.3/9.1$) is close to the nugget effect obtained by the numerical optimisation.

37.6 Conclusion

After applying several different spatial models, we conclude that it is possible to detect the spatial trend of environmental factors in plant cover, even if the vegetation is spatially heterogeneous. Remote sensing data are important to estimate the spatial distribution of specific species and plant community features.

The process of spatial analysis involves many steps to find the best spatial model that describes the relationship between environmental factors and site vegetation (Figure 37.10). The process consists of data exploration, model creation based only on environmental factors, and model refinement to take fine-scale stochastic and local interactions due to auto-correlation into account. The AIC for the optimal linear regression model without an auto-correlation structure was 2507, and using the Bailey and Gatrell scheme, LM(ce), gave AIC = 2478. For the AM and the AM(ce) using the numerical optimisation gave AIC = 2622 and AIC = 2461 respectively. Hence, incorporating a spatially correlated, stochastic component of vegetation variability (presented as residuals) into each model improve its measures of fit.

Information about anisotropy (obtained by variogram analysis of residuals) can help us to make some inferences on the local spatial behaviour of the investigated variable. The anisotropy angle was 90°, which means that along this direction the spatial auto-correlation diminishes more slowly than in the 0 ° direction. Disregarding the anisotropy may overestimate or underestimate the variable in some directions, and as a result it may give an improper spatial picture of the phenomenon.

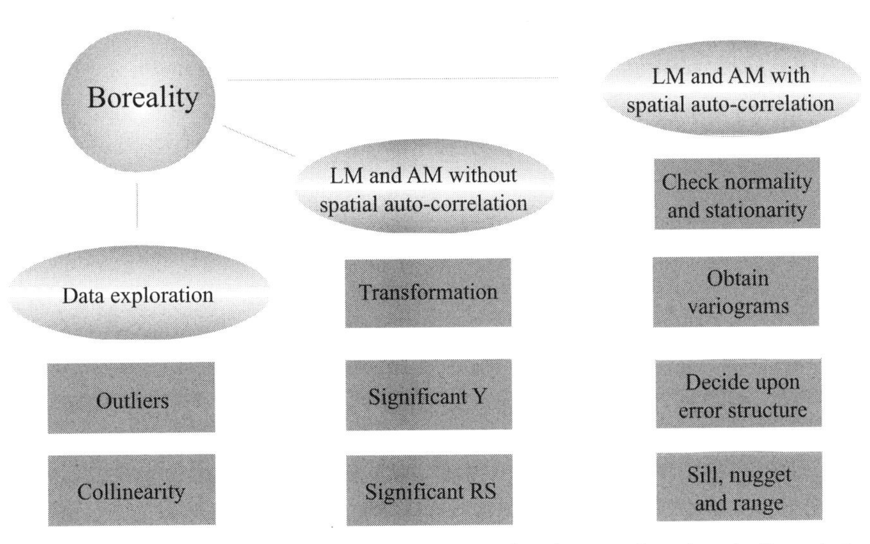

Figure 37.10. Outline of the data analyses. The data exploration indicated the presence of outliers and collinearity. We then applied linear regression and additive modelling and assumed that the error component was independently distributed. All explanatory variables in this model were significantly different from 0 at the 5% level, except for the *X* coordinate. We continued the analysis by allowing for a residual spatial correlation dependency. Using a sample variogram it was decided to use a spherical correlation structure. The resulting model can be written as: Boreality = *F*(Explanatory variables) + spatial correlated noise. The advantage of this model is that we are less likely to commit a type I error because we allow for spatial dependency.

Acknowledgement

We would like to thank Neil Campbell and Pam Sikkink for useful comments on an earlier draft.

References

Abaunza P (In Press) Using life history parameters as an important basis for the initial recognition of stock management units in horse mackerel (Trachurus trachurus). Fisheries Research

Abaunza P, Villamor B, Pérez JR (1995) Infestation by larvae of *Anisakis simplex* (Nematoda: Ascaridata) in horse mackerel, *Trachurus trachurus* and Atlantic mackerel, *Scomber scombrus*, in ICES Divisions VIIIb, VIIIc and IXa (N - NW of Spain). Sci Mar 59:223–233

Abaunza P, Murta A, Campbell N, Cimmaruta R, Comesaña S, Dahle G, Gallo E, García Santamaría MT, Gordo L, Iversen S, MacKenzie K, Magoulas A, Mattiucci S, Molloy J, Nascetti G, Pinto AL, Quinta R, Ramos P, Ruggi A, Sanjuan A, Santamaría MT, Santos AT, Stransky C, Zimmerman C (In Press) Considerations on sampling strategies for an holistic approach to stock identification: the example of the HOMSIR project. J Fish Bio

Abbott I, Marchant N, Cranfield R (2000) Long-term change in the floristic composition and vegetation structure of Carnac Island, Western Australia. J Biog 27:333–346

Alverson DL, Freeberg MH, Pope JG, Murawski SA (1994) A global assessment of fisheries bycatch and discards. FAO Fish. (Tech Paper No. 339, Rome, FAO)

Alward RD, Detling JK, Milchunas DG (1999) Grassland vegetation changes and nocturnal global warming. Science 283:229–231

Amaral V, Cabral HN (2004) Ecology of *Monochirus hispidus* in the Sado estuary. J Fish Bio 64:460–474

Amemiya T (1985) Advanced Econometrics, Cambridge, MA: Harvard University Press

Anscombe FJ (1973) Graphs in statistical analysis. Am Statist 27:17–21

Anselin L (1988) Spatial Econometrics: Methods and Models, Dordrecht The Netherlands: Kluwer Academic Publishers

Anselin L, Florax RJGM, Rey SJ (eds) (2004) Advances in Spatial Econometrics, Methodology, Tools and Applications. Series: Advances in Spatial Science, XXII, 513 pp

Austin MP (2002) Spatial prediction of species distribution: an interface between ecological theory and statistical modeling. Ecolog Model 157:101–118

Austin MP, Gaywood MJ (1994) Current problems of environmental gradients and species response curves in relation to continuum theory. J Veget Scie 5:473–482

Austin MP, Nicholls AO, Doherty MD, Meyers JA (1994) Determining species response functions to an environmental gradient by means of a b-function. J Veget Scie 5:215–228

Bailey TC, Gatrell AC (1995) Interactive Spatial Data Analysis, Harlow, UK: Longman Higher Education

Bakin OV, Rogova TV, Sitnikov AP (2000) Vascular Plans of Tatarstan. Kazan, Russia: Kazan State University Publishing, 496 pp (in Russian)

Bakker JP, Bunje J, Dijkema KS, Frikke J, Hecker N, Kers AS, Korber P, Kohlus J, Stock M (2005) QSR Update 2004. In: Salt Marshes. Chapter 7. Wilhemshaven, Germany: CWSS 165–180 pp. http://www.waddensea-secretariat.org/

Banko WE (1987a) History of endemic Hawaiian birds. Part 1. Population Histories—Species Accounts: Freshwater Birds: Hawaiian Coot `Alae`ke`oke`o. Cooperative National Park Resources Studies Unit, Honolulu, HI: University of Hawaii at Manoa

Banko WE (1987b) History of endemic Hawaiian birds. Part 1. Population Histories—Species Accounts: Freshwater Birds: Hawaiian Gallinule 'Alae'ula. Cooperative National Park Resources Studies Unit, Honolulu, HI: University of Hawaii at Manoa

Basualdo M, Bedascarrasbure EL, De Jong D (2000) Africanized Honey Bees (Hymenoptera: Apidae) have a greater fidelity to sunflowers than European bees. J Econ Entomol 93:304–307

Baum JK, Myers RA, Kehler DG, Worm B, Harley SJ, Doherty PA (2003) Collapse and conservation of shark populations in the Northwest Atlantic. Science 299:389–392

Bax NJ (1998) The significance and prediction of predation in marine fisheries. ICES J Mar Sci 55:997–1030

Baxter MA (1994) “A guide to seasonal adjustment of monthly data with X-11,” 3rd edn. Central Statistical Office, United Kingdom

Beals EW (1984) Bray-Curtis-ordination: an effective strategy for analysis of multivariate ecological data. Adv Ecol Res 14:1–55

Beeftink WG (1987) Vegetation responses to changes in tidal inundation of salt marshes. In: Van Andel J, Bakker JP, Snaydon RW (eds) Disturbance in grasslands, Dordrecht, The Netherlands: Junk Publishers, 97–117 pp

Belkin IM (2004) Propagation of the “Great Salinity Anomaly” of the 1990s around the northern North Atlantic. Geophys Res Lett: 31, L08306, doi: 10.1029/2003GL019334

Bell MA, Legendre P (1987) Multicharacter chronological clustering in a sequence of fossil sticklebacks. Syst Zool 36:52–61

Bemvenuti CE, Capitoli RR, Gianuca NM (1978) Estudos de ecologia bentônica na região estuarial da Lagoa dos Patos. II – Distribuição quantitativa do macrobentos infralitoral. Atlântica 3:23–32

Bértola GR, Ferrante A (1996) Dinámica y transporte de un canal de mareas (Bahía Samborombón, Argentina). Thalassas 12:107–119

Beukema JJ (1974) Seasonal changes in the biomass of the macro-benthos of a tidal flat area in the Dutch Wadden Sea. Neth J Sea Res 8:94–107

Beukema JJ (1979) Biomass and species richness of the macrobenthic animals living on a tidal flat area in the Dutch Wadden Sea; effects of a severe winter. Neth J Sea Res 13:203–223

Beukema JJ (1992) Long-term and recent changes in the benthic macrofauna living on tidal flats in the western part of the Wadden Sea. Neth J Sea Res 20:135–141

Beukema JJ (2002) Expected changes in the benthic fauna of the Wadden Sea tidal flats as a result of sea-level rise or bottom subsidence. J Sea Res 47:25–39

Bianchi G, Gislason H, Graham K, Hill L, Jin X, Koranteng K, Manickchand-Heileman S, Payá I, Sainsbury K, Sanchex F, Zwanenburg K (2000) Impact of fishing on size composition and diversity of demersal fish communities. ICES J Mar Sci 57:558–571

Blanco DE (1998) Uso de hábitat por tres especies de aves playeras (*Pluvialis dominica*, *Limosa haemastica* y *Calidris fuscicollis*) en relación con la marea en Punta Rasa, Argentina. Revista Chilena Hist Nat 71:87–94

Bookstein FL (1989) `Size and shape': A comment on semantics. Syst Zool 38:173–180

Bookstein FL (1991) Morphometric tools for landmark data: geometry and biology, New York: Cambridge Univ Press

Borcard D, Legendre P, Drapeau P (1992) Partialling out the spatial component of ecological variation. Ecol 73:1045–1055

Boreman J, Nakashima BS, Wilson JA, Kendall RL (eds) (1997) Northwest Atlantic groundfish: perspectives on a fishery collapse, Bethesda, MD: American Fisheries Society

Borges MF, Turner RJ, Casey J (1993) Plasma transferring polymorphisms in scad (*Trachurus trachurus* L.) populations from the north-east Atlantic. ICES J Mar Sci 50:299–301

Borges MF, Santos AMP, Crato N, Mendes H, Mota B (2003) Sardine regime shifts off Portugal: a time series analysis of catches and wind conditions. Sci Mar 67 (Suppl. 1):235–244

Bouyoucos GJ (1936) Directions for making mechanical analysis of soils by the hydrometer method. Soil Scie 42:225–228

Bowering WR, Morgan MJ, Brodie WB (1997) Changes in the population of American plaice (*Hippoglossoides platessoides*) off Labrador and northeastern Newfoundland: a collapsing stock with low exploitation. Fish Res 30:199–216

Bowman A, Azzalini A (1997) Applied smoothing techniques for data analysis: the Kernel approach with S-Plus illustrations. Oxford, UK: Oxford University Press

Box GEP, Pierce DA (1970) The distribution of residual autocorrelations in autoregressive-integrated moving average time series models. J Amer Stat Assoc 65:1509–1526

Breiman L, Friedman JH, Olshen RA, Stone CJ (1983) Classification and Regression Trees, Belmont, CA: Wadsworth

Brockwell PJ, Davis RA (2002) Introduction to time series and forecasting, 2nd edn. New York: Springer-Verlag

Brown AC, McLahlan A (1990) Ecology of Sandy Shores. Amsterdam, The Netherlands: Elsevier Science Publishers

Brown DJ, Boyd IL, Cripps GC, Butler PJ (1999) Fatty acid signature analysis from the milk of Antarctic fur seals and southern elephant seals from South Georgia: implications for diet determination. Marine Ecol Progr Ser 187:251–263

Brown HK, Prescott RJ (1999) Applied Mixed Models in Medicine. New York: John Wiley and Sons

Brown SK, Mahon R, Zwanenburg KCT, Buja KR, Claflin LW, O'Boyle RN, Atkinson B, Sinclair M, Howell G, Monaco ME (1996) East Coast of North America groundfish: initial explorations of biogeography and species assemblages. National Oceanic and Atmospheric Administration and Dartmouth, NS, Department of Fisheries and Oceans. 111 pp

Brzeziecki B, Kienast F, Wildi O (1995) Modelling potential impacts of climate change on the spatial distribution of zonal forest communities in Switzerland. J Veg Sci 6:257–268

Burkart R, Bárbaro NO, Sánchez RO, Gómez DA (1999) Eco-Regiones de la Argentina. Reporte de la Secretaría de Recursos Naturales y Desarrollo Sustentable, Argentina, 42 pp

Burnham KP, Anderson DR (2002) Model Selection and Multimodel Inference. Springer

Cadima JFCL, Jolliffe IT (1995) Loadings and correlations in the interpretation of principal components. J Appl Stat 22:203–214

Cadima JFCL, Jolliffe IT (1996) Size and shape-related principal component analysis. Biomet 52:710–716

Cadrin SX, Friedland KD, Waldman JR (eds) (2005) Stock Identification Methods: Applications in Fishery Science. Amsterdam, The Netherlands: Academic Press

Casetti E (1972) Generating models by the spatial expansion method: applications to geographical research. Geogr Anal 4:81–91

Casetti E (1982) Drift analysis of regression parameters: an application to the Investigation of fertility development relations. Modeling Simulation 13:961–966

Chambers JM, Hastie TJ (1992) Statistical Models in S. Pacific Grove, CA: Wadsworth & Brooks/Cole Computer Science Series

Chambers JM, Cleveland WS, Kleiner B, Tukey PA (1983) Graphical Methods for Data Analysis. Belmont, CA: Duxbury Press

Chang PR (1990) Strategies for Managing Endangered Waterbirds on Hawaiian National Wildlife Refuges. Unpubl. MS. Thesis, Amherst, MA: Univ. of Massachusetts

Chatfield C (2003) The analysis of time series: an introduction, 6th edn. London, UK: Chapman and Hall, Ltd

Chen CS, Pierce GJ, Wang J, Robin JP, Poulard JC, Pereira J, Zuur AF, Boyle PR, Bailey N, Beare DJ, Jereb P, Ragonese S, Mannini A, Orsi-Relini L (2006) The apparent disappearance of *Loligo forbesi* from the south of its range in the 1990s: trends in *Loligo* spp. abundance in the northeast Atlantic and possible environmental influences. Fish Res 78:44–54

Cheverud J, Dow M, Leutenegger W (1985) The quantitative assessment of phylogenetic constraints in comparative analyses: sexual dimorphism in body weight among primates. Evolution 39:1335–1351

Chiles JP, Delfiner P (1999) Geostatistics. Modeling Spatial Uncertainty, New York: J Wiley & Sons

Christensen V, Guenette S, Heymans JJ, Walters C, Watson R, Zeller D, Pauly D (2003) Hundred-year decline of North Atlantic predatory fishes. Fish Fisheries 4:1–24

Clarke KR (1993) Non-parametric multivariate analyses of changes in community structure. Aust J Ecol 18:117–143 [1294]

Clarke KR, Ainsworth M (1993) A method of linking multivariate community structure to environmental variables. Mar Ecol Prog Ser 92:205–219

Clarke KR, Warwick RM (1994) Change in Marine Communities: An Approach to Statistical Analysis and Interpretation. 1st edn, Plymouth, UK: Plymouth Marine Laboratory, 144 pp, Plymouth, UK: PRIMER-E, 172 pp

Clarke ED, Spear LB, Mccracken ML, Marques FFC, Borchers DL, Buckland ST, Ainley DG (2003) Validating the use of generalized additive models and at-sea surveys to estimate size and temporal trends of seabird populations. J Appl Ecol 40:278–292

Claude J, Tong H, Paradis E, Auffray JC (2003) A geometric morphometric assessment of the effects of environment and cladogenesis on the evolution of the turtle shell. Biol Journal of the Linnean Society 79:485–50

Claude J P, Pritchard CH, Tong HY, Paradis E, Auffray JC (2004) Ecological correlates and evolutionary divergence in the skull of turtles: a geometric morphometric assessment. Sys Biol 53:933–948

Cleveland WS (1985) The Elements of Graphing Data, Monterey, CA: Wadsworth

Cleveland WS (1993) Visualizing Data, Summit, NJ: Hobart Press, 330 pp

Cliff AD, Ord JK (1973) Spatial Autocorrelation, London, UK: Pion

Cliff AD, Ord JK (1981) Spatial Processes - Models and Applications, London, UK: Pion

Coles S (2004) An Introduction to Statistical Modeling of Extreme Values, London, UK: Limited Springer-Verlag

Costa MJ, Bruxelas A (1989) The structure of fish communities in the Tagus estuary, Portugal, and its role as a nursery for commercial fish species. Scient Mar 53:561–566

Costlow JD, Bookhout CG (1969) Temperature and meroplankton. Chesapeake Sci 10:252–257

Crawley MJ (2002) Statistical Computing. An Introduction to Data Analysis Using S-Plus, New York: Wiley

Crawley MJ (2005) Statistics. An introduction using R. New York: Wiley

Cressie NAC (1993) Statistics for Spatial Data. New York: Wiley

Cressie NAC (1996) Change of support and the modifiable areal unit problem, Geogra Syst 3:159–180

Dalgaard P (2002) Introductory Statistics with R. Berlin, Germany: Springer

Darroch JN Mosimann, JE (1985) Canonical and principal components of shape. Biometrika 72:241–252

Davison AC, Hinkley DV (1997) Bootstrap Methods and their Applications. Cambridge, UK: Cambridge University Press

Dayhoff JE (1990) Neural Network Architectures: An Introduction. New York: Van Nostrand Reinhold, 259 pp

De Jong DJ, Van der Pluijm AM (1994) Consequences of a tidal reduction for the salt-marsh vegetation in the Oosterschelde estuary (The Netherlands). Hydrobiologia 282/283:317–333

De'Ath G (2002) Multivariate regression trees: a new technique for modeling species environment relationships. Ecol 83:1105–1117

De'Ath G, Fabricus KA (2000) Classification and regression trees: a powerful yet simple technique for ecological data analysis. Ecol 81:3178–3192

Dedio W, Putt ED (1980) Sunflower. In: Fehr WR, Hadley HH (eds) Hybridizations of crops plants. Madison, WI: American Society of Agronom and Crop Science Society of America, 631–644 pp

DeGrandi-Hoffman G, Martin JH (1993) The size and distribution of the honey bee (*Apis mellifera* L.) cross-pollinating population on male-sterile sunflowers (*Helianthus annuus* L.). J Apic Res 32:159–165

Delaude A, Tasei JN, Rollier M (1978) Pollinator insects of sunflower (*Helianthus annuus* L.) in France. Pollination of male-sterile lines for hybrid seed production. In: IVth. Symposium on Pollination. Md. Agric. Exp. Sta. Spec. Misc. Publ, 29–40 pp

Diggle PJ (1990) Time Series: A Biostatistical Introduction, London, UK: Oxford University Press

Diggle PJ, Heagerty P, Liang KY, Zeger SL (2002) The Analysis of Longitudinal Data, 2nd Edition. Oxford, UK: Oxford University Press

Dijkema KS (1990) Salt- and brackish marshes around the Baltic Sea and adjacent parts of the North Sea, their vegetation and management. Biol Conserv 51:191–209

Dijkema KS (1997) Impact prognosis for salt marshes from subsidence by gas extraction in the Wadden Sea. Jo Coastal Rese 13:1294–1304

Dijkema KS, Bossinade JH, Bouwsema P, de Glopper RJ (1990) Salt marshes in the Netherlands Wadden Sea: rising high tide levels and accretion enhancement. In: Beukema JJ,Wolff WJ, Brouns JJWM (eds) Expected Effects of Climatic Change on Marine Coastal Ecosystems. Dordrecht, The Netherlands: Kluwer Academic Publishers 173–188 pp

Dobson AJ (2002) Introduction to Generalized Linear Models 2nd edn. Boca Raton, FL: Chapman & Hall/ CRC Press

Donald PF (2004) Biodiversity impacts of some agricultural commodity production systems. Conser Biol 18:17–38

Dos Santos A, Gonzalez-Gordillo JI (2004) Illustrated keys for the identification of the Pleocyemata (Crustacea, Decapoda) zoeal stages, from the coastal region of south western Europe. J Mar Biol Ass U.K. 84:205–227

Drane D, Macpherson R, White K (1982) Pollination studies in hybrid sunflower seed production. In: 10 th International Sunflower Conference, Paradise, Australia. 95–100 pp

Draper N, Smith H (1998) Applied Regression Analysis, 3rd edn. New York: Wiley

Drinkwater KR (2002) A review of the role of climate variability in the decline of northern cod. In Mcginn A (ed) Fisheries in a Changing Climate. Bethesda, MD: American Fisheries Society, 113–130 pp

Drinkwater Kf, Harding GC (2001) Effects of the Hudson Strait outflow on the biology of the Labrador Shelf. Can J Fish Aquat Sci 58:171–184.

Drinkwater KF, Mountain DG (1997) Climate and Oceanography. In Boreman J, Nakashima BS, Wilson JA, Kendall RL (eds) Northwest Atlantic groundfish: Perspectives on a Fishery Collapse. Bethesda, MD: American Fisheries Society, 3–25 pp

Drinkwater KF, Loeng H, Megrey BA, Bailey N, Cook RM (eds) (2005) The influence of climate change on North Atlantic fish stocks. ICES J Mar Sci 62:1203–1542

Dryden I, Mardia K (1998) Statistical Shape Analysis, John Wiley and Sons, New York

Durbin J, Koopman SJ (2001) Time Series Analysis by State Space Methods, Oxford, UK: Oxford University Press

Eastwood PD, Meaden GJ, Carpentier A, Rogers SL (2003) Estimating limits to the spatial extent and suitability of sole (*Solea solea*) nursery grounds in the Dover Strait. J Sea Res 50:151–165

Edwards H O (1957) Departmental letter to Yellowstone National Park personnel on exclosure construction and rationale. 1 p

Efron B, Tibshirani RJ (1993) An Introduction to the Boostrap. New York: Chapman and Hall

Eiríksson H (1999) Spatial variability of CPUE and mean size as possible criteria for unit stock demarcations in analytical assessments of Nephrops at Iceland. Rit Fiskideildar 16:239–245

Elías R (1992b) Quantitative benthic community structure in Blanca Bay and its relationship with organic enrichment. Marine Ecol 13:189–201

Elías R, Bremec C (1994) Biomonitoring of water quality using benthic communities in Blanca Bay (Argentina). Scie Total Environ 158:45–49

Elphick CS, Oring LW (1998) Winter management of California rice fields for waterbirds. J Appl Ecol 35:95–108

Elphick CS, Oring LW (2003) Effects of rice field management on winter waterbird communities: conservation and agronomic implications. Agric Ecosys Environ 94:17–29

Eltink ATGW (1992) Horse mackerel egg production and spawning stock size in the North Sea in 1991. ICES CM 1992/H:21

Engelhaard GH, Heino M (2004) Maturity changes in Norwegian spring-spawning herring before, during, and after a major population collapse. Fish Res 66:299–310

Engilis AJ, Pratt TK (1993) Status and population trends of Hawaii's native waterbirds, 1977–1987. Wilson Bull 105:142–158

Erchinger HF (1985) Dünen, Watt und Salzwiesen (in German). Der Niedersächsische Ministerie für Ernährung, Landwirtschaft und Forsten, Hannover. 59 pp

Erchinger HF (1995) Intaktes Deichvorland für Küstenschutz unverzichtbar (in German). Wasser und Boden 47:48–53

Erzini K (2005) Trends in NE Atlantic landings (southern Portugal): identifying the relative importance of fisheries and environmental variables. Fish Oceanogr 14:195–209

Erzini K, Inejih CAO, Stobberup KA (2005) An application of two techniques for the analysis of short, multivariate non-stationary time series of Mauritanian trawl survey data. ICES J of Mar Scie 62:353–359

Everitt B (2005) An R and S-Plus Companion to Multivariate Analysis. London, UK: Springer-Verlag

FAO (2000) FAO Yearbook on Fishery Statistics. Rome, Italy: UN Food and Agriculture Organisation

Faraway JJ (2006) Linear Models with R. London, UK: Chapman and Hall/CRC Press

Fay PA, Carlisle JD, Danner BT, Lett MS, McCarron JK, Stewart C, Knapp AK, Blair MJ, Collins SL (2002) Altered rainfall patterns, gas ex-change, and growth in grasses and forbs. Int Jo Plant Sci 163:549–557

Felsenstein J (1985) Phylogenies and the comparative method. Amer Naturalist 125:1–12

Fernandes TF, Elliott M, Da Silva MC (1995) The management of European estuaries: a comparison of the features, controls and management framework of the Tagus (Portugal) and the Huber (England). Neth J Aquat Ecol 29:459–468

Fincham A, Williamson DI (1978) Decapoda larvae, VI. Caridea. Families: Palaemonidae and Processidae. ICES, Fiches d'identification du zooplankton, fiche no.159/160

Findley DF, Monsell BC, Bell WR, Otto MC, Chen BC (1997) New capabilities and methods of the X-12-ARIMA seasonal adjustment program, Jo of Busi Econom Stati

Fitzmaurice GN, Laird NM, Ware J (2004) Applied longitudinal analysis. Washington, DC: Wiley-IEEE

Fotheringham AS, Wong DWS (1991) The modifiable areal unit problem in multivariate statistical analysis. Envir Plann A 23:1025–1044.

Fowler J, Cohen L, Jarvis P (1998) Practical statistics for field biology. 2nd edn, New York: Wiley.

Fox J (2000) Nonparametric Simple Regression. Smoothing Scatterplots. Thousand Oaks, CA: Sage Publications.

Fox J (2002a) An R and S-Plus Companion to Applied Regression. Thousand Oaks, CA: Sage Publications

Fox J (2002b) Multiple and Generalized Nonparametric Regression (Quantitative Applications in the Social Sciences S.), Thousand Oaks, CA: Sage Publications

França S, Vinagre C, Costa MJ, Cabral HN (2004) Use of the coastal areas adjacent to the Douro estuary as a nursery area for pouting, *Trisopterus luscus* Linnaeus, 1758. J Appl Ichthyol 20:99–104

Fuhlendorf SD, Briske DD, Smeins FE (2001) Herbaceous vegetation change in variable rangeland environments: the relative contribution of grazing and climatic variability. Appl Vegetat Sci 4:177–188

Gabriel KR (1995) In: Krzanowski WJ (ed.), Recent Advances in Descriptive Multivariate analysis, Oxford, UK: Oxford University Press

Gabriel KR, Odoroff CL (1990) Biplots in biomedical research. Stati Medici 9:469–485

Gaffney ES (1975) A phylogeny and classification of the higher categories of turtles. Bull Amer Mus Nat Hist 155:387–436

Garthwaite PH, Jolliffe IT, Jones B (1995) Statistical Inference, Englewood Cliggs, NJ: Prentice Hall

Gauch HG (1982) Multivariate Analysis in Community Ecology. Cambridge, UK: Cambridge Univ Press

Geiger EL, McPherson GR (2005) Response of semi-desert grasslands invaded by non-native grasses to altered disturbance regimes. J Biogeogr 32:895–902

Getis, A. (1991) Spatial interaction and spatial autocorrelation: a cross product approach, Environ Plann A 23:1269–1277

Giménez L, Borthagaray AI, Rodríguez M, Brazeiro A, Dimitriadis C (2005) Scale-dependent patterns of macrofaunal distribution in soft-sediment intertidal habitats along a large-scale estuarine gradient. Helgol Mar Res 59:224–236

Glantz MH (ed) (1992) Climate variability, climate change and fisheries. Cambridge, UK: Cambridge University Press, 450 pp

Gomes MC, Haedrich RL, Villagarcía MG (1995) Spatial and temporal changes in the groundfish assemblages on the Northeast Newfoundland/Labrador Shelf, Northwest Atlantic, 1978–1991. Fish Oceanogr 4:85–101

Goovaerts P (1997) Geostatistics for Natural Resources Evaluation. Oxford, UK: Oxford Univ. Press

Gower JC, Hand DJ (1996) Biplots. London: Chapman and Hall

Grafen A (1989) The phylogenetic regression. Philos. Trans R Soc Lond B Biol Sci 326:119–157

Greene WH (2000) Econometric Analysis, 4th edn. New York: Macmillan

Greenacre JM (1984) Theory and Application of Correspondence Analysis. NewYork: Academic Press

Greenacre JM (1993) Biplots in corresponcence analysis. J Appl Statist 2:251–269

Griffin CR, Shallenberger RJ, Fefer SI (1989) Hawaii's endangered waterbirds: a resource management challenge, 1165–1175 pp in Freshwater Wetlands and Wildlife

Haddon M (2001) Modelling and Quantitative Methods in Fisheries. Boca Raton, FL: Chapman Hall/CRC Press

Haedrich RL (1995) Structure over time of an exploited deep-water fish assemblage. In: Hopper AG (ed) Deep-water fisheries of the North Atlantic oceanic slope. Dordrecht, The Netherlands: Kluwer Academic Publishers, 27–50 pp

Haedrich RL, Barnes SM (1997) Changes over time of the size structure in an exploited shelf fish community. Fish Res 31:229–23

Haining R (1990) Spatial data analysis in the social and environmental sciences. Cambridge, UK: Cambridge Univ. Press

Hair JF, Anderson RE, Tatham RL, Black WC (1998) Multivariate data analysis, 5th edn. Englewood Cliffs, NY: Prentice Hall

Hamilton LC, Haedrich RL, Duncan CM (2004) Above and below the water: social/ecological transformation in the northwest Newfoundland. Popula Environ 25:195–215

Hammill MO, Stenson GB (2000) Estimated prey consumption by harp seals (*Phoca groenlandica*), hooded seals (*Cystophora cristata*), grey seals (*Halichoerus grypus*) and harbour seals (*Phoca vitulina*) in Atlantic Canada. J Northw Atl Fish Sci 26:1–23

Hammond PS, Berggren P, Benke H, Borchers DL, Collet A, Heide-Jorgensen MP, Heimlich S, Hiby AR, Leopold MF, Oien N (2002) Abundance of harbour porpoise and other cetaceans in the North Sea and adjacent waters. J Appl Ecol 39:361–376

Hanson PJ, Koenig CC, Zdanowicz VS (2004) Elemental composition of otoliths used to trace estuarine habitats of juvenile gag, *Mycteroperca microlepis*, along the west coast of Florida. Mar Eco Prog Ser 267:253–265

Hardin JW, Hibe JM (2003) Generalized estimation equations. Boca Raton, FL: Chapman and Hall/CRC Press

Hare SR, Mantua NJ (2000) Empirical evidence for North Pacific regime shifts in Pacific North America. Prog Oceanog 47:103–145

Harvey AC (1989) Forecasting, structural time series models and the Kalman filter. Cambridge, UK: Cambridge University Press

Hastie T, Tibshirani RJ (1990) Generalized Additive Models. London, UK: Chapman and Hall

Heitmeyer ME, Connelly DP, Pederson RL (1989) The central, imperial, and coachella valleys of California. In: Smith LM, Pederson R, Kaminski RH (eds) Habitat Management for Migrating and Wintering Waterfowl in North America. Lubbock, TX: Texas Tech University Press, 475–505 pp

Helbig J, Mertz G, Pepin P (1992) Environmental influences on the recruitment of Newfoundland/Labrador cod. Fish Oceanog 1:39–56

Hernandez MJ (2003) Database design for mere mortals. Boston, MA: Addison Wesley

Herrington WC, Bearse HM, Firth FE (1939) Observations on the life history, occurrence and distribution of the redfish parasite, *Sphyrion* lumpi. US Bureau of Fisheries Special Report 5:1–18

Hill MO (1973) Reciprocal Averaging: An eigenvector method of Ordination. J Ecol 61:237–49

Hill MO (1974) Correspondence analysis: a neglected multivariate method. Appl Stat 23:340–354

Hill MO (1979) DECORANA_A FORTRAN program for detrended correspondence analysis and reciprocal averaging. Section of Ecology and Systematics. Ithaca, NY: Cornell University, 468–470 pp

Holme NA, Mc Intyre AD (1984) Methods for the Study of Marine Benthos. Oxford, UK: Blackwell Scientific Publ

Hooker SK, Iverson SJ, Ostrom P, Smith SC (2001) Diet of northern bottlenose whales inferred from fatty-acid and stable-isotope analyses of biopsy samples. Canadian J Zool 79:1442–1454

Hosmer DW, Lemeshow S (2000) Applied Logistic Regression. New York: Wiley Serie in Probability and Statistics

Huberty CJ (1994) Applied Discriminant Analysis. New York: Wiley

Hubert L J, Golledge R G, Constanzo C M (1981) Generalized procedures for evaluating spatial autocorrelation. Geograph Anal 13:224–233

Hurrell JW (1995) Decadal trends in the North Atlantic Oscillation: regional temperatures and precipitation. Sci 269:676–679

Hurrell JW, Kushnir Y, Ottersen G, Visbeck M (2003) The North Atlantic Oscillation: Climatic Significance and Environmental Impact. Geophysical Monograph Series. Washington, DC: American Geophysical Union, 1–35 pp

Huse G, Gjøsæter H (1999) A neural network approach for predicting stock abundance of the Barents Sea capelin. Sarsia 84:457–464

ICES (1992) Report of the study group on the stock identity of mackerel and horse Mackerel. ICES CM 1992/H:4

ICES (2004) Report of the working group on the assessment of mackerel, horse mackerel, sardine and anchovy. ICES CM 2004/ACFM:08

Ieno E, Bastida R (1998) Spatial and temporal patterns in coastal macrobenthos of Samborombón Bay, Argentina: a case study of very low diversity. Estuaries 21:690–699

Ieno E, Alemany D, Blanco DE, Bastida R (2004) Prey size selection by red knot feeding on mud snails at Punta Rasa (Argentina) during migration. Waterbirds 27:493–498

Isaaks EH, Srivastava RM (1989) An Introduction to Applied Geostatistics. Oxford, UK: Oxford University press

Iverson SJ, Arnould JPY, Boyd IL (1997) Milk fatty acid signature indicate both major and minor shifts in the diet of lactating Antarctic fur seals. Canadian J Zool 75:188–197

Iverson SJ, Oftedal OT, Bowen WD, Boness DJ, Sampugna J (1995) Prenatal and postnatal transfer of fatty acids from mother to pup in the hooded seal. J Compar Physiol B 165:1–12

Iverson SJ, Field C, Bowen WD, Blanchard, W (2004) Quantitative fatty acid signature analysis: a new method of estimating predator diets. Ecol Monogr 74:211–235

Jacobs MJ, Schloeder CA (2002) Fire frequency and species associations in perennial grasslands of south-west Ethiopia. African J Ecol 40:1–9

Jager ZH, Kleef L, Tydeman P (1993) The distribution of 0-group flatfish in relation to abiotic factors on the tidal flats in the brackish Dolard (Ems Estuary, Wadden Sea). J Fish Biol 43:31–43

Jambu M (1991) Exploratory and Multivariate Data Analysis. Boston, MA: Academic Press

Janssen GM, Mulder S (2004) De ecologie van de zandige kust van Nederland (in Dutch). Ministerie van Verkeer en Waterstaat, Rijkswaterstaat / RIKZ / 033

Janssen GM, Mulder S (2005) Zonation of macrofauna across sandy beaches and surf zones along the Dutch coast. Oceanologia 47:265–282

Jolicoeur P (1963) The multivariate generalization of the allometry equation. Biometrics 19: 497–499

Jolliffe IT (2002) Principal Component Analysis. New York: Springer

Jolliffe IT, Trendafilov NT, Uddin M (2003) A modified principal component technique based on the LASSO. J of Computat Graph Stat 12:531–547

Jolly WM, Graham JM, Michaelis A, Nemani R, Running SW (2004) A flexible, integrated system for generating meterological surfaces derived from point sources across multiple geographic scales. Environ Model Software 20:873–882

Jones JP, Casetti E (1992) Applications of the Expansion Method. London: Routledge

Jongman RHG, Braak CJF, van Ter Tongeren OFR (1995) Data Analysis in Community and Landscape Ecology. Cambridge, UK: Cambridge University Press

Jungers WL, Falsetti AB, Wall CE (1995) Shape, relative size, and size-adjustments in morphometrics. Yearbook of Physi Anthropol 38:137–161

Kaiser J (2001) How rain pulses drive biome growth. Science 291:413–414

Kent M, Coker P (1992) Vegetation Description and Analysis. A Practical Approach. New York: John Wiley and Son, X+ 363 pp

Kimmerer WJ (2002a) Physical, biological and management responses to variable freshwater flow into the San Fransisco estuary. Estuaries 25:1275–1290

Kimmerer WJ (2002b) Effects of freshwater flow on abundance of estuarine organisms: physical effects or trophic linkages? Mar Ecol Prog Ser 243:39–55

Kirsch PE, Iverson SJ, Bowen WD (2000) Effect of a low-fat diet on body composition and blubber fatty acids of captive juvenile harp seals (*Phoca groenlandica*). Physiol Biochem Zool 73:45–59

Kirsch PE, Iverson SJ, Bowen WD, Kerr SR, Ackman RG (1998) Dietary effects on the fatty acid signature of whole Atlantic cod (*Gadus morhua*). Canadian Jo Fisheries Aquatic Scie 55:1378–1386

Klecka WR (1980) Discriminant Analysis. London: Sage

Kleinbaum DG Klein M (2002) Logistic Regression A Self-Learning Text. New York: Sringer-Verlag

Knapp AK, Smith MD (2001) Variation among biomes in temporal dynamics of aboveground primary production. Science 291:481–484

Kooijman SALM (1977) Species abundance with optimum relations to environmental factors. Ann Syst Res 6:123–138

Koopman H N (2001) The structure and function of the blubber of odontocetes. PhD thesis. Durham, NC: Duke University, 407 pp

Koranteng K, McGlade JM (2001) Evolution climatique a long terme du plateau continental du Ghana et du golfe de Guinee (1963–1992). Oceanol Acta 24:187–198

Koutsikopoulos C, Desaunay Y, Dorel D, Marchand J (1989) The role of coastal areas in the life history of sole (*Solea solea* L.) in the Bay of Biscay. Scient Mar 53:567–575

Krijgsveld KL, Lensink R, Schekkerman H, Wiersma P, Poot MJM, Meesters EHWG, Dirksen S (2005) Baseline studies North Sea wind farms: fluxe, flight paths and altitudes of flying birds 2003–2004. Report 05–041. Culemborg, The Netherlands: Bureau Waardenburg

Krzanowski WJ (1988) Principles of Multivariate Analysis. Oxford UK: Oxford University Press

Krzanowski WJ, Marriott FHC (1994) Multivariate Analysis Part 1: Distributions, Ordination and Inference. London: Wiley

Lana PC, Almeida MVO, Freitas CAF, Couto ECG, Conti LMP, Gonzáles-Peronti AL, Giles AG, Lopes MJS, Silva MHC, Pedroso LA (1989) Estrutura espacial de associações macrobênticas sublitorais da Gamboa Perequê (Pontal do Sul, Paraná). Nerítica 4:119–136

Leach MK, Givnish TJ (1996) Ecological determinants of species loss in remnant prairies. Science 273:1555–1558
Learmonth JA (2006) Life history and fatty acid analysis of harbour porpoises (*Phocoena phocoena*) from Scottish waters. PhD thesis. Aberdeen, Scotland: University of Aberdeen: 320 pp
Legendre P (1993) Spatial autocorrelation: trouble or new paradigm? Ecol 74:1659–1673
Legendre P, Anderson MJ (1999) Distance-based redundancy analysis: testing multi-species responses in multi-factorial ecological experiments. Ecolog Monogr 69:1–24
Legendre P, Gallagher E (2001) Ecologically meaningful transformation for ordination of species data. Oecologia 129: 271–280
Legendre P, Legendre L (1998) Numerical Ecology (2nd English edn). Amsterdam, The Netherlands: Elsevier, 853 pp
Legendre P, Dallot S, Legendre L (1985) Succession of species within a community: Chronological clustering, with application to marine and freshwater zooplankton. Am Nat 125:257–288
Leps J, Smilauer P (2003) Multivariate analysis of ecological data using CANOCO. Cambridge, UK: Cambridge University Press
Lester RJG (1990) Reappraisal of the use of parasites for fish stock identification. Aus J Mar Freshwater Res 41:855–864
Lester RJG, Barnes A, Habib G (1985) Parasites of Skipjack Tuna: fishery implications. Fish Bull 83:343–356
Liang KY, Zeger SL (1986) Longitudinal data analysis using generalized linear models. Biometrika 73:13–22
Lindsey JK (2004) Introduction to Applied Statistics. A Modelling Approach, 2nd edn. Oxford, UK: Oxford Univerisity Press
Lira-Noriega A (2003) La vegetación de los potreros del norte de la sierra de Los Tuxtlas. BSc.-Thesis (in Spanish). Universidad Nacional Autónoma de México. 98 pp
Ljung GM, Box GEP (1978) On a measure of lack of fit in time series models. Biometrika 65:297–303
Ljung L (1987) System Identification: Theory for the User (Prentice-Hall Information & System Sciences Series) Englewood Cliffs, NY: Prentice Hall PTR
Lloret J, Lleonart J, Solé I, Fromentin JM (2001) Fluctuations of landings and environmental conditions in the North-Western Mediterranean sea. Fish Oceanogr 10: 33–50
Lütkepohl H (1991) Introduction to multiple time series Analysis. Berlin, Germany: Springer-Verlag
MacDonald P (2005) Stock identification of the horse mackerel *Trachurus trachurus* L. from the North-west African coast using parasites as biological tags. MSc. thesis. Aberdeen, Scotland: University of Aberdeen
Macer CT (1977) Some aspects of the biology of the horse mackerel (*Trachurus trachurus* L.) in waters around Britain. J Fish Biol 10:51–62
MacKenzie K (1983) Parasites as biological tags in fish population studies. Adv App Bio 7:251–331
MacKenzie K, Campbell N, Mattiucci S, Ramos P, Pereira A, Abaunza P, (2004) A checklist of the protozoan and metazoan parasites reported from the Atlantic horse mackerel, *Trachurus trachurus* (L.). Bull Eur Ass Fish Pathol 24:180–184
MacKenzie K, Campbell N, Mattiucci S, Ramos P, Pinto AL, Abaunza P (In Press) Parasites as biological tags for stock identification of Atlantic horse mackerel *Trachurus trachurus* L. Fish Res
Magurran, AE (2004) measuring biological diversity. Oxford, UK: Blackwell Publishing

Maindonald J, Braun J (2003) Data Analysis and Graphics using R. Cambridge, UK: Cambridge University Press

Makridakis S, Wheelwright S, Hyndman R (1998) Forecasting: Methods and Applications, 3rd edn. New York: Wiley, 642 pp

Manly BFJ (2001) Statistics for Environmental Science and Management. London, UK: Chapman & Hall

Manly BFJ (2004) Multivariate Statistical Methods: A Primer, 3rd edn. Boca Raton, FL Chapman & Hall/CRC Press

Maravelias CD, Haralabous J, Papaconstantinou C (2003) Predicting demersal fish species distributions in the Mediterranean Sea using artificial neural networks. Mar Eco Prog Ser 25:249–258

Marchand J, Masson G (1988) Inshore migration, distribution, behaviour and feeding activity of sole, *Solea solea* (L.), postlarvae in the Vilaine estuary, France. J Fish Biol 33:227–228

Mardia K V, Coombes A, Kirkbride J, Linney A, Bowie J L (1996) On statistical problems with face identification from photographs. J Appl Statist 23:655–675

Marsden RF, Mysak LA, Myers RA (1991) Evidence for stability enhancement of sea ice in the Greenland and Labrador Seas. J Geophys Res 96:4783–4789

Martin JP (2002) Aspectos biológicos y ecológicos de los poliquetos de ambientes mixohalinos de la Provincia de Buenos Aires (In Spanish). Tesis Doctoral, Universidad Nacional de Mar del Plata, Argentina. 314 pp

Martins EP, Hansen TF (1997) Phylogenies and the comparative method: a general approach to incorporating phylogenetic information into the anlaysis of interspecific data. Amer Nat 149: 646–667. ERRATUM Am Nat 153:448

McCullagh P, Nelder J (1989) Generalized Linear Models. London, UK: Chapman and Hall, McCulloch WS, Pitts WH (1943) A logical calculus of the ideas immanent in nervous activity. Bull Math Biophysics 5:115–133

McCune B, Grace JB (2002) Analysis of ecological communities. Gleneden Beach, OR:MjM Software De-sign

McGinn NA (ed) (2002) Fisheries in a changing climate. American Fisheries Society Symposium 32, Bethesda, MD: American Fisheries Society

Mc. Gregor SE (1976) Insect pollination of cultivated crop plants. US Department of Agriculture Handbook Nº 496, 411pp

McNeely JA, Scherr SJ (2002) Ecoagriculture: Strategies to Feed the World and Save Wild Biodiversity. Washington, DC: Island Press

Mendelssohn R, Schwing F (1997) Application of State-Space Models to Ocean Climate Variability in the Northeast Pacific Ocean. Lecture Notes in Statistics: Application of Computer Aided Time Series Modeling, Masanao Aoki and Arthur M. Havenner, (eds) New York: Springer, 255–278 pp

Mendelssohn R, Schwing FB (2002) Common and uncommon trends in SST and wind stress in the California and Perú-Chile current systems. Prog Oceanogr 53:141–162

Molenaar PCM (1985) A dynamic factor model for the analysis of multivariate time series. Psychometrika 50:181–202

Molenaar PCM, de Gooijer JG, Schmitz B (1992) Dynamic factor analysis of nonstationary multivariate time series. Psychometrika 57:333–349

Moller J (1994) Lectures on Random Voronoi Tessellations, Lecture Notes in Statistics 87, NewYork: Springer-Verlag

Montgomery DC, Peck EA (1992) Introduction to linear regression analysis. New York: Wiley, 504 pp

Moore BR, Buckworth RC, Moss H, Lester RJG (2003) Stock discrimination and movements of narrow-barred Spanish mackerel across northern Australia as indicated by parasites. J Fish Biol 63:765–779

Moran PAP (1950) Notes on continuous stochastic phenomena. Biometrika 37:17–23

Morissette L, Hammill MO, Savenkoff C (2006) The trophic role of marine mammals in the Northern Gulf of St. Lawrence. Mar Mamm Sci 22:74–103

Mosquera J, Gomez-Gesteira M, Perez-Villar V (2000) Using parasites as biological tags of fish populations: a dynamical model. Bull Math Biol 62:87–99

Mosteller F, Tukey JW (1977) Data analysis and regression: a second course in statistics, Reading, MA: Addison Wesley

Mulder S (2000) Ecologie van de zandige kust (In Dutch). Werkdocument. RIKZ/OS/2000.617x

Muñoz-Carpena R, Ritter A, Li YC (2005) Dynamic factor analysis of groundwater quality trends in an agricultural area adjacent to Everglades National Park. J Contam Hydro 80:49–70

Murillo M de las NM (2001) Size structure and production in a demersal fish community. Deep-sea Fisheries Symposium (NAFO conference). Havana, Cuba. 12–14 September 2001. Published as: Martinez M MS 2001 NAFO SCR Doc. No. 134, Serial No. N4529, 16 pp

Murta AG (2000) Morphological variation of horse mackerel (*Trachurus trachurus*) in the Iberian and North African Atlantic: implications for stock identification. ICES J Mar Sci 57:1240–1248

Murta AG, Borges MF, Silveiro ML (1995) Infestation of the horse mackerel, *Trachurus trachurus* (L.) by *Anisakis simplex* (Rudolphi, 1809) (Nematoda, Ascaridida) larvae in Portuguese waters (ICES Div. Ixa). ICES CM 1995/H:19

Musgrove AJ, Pollitt MS, Hall C, Hearn RD, Holloway SJ, Marshall PE, Robinson JA. Cranswick PA (2002) The Wetland Bird Survey 2000–2001: Wildfowl and Wader Counts. Slimbridge, UK: BTO/ WWT/ RSPB/ JNCC

Myers JP, Myers LP (1979) Shorebirds of coastal Buenos Aires Province. Ibis 121:186–200

Myers RA, Worm B (2003) Rapid worldwide depletion of predatory fish communities. Nature 423:280–283

Neat FC, Wright PJ, Zuur AF, Gibb IM, Gibb FM, Tulett D, Righton DA and Turner RJ (2006) Residency and depth movements of a coastal group of Atlantic cod (*Gadus morhua* L.). Marine Biol 148:643–654

Nefedov GN, Alferova NM, Chuksin UV (1978) Polymorphic esterases of horse mackerel in the north-east Atlantic. Biologia Morya 2:64–74

Neis B, Schneider DC, Felt L, Haedrich RL, Fischer J, Hutchings JA (1999) Northern cod stock assessment: what can be learned from interviewing resource users? Can J Fish Aq Sci 56:1944–1963

Nelson DW, Sommers LE (1982) Total carbon, organic carbon, and organic matter. (In: Page, Miller, Keeney (ed). Methods of Soil Analysis, Part 2. Madison, WI: ASA Press, 539–578 pp

Newton RR, Rudestam KE (1999) Your Statistical Consultant. London,UK: Sage

NRC (1999) Sustaining Marine Fisheries. Washington, DC: National Academy Press Núñez JA (1982) Foraging pressure and its annual variation: a method of evaluation using artificial food sources. J Apic Res 21:134–138

O'Beirne P (2005) Spreadsheet Check and Control: 47 key practices to detect and prevent errors. Wexford, Ireland: Systems Publishing

Olff H, Bakker JP, Fresco LFM (1988) The effect of fluctuations in tidal inundation frequency on a salt marsh vegetation. Vegetatio 78:13–19

Openshaw S, Taylor PJ (1979) A million or so correlation coefficients: three experiments on the modifiable areal unit problem. In: Wrigley N (ed) Statistical Applications in the Spatial Sciences, London: Pion 127–144 pp

Ord K (1975) Estimation Methods for Models of Spatial Interaction, J Stat Assoc 70:120–126

Ostrom CW (1990) Time series analysis: Regression Techniques. 2nd edn. Thousand Oaks, CA: Sage Publications Inc

Otto L, Zimmerman JTF, Furnes GK, Mork M, Saetre R, Becker G (1990) Review of the physical oceanography of the North Sea. Neth J Sea Res 26:161–238

Palmer MW (1993) Putting things in even better order: the advantages of canonical correspondence analysis. Ecol: 74:2215–2230.

Pampel F (2000) Logistic Regression: A Primer. Sage Series on Quantitative Applications in the Social Sciences. Thousand Oaks, CA: Sage

Pannatier Y (1996) Variowin. Software for Spatial Data Analysis in 2D. New York: Springer

Paradis E, Claude J (2002) Analysis of comparative data using generalized estimating equations. J Theoret Biol 218:175–185

Parsons LS, Lear WH (2001) Climate variability and marine ecosystem impacts: a North Atlantic perspective. Prog Oceanogr 49:167–188

Pauly D, Maclean J (2003) In a perfect ocean: the state of fisheries and ecosystems in the North Atlantic Ocean. Sci 22:137:154

Pike RB, Williamson DI (1958) (revised 1959). Crustacea Decapoda: larvae. XI. Paguridea, Coenobitidea, Dromiidea, and Homolidea. ICES, Fiches d'identification du zooplankton, sheet 81

Pike RB, Williamson DI (1964) The larvae of some species of Pandalidae (Decapoda). Crustaceana 6:226–284

Pike RB, Williamson DI (1972) Crustacea Decapoda: larvae. X. Galatheidea. ICES, Fiches d'identification du zooplankton, sheet 139

Pinheiro JC, Bates DM (2000) Mixed-effects models in S and S-Plus. New York: Springer Verlag

Porfiriev VS (1968) Vegetation of Raifa. Proceedings of Volga-Kama State Reserve. Kazan, Russia: publishing of Kazan State University. 247 pp, (in Russian)

Quinn GP Keough MJ (2002) Experimental design and data analysis for biologists. Cambridge, UK: Cambridge University Press

R Development Core Team (2005) R: A Language and Environment for statistical Computing. Vienna, Austria: R Foundation for Statistical Computing, http://www.R-project.org

Rabus B, Eineder M, Roth A, Bamler R (2003) The shuttle radar topography mission - a new class of digital elevation models acquired by space borne radar, ISPRS. J Photogram Remote Sens 57:241–262

Radford BJ, Rhodes JW (1978) Effect of honeybee activity on the cross-pollination of male sterile sunflowers. Queensland J Agric Anim Sci 35:153–157

Raffaelli D, Hawkins S (1996) Intertidal ecology. London, UK: Chapman & Hall

Raftery AE (1995) Bayesian model selection in social research. In: Marsden PV (ed) London: Tavistock. Sociological Methodology 111–163 pp

Ranatunga C (1989) Methods of removing size from a data set. Unpublished M.Sc. dissertation. Cantebury, UK: University of Kent at Canterbury

Ravier C, Fromentin J (2001) Long-term fluctuations in the eastern Atlantic and Mediterranean bluefin tuna population. ICES J Mar Sci 58:1299–1317

Raymont JEG (1980) Plankton and productivity in the oceans. Vol. 2. Zooplankton. 2nd edn. Oxford UK: Pergamon Press 350 pp

Reed JM, Oring LW (1993) Long-term population trends of the endangered Ae`o (Hawaiian stilt, *Himantopus mexicanus knudseni*). Transactions of the Western Regional Wildlife Society 29:54–60

Reed JM, Silbernagle MD, Evans K, Engilis A, Jr, Oring LW (1998) Subadult movement patterns of the endangered Hawaiian stilt (*Himantopus mexicanus knudseni*). Auk 115:791–797

Ribbands CR (1964) The behavior and social life of honeybees. New York: Dover Publications Inc, 352 pp

Rijnsdorp AD, van Stralen M, van der Veer HW (1985) Selective tidal transport of North Sea plaice larvae *Pleuronectes platessa* in coastal nursery areas. Trans Amer Fish Soc. 114:461–470

Riley JD, Symonds DJ, Woolner L (1981) On the factors influencing the distribution of 0-group demersal fish in coastal waters. Rapp. P. -v. Réun Cons int Explor Mer 178:223–228

Ritter A, Muñoz-Carpena R (2006) Dynamic factor modeling of ground and surface water levels in an agricultural area adjacent to Everglades National Park. J Hydrol 317:340–354

Rochet MJ, TrenkelV, Bellail R, Coppin F, Le Pape O, Mahe JC, Morin J, Poulard JC, Schlaich I, Souplet A, Verin Y, Bertrand J (2005) Combining indicator trends to assess ongoing changes in exploited fish communities: diagnostic of communities off the coasts of France. ICES J Mar Sci 62: 1647–1664

Ruppert D, Wand MP, Carroll RJ (2003) Semiparametric Regression. Cambridge, UK: Cambridge Univerity Press

Sakamoto Y, Ishiguro M, Kitagawa G (1986) Akaike Information Criterion Statistics. Dordrecht, The Netherlands: D. Reidel Publishing Company

Salia SB (2005) Neural networks used in classification with emphasis on biological populations in stock identification methods: Applications in Fishery Science. Cadrin SX, Friedland KD, Waldman JR (eds) Amsterdam, The Netherlands: Academic Press 253–270 pp

Santos AMP, Borges M de F, Groom S (2001) Sardine and horse mackerel recruitment and upwelling off Portugal. ICES J Mar Sci 58:589–596

Santos MB, Pierce GJ, Learmonth JA, Reid RJ, Ross HM, Patterson IAP, Reid DG, Beare D (2004) Variability in the diet of harbor porpoises (*Phocoena phocoena*) in Scottish waters 1992–2003. Marine Mammal Sci 20:1–27

Schabenberger O, Pierce FJ (2002) Contemporary statistical models for the plant and soil sciences. Boca Raton, FL: CRC Press

Schimek MG (ed) 2000. Smoothing and regression: approaches, computation and application, New York: Wiley

Shapiro DE, Switzer P (1989) Extracting time trends from multiple monitoring sites. Technical report No. 132. Department of Statistics, Stanford, CA: Stanford University

Sharp GD, Csirke J (eds) (1983) Proc. Expert Consultation on the Changes in Abundance and Species Composition of Neritic Fish Resources, 4 April 18–29, 1983, San Jose, Costa. FAO Fish. Rep. 291

Sharitz RR, JW Gibbons (eds) DOE Symposium Series No. 61, Oak Ridge, TN: USDOE Office of Scientific and Technical Information

Shaw PJA (2003) Multivariate Statistics for the Environmental Sciences. London, UK: Hodder Arnold

Shelton PA, Sinclair AF, Chouinard GA, Mohn R, Duplisea DE (2006) Fishing under low productivity conditions is further delaying recovery of Northwest Atlantic cod (*Gadus morhua*). Canadian J Fisheries and Aquatic Scie 63:235–238

Shumway RH, Stoffer DS (1982) An approach to time series smoothing and forecasting using the EM algorithm. J Time Series Anal 3:253–264

Shumway RH, Stoffer DS (2000) Time Series Analysis and Its Applications. New York: Springer Verlag

Skinner JA (1987) Abundance and spatial distribution of bees visiting male-sterile and male-fertile sunflower cultivars in California. Environ Entomol 16:922–927

Slutz RJ, Lubker SJ, Hiscox JD, Woodruff SD, Jenne RL, Joseph DH, Steurer PM, Elms JD (1985): Comprehensive Ocean-Atmosphere Data Set; Release 1. NOAA Environmental Research Laboratories, Climate Research Program, Boulder, CO, 268 pp (NTIS PB86-105723)

Smith M (1993) Neural Networks for Statistical Modelling. New York: Van Nostrand Reinhold, 256 pp

Smith MT, Addison JT (2003) Methods for stock assessment of crustacean fisheries. Fish Res 65:231–256

Smith SJ, Iverson SJ, Bowen WD (1997) Fatty acid signatures and classification trees: new tools for investigating the foraging ecology of seals. Canadian J Fisher Aquat Sci 54:1377–1386

Snijders T, Bosker R (1999) An introduction to basic and advanced multilevel modelling. Thousand Oaks, CA: SAGE Publications Ltd

Sobrino I, Silva L, Bellido JM, Ramos F (2002) Rainfall, river discharges and sea temperature as factors affecting abundance of two coastal benthic cephalopod species in the Gulf of Cádiz (SW Spain). Bull Mar Sci **71**:851–865

Sokal RR , Rohlf FJ (1995) Biometry, 3rd edn. 887 pp, New York: Freeman

Solow AR (1994) Detecting changes in the composition of a multispecies community. Biomet 50:556–565

Somers KM (1986) Multivariate allometry and removal of size with principal components analysis. Syst Zool 35:359–368

Somers KM (1989) Allometry, isometry and shape in principal component analysis. Syst Zool 38:169–173

Sousa Reis C, Dornelas M, Lemos R, Santos R (2002) Chapter 11. Fisheries. In: Santos FD, Forbes K, Moita R (eds) Climate Change in Portugal: Scenarios, Impacts and Adaptation Measures - SIAM Project. Lisbon, Portugal, Gravida, 415–452 pp

Stenson GB, Hammill MO, Lawson JW (1997) Predation by harp seals in Atlantic Canada: preliminary consumption estimates for Arctic cod, capelin and Atlantic cod. J Northw Atl Fish Sci 22:137–154

Stern RD, Coe R, Allan EF, Dale IC (2004) Good Statistical Practice for Natural Resources Research. Wallingford, UK: CABI Publishing

Stowasser G, Pierce GJ, Moffat CF, Collins MA, Forsythe JW (2006) Experimental study on the effect of diet on fatty acid and stable isotope profiles of the squid *Lolliguncula brevis*. J Exper Mar Bio Ecol 333:97–114

Sundberg P (1989) Shape and size-constrained principal components analysis. Syst Zool 38:166–168

Svendsen E, Saetre R, Mork M (1991) Features of the northern North Sea circulation. Continen Shelf Res 11:493–508

Switzer P, Green AA (1984) Min/max autocorrelation factors for multivariate spatial imagery. Technical Report 6, Department of Statistics, Stanford, CA: Stanford University

Tabachnich B, Fidell LS (2001) Using Multivariate Statistics 4th edn, Boston MA: Allyn & Bacon

Tajsin AC (1969) Relief and Waters. Volga-Kama State Reserve. Kazan, Russia: Tatar book publishing house, 152 pp (in Russian)

Tajsin AC (1972) About relief influence on Raifa's natural landscape complexes. Proceedings of Volga-Kama State Reserve. Issue 2. Kazan, Russia: publishing of Kazan State University. 184 pp (in Russian)
Ter Braak CJF (1985) Correspondence analysis of incidence and abundance data: properties in terms of a unimodal response model. Biometrics 41:859–873
Ter Braak CJF (1986) Canonical correspondence analysis: a new eigenvector technique for multivariate direct gradient analysis. Ecol 67:1167–1179
Ter Braak CJF (1994) Canonical community ordination. Part I: Basic theory and linear methods. Ecosc 1:127–40
Ter Braak CJF, Looman CWN (1986) Weighted averaging, logistic regression and the Gaussian response model Vegetatio 65:3–11
Ter Braak CJF, Prentice IC (1988) A theory of gradient analysis. Adv Ecolo Res, 18:271–317
Ter Braak CJF, Verdonschot PFM (1995) Canonical correspondence analysis and related multivariate methods in aquatic ecology. Aquatic Sci 5/4:1–35
Therneau TM (1983) A short introduction to recursive partitioning. Orion Technical. Report 21, Stanford, CA: Stanford University, Department of Statistics
Thiel R, Cabral HN, Costa MJ (2003) Composition, temporal changes and ecological guild classification of the ichthyofaunas of large European estuaries – a comparison between the Tagus (Portugal) and the Elbe (Germany). J Applied Ichthyol 19:330–342
Thompson WL (ed) (2004) Sampling rare or elusive species. Washington, DC: Island Press
Thorrington-Smith M (1971) West Indian Ocean Phytoplankton: a numerical investigation of phytohydrographic regions and their characteristics phytoplankton associations. Mar Biol (Berl) 9:115–137
Tibshirani R (1996) Regression shrinkage and selection via the lasso. J Royal Statisti Soci, B 58:267–288
Tobler W (1979) Cellular geography, In: Gale S, Olsson G (eds) Philosophy in Geography, 379–386 pp
Tu W (2002) Zero-inflated data. Encyclop Environmet 4:2387–2391
Twisk Jos WR (2006) Applied Multilevel Analysis. A Practical Guide for Medical Researchers. Cambridge, UK: Cambridge University Press
Upton G, Fingleton B (1985) Spatial Data Analysis by Example, Chichester, UK: Wiley
Van der Meer J (1997) Sampling design of monitoring programmes for marine benthos: a comparison between the use of fixed versus randomly selected stations. J Sea Res 37:167–179
Venables WN, Ripley BD (2002) Modern Applied Statistics with S. 4th edn. New York: Springer
Verzani J (2005) Using R for introductory statistics. Boca Raton, FL: CRC Press
Villagarcía MG, Haedrich RL, Fischer J (1999) Groundfish assemblages of eastern Canada examined over two decades. In: Newell D, Ommer RE (eds), Fishing Places, Fishing People. Toronto, Ontario: University of Toronto Press, 239–259 pp
Wahle, RA (2003) Revealing stock-recruitment relationships in lobsters and crabs: is experimental ecology the key? Fish Res 65:3–32
Waite MB (1895) The pollination of pear flowers. US Dept Agr Div Path Bul 5
Walkley A, Black A (1965) Chapter 4. In: Black A, Evans J (eds) Methods of soil analysis: 155–203. Madison WI: American Society of Agronomy
Walton M, Henderson RJ, Pomeroy PP (2000) Use of blubber fatty acid profiles to distinguish dietary differences between grey seals *Halichoerus grypus* from two UK breeding colonies. Marine Ecology Progress Series 193:201–208
Wentworth C (1922) A scale for grade and class terms for clastic sedimentes. J Geol 30:377–392

Western Regional Climatic Center (2002) Western region climate summaries Western Regional Climatic Center. http://www.wrcc.dri.edu/climsum.html
Whitaker RH (1978) Classification of plant communities. Junk, The Hague
Whitehorn M, Marklyn B (2001) Inside relational databases. London, UK: Springer
Williamson DI (1957a) Crustacea Decapoda: larvae. I. General. ICES, Fiches d'identification du zooplankton, sheet 67
Williamson DI (1957b) Crustacea Decapoda: larvae. V. Caridea, Family Hippolyidae. ICES, Fiches d'identification du zooplankton, sheet 68
Williamson DI (1960) Crustacea Decapoda: larvae. VII. Caridea, Family Crangonidae. Stenopoidea. ICES, Fiches d'identification du zooplankton, sheet 90
Williamson DI (1962) Crustacea Decapoda: larvae. III. Caridea, Families Oplophoridae, Nematocarcinidae and Pasiphaeidae. ICES, Fiches d'identification du zooplankton, sheet 92
Williamson DI (1967) Crustacea Decapoda: larvae. IV. Caridea. Families: Pandalidae and Alpheridae. ICES, Fiches d'identification du zooplankton, sheet 109
Williamson DI (1983) Crustacea Decapoda: larvae. VIII. Nephropidea, Palinuridea, and Eryonidea. ICES, Fiches d'identification du zooplankton, fiche no. 167/168
Wood SN (2004) Stable and efficient multiple smoothing parameter estimation for generalized additive models. J the Amer Stat Assoc 99:673–686
Wood SN (2006) Generalized Additive Models. An Introduction with R. Boca Raton, FL: Chapman & Hall/CRC
Woodruff SD, Slutz RJ, Jenne RL, Steurer PM (1987) A comprehensive ocean-atmosphere data set. Bull Amer Meteor Soc 68:1239–1250.
Yanez E, Barbieri MS, Silva C, Nieto K, Espindola F (2001) Climate variability and pelagic fisheries in northern Chile. Prog Oceanogr 49:581–596
Zar JH (1999) Biostatistical Analysis, 4th edn. Prentice-Hall, Upper Saddle River, NJ
Zelditch ML, Swiderski D, Sheets D, Fink WL (2004) Geometric Morphometrics for Biologists: A Primer. London, UK: Elsevier
Zuur AF (1999) Dimension reduction techniques in community ecology with applications to spatio-temporal marine ecological data. PhD thesis, Aberdeen, Scotland: University of Aberdeen, 299 pp
Zuur AF, Pierce GJ (2004) Common trends in Northeast Atlantic squid time series. J Sea Res 52:57–72
Zuur AF, Fryer RJ, Jolliffe IT, Dekker R, Beukema JJ (2003a) Estimating common trends in multivariate time series using dynamic factor analysis. Environmet 14:665–685
Zuur AF, Tuck ID, Bailey N (2003b) Dynamic factor analysis to estimate common trends in fisheries time series. Can J Fish Aquat Sci 60:542–552
Zwanenburg KCT (2000) The effects of fishing on demersal fish communities of the Scotian Shelf. ICES J Mar Sci 57:503–509

Index

Printed in the United States of America.